D0984675

THE SINGLE SERVER QUEUE

NORTH-HOLLAND SERIES IN

APPLIED MATHEMATICS AND MECHANICS

VOLUME 8

NORTH-HOLLAND PUBLISHING COMPANY — AMSTERDAM · LONDON

THE SINGLE SERVER QUEUE

BY

J. W. COHEN

Professor of Mathematics,
Technological University, Delft

1969

NORTH-HOLLAND PUBLISHING COMPANY — AMSTERDAM · LONDON

PUBLISHERS:

NORTH-HOLLAND PUBLISHING CO. - AMSTERDAM
NORTH-HOLLAND PUBLISHING COMPANY LTD. - LONDON

This book is published and distributed
in the United States by
AMERICAN ELSEVIER PUBLISHING COMPANY, INC.
52 Vanderbilt Avenue, New York, N.Y. 10017

Library of Congress Catalog Card Number: 69-18389
Standard Book Number: 7204 2358 9

PRINTED IN THE NETHERLANDS

To Annette B.

EDITORIAL NOTE

The enormous increase in the amount of research information, published in an ever-growing number of scientific journals, has stimulated the demand for authoritative monographs on well-defined topics. Such monographs have become virtually indispensable to young research workers and students working in a particular field, who are either bewildered by the vast quantity of literature in existence, or are in danger of duplicating research that has already been published, but is not readily accessible. Specialists also may benefit from the availability of competent surveys by fellow experts in their own field.

The North-Holland Series in Applied Mathematics and Mechanics is intended to help meet this demand. The editors believe that a continuing close relationship between applied mathematics and mechanics, having proved so fruitful in the past, will continue to benefit both subjects in the future. The series will include original monographs as well as translations of outstanding works which would otherwise have remainded inaccessible to many readers.

PREFACE

Queueing theory is one of the most important branches of modern probability theory applied in technology and management. As far as the future development of technology and management may be extrapolated from the past and present state of affairs the need for a deeper insight into queueing theory will increase rapidly. It is hardly necessary to point out the many actual queueing situations encountered in every-day life. Production lines, the theory of scheduling and transportation (both surface and air traffiic), the design of automatic equipment such as telephone and telegraph exchanges, and particularly the rapidly growing field of information handling and data processing are but a few fields in which queueing situations are encountered. To characterize: it arises in every situation, in which a facility for common use is provided, where waiting and queueing may and usually do arise. The organisation and the performance of the facility on the one hand and the behaviour of the users on the other determine the queueing system.

Although the broad field of applications amply justifies an intensive study of queueing models, it turned out that these models are also of great use outside the field of queueing theory, e.g. in inventory and maintenance theory. Moreover, the analytical problems encountered in the study of queueing models are often very interesting from a mathematical point of view and, consequently, obtained much attention from probability theorists.

As with so many branches of applied mathematics the task of queueing theory consists of the classification of the more fundamental models, of the design of analytical methods for the study of these models and of the mathematical analysis of those quantities which describe the essential features and properties of the model. On the basis of this knowledge, eventually supplemented by experimental studies usually performed by simulation techniques, the designer of systems with queueing situations should obtain the information and feeling needed to predict the behaviour of the actual queueing situation. The step from the model and its mathematical description to the actual situation is usually the most difficult one, the more so if the actual situation is too complicated to allow a fruitful theoretical or experimental investigation of its model. A sound knowledge of the fundamental models and their properties is often the best guide in making this step.

The present book concentrates on the most basic model of queueing theory, i.e. the single server model. Its aim is two-fold. Firstly, a description of those mathematical techniques which have been proved to be the most fruitful for the investigation of queueing models, and secondly, an extensive analysis of the single server queue and its most important variants. Even within this limited range restrictions had to be made, but the author's goal will be reached if the reader acquires an understanding of the models, purposes and methods of the theory of queues.

The book is divided into three parts. The first part deals with those topics of the theory of stochastic processes which have been successfully used in queueing theory. Part II is devoted to the simplest single server model. A number of analytical techniques for this model is discussed. The most powerful one is that based on Pollaczek's approach, which is closely related to the Wiener-Hopf technique, combined with renewal theory. This technique is often very intricate and in simple cases it is possible to use a much more elegant argument, which is often shorter. However, it is the author's conviction that the primary duty of applied mathematics is the development of sharp and powerful tools leading to useful results. For this reason in the derivations elegance is often sacrificed to utility. In part III several of the more important variants of the single server model are discussed. It concludes with a chapter on asymptotic relations and limit theorems.

The book is intended for applied mathematicians concerned with system design or interested in applied probability theory. A sound knowledge of Volume I of Feller's "An Introduction to Probability Theory and its Applications" is needed, while some knowledge of advanced probability theory, e.g. subjects discussed in Feller's Volume II, is desirable; the Laplace-Stieltjes transform and the theory of functions are tools extensively used in the text. A short review of literature is added to each chapter. Completeness of these reviews is not claimed.

Relations which are not separated by text have only one reference number placed at the end of the first one; a reference to relation (5.24) in a section of part I refers to relation 5.24 of that part, whereas references to relations outside a part are prefixed by a roman numeral indicating the part, so (I.5.24) means relation (5.24) of chapter 5 of part I. The name of an author followed by a date refers to the list of references. Much trouble has been taken to give every symbol a unique meaning. Occasionally, deviations from this ideal could not be avoided, but sufficient provisions have been made to avoid confusion. The symbol '$\stackrel{\text{def}}{=}$' stands for the defining equality sign. All symbols indicating stochastic variables are printed in bold type.

Thanks are due to a number of colleagues, collaborators and students for reading chapters and suggesting improvements, in particular to Mr. S. J. de Lange, Mr. P. B. M. Roes and Mr. J. H. A. de Smit. I express my gratitude to Mrs. N. Zuidervaart and Miss M. Berenschot for their efficient typing of the manuscript and to Mr. B. Broere for his help with drawing the figures. Special thanks are also due to my friend Richard Syski. The stimulating discussions and correspondence I had with him about the writing of the book were a real contribution to its completion.

The Hague, 1968 J. W. COHEN

CONTENTS

Part I. Stochastic Processes

PART II. THE SINGLE SERVER QUEUE

Part III. Some Variants of the Single Server Queue

PART I

STOCHASTIC PROCESSES

I.1. INTRODUCTION

The theory of stochastic processes is a rather recently developed branch of probability theory. One of the first processes investigated was that of the Brownian motion; this process has been extensively studied and from its investigation fundamental contributions to probability theory have originated. The first important results concerning this process date back to the beginning of this century. At the same time telephone engineers were confronted with a type of stochastic process, today called a birth and death process, which turned out to be of fundamental importance for designing telephone exchanges. The theory of stochastic processes originating from needs in physics and technology is at present a rather well developed theory; a large number of basic processes have been classified and the most important properties of these processes are known (cf. DOOB [1953], LOÈVE [1960], BLANC-LAPIERRE et FORTET [1953], PARZEN [1962]).

The concept of "stochastic variable" or "random variable" is fundamental in modern probability theory and its applications; we assume that the reader is acquainted with it. Let t denote a parameter assuming values in a set T, and let $\boldsymbol{x}_t$ represent a stochastic variable for every $t \in T$. We thus obtain a family $\{\boldsymbol{x}_t, t \in T\}$ of stochastic variables. Such a family will be called a stochastic process if the parameter t stands for time, and from now on t will be interpreted as such. The elements of T are hence time points, and T will be a linear set, denumerable or non-denumerable. For queueing theory the most important cases are that T is the interval $[0, \infty)$ or that $T = \{t\colon t = t_n, n = 0, 1, 2, \ldots\}$ with $t_0 = 0$, and particularly with $t_n = ne$, where e is the time unit. If $T = \{t\colon t \in [0, \infty)\}$ then the stochastic process $\{\boldsymbol{x}_t, t \in T\}$ is said to be a process with continuous time parameter, whereas in the second case the process is said to have a discrete time parameter.

For arbitrary $t \in T$ the set of all possible realisations of the stochastic variable $\boldsymbol{x}_t$ is the sample space or state space of $\boldsymbol{x}_t$. From now on it will be assumed that all $\boldsymbol{x}_t$, $t \in T$ have the same state space. The state space may be a denumerable or non-denumerable set. For instance $\boldsymbol{x}_t$ may represent the number of customers waiting at a service station at time t; or $\boldsymbol{x}_t$ represents, at time t, the total time a server has been busy since $t=0$. However, it may also happen that the state space of the process is a vector space. For instance $\boldsymbol{x}_t$ is the vector variable $(\boldsymbol{\xi}_t, \boldsymbol{\tau}_t)$, where at time t the variable $\boldsymbol{\xi}_t$ stands for the number of customers waiting, and $\boldsymbol{\tau}_t$ for the length of time between t and the arrival of the next customer.

What information should be known to describe a stochastic process $\{\boldsymbol{x}_t, t \in T\}$ completely? We shall not study this question here but refer the reader to the existing literature on this subject (cf. e.g. DOOB [1953]). For our purpose it is (in general) sufficient to know the n-dimensional joint distribution of the variables $\boldsymbol{x}_{t_1}, \boldsymbol{x}_{t_2}, \ldots, \boldsymbol{x}_{t_n}$,

$$F_{t_1, \ldots, t_n}(x_1, \ldots, x_n),$$

for every finite positive integer n, for every point set $(t_1, t_2, \ldots, t_n)$ belonging to T and for all $x_1, x_2, \ldots, x_n$ belonging to the state space of the process. When investigating a stochastic process our aim will be to find these n-dimensional joint distributions. It should be noticed that the functions $F_{\ldots}(., \ldots, .)$ just mentioned cannot be completely arbitrary multidimensional distribution functions. As has been shown by Kolmogorov, these distribution functions have to satisfy two consistency conditions, viz. (*i*) if $m_1, \ldots, m_n$ is a permutation of $1, 2, \ldots, n$ then

$$F_{t_1, \ldots, t_n}(x_1, \ldots, x_n) = F_{t_{m_1}, \ldots, t_{m_n}}(x_{m_1}, \ldots, x_{m_n}),$$

and (*ii*) for $m = 1, 2, \ldots, n-1$,

$$F_{t_1, \ldots, t_m}(x_1, \ldots, x_m) = \lim_{x_{m+1} \to \infty} \ldots \lim_{x_n \to \infty} F_{t_1, \ldots, t_n}(x_1, \ldots, x_n).$$

That there are questions concerning a stochastic process which cannot be answered from the mere knowledge of all finite dimensional distributions may be illustrated by the following example. For the stochastic process $\{\boldsymbol{x}_t, t \in [0, \infty)\}$ with state space the real line the event

$$\max_{0<t<\tau} \boldsymbol{x}_t < \alpha,$$

where α and τ are given numbers, is an event involving more than a finite, even more than a denumerably infinite, number of stochastic variables. The probability of such an event cannot be defined by the finite dimensional distributions of the process, and additional information about the process is needed. Such information is usually provided by describing the existence of a limit procedure which enables the definition and calculation of the probability of the event above from the probabilities of the events

$$\max\{x_{t_i},\ i=1, ..., m\}<\alpha,\ 0<t_i<\tau,\ i=1, 2, ..., m;\ m=1, 2, ...$$

Roughly speaking, a stochastic process may be considered as the probabilistic counterpart of a deterministic process, and akin to the latter type of process one is interested in its time dependent structure and its behaviour for large values of the time parameter. The general theory of stochastic processes is, however, far more complicated than that of deterministic processes, since a stochastic process is, if one considers all its possible realisations or sample functions, in fact a family of time processes. From the viewpoint of the applications, however, the study of both types of processes is often directed to the same goals. A deterministic process may also be considered as a degenerate stochastic process; it is a stochastic process in which there is no randomness at all. On the other hand a stochastic process $\{x_t,\ t\in[0, \infty)\}$ for which the randomness is present in its most extreme form, for instance if all $x_{t_1}, ..., x_{t_n}$ are independent variables for every n and every subset $\{t_1, t_2, ..., t_n\}$ of T, is hardly of any interest, theoretically nor practically. Such a process is much too irregular and does not possess any link at all between past, present, and future in its development. At least some relation between the past, present and future of the process is desirable to make a process interesting for theoretical investigation and to be of any use as a model for applications.

Stochastic processes are mainly classified by properties of and relations between their finite dimensional distribution functions. For an extensive review of such a classification the reader is referred to DOOB [1953].

I.2. MARKOV CHAINS WITH A DISCRETE TIME PARAMETER

I.2.1. Definition of the process

In this chapter stochastic processes with a discrete time parameter will be considered. The state space, to be denoted by S, will be a denumerable set. The time parameter set T will be the set of non-negative numbers $\{0, 1, 2, \ldots\}$.

It is supposed that for this process

$$\{\boldsymbol{x}_n, n=0, 1, 2, \ldots\}$$

all finite dimensional joint distributions are given, so that for every finite integer $m=1, 2, \ldots$,

$$\Pr\{\boldsymbol{x}_{k_1}=i_1, \boldsymbol{x}_{k_2}=i_2, \ldots, \boldsymbol{x}_{k_m}=i_m\}$$

is known for every $i_h \in S$, $h=1, \ldots, m$, and every subset $\{k_1, \ldots, k_m\}$ of $\{0, 1, 2, \ldots\}$.

This process $\{\boldsymbol{x}_n, n=0, 1, 2, \ldots\}$ is in fact a sequence of stochastic variables $\boldsymbol{x}_0, \boldsymbol{x}_1, \boldsymbol{x}_2, \ldots$. Sequences of stochastic variables have been investigated in probability theory for a long time, especially sequences of independent and identically distributed variables. For these types of sequences the interest is mainly centered on properties of the partial sums $\boldsymbol{x}_0+\boldsymbol{x}_1+\ldots+\boldsymbol{x}_n$, $n=0, 1, 2, \ldots$. Sequences of independent variables are processes without any dependency between past, present and future. Due to the fact that the time parameter set is discrete, however, they are of great interest. In this chapter we shall consider a first generalisation of these processes by introducing dependence between present and future in the development of the process. Processes of this type were first considered by A. A. Markov at the beginning of this century; today, they form a very important chapter of the theory of probability.

DEFINITION. The process $\{x_n, n=0, 1, \ldots\}$ is a *discrete time parameter Markov chain* if for every integer $n \geq 0$ and every $i_k \in S$, $k=0, \ldots, n$,

$$\Pr\{x_{n+1}=i_{n+1} \mid x_k=i_k,\ k=0, \ldots, n\}=\Pr\{x_{n+1}=i_{n+1} \mid x_n=i_n\},$$

whenever the left-hand side is defined.

Evidently, the definition states that if in the sequence $\{x_n, n=0, 1, \ldots\}$ at time n, the realisation of x_n is known then the conditional probability of the future realisation of x_{n+1} depends only on x_n, i.e. on the present, and not on x_k, $k=0, 1, \ldots, n-1$. Hence the future realisation of x_{n+1} is independent of the past of the process up to and including time $n-1$. We shall show below that for any future event after time n the conditional probability given x_k, $k=0, \ldots, n$ is only dependent on x_n.

In the following $\{x_n, n=0, 1, \ldots\}$ will denote a discrete time parameter Markov chain. Using elementary properties of conditional probabilities it is easily seen that the definition above implies

$$\begin{aligned}
&\Pr\{x_{n+2}=i_{n+2}, x_{n+1}=i_{n+1} \mid x_k=i_k,\ k=0, \ldots, n\} \Pr\{x_k=i_k,\ k=0, \ldots, n\}\\
&\quad=\Pr\{x_k=i_k,\ k=0, \ldots, n+2\}\\
&\quad=\Pr\{x_{n+2}=i_{n+2} \mid x_k=i_k,\ k=0, \ldots, n+1\} \Pr\{x_k=i_k,\ k=0, \ldots, n+1\}\\
&\quad=\Pr\{x_{n+2}=i_{n+2} \mid x_{n+1}=i_{n+1}\} \Pr\{x_{n+1}=i_{n+1} \mid x_n=i_n\}\\
&\quad\cdot \Pr\{x_k=i_k,\ k=0, \ldots, n\},
\end{aligned}$$

so that, whenever conditional probabilities exist (which will be assumed throughout)

$$\begin{aligned}
&\Pr\{x_{n+2}=i_{n+2}, x_{n+1}=i_{n+1} \mid x_k=i_k,\ k=0, \ldots, n\}\\
&\qquad=\Pr\{x_{n+2}=i_{n+2} \mid x_{n+1}=i_{n+1}\} \Pr\{x_{n+1}=i_{n+1} \mid x_n=i_n\}.
\end{aligned}$$

The right-hand side of the last relation is independent of $x_k=i_k$, $k=0, 1, \ldots, n-1$, hence

$$\begin{aligned}
&\Pr\{x_{n+2}=i_{n+2}, x_{n+1}=i_{n+1} \mid x_k=i_k,\ k=0, \ldots, n\}\\
&\qquad=\Pr\{x_{n+2}=i_{n+2}, x_{n+1}=i_{n+1} \mid x_n=i_n\}.
\end{aligned}$$

The generalisation of the formulas above is

$$\begin{aligned}
&\Pr\{x_h=i_h,\ h=n+1, \ldots, n+m \mid x_k=i_k,\ k=0, 1, \ldots, n\} \qquad (2.1)\\
&\qquad= \Pr\{x_h=i_h,\ h=n+1, \ldots, n+m \mid x_n=i_n\}\\
&\qquad= \prod_{k=1}^{m} \Pr\{x_{n+k}=i_{n+k} \mid x_{n+k-1}=i_{n+k-1}\},
\end{aligned}$$

for all integers $n \geqq 0$, $m \geqq 1$, and all $i_k \in S$, $k=0, \ldots, n+m$. The proof of (2.1) will be omitted since it is an immediate extension of the two preceding relations.

The first part of relation (2.1) shows that for a discrete time parameter Markov chain the conditional probability of any future event after time n, given $\boldsymbol{x}_k$, $k=0, \ldots, n$, is indeed independent of $\boldsymbol{x}_k$, $k=0, \ldots, n-1$. The second part of relation (2.1) permits the following important conclusion. If the distribution function of $\boldsymbol{x}_0$ is known and if for all integers $n \geqq 0$ the *one-step transition probabilities* at time n, i.e.

$$\Pr\{\boldsymbol{x}_{n+1}=i_{n+1} \mid \boldsymbol{x}_n=i_n\}$$

are known for all i_n, $i_{n+1} \in S$, then every finite dimensional joint distribution of the discrete time parameter Markov chain $\{\boldsymbol{x}_n, n=0, 1, \ldots\}$ can be calculated. To see this take $n=0$ in (2.1) then

$$\begin{aligned}\Pr\{\boldsymbol{x}_h=i_h, h=0, \ldots, m\} &= \Pr\{\boldsymbol{x}_h=i_h, h=1, \ldots, m \mid \boldsymbol{x}_0=i_0\} \Pr\{\boldsymbol{x}_0=i_0\} \\ &= \Pr\{\boldsymbol{x}_0=i_0\} \prod_{k=1}^{m} \Pr\{\boldsymbol{x}_k=i_k \mid \boldsymbol{x}_{k-1}=i_{k-1}\},\end{aligned}$$

and from this relation the statement above is immediately obvious. The conclusion above shows that for the description of a discrete time parameter Markov chain it is sufficient to give the *initial distribution*, i.e. the distribution function of $\boldsymbol{x}_0$, and the conditional distributions of $\boldsymbol{x}_{n+1}$ given $\boldsymbol{x}_n$ for all integers $n \geqq 0$.

Since for any integer $n \geqq 0$,

$$\bigcup_{i \in S} \{\boldsymbol{x}_n=i\} = \Omega,$$

where Ω is the sure event, we have

$$\begin{aligned}&\Pr\{\boldsymbol{x}_{n+2}=i_{n+2} \mid \boldsymbol{x}_k=i_k, k=0, \ldots, n\} \\ &\quad= \Pr\{\boldsymbol{x}_{n+2}=i_{n+2}, \bigcup_{i \in S} \boldsymbol{x}_{n+1}=i \mid \boldsymbol{x}_k=i_k, k=0, \ldots, n\} \\ &\quad= \sum_{i \in S} \Pr\{\boldsymbol{x}_{n+2}=i_{n+2}, \boldsymbol{x}_{n+1}=i \mid \boldsymbol{x}_k=i_k, k=0, \ldots, n\} \\ &\quad= \sum_{i \in S} \Pr\{\boldsymbol{x}_{n+2}=i_{n+2}, \boldsymbol{x}_{n+1}=i \mid \boldsymbol{x}_n=i_n\} \\ &\quad= \Pr\{\boldsymbol{x}_{n+2}=i_{n+2}, \bigcup_{i \in S} \boldsymbol{x}_{n+1}=i \mid \boldsymbol{x}_n=i_n\} = \Pr\{\boldsymbol{x}_{n+2}=i_{n+2} \mid \boldsymbol{x}_n=i_n\}.\end{aligned}$$

By the same argument we obtain

$$\Pr\{\boldsymbol{x}_{n+m}=i_{n+m} \mid \boldsymbol{x}_k=i_k, k=0, \ldots, n\}=\Pr\{\boldsymbol{x}_{n+m}=i_{n+m} \mid \boldsymbol{x}_n=i_n\},$$

for all integers $n \geqq 0$, $m \geqq 1$ and i_{n+m}, $i_k \in S$; the conditional probability

$$\Pr\{\boldsymbol{x}_{n+m}=j \mid \boldsymbol{x}_n=i\}$$

is called the *m-step transition probability* of the chain at time n.

An elementary property of conditional probabilities shows that

$$\begin{aligned}\Pr\{&\boldsymbol{x}_{n+h+m}=i_{n+h+m},\ \boldsymbol{x}_{n+m}=i_{n+m} \mid \boldsymbol{x}_n=i_n\}\\ &=\Pr\{\boldsymbol{x}_{n+h+m}=i_{n+h+m} \mid \boldsymbol{x}_{n+m}=i_{n+m},\ \boldsymbol{x}_n=i_n\}\Pr\{\boldsymbol{x}_{n+m}=i_{n+m} \mid \boldsymbol{x}_n=i_n\}.\end{aligned}$$

Since

$$\Pr\{\boldsymbol{x}_{n+h+m}=i_{n+h+m} \mid \boldsymbol{x}_k=i_k,\ k=0, \ldots, n+m\}$$

is independent of the event $\{\boldsymbol{x}_k=i_k,\ k=0, \ldots, n+m-1\}$ it follows that

$$\Pr\{\boldsymbol{x}_{n+h+m}=i_{n+h+m} \mid \boldsymbol{x}_{n+m}=i_{n+m},\ \ \boldsymbol{x}_n=i_n\}=\Pr\{\boldsymbol{x}_{n+h+m}=i_{n+h+m} \mid \boldsymbol{x}_{n+m}=i_{n+m}\}$$

and, therefore, that

$$\begin{aligned}\Pr\{&\boldsymbol{x}_{n+h+m}=i_{n+h+m},\ \boldsymbol{x}_{n+m}=i_{n+m} \mid \boldsymbol{x}_n=i_n\}\\ &=\Pr\{\boldsymbol{x}_{n+h+m}=i_{n+h+m} \mid \boldsymbol{x}_{n+m}=i_{n+m}\}\Pr\{\boldsymbol{x}_{n+m}=i_{n+m} \mid \boldsymbol{x}_n=i_n\}.\end{aligned}$$

Summing the last relation over all $i_{n+m} \in S$ we obtain

$$\begin{aligned}\Pr\{&\boldsymbol{x}_{n+h+m}=j \mid \boldsymbol{x}_n=i\}\\ &=\sum_{k\in S}\Pr\{\boldsymbol{x}_{n+h+m}=j \mid \boldsymbol{x}_{n+m}=k\}\Pr\{\boldsymbol{x}_{n+m}=k \mid \boldsymbol{x}_n=i\}, \qquad (2.2)\end{aligned}$$

for all integers $n \geqq 0$, $h \geqq 1$, $m \geqq 1$, and all $i, j, k \in S$.

The last relation is an important one, it is in fact an immediate consequence of the property that for a Markov chain with discrete time parameter the conditional property of any future event given the present and the past depends only on the present and not on the past. Relation (2.2) permits the direct calculation of the higher step transition probabilities from the one-step transition probabilities. Obviously, we also have for these transition probabilities that, for all integers $n \geqq 0$, $m \geqq 1$,

$$\sum_{j\in S}\Pr\{\boldsymbol{x}_{n+m}=j \mid \boldsymbol{x}_n=i\}=1 \qquad \text{for all } i \in S. \qquad (2.3)$$

A realisation or sample function of a Markov chain with discrete time parameter is described by the successive values taken on by the stochastic variables $\boldsymbol{x}_n$, $n=0, 1, \ldots$. Such a sample function can be considered as a path described by a moving point. In this way it is often possible to obtain a useful heuristic description of a stochastic process. To give such an interpretation suppose there exists a one-to-one correspondence between the elements i of S and the elements E_i of a set $\{E_i, i \in S\}$, such that E_i corresponds with i and conversely i with E_i. Identifying the event $\{x_n = i\}$ with the event "the moving point or the system is in state E_i at time n" we obtain an interpretation of the realisations of the Markov chain in terms of paths described by a point moving over the elements of $\{E_i, i \in S\}$. In this terminology the conditional probability

$$\Pr\{x_{n+m} = j \mid x_n = i\}$$

is called the *transition probability from state E_i at time n to state E_j at time $n+m$.*

From the above discussion it is obvious that the mechanism which controls the development of a Markov chain with discrete time parameter is described by the transition probabilities of the process. These transition probabilities may be time-dependent or time-independent. A Markov chain with time dependent transition probabilities will have a much more complicated structure than a chain with time independent transition probabilities. In their most general form processes of the first type are far too irregular to permit fruitful theoretical investigation. Processes of the latter type, however, are very important, theoretically as well as practically.

DEFINITION. The Markov chain $\{x_n, n=0, 1, \ldots\}$ has *stationary* transition probabilities if for all $i, j \in S$ and all integers $m \geqq 1$, $n \geqq 0$,

$$\Pr\{x_{n+m} = j \mid x_n = i\}$$

is independent of n.

A necessary and sufficient condition for the Markov chain $\{x_n, n=0,1,\ldots\}$ to have stationary transition probabilities is that the one-step transition probabilities

$$\Pr\{x_{n+1} = j \mid x_n = i\}, \qquad i, j \in S,$$

are all independent of n. The necessity of the condition is evident; the sufficiency of it follows immediately from (2.2) by complete induction with respect to h and m starting with $h=1$, $m=1$.

From now on we shall only consider Markov chains with stationary transition probabilities.

Denote by

$$p_{ij}^{(m)} \overset{\text{def}}{=} \Pr\{x_{n+m} = j \mid x_n = i\}, \qquad i, j \in S, \quad m = 1, 2, \ldots,$$

the m-step transition probability for such a Markov chain. The matrix $P^{(m)}$ with these m-step transition probabilities as elements

$$P^{(m)} \stackrel{\text{def}}{=} (p_{ij}^{(m)}), \qquad P = (p_{ij}) \stackrel{\text{def}}{=} P^{(1)},$$

i being the row index and j the column index, has the following properties:

(i) its elements are non-negative;

(ii) its row sums are equal to one;

(iii) for all $m, n=1, 2, \ldots,$

$$P^{(m+n)} = P^{(m)}\, P^{(n)} = P^{(n)}\, P^{(m)}. \tag{2.4}$$

The second property follows from (2.3), whereas the third one is a restatement of (2.2) by the rule of matrix multiplication.

We further define for all $i, j \in S$,

$$p_{ij}^{(0)} \stackrel{\text{def}}{=} \delta_{ij}, \qquad P^{(0)} \stackrel{\text{def}}{=} (p_{ij}^{(0)}) = I,$$

where δ_{ij} is Kronecker's symbol, and I the unit matrix.

Since

$$\Pr\{\boldsymbol{x}_h = i_h,\ h=0, 1, \ldots, m\} = \Pr\{\boldsymbol{x}_0 = i_0\} \prod_{k=1}^{m} p_{i_{h-1} i_h},$$

it follows that, for a Markov chain $\{\boldsymbol{x}_n,\ n=0, 1, \ldots\}$ with stationary transition probabilities, all finite dimensional joint distributions can be calculated if the matrix P of one-step transition probabilities and the distribution function of $\boldsymbol{x}_0$ are known. Of particular interest are the absolute distributions. Define

$$a_j^{(n)} \stackrel{\text{def}}{=} \Pr\{\boldsymbol{x}_n = j\}, \qquad j \in S, \quad n=0, 1, \ldots,$$

then, evidently,

$$\sum_{j \in S} a_j^{(n)} = 1, \qquad n=0, 1, \ldots .$$

From the definition of $p_{ij}^{(m)}$ we obtain

$$a_j^{(n+m)} = \sum_{i \in S} a_i^{(n)}\, p_{ij}^{(m)}, \qquad j \in S, \quad n, m=0, 1, \ldots . \tag{2.5}$$

DEFINITION. A Markov chain $\{\boldsymbol{x}_n,\ n=0, 1, \ldots\}$ with stationary transition probabilities is *stationary* if $a_j^{(n)}$, $j \in S$, $n=0, 1, \ldots$, is independent of n.

Denoting by

$$a_j \stackrel{\text{def}}{=} \Pr\{\boldsymbol{x}_n = j\}, \qquad j \in S, \quad n=0, 1, \ldots,$$

a *stationary distribution* if the Markov chain is stationary then it follows from (2.5) that the a_j, $j \in S$ satisfy

$$a_j = \sum_{i \in S} a_i p_{ij}, \qquad j \in S, \tag{2.6}$$

and, generally,

$$a_j = \sum_{i \in S} a_i p_{ij}^{(m)}, \qquad m=0, 1, \ldots, j \in S. \tag{2.7}$$

Further, it is easily proved that for a stationary Markov chain the finite dimensional joint distribution

$$\Pr\{\boldsymbol{x}_h = i_h,\ h=n, n+1, \ldots, n+m\}$$

is independent of n for every $m=0, 1, \ldots$.

An important problem in the theory of Markov chains and also in applications is whether the chain possesses a stationary distribution. Particularly in applications the necessary and sufficient conditions for the existence of a stationary distribution are usually of essential importance. From what has been said above, it may be conjectured that the existence of a stationary distribution is related to the existence of a non-negative, non-null solution of the (finite or infinite) set of linear equations

$$x_j = \sum_{i \in S} x_i p_{ij}, \qquad j \in S.$$

A second important question, partially related to the one just mentioned, concerns the behaviour of $p_{ij}^{(n)}$ for $n \to \infty$. Both questions will be discussed in the following sections.

I.2.2. Classification of states

For the Markov chain $\{\boldsymbol{x}_n, n=0, 1, \ldots\}$ with stationary transition probabilities we define for $i, j \in S$,

$$f_{ij}^{(m)} \stackrel{\text{def}}{=} \begin{cases} 0, & m=0, \\ \Pr\{\boldsymbol{x}_{n+1}=j \mid \boldsymbol{x}_n=i\}, & m=1, \\ \Pr\{\boldsymbol{x}_{n+m}=j,\ \boldsymbol{x}_{n+h} \neq j,\ h=1, \ldots, m-1 \mid \boldsymbol{x}_n=i\}, & m=2, 3, \ldots . \end{cases}$$

Since the transition probabilities are stationary it is easily verified that $f_{ij}^{(m)}$ as defined above is independent of n, $n=0, 1, \ldots$. Evidently, $f_{ij}^{(m)}$ is the conditional probability that the chain reaches state E_j at time $n+m$ for the first time after time n whenever it was at this moment in state E_i; i.e. $f_{ij}^{(m)}$,

$m=1, 2, \ldots$, is the distribution of the *entrance time* into state E_j when starting in state E_i. By considering all disjoint possibilities of reaching E_j from E_i in exactly m steps it is easily seen that

$$p_{ij}^{(m)} = f_{ij}^{(m)} + f_{ij}^{(m-1)} p_{jj}^{(1)} + f_{ij}^{(m-2)} p_{jj}^{(2)} + \ldots + f_{ij}^{(1)} p_{jj}^{(m-1)}, \qquad m=1, 2, \ldots . \tag{2.8}$$

In particular for $j=i$,

$$p_{ii}^{(m)} = f_{ii}^{(m)} + f_{ii}^{(m-1)} p_{ii}^{(1)} + f_{ii}^{(m-2)} p_{ii}^{(2)} + \ldots + f_{ii}^{(1)} p_{ii}^{(m-1)}, \qquad m=1, 2, \ldots . \tag{2.9}$$

From (2.9) it is easily seen that once the transition probabilities $p_{ii}^{(m)}$, $m=0, 1, 2, \ldots$, are known $f_{ii}^{(m)}$ can be calculated recursively for all $m=1, 2, \ldots$, and are uniquely determined by (2.9).

For the following considerations it is no loss of generality to suppose that

$$\Pr\{x_0=i\}=1 \qquad \text{for a fixed } i\in S,$$

i.e. the Markov chain starts in state E_i.

Define the sequence of stochastic variables y_n, $n=1, 2, \ldots$, as follows: $\{y_n=k\}$ is the event realized whenever the nth occurrence of E_i happens at time k after the preceding occurrence of E_i, $k=1, 2, \ldots$. Evidently,

$$\Pr\{y_n=k\}=f_{ii}^{(k)}, \qquad k=1, 2, \ldots; \quad n=1, 2, \ldots .$$

Since

$$\Pr\{x_{n+h}\neq i,\ h=1, \ldots, k-1,\ x_{n+k}=i \mid x_n=i,\ x_m=i_m,\ m=0, 1, \ldots, n-1\}$$

is independent of i_m, $m=0, 1, \ldots, n-1$, it is easily verified that the stochastic variables of the sequence y_n, $n=1, 2, \ldots$, are independent and identically distributed. The distribution of y_n will be called the *return time distribution* of state E_i. The process $\{y_n,\ n=1, 2, \ldots\}$ is a discrete renewal process; here it is for the Markov chain $\{x_n,\ n=0, 1, \ldots\}$ with initial state E_i the *imbedded renewal process* for the state E_i. For a discussion of renewal theory the reader is referred to chapter 6.

By writing

$$f_{ii} \stackrel{\text{def}}{=} \sum_{k=1}^{\infty} f_{ii}^{(k)} = \Pr\{y_n<\infty\},$$

it is seen that f_{ii} represents the probability of a return to E_i when starting in E_i. Now two cases are to be distinguished, viz.

$$f_{ii}=1 \quad \text{and} \quad f_{ii}<1.$$

Only in the first case $\{f_{ii}^{(k)},\ k=1, 2, \dots\}$ is a proper probability distribution. Defining for the second case

$$\Pr\{y_n=\infty\} \overset{\text{def}}{=} 1 - f_{ii}, \qquad n=1, 2, \dots,$$

it is seen that y_n is a so called *generalized stochastic variable.* If $f_{ii}=1$ a return to E_i when starting in E_i is certain and E_i is called a *recurrent* state. Otherwise it is denoted as a *transient* state, and for such a state a return is not certain since the probability of a return f_{ii} is less than one.

The probability that starting from E_i the system passes state E_i exactly m times is given by $f_{ii}^m(1-f_{ii})$ so that

$$1 - f_{ii}^{m+1} = \sum_{k=0}^{m} f_{ii}^k(1-f_{ii})$$

is the probability that E_i is reached at most m times, $m=1, 2, \dots$; consequently, for $m\to\infty$,

$$\Pr\{E_i, \text{i.o.}\} \equiv \Pr\{E_i, \text{infinitely often}\} = \begin{cases} 1 & \text{if } f_{ii}=1, \\ 0 & \text{if } f_{ii}<1. \end{cases}$$

The average return time μ_i of a recurrent state E_i is defined by

$$\mu_i = \sum_{k=1}^{\infty} k f_{ii}^{(k)}.$$

If μ_i is finite the state E_i is called *persistent* or *positive recurrent*, whereas E_i is called *null recurrent* if μ_i is infinite.

Whenever $f_{ii}^{(k)}>0$ for some k then the greatest common divisor λ_i of those k for which $f_{ii}^{(k)}>0$ will be called the *period* of state E_i if its value is greater than one, E_i is then called *periodic.* If $\lambda_i=1$ then E_i is *aperiodic.* It should be noted that whenever E_i is a periodic state with period λ_i in the Markov chain $\{x_n,\ n=0, 1, \dots\}$, then it is an aperiodic state of the Markov chain $\{x_{n\lambda_i},\ n=0, 1, \dots\}$, the latter chain having $(p_{ij}^{(\lambda_i)})$ as one-step transition matrix.

From what has been said above it is seen that a state is transient or null recurrent or positive recurrent and in all three cases it may be periodic or aperiodic. An aperiodic positive recurrent state is often called *ergodic.* The classification of states introduced above has been defined by starting from the distribution of the return time. This distribution is uniquely determined by the probabilities $p_{ii}^{(m)}$, $m=1, 2, \dots$, (cf. (2.9)). Since these probabilities are usually given, it is desirable to express necessary and sufficient conditions for a state to be transient, null recurrent or positive recurrent, in terms of the $p_{ii}^{(m)}$, $m=1, 2, \ldots$. The two following theorems describe these criteria.

THEOREM 2.1. *State E_i is transient if and only if $\sum_{n=0}^{\infty} p_{ii}^{(n)}$ converges.*

Proof. Define the generating functions

$$P_{ii}(s) \stackrel{\text{def}}{=} \sum_{n=0}^{\infty} p_{ii}^{(n)} s^n, \qquad |s|<1,$$

$$F_{ii}(s) \stackrel{\text{def}}{=} \sum_{n=0}^{\infty} f_{ii}^{(n)} s^n, \qquad |s|\leqq 1.$$

Since by definition $p_{ii}^{(0)}=1$ and $f_{ii}^{(0)}=0$ it follows from (2.9) for $|s|<1$, that

$$\begin{aligned} P_{ii}(s) - 1 = \sum_{m=0}^{\infty} p_{ii}^{(m)} s^m - 1 &= \sum_{m=0}^{\infty} s^m \sum_{k=0}^{m} f_{ii}^{(m-k)} p_{ii}^{(k)} \\ &= \sum_{k=0}^{\infty} \sum_{m=k}^{\infty} f_{ii}^{(m-k)} s^{m-k} p_{ii}^{(k)} s^k = P_{ii}(s)\, F_{ii}(s), \end{aligned}$$

and hence for $|s|<1$,

$$P_{ii}(s) = \frac{1}{1-F_{ii}(s)}. \tag{2.10}$$

Since (cf. app. 1)

$$f_{ii} = \lim_{s\uparrow 1} F_{ii}(s),$$

we have for $f_{ii}<1$,

$$\lim_{s\uparrow 1} P_{ii}(s) = \frac{1}{1-f_{ii}} < \infty.$$

Hence, since $p_{ii}^{(n)}\geqq 0$, $n=0, 1, \ldots$, it follows (cf. app. 1) that

$$\sum_{n=0}^{\infty} p_{ii}^{(n)} = \frac{1}{1-f_{ii}} < \infty.$$

Conversely, if $\sum_{n=0}^{\infty} p_{ii}^{(n)} < \infty$, then $\lim_{s\uparrow 1} (1-F_{ii}(s))>0$ (cf. app. 1) so that $f_{ii}<1$. The proof is complete.

THEOREM 2.2. *For an aperiodic, recurrent state E_i,*

$$\lim_{n\to\infty} p_{ii}^{(n)} = \begin{cases} \dfrac{1}{\mu_i} & \text{if } \mu_i<\infty, \\ 0 & \text{if } \mu_i=\infty. \end{cases}$$

For a periodic, recurrent state E_i with period λ_i,

$$\lim_{n\to\infty} p_{ii}^{(n\lambda_i)} = \begin{cases} \dfrac{\lambda_i}{\mu_i} & \text{if } \mu_i < \infty, \\ 0 & \text{if } \mu_i = \infty. \end{cases}$$

Theorem 2.2 is in fact the main theorem of discrete renewal theory; its proof will be omitted here and the reader is referred for it to FELLER [1957] (cf. also chapter 6.4 where renewal theory will be discussed). The main point of the proof of the theorem is to show that $p_{ii}^{(n)}$ has a limit for $n\to\infty$ in the aperiodic case. Whenever $F_{ii}(s)$ is a rational function of s then μ_i is always finite and the proof can be easily deduced from (2.10) by using a partial fraction expansion for $\{1-F_{ii}(s)\}^{-1}$ and noting that the only zero of $1-F_{ii}(s)$ in the region $|s|\leqq 1$ is $s=1$. A general proof can also be given by applying a Tauberian theorem. Once the first part of the theorem has been proved the second part follows easily from it by applying it to the Markov chain $\{x_{n\lambda_i},\ n=0, 1, \ldots\}$, in the latter chain E_i is then an aperiodic state.

Until now we have only considered properties related to the return of a state E_i when starting in E_i. In the following we shall pay attention to questions which are related to the possibility of passage from one state to another.

DEFINITION. A state E_i *leads to* state E_j if a positive integer n exists such that $p_{ij}^{(n)}>0$. If E_i leads to E_j and vice versa then E_i and E_j are called *communicating* states.

Suppose E_i is a recurrent state and that it leads to E_j, so that $p_{ij}^{(n)}>0$ for at least one n. Let N be the smallest integer of those n for which $p_{ij}^{(n)}>0$, i.e. a passage from E_i to E_j takes at least N steps. Put

$$\alpha \overset{\text{def}}{=} p_{ij}^{(N)}.$$

We prove that E_j leads to E_i. Suppose this is not true, i.e. $p_{ji}^{(n)}=0$ for all $n=1, 2, \ldots$. Then the probability of no return to E_i when starting in E_i is at least α, since a passage from E_i to E_j implies by assumption no return. Hence the probability of a return to E_i when starting in E_i is at most $1-\alpha$, so that $f_{ii}\leqq 1-\alpha<1$. However, this is a contradiction since E_i is recurrent and hence $f_{ii}=1$. Consequently, $p_{ji}^{(m)}>0$ for at least one integer m.

Let E_i be recurrent and leading to E_j; by M we denote the smallest integer of all those m for which $p_{ji}^{(m)}>0$. Put

$$\beta \overset{\text{def}}{=} p_{ji}^{(M)}.$$

From (2.4) it follows that

$$P^{(N+n+M)} = P^{(N)}\, P^{(n)}\, P^{(M)},$$

and hence

$$p_{ii}^{(N+n+M)} \geqq p_{ij}^{(N)}\, p_{jj}^{(n)}\, p_{ji}^{(M)} = \alpha\beta p_{jj}^{(n)},$$

$$p_{jj}^{(M+n+N)} \geqq p_{ji}^{(M)}\, p_{ii}^{(n)}\, p_{ij}^{(N)} = \alpha\beta p_{ii}^{(n)}.$$

Since $f_{ii}=1$ implies that $\sum_{n=\infty}^{\infty} p_{ii}^{(n)}$ diverges (cf. theorem 2.1) it follows from the second inequality that $\sum_{n=0}^{\infty} p_{jj}^{(n)}$ diverges and hence by the same theorem that $f_{ij}=1$, i.e. E_j is a recurrent state. If E_i is periodic with period λ_i then necessarily $N+M=0 \bmod \lambda_i$, so that $p_{jj}^{(n)}=0$ for $n\neq 0 \bmod \lambda_i$ because of the first inequality above, whereas the second inequality shows that $p_{jj}^{(N+n+M)}>0$ for $n=0 \bmod \lambda_i$. Hence E_j has the same period as E_i. Next, it is seen that if E_i is null recurrent so that $\lim_{n\to\infty} p_{ii}^{(n\lambda_i)}=0$ for a $\lambda_i \geqq 1$, then the inequalities above imply that for the same λ_i also $\lim_{n\to\infty} p_{jj}^{(n\lambda_i)}=0$, i.e. E_j is null recurrent. Since the converse statement is also true it follows that if E_i is positive recurrent then E_j is also positive recurrent. Hence the following theorem has been proved.

THEOREM 2.3. *If in a Markov chain with stationary transition probabilities a recurrent state E_i leads to another state E_j then these states are communicating and of the same type; i.e. both are positive recurrent or null recurrent and in case of periodicity they have the same period.*

Next, we consider classes of states of a Markov chain.

DEFINITION. A class $\mathscr{E}$ of states of a Markov chain is said to be *closed* if $p_{ij}^{(n)}=0$ for all $n=1, 2, \ldots$, for every $E_i \in \mathscr{E}$ and every $E_j \notin \mathscr{E}$.
A closed class is *minimal* if it does not contain a proper subclass which is closed.

Since by definition from a state of a closed class no transition to a state outside this class is possible, it follows that if $\mathscr{E}$ is a closed class then $\sum_{E_j \in \mathscr{E}} p_{ij}^{(n)}=1$ for every $E_i \in \mathscr{E}$, and all $n=1, 2, \ldots$. Evidently, the converse statement is also true. A closed class may consist of only one state E_i and it is called an *absorbing* state if another state leads to it. In that case E_i is aperiodic and positive recurrent since necessarily $p_{ii}^{(n)}=1$ for all $n=1, 2, \ldots$.
For a closed class containing more than one state we have

THEOREM 2.4. *If $\mathscr{E}$ is a minimal class then every pair of its states is communicating and all its states are of the same type.*

Proof. Let E_i and E_j belong to $\mathscr{E}$ and suppose E_j cannot be reached from E_i. The class $\mathscr{E}(E_i)$ of those states which can be reached from E_i is contained in $\mathscr{E}$ and is closed. By hypothesis $E_j \notin \mathscr{E}(E_i)$ so $\mathscr{E}(E_i)$ is a closed proper subclass of $\mathscr{E}$, but since $\mathscr{E}$ is minimal we have a contradiction. Therefore, any state of $\mathscr{E}$ leads to any other state of $\mathscr{E}$, and hence every two states of $\mathscr{E}$ are communicating. If a state of $\mathscr{E}$ is recurrent then on behalf of theorem 2.3 they are all of the same type. Consequently, if a state of $\mathscr{E}$ is transient then all other states must be transient; moreover for this case it is easily seen from the inequalities above that they are all aperiodic or all periodic with the same period. The proof is complete.

From theorem 2.4 the following corollary is easily proved.

COROLLARY 2.4. *A recurrent state is always element of a minimal class. A state not belonging to a minimal class is transient.*

DEFINITION. A Markov chain with stationary transition probabilities is called *irreducible* if its state space S is a minimal class. An irreducible Markov chain is transient or null recurrent or positive recurrent if all its states are transient, null recurrent or positive recurrent, respectively; it is ergodic if it is positive recurrent and aperiodic.

It should be noted that if the Markov chain $\{x_n, n=0, 1, \ldots\}$ is irreducible and periodic with period λ, then the Markov chain $\{x_{n\lambda}, n=0, 1, \ldots\}$ is not irreducible.

I.2.3. Ergodic properties of aperiodic irreducible Markov chains

In this section we consider an aperiodic and irreducible Markov chain with stationary transition probabilities. For this type of Markov chains we formulate and prove a theorem which is of great importance for applications.

THEOREM 2.5. *For an irreducible and aperiodic Markov chain with stationary transition probabilities:*

(*i*) *if the states are positive recurrent then*

$$\lim_{n\to\infty} p_{ij}^{(n)} = u_j > 0 \qquad \textit{for all } i, j \in S, \tag{2.11}$$

where

$$u_j \overset{\text{def}}{=} \mu_j^{-1},$$

$\{u_j,\ j \in S\}$ is a probability distribution, uniquely determined by the set of equations

$$z_j = \sum_{i \in S} z_i p_{ij}, \qquad j \in S, \tag{2.12}$$

it is the only stationary distribution of the Markov chain, and

$$\lim_{n \to \infty} a_j^{(n)} = u_j \qquad \textit{for all } j \in S;$$

(*ii*) *if the states are null recurrent or transient then*

$$\lim_{n \to \infty} p_{ij}^{(n)} = 0 \qquad \textit{for all } i, j \in S,$$

$$\lim_{n \to \infty} a_j^{(n)} = 0 \qquad \textit{for all } j \in S,$$

and no stationary distribution exists.

Proof. From theorem 2.2 the relation (2.11) follows immediately for $i=j$. Now let $i \neq j$. The probability of reaching E_j when starting in E_i is given by $\sum_{n=1}^{\infty} f_{ij}^{(n)}$ and it should be equal to one, otherwise there is a positive probability to pass from E_j to E_i and never to return, which is impossible since E_j is recurrent and E_j leads to E_i because of theorem 2.4. Hence for every $\varepsilon > 0$ an integer N exists such that $f_{ij}^{(1)} + f_{ij}^{(2)} + \ldots + f_{ij}^{(N)} > 1 - \varepsilon$. Applying theorem 2.2 it follows that in (2.8) for m sufficiently large the sum of the last N terms will differ arbitrarily little from $u_j\{f_{ij}^{(1)} + \ldots + f_{ij}^{(N)}\}$ and hence from u_j. The sum of the first $m-N$ terms in (2.8) is majorized by $f_{ij}^{(m)} + \ldots + f_{ij}^{(N+1)}$, and hence by $\sum_{n=N+1}^{\infty} f_{ij}^{(n)} < \varepsilon$. This proves (2.11).

From (2.11) and

$$p_{ij}^{(m+1)} = \sum_{h \in S} p_{ih}^{(m)} p_{hj}, \qquad m = 1, 2, \ldots,$$

it follows, since all terms are non-negative, that

$$u_j \geqq \sum_{k \in S} u_k p_{kj}.$$

If the sign "$>$" would apply then

$$\sum_{j \in S} u_j > \sum_{j \in S} \sum_{k \in S} u_k p_{kj}$$

$$= \sum_{k \in S} u_k \sum_{j \in S} p_{kj} = \sum_{k \in S} u_k,$$

which is a contradiction, hence we have

$$u_j = \sum_{k \in S} u_k p_{kj} \qquad \text{for all } j \in S,$$

so that $\{u_j, j \in S\}$ is a solution of (2.12). If $\{v_j, j \in S\}$ with $\sum_{j \in S} |v_j| < \infty$ is a solution of (2.12) then

$$v_j = \sum_{i \in S} v_i p_{ij} = \sum_{i \in S} \sum_{k \in S} v_k p_{ki} p_{ij} = \sum_{k \in S} v_k p_{kj}^{(2)},$$

and generally for all $n = 1, 2, \ldots$,

$$v_j = \sum_{i \in S} v_i p_{ij}^{(n)}, \qquad j \in S.$$

From (2.11) and the absolute convergence of $\sum_{i \in S} v_i$ it follows by letting $n \to \infty$ that

$$v_j = u_j \sum_{i \in S} v_i, \qquad j \in S,$$

i.e. the ratio of v_j and u_j is independent of j. Hence, if $\{v_j, j \in S\}$ is a probability distribution, i.e. $v_j \geqq 0$, $\sum_{j \in S} v_j = 1$, then $u_j = v_j$ for all $j \in S$. Consequently $\{u_j, j \in S\}$ is a probability distribution, and as such it is uniquely determined by (2.12). It follows from (2.6) that it is a stationary distribution, and obviously it is the only one.

Finally, since from (2.5),

$$a_j^{(n)} = \sum_{i \in S} a_i^{(0)} p_{ij}^{(n)}, \tag{2.13}$$

it follows easily from (2.11) that $a_j^{(n)}$ has a limit for $n \to \infty$ and

$$\lim_{n \to \infty} a_j^{(n)} = u_j, \qquad j \in S.$$

The first part of the theorem has now been proved. To prove the second statement, it is noted that for an aperiodic transient or null recurrent state $p_{jj}^{(n)} \to 0$ for $n \to \infty$ (cf. theorems 2.1 and 2.2). Since $\sum_{n=1}^{\infty} f_{ij}^{(n)} \leqq 1$ it follows from (2.8) that

$$\lim_{n \to \infty} p_{ij}^{(n)} = 0 \qquad \text{for all } i, j \in S.$$

From this relation and (2.13) it is further seen that $a_j^{(n)}$ tends to zero for $n \to \infty$, $j \in S$. That no stationary distribution exists is now easily seen from (2.7) for $m \to \infty$. The proof is complete.

It should be noted that if the state space S is a minimal class and its states are null recurrent or transient then S cannot be a finite set of states.

This follows easily from the second part of theorem 2.5 and (cf. (2.3)) from

$$\sum_{j \in S} p_{ij}^{(n)} = 1, \qquad n = 1, 2, \ldots, \quad i \in S.$$

The case in which S is a minimal class consisting of aperiodic positive recurrent states is important for applications. For this situation the above theorem shows that, whatever the initial distribution of the process, the absolute distribution of $\boldsymbol{x}_n$ tends always to the same limit as $n \to \infty$. In physics and telephone traffic theory it is then often said that such a system tends, for $n \to \infty$, to the situation of statistical equilibrium, meaning that for large n the influence of the initial situation will disappear and that the system reaches a situation for which the absolute distribution of $\boldsymbol{x}_n$ is time independent, i.e. stationary.

I.2.4. Foster's criteria

Various methods used in the investigation of queueing situations are based on the theory of Markov chains. When applying these methods it is important to investigate the conditions which guarantee that a state of statistical equilibrium exists. Therefore, we need criteria to decide whether a state of a Markov chain is positive recurrent, null recurrent or transient. The criteria given in the preceding section are often rather difficult to apply. For the special type of Markov chains occurring in queueing theory Foster has given useful criteria and we now discuss these criteria.

The Markov chain $\{\boldsymbol{x}_n, n = 0, 1, 2, \ldots\}$ is assumed to be an aperiodic chain with stationary transition probabilities. It is further assumed that its state space is the union of a minimal class C and a class T of transient states. Hence a state belonging to T can never be reached from a state of C. When starting in a state belonging to T it is possible that at some time the system enters C, i.e. reaches a state of C for the first time; on the other hand it may be that a state of C is never reached, i.e. the system passes always through states of T. We shall investigate first the latter case.

Let $y_j^{(n)}$ denote the conditional probability that when starting in E_j, $j \in T$, the system is at the nth step in a state belonging to T. It is now easily seen that for all $j \in T$,

$$y_j^{(1)} = \sum_{i \in T} p_{ji}, \tag{2.14}$$

$$y_j^{(n+1)} = \sum_{i \in T} p_{ji} y_i^{(n)}, \qquad n = 1, 2, \ldots .$$

These relations show that, once the transition matrix P is known, the $y_j^{(n)}$, $n=1, 2, \dots; j\in T$, can be calculated recursively. It is easily proved that $y_j^{(n)}$ has a limit for $n\to\infty$. To show this, it is first noted that

$$0 \leqq y_j^{(1)} \leqq 1 \quad \text{and} \quad y_j^{(2)} \leqq \sum_{i\in T} p_{ji} = y_j^{(1)}, \qquad \text{for all } j\in T.$$

If for all $i\in T$ and some $m\geqq 2$,

$$y_i^{(m)} \leqq y_i^{(m-1)},$$

then the above relations show that

$$y_j^{(m+1)} \leqq \sum_{i\in T} p_{ji} y_i^{(m-1)} = y_j^{(m)} \qquad \text{for all } j\in T. \tag{2.15}$$

Hence, it follows by induction that for every $j\in T$ the sequence $y_j^{(n)}$, $n=1, 2, \dots$, is non-increasing; so that since the $y_j^{(n)}$ are bounded $\lim y_j^{(n)}$ for $n\to\infty$ exists. We write

$$y_j \overset{\text{def}}{=} \lim_{n\to\infty} y_j^{(n)}, \qquad j\in T.$$

Note that y_j is the conditional probability that when starting in E_j, $j\in T$ the system always passes through states of T; obviously y_j is also the probability of never entering class C when starting in E_j, $j\in T$.

From (2.14) and (2.15) we have for $j\in T$,

$$y_j \geqq \sum_{i\in T} p_{ji} y_i \quad \text{and} \quad y_j \leqq \sum_{i\in T} p_{ji} y_i,$$

hence $\{y_j,\ j\in T\}$ is a solution bounded by one of

$$z_j = \sum_{i\in T} p_{ji} z_i, \qquad j\in T. \tag{2.16}$$

The system of equations (2.16) is homogeneous and may have more than one solution. To determine which of all possible solutions of (2.16) represents the limit of $y_j^{(n)}$, $j\in T$ for $n\to\infty$ we need the following theorem.

THEOREM 2.6. *(i) If and only if the system* (2.16) *has no bounded solution except the null solution, then* $y_j=0$, $j\in T$. *(ii) Any non-null solution* $\{z'_j,\ j\in T\}$ *of* (2.14) *and bounded by one satisfies*

$$|z'_j| \leqq y_j \qquad \textit{for all } j\in T,$$

i.e. $\{y_j,\ j\in T\}$ *is the maximal solution of all solutions of* (2.16) *which are bounded by one.*

Proof. Note that any bounded solution of (2.16) can be transformed into a solution bounded by one. If no bounded solution exists except the null solution then y_j must be zero for all $j \in T$, since $\{y_j, j \in T\}$ satisfies (2.16). To prove the converse suppose the second part of the theorem is true, then evidently if $y_j = 0$ for all $j \in T$, no bounded non-null solution can exist. It remains to prove the second statement. Suppose, therefore, that $\{z'_j, j \in T\}$ is a solution of (2.16) bounded by one with at least one of the z'_j different from zero. Then it follows from (2.16) that

$$|z'_j| \leqq \sum_{i \in T} p_{ji} |z'_i| \leqq \sum_{i \in T} p_{ji} = y_j^{(1)}, \qquad \text{all } j \in T.$$

Hence

$$|z'_j| \leqq \sum_{i \in T} p_{ji} y_i^{(1)} = y_j^{(2)}, \qquad \text{all } j \in T;$$

and by induction we obtain

$$|z'_j| \leqq y_j^{(n)} \qquad \text{for } n = 1, 2, \ldots, \text{ and all } j \in T.$$

Consequently,

$$|z'_j| \leqq y_j \qquad \text{for all } j \in T.$$

This proves the theorem.

So far we considered only the situation in which the chain when starting in a state of T passes through states belonging to T. Next, we shall consider some aspects of entrance into C when starting in a state of T. By η_j we shall denote the entrance time from state E_j, $j \in T$ into class C; i.e. η_j is the time at which for the first time a state belonging to C is reached when starting in E_j, $j \in T$, so that for $n = 1, 2, \ldots$, and $j \in T$,

$$\Pr\{\eta_j = n\} = \Pr\{\boldsymbol{x}_{n+m} \in C, \ \boldsymbol{x}_{m+k} \in T, \ k = 1, \ldots, n-1 \mid \boldsymbol{x}_m = j\},$$

where the right-hand side is independent of m since the transition probabilities are stationary.

Putting

$$x_j^{(n)} \overset{\text{def}}{=} \Pr\{\eta_j = n\}, \qquad n = 1, 2, \ldots; \ j \in T,$$

then

$$x_j^{(1)} = \sum_{i \in C} p_{ji}, \qquad j \in T, \tag{2.17}$$

$$x_j^{(n+1)} = \sum_{i \in T} p_{ji} x_i^{(n)}, \qquad n = 1, 2, \ldots; \ j \in T.$$

This set of equations shows how the $x_j^{(n)}$, $j \in T$, $n = 1, 2, \ldots$, can be calculated recursively. Evidently

$$x_j \overset{\text{def}}{=} \Pr\{\eta_j < \infty\} = \sum_{n=1}^{\infty} x_j^{(n)}, \qquad j \in T,$$

represents the probability of ultimately reaching a state of class C when starting in E_j, $j \in T$; x_j is called the *absorbtion* or *entrance* probability from E_j into C. It follows readily that $\{x_j, j \in T\}$ is a non-negative and bounded solution of

$$z_j = \sum_{i \in T} p_{ji} z_i + \sum_{i \in C} p_{ji}, \qquad j \in T. \tag{2.18}$$

THEOREM 2.7. (*i*) *If and only if* $y_j = 0$ *for all* $j \in T$ *then* $\{x_j, j \in T\}$ *is the only bounded solution of* (2.16). (*ii*) *Any non-negative solution* $\{z_j'', j \in T\}$ *of* (2.16) *satisfies*

$$x_j \leqq z_j'' \qquad \textit{for all } j \in T,$$

i.e. $\{x_j, j \in T\}$ *is the minimal solution of all non-negative solutions of* (2.18).

Proof. Since the homogeneous equation corresponding to (2.18) is given by (2.16) the first statement follows immediately from theorem 2.6. To prove the second statement let $\{z_j'', j \in T\}$ be a non-negative solution of (2.18), so that

$$z_j'' = \sum_{i \in T} p_{ji} z_i'' + \sum_{i \in C} p_{ji}.$$

Hence from (2.17),

$$z_j'' \geqq x_j^{(1)}, \qquad j \in T;$$

therefore, again by (2.17),

$$z_j'' \geqq \sum_{i \in T} p_{ji} x_i^{(1)} + x_j^{(1)} = x_j^{(2)} + x_j^{(1)}.$$

Proceeding by induction it is seen that

$$z_j'' \geqq x_j^{(n)} + x_j^{(n-1)} + \ldots + x_j^{(1)}, \qquad n = 1, 2, \ldots; \quad j \in T.$$

Hence for $n \to \infty$,

$$z_j'' \geqq x_j \qquad \text{for all } j \in T.$$

This proves the theorem.

In deriving the above results no use was made of the fact that the state space is here assumed to be the union of C and T; so that these results are independent of this assumption. Evidently, in this particular case we have

$$x_j^{(n)} = y_j^{(n-1)} - y_j^{(n)}, \qquad n = 1, 2, \ldots; \quad j \in T,$$

if we put

$$y_j^{(0)} \overset{\text{def}}{=} 1;$$

and hence

$$x_j = 1 - y_j, \qquad j \in T.$$

Suppose, for the moment, that $y_j = 0$ for all $j \in T$, then when starting in a state of T the system will ultimately reach a state belonging to C. Since $x_j = 1$, $j \in T$, it follows that for every $j \in T$ the sequence $x_j^{(n)}$, $n = 1, 2, \ldots$, is a proper probability distribution. The mean ν_j of this distribution

$$\nu_j = \sum_{n=1}^{\infty} n x_j^{(n)}, \qquad j \in T,$$

is the *mean entrance time* from E_j, $j \in T$ into C. From (2.17) it is now easily found that $\{\nu_j, j \in T\}$ satisfies

$$\nu_j = \sum_{i \in T} p_{ji} \nu_i + 1, \qquad j \in T. \tag{2.19}$$

We shall now discuss Foster's criteria. These criteria relate to a Markov chain with stationary transition probabilities, with state space $E_0, E_1, E_2, \ldots$, and with all states communicating and aperiodic. The assumption of aperiodicity is not essential and the criteria can be easily extended to periodic chains (cf. KANTERS [1965]).

Foster's criteria are (cf. FOSTER [1953])

(*i*) The chain is ergodic if

$$z_j = \sum_{i=0}^{\infty} z_i p_{ij}, \qquad j = 0, 1, \ldots,$$

possesses a non-null solution such that $\sum_{i=0}^{\infty} |z_j| < \infty$; and only if this property is possessed by any non-negative solution of the inequalities

$$z_j \geqq \sum_{i=0}^{\infty} z_i p_{ij}, \qquad j = 0, 1, \ldots .$$

(*ii*) The chain is ergodic if an $\varepsilon > 0$ and an integer $i_0 > 0$ exist such that the inequalities

$$z_i - \varepsilon \geqq \sum_{j=0}^{\infty} p_{ij} z_j, \qquad i = i_0 + 1, i_0 + 2, \ldots,$$

$$\lambda_i \overset{\text{def}}{=} \sum_{j=0}^{\infty} p_{ij} z_j < \infty, \qquad i = 0, 1, \ldots, i_0,$$

have a non-negative solution.

(*iii*) If the chain is ergodic then the finite mean first passage times ν_j from E_j to E_0 satisfy

$$\nu_i = \sum_{j=1}^{\infty} p_{ij}\nu_j + 1, \qquad i=1, 2, \ldots,$$

$$\sum_{j=1}^{\infty} p_{0j}\nu_j < \infty.$$

(*iv*) The chain is transient if and only if

$$z_i = \sum_{j=0}^{\infty} p_{ij}z_j, \qquad i=1, 2, \ldots,$$

has a bounded non-constant solution.

(*v*) The chain is recurrent if the set of inequalities

$$z_i \geqq \sum_{j=0}^{\infty} p_{ij}z_j, \qquad i=1, 2, \ldots,$$

has a solution such that $z_i \to \infty$ for $i \to \infty$.

(*vi*) The chain is transient if and only if

$$z_i \geqq \sum_{j=0}^{\infty} p_{ij}z_j, \qquad i=1, 2, \ldots,$$

has a bounded solution such that $z_i < z_0$ for some i.

Below we shall give the proofs of these criteria.

Proof of (*i*). Since the chain is aperiodic and irreducible it follows from theorem 2.5 that

$$u_j = \lim_{n \to \infty} p_{ij}^{(n)}$$

exists for all j and that for all j either $u_j > 0$ or $u_j = 0$. If z'_j, $j=0, 1, \ldots$, is a non-null solution with $\sum_{j=0}^{\infty} |z'_j| < \infty$ then

$$z'_j = \sum_{i=0}^{\infty} z'_i p_{ij}, \qquad j=0, 1, \ldots,$$

implies for all $n=1, 2, \ldots$,

$$z'_j = \sum_{i=0}^{\infty} z'_i p_{ij}^{(n)}, \qquad j=0, 1, \ldots,$$

and hence for $n \to \infty$,

$$z'_j = \sum_{i=0}^{\infty} z'_i u_j, \qquad j=0, 1, \ldots;$$

therefore $u_j > 0$, since z_j, $j=0, 1, \ldots$, is a non-null solution. Hence all states aree rgodic. Conversely, suppose $u_j > 0$ for all $j=0, 1, \ldots$. If z_j', $j=0, 1, \ldots$, is a non-negative solution of

$$z_j \geqq \sum_{i=0}^{\infty} z_i p_{ij}, \qquad j=0, 1, \ldots,$$

then the relations above also apply with the inequality sign, i.e.

$$z_j' \geqq \sum_{i=0}^{\infty} z_j' p_{ij}^{(n)}, \qquad n=1, 2, \ldots, \quad j=0, 1, \ldots,$$

and hence for $n \to \infty$,

$$z_j' \geqq \sum_{i=0}^{\infty} z_i' u_j, \qquad j=0, 1, \ldots,$$

consequently,

$$\sum_{i=0}^{\infty} z_i' < \infty.$$

The proof of (*i*) is now complete.

Proof of (*ii*). Let $z_j^{(1)}$, $j=0, 1, \ldots$, denote a non-negative solution satisfying the conditions of the second criterion. Define

$$z_i^{(n+1)} \stackrel{\text{def}}{=} \sum_{j=0}^{\infty} p_{ij}^{(n)} z_j^{(1)}, \qquad i=0, 1, \ldots .$$

Then

$$\begin{aligned}
z_i^{(n+2)} &= \sum_{j=0}^{\infty} p_{ij}^{(n+1)} z_j^{(1)} = \sum_{j=0}^{\infty} \sum_{k=0}^{\infty} p_{ik}^{(n)} p_{kj} z_j^{(1)} \\
&= \sum_{k=0}^{i_0} p_{ik}^{(n)} \sum_{j=0}^{\infty} p_{kj} z_j^{(1)} + \sum_{k=i_0+1}^{\infty} p_{ik}^{(n)} \sum_{j=0}^{\infty} p_{kj} z_j^{(1)} \\
&\leqq \sum_{k=0}^{i_0} p_{ik}^{(n)} \lambda_k + \sum_{k=i_0+1}^{\infty} p_{ik}^{(n)} (z_k^{(1)} - \varepsilon) \\
&= \sum_{k=0}^{i_0} p_{ik}^{(n)} \lambda_k - \varepsilon \{1 - \sum_{k=0}^{i_0} p_{ik}^{(n)}\} + \sum_{k=0}^{\infty} p_{ik}^{(n)} z_k^{(1)} - \sum_{k=0}^{i_0} p_{ik}^{(n)} z_k^{(1)} \\
&\leqq \sum_{k=0}^{i_0} p_{ik}^{(n)} (\lambda_k + \varepsilon) - \varepsilon + z_i^{(n+1)}.
\end{aligned}$$

Hence

$$z_i^{(3)} - z_i^{(2)} \leqq \sum_{k=0}^{i_0} p_{ik}(\lambda_k+\varepsilon) - \varepsilon, \qquad i=0, 1, \ldots,$$

. . . .

$$z_i^{(n+2)} - z_i^{(n+1)} \leqq \sum_{k=0}^{i_0} p_{ik}^{(n)}(\lambda_k+\varepsilon) - \varepsilon,$$

so that, since $z_i^{(1)} < \infty$ implies $z_i^{(n)} < \infty$, $i=0, 1, \ldots,$

$$0 \leqq \frac{1}{n} z_i^{(n+2)} \leqq \sum_{k=0}^{i_0} \frac{\lambda_k+\varepsilon}{n} \sum_{m=1}^{n} p_{ik}^{(m)} - \varepsilon + \frac{1}{n} z_i^{(2)}.$$

Hence for $n \to \infty$,

$$\liminf_{n\to\infty} \sum_{k=0}^{i_0} \frac{\lambda_k+\varepsilon}{n} \sum_{m=1}^{n} p_{ik}^{(m)} \geqq \varepsilon, \qquad i=0, 1, \ldots .$$

Since all states are aperiodic $p_{ik}^{(n)}$, $n=1, 2, \ldots$, has a limit for $n \to \infty$, and therefore the Césaro 1-limit of this sequence exists and is equal to it;

$$u_k = \lim_{n\to\infty} p_{ik}^{(n)} = \lim_{n\to\infty} \frac{1}{n} \sum_{m=1}^{n} p_{ik}^{(m)}.$$

Therefore,

$$\sum_{k=0}^{i_0} (\lambda_k+\varepsilon)\, u_k \geqq \varepsilon > 0,$$

so that at least one of the u_k, $k=0, 1, \ldots, i_0$, is positive since $\lambda_k \geqq 0$. Consequently, one state is positive recurrent and hence all states are positive recurrent since they are all communicating. This proves the second criterion.

Proof of (iii). Consider the modified Markov chain with state space E_0, E_1, ..., and with stationary transition probabilities $\bar{p}_{ij}$ defined by

$$\bar{p}_{ij} \stackrel{\text{def}}{=} p_{ij} \qquad \text{for} \quad i=1, 2, \ldots; \quad j=0, 1, 2, \ldots,$$

$$\bar{p}_{0j} \stackrel{\text{def}}{=} \begin{cases} 0 & \text{for} \quad j=1, 2, \ldots, \\ 1 & \text{for} \quad j=0. \end{cases}$$

In this modified chain E_0 is an absorbing state and it is easily verified that E_0 is a positive recurrent state for this chain. For this chain the states E_1, E_2, ..., are now all transient states, since in the original chain all states are positive recurrent, so that in this chain E_0 is reached with probability one from any E_j, $j=1, 2, \ldots$. This remains true for the modified chain as it is seen from the definition of its one-step transition matrix $(\bar{p}_{ij})$. Defining the classes C and T by

$$C \stackrel{\text{def}}{=} \{E_0\}, \qquad T \stackrel{\text{def}}{=} \{E_1, E_2, \ldots\},$$

it is obvious that the entrance probability x_j from a state E_j of T into C is equal to one. Moreover, for the modified chain the mean entrance time ν_j from $E_j \in T$ into C is identical with the mean first passage time from E_j to E_0 in the original chain. In the original chain all states are ergodic so the mean first passage times are all finite. Since for the original chain $f_{j0}^{(n)}$ is the probability of a first passage from E_j to E_0 in exactly n steps (cf. section 2.2) it follows that

$$\nu_j = \sum_{n=1}^{\infty} n f_{j0}^{(n)}.$$

From the relations stated between the original chain and the modified chain and from (2.19) it follows that the ν_j, $j=1, 2, \ldots$, satisfy

$$\sum_{j=1}^{\infty} p_{ij}\nu_j = \nu_i - 1, \qquad i=1, 2, \ldots .$$

For the original chain

$$1 + \sum_{j=1}^{\infty} p_{0j}\nu_j$$

is the average return time μ_0 of state E_0 and μ_0 is evidently finite since E_0 is positive recurrent for the original chain. The proof is complete.

Proof of (*iv*). For the modified chain introduced in the proof of (*iii*), T is a class of transient states. Suppose that z_j', $j=0, 1, \ldots$, is a non-constant solution bounded by one of

$$z_i = \sum_{j=0}^{\infty} p_{ij} z_j, \qquad i=1, 2, \ldots,$$

then $z_0'' \stackrel{\text{def}}{=} 0$, $z_j'' \stackrel{\text{def}}{=} z_j' - z_0'$, $j=1, 2, \ldots$, is a bounded non-null solution of

$$z_i = \sum_{j=1}^{\infty} p_{ij} z_j, \qquad i=1, 2, \ldots .$$

From theorem 2.6 it now follows that a state $E_j \in T$ exists such that the probability of always remaining in T when starting in E_j is positive for the modified chain and hence also for the original chain. Hence for the latter chain of which every pair of states is communicating

$$f_{j0} \stackrel{\text{def}}{=} \sum_{n=1}^{\infty} f_{j0}^{(n)} < 1, \qquad E_j \in T,$$

and this can be only true if E_0 is transient in the original chain (cf. theorem 2.3). Hence the original chain is transient.

To prove the converse, write

$$\bar{\pi}_{ij} \overset{\text{def}}{=} \lim_{n\to\infty} \bar{p}_{ij}^{(n)}, \qquad i,j=0,1,\ldots;$$

the limits exist since the modified chain is also aperiodic. Then $\bar{\pi}_{00}=1$, $\bar{\pi}_{ij}=0$ for all $E_j \in T$. From (2.8) it now follows that $\bar{\pi}_{i0}$, $i \neq 0$, represents the probability of reaching E_0 when starting in $E_i \in T$ for the original chain. If the original chain is transient then $\bar{\pi}_{i0}<1$ for at least one $E_i \in T$. For if $\bar{\pi}_{i0}=1$ for all $E_i \in T$ then E_0 would be recurrent in the original chain, and hence this chain would be recurrent. Since

$$\bar{\pi}_{i0} = \sum_{j=0}^{\infty} \bar{p}_{ij}\bar{\pi}_{j0}, \qquad i=0,1,\ldots,$$

it follows that $z_0'=\bar{\pi}_{00}=1$, $z_i'=\bar{\pi}_{i0}$, $i=1,2,\ldots$, is a bounded non-constant solution of

$$z_i = \sum_{j=0}^{\infty} p_{ij}z_j, \qquad i=1,2,\ldots.$$

The proof is complete.

Proof of (*v*). Suppose z_i', $i=0,1,\ldots$, satisfies the conditions of criterion (*v*) then we have for the modified chain

$$\sum_{j=0}^{\infty} \bar{p}_{ij}z_j' \leqq z_i', \qquad i=0,1,\ldots,$$

from which it follows that for $n=1,2,\ldots$,

$$\sum_{j=0}^{\infty} \bar{p}_{ij}^{(n)}z_j' \leqq z_i', \qquad i=1,2,\ldots.$$

Since for any constant $\alpha>0$, $z_i'+\alpha$, $i=0,1,\ldots$, also satisfies the conditions of the theorem and $z_i' \to \infty$ for $i \to \infty$ we may assume that all $z_i' \geqq 0$. Hence for any positive integer m,

$$\sum_{j=m+1}^{\infty} p_{ij}^{(n)}z_j' \leqq z_i', \qquad i=1,2,\ldots;\quad n=1,2,\ldots.$$

Putting

$$v_m \overset{\text{def}}{=} \min_{i\geqq m+1} z_i',$$

it follows that

$$v_m \sum_{j=m+1}^{\infty} \bar{p}_{ij}^{(n)} \leqq z_i', \qquad i=1, 2, \ldots,$$

and

$$\sum_{j=0}^{m} \bar{p}_{ij}^{(n)} \geqq 1 - \frac{z_i'}{v_m}, \qquad i=1, 2, \ldots .$$

Since for the modified chain $\bar{\pi}_{ij}=0$ for $j \neq 0$ it follows from the last inequality by letting $n \to \infty$,

$$\bar{\pi}_{i0} \geqq 1 - \frac{z_i'}{v_m}, \qquad i=1, 2, \ldots .$$

Since $v_m \to \infty$ for $m \to \infty$ it follows that for every fixed $i \neq 0$,

$$\bar{\pi}_{i0} \geqq 1, \quad \text{so} \quad \bar{\pi}_{i0}=1, \qquad i=1, 2, \ldots .$$

Hence, in the modified chain E_0 is reached from any $E_i \in T$ with probability one. Consequently, this is also true for the original chain and hence for this chain a return to E_0 when starting in E_0 is certain; so E_0 is recurrent and hence the original chain is recurrent.
The proof is complete.

Proof of (vi). The necessary part is proved as in criterion (*iv*). Let z_i', $i=0, 1, \ldots$, satisfy the conditions of (*vi*), then this is also true for $z_i''=\alpha+\beta z_i'$, $i=0, 1, \ldots$, with constant α and $\beta>0$. Hence, we may and do assume that all $z_i' \geqq 0$ and $z_0'=1$. Again we have

$$\sum_{j=0}^{m} \bar{p}_{ij}^{(n)} z_i' \leqq z_i', \qquad i=1, 2, \ldots; \quad n=1, 2, \ldots .$$

So for $n \to \infty$, and since $\bar{\pi}_{ij}=0$ for $j=1, 2, \ldots$, it follows that

$$\bar{\pi}_{i0} \leqq z_i', \qquad i=1, 2, \ldots .$$

Since $z_i' < z_0' = 1$ for at least one $i=1, 2, \ldots$, it follows $\bar{\pi}_{i0}<1$ for at least one $i=1, 2, \ldots$. Hence, for at least one i the probability that E_0 is reached from E_i in the original chain is less than one. So the original chain is transient. The last criterion is proved.

I.2.5. Taboo probabilities

For a Markov chain $\{x_n, n=0, 1, \ldots\}$ with stationary transition probabilities we define for $i, j, k \in S$, $k \neq j$,

$$p_{k;ij}^{(n)} \overset{\text{def}}{=} \begin{cases} \delta_{ij} & \text{for} \quad n=0, \\ p_{ij} & \text{for} \quad n=1, \\ \Pr\{x_n=j,\ x_m \neq k,\ m=1,\dots,n-1 \mid x_0=i\}, & n=2,3,\dots; \end{cases}$$

$$f_{k;ij}^{(n)} \overset{\text{def}}{=} \begin{cases} 0 & \text{for} \quad n=0, \\ f_{ij} & \text{for} \quad n=1, \\ \Pr\{x_n=j,\ x_{m_1} \neq k,\ x_{m_2} \neq j;\ m_1,m_2=1,\dots,n-1 \mid x_0=i\}, & n=2,3,\dots . \end{cases}$$

Evidently, $p_{k;ij}^{(n)}$ is the probability that the system does not pass through E_k and that it is in E_j at time n when starting in E_i; $f_{k;ij}^{(n)}$ is the probability of a first entrance in E_j from E_i in exactly n steps without having passed through E_k. Probabilities as defined above are called *taboo probabilities* and E_k is here the *taboo state*. These taboo probabilities are an important tool for the investigation of queueing systems. It is easily verified that

$$p_{ij}^{(n)} = p_{k;ij}^{(n)} + \sum_{m=0}^{n-1} f_{ik}^{(n)} p_{kj}^{(n-m)}, \qquad n=1, 2, \dots,$$

$$p_{k;ij}^{(n)} = f_{k;ij}^{(n)} + \sum_{m=0}^{n-1} f_{k;ij}^{(m)} p_{k;jj}^{(n-m)}, \qquad n=1, 2, \dots .$$

From the latter relations and from (2.8) and (2.9) the taboo probabilities can be calculated if the one step transition probabilities are known.

Define for $|s|<1$,

$$P_{ij}(s) \overset{\text{def}}{=} \sum_{n=0}^{\infty} p_{ij}^{(n)} s^n, \qquad F_{ij}(s) \overset{\text{def}}{=} \sum_{n=0}^{\infty} f_{ij}^{(n)} s^n.$$

From the relations above with $i=j \neq k$ we obtain for $|s|<1$,

$$\sum_{n=0}^{\infty} s^n p_{k;ii}^{(n)} = P_{ii}(s) - F_{ik}(s)\, P_{ki}(s),$$

$$\sum_{n=0}^{\infty} s^n f_{k;ii}^{(n)} = \frac{\sum_{n=0}^{\infty} p_{k;ii}^{(n)} s^n - 1}{\sum_{n=0}^{\infty} p_{k;ii}^{(n)} s^n}.$$

Since from (2.8) and (2.9) we have for $|s|<1$,

$$P_{ii}(s) = \frac{1}{1-F_{ii}(s)}, \qquad P_{ki}(s) = \frac{F_{ki}(s)}{1-F_{ii}(s)}, \qquad k \neq i,$$

we obtain

$$\sum_{n=0}^{\infty} s^n p_{k;ii}^{(n)} = \frac{1-F_{ik}(s)\,F_{ki}(s)}{1-F_{ii}(s)}, \qquad |s|<1.$$

Suppose that the state space S is irreducible and that all its states are ergodic. The average entrance times

$$\nu_{ij} \stackrel{\text{def}}{=} \sum_{n=1}^{\infty} n f_{ij}^{(n)}, \qquad i,j \in S,$$

are now all finite. From the relations above (cf. app. 1) it follows easily that in this particular case

$$\sum_{n=0}^{\infty} p_{k;ii}^{(n)} = \frac{\nu_{ik}+\nu_{ki}}{\nu_{ii}},$$

$$\sum_{n=0}^{\infty} f_{k;ii}^{(n)} = 1 - \frac{\nu_{ii}}{\nu_{ik}+\nu_{ki}}.$$

Note that $\sum_{n=0}^{\infty} f_{k;ii}^{(n)}$ is the probability of a return to E_i without passing E_k.

I.3. MARKOV CHAINS WITH A CONTINUOUS TIME PARAMETER

I.3.1. Definition of the process

In this chapter stochastic processes with a continuous time parameter will be considered. The time parameter set T will be the set of non-negative numbers. The state space S will be a denumerable set, finite or infinite.

It is supposed that for the process $\{x_t,\ t\in[0, \infty)\}$ all finite dimensional joint distributions are given, i.e. for every finite integer $m=1, 2, \ldots,$

$$\Pr\{x_{t_1}=i_1,\ x_{t_2}=i_2,\ \ldots,\ x_{t_m}=i_m\}$$

is known for all $i_1, i_2, \ldots, i_m\in S$ and every subset $\{t_1, \ldots, t_m\}$ of $[0, \infty)$.

The definition of a Markov chain with a continuous time parameter is quite similar to that with a discrete time parameter.

DEFINITION. The class of stochastic variables $\{x_t,\ t\in[0, \infty)\}$ is a continuous time parameter Markov chain if for every integer $n\geqq 0$, for $i_k\in S$, $k=0, 1, \ldots,$ and any sequence $t_0, \ldots, t_{n+1}$ with $0\leqq t_0<t_1<\ldots<t_{n+1}$,

$$\Pr\{x_{t_{n+1}}=i_{n+1} \mid x_{t_k}=i_k,\ k=0, 1, \ldots, n\}=\Pr\{x_{t_{n+1}}=i_{n+1} \mid x_{t_n}=i_n\},$$

whenever the left-hand side is defined.

The definition states that if of the process $\{x_t,\ t\in[0, \infty)\}$ the realisation of x_t is known at a time $t=t_n$ then the conditional probability of any future realisation of a x_t depends only on the realisation of x_{t_n} and is independent of $x_{t_0}, \ldots, x_{t_{n-1}}$ for every integer $n\geqq 1$ and every sequence $t_0, t_1, \ldots, t_{n-1}$ with $t_k<t_n$, $k=0, \ldots, n-1$.

As in section 2.1 it follows from the definition above that

$$\Pr\{x_{t_h}=i_h,\ h=n+1, \ldots, m \mid x_{t_k}=i_k,\ k=1, \ldots, n\} \tag{3.1}$$
$$= \Pr\{x_{t_h}=i_h,\ h=n+1, \ldots, m \mid x_{t_n}=i_n\}$$
$$= \prod_{k=1}^{m} \Pr\{x_{t_{n+k}}=i_{n+k} \mid x_{t_{n+k-1}}=i_{n+k-1}\},$$

for all $t_1, \ldots, t_m$ with $0 \leqq t_1 < \ldots < t_m$.
More generally, it may be deduced from the definition (cf. e.g. DOOB [1953]) that for $t_n < t_h$, $h=n+1, \ldots, m$,

$$\Pr\{x_{t_h}=i_h,\ h=n+1, \ldots, m \mid x_s=i_s,\ 0 \leqq s \leqq t_n\}$$
$$=\Pr\{x_{t_h}=i_h,\ h=n+1, \ldots, m \mid x_{t_n}=i_n\}. \tag{3.2}$$

The latter relation shows more clearly that the conditional probability of the future event $\{x_{t_h}=i_h,\ h=n+1, \ldots, m\}$ after time $t=t_n$ given the past $x_s=i_s$, $0 \leqq s < t_n$ and the present $x_{t_n}=i_n$ is independent of the past.
In the same way as (2.2) has been derived it follows

$$\Pr\{x_{t_1+t_2+t_3}=j \mid x_{t_1}=i\}$$
$$= \sum_{h \in S} \Pr\{x_{t_1+t_2+t_3}=j \mid x_{t_1+t_2}=h\} \Pr\{x_{t_1+t_2}=h \mid x_{t_1}=i\}.$$

Putting

$$P_{ij}(t, s) \overset{\text{def}}{=} \Pr\{x_s=j \mid x_t=i\}, \qquad s>t \geqq 0, \qquad i, j \in S,$$

the last relation may be written as

$$P_{ij}(t, s) = \sum_{h \in S} P_{ih}(t, \tau) P_{hj}(\tau, s), \qquad s>\tau>t \geqq 0, \qquad i, j \in S. \tag{3.3}$$

This relation is known as the *Chapman-Kolmogorov relation* and it plays an important role in the theory of Markov chains with a continuous time parameter. Evidently, for $s>t \geqq 0$, $i, j \in S$,

$$P_{ij}(t, s) \geqq 0, \qquad \sum_{j \in S} P_{ij}(t, s) = 1. \tag{3.4}$$

As before (cf. section 2.2) it is said that the Markov chain is in state E_i at time t if the realisation of x_t is i. Further $P_{ij}(t, s)$ with $s>t$ is called the *transition probability* from state E_i at time t to state E_j at time s.
The *transition matrix* with parameters t and s, is defined by

$$P(t, s) \overset{\text{def}}{=} (P_{ij}(t, s)), \qquad 0 \leqq t < s, \quad i, j \in S.$$

Hence in matrix notation the Chapman-Kolmogorov relation reads

$$P(t, s)=P(t, \tau)\, P(\tau, s), \qquad 0 \leqq t < \tau < s.$$

Defining for all $t \in [0, \infty)$,

$$P(t, t) \stackrel{\text{def}}{=} I,$$

where I is the unit matrix, it is seen that the Chapman-Kolmogorov relation holds for $0 \leqq t \leqq \tau \leqq s$.

Putting

$$a_j(t) \stackrel{\text{def}}{=} \Pr\{x_t = j\}, \qquad t \in [0, \infty), \quad j \in S,$$

so that $\{a_j(t), j \in S\}$ is the absolute distribution of x_t, then it follows easily from the definition of $P_{ij}(t, s)$ that

$$a_j(s) = \sum_{i \in S} a_i(t)\, P_{ij}(t, s), \qquad t \leqq s, \quad j \in S. \tag{3.5}$$

From (3.1) and (3.5) it is seen that all finite dimensional joint distributions of the process $\{x_t, t \in [0, \infty)\}$ can be calculated whenever the initial distribution $\{a_i(0), i \in S\}$ and the transition matrix $P(t, s)$ for all s and t with $s > t \geqq 0$ are known.

From the last statement it is seen that the mechanism which controls the development in time of the Markov chain $\{x_t, t \in [0, \infty)\}$ is described by the transition matrix function $P(.,.)$. As with discrete time parameter Markov chains we shall restrict here the discussion to Markov chains for which $P(.,.)$ is independent of the absolute time.

DEFINITION. The Markov chain $\{x_t, t \in [0, \infty)\}$ has *stationary* transition probabilities if

$$\Pr\{x_{t+s} = j \mid x_t = i\}$$

is independent of t for all $i, j \in S$ and all $s \in [0, \infty)$.

In this chapter we shall from now on consider only Markow chains with stationary transition probabilities.

From the definition it follows that $P_{ij}(t, s)$ is for all $i, j \in S$ a function of $s - t$; we define

$$p_{ij}(s) \stackrel{\text{def}}{=} \begin{cases} P_{ij}(t, t+s), & s > 0, \quad t \geqq 0, \\ \delta_{ij}, & s = 0, \end{cases}$$

and

$$P(s) \stackrel{\text{def}}{=} P(t, t+s), \qquad s \geqq 0, \quad t \geqq 0.$$

From (3.4) it follows for all $i, j \in S$, that

$$p_{ij}(s) \geqq 0, \qquad \sum_{j \in S} p_{ij}(s) = 1, \quad s \geqq 0, \tag{3.6}$$

and the Chapman-Kolmogorov relation now reads

$$P(t+s)=P(t)\,P(s), \qquad s\geqq 0, \quad t\geqq 0. \tag{3.7}$$

DEFINITION. A Markov chain with stationary transition probabilities is *stationary* if

$$a_i(t)=\Pr\{x_t=i\}, \qquad i\in S,$$

is independent of t.

Denoting the *stationary distribution* for a stationary Markov chain by $\{a_i,\ i\in S\}$ it follows from (3.5) that

$$a_j = \sum_{i\in S} a_i p_{ij}(t), \qquad j\in S, \quad \text{all } t\geqq 0.$$

I.3.2. The Q-matrix

From the preceding section it is seen that for a Markov chain with stationary transition probabilities all finite dimensional joint distributions can be calculated if the initial distribution $\{a_i(0),\ i\in S\}$ and the transition matrix function $P(t)$, $0\leqq t<\infty$ are known. Now $P(.)$ has the property (3.7) showing that relations exist between the various "values" of $P(.)$, and it is seen from (3.7) that $P(t)$ can be calculated for all $t>\delta$ whenever $P(t)$ is known for all $t\in(0,\delta]$ if $\delta>0$. Hence, by letting $\delta\to 0$ it seems plausible that if $P(.)$ has a "derivative" at $t=0$, then it would be sufficient to know this derivative to calculate $P(t)$ for all $t>0$. Under some rather weak restrictions this is indeed true.

From now on *it will be assumed* that for $t\downarrow 0$ the following limits exist and are given by

$$\lim_{t\downarrow 0} p_{ij}(t)=\delta_{ij} \qquad \text{for all } i,j\in S,$$

or in matrix notation

$$\lim_{t\downarrow 0} P(t) \overset{\text{def}}{=} (\lim_{t\downarrow 0} p_{ij}(t)) = I. \tag{3.8}$$

This assumption has far reaching consequences, and a transition matrix function having the property (3.8) is therefore called a *standard transition matrix function.* From (3.8) and (3.7) it can now be deduced (cf. CHUNG [1960]) that the right-hand derivative of $p_{ij}(t)-\delta_{ij}$ at $t=0$ exists for all

$i, j \in S$ and that it is always finite for $i \neq j$;

$$-q_i \overset{\text{def}}{=} q_{ii} \overset{\text{def}}{=} \lim_{t \downarrow 0} \frac{p_{ii}(t)-1}{t}, \qquad i \in S, \tag{3.9}$$

$$q_{ij} \overset{\text{def}}{=} \lim_{t \downarrow 0} \frac{p_{ij}(t)}{t}, \qquad i \neq j, \quad i, j \in S.$$

Obviously, we have from (3.9) that

$$q_i \geqq \sum_{\substack{j \in S \\ j \neq i}} q_{ij}, \qquad i \in S.$$

By defining

$$Q \overset{\text{def}}{=} (q_{ij}), \qquad i, j \in S,$$

it is seen that

$$Q = \lim_{t \downarrow 0} \frac{P(t)-I}{t} \overset{\text{def}}{=} \left(\lim_{t \downarrow 0} \frac{p_{ij}(t)-\delta_{ij}}{t} \right).$$

The matrix Q is the so called Q-matrix of the Markov chain or the *infinitesimal generator* of the transition matrix function $P(.)$. We now introduce the matrix $R=(r_{ij})$, in the case when all the q_i are finite, by writing

$$r_{ij} \overset{\text{def}}{=} \begin{cases} (1-\delta_{ij}) \dfrac{q_{ij}}{q_i}, & q_i > 0, \quad i, j \in S, \\ \delta_{ij}, & q_i = 0, \quad i, j \in S. \end{cases}$$

The matrix R is called the *jump matrix* of the Markov chain. Its generic element r_{ij} is the conditional probability of a jump from i to j given that a jump from i occurs. From (3.9) it follows that if $q_i < \infty$,

$$\begin{aligned} 1-p_{ii}(t) &= q_i t + o(t) \qquad \text{for } t \downarrow 0, \\ p_{ij}(t) &= q_{ij} t + o(t) \qquad \text{for } t \downarrow 0, \quad i \neq j. \end{aligned} \tag{3.10}$$

Hence for small positive values of Δt we may interpret $q_{ij}\Delta t$ as the elementary conditional probability that whenever the system is at time t in state E_i it will make a transition from E_i to E_j during $t \div t+\Delta t$; similarly, if q_i is finite, $q_i \Delta t$ is the elementary conditional probability that whenever the system is at time t in E_i it will leave this state during $t \div t+\Delta t$.

To elucidate the meaning of q_i further we shall calculate the conditional probability that whenever at time s the system is in state E_i it remains during $s \div s+t$ in E_i, i.e. during $(s, s+t)$ no transition occurs. This probability may be expressed as

$$\Pr\{x_\tau = i,\ s<\tau<s+t \mid x_s = i\}, \qquad s \geqq 0,\quad t>0.$$

For the evaluation of this probability we encounter one of the principal difficulties of the theory of stochastic processes. As has been stated above, all finite dimensional joint distributions can be calculated if the initial distribution and the transition matrix function $P(.)$ are given. However, the probability mentioned above refers to a non-denumerable set of time points and it will be evident that it cannot be found from (or even defined by) the finite dimensional distributions. However, from the assumption that the transition matrix function is standard (cf. (3.8)) the probability mentioned above may be calculated by applying a limit procedure to the finite dimensional distributions.

Let M be any denumerable set which is dense in $[0, \infty)$, then, since the transition matrix function is standard, it can be shown (cf. CHUNG [1960]) that

$$\Pr\{x_\tau = i,\ s<\tau<s+t \mid x_s = i\} = \Pr\{x_\tau = i,\ \tau \in M \cap (s, s+t) \mid x_s = i\}.$$

Since M is denumerable and dense in $[0, \infty)$ it is possible to find for each $n = 1, 2, \ldots,$ a set of points $M_n = \{s_\nu^{(n)},\ \nu = 1, \ldots, n+1\} \subset M \cap (s, s+t)$ with $s_\nu^{(n)} < s_{\nu+1}^{(n)},\ \nu = 1, \ldots, n$ and such that

$$\max_{0 \leqq \nu \leqq n+1} \delta_\nu^{(n)} \to 0 \quad \text{and} \quad M_n \uparrow M \qquad \text{for } n \to \infty,$$

where

$$\delta_0^{(n)} \stackrel{\text{def}}{=} s_1^{(n)} - s, \quad \delta_\nu^{(n)} \stackrel{\text{def}}{=} s_{\nu+1}^{(n)} - s_\nu^{(n)}, \quad \nu = 1, \ldots, n; \quad \delta_{n+1}^{(n)} = s + t - s_{n+1}^{(n)}.$$

Hence, applying a well-known convergence theorem for measures (cf. LOÈVE [1960]) and using (3.10) then, if $q_i < \infty$,

$$\begin{aligned}
\Pr\{x_\tau = i,\ \tau \in M \cap (s, s+t) \mid x_s = i\} &= \lim_{n\to\infty} \Pr\{x_{s_\nu^{(n)}} = i,\ \nu = 1, \ldots, n+1 \mid x_s = i\} \\
&= \lim_{n\to\infty} \prod_{\nu=0}^{n} p_{ii}(\delta_\nu^{(n)}) = \lim_{n\to\infty} \prod_{\nu=0}^{n} \{1 - q_i \delta_\nu^{(n)} + o(\delta_\nu^{(n)})\} \\
&= \lim_{n\to\infty} \exp\{-q_i \sum_{\nu=0}^{n} \delta_\nu^{(n)}\} = e^{-q_i t}.
\end{aligned}$$

Consequently, for $q_i < \infty$,

$$\Pr\{x_\tau = i,\ s<\tau<s+t \mid x_s = i\} = e^{-q_i t}, \qquad t>0. \tag{3.11}$$

From (3.11) it follows that if the system is in state E_i at some moment s

then $1-e^{-q_i t}$, $t \geqq 0$ is the probability that the system will leave this state before time $t+s$. We now introduce the stochastic variable ρ_i as the *exit time* from state E_i; suppose at a time s the system is in state E_i, then

$$\rho_i \stackrel{\text{def}}{=} \inf\{t: t>0, x_{t+s} \neq i\},$$

is the greatest lower bound of all those t for which the system is at time $t+s$ not in E_i. The above result shows that for $q_i < \infty$,

$$\Pr\{\rho_i \geqq t\} = e^{-q_i t}, \qquad t \geqq 0, \tag{3.12}$$

i.e. *if the transition matrix is standard and if* $0 < q_i < \infty$, *then the exit time of* E_i *is negative exponentially distributed with mean* q_i^{-1}.

Suppose again that the system is in state E_i at time s and that q_i and q_j are both non-zero and finite. Define

$$\tau_j \stackrel{\text{def}}{=} \inf\{t: t>0, x_{s+t}=j\}, \qquad \tau'_j \stackrel{\text{def}}{=} \inf\{t: t>\tau_j, x_{s+t} \neq j\},$$

then if τ_j and τ'_j are finite with probability one

$$\lambda_j \stackrel{\text{def}}{=} \tau'_j - \tau_j,$$

denotes the *sojourn time* of state E_j, i.e. λ_j represents the length of the uninterrupted time which the system spent in E_j. It can now be proved (cf. DOOB [1953]), that

$$\Pr\{\lambda_j \geqq t\} = e^{-q_j t}, \qquad t \geqq 0, \quad q_j < \infty, \tag{3.13}$$

and hence

$$E\{\lambda_j\} = \frac{1}{q_j}, \qquad 0 < q_j < \infty.$$

The relation (3.13) also holds if $q_j = \infty$. In this case E_j is called an *instantaneous* state, otherwise E_j is called *stable*. If $q_j = 0$ then E_j is an *absorbing state*, and it is then easily verified that $p_{jj}(t) = 1$ for all $t \geqq 0$. Obviously, a stable, non-absorbing state has a finite and non-zero mean sojourn time. This leads to the conjecture that whenever a transition occurs from a stable state to another stable state then the sample function or realisation of x_t has an ordinary discontinuity at the epoch of transition, i.e. a jump. The sample functions of the Markov chain $\{x_t, t \in [0, \infty)\}$ may have much more complicated discontinuities, and in general sample functions with discontinuities other than simple jumps may occur with non-zero probability. Discontinuities may be caused by the occurrence of instantaneous states; it is also possible that in a finite time interval an infinite number of transitions occur with non-zero probability, even if all states are stable (explosive processes). For instance, let S be the set of all non-negative integers, and suppose for the moment that all the q_j are finite and non-zero and that

$$\sum_{j=0}^{\infty} q_j^{-1} < \infty.$$

Then

$$\sum_{j=0}^{\infty} \sigma^2\{\boldsymbol{\lambda}_j\} = \sum_{j=0}^{\infty} E\{\boldsymbol{\lambda}_j\} < \infty,$$

and this leads to the conjecture that with non-zero probability the system may pass through all or nearly all states in a finite time. Evidently, such realisations of the process are not step-functions, i.e. functions with, at most, a finite number of jumps in any finite interval.

It can be proved (cf. CHUNG [1960]) that if infinitely many q_j are positive then

$$\sum_{j} q_j^{-1} = \infty,$$

the summation being performed over all those $j \in S$ for which $q_j \neq 0$, is a sufficient condition that with probability one all realisations of the Markov chain $\{\boldsymbol{x}_t,\ t \in [0, \infty)\}$ are step-functions.

From now on *it will be assumed* that

$$\alpha \stackrel{\text{def}}{=} \sup_{j \in S} q_j < \infty. \tag{3.14}$$

It will be shown below (cf. corollary 3.1) that (3.14) implies

$$q_i = \sum_{\substack{j \in S \\ j \neq i}} q_{ij}, \qquad i \in S. \tag{3.15}$$

From (3.14) it follows that all states are stable and that with probability one every realisation is a step-function. We shall always take these step-functions to be continuous from the left i.e. if t is a transition point then

$$\boldsymbol{x}_t \stackrel{\text{def}}{=} \lim_{s \uparrow t} \boldsymbol{x}_s.$$

The definition of a Markov chain with a continuous time parameter has been given in section 3.1. Heuristically formulated it implies that for any given time t (the present) any future event after t is independent of any past event before t given the state of the system at t. Here t has to be chosen independent of the evolution of the process $\{\boldsymbol{x}_t,\ t \in [0, \infty)\}$. The question which now arises is whether or not this defining property will be also true for a time point t which depends on the evolution of the process, e.g. if we choose for t the moment of entrance or exit of a given state. To give an example, let ${}_1p_{ij}(t)$ denote the probability that whenever the system is in a non-absorbing state E_i at time s it is at time $s+t$ in E_j and has reached E_j by just one transition from E_i:

$$\begin{aligned}
{}_1p_{ij}(t) &\overset{\text{def}}{=} \Pr\{x_{s+\rho_i+\sigma}=j,\ 0<\sigma\leqq t-\rho_i,\ \rho_i<t \mid x_s=i\} \\
&= \Pr\{\lambda_j\geqq t-\rho_i,\ x_{s+\rho_i+}=j,\ 0\leqq\rho_i<t \mid x_s=i\} \\
&= \int_0^t \Pr\{\lambda_j\geqq t-\sigma,\ x_{s+\sigma+}=j \mid \rho_i=\sigma,\ x_s=i\}\,\mathrm{d}_\sigma \Pr\{\rho_i<\sigma \mid x_s=i\} \\
&= \int_0^t \Pr\{\lambda_j\geqq t-\sigma \mid x_{s+\rho_i+}=j,\ \rho_i=\sigma,\ x_s=i\} \Pr\{x_{s+\rho_i+}=j \mid \rho_i=\sigma,\ x_s=i\} \\
&\quad\cdot \mathrm{d}_\sigma \Pr\{\rho_i<\sigma \mid x_s=i\}.
\end{aligned}$$

Supposing that the Markov property (3.2) also holds if the present is taken to be the moment of exit from E_i it follows from (3.2) and (3.13),

$$\Pr\{\lambda_j\geqq t-\sigma \mid x_{s+\rho_i+}=j,\ \rho_i=\sigma,\ x_s=i\}=\Pr\{\lambda_j\geqq t-\sigma \mid x_{s+\rho_i+}=j\}=\mathrm{e}^{-q_j(t-\sigma)},$$

$$\Pr\{x_{s+\rho_i+}=j \mid \rho_i=\sigma,\ x_s=i\}=\Pr\{x_{s+\rho_i+}=j \mid x_{s+\rho_i}=i\}.$$

The latter probability is the conditional probability that whenever a transition from E_i occurs this transition leads to E_j. Since $0<q_i<\infty$ and all realisations of the process are step-functions with probability one it follows from (3.10), (3.15) and the assumption above that

$$\Pr\{x_{s+\rho_i+}=j \mid x_{s+\rho_i}=i\} = \frac{q_{ij}}{q_i}. \tag{3.16}$$

Hence, using (3.12),

$${}_1p_{ij}(t) = \int_0^t \mathrm{e}^{-q_j(t-\sigma)}\,\frac{q_{ij}}{q_i}\,q_i\,\mathrm{e}^{-q_i\sigma}\,\mathrm{d}\sigma = q_{ij}\int_0^t \mathrm{e}^{-q_i\sigma-q_j(t-\sigma)}\,\mathrm{d}\sigma.$$

The validity of this result may be shown as follows. Consider for $n=4, 5, \dots,$ the probability

$$p_{ij}(t;n) \overset{\text{def}}{=}$$

$$\Pr\left\{\bigcup_{\nu=0}^{n-2}\left(x_{s+\sigma_1}=j,\ x_{s+\sigma_2}=i,\ \frac{\nu+1}{n}t<\sigma_1\leqq t,\ 0\leqq\sigma_2\leqq\frac{\nu}{n}t\right)\middle|\ x_s=i\right\}.$$

It may be proved (cf. the derivation of (3.12)) that

$$ {}_1p_{ij}(t) = \lim_{n\to\infty} p_{ij}(t; n). $$

Applying definition (3.1) it follows that

$$ p_{ij}(t; n) = \sum_{\nu=0}^{n-2} \exp\left(-q_i \frac{\nu}{n} t\right) p_{ij}\left(\frac{t}{n}\right) \exp\left(-q_j \frac{n-\nu-1}{n} t\right). $$

For large values of n and fixed t it follows from (3.10) that

$$ p_{ij}\left(\frac{t}{n}\right) = q_{ij}\frac{t}{n} + \mathrm{o}\left(\frac{t}{n}\right), \qquad i \neq j, $$

and hence letting $n\to\infty$ in the above relation

$$ \lim_{n\to\infty} p_{ij}(t; n) = \int_0^t q_{ij} \exp\{-q_i s - q_j(t-s)\}\, \mathrm{d}s, $$

so that the same expression as above is found for ${}_1p_{ij}(t)$.

This result leads to the conjecture that the Markov property (3.2) holds even if for the present an exit moment is taken. For Markov chains as considered in this chapter this conjecture is correct; and a similar property holds for the entrance moment (cf. CHUNG [1960]).

Let ${}_np_{ij}(t)$ denote the probability that whenever the system is in state E_i at time s, it is at time $t+s$ in E_j and has reached E_j from E_i in exactly n transitions, i.e. the sample function of the process has exactly n jumps during $[s, s+t)$, $n=0, 1, \ldots$. From (3.12) it follows immediately that

$$ {}_0p_{ij}(t) = \delta_{ij}\, \mathrm{e}^{-q_i t} = \delta_{ij}\, \mathrm{e}^{-q_j t}, \qquad i, j \in S, \quad t>0. \tag{3.17} $$

An expression for ${}_1p_{ij}(t)$ has already been obtained above. We now derive two recursive relations for ${}_np_{ij}(t)$; the first relation by considering the first jump after s, the second relation by considering the last jump before $s+t$. By a probabilistic argument similar to that used above in the derivation of the expression for ${}_1p_{ij}(t)$ we obtain for $i, j \in S$,

$$ {}_{n+1}p_{ij}(t) = \sum_{\substack{k\in S\\ k\neq i}} \int_0^t \mathrm{e}^{-q_i\sigma}\, q_{ik}\, {}_np_{kj}(t-\sigma)\, \mathrm{d}\sigma, \qquad n=0, 1, \ldots . \tag{3.18} $$

This relation is also true if $q_i=0$, since then all $q_{ik}=0$ (cf. (3.15)) and ${}_np_{ij}(t) = 0$ for all $j \in S$, $n=1, 2, \ldots$, because of (3.12).

Supposing that the last jump before $t+s$ occurs at time $s+\sigma \div s+\sigma+\mathrm{d}\sigma$ it follows that

$$\int_{\sigma=0}^{t} {}_n p_{ik}(\sigma)\, q_{kj}\, \mathrm{d}\sigma\, \mathrm{e}^{-q_j(t-\sigma)},$$

is the probability of going from E_i to E_j in $n+1$ steps and that E_k is the last state preceding E_j. Hence for $i, j \in S$,

$$ {}_{n+1}p_{ij}(t) = \sum_{\substack{k\in S\\ k\neq j}} \int_{\sigma=0}^{t} {}_n p_{ik}(\sigma)\, q_{kj}\, \mathrm{e}^{-q_j(t-\sigma)}\, \mathrm{d}\sigma, \qquad n=0, 1, \ldots . \tag{3.19}$$

Since with probability one all realisations of the process are step-functions it follows that

$$p_{ij}(t) = \sum_{n=0}^{\infty} {}_n p_{ij}(t), \qquad t \geqq 0, \quad i, j \in S. \tag{3.20}$$

From (3.17) and (3.18) and also from (3.17) and (3.19) it is seen that ${}_n p_{ij}(t)$, $n=1, 2, \ldots;\ t \geqq 0,\ i, j \in S$, can be calculated recursively, that they are uniquely determined by these relations and that they are continuous functions of t for $t>0$. As the series in (3.20) converges uniformly in every finite t-interval it is seen that $p_{ij}(t)$ is a continuous function of t for $t>0$.

It is readily verified by induction that the following integrals exist and that

$$\int_0^{\infty} {}_0 p_{ij}(t)\, q_j\, \mathrm{d}t = \delta_{ij},$$

$$\int_0^{\infty} {}_{n+1}p_{ij}(t)\, q_j\, \mathrm{d}t = \sum_{\substack{k\in S\\ k\neq j}} \left\{ \int_0^{\infty} {}_n p_{ik}(\sigma)\, q_k\, \mathrm{d}\sigma \right\} \frac{q_{kj}}{q_k}, \qquad n=0, 1, \ldots,$$

from which it follows that

$$\int_0^{\infty} {}_m p_{ij}(t)\, q_j\, \mathrm{d}t = r_{ij}^{(m)}, \qquad m=0, 1, \ldots; \quad i, j \in S,$$

where $r_{ij}^{(m)}$ is the generic element of the mth power of the jump matrix (r_{ij}).

THEOREM 3.1. *If $(p_{ij}(.))$ is a standard transition matrix function and if*

$$\alpha \stackrel{\text{def}}{=} \sup_{j\in S} q_j < \infty,$$

then $p_{ij}(t)$, $t \geqq 0$, $i, j \in S$ is a solution of the system of integral equations

$$z_{ij}(t) = \delta_{ij}\, \mathrm{e}^{-q_i t} + \sum_{\substack{k\in S\\ k\neq i}} \int_0^{t} \mathrm{e}^{-q_i \sigma}\, q_{ik}\, z_{kj}(t-\sigma)\, \mathrm{d}\sigma, \qquad i, j \in S, \tag{3.21}$$

and also of the system

$$z_{ij}(t) = \delta_{ij}\, e^{-q_j t} + \sum_{\substack{k\in S\\ k\neq j}} \int_0^t z_{ik}(\sigma)\, q_{kj}\, e^{-q_j(t-\sigma)}\, d\sigma, \qquad i,j\in S. \tag{3.22}$$

Any non-negative solution $z_{ij}(t)$ of the system (3.21) *satisfies for all $t \geqq 0$,*

$$z_{ij}(t) \geqq p_{ij}(t), \qquad i,j\in S;$$

and if for all $t\geqq 0$,

$$\sum_{j\in S} z_{ij}(t) = 1, \qquad i\in S,$$

then for all $t\geqq 0$,

$$z_{ij}(t) = p_{ij}(t), \qquad i,j\in S;$$

a similar statement holds for the system (3.22).

Proof. It follows from (3.17), (3.18) and (3.20) that

$$p_{ij}(t) = \delta_{ij}\, e^{-q_i t} + \sum_{\substack{k\in S\\ k\neq i}} \int_0^t e^{-q_i\sigma}\, q_{ik} p_{kj}(t-\sigma)\, d\sigma, \tag{3.23}$$

and from (3.17), (3.19) and (3.20) that

$$p_{ij}(t) = \delta_{ij}\, e^{-q_j t} + \sum_{\substack{k\in S\\ k\neq j}} \int_0^t p_{ik}(\sigma)\, q_{kj}\, e^{-q_j(t-\sigma)}\, d\sigma, \tag{3.24}$$

so that the first assertion is proved.

If $z_{ij}(t)$ is a non-negative solution, i.e. it satisfies (3.21) and

$$z_{ij}(t) \geqq 0 \qquad \text{for all } t\geqq 0, \quad i,j\in S,$$

then it follows from (3.21) and (3.17) that

$$z_{ij}(t) \geqq \delta_{ij}\, e^{-q_i t} = {}_0p_{ij}(t).$$

Suppose that for some integer $N\geqq 0$,

$$z_{ij}(t) \geqq \sum_{n=0}^{N} {}_np_{ij}(t), \qquad t\geqq 0, \quad i,j\in S,$$

then from (3.21),

$$z_{ij}(t) \geqq \delta_{ij}\, e^{-q_j t} + \sum_{\substack{k\in S\\ k\neq i}} \int_0^t e^{-q_i\sigma}\, q_{ik} \sum_{n=0}^{N} {}_np_{kj}(t-\sigma)\, d\sigma,$$

and hence by (3.18),

$$z_{ij}(t) \geqq \sum_{n=0}^{N+1} {}_np_{ij}(t), \qquad t\geqq 0, \quad i,j\in S.$$

Hence, by induction, it is seen that the last relation is valid for all $N=0, 1, 2, \dots$, and therefore by (3.20) $z_{ij}(t) \geqq p_{ij}(t)$, for all $t \geqq 0$, $i, j \in S$. Consequently, if

$$\sum_{j \in S} z_{ij}(t) = 1, \qquad t \geqq 0, \quad i \in S,$$

then since

$$1 = \sum_{j \in S} z_{ij}(t) \geqq \sum_{j \in S} p_{ij}(t) = 1, \qquad t \geqq 0, \quad i \in S,$$

and the fact that $p_{ij}(t)$ and $z_{ij}(t)$ are non-negative

$$z_{ij}(t) = p_{ij}(t), \qquad t \geqq 0, \quad i, j \in S.$$

The statements for the system (3.22) are established in the same way; hence the theorem is proved.

COROLLARY 3.1. *Under the conditions of theorem* 3.1,

$$q_i = \sum_{\substack{j \in S \\ j \neq i}} q_{ij}, \qquad i \in S.$$

Proof. For $q_i=0$ the statement is evident (cf. (3.9)), suppose therefore that $q_i>0$. Since $p_{ij}(t)$ is a solution of (3.21) it follows from (3.6) and (3.21) that

$$\begin{aligned} 1 = \sum_{j \in S} p_{ij}(t) &= \mathrm{e}^{-q_i t} + \sum_{\substack{k \in S \\ k \neq i}} \int_0^t \mathrm{e}^{-q_i \sigma} q_{ik} \, \mathrm{d}\sigma \\ &= \mathrm{e}^{-q_i t} + \frac{1 - \mathrm{e}^{-q_i t}}{q_i} \sum_{\substack{k \in S \\ k \neq i}} q_{ik} \qquad \text{for all } t \geqq 0, \end{aligned}$$

from which the statement follows.

The integral relation (3.23) has a direct probabilistic interpretation if one considers the totality of possibilities at the first jump in the interval $[s, s+t)$, whereas (3.24) may be based on considering the last jump before $s+t$. In this way another derivation of (3.23) and (3.24) is obtained. For these reasons (3.23) is called the backward and (3.24) the forward integral relation.

THEOREM 3.2. *Under the conditions of theorem* 3.1, $p_{ij}(t)$ *is given by*

$$p_{ij}(t) = \mathrm{e}^{-\beta t} \sum_{n=0}^{\infty} \frac{\beta^n t^n}{n!} p_{ij}^{(n)}, \qquad t \geqq 0, \quad i, j \in S, \tag{3.25}$$

where $P=(p_{ij})$, $i,j\in S$ is a stochastic matrix defined by

$$P\overset{\text{def}}{=}I+\frac{1}{\beta}Q,\qquad P^0\overset{\text{def}}{=}I,\qquad \beta\geqq\alpha\ (cf.\ (3.14)).$$

Proof. From (3.9), (3.14) and (3.15) it follows that

$$p_{ij}\geqq 0,\qquad \sum_{j\in S}p_{ij}=1,\qquad i,j\in S,$$

so that P is a stochastic matrix.

Denoting the right-hand side of (3.25) by $\bar{p}_{ij}(t)$ then by using Laplace transforms it is easily verified by direct substitution into (3.21) that $\bar{p}_{ij}(t)$ is a non-negative solution of this integral equation. Since

$$\begin{aligned}\sum_{j\in S}\bar{p}_{ij}(t)&=\mathrm{e}^{-\beta t}\sum_{n=0}^{\infty}\frac{\beta^n t^n}{n!}\sum_{j\in S}p_{ij}^{(n)}\\&=\mathrm{e}^{-\beta t}\sum_{n=0}^{\infty}\frac{\beta^n t^n}{n!}=1,\end{aligned}$$

it follows from the preceding theorem that $\bar{p}_{ij}(t)=p_{ij}(t)$, $t\geqq 0$, $i,j\in S$. The proof is therefore complete.

Note. From (3.25) it is readily seen that

$$\begin{aligned}P(t)&=\mathrm{e}^{-\beta t}\sum_{n=0}^{\infty}\frac{\beta^n t^n}{n!}\left(I+\frac{1}{\beta}Q\right)^{(n)}=\mathrm{e}^{-\beta t}\sum_{n=0}^{\infty}\frac{\beta^n t^n}{n!}\sum_{k=0}^{n}\binom{n}{k}\beta^{-k}Q^{(k)}\\&=\sum_{k=0}^{\infty}\frac{t^k}{k!}Q^{(k)},\end{aligned}$$

so that $P(t)=(p_{ij}(t))$ is independent of β. Hence, in (3.25) and in the definition of P, β may be any real number not less than α.

From the Chapman-Kolmogorov relation (3.7) it follows that for $t\geqq 0$, $h\geqq 0$,

$$\frac{p_{ij}(t+h)-p_{ij}(t)}{h}=\frac{p_{ii}(h)-1}{h}p_{ij}(t)+\sum_{\substack{k\in S\\k\neq i}}\frac{p_{ik}(h)}{h}p_{kj}(t),$$

$$\frac{p_{ij}(t+h)-p_{ij}(t)}{h}=p_{ij}(t)\frac{p_{jj}(h)-1}{h}+\sum_{\substack{k\in S\\k\neq j}}p_{ik}(t)\frac{p_{kj}(h)}{h}.$$

From (3.25) it is seen that $p_{ij}(t)$ has a derivative for all $t>0$. Replacing $p_{ik}(h)$ and $p_{kj}(h)$ in the above relations by their expressions according to

(3.25) it is easily shown by letting $h\downarrow 0$ that

$$\frac{\mathrm{d}}{\mathrm{d}t}p_{ij}(t) = -q_i p_{ij}(t) + \sum_{\substack{k\in S\\ k\neq i}} q_{ik} p_{kj}(t), \tag{3.26}$$

$$\frac{\mathrm{d}}{\mathrm{d}t}p_{ij}(t) = -q_j p_{ij}(t) + \sum_{\substack{k\in S\\ k\neq j}} p_{ik}(t)\, q_{kj}, \tag{3.27}$$

for $i, j\in S$, $t>0$.

The system of equations (3.26) is denoted as the system of *backward* differential equations, the system (3.27) is known as the *forward* system of differential equations. For both systems the boundary conditions are given by

$$p_{ij}(0+) = \delta_{ij}. \tag{3.28}$$

The sum in (3.26) is a continuous function of t (cf. (3.25)) and consists of non-negative terms. Integrating the differential equation (3.26) for $p_{ij}(t)$ and using (3.28) we are led to the backward integral relation (3.23). Conversely, upon differentiating (3.23) with respect to t we obtain (3.26), and from (3.23) for $t\to 0$ the conditions (3.28) result. A similar statement is easily seen to hold for the forward differential relation (3.27) and the forward integral relation (3.24).

THEOREM 3.3. *Under the conditions of theorem* 3.1, $p_{ij}(t)$, $i, j\in S$, $t\geqq 0$, *is a solution of the system of differential equations*

$$\frac{\mathrm{d}}{\mathrm{d}t}z_{ij}(t) = -q_i z_{ij}(t) + \sum_{\substack{k\in S\\ k\neq i}} q_{ik} z_{kj}(t), \qquad t>0, \quad i, j\in S, \tag{3.29}$$

and also of the system

$$\frac{\mathrm{d}}{\mathrm{d}t}z_{ij}(t) = -q_j z_{ij}(t) + \sum_{\substack{k\in S\\ k\neq j}} z_{ik}(t)\, q_{kj}, \qquad t>0, \quad i, j\in S, \tag{3.30}$$

with boundary conditions for both systems

$$z_{ij}(0+) = \delta_{ij}, \qquad i, j\in S. \tag{3.31}$$

Any non-negative solution $z_{ij}(t)$ *of the system* (3.29) *satisfying* (3.31) *has the property:*

$$z_{ij}(t) \geqq p_{ij}(t), \qquad i, j\in S, \quad \textit{for all } t\geqq 0,$$

and if, for all $t \geqq 0$,

$$\sum_{j \in S} z_{ij}(t) = 1, \qquad i \in S,$$

then

$$z_{ij}(t) = p_{ij}(t), \qquad i, j \in S, \quad \text{for all } t \geqq 0.$$

A similar statement holds for the system (3.30).

Proof. That $p_{ij}(t)$ is a solution of (3.29) and (3.31) follows immediately from (3.26) and (3.28) or by direct substitution of the expression (3.25) into (3.29) and (3.31).

Suppose $z_{ij}(t)$ is a solution of (3.29) and (3.31) such that $z_{ij}(t) \geqq 0$ for all $t > 0$ and all $i, j \in S$. Then, from (3.29),

$$\frac{\mathrm{d}}{\mathrm{d}t} z_{ij}(t) \geqq -q_i z_{ij}(t),$$

so that using (3.31) and (3.17),

$$z_{ij}(t) \geqq \delta_{ij}\, \mathrm{e}^{-q_i t} = {}_0p_{ij}(t).$$

Hence from (3.29),

$$\frac{\mathrm{d}}{\mathrm{d}t} z_{ij}(t) \geqq -q_i z_{ij}(t) + \sum_{\substack{k \in S \\ k \neq i}} q_{ik}\, {}_0p_{kj}(t),$$

from which we obtain

$$z_{ij}(t) \geqq \delta_{ij}\, \mathrm{e}^{-q_i t} + \int_0^t \mathrm{e}^{-q_i(t-\sigma)} \sum_{\substack{k \in S \\ k \neq i}} q_{ik}\, {}_0p_{kj}(\sigma)\, \mathrm{d}\sigma,$$

or by (3.18),

$$z_{ij}(t) \geqq {}_0p_{ij}(t) + {}_1p_{ij}(t).$$

By complete induction we obtain

$$z_{ij}(t) \geqq \sum_{n=0}^{N} {}_np_{ij}(t),$$

for all $N = 0, 1, 2, \ldots$. Consequently, from (3.20) for all $t \geqq 0$,

$$z_{ij}(t) \geqq p_{ij}(t), \qquad i, j \in S.$$

The rest of the proof is similar to that of theorem 3.1. The proof is therefore complete.

Theorems 3.1 and 3.3 enable us to calculate $P(t)$ whenever we are given the Q-matrix. The solution obtained directly from (3.25) is often difficult to handle analytically. In most applications it is wise first of all to investigate the forward differential equations since, in this system, for fixed i only the unknowns $p_{ij}(t)$, $j \in S$ appear. Finally, it should be noted that in most applications the conditions mentioned in theorems 3.1 and 3.3 impose no serious restrictions.

I.3.3. Types of states and ergodic properties

Suppose that at time s the system is in a non-absorbing state E_i; as in the preceding section ρ_i is the exit time from E_i, so that at time $s+\rho_i$ the system leaves E_i. We define

$$\alpha_{ij} \overset{\text{def}}{=} \inf\{t\colon t>\rho_i,\ x_{s+t}=j\}, \qquad i, j \in S,$$

i.e. α_{ij} is the time between s and the moment of entrance into E_j for the first time after s; α_{ij} is therefore called the (first) *entrance time* from E_i into E_j if $i \neq j$. Similarly, α_{ii} represents the (first) *return time* from E_i to E_i, i.e. the time between s and the first re-entrance after $s+\rho_i$ into E_i. The distribution function of α_{ij} will be denoted by

$$F_{ij}(t) \overset{\text{def}}{=} \begin{cases} \Pr\{\alpha_{ij}<t\}, & t>0, \\ 0, & t \leqq 0. \end{cases}$$

If E_i is an absorbing state we define $F_{ii}(t)$ to be the unit step-function,

$$F_{ii}(t) \overset{\text{def}}{=} \begin{cases} 1 & \text{if } t>0, \\ 0 & \text{if } t \leqq 0, \end{cases}$$

and $F_{ij}(t)$ to be zero for all $j \in S$, all $t \in (-\infty, \infty)$.
Since the Markov property holds for a transition point, it follows that

$$p_{ij}(t) = \delta_{ij}\, \mathrm{e}^{-q_i t} + \int_0^t p_{jj}(t-\tau)\, \mathrm{d}F_{ij}(\tau), \qquad t \geqq 0, \quad i, j \in S. \tag{3.32}$$

In the Markov chain $\{x_t,\ t \in [0, \infty)\}$ a state E_i will be called *recurrent* if its return time is finite with probability one, i.e. if

$$\lim_{t\to\infty} F_{ii}(t) = 1,$$

otherwise E_i will be denoted as a *transient* state. A recurrent state E_i having a finite average return time is called *positive recurrent* or *persistent* or *ergodic*. It will be called a *null recurrent* state if its average return time is infinite. We now have

THEOREM 3.4. *Under the conditions of theorem* 3.1*:*
(*i*) $\lim_{t\to\infty} p_{ii}(t)$ *exists for all* $i \in S$*;*

(*ii*) E_i *is transient if and only if* $\int_0^\infty p_{ii}(t)\,\mathrm{d}t < \infty$*;*

(*iii*) *if* E_i *is positive recurrent then*

$$\lim_{t\to\infty} p_{ii}(t) = \frac{1}{q_i}\,\frac{1}{\mu_i} > 0, \qquad \mu_i \overset{\text{def}}{=} E\{\alpha_{ii}\};$$

(*iv*) *if* E_i *is null recurrent, then*

$$\lim_{t\to\infty} p_{ii}(t) = 0.$$

Proof. Let M denote a discrete time parameter Markov chain with state space S and with one step transition matrix $P=(p_{ij})$ as defined in theorem 3.2. Since $p_{ii}>0$, $i\in S$, all states of S are aperiodic for the chain M. Hence $p_{ii}^{(n)}$ has a limit for $n\to\infty$ (cf. theorems 2.1 and 2.2), say v_i. It follows from (3.25) that for $i\in S$, $t\geqq 0$, and N an arbitrary integer

$$p_{ii}(t) - v_i = \mathrm{e}^{-\beta t} \sum_{n=0}^{N} \frac{\beta^n t^n}{n!}\,(p_{ii}^{(n)} - v_i) + \mathrm{e}^{-\beta t} \sum_{n=N+1}^{\infty} \frac{\beta^n t^n}{n!}\,(p_{ii}^{(n)} - v_i).$$

For arbitrary $\varepsilon>0$ we may choose N such that $|p_{ii}^{(n)} - v_i| < \varepsilon$ for $n>N$. Since the first term of the right-hand side tends to zero for $t\to\infty$ and the second term is, in absolute value, always less than ε it follows that $p_{ii}(t)$ has a limit for $t\to\infty$, and

$$\lim_{t\to\infty} p_{ii}(t) = \lim_{n\to\infty} p_{ii}^{(n)}. \tag{3.33}$$

The first statement is proved.
To prove the second statement, we define for s real

$$\Pi_{ij}(s) \overset{\text{def}}{=} \int_0^\infty \mathrm{e}^{-st}\, p_{ij}(t)\,\mathrm{d}t, \qquad s>0,$$

$$\varphi_{ij}(s) \overset{\text{def}}{=} \int_{0-}^\infty \mathrm{e}^{-st}\,\mathrm{d}F_{ij}(t), \qquad s\geqq 0.$$

It follows from (3.32) that

$$\Pi_{ii}(s) = \frac{1}{s+q_i}\,\frac{1}{1-\varphi_{ii}(s)}, \qquad s>0. \tag{3.34}$$

If E_i is transient it follows (cf. app. 3) that

$$\lim_{s\downarrow 0} \varphi_{ii}(s) = \lim_{t\to\infty} F_{ii}(t) < 1.$$

Hence $\Pi_{ii}(s)$ has a finite limit for $s\downarrow 0$, so that (cf. app. 2) $\int_0^\infty p_{ii}(t)\,\mathrm{d}t<\infty$. Conversely, if $\int_0^\infty p_{ii}(t)\,\mathrm{d}t$ converges then $\Pi_{ii}(s)$ has a finite limit for $s\downarrow 0$, so that from (3.34) (cf. app. 2) $F_{ii}(+\infty)<1$, i.e. E_i is transient. The second statement is proved. Suppose E_i is recurrent, so that $\varphi_{ii}(s)\to 1$ for $s\downarrow 0$, the existence of the limit of $p_{ii}(t)$ for $t\to\infty$ implies (cf. app. 2),

$$\begin{aligned}\lim_{t\to\infty} p_{ii}(t) &= \lim_{s\downarrow 0} s\Pi_{ii}(s) = \frac{1}{q_i}\lim_{s\downarrow 0}\frac{s}{1-\varphi_{ii}(s)}\\ &= -\frac{1}{q_i}\lim_{s\downarrow 0}\left[\frac{\mathrm{d}}{\mathrm{d}s}\varphi_{ii}(s)\right]^{-1}.\end{aligned}$$

Since

$$-\frac{\mathrm{d}}{\mathrm{d}s}\varphi_{ii}(s) = \int_0^\infty \mathrm{e}^{-st}\,t\,\mathrm{d}F_{ii}(t), \qquad s>0,$$

it follows (cf. app. 3) that

$$\lim_{s\downarrow 0}\frac{\mathrm{d}}{\mathrm{d}s}\varphi_{ii}(s) = \begin{cases} -\mu_i & \text{if } \int_0^\infty t\,\mathrm{d}F_{ii}(t)<\infty,\\ -\infty & \text{if } \int_0^\infty t\,\mathrm{d}F_{ii}(t)=\infty.\end{cases}$$

From these relations the third and fourth statements are established.

DEFINITION. A class $\mathscr{E}$ of states is called *closed* if $p_{ij}(t)=0$ for all $t>0$ for every $E_i\in\mathscr{E}$ and every $E_j\notin\mathscr{E}$; it is called *mimimal* if it is closed and does not contain a proper subclass which is closed. If the state space is minimal then the Markov chain is called *irreducible*.

Before continuing the study of the continuous time parameter Markov chain $\{x_t,\ t\in[0,\infty)\}$ we introduce two discrete time parameter Markov chains M_2 and M_3 both with stationary transition probabilities and state space S. The continuous time parameter chain will be referred to as M_1.

The transition matrix of M_2 will be the jump matrix (r_{ij}) of M_1, that of M_3 is the transition matrix P of theorem 3.2. We derive first some relations between the chains M_1, M_2 and M_3.

We have already proved that

$$\int_0^\infty {}_m p_{ij}(t)\, q_j\, \mathrm{d}t = r_{ij}^{(m)}, \qquad m=0, 1, \ldots, \quad i, j \in S.$$

From (3.20), (3.25) and the above relations it follows that if $q_j>0$,

$$\frac{1}{\beta} \sum_{n=0}^{\infty} p_{ij}^{(n)} = \int_0^\infty p_{ij}(t)\, \mathrm{d}t$$

$$= \sum_{m=0}^{\infty} \int_0^\infty {}_m p_{ij}(t)\, \mathrm{d}t = \frac{1}{q_j} \sum_{m=0}^{\infty} r_{ij}^{(m)}, \tag{3.35}$$

in the sense that all terms are infinite if one of them is infinite or all are finite and then the equality signs are valid.

From (3.35) important conclusions can be drawn. It is first noted that if $q_j=0$ then E_j is an absorbing state in all three chains M_1, M_2 and M_3 and the class with E_j as only element is minimal for all these chains. Taking in (3.35) $i=j$ then it is seen from theorems 2.1 and 3.4 that *if E_i is a nonrecurrent state in one chain then it is nonrecurrent in the other chains*, and similarly *if E_i is recurrent in one chain it is recurrent in the other ones*. Since $p_{ij}(t)>0$ for some $t>0$ implies $p_{ij}(t)>0$ for all $t>0$ (cf. (3.25)) it results from (3.35) that the three statements "$p_{ij}(t)>0$ for some $t>0$", "$p_{ij}^{(n)}>0$ for some integer $n>0$" and "$r_{ij}^{(m)}>0$ for some integer $m>0$" are equivalent. Hence, *if $\mathscr{E}$ is a (minimal) closed class in one of the chains it is a (minimal) closed class in the other chains.*

From (3.33) and theorems 2.2 and 3.4 it is seen that *if E_i is positive recurrent for M_1 then it is also positive recurrent for M_3 and conversely*. We now prove a similar statement for M_2 and M_3. Let $\mathscr{E}$ be a minimal class, and consider the two systems of equations

$$u_i = \sum_{j \in \mathscr{E}} u_j p_{ji}, \qquad i \in \mathscr{E}, \tag{3.36}$$

$$v_i = \sum_{j \in \mathscr{E}} v_j r_{ji}, \qquad i \in \mathscr{E}, \tag{3.37}$$

in the unknowns u_i, $i \in \mathscr{E}$, and v_i, $i \in \mathscr{E}$, respectively. From the definition of

P and R it follows that the system (3.36) is equivalent to

$$q_i u_i = \sum_{\substack{j \in \mathscr{E} \\ j \neq i}} u_j q_{ji}, \qquad i \in \mathscr{E},$$

and system (3.37) to

$$v_i = \sum_{\substack{j \in \mathscr{E} \\ j \neq i}} v_j \frac{q_{ji}}{q_j}, \qquad i \in \mathscr{E},$$

if $q_j > 0$, i.e. E_j non-absorbing. Since the case $q_j = 0$ is trivial ($\mathscr{E}$ contains then only one state) we suppose $q_j > 0$.

Since $\mathscr{E}$ is a minimal class (p_{ij}), $i, j \in \mathscr{E}$ and (r_{ij}), $i, j \in \mathscr{E}$ are stochastic matrices. Supposing for the sake of simplicity that M_2 has no periodic states in $\mathscr{E}$ (it is easily verified that this assumption is not at all essential) it follows from theorem2.5, since all states of M_3 are aperiodic, that for all $i, j \in \mathscr{E}$,

$$\text{if } \sum_{i \in \mathscr{E}} u_i = 1 \qquad \text{then } u_i = \lim_{n \to \infty} p_{ji}^{(n)},$$

and

$$\text{if } \sum_{i \in \mathscr{E}} v_i = 1 \qquad \text{then } v_i = \lim_{m \to \infty} r_{ji}^{(m)}.$$

Since $q_i u_i$, $i \in \mathscr{E}$, is a solution of (3.37) if u_i, $i \in \mathscr{E}$ is a solution of (3.36) and conversely, it follows

$$q_i \lim_{n \to \infty} p_{ji}^{(n)} = \lim_{m \to \infty} r_{ji}^{(m)}, \qquad i, j \in \mathscr{E}. \tag{3.38}$$

Hence, if the right-hand side of (3.38) is positive then $\lim_{n \to \infty} p_{ji}^{(n)} > 0$ and conversely, as q_i is assumed to be non-zero. Consequently, *if $E_i \in \mathscr{E}$ is positive recurrent for M_2 it is positive recurrent for M_3 and conversely.*

Since a recurrent state is always an element of a minimal class the results obtained above show that the following lemma has been proved.

LEMMA. *A class of states which is (minimal) closed in one of the chains M_1, M_2 or M_3 is (minimal) closed in the other chains. A state E_i is of the same type in all three chains.*

We now formulate and prove for continuous time parameter Markov chains the analogue of theorem 2.5.

THEOREM 3.5. *Under the conditions of theorem* 3.1*: if the Markov chain $\{x_t, t \in [0, \infty)\}$ is irreducible then:*

(*i*) *all its states are of the same type;*

(*ii*) *if all states are positive recurrent then* $p_{ij}(t)$ *has a limit for* $t \to \infty$,

$$u_j \stackrel{\text{def}}{=} \lim_{t\to\infty} p_{ij}(t) > 0, \qquad i, j \in S,$$

and $\{u_j, j \in S\}$ *is a probability distribution uniquely determined by the system of equations*

$$q_j z_j = \sum_{\substack{i\in S\\ i\neq j}} z_i q_{ij}, \qquad i \in S;$$

$\{u_j, j \in S\}$ *is the only stationary distribution of the Markov chain and*

$$\lim_{t\to\infty} a_j(t) = u_j, \qquad j \in S;$$

(*iii*) *if all states are null recurrent or transient then*

$$\lim_{t\to\infty} p_{ij}(t) = 0, \qquad i, j \in S,$$

$$\lim_{t\to\infty} a_j(t) = 0, \qquad j \in S.$$

Proof. If $q_j = 0$ for some $j \in S$, then E_j is an absorbing state and hence since S is minimal, S contains only one element. The theorem is then trivial, and we therefore discard this case.
Since S is irreducible theorem 2.5 and the above lemma imply that all states of M_2 are of the same type and hence by the same lemma this is also true for M_1. This proves the first statement. In the following we again suppose for the sake of simplicity that M_2 has no periodical states. From theorem 2.5 applied to M_3 it follows that $p_{ij}^{(n)}$ has for $n \to \infty$ a limit which is positive. As in the first part of the proof of theorem 2.4 it now follows from (3.25) that $p_{ij}(t)$ has a limit for $t \to \infty$ and

$$u_j = \lim_{t\to\infty} p_{ij}(t) = \lim_{n\to\infty} p_{ij}^{(n)} > 0, \qquad i, j \in S.$$

Theorem 2.5 applied to M_3 shows that $\{u_j, j \in S\}$ is a probability distribution, which is uniquely determined by

$$u_i = \sum_{j\in S} u_j p_{ji}, \qquad i \in S,$$

and this system of equations is equivalent to

$$q_i u_i = \sum_{\substack{j\in S\\ j\neq i}} u_j q_{ji}, \qquad i \in S.$$

From (3.25) it is easily verified that

$$u_i = \sum_{j\in S} u_j p_{ji}(t), \qquad t \geqq 0, \quad i \in S,$$

so that $\{u_i,\ i\in S\}$ is the only stationary distribution of M_1. Since (3.5) implies the relation

$$a_j(t+s) = \sum_{i\in S} a_i(s)\, p_{ij}(t), \qquad j\in S, \quad s>0, \quad t>0,$$

it follows easily that

$$\lim_{t\to\infty} a_j(t) = u_j, \qquad j\in S.$$

The second statement has now been proved.
The third statement is proved by applying (3.25), the argument in the proof of (*i*) of theorem 3.4, the lemma above and theorem 2.5 applied to M_3. The proof is complete.

In many applications the state space S is a minimal class of positive recurrent states. As in the discrete parameter case the theorem above shows that in this case the absolute distribution of $\boldsymbol{x}_t$ tends for $t\to\infty$ to a probability distribution which is independent of the initial distribution. We also say here that as $t\to\infty$ the system tends to the state of statistical equilibrium.

I.3.4. Foster's criteria

In section 2.4 we discussed Foster's criteria for discrete time parameter Markov chains. In this section we shall treat these criteria for continuous time parameter Markov chains. Before discussing these criteria we derive first a property of the mean first entrance time from a state E_i to a state E_j or into a class of states.

Suppose the state space S is the union of a minimal class C and a set T of transient states. Let $y_j(t)$, $j\in T$, denote for the continuous time parameter Markov chain M_1 the conditional probability that when starting in $E_j \in T$ at time s the system will still be in T at time $s+t$.
It now follows from (3.25) that

$$y_j(t) = \sum_{i\in T} p_{ji}(t)$$

$$= \mathrm{e}^{-\beta t} \sum_{n=0}^{\infty} \frac{\beta^n t^n}{n!} \sum_{i\in T} p_{ji}^{(n)}, \qquad t>0, \quad j\in T.$$

Since

$$y_j^{(n)} \stackrel{\text{def}}{=} \sum_{i \in T} p_{ji}^{(n)}, \qquad n=0, 1, \ldots,$$

represents for the discrete time parameter Markov chain M_3 with m-step transition matrix P^m the conditional probability that when starting in $E_j \in T$ the system is still in T after n transitions, it follows that

$$y_j^{(n)} \geqq y_j^{(n+1)}, \qquad n=1, 2, \ldots; \; j \in T.$$

Putting

$$y_j \stackrel{\text{def}}{=} \lim_{n \to \infty} y_j^{(n)}, \qquad j \in T,$$

then for M_3 the conditional probability of starting in E_j, $j \in T$ and always remaining in T is given by y_j. Applying the argument in the proof of (*i*) of theorem 3.4 to the expression for $y_j(t)$ above shows that

$$y_j = \lim_{t \to \infty} y_j(t) = \lim_{n \to \infty} \sum_{i \in T} p_{ji}^{(n)}, \qquad j \in T,$$

and hence y_j is also the probability for the Markov chain M_1 when starting in $E_j \in T$ of remaining always in T.

Defining

$$x_j(t) \stackrel{\text{def}}{=} 1 - y_j(t), \qquad j \in T, \quad t \geqq 0,$$

then $x_j(t)$ is for M_1 the probability when starting in $E_j \in T$ at time s that at time $s+t$ the system has reached a state belonging to C, i.e. $x_j(t)$ is the probability that the entrance time from E_j, $j \in T$ into C is less than t. Obviously, $x_j(t)$, $t \geqq 0$, is a proper probability distribution if and only if $y_j = 0$. Since y_j, $j \in T$ satisfy

$$y_j = \sum_{i \in T} p_{ji} y_i, \qquad j \in T,$$

as it follows from (2.16) applied to M_3, it follows that if $y_j = 0$ for some $j \in T$, then $y_j = 0$ for all $j \in T$. Suppose $y_j = 0$ for all $j \in T$, then

$$v_j \stackrel{\text{def}}{=} \int_0^\infty t \, \mathrm{d}x_j(t), \qquad j \in T,$$

represents the mean first entrance time from E_j, $j \in T$ into C. It follows easily that

$$v_j = \int_0^\infty y_j(t) \, \mathrm{d}t = \frac{1}{\beta} \sum_{n=0}^\infty y_j^{(n)}, \qquad j \in T.$$

Since

$$y_j^{(n+1)} = \sum_{i \in T} p_{ji} y_i^{(n)}, \qquad n=0, 1, \ldots, \quad j \in T,$$

we have

$$v_j - \frac{1}{\beta} = \sum_{i \in T} p_{ji} v_i, \qquad j \in T,$$

or

$$v_j - \frac{1}{\beta} = \left(1 - \frac{q_j}{\beta}\right) v_j + \sum_{\substack{i \in T \\ i \neq j}} \frac{q_{ji}}{\beta} v_i .$$

Hence, for the continuous time parameter Markov chain M_1 the mean first entrance times v_j, $j \in T$, satisfy

$$v_j = \sum_{\substack{i \in T \\ i \neq j}} \frac{q_{ji}}{q_j} v_i + \frac{1}{q_j}, \qquad j \in T, \tag{3.39}$$

if $y_j = 0$, $j \in T$.

We shall now discuss Foster's criteria for the continuous time parameter Markov chain $\{\boldsymbol{x}_t,\ t \in [0, \infty)\}$ for which the conditions of theorem 3.1 hold and for which the state space is irreducible and is the set of non-negative integers.

In section 2.4 Foster's criteria are described for a discrete time parameter Markov chain in case all its states are aperiodic. Hence, on account of the lemma of the preceding section we obtain the Foster criteria for the continuous parameter case if we apply to the discrete time parameter Markov chain M_2 with transition matrix R the criteria of section 2.4, in case all states of M_2 are aperiodic. It can be easily verified that Foster's criteria for the discrete time parameter case also hold if the states are periodic (cf. KANTERS [1965]).

An exception should be made for Foster's criterion (*iii*). If this is applied to the chain M_2 then the v_i, $i = 1, 2, \ldots$, represent the mean passage times from E_i to E_0 for the chain M_2 and these mean passage times differ from those of M_1 (cf. (3.39)). It is easily seen how this criterion should read for the continuous parameter case (the proof is analogous). For applications it is desirable to have a somewhat more general formulation of Foster's third criterion. This is given below and the proof is along the same lines as that of the original formulation.

From what has been said above it now follows that if the continuous time parameter Markov chain satisfies the conditions of theorem 3.1, and its state space is irreducible and consists of the set of non-negative integers then Foster's criteria for this chain read as follows:

(*i*) The chain is ergodic if

$$z_j = \sum_{\substack{i=0\\i\neq j}}^{\infty} z_i \frac{q_{ij}}{q_i}, \qquad j=0, 1, \ldots,$$

possesses a non-null solution such that $\sum_{j=0}^{\infty} |z_j| < \infty$; and only if this property is possessed by any non-negative solution of the inequalities

$$z_j \geqq \sum_{\substack{i=0\\i\neq j}}^{\infty} z_i \frac{q_{ij}}{q_i}, \qquad j=0, 1, \ldots .$$

Note: The theorem remains true if in the above relations q_i is replaced by q_j as can be seen by replacing z_i by $x_i q_i$. For applications the latter form is often easier to handle.

(*ii*) The chain is ergodic if an $\varepsilon > 0$ and an integer $i_0 > 0$ exist such that the inequalities

$$z_i - \varepsilon \geqq \sum_{\substack{j=0\\j\neq i}}^{\infty} \frac{q_{ij}}{q_i} z_j, \qquad i=i_0+1, i_0+2, \ldots,$$

$$\lambda_i \overset{\text{def}}{=} \sum_{\substack{j=0\\j\neq i}}^{\infty} \frac{q_{ij}}{q_i} z_j < \infty, \qquad i=0, 1, \ldots, i_0,$$

have a non-negative solution.

(*iii*) If the chain is ergodic then the finite mean first passage time ν_j from E_j into $\{E_0, E_1, \ldots, E_{i_0}\}$ satisfies

$$\nu_i = \sum_{\substack{j=i_0+1\\j\neq i}}^{\infty} \frac{q_{ij}}{q_i} \nu_j + \frac{1}{q_i}, \qquad i=i_0+1, i_0+2, \ldots,$$

$$\sum_{j=i_0+1}^{\infty} \frac{q_{ij}}{q_i} \nu_j < \infty, \qquad i=0, 1, \ldots, i_0,$$

where $i_0 \geqq 0$ is an arbitrary integer.

(*iv*) The chain is transient if and only if

$$z_i = \sum_{\substack{j=0\\j\neq i}}^{\infty} \frac{q_{ij}}{q_i} z_j, \qquad i=1, 2, \ldots,$$

has a bounded non-constant solution.

(*v*) The chain is recurrent if the inequalities

$$z_i \geqq \sum_{\substack{j=0 \\ j \neq i}}^{\infty} \frac{q_{ij}}{q_i} z_j, \qquad i=1, 2, \ldots,$$

have a solution such that $z_i \to \infty$ for $i \to \infty$.

(*vi*) The chain is transient if and only if

$$z_i \geqq \sum_{\substack{j=0 \\ j \neq i}}^{\infty} \frac{q_{ij}}{q_i} z_j, \qquad i=1, 2, \ldots,$$

has a bounded solution such that $z_i < z_0$ for some i.

I.4. BIRTH AND DEATH PROCESSES

I.4.1. Introduction

An important sub-class of the class of all Markov chains with continuous time parameter is formed by the birth and death processes. These processes are characterized by the property that if a transition occurs then this transition leads to a neighbouring state. In this chapter we shall only consider birth and death processes for which the state space is a linear ordered set and in nearly all cases it will be the set of non-negative integers. However, the reader will have no difficulties in extending the theory to more complex state spaces as for instance to a state space consisting of all pairs of non-negative integers if he bears in mind that a birth and death process is characterized by the property that only transitions to neighbouring states are possible. For the latter state space the neighbouring states of the generic state (i, j) are $(i+1, j)$, $(i-1, j)$, $(i, j+1)$ and $(i, j-1)$.

DEFINITION. A Markov chain $\{\boldsymbol{x}_t,\ t\in[0, \infty)\}$ with state space the set of non-negative integers and with

$$\begin{aligned} \lambda_i &\overset{\text{def}}{=} q_{i,i+1}, \qquad i=0, 1, \ldots, \\ \mu_i &\overset{\text{def}}{=} q_{i,i-1}, \qquad i=1, 2, \ldots, \\ q_{ij} &= 0 \qquad \text{for } j\neq i \text{ and } j\neq i\pm 1, \quad i=0, 1, \ldots, \\ q_i &= \lambda_i + \mu_i, \qquad i=0, 1, \ldots; \qquad \mu_0 \overset{\text{def}}{=} 0, \end{aligned} \tag{4.1}$$

is called: (i) a *birth process* if all μ_i, $i=1, 2, \ldots$, are zero; (ii) a *death process* if all λ_i, $i=0, 1, \ldots$, are zero; (iii) a *birth and death process* if at least some of the λ_i and some of the μ_i are positive.

The elementary probability $\lambda_i \, \mathrm{d}t$ is the probability that whenever at time t the system is in state E_i, i.e. $x_t = i$, the system will leave E_i during $t \div t + \mathrm{d}t$ and go to E_{i+1}; λ_i is denoted as the *birth rate* when the system is in state E_i, and similarly μ_i is the *death rate* in this state. The name "birth and death process" stems from the application of these processes in the study of biological processes such as the growth of bacteria populations (cf. BAILEY [1964]). For instance if the population consists of i bacteria at time t then the system is said to be at time t in state E_i and λ_i is the probability density that these i bacteria produce during $t \div t + \mathrm{d}t$ a new one, whereas $\mu_i \, \mathrm{d}t$ is the elementary probability that during $t \div t + \mathrm{d}t$ one of these i bacteria dies.

From what has been said above it is seen that we implicitly assumed that λ_i and μ_i are independent of time, or in other words that we only consider here Markov chains with stationary transition probabilities. For applications in queueing theory this is in general sufficient; however, it is easily seen how to generalize to the non-stationary situation. In this chapter *we shall again assume* that the condition (3.14) is valid. For applications in queueing theory this is hardly a restriction; in biological applications, however, the condition (3.14) is often too restrictive.

From (4.1) and (3.26) it follows that for the birth and death process the backward equations read for $j = 0, 1, \ldots,$

$$\frac{\mathrm{d}}{\mathrm{d}t} p_{0j}(t) = -\lambda_0 p_{0j}(t) + \lambda_0 p_{1j}(t), \tag{4.2}$$

$$\frac{\mathrm{d}}{\mathrm{d}t} p_{ij}(t) = -(\lambda_i + \mu_i)\, p_{ij}(t) + \lambda_i p_{i+1,j}(t) + \mu_i p_{i-1,j}(t), \qquad i = 1, 2, \ldots,$$

whereas the forward equations are given by: for $i = 0, 1, \ldots,$

$$\frac{\mathrm{d}}{\mathrm{d}t} p_{i0}(t) = -\lambda_0 p_{i0}(t) + \mu_1 p_{i1}(t), \tag{4.3}$$

$$\frac{\mathrm{d}}{\mathrm{d}t} p_{ij}(t) = -(\lambda_j + \mu_j)\, p_{ij}(t) + \lambda_{j-1} p_{i,j-1}(t) + \mu_{j+1} p_{i,j+1}(t), \qquad j = 1, 2, \ldots .$$

For both systems the boundary conditions are

$$p_{ij}(0+) = \delta_{ij}, \qquad i, j = 0, 1, \ldots . \tag{4.4}$$

The systems (4.3) and (4.4) have been investigated by many authors. Particularly we mention here the studies of KARLIN and MCGREGOR [1957, 1958, 1960]. These authors show that the general solution of (4.2) subjected to the boundary conditions (4.4) can be written as

$$p_{ij}(t) = \int_0^\infty e^{-xt}\, Q_i(x)\, Q_j(x)\, d\Psi(x), \qquad i, j=0, 1, \ldots, \tag{4.5}$$

where $Q_n(x)$ are polynomials in x, defined recursively by

$$\begin{aligned} &Q_0(x)=1, \\ -&xQ_0(x)=-(\lambda_0+\mu_0)\, Q_0(x)+\lambda_0 Q_1(x), \\ -&xQ_n(x)=\mu_n Q_{n-1}(x)-(\lambda_n+\mu_n)\, Q_n(x)+\lambda_n Q_{n+1}(x), \qquad n\geqq 1, \end{aligned} \tag{4.6}$$

and $\Psi(x)$ is a positive regular measure on $[0, \infty)$ such that

$$\int_0^\infty Q_i(x)\, Q_j(x)\, d\Psi(x) = \frac{\delta_{ij}}{\pi_j}, \qquad i, j=0, 1, \ldots, \tag{4.7}$$

with

$$\pi_0 \stackrel{\text{def}}{=} 1, \qquad \pi_n \stackrel{\text{def}}{=} \frac{\lambda_0 \lambda_1 \ldots \lambda_{n-1}}{\mu_1 \mu_2 \ldots \mu_n}, \qquad n=1, 2, \ldots . \tag{4.8}$$

Denoting by $P_j(t)$ for birth and death processes the distribution of $\boldsymbol{x}_t$, i.e.

$$P_j(t) \stackrel{\text{def}}{=} \Pr\{\boldsymbol{x}_t=j\}, \qquad j=0, 1, \ldots;\quad t\geqq 0,$$

then if at $t=0$ the system starts in state E_i, so that

$$P_j(0) = \Pr\{\boldsymbol{x}_0=j\} = \delta_{ij}, \tag{4.9}$$

it follows from

$$P_j(t) = p_{ij}(t), \qquad t\geqq 0,$$

and the forward equations (4.3) that

$$\begin{aligned} \frac{d}{dt} P_0(t) &= -\lambda_0 P_0(t) + \mu_1 P_1(t), \\ \frac{d}{dt} P_j(t) &= -(\lambda_j+\mu_j)\, P_j(t) + \lambda_{j-1} P_{j-1}(t) + \mu_{j+1} P_{j+1}(t), \qquad j=1,2, \ldots . \end{aligned} \tag{4.10}$$

In practical situations we shall frequently start our investigations from (4.9) and (4.10).

We shall now apply Foster's criteria to the general birth and death process with all the λ_i and μ_i non-zero (cf. section 3.4) to obtain necessary and sufficient conditions for the states to be positive recurrent, null recurrent or transient, respectively. Since all λ_i and μ_i are supposed to be non-zero it is easily seen that the state space $\{0, 1, 2, ...\}$ of the birth and death process is irreducible.

Application of Foster's first criterion with z_i replaced by $q_i x_i$ leads because of (4.1) to

$$\lambda_0 x_0 = \mu_1 x_1, \tag{4.11}$$

$$(\lambda_j + \mu_j) x_j = \lambda_{j-1} x_{j-1} + \mu_{j+1} x_{j+1}, \qquad j = 1, 2, \dots .$$

Rewriting this set of equations as

$$\lambda_0 x_0 - \mu_1 x_1 = 0,$$

$$\lambda_j x_j - \mu_{j+1} x_{j+1} = \lambda_{j-1} x_{j-1} - \mu_j x_j, \qquad j = 1, 2, \dots,$$

it is easily verified that the solution is given by (cf. (4.8)),

$$x_j = \pi_j x_0, \qquad j = 1, 2, \dots .$$

Hence since q_i, $i = 0, 1, \dots$, is uniformly bounded Foster's first criterion implies that

$$\sum_{j=1}^{\infty} \pi_j < \infty,$$

is a sufficient condition for the birth and death process to have all its states positive recurrent.

Next we apply Foster's third criterion. From (4.1) and

$$\nu_i = \sum_{\substack{j=1 \\ j \neq i}}^{\infty} \frac{q_{ij}}{q_i} \nu_j + \frac{1}{q_i}, \qquad i = 1, 2, \dots, \tag{4.12}$$

with $\nu_0 = 0$ it follows that

$$\nu_{i+1} - \nu_i = -\frac{1}{\lambda_i} + \frac{\mu_i}{\lambda_i} (\nu_i - \nu_{i-1}), \qquad i = 1, 2, \dots .$$

From the last relation we obtain

$$\nu_{i+1} - \nu_i = -\frac{1}{\lambda_i \pi_i} \sum_{j=1}^{i} \pi_j + \frac{\lambda_0}{\lambda_i \pi_i} \nu_1, \qquad i = 1, 2, \dots . \tag{4.13}$$

Since ν_i is the mean first passage time from state E_i to E_0 and since a passage from E_{i+1} to E_0 always leads via E_i it follows that $\nu_{i+1} \geqq \nu_i$. Consequently, from the above relation

$$\lambda_0 \nu_1 \geqq \sum_{j=1}^{i} \pi_j, \qquad i=1,2,\ldots .$$

Hence if $\lambda_0 \nu_1 < \infty$, $\sum_{j=1}^{i} \pi_j$ is uniformly bounded in $i=1,2,\ldots$. Consequently, from Foster's third criterion: if the chain is ergodic then

$$\sum_{j=1}^{\infty} \pi_j < \infty.$$

Consider the set of equations

$$z_i = \sum_{\substack{j=0 \\ j \neq i}}^{\infty} \frac{q_{ij}}{q_i} z_j, \qquad i=1,2,\ldots,$$

which (cf. (4.1)) are equivalent to

$$\lambda_i(z_{i+1}-z_i) = \mu_i(z_i - z_{i-1}), \qquad i=1,2,\ldots .$$

It follows that

$$z_{i+1} = z_1 + (z_1 - z_0) \sum_{j=1}^{i} \frac{\lambda_0}{\lambda_j \pi_j}, \qquad i=1,2,\ldots .$$

Hence, if we take $z_1 \neq z_0$, the above set of equations has a non-constant and bounded solution if

$$\sum_{j=1}^{\infty} \frac{\lambda_0}{\pi_j \lambda_j} < \infty.$$

Consequently, due to Foster's fourth criterion the above condition is sufficient for the states of the chain to be transient. If this condition is not satisfied, i.e. if the sum is divergent, the above set of equations has a solution such that $z_i \to \infty$ for $i \to \infty$ if we take $z_1 > z_0$, and hence Foster's fifth criterion guarantees that the states of the chain are all null recurrent if

$$\sum_{j=1}^{\infty} \frac{\lambda_0}{\lambda_j \pi_j} = \infty.$$

Hence, it has been shown that the birth and death process is (cf. (4.8)):

ergodic if and only if $\qquad \sum_{j=1}^{\infty} \pi_j < \infty;$

null recurrent if and only if $\sum_{j=1}^{\infty} \pi_j = \infty$ and $\sum_{j=1}^{\infty} \frac{\lambda_0}{\lambda_j \pi_j} = \infty$;

transient if and only if $\sum_{j=1}^{\infty} \frac{\lambda_0}{\lambda_j \pi_j} < \infty$.

Suppose from now on that the birth and death process is ergodic. From theorem 3.5 it then follows that the stationary distribution $\{u_j, j=0, 1, \ldots\}$ is uniquely determined by

$$q_j u_j = \sum_{\substack{i=0 \\ i \neq j}}^{\infty} u_i q_{ij}, \qquad j=0, 1, \ldots,$$

$$\sum_{j=0}^{\infty} u_j = 1.$$

From (4.1) and (4.11) it is easily verified that the solution of the above equations is given by

$$u_0 = \{\sum_{j=0}^{\infty} \pi_j\}^{-1}, \qquad u_j = \pi_j u_0, \qquad j=0, 1, \ldots . \tag{4.14}$$

Since theorem 3.4 implies

$$u_j = \lim_{t \to \infty} p_{jj}(t) = \frac{1}{q_j \nu_{jj}}, \qquad j=0, 1, \ldots,$$

where ν_{jj} represents the average first return time from E_j to E_j it follows that

$$\nu_{jj} = \frac{\sum_{k=0}^{\infty} \pi_k}{(\lambda_j + \mu_j)\pi_j}, \qquad j=0, 1, \ldots . \tag{4.15}$$

Next, we shall determine the average first passage time ν_{j0} from E_j to E_0. From the definition of ν_j in Foster's third criterion it is seen that $\nu_j = \nu_{j0}$ (section 3.4).

From the derivation of (3.39) it is seen that ν_j, $j=1, 2, \ldots$, is a non-negative solution of

$$z_j = \sum_{i=1}^{\infty} p_{ji} z_i + \frac{1}{\beta}, \qquad j=1, 2, \ldots, \tag{4.16}$$

with

$$p_{ij} = \delta_{ij} - \frac{q_{ij}}{\beta}, \qquad i, j=0, 1, \ldots .$$

Let z'_j, $j=1, 2, \ldots$, denote a non-negative solution of (4.16); then it is easily deduced (cf. the proof of theorem 2.7) using the notation of sections 2.5 and 3.4 that

$$z'_j \geqq \frac{1}{\beta} \sum_{n=0}^{\infty} \sum_{i=1}^{\infty} p_{0;ji}^{(n)} = \frac{1}{\beta} \sum_{n=0}^{\infty} y_j^{(n)} = \nu_j, \qquad j=1, 2, \ldots;$$

so that of all non-negative solutions of (4.16) ν_j, $j=1, 2, \ldots$, is the minimal solution. Since (4.16) is equivalent to (4.12), it follows from (4.13) that

$$z_{i+1} = z_1 + \sum_{k=1}^{i} \frac{1}{\lambda_k \pi_k} \{\lambda_0 z_1 - \sum_{j=1}^{k} \pi_j\}, \qquad i=1, 2, \ldots .$$

Hence, the minimal non-negative solution of (4.16) is obtained by taking

$$\lambda_0 z_1 = \sum_{k=1}^{\infty} \pi_k.$$

From the latter relations it follows that the average first passage time ν_{i0} from E_i to E_0 is given by

$$\nu_{i0} = \sum_{n=0}^{i-1} \frac{1}{\lambda_n \pi_n} \sum_{k=n+1}^{\infty} \pi_k, \qquad i=1, 2, \ldots .$$

The mean first passage time ν_{ij} from E_i to E_j for $i>j$ can be obtained in the same way as above. However, it can be obtained immediately from the right-hand side of the expression for ν_{i0} by replacing in this expression i by $i-j$, μ_h by μ_{h+j} and λ_h by λ_{h+j}, since in a birth and death process a transition leads always to a neighbouring state. After some simple calculations we find that

$$\nu_{ij} = \sum_{n=j}^{i-1} \frac{1}{\lambda_n \pi_n} \sum_{k=n+1}^{\infty} \pi_k, \qquad i=j+1, \ldots; \quad j=0, 1, \ldots . \tag{4.17}$$

To determine ν_{ij} for $i<j$ we start from the third Foster criterion (cf. section 3.4),

$$q_i \nu_{ij} = \sum_{\substack{h=0 \\ h\neq i}}^{j-1} q_{ih} \nu_{hj} + 1, \qquad i=0, \ldots, j-1.$$

From (4.1) it is easily found that the above set of equations is equivalent to

$$\nu_{i+1,j} = \nu_{0j} - \sum_{n=0}^{i} \frac{1}{\lambda_n \pi_n} \sum_{k=0}^{n} \pi_k, \qquad i=0, 1, \ldots, j-2.$$

Since the process is a birth and death process it follows from the definition of the average first passage time that

$$\nu_{i,i+1} + \nu_{i+1,j} = \nu_{ij}, \qquad i+1<j.$$

From the latter relations ν_{ij} can be easily calculated and the result reads

$$\nu_{ij} = \sum_{n=i}^{j-1} \frac{1}{\lambda_n \pi_n} \sum_{k=0}^{n} \pi_k, \qquad i=0, 1, \ldots, j-1; \quad j=1, 2, \ldots . \tag{4.18}$$

I.4.2. The Poisson process

For a large class of stochastic phenomena the Poisson process is an extremely useful model and therefore plays a prominent role in applications. The great applicability of the Poisson process makes it desirable to discuss this process from various viewpoints, and therefore we shall consider in this section a number of stochastic models which all lead to the Poisson process. We shall first discuss the Poisson process as a birth process.

DEFINITION. A birth process (with state space the set of non-negative integers) with constant birth rate, i.e.

$$\lambda_i = \lambda, \qquad i=0, 1, \ldots,$$

is called a (homogeneous) Poisson process with parameter λ.

To investigate this process we first determine the transition function $p_{ij}(t)$, $t \geqq 0$. The easiest way to obtain these functions is to apply theorem 3.2. From the definition of a birth process (see preceding section) it follows that (cf. (3.14)),

$$\alpha = q_i = q_{i,i+1} = \lambda, \qquad i=0, 1, \ldots,$$

$$q_{i,j} = 0 \quad \text{for} \quad i \neq j, \quad i \neq j-1; \quad i, j=0, 1, \ldots .$$

Hence, from theorem 3.2 with $\beta = \alpha$, we have

$$p_{ij} = \delta_{ij} + \frac{1}{\alpha} q_{ij} = \begin{cases} 1 & \text{for} \quad j=i+1; \quad i=0, 1, \ldots, \\ 0 & \text{for} \quad j \neq i+1; \quad i, j=0, 1, \ldots, \end{cases}$$

so that

$$p_{ij}^{(n)} = \begin{cases} 1 & \text{for} \quad j=i+n, \quad i=0, 1, \ldots; \quad n=1, 2, \ldots, \\ 0 & \text{for} \quad j \neq i+n, \quad i=0, 1, \ldots; \quad n=1, 2, \ldots, \end{cases}$$

and from (3.25) that, for $t \geqq 0$,

$$p_{ij}(t) = \begin{cases} \dfrac{(\lambda t)^{j-i}}{(j-i)!} \, \mathrm{e}^{-\lambda t}, & j=i, i+1, \ldots, \\ 0, & j=0, 1, \ldots, i-1. \end{cases} \tag{4.19}$$

Since, by definition, the Poisson process is a Markov chain $\{\boldsymbol{x}_t, \ t \in [0, \infty)\}$ with stationary transition probabilities it follows from (4.19) that

$$\begin{aligned} \Pr\{\boldsymbol{x}_{t+s} - \boldsymbol{x}_s = k \mid \boldsymbol{x}_s = i\} &= \Pr\{\boldsymbol{x}_{t+s} = i+k \mid \boldsymbol{x}_t = i\} \\ &= \frac{(\lambda t)^k}{k!} \, \mathrm{e}^{-\lambda t}, \qquad i, k=0, 1, \ldots; \quad t, s \geqq 0. \end{aligned} \tag{4.20}$$

The right-hand side of (4.20) is independent of i for all $i=0, 1, \ldots$; using this property it is easily shown that $\boldsymbol{x}_t - \boldsymbol{x}_s$, $t > s$, is a non-negative stochastic variable and that $\boldsymbol{x}_t - \boldsymbol{x}_s$ and $\boldsymbol{x}_s$ are independent variables for every fixed t and s with $t > s \geqq 0$. Writing

$$k_h \overset{\text{def}}{=} i_{n+h} - i_{n+h-1}, \qquad h=1, \ldots, m,$$

and using this property, we see that it follows from the relation (3.1) that

$$\begin{aligned} \Pr\{\boldsymbol{x}_{t_{n+h}} = i_{n+h}, \ h=1, \ldots, m \mid \boldsymbol{x}_{t_n} = i_n\} \\ &= \Pr\{\boldsymbol{x}_{t_{n+h}} - \boldsymbol{x}_{t_{n+h-1}} = k_h, \ h=1, \ldots, m \mid \boldsymbol{x}_{t_n} = i_n\} \\ &= \prod_{h=1}^{m} \Pr\{\boldsymbol{x}_{t_{n+h}} = i_{n+h} \mid \boldsymbol{x}_{t_{n+h-1}} = i_{n+h-1}\} \\ &= \prod_{h=1}^{m} \Pr\{\boldsymbol{x}_{t_{n+h}} - \boldsymbol{x}_{t_{n+h-1}} = k_h \mid \boldsymbol{x}_{t_{n+h-1}} = i_{n+h-1}\} \\ &= \prod_{h=1}^{m} \Pr\{\boldsymbol{x}_{t_{n+h}} - \boldsymbol{x}_{t_{n+h-1}} = k_h\}. \end{aligned}$$

Consequently,

$$\begin{aligned} \Pr\{\boldsymbol{x}_{t_{n+h}} - \boldsymbol{x}_{t_{n+h-1}} = k_h, \ h=1, \ldots, m \mid \boldsymbol{x}_{t_n} = i_n\} \\ = \Pr\{\boldsymbol{x}_{t_{n+h}} - \boldsymbol{x}_{t_{n+h-1}} = k_h, \ h=1, \ldots, m\}. \end{aligned}$$

These relations show that the stochastic variables $\boldsymbol{x}_{t_{n+h}} - \boldsymbol{x}_{t_{n+h-1}}$, $h=1, \ldots, m$, are independent. Hence, denoting by

$$\boldsymbol{y}_{s,t} \overset{\text{def}}{=} \boldsymbol{x}_t - \boldsymbol{x}_s, \qquad t > s \geqq 0,$$

the increment of the Poisson process during the time interval $[s, t)$ (note that we defined the sample functions to be continuous from the left, see section 3.2) the results obtained above imply the following theorem.

THEOREM 4.1. *In a Poisson process the increments* y_{s_i,t_i}, $i=1, \ldots, m$ *of disjoint intervals* $[s_i, t_i)$, $0 \leqq s_1 < t_1 \leqq s_2 < \ldots \leqq s_m < t_m$, *are independent variables; they all have Poisson distributions and*

$$\Pr\{y_{s,t}=k\} = \frac{\{\lambda(t-s)\}^k}{k!} e^{-\lambda(t-s)}, \qquad k=0, 1, \ldots; \quad t>s \geqq 0.$$

The Poisson process has been introduced here as a birth process; with this interpretation theorem 4.1 states that the numbers of births in disjoint time intervals are independent Poisson distributed variables. It follows further that $e^{-\lambda t}$ is the probability that during an interval of length t no birth occurs.

Next we consider the joint distribution of the time points of birth during an interval of length t when it is known that exactly k births occurred during the interval t.

THEOREM 4.2. *If an interval of length t contains exactly k birth points then the joint distribution of the instants at which births occurred is that of k points uniformly distributed over an interval of length t.*

Proof. Let $[t_i, t_i+h_i)$, $h_i>0$, $i=1, \ldots, k$, $0 \leqq t_1 < t_2 < \ldots < t_k < t$, be k non-overlapping intervals all contained in $[0, t]$, and let $O_1, \ldots, O_k$ denote k points placed at random and independent of one another in $[0, t]$, so that the abscissa of each point is uniformly distributed over $[0, t]$. Then the joint probability that each of the intervals $[t_i, t_i+h_i)$, $i=1, \ldots, k$, contains exactly one point is equal to

$$\frac{k!}{t^k} h_1 h_2 \ldots h_k,$$

since no distinction is made between the $k!$ permutations of the k points $O_1, \ldots, O_k$ over the k intervals $[t_i, t_i+h_i)$.

The conditional probability that whenever the interval $[0, t)$ contains k births each of the intervals $[t_i, t_i+h_i)$ contains exactly one birth is given by

$$\Pr\{y_{t_i,t_i+h_i}=1,\ y_{t_{j-1}+h_{j-1},t_j}=0,\ i=1, \ldots, k;\ j=1, \ldots, k+1 \mid y_{0,t}=k\}$$
$$= \frac{\Pr\{y_{t_i,t_i+h_i}=1,\ y_{t_{j-1}+h_{j-1},t_j}=0,\ i=1, \ldots, k;\ j=1, \ldots, k+1\}}{\Pr\{y_{0,t}=k\}},$$

where $t_0 + h_0 \stackrel{\text{def}}{=} 0$, $t_{k+1} \stackrel{\text{def}}{=} t$. From theorem 4.1 it follows that the above probability is equal to

$$\frac{\prod_{i=1}^{k} \lambda h_i \, e^{-\lambda h_i} \prod_{j=1}^{k+1} \exp\{-\lambda[t_j-(t_{j-1}-h_{j-1})]\}}{\{(\lambda t)^k/k!\}\, e^{-\lambda t}} = \frac{k!}{t^k} h_1 h_2 \dots h_k.$$

This proves the theorem.

Above, the Poisson process has been defined as a birth process with constant birth rate. Next, we shall introduce the Poisson process as a renewal process (see for the definition of renewal processes section 6.1).

THEOREM 4.3. *Let z_n, $n=1, 2, \dots$, denote a sequence of independent non-negative stochastic variables each being negative exponentially distributed with parameter λ. The stochastic process $\{\boldsymbol{x}_t, t\in[0, \infty)\}$ with the integer valued stochastic variable $\boldsymbol{x}_t$ for every fixed t defined by*

$$\boldsymbol{x}_0 \stackrel{\text{def}}{=} 0, \qquad \boldsymbol{x}_t = \max\{n: z_1+\dots+z_n<t\}, \qquad t>0,$$

is a Poisson process.

Proof. Since for all $i=1, 2, \dots$,

$$\Pr\{z_i<t\} = \begin{cases} 1 - e^{-\lambda t}, & t\geqq 0, \\ 0, & t<0, \end{cases}$$

and all z_i, $i=1, 2, \dots$, are independent variables it follows for all $k=0, 1, \dots$,

$$\Pr\{z_{1+k}+\dots+z_{n+k}<t\} = \int_0^t \lambda \frac{(\lambda\tau)^{n-1}}{(n-1)!} e^{-\lambda\tau} \, d\tau, \qquad n=1, 2, \dots .$$

The definition of $\boldsymbol{x}_t$ implies that, for $t>0$,

$$\boldsymbol{x}_t = \begin{cases} 0 & \text{if and only if} \quad z_1\geqq t, \\ n & \text{if and only if} \quad z_1+\dots+z_n<t\leqq z_1+\dots+z_{n+1}, \quad n=1, 2, \dots . \end{cases}$$

Hence, with $t_0 \stackrel{\text{def}}{=} 0$, we have

$$\Pr\{\boldsymbol{x}_t-\boldsymbol{x}_0=n \mid \boldsymbol{x}_0=0\} = \int_0^t \Pr\{z_{n+1}>t-\sigma\} \, d\Pr\{z_1+\dots+z_n<\sigma\}$$

$$= \frac{(\lambda t)^n}{n!} e^{-\lambda t}, \qquad t\geqq 0, \quad n=0, 1, \dots .$$

Putting

$$s_n \overset{\text{def}}{=} z_1 + \dots + z_n, \qquad n=1, 2, \dots,$$

we see that for $u \geqq 0$,

$$\Pr\{x_t - x_0 = n,\ s_{n+1} - t > u \mid x_0 = 0\} = \int_0^t \Pr\{z_{n+1} > t+u-\sigma\}\ \mathrm{d}\Pr\{s_n < \sigma\}$$

$$= \frac{(\lambda t)^n}{n!}\ \mathrm{e}^{-\lambda(t+u)} = \Pr\{x_t - x_0 = n \mid x_0 = 0\}\ \mathrm{e}^{-\lambda u}, \qquad n=0, 1, \dots .$$

With $t_2 > t_1$, $u > 0$, we have

$$\Pr\{x_{t_1} = n_1,\ x_{t_2} = n_1 + n_2,\ s_{n_1+n_2+1} - t_2 > u \mid x_0 = 0\}$$

$$= \int_{\sigma=t_1}^{t_2} \mathrm{d}_\sigma \Pr\{x_{t_1} - x_{t_0} = n_1,\ s_{n_1+1} - t_1 + s_{n_1+n_2} - s_{n_1+1} < \sigma,\ z_{n_1+n_2+1} > t_2 - t_1 + u - \sigma \mid x_0 = 0\}$$

$$= \int_{\sigma=t_1}^{t_2} \frac{\{\lambda(t_1 - t_0)\}^{n_1}}{n_1!}\ \mathrm{e}^{-\lambda(t_2-t_1)}\ \mathrm{e}^{-\lambda(t_2-t_1+u-\sigma)}\ \frac{(\lambda\sigma)^{n_2-1}}{(n_2-1)!}\ \mathrm{e}^{-\lambda\sigma}\ \lambda\ \mathrm{d}\sigma$$

$$= \frac{\{\lambda(t_1 - t_0)\}^{n_1}}{n_1!}\ \mathrm{e}^{-\lambda(t_1-t_0)}\ \frac{\{\lambda(t_2 - t_1)\}^{n_2}}{n_2!}\ \mathrm{e}^{-\lambda(t_2-t_1)}\ \mathrm{e}^{-\lambda u},$$

where we used the fact that s_{n_1+1} and $s_{n_1+n_2} - s_{n_1+1}$ are independent and assumed that $n_2 \geqq 1$. However, it is easily verified that this result is also true for $n_2 = 0$. By complete induction with respect to $n = 1, 2, \dots$, it is now easily proved that for any set of non-negative integers i_k, $k = 1, \dots, n$ and any set of time points $t_n > t_{n-1} > \dots > t_1 > t_0 = 0$,

$$\Pr\{x_{t_k} = \sum_{h=1}^{k} i_h,\ k=1, \dots, n,\ s_{i_1+\dots+i_n+1} - t_n > u \mid x_0 = 0\}$$

$$= \mathrm{e}^{-\lambda u} \prod_{k=1}^{n} \frac{\{\lambda(t_k - t_{k-1})\}^{i_k}}{i_k!}\ \mathrm{e}^{-\lambda(t_k - t_{k-1})}, \qquad u \geqq 0,$$

and hence

$$\Pr\{x_{t_k} - x_{t_{k-1}} = i_k,\ k=1, \dots, n \mid x_0 = 0\}$$

$$= \Pr\{x_{t_k} = \sum_{h=1}^{k} i_h,\ k=1, \dots, n \mid x_0 = 0\}$$

$$= \prod_{k=1}^{n} \frac{\{\lambda(t_k - t_{k-1})\}^{i_k}}{i_k!}\ \mathrm{e}^{-\lambda(t_k - t_{k-1})}.$$

From this relation it follows immediately that for $t>s\geqq 0$,

$$\Pr\{x_t-x_s=k\} = \frac{\{\lambda(t-s)\}^k}{k!}\,\mathrm{e}^{-\lambda(t-s)}, \qquad k=0, 1, \ldots,$$

and that the increments $x_{t_h}-x_{t_{h-1}}$, $h=1, 2, \ldots$, are independent variables, each one being Poisson distributed. It is further seen from the above relation that

$$\Pr\{x_{t_n} = \sum_{k=1}^{n} i_k \mid x_{t_h} = \sum_{k=1}^{h} i_k,\ h=1, \ldots, n-1,\ x_0=0\}$$

$$= \Pr\{x_{t_n} = \sum_{k=1}^{n} i_k \mid x_{t_{n-1}} = \sum_{k=1}^{n-1} i_k\},$$

so that $\{x_t,\ t\in[0, \infty)\}$ is a Markov chain with continuous time parameter and state space the set of non-negative integers. Since for $t>s$ and $i=0, 1, \ldots$,

$$\Pr\{x_t=j \mid x_s=i\} = \begin{cases} \dfrac{\{\lambda(t-s)\}^{j-i}}{(j-i)!}\,\mathrm{e}^{-\lambda(t-s)}, & j=i, i+1, \ldots, \\ 0, & j=0, 1, \ldots, i-1, \end{cases}$$

it follows immediately from the definition of section 4.1 that $\{x_t,\ t\in[0, \infty)\}$ is a birth process with constant birth rate λ, and hence $\{x_t,\ t\in[0, \infty)\}$ is a Poisson process. This completes the proof.

A third possibility of defining a Poisson process is described in the following theorem.

THEOREM 4.4. *For every* $t\geqq 0$ *let* x_t *be a non-negative integer valued stochastic variable with* $x_0 \overset{\text{def}}{=} 0$. *Then* $\{x_t,\ t\in[0, \infty)\}$ *is a Poisson process if: (i) for all* $t\geqq 0, s\geqq 0$, *the distribution of* $x_{t+s}-x_s$ *is independent of* s; *(ii)* $\{x_t,\ t\in[0, \infty)\}$ *has independent increments, i.e. for all sets of time points* $t_1, \ldots, t_n$ *with* $t_n>t_{n-1}>\ldots>t_0\geqq 0$, $n=2, 3, \ldots$, *the stochastic variables* $x_{t_h}-x_{t_{h-1}}$, $h=1, \ldots, n$ *are independent; (iii)* $\Pr\{x_{t+h}-x_t\geqq 2\}=\mathrm{o}(h)$ *for* $h\downarrow 0$.

Proof. Because of (*i*) we write

$$p_k(t) \overset{\text{def}}{=} \Pr\{x_{t+s}-x_s=k\}, \qquad k=0, 1, \ldots;\quad s\geqq 0.$$

From (*i*) and (*ii*) it follows that for every $\tau>0$,

$$p_0(n\tau) = \Pr\{x_{n\tau}-x_0=0\} = \Pr\{x_{h\tau}-x_{(h-1)\tau}=0,\ h=1, \ldots, n\}$$

$$= \prod_{h=1}^{n} \Pr\{x_{h\tau}-x_{(h-1)\tau}=0\} = p_0^n(\tau), \qquad n=1, 2, \ldots,$$

hence with $s=n\tau$,

$$p_0(s)=\left\{p_0\left(\frac{s}{n}\right)\right\}^n \quad \text{and} \quad \{p_0(s)\}^{k/n}=p_0\left(\frac{k}{n}s\right), \qquad k=1, 2, \ldots .$$

For a given $\sigma>0$ and a given integer $n\geqq 1$, let the integer m be defined by

$$\frac{m-1}{n}\tau\leqq\sigma\leqq\frac{m}{n}\tau.$$

Since

$$\begin{aligned}p_0(t)&=\Pr\{x_t-x_s=0,\ x_s-x_0=0\}\\&=\Pr\{x_t-x_s=0\}\,p_0(s)\leqq p_0(s), \qquad t>s>0,\end{aligned}$$

it follows that

$$\{p_0(\tau)\}^{m/n}\leqq p_0(\sigma)\leqq\{p_0(\tau)\}^{(m-1)/n}.$$

If $n\to\infty$, then $m\to\infty$ and from the definition of m it follows that $m/n\to\sigma/\tau$ and therefore that

$$p_0(\sigma)=\{p_0(\tau)\}^{\sigma/\tau}.$$

Consequently, a non-negative constant λ exists such that

$$p_0(\sigma)=\mathrm{e}^{-\lambda\sigma}, \qquad \sigma>0.$$

We may exclude the cases $\lambda=0$ and $\lambda=\infty$. For if $\lambda=0$ then $p_0(\sigma)=1$ for every $\sigma>0$, and consequently $\Pr\{x_t=0\}$ for all $t>0$. It is now easily proved that $\{x_t,\ t\in[0,\ \infty)\}$ is indeed a Poisson process, but it is degenerate and not of any interest. If $\lambda=\infty$ then $p_0(\sigma)=0$ for all $\sigma>0$, and hence

$$\Pr\{x_t-x_s\geqq 1,\ x_s-x_0\geqq 1\}=\{1-p_0(t-s)\}\{1-p_0(s)\}=1, \qquad t>s>0,$$

so that with $t=2s$,

$$\Pr\{x_{2s}-x_0\geqq 2\}=1 \qquad \text{for all } s>0,$$

and this result contradicts condition (*iii*) of the theorem. From now on we take $0<\lambda<\infty$.
From (*i*) and (*ii*) it now follows that

$$\begin{aligned}p_k(t+\tau)&=\sum_{h=0}^{k}\Pr\{x_{t+\tau}-x_t=h\}\Pr\{x_t-x_0=k-h\}\\&=\sum_{h=0}^{k}p_h(\tau)\,p_{k-h}(t), \qquad k=0, 1, \ldots .\end{aligned}$$

Hence

$$p_0(t+\tau) - p_0(t) = -\{1-p_0(\tau)\}\, p_0(t),$$

$$p_1(t+\tau) - p_1(t) = -\{1-p_0(\tau)\}\, p_1(t) + p_1(\tau)\, p_0(t),$$

$$p_k(t+\tau) - p_k(t) = -\{1-p_0(\tau)\}\, p_k(t) + p_1(\tau)\, p_{k-1}(t) + \sum_{h=2}^{k} p_h(\tau)\, p_{k-h}(t),$$

$$k=2, \dots .$$

We have from $p_0(\tau)=\exp(-\lambda\tau)$,

$$1 - p_0(\tau) = \lambda\tau + \mathrm{o}(\tau), \qquad \tau\downarrow 0,$$

and from (*iii*)

$$p_1(\tau) = 1 - p_0(\tau) - \sum_{k=2}^{\infty} p_k(\tau)$$

$$= \lambda\tau + \mathrm{o}(\tau), \qquad \tau\downarrow 0,$$

and

$$\sum_{h=2}^{k} p_h(\tau)\, p_{k-h}(t) \leqq \sum_{h=2}^{k} p_h(\tau) = \mathrm{o}(\tau), \qquad \tau\downarrow 0, \quad k=2, 3, \dots .$$

Hence, we obtain for $\tau\downarrow 0$,

$$\frac{p_k(t+\tau)-p_k(t)}{\tau} = \begin{cases} -\lambda p_0(t) + \dfrac{\mathrm{o}(\tau)}{\tau}, & k=0, \\[2ex] -\lambda p_k(t) + \lambda p_{k-1}(t) + \dfrac{\mathrm{o}(\tau)}{\tau}, & k=1, 2, \dots . \end{cases}$$

Consequently, $p_k(t)$ has a right-hand derivative for $t>0$. Replacing in the relations above $t+\tau$ by s and t by $s-\sigma$, it is seen that for $\sigma\downarrow 0$ the left-hand derivative of $p_k(s)$ also exists; these derivatives appear to be equal since it also follows from the relations above that $p_k(t)$ is continuous in $t>0$. Hence, we obtain for $t>0$,

$$\frac{\mathrm{d}}{\mathrm{d}t} p_0(t) = -\lambda p_0(t),$$

$$\frac{\mathrm{d}}{\mathrm{d}t} p_k(t) = -\lambda p_k(t) + \lambda p_{k-1}(t), \qquad k=1, 2, \dots .$$

Since $\boldsymbol{x}_0=0$, it follows that

$$p_k(0+) = \begin{cases} 1 & \text{for } k=0, \\ 0 & \text{for } k=1, 2, \dots . \end{cases}$$

It is immediately obvious that this set of differential equations together with the boundary conditions determine $p_k(t)$, $k=0, 1, \ldots; t>0$, uniquely, and it is easily found that

$$p_k(t) = \Pr\{x_{t+s} - x_s = k\} = \frac{(\lambda t)^k}{k!} e^{-\lambda t}, \qquad t>0, \quad s \geqq 0, \quad k=0, 1, \ldots .$$

Since $\{x_t,\ t \in [0, \infty)\}$ is a process with independent and stationary increments and the above results show that these increments all have the Poisson distribution we have now reached a result occurring also in the proof of theorem 4.3. Hence the proof now proceeds as does that of theorem 4.3. The proof is now complete.

I.4.3. Death process with constant death rate

The birth process with constant birth rate turned out to be the Poisson process. We shall now consider the death process with constant death rate; this process is also important in queueing theory.

Since the death rate is constant we write (cf. (4.1)),

$$\mu_i = \mu, \qquad i=1, 2, \ldots .$$

Suppose the system starts at time $t=0$ in state E_i, i.e. that

$$\Pr\{x_0 = i\} = 1.$$

From the definition of a death process it is evident that for all $t \geqq 0$,

$$p_{jh}(t) = \begin{cases} 1 & \text{for } h=j=0, \\ 0 & \text{for } h=j+1, j+2, \ldots; \quad j=0, 1, \ldots; \end{cases}$$

a result which is also easily deduced from the backward differential equations (4.2). From the forward differential equations (4.3) and the relations above we now obtain, for $t>0$,

$$\begin{aligned} \frac{\mathrm{d}}{\mathrm{d}t} p_{i0}(t) &= \mu p_{i1}(t), \\ \frac{\mathrm{d}}{\mathrm{d}t} p_{ij}(t) &= -\mu p_{ij}(t) + \mu p_{i,j+1}(t), \qquad j=1, \ldots, i-1, \\ \frac{\mathrm{d}}{\mathrm{d}t} p_{ii}(t) &= -\mu p_{ii}(t), \end{aligned}$$

and

$$p_{ij}(0+) = \delta_{ij}.$$

The solution of this set of differential equations can be found directly or by applying theorem 3.2. It is easily verified that the solution is given by

$$p_{ij}(t) = \begin{cases} \int_0^t \mu \frac{(\mu\sigma)^{i-1}}{(i-1)!} e^{-\mu\sigma} d\sigma, & j=0; \quad i=1, 2, \ldots; \quad t>0, \\ \frac{(\mu t)^{i-j}}{(i-j)!} e^{-\mu t}, & j=1, 2, \ldots, i; \quad i=1, 2, \ldots; \quad t>0. \end{cases}$$

I.4.4. Birth and death process with constant birth and death rates

The birth and death process with constant birth and death rates was one of the first queueing processes investigated. Here, we shall not discuss this process as a queueing process but postpone such an interpretation to section II.2.1.

According to the definition of a birth and death process (cf. (4.1)) we write

$$\lambda \stackrel{\text{def}}{=} \lambda_i, \qquad i=0, 1, \ldots,$$

$$\mu \stackrel{\text{def}}{=} \mu_i, \qquad i=1, 2, \ldots,$$

and define

$$a \stackrel{\text{def}}{=} \lambda/\mu.$$

Hence it follows from (4.8) that

$$\pi_n = a^n, \qquad n=0, 1, \ldots .$$

Applying the criteria developed in section 4.1 it is immediately found that the birth and death process with constant birth rate λ and constant death rate μ is

ergodic if and only if $\lambda/\mu = a < 1$,

null recurrent if and only if $a=1$,

transient if and only if $a>1$.

If the process is ergodic, so that $a<1$, then it follows from (4.14) that the stationary distribution $\{u_j, j=0, 1, \dots\}$ of the process is given by

$$u_j = a^j(1-a), \qquad j=0, 1, \dots, \tag{4.21}$$

and that the average first passage times from E_i into E_0, and into E_1, respectively, are given by (cf. (4.17)),

$$\nu_{i0} = \frac{i}{\mu(1-a)}, \qquad i=1, 2, \dots,$$

$$\nu_{i1} = \frac{i-1}{\mu(1-a)}, \qquad i=2, 3, \dots .$$

Next, we shall determine $p_{ij}(t)$, $t>0$, i.e. the transition probabilities of the process. Our starting point will be the forward differential equations with the appropriate boundary conditions as given by (4.9) and (4.10). Introducing a new time parameter

$$\tau \stackrel{\text{def}}{=} \mu t,$$

and writing

$$p_j(\tau) \stackrel{\text{def}}{=} P_j(\tau/\mu), \qquad j=0, 1, \dots; \quad t \geqq 0,$$

the differential equations and boundary conditions read

$$\frac{\mathrm{d}}{\mathrm{d}\tau} p_0(\tau) = -ap_0(\tau) + p_1(\tau), \tag{4.22}$$

$$\frac{\mathrm{d}}{\mathrm{d}\tau} p_j(\tau) = -(a+1)p_j(\tau) + ap_{j-1}(\tau) + p_{j+1}(\tau), \qquad j=1, 2, \dots,$$

$$p_j(0+) = \delta_{ij}.$$

To obtain the solution of the system (4.22) we introduce the Laplace transform of $p_j(\tau)$. Since theorem 3.2 implies that $p_j(\tau)$ is a continuous and bounded function for $\tau>0$ its Laplace transform will exist for $\operatorname{Re} \rho>0$, i.e.

$$\pi_j(\rho) \stackrel{\text{def}}{=} \int_0^\infty \mathrm{e}^{-\rho\tau} p_j(\tau)\,\mathrm{d}\tau, \qquad \operatorname{Re} \rho > 0. \tag{4.23}$$

From (4.22) and (4.23) it follows

$$-(a+\rho)\,\pi_0(\rho) + \pi_1(\rho) = -\delta_{i0},$$

$$-(1+a+\rho)\,\pi_j(\rho) + a\pi_{j-1}(\rho) + \pi_{j+1}(\rho) = -\delta_{ij}, \qquad j=1, 2, \dots .$$

Since $|p_j(\tau)| \leqq 1$ for $\tau \geqq 0, j=0, 1, \ldots$, it follows that we may define the generating function $P(p, \rho)$ of the sequence $\pi_j(\rho), j=0, 1, \ldots$,

$$P(p, \rho) \stackrel{\text{def}}{=} \sum_{j=0}^{\infty} \pi_j(\rho)\, p^j, \qquad |p|<1. \tag{4.24}$$

From the set of equations for $\pi_j(\rho)$ it is now found that

$$-(1+a+\rho)\,P(p,\rho) + \pi_0(\rho) + apP(p,\rho) + \frac{1}{p}\{P(p,\rho) - \pi_0(\rho)\} = -p^i,$$

so that

$$P(p, \rho) = \frac{(1-p)\,\pi_0(\rho) - p^{i+1}}{ap^2 - (1+a+\rho)\,p + 1}.$$

To determine $\pi_0(\rho)$ we first investigate the equation in x,

$$ax^2 - (1+a+\rho)\,x + 1 = 0, \qquad a>0, \quad \operatorname{Re}\rho>0.$$

The roots x_1 and x_2 are given by

$$x_1 = \frac{1+a+\rho}{2a} + \frac{1}{2a}\sqrt{\{(1+a+\rho)^2 - 4a\}}, \tag{4.25}$$

$$x_2 = \frac{1+a+\rho}{2a} - \frac{1}{2a}\sqrt{\{(1+a+\rho)^2 - 4a\}}.$$

We shall prove that

$$|x_1|<1, \qquad |x_2|>1 \qquad \text{for } \operatorname{Re}\rho>0.$$

To prove this we apply Rouché's theorem (see app. 6) with D the unit circle and

$$f(x) \stackrel{\text{def}}{=} 1 - (1+a+\rho)\,x, \qquad g(x) \stackrel{\text{def}}{=} ax^2.$$

Obviously, $f(x)$ and $g(x)$ are regular inside and on D. On this contour $|x|=1$ and

$$|1-(1+a+\rho)x| \geqq ||(1+a+\rho)x|-1| = |1+a+\rho|-1 > a = a\,|x^2|,$$

so that since $f(z)$ has exactly one root inside D our statement has been proved.

For every fixed ρ with $\operatorname{Re}\rho>0$ the generating function $P(p, \rho)$ should be an analytic function of p for $|p|<1$, since $\pi_j(\rho)$ is uniformly bounded in j. Hence, $p=x_2$ should be a zero of the denominator in the expression for

$P(p, \rho)$. Therefore we have

$$\pi_0(\rho) = \frac{x_2^{i+1}}{1-x_2}. \tag{4.26}$$

Consequently, we have

$$\begin{aligned} P(p, \rho) &= \frac{(1-p)\, x_2^{i+1} - (1-x_2)\, p^{i+1}}{\{ap^2-(1+a+\rho)\,p+1\}(1-x_2)} \\ &= \frac{(1-p)\, x_2^{i+1} - (1-x_2)\, p^{i+1}}{a(1-x_2)(x_1-x_2)} \left\{ \frac{1}{x_2-p} - \frac{1}{x_1-p} \right\}, \\ & \qquad |p|<1, \quad \mathrm{Re}\, \rho>0. \end{aligned} \tag{4.27}$$

Expanding the right-hand side of the above expression in powers of p we see from the definition of $P(p, \rho)$ that the coefficient of p^j represents $\pi_j(\rho)$. Hence we have

$$\begin{aligned} \pi_j(\rho) &= \frac{x_2^{i+1}}{a(1-x_2)(x_1-x_2)} \{x_2^{-j-1} - x_2^{-j} - (x_1^{-j-1} - x_1^{-j})\} \\ &\quad - \frac{\lambda_{ij}}{a(x_1-x_2)} \{x_2^{-(j-i)} - x_1^{-(j-i)}\} \\ &= (1-\lambda_{ij}) \frac{x_2^{-(j-i)}}{a(x_1-x_2)} + \lambda_{ij} \frac{a^{j-i} x_2^{j-i}}{a(x_1-x_2)} \\ &\quad + \frac{a^j}{\rho} \frac{x_2^{j+i} - 2a x_2^{j+i+1} + a^2 x_2^{j+i+2}}{a(x_1-x_2)}, \qquad j=0, 1, \ldots, \end{aligned} \tag{4.28}$$

where

$$\lambda_{ij} \overset{\text{def}}{=} \begin{cases} 0 & \text{for } j=0, 1, \ldots, i, \\ 1 & \text{for } j=i+1, i+2, \ldots . \end{cases}$$

Since (cf. ERDÉLYI [1954]),

$$\int_0^\infty \mathrm{e}^{-\rho t}\, b^j I_j(bt)\, \mathrm{d}t = \frac{\{\rho - \sqrt{(\rho^2-b^2)}\}^j}{\sqrt{(\rho^2-b^2)}}, \qquad \mathrm{Re}\, \rho > b, \quad j=0, 1, \ldots,$$

where

$$I_j(x) = I_{-j}(x) = \sum_{n=0}^{\infty} \frac{(\frac{1}{2}x)^{j+2n}}{(j+n)!\,n!}, \qquad j=0, 1, \ldots,$$

is the modified Bessel function of the first kind, it follows immediately from (4.25) that

$$\int_0^\infty e^{-\rho t}\, a^{-\frac{1}{2}j}\, e^{-(1+a)t}\, I_j(2t\sqrt{a})\, dt = \frac{x_2^j}{a(x_1-x_2)}, \qquad \operatorname{Re}\rho>0, \quad j=0, 1, \dots . \tag{4.29}$$

From (4.28), (4.23) and the above relations we now find that

$$\begin{aligned} p_j(\tau) &= a^{\frac{1}{2}(j-i)}\, e^{-(1+a)\tau}\, I_{j-i}(2\tau\sqrt{a}) \\ &\quad + a^{\frac{1}{2}(j-i)} \int_0^\tau e^{-(1+a)\sigma} \{I_{j+i}(2\sigma\sqrt{a}) - 2a^{\frac{1}{2}} I_{j+i+1}(2\sigma\sqrt{a}) \\ &\quad + a I_{j+i+2}(2\sigma\sqrt{a})\}\, d\sigma, \qquad j=0, 1, \dots . \end{aligned} \tag{4.30}$$

Since (cf. ERDÉLYI [1954]),

$$\int_0^\infty e^{-\rho t}\, \frac{k a^{-\frac{1}{2}k} I_k(2t\sqrt{a})}{t}\, e^{-(1+a)t}\, dt = x_2^k, \qquad k=1,2, \dots; \quad \operatorname{Re}\rho>0,$$

and since $|x_2|<1$, it follows from

$$\pi_0(\rho) = \frac{x_2^{i+1}}{1-x_2} = \sum_{k=0}^\infty x_2^{i+k+1},$$

that we also have

$$p_0(t) = \sum_{k=0}^\infty \frac{k+i+1}{t}\, a^{-\frac{1}{2}(k+i+1)}\, e^{-(1+a)t}\, I_{k+i+1}(2t\sqrt{a}), \qquad i=0, 1, \dots .$$

Consequently, it is seen that $p_0(t)\geqq 0$ for all $t\geqq 0$; this implies that

$$\begin{aligned} I_i(2\tau\sqrt{a})\, e^{-(1+a)\tau} + \int_0^\tau e^{-(1+a)\sigma}\{I_i(2\sigma\sqrt{a}) - 2a^{\frac{1}{2}} I_{i+1}(2\sigma\sqrt{a}) \\ + a I_{i+2}(2\sigma\sqrt{a})\}\, d\sigma \geqq 0, \end{aligned}$$

for $t\geqq 0$, and all $i=0, 1, \dots$. Replacing in this expression i by $i+j$ and noting that the well-known recurrence relation

$$I_n(t) = I_{n+2}(t) + \frac{2(n+1)}{t}\, I_{n+1}(t), \qquad n=0, 1, \dots; \quad t\geqq 0,$$

implies

$$I_{j-i}(2\tau\sqrt{a}) \geqq I_{j+i}(2\tau\sqrt{a}), \qquad \tau\geqq 0,$$

it follows that $p_j(\tau)$ as given by the above relation is non-negative for all $t \geqq 0$, $i, j=0, 1, \ldots$. Using this fact we have from (4.28) that

$$\int_0^\infty e^{-\rho t} \sum_{j=0}^\infty p_j(t)\, dt = \sum_{j=0}^\infty \pi_j(\rho) = \frac{1}{\rho},$$

and hence

$$\sum_{j=0}^\infty p_j(\tau) = 1 \qquad \text{for all } \tau \geqq 0.$$

Since for every fixed $\tau > 0$, $p_j(\tau)$, $j=0, 1, \ldots$, is a probability distribution and represents a solution of the forward differential equations of the birth and death process it follows from theorem 3.3 that the transition probabilities

$$p_{ij}(t) = \Pr\{\boldsymbol{x}_{t+s}=j \mid \boldsymbol{x}_s=i\} = p_j(\mu t),$$

are given by

$$\begin{aligned} p_{ij}(t) = {} & a^{\frac{1}{2}(j-i)}\, e^{-(1+a)\mu t}\, I_{j-i}(2\mu t\sqrt{a}) \\ & + a^{\frac{1}{2}(j-i)} \int_0^t e^{-(1+a)\mu\sigma} \{I_{j+i}(2\mu\sigma\sqrt{a}) - 2a^{\frac{1}{2}} I_{j+i+1}(2\mu\sigma\sqrt{a}) \\ & + a I_{j+i+2}(2\mu\sigma\sqrt{a})\}\, \mu\, d\sigma, \end{aligned} \tag{4.31}$$

for $t > 0$ and $i, j=0, 1, \ldots$.
Starting again from (4.28) and noting that

$$(1-x_2)(1-ax_2) = \rho x_2, \qquad |x_2| < 1,$$

we see that

$$\begin{aligned} \frac{a_j}{\rho} \frac{x_2^{j+i}(1-ax_2)^2}{a(x_1-x_2)} &= \frac{a^j x_2^{j+i+1}}{a(x_1-x_2)} \left\{ a + \frac{1-a}{1-x_2} \right\} \\ &= \frac{a^j x_2^{j+i+1}}{a(x_1-x_2)} \{1+(1-a) \sum_{k=1}^\infty x_2^k\}. \end{aligned}$$

Consequently, we obtain from (4.28) and the above relation the following series expansion for the transition probabilities:

$$\begin{aligned} p_{ij}(t) = {} & a^{\frac{1}{2}(j-i)}\, e^{-(1+a)\mu t}\, I_{j-i}(2\mu t\sqrt{a}) \\ & + a^{\frac{1}{2}(j-i-1)}\, e^{-(1+a)\mu t}\, I_{j+i+1}(2\mu t\sqrt{a}) \\ & + (1-a)\, a^j\, e^{-(1+a)\mu t} \sum_{k=i+j+2}^\infty a^{-\frac{1}{2}k} I_k(2\mu t\sqrt{a}), \qquad t \geqq 0, \quad i, j=0, 1, \ldots. \end{aligned} \tag{4.32}$$

The process studied in this section is a Markov chain with continuous time parameter. Theorem 3.5 shows that $p_{ij}(t)$ has a limit for $t\to\infty$. Hence by applying an Abelian theorem for the Laplace transform (see app. 2) and taking ρ to be real we have

$$\lim_{\rho\downarrow 0} \rho \int_0^\infty \mathrm{e}^{-\rho t}\, p_{ij}(t)\,\mathrm{d}t = \lim_{t\to\infty} p_{ij}(t).$$

Noting that

$$\lim_{\rho\downarrow 0} a(x_1-x_2) = |1-a|,$$

$$\lim_{\rho\downarrow 0} x_2 = \begin{cases} 1 & \text{for } a<1, \\ \dfrac{1}{a} & \text{for } a\geqq 1, \end{cases}$$

we see from (4.28) that

$$\lim_{t\to\infty} p_{ij}(t) = \begin{cases} a^j(1-a) & \text{for } a<1, \\ 0 & \text{for } a\geqq 1, \quad i,j=0,1,\ldots, \end{cases}$$

a result which agrees with that found in the beginning of this section (cf. (4.21)). Introducing the unit step-function $U(t)$ defined by

$$U(t) \overset{\text{def}}{=} \begin{cases} 0, & t<0, \\ \frac{1}{2}, & t=0, \\ 1, & t>0, \end{cases}$$

we may rewrite (4.30) as

$$\begin{aligned} p_{ij}(t) = {} & (1-a)\,a^j U(1-a) + a^{\frac{1}{2}(j-i)}\,\mathrm{e}^{-(1+a)\mu t}\, I_{j-i}(2\mu t\sqrt{a}) \\ & - a^{\frac{1}{2}(j-i)} \int_t^\infty \mathrm{e}^{-(1+a)\mu\sigma} \{I_{i+j}(2\mu\sigma\sqrt{a}) - 2a^{\frac{1}{2}} I_{i+j+1}(2\mu\sigma\sqrt{a}) \\ & + a I_{i+j+2}(2\mu\sigma\sqrt{a})\}\,\mu\,\mathrm{d}\sigma, \qquad t\geqq 0,\quad i,j=0,1,\ldots . \end{aligned} \tag{4.33}$$

To investigate the behaviour of $p_{ij}(t)$ for large values of t we need the following asymptotic relations (cf. ERDÉLYI [1953]),

$$I_m(t) = \frac{\mathrm{e}^t}{\sqrt{2\pi t}} \left\{1 - \frac{4m^2-1}{8t} + \mathrm{O}\left(\frac{1}{t^2}\right)\right\}, \qquad t\to\infty,$$

$$\int_t^\infty \mathrm{e}^{-x}\, x^{\alpha-1}\,\mathrm{d}x = t^{\alpha-1}\,\mathrm{e}^{-t}\left\{1 - \frac{1-\alpha}{t} + \mathrm{O}\left(\frac{1}{t^2}\right)\right\}, \qquad t\to\infty.$$

From these relations it follows for $t \to \infty$ and $a \neq 1$,

$$\int_t^\infty \mathrm{e}^{-(1+a)\mu\sigma}\, I_m(2\mu\sigma\sqrt{a})\, \mu\, \mathrm{d}\sigma$$

$$= \frac{\exp\{-(1-\sqrt{a})^2\,\mu t\}}{(1-\sqrt{a})^2 \cdot 2\sqrt{(\pi\mu t\sqrt{a})}} \left\{1 - \frac{1}{2\mu t}(1-\sqrt{a})^{-2} - \frac{4m^2-1}{16\mu t\sqrt{a}} + \mathrm{O}\left(\frac{1}{t^2}\right)\right\}.$$

From (4.33) and the above asymptotic relations it is now found for $i, j = 0, 1, \ldots,$ and $t \to \infty$,

$$p_{ij}(t) = \begin{cases} (1-a)\, a^j U(1-a) + a^{\frac{1}{2}(j-i)} \dfrac{\exp\{-(1-\sqrt{a})^2\,\mu t\}}{2(\sqrt{\pi})(\mu t\sqrt{a})^{\frac{3}{2}}} \\ \cdot\left\{\left(j - \dfrac{\sqrt{a}}{1-\sqrt{a}}\right)\left(i - \dfrac{\sqrt{a}}{1-\sqrt{a}}\right) + \mathrm{O}\left(\dfrac{1}{t}\right)\right\} & \text{for } a \neq 1, \\ \dfrac{1}{\sqrt{\pi\mu t}}\left\{1 + \mathrm{O}\left(\dfrac{1}{t}\right)\right\} & \text{for } a = 1. \end{cases} \tag{4.34}$$

The expressions for $p_{ij}(t)$ have been found above. By using the integral relation (cf. (3.32)),

$$p_{ij}(t) = \delta_{ij}\, \mathrm{e}^{-q_i t} + \int_0^t p_{jj}(t-\tau)\, \mathrm{d}F_{ij}(\tau), \qquad t \geqq 0,$$

it is now possible to determine the distribution $F_{ij}(.)$ of the first entrance time from state E_i to E_j. As an example and also with a view to later application we shall determine $F_{10}(t)$.

Since

$$p_{10}(t) = \int_0^t p_{00}(t-\tau)\, \mathrm{d}F_{10}(\tau),$$

it follows by taking Laplace transforms and applying the expressions for the Laplace transforms of $p_{ij}(t)$ that

$$\int_0^\infty \mathrm{e}^{-\rho t}\, \mathrm{d}F_{10}(t) = \frac{1 + a + \rho/\mu - \sqrt{\{(1+a+\rho/\mu)^2 - 4a\}}}{2a}, \qquad \mathrm{Re}\, \rho > 0.$$

From this relation it is easily deduced that

$$F_{10}(t) = \begin{cases} 0, & t<0, \\ \int_0^t \frac{a^{-\frac{1}{2}} I_1(2\mu\sigma\sqrt{a})\, \mathrm{e}^{-(1+a)\mu\sigma}}{\sigma}\, \mathrm{d}\sigma, & t \geqq 0. \end{cases} \tag{4.35}$$

From (4.35) it follows that

$$\lim_{t\to\infty} F_{10}(t) = \lim_{\rho\downarrow 0} \int_0^\infty \mathrm{e}^{-\rho t}\, \mathrm{d}F_{10}(t) = \begin{cases} 1 & \text{if } a<1, \\ \frac{1}{a} & \text{if } a \geqq 1. \end{cases}$$

Hence, if $a \leqq 1$, $F_{10}(.)$ is a proper distribution whereas if $a>1$ it is improper. Therefore, if $a \geqq 1$ then $1-1/a$ is the probability of never reaching state E_0 when starting in E_1. Finally, it is noted that from the Laplace-Stieltjes transform of $F_{10}(t)$ the moments of this distribution are easily obtained. For instance

$$\nu_{10} = \lim_{\rho\downarrow 0} -\frac{\mathrm{d}}{\mathrm{d}\rho} \int_0^\infty \mathrm{e}^{-\rho t}\, \mathrm{d}F_{10}(t) = \frac{1}{\mu}\, \frac{1}{1-a},$$

a result which agrees with that found in the beginning of this section.

I.5. DERIVED MARKOV CHAINS

I.5.1. Introduction

Some years ago COHEN [1962[a], 1962[b], 1963] investigated a method by which from a given Markov chain, new Markov chains – the so called derived Markov chains – can be constructed. It turned out that some special derived Markov chains are very useful for obtaining solutions of queueing problems. In this chapter we shall describe a short outline of the theory of derived Markov chains, and consider some special derived processes which have been used in queueing theory (cf. COHEN [1963], ROES [1966] and section III.2.7). Derived Markov chains belong to the class of subordinated processes (cf. FELLER [1966]) and have been introduced by BOCHNER [1955], see also DYNKIN [1965].

Consider a class of probability distribution functions $b(s, t)$, $s \geqq 0$, $-\infty < t < \infty$, such that:
for every fixed $s \geqq 0$,

$$b(s,t) \begin{cases} = 0 & \text{for } t \leqq 0, \\ \geqq 0 & \text{for } t > 0, \end{cases}$$

$$\lim_{t\to\infty} b(s, t) = 1, \qquad b(s, t-) = b(s, t);$$

and for every fixed $t \geqq 0$,

$$b(s_1+s_2, t) = \int_{0-}^{t} b(s_1, t-\tau)\, \mathrm{d}_\tau b(s_2, \tau), \qquad s_1, s_2 \geqq 0, \tag{5.1}$$

so that for every $s \geqq 0$, $b(s, .)$ is a probability distribution of a non-negative stochastic variable. By taking the characteristic function of $b(s, .)$ it follows from (5.1) that for every fixed s, $b(s, .)$ is an infinitely divisible distribution function. Note: a probability distribution function is called *infinitely divisible* if for every natural number n its characteristic function is the nth power of a non-degenerate characteristic function (cf. LUKACS [1960]).

Let m denote a finite, non-negative constant and let $\Psi(t)$, $t\in(0, \infty)$ represent a non-decreasing function of t bounded above with the properties

$$\lim_{t\to\infty} \Psi(t) = 0, \qquad \int_0^1 t \, \mathrm{d}\Psi(t) < \infty.$$

It can be shown (cf. HILLE and PHILLIPS [1957], COHEN [1962b]) that every function $b(.,.)$ of the type described above determines uniquely a constant m and function $\Psi(.)$ as described above and conversely. The relation between $b(.,.)$ and the pair m, $\Psi(.)$ is given by

$$\int_{-\infty}^{\infty} \mathrm{e}^{-t\xi} \, \mathrm{d}_t b(s,t) = \exp\{-ms\xi - s\int_0^{\infty} (1-\mathrm{e}^{-t\xi}) \, \mathrm{d}\Psi(t)\}, \qquad \mathrm{Re}\,\xi \geqq 0, \quad s\geqq 0. \tag{5.2}$$

From (5.2) it follows that

$$\int_{-\infty}^{\infty} t \, \mathrm{d}_t b(s, t) = ms + s\int_0^{\infty} t \, \mathrm{d}\Psi(t), \qquad s\geqq 0,$$

in the sense that both members are infinite, or are finite and then equal.

Let ${}_1M$ denote a continuous time parameter Markov chain with state space S and with stationary transition matrix, which is assumed to be a standard transition matrix,

$${}_1P(t) \equiv ({}_1p_{ij}(t)), \qquad t\geqq 0, \quad i, j\in S.$$

The Q-matrix of ${}_1P(t)$ is

$${}_1Q \equiv ({}_1q_{ij}), \qquad i, j \in S,$$

and it is again assumed that ${}_1Q$ satisfies the conditions (3.14).

We define

$${}_1P(t) \stackrel{\text{def}}{=} I \qquad \text{for } t<0,$$

and introduce the matrix function

$${}_2P(s) \equiv ({}_2p_{ij}(s)), \qquad i, j\in S,$$

defined by

$${}_2p_{ij}(s) \stackrel{\text{def}}{=} \int_{-\infty}^{\infty} {}_1p_{ij}(t) \, \mathrm{d}_t b(s, t), \qquad s\geqq 0, \quad i, j\in S. \tag{5.3}$$

Since ${}_1p_{ij}(t)$ is bounded and continuous on $(-\infty, \infty)$ the left-hand side of (5.3) is well-defined. It is easily verified that

$$ {}_2p_{ij}(s) \geqq 0, \qquad \sum_{j\in S} {}_2p_{ij}(s) = 1, \qquad s \geqq 0, \quad i, j \in S. $$

Taking $\xi = -i\omega$ in (5.2) the complete convergence theorem (cf. LOÈVE [1960]) shows that $b(s, t)$ converges for $s \to 0$ completely to the unit step-function $U(t)$. Hence, since ${}_1p_{ij}(t)$ is bounded and continuous on $(-\infty, \infty)$, it follows from Helly-Bray's theorem (cf. LOÈVE [1960]) that

$$ \lim_{s\to 0} {}_2p_{ij}(s) = \delta_{ij}, \qquad i, j \in S. $$

Applying the monotone convergence theorem and Fubini's theorem (cf. LOÈVE [1960]) it is seen that

$$ \begin{aligned} \sum_{k\in S} {}_2p_{ik}(s_1)\, {}_2p_{kj}(s_2) &= \sum_{k\in S} \int_{-\infty}^{\infty} \int_{-\infty}^{\infty} {}_1p_{ik}(t_1)\, {}_1p_{kj}(t_2)\, \mathrm{d}_{t_1} b(s_1, t_1)\, \mathrm{d}_{t_2} b(s_2, t_2) \\ &= \int_{-\infty}^{\infty} \int_{-\infty}^{\infty} {}_1p_{ij}(t_1 + t_2)\, \mathrm{d}_{t_1} b(s_1, t_1)\, \mathrm{d}_{t_2} b(s_2, t_2) \\ &= \int_{-\infty}^{\infty} {}_1p_{ij}(t)\, \mathrm{d}_t \int_{0-}^{t} b(s_1, t - t_2)\, \mathrm{d}_{t_2} b(s_2, t_2) \\ &= \int_{-\infty}^{\infty} {}_1p_{ij}(t)\, \mathrm{d}_t b(s_1 + s_2, t) = {}_2p_{ij}(s_1 + s_2), \end{aligned} $$

$$ s_1, s_2 \geqq 0, \quad i, j \in S. $$

Consequently, ${}_2P(s)$, $s \geqq 0$, is a standard transition matrix function. It can be proved (cf. COHEN [1962[b]], STAM [1962]) that its Q-matrix

$$ {}_2Q \equiv ({}_2q_{ij}), $$

is given by

$$ {}_2q_{ij} = m_1 q_{ij} + \int_0^{\infty} \{{}_1p_{ij}(t) - \delta_{ij}\}\, \mathrm{d}\Psi(t), \qquad i, j \in S. \tag{5.4} $$

Consequently, a continuous time parameter Markov chain ${}_2M$ with state space S and with stationary standard transition matrix function ${}_2P(s)$ exists. Such a Markov chain ${}_2M$ will be said to be a Markov chain *derived* from ${}_1M$; the function $b(.,.)$ will be called the *deriving function.*

It should be noted that if $\{\boldsymbol{x}_t, t \in [0, \infty)\}$ represents the Markov chain ${}_1M$ and if for every fixed $s \geqq 0$ the non-negative stochastic variable $\boldsymbol{\tau}_s$, with state space the real line, is distributed according to $b(s, t)$, $-\infty < t < \infty$, then the stochastic process $\{\boldsymbol{y}_s, s \in [0, \infty)\}$ with

$$\boldsymbol{y}_s \stackrel{\text{def}}{=} \boldsymbol{x}_{\tau_s}, \qquad s \geqq 0.$$

is a Markov chain, and can be interpreted as the Markov chain ${}_2M$ described above.

I.5.2. Derived Poisson process

In this section the Poisson process with parameter λ, as described in section 4.2, is used as original chain ${}_1M$:

$${}_1p_{ij}(t) = \begin{cases} \dfrac{(\lambda t)^{j-i}}{(j-i)!} \mathrm{e}^{-\lambda t}, & t \geqq 0, \quad j=i, i+1, \ldots; \quad i=0, 1, \ldots, \\ 0, & j=0, 1, \ldots, i-1. \end{cases}$$

Denoting by ${}_2M$ the continuous time parameter Markov chain derived from ${}_1M$ by the deriving function $b(.,.)$ we have for $s \geqq 0$,

$${}_2p_{ij}(s) = \begin{cases} \displaystyle\int_{0-}^{\infty} \frac{(\lambda t)^{j-i}}{(j-i)!} \mathrm{e}^{-\lambda t} \, \mathrm{d}_t b(s, t), & j=i, i+1, \ldots; \quad i=0, 1, \ldots, \\ 0, & j=0, 1, \ldots, i-1, \end{cases} \tag{5.5}$$

the Q-matrix of this process is given by

$${}_2q_{ij} = \begin{cases} \displaystyle\int_{0}^{\infty} \left\{ \frac{(\lambda t)^{j-i}}{(j-i)!} \mathrm{e}^{-\lambda t} - \delta_{ij} \right\} \mathrm{d}\Psi(t), & j=i, i+1, \ldots; \quad i=0, 1, \ldots, \\ 0, & j=0, 1, \ldots, i-1, \end{cases} \tag{5.6}$$

if we take $m=0$.

The process ${}_2M$ will be called a *derived Poisson process*. Denoting by $\{{}_2\boldsymbol{x}_s, s \in [0, \infty)\}$ a derived Poisson process, it is easily proved (cf. section 4.2) that it has independent increments. The distribution of the increment ${}_2\boldsymbol{x}_\tau - {}_2\boldsymbol{x}_\sigma$ over the interval $[\sigma, t)$, $0 \leqq \sigma < \tau$, being given by

$$\Pr\{{}_2\boldsymbol{x}_\tau - {}_2\boldsymbol{x}_\sigma = k\} = \int_0^\infty \frac{(\lambda t)^k}{k!} \mathrm{e}^{-\lambda t} \, \mathrm{d}_t b(\tau - \sigma, t), \qquad k=0, 1, \ldots .$$

In general this derived Poisson process is not a birth process. From the expression for ${}_2q_{ij}$ it is seen that whenever a jump occurs from state E_i this jump may lead to any state E_j with $j>i$.

We shall treat here an example. Let

$$m=0, \qquad \Psi(t) = -\int_t^\infty \tau^{-1}\,\mathrm{e}^{-\mu\tau}\,\mathrm{d}\tau, \qquad \mu>0,$$

it follows from (5.2) and a table for Laplace transforms (cf. ERDÉLYI [1954]) that

$$b(s,t) = \frac{\mu^s}{\Gamma(s)}\int_0^t \tau^{s-1}\,\mathrm{e}^{-\mu\tau}\,\mathrm{d}\tau, \qquad t\geqq 0, \quad s\geqq 0.$$

Hence from (5.5) we obtain

$${}_2p_{ij}(s) = \begin{cases} \dfrac{\Gamma(j-i+s)}{(j-i)!\,\Gamma(s)}\,\dfrac{\mu^s\lambda^{j-i}}{(\lambda+\mu)^{j-i+s}}, & j=i,\, i+1,\, \dots; \quad i=0,1,\dots, \\ 0, & j=0,1,\dots,i-1, \end{cases}$$

and from (5.6),

$${}_2q_{ij} = \begin{cases} \dfrac{1}{(j-i)}\,\dfrac{1}{\lambda+\mu}\left(\dfrac{\lambda}{\lambda+\mu}\right)^{j-i}, & j=i+1,\ i+2,\ \dots; \quad i=0,1,\dots, \\ \log\dfrac{\mu}{\lambda+\mu}, & j=i=0,1,\dots, \\ 0, & j=0,1,\dots,i-1. \end{cases}$$

Usually, it is difficult to find an explicit form for $b(s,t)$ if m and $\Psi(t)$ are given; however in the applications it is rather the Laplace-Stieltjes transform of $b(s,t)$ than $b(s,t)$ itself which is needed. This Laplace-Stieltjes transform is easily obtained from (5.3) if m and $\Psi(t)$ are known (see also section III.2.7).

I.5.3. Derived death process

In this section ${}_1M$ stands for the death process with constant death rate μ as described in section 4.3; so we have for all $t\geqq 0$,

$$
{}_1p_{jh}(t) = \begin{cases} 1, & h=j=0, \\ 0, & h=j+1, j+2, \dots; \quad j=0, 1, \dots, \end{cases}
$$

$$
{}_1p_{ij}(t) = \begin{cases} \displaystyle\int_0^t \mu \frac{(\mu\sigma)^{i-1}}{(i-1)!} \mathrm{e}^{-\mu\sigma} \,\mathrm{d}\sigma; & j=0; \quad i=1, 2, \dots, \\ \displaystyle\frac{(\mu t)^{i-j}}{(i-j)!} \mathrm{e}^{-\mu t}, & j=1, 2, \dots, i; \quad i=1, 2, \dots . \end{cases}
$$

Denoting by ${}_2M$ the Markov chain derived from ${}_1M$ by the deriving function $b(s, t)$, we have

$$
{}_2p_{jh}(s) = \begin{cases} 1, & h=j=0, \\ 0, & h=j+1, j+2, \dots; \quad j=0, 1, \dots, \end{cases}
$$

$$
{}_2p_{ij}(s) = \begin{cases} \displaystyle\int_0^\infty \left\{ \int_0^t \mu \frac{(\mu\sigma)^{i-1}}{(i-1)!} \mathrm{e}^{-\mu\sigma} \,\mathrm{d}\sigma \right\} \mathrm{d}_t b(s, t), & j=0; \quad i=1, 2, \dots, \\ \displaystyle\int_0^\infty \frac{(\mu t)^{i-j}}{(i-j)!} \mathrm{e}^{-\mu t} \,\mathrm{d}_t b(s, t), & j=1, 2, \dots, i; \quad i=1, 2, \dots, \end{cases}
$$

and, assuming $m=0$, the Q-matrix is given by

$$
{}_2q_{jh} = \begin{cases} 0, & h=j=0, \\ 0, & h=j+1, j+2, \dots; \quad j=0, 1, \dots, \end{cases}
$$

$$
{}_2q_{ii} = \int_0^\infty (\mathrm{e}^{-\mu t} - 1) \,\mathrm{d}\Psi(t), \qquad i=1, 2, \dots,
$$

$$
{}_2q_{ij} = \begin{cases} \displaystyle\int_0^\infty \left\{ \int_0^t \frac{(\mu\tau)^{i-1}}{(i-1)!} \mathrm{e}^{-\mu\tau} \,\mathrm{d}(\mu\tau) \right\} \mathrm{d}\Psi(t), & j=0; \quad i=1, 2, \dots, \\ \displaystyle\int_0^\infty \frac{(\mu t)^{i-j}}{(i-j)!} \mathrm{e}^{-\mu t} \,\mathrm{d}\Psi(t), & j=1, 2, \dots, i-1; \quad i=2, 3, \dots . \end{cases}
$$

This Markov chain ${}_2M$ will play a role in queueing theory. Obviously, it is not a death process since if the process leaves state E_i then it may jump to any state E_j with $j<i$. The transition probabilities ${}_2p_{ij}(s)$ satisfy the forward differential equations (3.27). These equations now read

$$\frac{\mathrm{d}}{\mathrm{d}s}\, {}_2p_{ii}(s) = {}_2q_{ii}\, {}_2p_{ii}(s), \qquad i=1, 2, \ldots, \tag{5.7}$$

$$\frac{\mathrm{d}}{\mathrm{d}s}\, {}_2p_{ij}(s) = {}_2q_{jj}\, {}_2p_{ij}(s) + \sum_{k=j+1}^{i} {}_2p_{ik}(s)\, {}_2q_{kj},$$
$$j=1, 2, \ldots, i-1; \quad i=2, 3, \ldots, \tag{5.8}$$

$$\frac{\mathrm{d}}{\mathrm{d}s}\, {}_2p_{i0}(s) = \sum_{k=1}^{i} {}_2p_{ik}(s)\, {}_2q_{k0}, \qquad i=1, 2, \ldots, \tag{5.9}$$

$$\frac{\mathrm{d}}{\mathrm{d}s}\, {}_2p_{00}(s) = 0. \tag{5.10}$$

Next, we construct another continuous time parameter Markov chain ${}_3M$ with the aid of the results obtained above for ${}_2M$. The state space of the Markov chain ${}_3M$ will be the set of integers $\{-1, 0, 1, \ldots\}$. The Q-matrix of this chain is defined by

$${}_3q_{ij} \stackrel{\mathrm{def}}{=} {}_2q_{ij}, \qquad i=1, 2, \ldots; \quad j=0, \tag{5.11}$$

$${}_3q_{-1,j} \stackrel{\mathrm{def}}{=} 0, \qquad j=-1, 0, 1, \ldots,$$

$${}_3q_{0,-1} = -{}_3q_{00} \stackrel{\mathrm{def}}{=} \int_0^\infty \{1 - \mathrm{e}^{-\mu t}\}\, \mathrm{d}\Psi(t),$$

$${}_3q_{0,j} \stackrel{\mathrm{def}}{=} 0, \qquad j=1, 2, \ldots,$$

$${}_3q_{i,-1} \stackrel{\mathrm{def}}{=} 0, \qquad i=1, 2, \ldots .$$

The stationary transition probabilities ${}_3p_{ij}(s)$ of ${}_3M$ satisfy

$${}_3p_{ij}(0+) = \delta_{ij}, \tag{5.12}$$

and the forward differential equations (cf. (3.27)) are

$$\frac{\mathrm{d}}{\mathrm{d}s}\, {}_3p_{ii}(s) = {}_3q_{ii}(s)\, {}_3p_{ii}(s), \qquad i=-1, 0, 1, \ldots, \tag{5.13}$$

$$\frac{\mathrm{d}}{\mathrm{d}s}\, {}_3p_{ij}(s) = {}_3q_{jj}(s)\, {}_3p_{ij}(s) + \sum_{k=j+1}^{i} {}_3p_{ik}(s)\, {}_3q_{kj},$$
$$j=1, 2, \ldots, i-1; \quad i=2, 3, \ldots, \tag{5.14}$$

$$\frac{\mathrm{d}}{\mathrm{d}s}\,{}_3p_{i0}(s) = {}_3q_{00}\,{}_3p_{i0}(s) + \sum_{k=1}^{i} {}_3p_{ik}(s)\,{}_3q_{k0}, \qquad i=1, 2, \dots, \tag{5.15}$$

$$\frac{\mathrm{d}}{\mathrm{d}s}\,{}_3p_{i,-1}(s) = \sum_{k=0}^{i} {}_3p_{ik}(s)\,{}_3q_{k,-1}, \qquad i=0, 1, \dots, \tag{5.16}$$

$$\frac{\mathrm{d}}{\mathrm{d}s}\,{}_3p_{ij}(s) = 0, \qquad i=-1, 0, 1, \dots; \qquad j=i+1, i+2, \dots . \tag{5.17}$$

From (5.12), (5.13) and (5.17) it follows immediately that, for all $s \geqq 0$,

$$\begin{aligned} {}_3p_{ij}(s) &= 0, \qquad j=i+1, i+2, \dots; \quad i=-1, 0, 1, \dots, \\ {}_3p_{-1,-1}(s) &= 1. \end{aligned} \tag{5.18}$$

Defining

$$\mu(x) \stackrel{\text{def}}{=} \int_0^\infty \{1 - \mathrm{e}^{-\mu t(1-x)}\}\,\mathrm{d}\Psi(t), \qquad |x|<1, \tag{5.19}$$

we see that it follows from (5.12) and (5.13) that

$${}_3p_{ii}(s) = \mathrm{e}^{-s\mu(0)}, \qquad i=0, 1, \dots; \quad s \geqq 0. \tag{5.20}$$

From (5.8), (5.11), (5.14), (5.19), (5.20) and the expression for ${}_2p_{ij}(s)$ it is readily seen that

$${}_3p_{ij}(s) = {}_2p_{ij}(s), \qquad s \geqq 0, \quad j=1, 2, \dots, i-1; \quad i=2, 3, \dots . \tag{5.21}$$

Further, by (5.12) and (5.16) for $i=0$,

$${}_3p_{0,-1}(s) = 1 - \mathrm{e}^{-s\mu(0)}, \qquad s \geqq 0. \tag{5.22}$$

Comparing (5.15) and (5.9) it follows from (5.21) that

$$\frac{\mathrm{d}}{\mathrm{d}s}\,{}_3p_{i0}(s) = -\mu(0)\,{}_3p_{i0}(s) + \frac{\mathrm{d}}{\mathrm{d}s}\,{}_2p_{i0}(s), \qquad i=1, 2, \dots; \tag{5.23}$$

and (5.16) may be rewritten as

$$\frac{\mathrm{d}}{\mathrm{d}s}\,{}_3p_{i,-1}(s) = {}_3p_{i0}(s)\,\mu(0), \qquad i=1, 2, \dots . \tag{5.24}$$

It is not difficult to obtain from (5.23) and (5.24) explicit expressions for ${}_3p_{i0}(s)$ and ${}_3p_{i,-1}(s)$. In the application of this process ${}_3M$, however, it suffices to know their generating functions

$$\begin{aligned} {}_3P_0(s; x) &\stackrel{\text{def}}{=} \sum_{i=1}^\infty x^i\,{}_3p_{i0}(s), \qquad |x|<1, \\ {}_3P_{-1}(s; x) &\stackrel{\text{def}}{=} \sum_{i=1}^\infty x^i\,{}_3p_{i,-1}(s), \qquad |x|<1. \end{aligned}$$

Since

$$\sum_{i=1}^{\infty} x^i \, {}_2p_{i0}(s) = \frac{x}{1-x} \{1 - \mathrm{e}^{-s\mu(x)}\}, \qquad |x|<1,$$

it follows from (5.23), (5.24) and (5.12) that

$$\begin{aligned} {}_3P_0(s;x) &= \frac{x}{1-x} \frac{\mu(x)}{\mu(x)-\mu(0)} \{\mathrm{e}^{-s\mu(0)} - \mathrm{e}^{-s\mu(x)}\}, \qquad (5.25) \\ {}_3P_{-1}(s;x) &= \frac{x}{1-x}\left\{1 - \frac{\mu(x)\,\mathrm{e}^{-s\mu(0)} - \mu(0)\,\mathrm{e}^{-s\mu(x)}}{\mu(x)-\mu(0)}\right\}. \end{aligned}$$

It is easily verified that the solution ${}_3p_{ij}(s)$, $i,j=-1, 0, 1, \ldots$, of the set of differential equations (5.12), ..., (5.17) constructed above is a non-negative solution. By summing these differential equations it is easily seen from (5.18) that for all $i=-1, 0, 1, \ldots$,

$$\sum_{j=-1}^{\infty} \frac{\mathrm{d}}{\mathrm{d}s} \, {}_3p_{ij}(s) = \sum_{j=-1}^{i} \frac{\mathrm{d}}{\mathrm{d}s} \, {}_3p_{ij}(s) = \frac{\mathrm{d}}{\mathrm{d}s} \sum_{j=-1}^{i} {}_3p_{ij}(s) = 0,$$

so that from (5.12),

$$\sum_{j=-1}^{i} {}_3p_{ij}(s) = \sum_{j=-1}^{\infty} {}_3p_{ij}(s) = 1, \qquad i=-1, 0, 1, \ldots; \quad s \geqq 0.$$

Consequently, theorem 3.3 shows that the solution found above for ${}_3p_{ij}(s)$, $i,j=-1, 0, 1, \ldots; s \geqq 0$ represents the transition probabilities of the Markov chain ${}_3M$ with Q-matrix given by (5.11). This Markov chain ${}_3M$ will play an interesting role in defining servicing processes in queueing theory (cf. section III.2.7). The Markov process ${}_2M$ defined above will be referred to as a *derived death process*, while ${}_3M$ will be called a *quasi-derived death process*.

I.6. RENEWAL THEORY AND REGENERATIVE PROCESSES

I.6.1. Introduction

Renewal theory is an important and interesting chapter of the theory of stochastic processes. It is still subject of present day research and an extensive literature is available. In this chapter we shall outline the basic ideas and theorems of renewal theory.

Let $z_1, z_2, \dots,$ denote a series of independent, non-negative stochastic variables with $z_2, z_3, \dots,$ identically distributed; their distribution functions are denoted by

$$F_1(t) \stackrel{\text{def}}{=} \begin{cases} \Pr\{z_1 < t\}, \\ 0, \end{cases} \qquad F(t) \stackrel{\text{def}}{=} \begin{cases} \Pr\{z_i < t\}, & t > 0, \quad i = 2, 3, \dots, \\ 0, & t \leqq 0. \end{cases}$$

It will always be assumed *) that

$$F_1(0+) = 0, \qquad F(0+) = 0;$$

the distribution functions will be considered to be continuous from the left. We further introduce

$$s_0 \stackrel{\text{def}}{=} 0, \qquad s_n \stackrel{\text{def}}{=} z_1 + \dots + z_n, \quad n = 1, 2, \dots .$$

DEFINITION. The stochastic process $\{\boldsymbol{\nu}_t, \ t \in [0, \infty)\}$ with

$$\boldsymbol{\nu}_t \stackrel{\text{def}}{=} \max\{n : s_n < t\}, \qquad \boldsymbol{\nu}_0 \stackrel{\text{def}}{=} 0,$$

will be called a *general renewal process* if $F_1(t)$ and $F(t)$ are not identical;

*) This assumption is not essential.

if $F_1(t) \equiv F(t)$ the process is called a *renewal process*. $F(t)$ will be called the *renewal distribution*, and $F_1(t)$ the distribution of the first renewal time.

From the definition it follows immediately that for $t>0$,

$$\begin{aligned} &\{\boldsymbol{\nu}_t=0\}=\{\boldsymbol{s}_1 \geqq t\}, && && (6.1)\\ &\{\boldsymbol{\nu}_t=n\}=\{\boldsymbol{s}_n<t,\ \boldsymbol{s}_{n+1}\geqq t\}, && n=1, 2, \ldots, \\ &\{\boldsymbol{\nu}_t<n\}=\{\boldsymbol{s}_n\geqq t\}, && n=1, 2, \ldots . \end{aligned}$$

To illustrate the meaning of the term 'renewal' suppose that $\boldsymbol{z}_{n+1}$ represents the lifetime of the $(n+1)$th electrical bulb placed in a lamp-post immediately after failure of the nth bulb. If all bulbs have the same lifetime distribution and if at time $t=0$ a new bulb was installed then $\boldsymbol{\nu}_t$ as defined above represents the number of replacements, i.e. *renewals*, during the interval $[0, t)$, and $\{\boldsymbol{\nu}_t,\ t\in[0, \infty)\}$ is a renewal process. If at time $t=0$ there was already a bulb functioning in the lamp-post then the distribution of the time from $t=0$ to the moment of failure of this bulb generally differs from its lifetime distribution so that now $\{\boldsymbol{\nu}_t,\ t\in[0, \infty)\}$ is a general renewal process.

Another example is obtained if one considers a stationary stream of arrivals of customers at a ticket window. Suppose the interarrival times of successive arriving customers are independent and identically distributed stochastic variables with distribution function $F(t)$. Let $F_1(t)$ denote the distribution of the time $\boldsymbol{z}_1$ at which the first customer arrives after $t=0$, and denote by $\boldsymbol{z}_i$ the ith *interarrival time*, i.e. the time between the moment of arrival of the $(i-1)$th customer and that of the ith arriving customer after $t=0$, $i=2, 3, \ldots$. Then $\boldsymbol{\nu}_t$ as defined above represents the number of customers who have arrived during $[0, t)$.

The *renewal function* $m(t)$, $t\geqq 0$, is defined by

$$m(t) \overset{\text{def}}{=} \mathrm{E}\{\boldsymbol{\nu}_t\}, \qquad t\geqq 0,$$

and represents the average number of renewals in $[0, t)$.

From (6.1) it follows that

$$m(t) = \sum_{n=1}^{\infty} \Pr\{\boldsymbol{\nu}_t\geqq n\} = \sum_{n=0}^{\infty} F_1(t) * F^{n*}(t), \tag{6.2}$$

where $F^{n*}(t)$ denotes the n-fold convolution of $F(t)$ with itself, $n=1, 2, \ldots$; $F^{0*}(t)$ is by definition the probability distribution degenerated at $t=0$ i.e.

$$U_1(t) = F^{0*}(t) = \begin{cases} 0, & t\leqq 0, \\ 1, & t>0. \end{cases}$$

For any integer $m \geqq 1$ we may write

$$m(t) = F_1(t) * \{U_1(t)+F(t)+\ldots+F^{(m-1)*}(t)\}$$

$$+ F^{m*}(t) * F_1(t) * \{F^{0*}(t)+\ldots+F^{(m-1)*}(t)\} * \sum_{n=0}^{\infty} F^{(nm)*}(t), \tag{6.3}$$

and hence for $t>0$,

$$m(t) = F_1(t) * \{U_1(t)+F(t)+\ldots+F^{(m-1)*}(t)\} + \int_0^t F^{m*}(t-\tau)\,\mathrm{d}m(\tau). \tag{6.4}$$

From the definition of $F_1(t)$ and $F(t)$ it follows that a positive number Δ exists such that $F(t)<1$ for $t<\Delta$. Consequently, for every finite $t>0$, it is possible to determine the integer m such that $F^{m*}(t)<1$ and $m\Delta>t$. Since

$$F^{(nm+m)*}(t) = \int_0^t F^{m*}(t-\tau)\,\mathrm{d}F^{(nm)*}(\tau) \leqq F^{m*}(t)F^{(nm)*}(t), \qquad t \geqq 0,$$

it follows that for $t<m\Delta$,

$$F^{(nm)*}(t) \leqq \{F^{m*}(t)\}^n < 1, \qquad n=1, 2, \ldots .$$

Hence, for $t<m\Delta$, we obtain from (6.3),

$$m(t) \leqq F_1(t) * \{U_1(t)+\ldots+F^{(m-1)*}(t)\} \frac{1}{1-F^{m*}(t)}.$$

From this inequality it follows that $m(t)$ is finite for every finite value of t.

Taking $m=1$ in (6.4) we see that the renewal function satisfies the integral relation

$$m(t) = F_1(t) + \int_0^t F(t-\tau)\,\mathrm{d}m(\tau), \qquad t \geqq 0. \tag{6.5}$$

It may be proved that (6.5) considered as an integral equation for $m(t)$, the so called *renewal equation*, has a unique non-decreasing solution which is given by (6.2) (cf. FELLER [1941, 1966]).

Introducing the Laplace-Stieltjes transforms

$$f_1(s) \stackrel{\text{def}}{=} \int_0^\infty \mathrm{e}^{-st}\,\mathrm{d}F_1(t); \qquad f(s) \stackrel{\text{def}}{=} \int_0^\infty \mathrm{e}^{-st}\,\mathrm{d}F(t), \qquad \operatorname{Re} s \geqq 0,$$

we obtain from (6.5) or from (6.2) that

$$\mu(s) \stackrel{\text{def}}{=} \int_0^\infty \mathrm{e}^{-st}\,\mathrm{d}m(t) = \frac{f_1(s)}{1-f(s)}, \qquad \operatorname{Re} s>0; \tag{6.6}$$

a relation which is often very useful for determining $m(t)$.

Since (cf. (6.1)) for $t>0$,

$$\Pr\{\nu_t=n\} = \int_0^t \{1-F(t-\tau)\}\, \mathrm{d}(F_1(\tau) * F^{(n-1)*}(\tau))$$

$$= F_1(t) * F^{(n-1)*}(t) * \{U_1(t)-F(t)\}, \qquad n=0, 1, \dots,$$

it follows with

$$m_2(t) \overset{\text{def}}{=} \mathrm{E}\{\nu_t^2\}, \qquad t \geqq 0,$$

that

$$\int_0^\infty \mathrm{e}^{-st}\, \mathrm{d}m_2(t) = \{1-f(s)\} f_1(s) \sum_{n=1}^\infty n^2 f^{n-1}(s)$$

$$= f_1(s) \frac{1+f(s)}{\{1-f(s)\}^2}, \qquad \operatorname{Re} s>0.$$

Hence, if $\{\nu_t,\ t\in[0, \infty)\}$ is a renewal process so that $F_1(t)\equiv F(t)$ then

$$\int_0^\infty \mathrm{e}^{-st}\, \mathrm{d}m_2(t) = \mu(s) + 2\mu^2(s).$$

Therefore,

$$m_2(t) = m(t) + 2\int_0^t m(t-\tau)\, \mathrm{d}m(\tau), \qquad t\geqq 0.$$

By the same method expressions for the higher moments of ν_t can be obtained and it is easily verified that all moments of ν_t are finite for finite t.

A general renewal process is called *stationary* if

$$\mu \overset{\text{def}}{=} \int_0^\infty t\, \mathrm{d}F(t) < \infty, \quad \text{and} \quad F_1(t) = \frac{1}{\mu}\int_0^t \{1-F(\tau)\}\, \mathrm{d}\tau, \quad t\geqq 0.$$

Since

$$\int_0^\infty \mathrm{e}^{-st}\, \frac{1-F(t)}{\mu}\, \mathrm{d}t = \frac{1}{\mu s}\{1-f(s)\}, \qquad \operatorname{Re} s\geqq 0,$$

it follows from (6.6) that for a stationary renewal process

$$\mu(s) = \frac{1}{\mu s}, \tag{6.7}$$

and hence

$$m(t) = \frac{t}{\mu}, \qquad t \geqq 0. \tag{6.8}$$

A renewal process will be called *discrete* if the renewal distribution $F(t)$ is a lattice distribution, i.e. if $F(t)$ can be written as

$$F(t) = \sum_{k=0}^{\infty} b_k U_1(t - k\tau), \tag{6.9}$$

with

$$\sum_{k=0}^{\infty} b_k = 1, \qquad b_k \geqq 0, \qquad k=1, 2, \dots; \quad b_0=0,$$

and with the greatest common divisor of those k for which $b_k>0$ being one, and where τ, the *period*, is a positive constant.

Let $\{\boldsymbol{\nu}_t, t \in [0, \infty)\}$ be a discrete renewal process with renewal distribution $F(t)$ given by (6.9). It then follows from (6.2) that

$$m(t) = \sum_{n=1}^{\infty} F^{n*}(t).$$

By writing

$$b_k^{1*} \overset{\text{def}}{=} b_k, \qquad b_k^{n*} = \sum_{h=0}^{k} b_h^{(n-1)*} b_{k-h}, \qquad k=0, 1, \dots; \quad n=2, 3, \dots,$$

we deduce that

$$F^{n*}(t) = \sum_{k=0}^{\infty} b_k^{n*} U_1(t-k\tau), \qquad n=1, 2, \dots,$$

and hence that

$$m(t) = \sum_{k=0}^{\infty} w_k U_1(t-k\tau), \tag{6.10}$$

where

$$w_k \overset{\text{def}}{=} \sum_{n=1}^{\infty} b_k^{n*}, \qquad k=0, 1, \dots . \tag{6.11}$$

From the definition of b_k^{n*} it follows that

$$w_k = b_k + \sum_{n=2}^{\infty} \sum_{h=0}^{k} b_h^{(n-1)*} b_{k-h},$$

so that

$$w_k = b_k + \sum_{h=0}^{k} w_h b_{k-h}, \qquad k=0, 1, \dots . \tag{6.12}$$

The relation (6.12) represents the *renewal equation* for a discrete renewal process.

From (6.10) it follows that

$$w_{k+1}=m((k+1)\,\tau+\sigma)-m(k\tau+\sigma),\qquad 0<\sigma\leqq\tau,\qquad k=0,1,\ldots,\qquad (6.13)$$

so that w_{k+1} represents the average number of renewals in the interval $[k\tau+\sigma,(k+1)\,\tau+\sigma)$, $0<\sigma\leqq\tau$, whereas $w_0=0$.

It should be noted that the renewal process discussed in section 2.2 is a discrete renewal process as defined above. In the notation used here we have for that process $\tau=1$, $b_0=0$, $b_k=f_{ii}^{(k)}$, $w_k=p_{ii}^{(k)}$, $k=1,2,\ldots$, and for $k=1$, $2,\ldots$, the relation (2.9) is equivalent to the renewal equation (6.12).

I.6.2. Renewal theorems

The central problem in renewal theory concerns the behaviour of the renewal function $m(t)$ for $t\to\infty$. This behaviour will be described in the following theorems. These theorems will be discussed for a renewal process first, and afterwards it will be shown that they also hold for a general renewal process.

THEOREM 6.1 (*Elementary renewal theorem*).

$$\lim_{t\to\infty}\frac{m(t)}{t}=\begin{cases}\dfrac{1}{\mu} & \text{if } \mu<\infty,\\[2ex] 0 & \text{if } \mu=\infty,\end{cases}$$

where

$$\mu\overset{\text{def}}{=}\int_0^\infty t\,\mathrm{d}F(t).$$

Proof. Suppose first $\mu<\infty$ then

$$\mu=\int_0^\infty t\,\mathrm{d}F(t)=\int_0^\infty \mathrm{d}F(t)\int_0^t \mathrm{d}y$$

$$=\int_0^\infty \mathrm{d}y\int_{t=y}^\infty \mathrm{d}F(t)=\int_0^\infty\{1-F(y)\}\,\mathrm{d}y.$$

Since

$$f(s)=\int_0^\infty \mathrm{e}^{-st}\,\mathrm{d}F(t),\qquad s\geqq 0,$$

implies that

$$\int_0^\infty e^{-st}\{1-F(t)\}\,dt = \frac{1-f(s)}{s}, \qquad s>0,$$

it follows that

$$\lim_{s\downarrow 0}\frac{1-f(s)}{s} = \int_0^\infty \{1-F(t)\}\,dt = \mu, \tag{6.14}$$

$$\lim_{s\downarrow 0} f(s) = 1,$$

so that, since we consider a renewal process, we deduce from (6.6) that

$$\mu(s) = \int_0^\infty e^{-st}\,dm(t) = \frac{f(s)}{1-f(s)} \approx \frac{1}{\mu s} \qquad \text{for} \quad s\downarrow 0.$$

Consequently, since $m(t)$ is non-decreasing, a Tauberian theorem for the Laplace-Stieltjes transform (cf. app. 4) yields

$$\lim_{t\to\infty}\frac{m(t)}{t} = \frac{1}{\mu} \qquad \text{if} \quad \mu<\infty.$$

If $\mu=\infty$ we consider the renewal process $\{\boldsymbol{\nu}_t^{(1)},\ t\in[0,\infty)\}$ with renewal distribution $F^{(1)}(t)$ defined by

$$F^{(1)}(t) = \begin{cases} F(t) & \text{for} \quad t\leqq \Delta, \\ 1 & \text{for} \quad t>\Delta, \end{cases}$$

where Δ is a positive constant. Obviously, $\boldsymbol{\nu}_t^{(1)} \geqq \boldsymbol{\nu}_t$ with probability one for every $t\geqq 0$, and hence $m^{(1)}(t)\geqq m(t)$, $t\geqq 0$, where $m^{(1)}(t)$ is the renewal function of the process $\{\boldsymbol{\nu}_t^{(1)},\ t\in[0,\infty)\}$.
Therefore,

$$\frac{m^{(1)}(t)}{t} \geqq \frac{m(t)}{t}, \qquad t>0,$$

and with

$$\mu^{(1)} \overset{\text{def}}{=} \int_0^\Delta t\,dF^{(1)}(t) < \infty,$$

it follows from the above results that

$$\lim_{t\to\infty}\frac{m^{(1)}(t)}{t} = \frac{1}{\mu^{(1)}} \geqq \limsup_{t\to\infty}\frac{m(t)}{t} \geqq 0.$$

Letting $\Delta \to \infty$ we see that $(\mu^{(1)})^{-1} \to 0$ leads to

$$\lim_{t\to\infty} \frac{m(t)}{t} = 0.$$

This completes the proof.

This theorem shows that $m(t)/t$ has a limit for $t \to \infty$. It turns out that the average number of renewals in $[0, t)$ is proportional to t for large values of t, a result which has an intuitive meaning. It is reasonable to conjecture that a similar statement holds for the average number of renewals in an interval $[t-h, t)$ for $t \to \infty$ and a fixed $h > 0$. However, from the discussion of a discrete renewal process in the preceding section it is seen that $m(t)$ can be a step-function so that in general $m(t)-m(t-h)$ does not necessarily possess a limit for $t \to \infty$. For this reason, it is necessary to distinguish in the investigation of the behaviour of $m(t)-m(t-h)$ as $t \to \infty$ between renewal distributions which are lattice distributions and those which are not.

THEOREM 6.2 (*The key renewal theorem of Smith*). *If the renewal distribution $F(t)$ is not a lattice distribution and $g(u)$ is a function of bounded variation on $0 \leqq u < \infty$ then, if $\int_0^\infty g(u)\,du$ exists,*

$$\lim_{t\to\infty} \int_0^t g(t-u)\,dm(u) = \begin{cases} \dfrac{1}{\mu} \displaystyle\int_0^\infty g(u)\,du & \text{if } \mu < \infty, \\[2ex] 0 & \text{if } \mu = \infty. \end{cases}$$

This theorem leads immediately to Blackwell's theorem if we take

$$g(u) \overset{\text{def}}{=} \frac{1}{h}\{1-U(t-h)\}, \qquad -\infty < t < \infty.$$

Conversely, theorem 6.2 can be deduced from Blackwell's theorem (see TAKÁCS [1962] p. 227).

THEOREM 6.3 (*Blackwell's theorem*). *If the renewal distribution $F(t)$ is not a lattice distribution then for every fixed $h > 0$,*

$$\lim_{t\to\infty} \frac{m(t)-m(t-h)}{h} = \begin{cases} \dfrac{1}{\mu} & \text{if } \mu < \infty, \\[2ex] 0 & \text{if } \mu = \infty. \end{cases}$$

For discrete renewal processes there are theorems equivalent to those above (cf. SMITH [1954]). We only mention here the equivalent of theorem 6.3.

THEOREM 6.4 (*Theorem of Erdös, Pollard and Feller*). *If the renewal distribution $F(t)$ is a lattice distribution with period τ and first moment μ, then* (*cf.* (6.11)),

$$\lim_{n\to\infty} w_n = \begin{cases} \tau/\mu & \text{if } \mu<\infty, \\ 0 & \text{if } \mu=\infty. \end{cases}$$

Theorem 6.3 leads to the conjecture that, if $m(t)$ is a differentiable function, the derivative $m'(t)$ will tend to $1/\mu$ for $t\to\infty$. Under certain conditions this conjecture is true. If the renewal distribution $F(t)$ is absolutely continuous so that for almost all $t\geqq 0$,

$$\varphi(t) \overset{\text{def}}{=} \frac{\mathrm{d}}{\mathrm{d}t} F(t),$$

is the density of the renewal distribution, then it is easily shown from (6.2) with $F_1(t)\equiv F(t)$ that $m(t)$ has a derivative $m'(t)$ given by

$$m'(t) = \sum_{n=1}^{\infty} \varphi^{n*}(t), \qquad t\geqq 0,$$

where

$$\varphi^{(n+1)*}(t) = \int_0^t \varphi(t-\tau)\, \varphi^{n*}(\tau)\, \mathrm{d}\tau, \qquad n=1, 2, \ldots; \quad t\geqq 0,$$

with

$$\varphi^{1*}(t) \overset{\text{def}}{=} \varphi(t), \qquad t\geqq 0.$$

The relevant theorem now reads:

THEOREM 6.5. *If for some $T>0$, $\varphi(t)$ is monotone for $t>T$, or if for some $\delta>0$,*

$$\int_0^{\infty} \{\varphi(t)\}^{1+\delta}\, \mathrm{d}t < \infty,$$

then $m'(t)$ has a limit for $t\to\infty$ and

$$\lim_{t\to\infty} m'(t) = \begin{cases} 1/\mu & \text{if } \mu<\infty, \\ 0 & \text{if } \mu=\infty. \end{cases}$$

We shall not discuss here the proofs of the theorems 6.2, 6.4 and 6.5, but refer the reader to the literature (cf. BLACKWELL [1948]; SMITH [1954];

FELLER [1957, 1966]). The proofs are rather intricate and the main difficulty is the proof of the existence of the limits mentioned in the theorems. Assuming that these limits exist, their values are easily calculated. For instance from (6.14) and

$$\int_0^\infty e^{-st}\, d\left\{\int_0^t g(t-u)\, dm(u)\right\} = \frac{f(s)}{\{1-f(s)\}/s}\, \frac{1}{s}\, g_0(s), \qquad s>0,$$

$$\int_0^\infty e^{-st}\, d\left\{\int_0^t g(u)\, du\right\} = \frac{1}{s}\, g_0(s), \qquad s>0,$$

with

$$g_0(s) \stackrel{\text{def}}{=} \int_0^\infty e^{-st}\, dg(t), \qquad s>0,$$

it follows immediately by applying an Abelian theorem for the Laplace-Stieltjes transform (cf. app. 3) that

$$\lim_{t\to\infty} \int_0^t g(t-u)\, dm(u) = \frac{1}{\mu} \int_0^\infty g(t)\, dt \qquad \text{if} \quad \mu<\infty,$$

once the existence of the limit has been established.

Let us consider an example. Suppose that the renewal distribution is not a lattice distribution and that its Laplace-Stieltjes transform $f(s)$ is a rational function of s, so that we may write

$$\int_0^\infty e^{-st}\, dF(t) = f(s) = \frac{u(s)}{v(s)}, \qquad \operatorname{Re} s \geqq 0,$$

where $u(s)$ and $v(s)$ are polynomials in s, the degree of $u(s)$ being at most equal to that of $v(s)$. It now follows that

$$\mu(s) = \frac{f(s)}{1-f(s)} = \frac{u(s)}{v(s)-u(s)}, \qquad \operatorname{Re} s>0.$$

From

$$\left|\frac{u(s)}{v(s)}\right| \leqq \int_0^\infty e^{-t\operatorname{Re} s}\, dF(t) < \int_0^\infty dF(t) = 1 \qquad \text{for} \quad \operatorname{Re} s>0,$$

and since it is impossible that $|f(s_0)|=1$ for a $s_0 \neq 0$, Re $s_0=0$, because $F(t)$ is not a lattice distribution (see LUKACS [1960] p. 25), it follows that

$$|u(s)|<|v(s)|, \qquad s\neq 0, \quad \text{Re}\, s \geqq 0.$$

Consequently, the poles of $\mu(s)$ all have a negative real part, except the pole at $s=0$. Since $f(s)$ is a rational function of s, all moments of $F(t)$ are finite and hence $s=0$ is a simple pole of $\mu(s)$. The residue at $s=0$ is $1/\mu$, therefore $\mu(s)-1/\mu s$ is the Laplace-Stieltjes transform of a function which has a finite limit for $t\to\infty$. Since for real values of s,

$$f(s) = 1 - \mu s + \tfrac{1}{2}s^2\mu_2 + o(s^2) \qquad \text{for} \quad s\downarrow 0,$$

so that

$$\mu(s) - \frac{1}{\mu s} = \frac{\mu_2}{2\mu^2} - 1 + O(s) \quad \text{for} \quad s\downarrow 0,$$

it follows by applying an Abelian theorem for the Laplace-Stieltjes transform (cf. app. 3) that

$$m(t) - \frac{t}{\mu} \to \frac{\mu_2}{2\mu^2} - 1 \qquad \text{for} \quad t\to\infty, \tag{6.15}$$

where

$$\mu_2 = \int_0^\infty t^2\, \mathrm{d}F(t).$$

This result applies whenever $F(t)$ is not a lattice distribution and $\mu_2<\infty$ (see SMITH [1954]).

Continuing the example above note that (see preceding section)

$$\int_0^\infty \mathrm{e}^{-st}\, \mathrm{d}m_2(t) = \frac{f(s)\{1+f(s)\}}{\{1-f(s)\}^2}.$$

This function has a pole with multiplicity two at $s=0$, while all the other poles have a negative real part. Since for real s,

$$\frac{f(s)\{1+f(s)\}}{\{1-f(s)\}^2} = \frac{2}{\mu^2 s^2} + \frac{1}{\mu^3 s}\{2\mu_2-3\mu^2\} + O(1), \qquad s\downarrow 0,$$

it follows by an Abelian theorem (cf. app. 3) that

$$m_2(t) = \mathrm{E}\{\boldsymbol{v}_t^2\} = \frac{t^2}{\mu^2} + \frac{2\mu_2-3\mu^2}{\mu^3}\, t + O(1) \qquad \text{for} \quad t\to\infty.$$

From this result and (6.15) we now obtain

$$\operatorname{var}\{\boldsymbol{\nu}_t\} = \frac{\mu_2-\mu^2}{\mu^3}\,t+O(1) \qquad \text{for} \quad t\to\infty,$$

a result which holds if $F(t)$ is not a lattice distribution and $\mu_2<\infty$.

Generally, it can be shown by starting from the relation (cf. (6.1)),

$$\Pr\{\boldsymbol{\nu}_t \geqq n\} = \Pr\{\boldsymbol{s}_n < t\},$$

and applying the central limit theorem to the sum $\boldsymbol{s}_n = \boldsymbol{x}_1 + \ldots + \boldsymbol{x}_n$ that if $\mu_2<\infty$ then for $t\to\infty$,

$$\Pr\left\{\boldsymbol{\nu}_t \geqq \frac{t}{\mu} - \alpha\,\frac{\sqrt{(\mu_2-\mu^2)}}{\mu}\sqrt{\frac{t}{\mu}}\right\} \to \frac{1}{\sqrt{2\pi}}\int_{-\infty}^{\alpha} e^{-\frac{1}{2}\eta^2}\,d\eta.$$

In the example discussed above we considered a renewal process. It will be evident how the calculations are to be performed if the process is a general renewal process. To show that the theorems 6.1, ..., 6.4 are also valid for general renewal processes let $m_g(t)$ and $m(t)$ denote the renewal functions of the general renewal process and the renewal process, respectively, both processes having the same renewal distribution. From the proof of theorem 6.1 it is easily seen that this theorem is also valid for $m_g(t)$.

Suppose the renewal distribution $F(t)$ is not a lattice distribution and that $g(t)$ is of bounded variation on $[0, \infty)$ with $\int_0^\infty g(t)\,dt<\infty$. Since

$$m_g(t) = F_1(t) + F_1(t) * m(t),$$

we have

$$g(t) * m_g(t) = g(t) * F_1(t) + g(t) * F_1(t) * m(t). \tag{6.16}$$

We first prove that $g(t) * F_1(t)$ has a limit for $t\to 0$ which equals zero. Since $g(t)$ is of bounded variation it can be expressed as the difference of two bounded non-increasing functions. It is therefore sufficient to prove the statement above for $g(t)$ a bounded non-increasing non-negative function with $g(t)\to 0$ for $t\to\infty$, since $\int_0^\infty g(t)\,dt<\infty$.

Hence for a given $\varepsilon>0$, a $t_0>0$ can be found such that

$$|g(t)|<\varepsilon, \qquad 1-F_1(t)<\varepsilon, \qquad \text{for} \quad t>t_0.$$

For $t>2t_0$,

$$\begin{aligned} \left|\int_0^t g(t-u)\,dF_1(u)\right| &\leqq \int_0^{t_0} |g(t-u)|\,dF_1(u) + \int_{t_0}^t |g(t-u)|\,dF_1(u) \\ &\leqq \varepsilon F_1(t_0) + \varepsilon g(0+), \end{aligned}$$

and, therefore, $g(t)*F_1(t)\to 0$ for $t\to\infty$. In the same way it is easily proved that

$$\lim_{t\to\infty}\int_0^t du \int_0^u g(u-v)\, dF_1(v) = \lim_{t\to\infty}\int_0^t dF_1(v) \int_{u=v}^t g(u-v)\, du = \int_0^\infty g(u)\, du.$$

Since $g(t)*F_1(t)$ is also of bounded variation on $[0, \infty)$ it follows from (6.16) and theorem 6.2 that $g(t)*m_g(t)$ has a limit for $t\to\infty$ and that this limit is equal to that of $g(t)*m(t)$. Theorem 6.3 follows as before from theorem 6.2.

Suppose next that the renewal distribution is a lattice distribution, then from (6.10) and

$$m_g(t) = F_1(t) + F_1(t) * m(t),$$

we have

$$m_g(t) = F_1(t) + \sum_{k=0}^{[t/\tau]} F_1(t-k\tau)\, w_k, \qquad t\geq 0,$$

where $[t/\tau]$ denotes the highest integer not exceeding t/τ. It follows that

$$m_g(t) - m_g(t-\tau) = F_1(t) - F_1(t-\tau) + \sum_{k=0}^{[t/\tau]} F_1(t-k\tau)\, w_k - \sum_{k=0}^{[t/\tau]-1} F_1(t-(k+1)\,\tau)\, w_k.$$

For $t\to\infty$ and fixed k we have

$$F_1(t-k\tau)-F_1(t-(k+1)\,\tau)\to 0.$$

We may choose the integer N so large that for $k\leq N+1$, w_k differs by an arbitrarily small amount from its limit value given by theorem 6.4, and hence for t sufficiently large with $h \overset{\text{def}}{=} [t/\tau]$,

$$\begin{aligned} m_g(t) - m_g(t-\tau) &\approx F_1(t) - F_1(t-\tau) \\ &\quad + \sum_{k=0}^{N} \{F_1(t-k\tau)-F_1(t-(k+1)\,\tau)\}\, w_k \\ &\quad + w_N \sum_{k=N+1}^{h-1} \{F_1(t-k\tau)-F_1(t-(k+1)\,\tau)\} \\ &\quad + F_1(t-h\tau)\, w_N \approx F_1(t-(N+1)\,\tau)\, w_N, \end{aligned}$$

so that for $t\to\infty$ and then $N\to\infty$ we find

$$\lim_{t\to\infty} \{m_g(t) - m_g(t-\tau)\} = \lim_{N\to\infty} w_N,$$

which is the analogue of the result of theorem 6.4.

I.6.3. Past lifetime and residual lifetime

The stochastic variables

$$\eta_t \overset{\text{def}}{=} t - s_{\nu_t}, \qquad t>0,$$

$$\zeta_t \overset{\text{def}}{=} s_{\nu_t+1} - t, \qquad t>0,$$

are called the *past lifetime* and the *residual lifetime* at time t, respectively. Evidently, η_t represents the time between t and the moment of the last renewal before t, if there is any, otherwise $\eta_t = t$, whereas ζ_t is the time between t and the first renewal after t.

To obtain the probability distribution of the past lifetime we note that η_t cannot exceed t; moreover for $0<\eta<t$,

$$\{\eta_t<\eta\} = \bigcup_{n=1}^{\infty} \{t-\eta<s_n<t,\ s_{n+1}\geqq t\},$$

so that

$$\Pr\{\eta_t<\eta\} = \sum_{n=1}^{\infty} \int_{u=(t-\eta)+}^{t} \Pr\{x_{n+1}\geqq t-u\}\,\mathrm{d}\Pr\{s_n<u\}, \qquad t>\eta>0.$$

Hence for $t>0$ we have from (6.2),

$$\Pr\{\eta_t<\eta\} = \begin{cases} 0, & \eta\leqq 0, \\ \displaystyle\int_{u=(t-\eta)+}^{t} \{1-F(t-u)\}\,\mathrm{d}m(u), & 0<\eta\leqq t, \\ 1, & \eta>t. \end{cases} \tag{6.17}$$

From the definition of ζ_t it follows that

$$\{\zeta_t<\zeta\} = \bigcup_{n=0}^{\infty} \{s_n<t,\ t\leqq s_{n+1}<t+\zeta\}, \qquad \zeta>0,\quad t>0,$$

so that for $\zeta>0$,

$$\begin{aligned}\Pr\{\zeta_t<\zeta\} &= \sum_{n=0}^{\infty} \int_{u=0}^{t} \Pr\{t-u\leqq x_{n+1}<t+\zeta-u\}\,\mathrm{d}\Pr\{s_n<u\} \\ &= F_1(t+\zeta) - F_1(t) + \int_{u=0}^{t} \{F(t-u+\zeta)-F(t-u)\}\,\mathrm{d}m(u).\end{aligned}$$

Noting that $m(t)$ satisfies (6.5) we have

$$\begin{aligned}\Pr\{\xi_t<\zeta\} &= m(t+\zeta) - \int_{u=0}^{t+\zeta} F(t+\zeta-u)\,\mathrm{d}m(u)\\ &\quad - m(t) + \int_{u=0}^{t} F(t+\zeta-u)\,\mathrm{d}m(u)\\ &= \int_{u=t}^{t+\zeta} \{1-F(t+\zeta-u)\}\,\mathrm{d}m(u), \qquad \zeta>0.\end{aligned}$$

Hence for $t\geqq 0$,

$$\Pr\{\xi_t<\zeta\} = \begin{cases} 0, & \zeta\leqq 0,\\ \displaystyle\int_{u=t}^{t+\zeta} \{1-F(t+\zeta-u)\}\,\mathrm{d}m(u), & \zeta>0. \end{cases} \tag{6.18}$$

The joint distribution of η_t and ξ_t can be determined from the following relations

$$\{\eta_t<\eta,\ \xi_t<\zeta\} = \begin{cases} \displaystyle\bigcup_{n=1}^{\infty} \{t-\eta<s_n<t,\ t\leqq s_{n+1}<t+\zeta\}, & 0<\eta<t,\quad \zeta>0,\\ \{\xi_t<\zeta\}, & \eta>t>0,\quad \zeta>0. \end{cases}$$

Further it follows that

$$\{\xi_t<\zeta\} = \{\eta_{t+\zeta}\leqq\zeta\}, \qquad \zeta>0, \tag{6.19}$$

$$\{\eta_t\geqq\eta,\ \xi_t\geqq\zeta\} = \{\eta_{t+\zeta}\geqq\eta+\zeta\}, \qquad \zeta\geqq 0,\quad \eta\geqq 0.$$

Let us consider the case in which $\{\boldsymbol{\nu}_t,\ t\in[0,\infty)\}$ is a renewal process with renewal distribution the negative exponential distribution with parameter λ, i.e.

$$F_1(t) = F(t) = \begin{cases} 1-\mathrm{e}^{-\lambda t}, & t>0,\\ 0, & t\leqq 0, \end{cases}$$

where $\lambda>0$.
It then follows from theorem 4.3 that $\{\boldsymbol{\nu}_t,\ t\in[0,\infty)\}$ is a Poisson process, and it is easily verified from (6.6), (6.17) and (6.18) that for $t\geqq 0$,

$$m(t) = \lambda t,$$

$$\Pr\{\eta_t<\eta\} = \begin{cases} 0, & \eta\leqq 0,\\ 1-\mathrm{e}^{-\lambda\eta}, & 0<\eta\leqq t,\\ 1, & 0<t<\eta; \end{cases}$$

$$\Pr\{\xi_t<\zeta\} = \begin{cases} 0, & \zeta\leqq 0,\\ 1-\mathrm{e}^{-\lambda\zeta}, & \zeta>0. \end{cases}$$

Application of the second relation of (6.19) yields

$$\Pr\{\eta_t \geqq \eta,\ \xi_t \geqq \eta\} = \Pr\{\eta_{t+\zeta} \geqq \eta + \zeta\}$$

$$= \begin{cases} e^{-\lambda(\eta+\zeta)}, & 0 \leqq \eta \leqq t, \quad \zeta \geqq 0, \\ 0, & t < \eta, \quad \zeta \geqq 0. \end{cases}$$

From these relations it is seen that η_t and ξ_t are independent variables for every fixed $t>0$, and that the probability distribution of ξ_t is independent of t.

Next, we consider a renewal process for which the past lifetime η_t and the residual lifetime ξ_t are independent variables for every fixed $t>0$ and for which the probability distribution of ξ_t is independent of t.

From (6.19) and the independence of η_t and ξ_t it follows that

$$\Pr\{\eta_t \geqq \eta\} \Pr\{\xi_t \geqq \zeta\} = \Pr\{\eta_{t+\zeta} \geqq \eta + \zeta\}, \qquad \eta \geqq 0, \quad \zeta \geqq 0. \tag{6.20}$$

Since

$$\Pr\{\eta_t \geqq t\} = \Pr\{\eta_t = t\} = 1 - F(t),$$

it follows from (6.20) with $\eta = t$, for those values of t for which $F(t) \neq 1$, that

$$\Pr\{\xi_t \geqq \zeta\} = \frac{1 - F(t+\zeta)}{1 - F(t)}, \qquad t > 0, \quad \zeta \geqq 0.$$

By definition of ξ_t we have

$$\Pr\{\xi_{0+} \geqq \zeta\} = 1 - F(\zeta), \qquad \zeta \geqq 0.$$

Since the probability distribution of ξ_t is supposed to be independent of t we obtain from the latter relations

$$1 - F(t+\zeta) = (1 - F(t)) \cdot (1 - F(\zeta)), \qquad t > 0, \quad \zeta \geqq 0. \tag{6.21}$$

From this relation it is seen that for $t>0$, $1-F(t)$ is a solution, bounded in every finite interval, of the functional equation

$$b(t+\sigma) = b(t)\, b(\sigma), \qquad t > 0, \quad \sigma \geqq 0.$$

The real solutions of this functional equation which are bounded in every finite interval are either identically zero or given by

$$b(t) = e^{-\lambda t}, \qquad t > 0, \quad -\infty < \lambda < \infty,$$

λ being a constant (see PARZEN [1962] p. 121). (*Note:* the proof of this statement has in fact already been given in the first part of the proof of theorem 4.4; to see this identify $b(t)$ and $p_0(t)$).

Hence, since $F(0+)=0$, we must have

$$1-F(t)=\mathrm{e}^{-\lambda t}, \qquad t>0, \quad \lambda>0.$$

Consequently, from theorem 4.3 we deduce that *if for a renewal process the past lifetime η_t and the residual lifetime ζ_t are independent variables for every $t>0$ and the distribution of ζ_t is independent of t then the distribution functions of η_t and ζ_t are the same as for the Poisson (renewal) process.*

In section 6.1 we defined a stationary renewal process and showed that for such a process (cf. (6.8)),

$$m(t)=t/\mu, \qquad t\geqq 0.$$

From (6.17) and (6.18) it is now easily verified that *for a stationary renewal process the distributions of the past lifetime and the residual lifetime are given, respectively, by*

$$\Pr\{\eta_t<\eta\}=\begin{cases}0, & \eta\leqq 0,\\ \dfrac{1}{\mu}\displaystyle\int_0^{\eta}\{1-F(u)\}\,\mathrm{d}u, & 0<\eta\leqq t,\\ 1, & \eta>t;\end{cases}$$

and

$$\Pr\{\zeta_t<\zeta\}=\begin{cases}0, & \zeta\leqq 0,\\ \dfrac{1}{\mu}\displaystyle\int_0^{\zeta}\{1-F(u)\}\,\mathrm{d}u, & \zeta>0.\end{cases}$$

It is seen that for a stationary renewal process the distribution of the residual lifetime ζ_t is independent of t. From

$$\frac{1}{\mu}\int_{y=0}^{\infty} y\{1-F(y)\}\,\mathrm{d}y=\frac{1}{\mu}\int_0^{\infty} y\,\mathrm{d}y\int_{t=y}^{\infty}\mathrm{d}F(t)$$

$$=\frac{1}{\mu}\int_0^{\infty}\mathrm{d}F(t)\int_0^{t} y\,\mathrm{d}y=\frac{1}{2\mu}\int_0^{\infty} t^2\,\mathrm{d}F(t),$$

it appears that *for a stationary renewal process the first moment of the distribution of ζ_t is finite if and only if the second moment of the renewal distribution is finite.*

From (6.19) and the above formulae it now follows that

$$\Pr\{\eta_t \geqq \eta,\ \xi_t \geqq \zeta\} = \Pr\{\eta_{t+\zeta} \geqq \eta+\zeta\}$$

$$= \frac{1}{\mu} \int_{\eta+\zeta}^{\infty} \{1-F(\tau)\}\, \mathrm{d}\tau, \qquad t \geqq \eta \geqq 0, \quad \zeta \geqq 0.$$

Starting from this relation we easily deduce that for a stationary renewal process

$$\Pr\{\eta_t + \xi_t < \alpha\} = \frac{1}{\mu} \int_0^{\alpha} u\, \mathrm{d}F(u), \qquad \alpha > 0, \quad t > 0.$$

It should be stressed that the distribution of $\eta_t + \xi_t$ is not identical with the renewal distribution and that the mean of $\eta_t + \xi_t$ is finite if and only if the second moment of $F(t)$ is finite.

Next, we shall investigate for the general renewal process the distribution functions of the past and residual lifetime for $t \to \infty$.

Consider first the case in which the renewal distribution $F(t)$ is not a lattice distribution. Introduce the function

$$g(u) \stackrel{\text{def}}{=} \{1-F(u)\}\{1-U(u-\chi)\}, \qquad \chi > 0,$$

where χ is some positive constant. It now follows from theorem 6.2 that

$$\lim_{t\to\infty} \int_{t-\chi}^{t} \{1-F(t-u)\}\, \mathrm{d}m(u) = \lim_{t\to\infty} \int_0^t g(t-u)\, \mathrm{d}m(u)$$

$$= \frac{1}{\mu} \int_0^{\infty} g(\sigma)\, \mathrm{d}\sigma = \begin{cases} \dfrac{1}{\mu} \displaystyle\int_0^{\chi} \{1-F(\sigma)\}\, \mathrm{d}\sigma & \text{if } \mu < \infty, \\ 0 & \text{if } \mu = \infty. \end{cases}$$

Further from (6.17) and the first relation of (6.19) we find that, if $F(t)$ is not a lattice distribution and $\mu < \infty$,

$$\lim_{t\to\infty} \Pr\{\eta_t < \chi\} = \lim_{t\to\infty} \Pr\{\xi_t < \chi\} = \begin{cases} 0, & \chi < 0, \\ \dfrac{1}{\mu} \displaystyle\int_0^{\chi} \{1-F(\sigma)\}\, \mathrm{d}\sigma, & \chi \geqq 0. \end{cases} \tag{6.22}$$

From the second relation of (6.19) we obtain for $t \to \infty$,

$$\lim_{t\to\infty} \Pr\{\eta_t \geqq \eta,\ \zeta_t \geqq \zeta\} = 1 - \frac{1}{\mu} \int_0^{\eta+\zeta} \{1-F(\sigma)\}\, d\sigma, \qquad \eta \geqq 0,\quad \zeta \geqq 0,$$

and hence from (6.22),

$$\lim_{t\to\infty} \Pr\{\eta_t < \eta,\ \zeta_t < \zeta\} = \begin{cases} 0, & \eta<0, \text{ or } \zeta<0, \\ \dfrac{1}{\mu}\displaystyle\int_0^{\eta} \{1-F(\sigma)\}\, d\sigma + \dfrac{1}{\mu}\displaystyle\int_0^{\zeta} \{1-F(\sigma)\}\, d\sigma & \\ \quad - \dfrac{1}{\mu}\displaystyle\int_0^{\eta+\zeta} \{1-F(\sigma)\}\, d\sigma, & \zeta \geqq 0,\quad \eta \geqq 0. \end{cases} \tag{6.23}$$

Suppose next that the distributions $F(t)$ and $F_1(t)$ are both lattice distributions with period τ. We again consider the general renewal process. To investigate the distribution of η_t for large values of t it is remarked that if $t = \rho\tau + \sigma_1$, with ρ an integer and $0 \leqq \sigma_1 < \tau$, then η_t is a discrete variable with sample space $\{\alpha\tau + \sigma_1,\ \alpha = 0, 1, \ldots, \rho\}$. Since for the general renewal process (see preceding section)

$$m(t) = F_1(t) + F_1(t) * \sum_{k=0}^{\infty} w_k U_1(t - k\tau),$$

we have from (6.17) that, for $0 < \eta \leqq t$,

$$\Pr\{\eta_t < \eta\} = \int_{(t-\eta)+}^{t} \{1 - F(t-u)\}\, dF_1(u)$$

$$+ \int_{(t-\eta)+}^{t} \{1-F(t-u)\}\, d_u\{F_1(u) * \sum_{k=0}^{\infty} w_k U_1(u - k\tau)\}.$$

It is easily verified that for $t \to \infty$ the first integral above tends to zero.

For $t = \rho\tau + \sigma_1$, $\eta = \alpha\tau + \sigma$, the second integral equals

$$\int_{0+}^{\alpha\tau+\sigma_1} \{1 - F(\alpha\tau + \sigma_1 - u)\}\, d_u \sum_{k=0}^{\infty} w_k F_1\{u + (\rho - \alpha - k)\,\tau\}.$$

To evaluate this integral for $\rho \to \infty$ it is first noted that for every fixed integer $N > 0$,

$$\int_{0+}^{\alpha\tau+\sigma_1} \{1-F(\alpha\tau+\sigma_1-u)\}\, \mathrm{d}_u \sum_{k=0}^{N} w_k F_1\{u+(\rho-\alpha-k)\tau\}$$

$$\leqq \sum_{k=0}^{N} w_k\{F_1\{(\rho-k)\tau+\sigma_1\} - F\{(\rho-\alpha-k)\tau+\}\}.$$

Obviously, the right-hand side of the last inequality tends to zero as $\rho\to\infty$. Further

$$\sum_{k=N+1}^{\infty} w_k \int_{0+}^{\alpha\tau+\sigma_1} \{1-F(\alpha\tau+\sigma_1-u)\}\, \mathrm{d}_u F_1\{u+(\rho-\alpha-k)\tau\}$$

$$\approx \frac{\tau}{\mu} \sum_{k=N+1}^{\infty} \sum_{h=1}^{\alpha} \{1-F\{(\alpha-h)\tau+\}\}\{F_1\{\sigma_1+(\rho-\alpha-k+h)\tau\}$$

$$- F_1\{\sigma_1+(\rho-\alpha-k+h-1)\tau\}\}$$

$$= \frac{\tau}{\mu} \sum_{i=0}^{\alpha-1} \{1-F(i\tau+)\}\, F_1\{\sigma_1+(\rho-N-1-i)\tau\},$$

for N sufficiently large, since from theorem 6.4,

$$\lim_{n\to\infty} w_n = \tau/\mu.$$

Letting $\rho\to\infty$ and then $N\to\infty$ it is easily found from the relations above that if $\mu<\infty$ and $0\leqq\sigma_1<\tau$ then

$$\lim_{\rho\to\infty} \Pr\{\eta_{\rho\tau+\sigma_1}<\alpha\tau+\sigma_1\} = \begin{cases} 0, & \alpha=0, \\ \dfrac{\tau}{\mu}\displaystyle\sum_{i=0}^{\alpha-1} \{1-F(i\tau+)\}, & \alpha=1, 2, \dots . \end{cases} \tag{6.24}$$

From (6.19) and (6.24) we obtain with $t=\rho\tau-\sigma_2$, ρ an integer and $0\leqq\sigma_2<\tau$, for $\rho\to\infty$, $\mu<\infty$,

$$\lim_{\rho\to\infty} \Pr\{\zeta_{\rho\tau-\sigma_2}<\alpha\tau+\sigma_2\} = \begin{cases} 0, & \alpha=0, \\ \dfrac{\tau}{\mu}\displaystyle\sum_{i=0}^{\alpha-1} \{1-F(i\tau+)\}, & \alpha=1, 2, \dots . \end{cases} \tag{6.25}$$

Further from (6.19), (6.24) and (6.25) for $\mu<\infty$ and noting that $\sigma_1+\sigma_2=\tau$,

$$\lim_{\rho\to\infty} \Pr\{\eta_{\rho\tau+\sigma_1}<\alpha\tau+\sigma_1,\ \zeta_{(\rho+1)\tau-\sigma_2}<\beta\tau+\sigma_2\}$$

$$=\begin{cases} 0, & \alpha=0 \quad\text{or}\quad \beta=0,\\ \dfrac{\tau}{\mu}\sum\limits_{i=0}^{\alpha-1}\{1-F(i\tau+)\}+\dfrac{\tau}{\mu}\sum\limits_{i=0}^{\beta-1}\{1-F(i\tau+)\} & \\ \quad -\dfrac{\tau}{\mu}\sum\limits_{i=0}^{\alpha+\beta-1}\{1-F(i\tau+)\}, & \alpha,\beta=1,2,\dots. \end{cases} \tag{6.26}$$

From the results obtained above it is seen that *if the renewal distribution $F(t)$ is not a lattice distribution then the distribution of η_t, of ζ_t and the joint distribution of η_t and ζ_t have limits for $t\to\infty$ and if $F(t)$ has a finite first moment then these limit distributions are identical with the corresponding distributions of the stationary renewal process.* If $F(t)$ is a lattice distribution then the distributions of η_t and ζ_t have no limits for $t\to\infty$. However, if we consider special subsequences, i.e. restrict t to vary over the set $\{\rho t+\sigma,\ \rho=1,2,\dots\}$ then these subsequences have limits.

I.6.4. Regenerative processes

Let ξ_n, $n=0,1,\dots$, be a sequence of non-negative, independent variables with $\xi_0 \stackrel{\text{def}}{=} 0$; for $n=1,2,\dots$, they are identically distributed with distribution function $F(t)$:

$$F(t)=\begin{cases}\Pr\{\xi_n<t\}, & t>0,\\ 0, & t\leqq 0.\end{cases}$$

In this section it will always be assumed that

$$F(0+)=0,\qquad \mu=\int_0^\infty t\,\mathrm{d}F(t)<\infty.$$

We consider the renewal process $\{\boldsymbol{\nu}_t,\ t\in[0,\infty)\}$ defined by

$$\boldsymbol{s}_n \stackrel{\text{def}}{=} \xi_0+\xi_1+\dots+\xi_n,\qquad n=0,1,\dots,$$

and

$$\boldsymbol{\nu}_t \stackrel{\text{def}}{=} \max\{n:\ \boldsymbol{s}_n<t\},\qquad t>0,\quad \boldsymbol{\nu}_0\stackrel{\text{def}}{=}0.$$

This renewal process will play a basic role in the definition of regenerative processes. We could have started with a general renewal process but for the sake of simplicity this will not be done. The definitions can be easily extended, however, for $\{\boldsymbol{v}_t,\ t\in[0,\infty)\}$ a general renewal process and the results to be obtained will usually hold true for such a renewal process.

The regenerative processes, which we shall consider here, will have a denumerable state space S for which we shall take the set of integers. This type of state space is chosen since for our purpose it is the most important one. The sample functions of these processes are continuous from the left by definition.

DEFINITION. The process $\{\boldsymbol{x}_t,\ t\in[0,\infty)\}$ with initial distribution

$$a_i^{(0)} \overset{\text{def}}{=} \Pr\{\boldsymbol{x}_0=i\}, \qquad i\in S,$$

is regenerative with respect to the renewal process $\{\boldsymbol{v}_t,\ t\in[0,\infty)\}$ if it has the following properties:

(*i*)
$$\Pr\{\boldsymbol{x}_{t_{h_1}}=j_{h_1},\ \boldsymbol{s}_{k_1}<s_{k_1},\ h_1=m+1,\dots,m_1;\ k_1=n+1,\dots,n_1 \mid$$
$$\boldsymbol{x}_{t_{h_2}}=j_{h_2},\ \boldsymbol{x}_t=i,\ \boldsymbol{s}_{k_2}=s_{k_2},\ \boldsymbol{s}_n=t,\ h_2=0,\dots,m;\ k_2=0,\dots,n\}$$
$$=\Pr\{\boldsymbol{x}_{t_{h_1}}=j_{h_1},\ \boldsymbol{s}_{k_1}<s_{k_1},\ h_1=m+1,\dots,m_1;\ k_1=n+1,\dots,n_1 \mid$$
$$\boldsymbol{x}_t=i,\ \boldsymbol{s}_n=t\},$$

for $t\geqq 0$, $n=0,1,\dots;\ n_1=n+1,\dots;\ m=0,1,\dots;\ m_1=m+1,\dots;$

$i, j_{h_1}, j_{h_2}\in S$, and $0\leqq t_1<\dots<t_m<t<t_{m+1}<\dots<t_{m_1};$

$0\leqq s_1<\dots<s_{n-1}<t<s_{n+1}<\dots<s_{n_1};$

(*ii*) for $n=0,1,\dots;\ v>u\geqq 0$,

$$\Pr\{\boldsymbol{x}_{t_h}=j_h,\ u<t_h\leqq v,\ h=1,\dots,m \mid \boldsymbol{x}_u=i,\ \boldsymbol{s}_n=u,\ \boldsymbol{s}_{n+1}=v\}$$

is only a function of the variables t_h-u, $h=1,\dots,m$ for fixed $i, j_h\in S$, and $p_{ij}(t)$ with

$$p_{ij}(t-u)\overset{\text{def}}{=}\begin{cases}\Pr\{\boldsymbol{x}_t=j \mid \boldsymbol{x}_u=i,\ \boldsymbol{s}_n=u,\ \boldsymbol{s}_{n+1}=v\}, & 0\leqq u\leqq t\leqq v,\\ 0, & t<u,\end{cases}$$

is integrable with respect to $F(t)$ over $[0,\infty)$ for all $i,j\in S$;

(*iii*) the variables $\boldsymbol{\xi}_{n+1}, \boldsymbol{\xi}_{n+2}, \dots,$ are conditionally independent of $\boldsymbol{x}_t$ for $\boldsymbol{s}_n=t$, $n=0,1,\dots;\ t>0$;

whenever the conditional probabilities above are defined.

It is obvious that the first defining property states that the future development of the process $\{\boldsymbol{x}_t,\ t\in[0,\infty)\}$ after a renewal, given the state of the process just before this renewal, is independent of the past of the process until this renewal. The second defining property shows that the development of the process between two successive renewals is independent of the renewal process and independent of the absolute time. The last defining property states that the renewal process $\boldsymbol{\nu}_t$ is not influenced in its development by the $\boldsymbol{x}_t$ process. The assumption that the renewal distribution $F(t)$ has a finite first moment is not essential in the definition above, it has been introduced for the sake of simplicity.

Roughly speaking, we may say that the process $\{\boldsymbol{x}_t,\ t\in[0,\infty)\}$ is stopped or interrupted in its development at a renewal epoch in the renewal process $\{\boldsymbol{\nu}_t,\ t\in[0,\infty)\}$, and then the process $\{\boldsymbol{x}_t,\ t\in[0,\infty)\}$ is restarted in the state reached just before this renewal without its future development being influenced by the past before this renewal.

Regenerative processes have been studied by SMITH [1955], but his definition differs from the one given above; he assumes that the future development of the process after a renewal is also independent of the state of the process reached at the renewal point. In two interesting papers PYKE [1961[a], 1961[b]] has treated processes which are related to the regenerative process introduced here.

For $t>0$ the stochastic variable $\boldsymbol{x}_{\boldsymbol{s}_{\boldsymbol{\nu}_t}}$ represents the state of the $\boldsymbol{x}_t$ process just before the last renewal occurring before time t if $\boldsymbol{\nu}_t>0$, since the sample functions are supposed to be continuous from the left. Consider the stochastic process $\{\boldsymbol{x}_{\boldsymbol{s}_n},\ n=0,1,\dots\}$ or $\{\boldsymbol{y}_n,\ n=0,1,\dots\}$ with

$$\boldsymbol{y}_n \stackrel{\text{def}}{=} \boldsymbol{x}_{\boldsymbol{s}_n}, \qquad n=0,1,\dots, \qquad \boldsymbol{y}_0 \stackrel{\text{def}}{=} \boldsymbol{x}_0.$$

This discrete time parameter process is an imbedded process of the regenerative process $\{\boldsymbol{x}_t,\ t\in[0,\infty)\}$. We shall show that this *imbedded process is a discrete time parameter Markov chain with state space S and stationary transition probabilities given by*

$$\Pr\{\boldsymbol{y}_{n+1}=j \mid \boldsymbol{y}_n=i\} = \int_0^\infty p_{ij}(t)\,\mathrm{d}F(t), \qquad n=0,1,\dots;\quad i,j\in S.$$

From the definition above it follows that

$$p_{ij}(t) \geqq 0, \qquad \sum_{j\in S} p_{ij}(t) = 1 \quad \text{for} \quad t>0,\quad i,j\in S,$$

and consequently that

$$\sum_{j\in S}\int_0^\infty p_{ij}(t)\,\mathrm{d}F(t) = 1, \qquad i\in S.$$

By taking $m_1=n_1=n+1$, $t_{n+1}=s_{n+1}$, $t_{h_2}=s_{k_2}$, $h_2=k_2=0, 1, \ldots, n$, we obtain from the defining properties

$$\Pr\{y_{n+1}=j_{n+1}, \mathbf{s}_{n+1}<s_{n+1} \mid y_k=j_k, \mathbf{s}_k=s_k, s_n=t, k=0, \ldots, n\}$$

$$= \Pr\{y_{n+1}=j_{n+1}, \mathbf{s}_{n+1}<s_{n+1} \mid y_n=j_n, \mathbf{s}_n=t\}$$

$$= \int_{v=t}^{s_{n+1}} \Pr\{y_{n+1}=j_{n+1} \mid \mathbf{s}_{n+1}=v, y_n=j_n, \mathbf{s}_n=t\} \, \mathrm{d}_v \Pr\{\mathbf{s}_{n+1}<v \mid y_n=j_n, \mathbf{s}_n=t\}$$

$$= \int_{v=t}^{s_{n+1}} p_{j_n j_{n+1}}(v-t) \, \mathrm{d}_v F(v-t).$$

Hence for $t \geqq 0$, $n=0, 1, \ldots$; $0 \leqq s_0<s_1<\ldots<s_n=t$,

$$\Pr\{y_{n+1}=j_{n+1} \mid y_k=j_k, \mathbf{s}_k=s_k, s_n=t, k=0, \ldots, n\} = \int_{v=0}^{\infty} p_{j_n j_{n+1}}(v) \, \mathrm{d}F(v),$$

so that, since the right-hand side is independent of j_k, s_k, $k=0, \ldots, n-1$, it follows that

$$\Pr\{y_{n+1}=j \mid y_n=i, y_k=j_k, k=0, \ldots, n-1\}$$

$$= \Pr\{y_{n+1}=j \mid y_n=i\} = \int_0^{\infty} p_{ij}(v) \, \mathrm{d}F(v), \qquad n=0, 1, \ldots; \quad i, j \in S.$$

Hence, it is seen from section 2.1 that $\{y_n, n=0, 1, \ldots\}$ is indeed a discrete time parameter Markov chain with stationary transition probabilities.

We shall consider for this Markov chain the entrance time

$$\boldsymbol{\sigma}_{ij} \stackrel{\text{def}}{=} \min\{k: k>0, y_k=j \mid y_0=i\}, \qquad i, j \in S;$$

hence (cf. section 2.2) for $k=1, 2, \ldots$,

$$f_{ij}^{(k)} \stackrel{\text{def}}{=} \Pr\{\boldsymbol{\sigma}_{ij}=k\} = \Pr\{y_k=j, y_h \neq j, h=1, \ldots, k-1 \mid y_0=i\}.$$

Define for $y_0=i$, $t>0$,

$$\boldsymbol{\sigma}_{ij}(t) \stackrel{\text{def}}{=} \begin{cases} 0 & \text{if } \boldsymbol{\nu}_t=0 \text{ or } y_k \neq j, \mathbf{s}_k<t, k=1, \ldots, \boldsymbol{\nu}_t, \\ k & \text{if } y_k=j, \mathbf{s}_k<t, y_h \neq j, h=1, \ldots, k-1; k=1, \ldots, \boldsymbol{\nu}_t, \end{cases}$$

so that, for $t>0$, $k=1, 2, \ldots$; $i, j \in S$,

$$F_{ij}^{(0)}(t) \stackrel{\text{def}}{=} \Pr\{\boldsymbol{\sigma}_{ij}(t)=0\} = \Pr\{\mathbf{s}_h<t, y_h \neq j, h=1, \ldots, \boldsymbol{\nu}_t \mid y_0=i\}$$
$$+ \Pr\{\boldsymbol{\nu}_t=0 \mid y_0=i\},$$

$$F_{ij}^{(k)}(t) \stackrel{\text{def}}{=} \Pr\{\boldsymbol{\sigma}_{ij}(t)=k\} = \Pr\{y_k=j, \mathbf{s}_k<t, y_h \neq j, h=1, \ldots, k-1 \mid y_0=i\}.$$

We also introduce the stochastic variable $\boldsymbol{\tau}_{ij}$ which will denote the time between $t=0$ with $\boldsymbol{y}_0=i$ and the moment ρ of the first renewal after $t=0$ with $\boldsymbol{x}_\rho=j$, so that

$$\{\boldsymbol{\tau}_{ij}<s\} = \bigcup_{k=1}^{\infty} \{\boldsymbol{\sigma}_{ij}(s)=k\} = \bigcup_{k=1}^{\infty} \{\boldsymbol{\xi}_1+\dots+\boldsymbol{\xi}_k<s, \boldsymbol{\sigma}_{ij}=k\}$$

$$= \{\boldsymbol{\xi}_1+\dots+\boldsymbol{\xi}_{\boldsymbol{\sigma}_{ij}}<s\}, \qquad s>0;\ i,j\in S.$$

Obviously, we now have for all $i,j\in S$,

$$F_{ij}(t) \overset{\text{def}}{=} \Pr\{\boldsymbol{\tau}_{ij}<t\} = \begin{cases} \sum_{k=1}^{\infty} F_{ij}^{(k)}(t) = \Pr\{\boldsymbol{\sigma}_{ij}(t)>0\}, & t>0, \\ 0, & t\leqq 0, \end{cases}$$

and

$$F_{ij}(0+) = \Pr\{\boldsymbol{\tau}_{ij}<0+\} = 0,$$

since $F(0+)=0$. Moreover, for all $i,j\in S$,

$$F_{ij}^{(0)}(t) + F_{ij}(t) = 1, \qquad t > 0, \qquad \sum_{k=1}^{\infty} f_{ij}^{(k)} \leqq 1,$$

and as $t\to\infty$, $F_{ij}^{(k)}(t)$ tends monotonically to $f_{ij}^{(k)}$. Hence

$$1 \geqq 1 - F_{ij}^{(0)}(\infty) = F_{ij}(\infty) = \lim_{t\to\infty} \sum_{k=1}^{\infty} F_{ij}^{(k)}(t)$$

$$= \sum_{k=1}^{\infty} F_{ij}^{(k)}(\infty) = \sum_{k=1}^{\infty} f_{ij}^{(k)}.$$

If, at some renewal point ρ, $\boldsymbol{x}_\rho=j$ then we shall say that the regenerative process $\{\boldsymbol{x}_t,\ t\in[0,\infty)\}$ is at time ρ in the renewal state j. A renewal state will be called *transient* if $F_{jj}(\infty)<1$, and *recurrent* otherwise. It is *positive recurrent* if $F_{jj}(\infty)=1$ and

$$\mu_{jj} \overset{\text{def}}{=} \int_0^\infty t\,\mathrm{d}F_{jj}(t) < \infty,$$

and *null recurrent* if $F_{jj}(\infty)=1$, $\mu_{jj}=\infty$.

From the above inequality it is seen that $F_{ij}(\infty)<1$ implies $\sum_{k=1}^{\infty} f_{ij}^{(k)}<1$ and conversely. Moreover, by taking $i=j$ it is seen that if state j is recurrent in the Markov chain $\{\boldsymbol{y}_n,\ n=0,1,\dots\}$ then the renewal state j is also recurrent and conversely.

Since

$$F_{jj}(t)=\Pr\{\boldsymbol{\xi}_1+\dots+\boldsymbol{\xi}_{\boldsymbol{\sigma}_{jj}}<t\}, \qquad t>0,$$

it follows that

$$\mu_{jj} = \mathrm{E}\{\xi_1 + \ldots + \xi_{\sigma_{jj}}\}, \qquad j \in S.$$

The third defining property shows that the event $\{\boldsymbol{\sigma}_{jj} = k\}$ and the variables $\xi_{k+1}, \xi_{k+2}, \ldots$, are independent, hence by Wald's theorem (cf. app. 7) we have

$$\mu_{jj} = \mathrm{E}\{\xi_1\}\, \mathrm{E}\{\boldsymbol{\sigma}_{jj}\} = \mu \mathrm{E}\{\boldsymbol{\sigma}_{jj}\}. \tag{6.27}$$

$\mathrm{E}\{\boldsymbol{\sigma}_{jj}\}$ represents the average number of steps between two successive occurrences of state j in the Markov chain $\{\boldsymbol{y}_n,\ n=0, 1, \ldots\}$, therefore, it is also the average number of renewals of the renewal process $\{\boldsymbol{v}_t,\ t \in [0, \infty)\}$ between two successive occurrences of the renewal state j. If j is a recurrent state of the discrete time parameter Markov chain and $\mathrm{E}\{\boldsymbol{\sigma}_{jj}\} < \infty$ then the renewal state j is recurrent and in fact positive recurrent since we have assumed μ to be finite; if state j is null recurrent, however, then the renewal state j is always null recurrent irrespective of the value of μ. If we allow μ to be infinite, a renewal state can be null recurrent while it is positive recurrent in the Markov chain $\{\boldsymbol{y}_n,\ n=0, 1, \ldots\}$.

It follows from the above relations that whenever $\sum_{k=1}^{\infty} f_{ij}^{(k)} = 1$ then $F_{ij}(\infty) = 1$, so that $F_{ij}(t)$ is a proper probability distribution. Evidently, $\sum_{k=1}^{\infty} f_{ij}^{(k)} = 1$ for all $i, j \in S$ if the imbedded Markov chain is irreducible and has all its states recurrent.

Let us assume for the present that $\sum_{k=1}^{\infty} f_{ij}^{(k)} = 1$ for all $i, j \in S$, so that $F_{ij}(t)$ is a proper probability distribution for all $i, j \in S$. We now consider some relations between the quantities introduced above.

Obviously, we have for $t > 0$, $i, j \in S$,

$$F_{ij}^{(1)}(t) = \int_0^t p_{ij}(\sigma)\, \mathrm{d}F(\sigma),$$

$$F_{ij}^{(n)}(t) = \sum_{\substack{k \in S \\ k \neq j}} \int_0^t p_{ik}(u)\, F_{kj}^{(n-1)}(t-u)\, \mathrm{d}F(u), \qquad n = 2, 3, \ldots .$$

Summing these relations over $n = 1, 2, \ldots$, we find that for $i, j \in S$, $t > 0$,

$$F_{ij}(t) = \sum_{\substack{k \in S \\ k \neq j}} \int_0^t p_{ik}(u)\, F_{kj}(t-u)\, \mathrm{d}F(u) + \int_0^t p_{ij}(u)\, \mathrm{d}F(u).$$

Consider the set of integral equations

$$z_{ij}(t) = \sum_{\substack{k \in S \\ k \neq j}} \int_0^t p_{ik}(u)\, z_{kj}(t-u)\, \mathrm{d}F(u) + \int_0^t p_{ij}(u)\, \mathrm{d}F(u), \qquad t > 0, \quad i, j \in S,$$

then *any non-negative solution* $z_{ij}^{(1)}(t)$, $t>0$, $i, j \in S$, *of this set of integral relations satisfies*

$$z_{ij}^{(1)}(t) \geqq F_{ij}(t) \qquad \text{for all} \quad t>0, \quad i,j\in S.$$

To prove this we first note that

$$z_{ij}^{(1)}(t) \geqq \int_0^t p_{ij}(u)\,\mathrm{d}F(u), \qquad t>0, \quad i,j\in S,$$

so that for all $t>0$,

$$z_{ij}^{(1)}(t) \geqq F_{ij}^{(1)}(t), \qquad i,j\in S.$$

Hence

$$z_{ij}^{(1)}(t) \geqq \sum_{\substack{k\in S\\ k\neq j}} \int_0^t p_{ik}(u)\,F_{kj}^{(1)}(t-u)\,\mathrm{d}F(u) + \int_0^t p_{ij}(u)\,\mathrm{d}F(u),$$

so that for all $t>0$,

$$z_{ij}^{(1)}(t) \geqq F_{ij}^{(2)}(t) + F_{ij}^{(1)}(t), \qquad i,j\in S.$$

Repeating this procedure we obtain for every integer $N>0$, all $t>0$,

$$z_{ij}^{(1)}(t) \geqq \sum_{n=1}^{N} F_{ij}^{(n)}(t), \qquad i,j\in S.$$

Therefore,

$$z_{ij}^{(1)}(t) \geqq F_{ij}(t), \qquad i,j\in S, \quad \text{all } t>0.$$

This establishes our statement.

We next consider *a general renewal process with renewal distribution* $F_{jj}(t)$, *and with* $F_{ij}(t)$ *as the distribution function of the first renewal time.* Denoting by $m_{ij}(t)$ the renewal function of this general renewal process then since $F_{ij}(0+)=0$, $i,j\in S$,

$$m_{ij}(t) = F_{ij}(t) + \int_0^t F_{jj}(t-u)\,\mathrm{d}m_{ij}(t), \qquad t\geqq 0, \quad i,j\in S,$$

and

$$m_{ij}(t) = F_{ij}(t) + \sum_{n=1}^{\infty} F_{ij}(t) * F_{jj}^{n*}(t), \qquad t\geqq 0, \quad i,j\in S.$$

It follows from theorem 6.1 that

$$\lim_{t\to\infty} \frac{m_{ij}(t)}{t} = \begin{cases} \dfrac{1}{\mu_{jj}} & \text{if } \mu_{jj}<\infty, \\[2ex] 0 & \text{if } \mu_{jj}=\infty. \end{cases}$$

From the definitions above we have

$$\Pr\{y_1=j,\ s_1<t \mid y_0=i\} = F_{ij}^{(1)}(t),$$

$$\Pr\{y_n=j,\ s_n<t \mid y_0=i\} = \sum_{k=1}^{n-1} \int_0^t F_{jj}^{(n-k)}(t-u)\, \mathrm{d}_u \Pr\{y_k=j,\ s_k<u \mid y_0=i\}$$
$$+ F_{ij}^{(n)}(t), \qquad t \geqq 0, \quad n=2, 3, \ldots .$$

Summing these relations over $n=1, 2, \ldots$, yields for $t \geqq 0$, $i, j \in S$,

$$\sum_{n=1}^{\infty} \Pr\{y_n=j,\ s_n<t \mid y_0=i\}$$
$$= F_{ij}(t) + \int_0^t F_{jj}(t-u)\, \mathrm{d}_u \sum_{n=2}^{\infty} \Pr\{y_n=j,\ s_n<u \mid y_0=i\}.$$

This relation represents *the renewal equation for the general renewal process* with renewal distribution $F_{jj}(t)$ and distribution of the first renewal time $F_{ij}(t)$; hence (cf. (6.5)),

$$\sum_{n=1}^{\infty} \Pr\{y_n=j,\ s_n<t \mid y_0=i\} = m_{ij}(t), \qquad t \geqq 0, \quad i, j \in S. \tag{6.28}$$

This relation is obvious; the left-hand side equals the average number of occurrences of the renewal state j during the interval $[0, t)$.

From the latter relation we derive *a set of integral equations for* $m_{ij}(t)$. From the defining properties (*i*), (*ii*), (*iii*) we have

$$m_{ij}(t) = \int_0^t p_{ij}(\sigma)\, \mathrm{d}F(\sigma) + \sum_{n=2}^{\infty} \Pr\{y_n=j,\ s_n<t \mid y_0=i\}, \tag{6.29}$$

and

$$\sum_{n=2}^{\infty} \Pr\{y_n=j,\ s_n<t \mid y_0=i\}$$
$$= \sum_{h \in S} \sum_{n=2}^{\infty} \int_{u=0}^{t} \Pr\{y_n=j,\ s_n<t \mid y_{n-1}=h,\ s_{n-1}=u,\ y_0=i\}$$
$$\cdot \mathrm{d}_u \Pr\{y_{n-1}=h,\ s_{n-1}<u \mid y_0=i\}.$$

Since for $t>u$,

$$\Pr\{y_n=j,\ s_n<t \mid y_{n-1}=h,\ s_{n-1}=u,\ y_0=i\}$$
$$= \int_{v=u}^{t} \Pr\{y_n=j \mid y_{n-1}=h,\ s_n=v,\ s_{n-1}=u,\ y_0=i\}$$
$$\cdot \mathrm{d}_v \Pr\{s_n<v \mid y_{n-1}=h,\ s_{n-1}=u,\ y_0=i\}$$
$$= \int_{v=u}^{t} p_{hj}(v-u)\, \mathrm{d}_v F(v-u) = \int_0^{t-u} p_{hj}(\sigma)\, \mathrm{d}F(\sigma),$$

it follows from the relations above that for $t \geqq 0$, $i, j \in S$,

$$m_{ij}(t) = \int_0^t p_{ij}(\sigma)\, \mathrm{d}F(\sigma) + \sum_{h \in S} \int_0^t \{ \int_0^{t-u} p_{hj}(\sigma)\, \mathrm{d}F(\sigma)\}\, \mathrm{d}_u m_{ih}(u),$$

which represents the set of integral equations for the renewal functions $m_{ij}(t)$. The obvious method to solve this set of integral equations is by applying the Laplace-Stieltjes transform.

Before applying the results of renewal theory it should be noted that if $F(t)$ is not a lattice distribution, $F_{ij}(t)$ is not a lattice distribution either and conversely.

Since

$$\Pr\{\boldsymbol{y}_n = j,\, \boldsymbol{s}_n < t \mid \boldsymbol{y}_0 = i\} = \Pr\{\boldsymbol{y}_n = j,\, \boldsymbol{\nu}_t \geqq n \mid \boldsymbol{y}_0 = i\}, \qquad n = 1, 2, \ldots,$$

it follows from theorem 6.1 applied to the general renewal process with renewal distribution $F_{jj}(t)$ that for $t \to \infty$,

$$\sum_{n=1}^{\infty} \Pr\{\boldsymbol{y}_n = j,\, \boldsymbol{\nu}_t \geqq n \mid \boldsymbol{y}_0 = i\} \to \begin{cases} 0 & \text{if } \mu_{jj} = \infty, \\ \dfrac{t}{\mu_{jj}} & \text{if } \mu_{jj} < \infty. \end{cases}$$

Since $\mu_{jj} < \infty$ implies $\mathrm{E}\{\boldsymbol{\sigma}_{jj}\} < \infty$ and conversely, it follows from theorem 2.2 and the latter result that, if state j is aperiodic in the Markov chain $\{\boldsymbol{y}_n,\, n = 0, 1, \ldots\}$, then

$$\lim_{t \to \infty} \frac{1}{t} \sum_{n=1}^{\infty} \Pr\{\boldsymbol{y}_n = j,\, \boldsymbol{\nu}_t \geqq n \mid \boldsymbol{y}_0 = i\} = \lim_{n \to \infty} \Pr\{\boldsymbol{y}_n = j\} \lim_{t \to \infty} \frac{1}{t} \sum_{n=1}^{\infty} \Pr\{\boldsymbol{\nu}_t \geqq n\}.$$

From the defining properties we further have

$$\begin{aligned} \Pr\{\boldsymbol{y}_n = j,\, \boldsymbol{\nu}_t = n \mid \boldsymbol{y}_0 = i\} &= \Pr\{\boldsymbol{y}_n = j,\, \boldsymbol{s}_n < t,\, \boldsymbol{s}_{n+1} \geqq t \mid \boldsymbol{y}_0 = i\} \\ &= \int_{u=0}^{t} \Pr\{\boldsymbol{s}_{n+1} \geqq t \mid \boldsymbol{y}_n = j,\, \boldsymbol{s}_n = u,\, \boldsymbol{y}_0 = i\}\, \mathrm{d}_u \Pr\{\boldsymbol{y}_n = j,\, \boldsymbol{s}_n < u \mid \boldsymbol{y}_0 = i\} \\ &= \int_0^t \{1 - F(t-u)\}\, \mathrm{d}_u \Pr\{\boldsymbol{y}_n = j,\, \boldsymbol{s}_n < u \mid \boldsymbol{y}_0 = i\}, \qquad t > 0. \end{aligned}$$

Hence by theorem 6.2 if $F(t)$ is not a lattice distribution

$$\lim_{t \to \infty} \sum_{n=1}^{\infty} \Pr\{\boldsymbol{y}_n = j,\, \boldsymbol{\nu}_t = n \mid \boldsymbol{y}_0 = i\} = \begin{cases} \dfrac{1}{\mu_{jj}} \displaystyle\int_0^{\infty} \{1 - F(v)\}\, \mathrm{d}v = \dfrac{\mu}{\mu_{jj}}, & \mu_{jj} < \infty, \\ 0, & \mu_{jj} = \infty. \end{cases} \tag{6.30}$$

A similar result is obtained if $F(t)$ is a lattice distribution.

THEOREM 6.6. *If for the regenerative process* $\{\boldsymbol{x}_t,\ t\in[0,\infty)\}$ *the imbedded Markov chain* $\{\boldsymbol{y}_n,\ n=0,1,\ldots\}$ *is irreducible and aperiodic then:*

(i) *if this Markov chain is ergodic*

$$u_i \overset{\text{def}}{=} \lim_{n\to\infty} \Pr\{\boldsymbol{y}_n=i\} = \lim_{t\to\infty} \Pr\{\boldsymbol{x}_{s_{\nu_t}}=i\} = \frac{\mu}{\mu_{ii}}, \qquad i\in S, \tag{6.31}$$

$$u_j = \sum_{i\in S} u_i \int_0^\infty p_{ij}(\sigma)\,\mathrm{d}F(\sigma), \qquad j\in S;\quad \sum_{i\in S} u_i = 1, \tag{6.32}$$

$$\lim_{t\to\infty} \Pr\{\boldsymbol{x}_t=j\} = \sum_{i\in S} \frac{u_i}{\mu} \int_0^\infty p_{ij}(\sigma)\{1-F(\sigma)\}\,\mathrm{d}\sigma, \qquad j\in S, \tag{6.33}$$

if $F(t)$ *is not a lattice distribution, whereas if* $F(t)$ *is a lattice distribution with period* τ *and* $0\leqq\sigma<\tau$, $\rho=0,1,\ldots,$

$$\lim_{\rho\to\infty} \Pr\{\boldsymbol{x}_{\rho\tau+\sigma}=j\} = \frac{\tau}{\mu} \sum_{i\in S} u_i \sum_{n=0}^\infty p_{ij}(n\tau+\sigma)\{1-F(n\tau+)\}, \qquad j\in S; \tag{6.34}$$

the limit distributions above are proper probability distributions;

(ii) *if the Markov chain is not ergodic then the limits above exist but are equal to zero.*

Proof. Since the discrete parameter Markov chain $\{\boldsymbol{y}_n,\ n=0,1,\ldots\}$ is ergodic with stationary transition probabilities given by

$$\Pr\{\boldsymbol{y}_{n+1}=j \mid \boldsymbol{y}_n=i\} = \int_0^\infty p_{ij}(v)\,\mathrm{d}F(v), \qquad i,j\in S,\quad n=0,1,\ldots,$$

(6.32) and the first part of (6.31) immediately follow from theorem 2.5. To prove the second part of (6.31) we refer to (6.30) and

$$\Pr\{\boldsymbol{x}_{s_{\nu_t}}=i\} = \sum_{n=0}^\infty \Pr\{\boldsymbol{y}_n=i,\ \boldsymbol{\nu}_t=n\}$$

$$= \sum_{n=0}^\infty \sum_{j\in S} \Pr\{\boldsymbol{y}_n=i,\ \boldsymbol{\nu}_t=n \mid \boldsymbol{y}_0=j\} \Pr\{\boldsymbol{y}_0=j\},$$

from which the statements follow immediately.

To prove (6.33) consider first

$$\Pr\{x_t=j,\ x_{s_{\nu_t}}=i \mid y_0=h\} = \sum_{n=0}^{\infty} \Pr\{x_t=j,\ y_n=i,\ \nu_t=n \mid y_0=h\}$$

$$= \sum_{n=0}^{\infty} \Pr\{x_t=j,\ y_n=i,\ s_n<t,\ s_{n+1}\geqq t \mid y_0=h\}$$

$$= \sum_{n=0}^{\infty} \int_{u=0}^{t} \int_{v=t}^{\infty} \Pr\{x_t=j \mid y_n=i,\ s_n=u,\ s_{n+1}=v,\ y_0=h\}$$

$$\cdot \mathrm{d}_v \Pr\{s_{n+1}<v \mid y_n=i,\ s_n=u,\ y_0=h\}\ \mathrm{d}_u \Pr\{y_n=i,\ s_n<u \mid y_0=h\}$$

$$= \sum_{n=0}^{\infty} \int_{u=0}^{t} \int_{v=t-u}^{\infty} p_{ij}(t-u)\,\mathrm{d}F(v)\,\mathrm{d}_u \Pr\{y_n=i,\ s_n<u \mid y_0=h\}$$

$$= \int_{u=0}^{t} p_{ij}(t-u)\{1-F(t-u)\}\ \mathrm{d}_u \sum_{n=0}^{\infty} \Pr\{y_n=i,\ s_n<u \mid y_0=h\}. \tag{6.35}$$

If $F(t)$ is not a lattice distribution, so that $F_{ii}(t)$ is not a lattice distribution either, it follows from (6.35) and (6.29) by applying theorem 6.2 to the general renewal process with renewal distribution $F_{ii}(t)$,

$$\lim_{t\to\infty} \Pr\{x_t=j,\ x_{s_{\nu_t}}=i \mid y_0=h\} = \lim_{t\to\infty} \Pr\{x_t=j,\ x_{s_{\nu_t}}=i\}$$

$$= \frac{1}{\mu_{ii}} \int_0^{\infty} p_{ij}(\sigma)\{1-F(\sigma)\}\ \mathrm{d}\sigma = \frac{u_i}{\mu} \int_0^{\infty} p_{ij}(\sigma)\{1-F(\sigma)\}\ \mathrm{d}\sigma.$$

Let S_1 be a finite subset of S then

$$\Pr\{x_t=j\} = \sum_{i\in S_1} \Pr\{x_t=j,\ x_{s_{\nu_t}}=i\} + \sum_{i\in S-S_1} \Pr\{x_t=j,\ x_{s_{\nu_t}}=i\},$$

and

$$\sum_{i\in S-S_1} \Pr\{x_t=j,\ x_{s_{\nu_t}}=i\} \leqq \sum_{i\in S-S_1} \Pr\{x_{s_{\nu_t}}=i\}.$$

From (6.31) and (6.32) it now follows that

$$1 = \lim_{t\to\infty} \sum_{i\in S} \Pr\{x_{s_{\nu_t}}=i\} = \sum_{i\in S_1} u_i + \lim_{t\to\infty} \sum_{i\in S-S_1} \Pr\{x_{s_{\nu_t}}=i\},$$

so that for every $\delta>0$ the subset S_1 can be determined such that

$$\lim_{t\to\infty} \sum_{i\in S-S_1} \Pr\{x_{s_{\nu_t}}=i\} = 1 - \sum_{i\in S_1} u_i < \delta.$$

From the relations above it is seen that $\Pr\{x_t=j\}$ has a limit for $t\to\infty$ and that (6.33) holds if $F(t)$ is not a lattice distribution. If $F(t)$ is a lattice distribution with period τ then it is easily verified from the definition of $F_{ii}(t)$, that this distribution is a lattice distribution with the same period.

Application of theorem 6.4 to the relation (6.35) leads as above to (6.34). By adding the limit distributions mentioned in (6.31), ..., (6.34) over i and j and noting that $u_i > 0$ it is easily verified that these distributions are proper probability distributions. The first part of the theorem has now been proved. If the imbedded Markov chain is not ergodic then $u_i = 0$ for all $i \in S$ and the calculations above show that all limits in (6.31), ..., (6.34) exist but all equal zero. The proof is complete.

Next, we shall consider *the joint distribution of* $\boldsymbol{x}_t$, $\boldsymbol{\eta}_t$ *and* $\boldsymbol{\xi}_t$, where $\boldsymbol{\eta}_t$ and $\boldsymbol{\xi}_t$ denote the past lifetime and the residual lifetime at time t for the renewal process $\{\boldsymbol{v}_t, t \in [0, \infty)\}$. For $t > \eta > 0$, $\zeta > 0$ we have

$$\Pr\{\boldsymbol{x}_t = j, \boldsymbol{\eta}_t < \eta, \boldsymbol{\xi}_t < \zeta, \boldsymbol{v}_t = n, \boldsymbol{y}_n = i \mid \boldsymbol{y}_0 = h\}$$

$$= \Pr\{\boldsymbol{x}_t = j, \boldsymbol{y}_n = i, t - \eta < \boldsymbol{s}_n < t, t \leqq \boldsymbol{s}_{n+1} < t + \zeta \mid \boldsymbol{y}_0 = h\}$$

$$= \int_{u=(t-\eta)+}^{t} \int_{v=t}^{t+\zeta} \Pr\{\boldsymbol{x}_t = j \mid \boldsymbol{y}_n = i, \boldsymbol{s}_n = u, \boldsymbol{s}_{n+1} = v, \boldsymbol{y}_0 = h\}$$

$$\cdot \mathrm{d}_v \Pr\{\boldsymbol{s}_{n+1} < v \mid \boldsymbol{y}_n = i, \boldsymbol{s}_n = u, \boldsymbol{y}_0 = h\} \, \mathrm{d}_u \Pr\{\boldsymbol{y}_n = i, \boldsymbol{s}_n < u \mid \boldsymbol{y}_0 = h\}$$

$$= \int_{u=(t-\eta)+}^{t} p_{ij}(t-u)\{F(t+\zeta-u) - F(t-u)\} \, \mathrm{d}_u \Pr\{\boldsymbol{y}_n = i, \boldsymbol{s}_n < u \mid \boldsymbol{y}_0 = h\}.$$

Hence for $t > \eta > 0$, $\zeta > 0$,

$$\Pr\{\boldsymbol{x}_t = j, \boldsymbol{\eta}_t < \eta, \boldsymbol{\xi}_t < \zeta, \boldsymbol{x}_{\boldsymbol{s}_{\boldsymbol{v}_t}} = i \mid \boldsymbol{y}_0 = h\}$$

$$= \int_{u=(t-\eta)+}^{t} p_{ij}(t-u)\{F(t+\zeta-u) - F(t-u)\} \, \mathrm{d}_u \sum_{n=1}^{\infty} \Pr\{\boldsymbol{y}_n = i, \boldsymbol{s}_n < u \mid \boldsymbol{y}_0 = h\}.$$

If $F(t)$ is not a lattice distribution and the imbedded Markov chain is ergodic, it follows after summation over all $i \in S$ that as $t \to \infty$,

$$\lim_{t\to\infty} \Pr\{\boldsymbol{x}_t = j, \boldsymbol{\eta}_t < \eta, \boldsymbol{\xi}_t < \zeta \mid \boldsymbol{y}_0 = h\} = \lim_{t\to\infty} \Pr\{\boldsymbol{x}_t = j, \boldsymbol{\eta}_t < \eta, \boldsymbol{\xi}_t < \zeta\}$$

$$= \sum_{i \in S} \frac{u_i}{\mu} \int_0^{\eta} p_{ij}(\sigma)\{F(\sigma+\zeta) - F(\sigma)\} \, \mathrm{d}\sigma, \qquad \eta > 0, \quad \zeta > 0. \tag{6.36}$$

From this relation we immediately obtain for $\zeta \geqq 0$, $\eta \geqq 0$,

$$\lim_{t\to\infty} \Pr\{\boldsymbol{x}_t=j,\ \zeta_t<\zeta\} = \sum_{i\in S} \frac{u_i}{\mu} \int_0^\infty p_{ij}(\sigma)\{F(\sigma+\zeta)-F(\sigma)\}\,\mathrm{d}\sigma, \tag{6.37}$$

$$\lim_{t\to\infty} \Pr\{\boldsymbol{x}_t=j,\ \eta_t<\eta\} = \sum_{i\in S} \frac{u_i}{\mu} \int_0^\eta p_{ij}(\sigma)\{1-F(\sigma)\}\,\mathrm{d}\sigma.$$

If $F(t)$ is a lattice distribution with period τ the relations analogous to (6.37) read with $t=\rho\tau+\sigma$, $0\leqq\sigma<\tau$, ρ an integer,

$$\lim_{\rho\to\infty} \Pr\{\boldsymbol{x}_{\rho\tau+\sigma}=j,\ \zeta_{\rho\tau+\sigma}<\alpha\tau+\sigma\}$$

$$=\begin{cases} \dfrac{\tau}{\mu}\sum\limits_{i\in S} u_i \sum\limits_{n=0}^{\infty} p_{ij}(n\tau+\sigma)\{F((n+\alpha)\tau+)-F(n\tau+)\}, & \alpha=1,2,\ldots, \\ 0, & \alpha=0; \end{cases}$$

$$\lim_{\rho\to\infty} \Pr\{\boldsymbol{x}_{\rho\tau+\sigma}=j,\ \eta_{\rho\tau+\sigma}<\alpha\tau+\sigma\}$$

$$=\begin{cases} \dfrac{\tau}{\mu}\sum\limits_{i\in S} u_i \sum\limits_{n=0}^{\alpha-1} p_{ij}(n\tau+\sigma)\{1-F(n\tau+)\}, & \alpha=1,2,\ldots, \\ 0, & \alpha=0. \end{cases}$$

An interesting quantity for the regenerative process is the conditional probability of $\boldsymbol{x}_{\boldsymbol{s}_n+}$ given $\boldsymbol{x}_{\boldsymbol{s}_n}$. Obviously,

$$\Pr\{\boldsymbol{x}_{\boldsymbol{s}_n+}=j,\ \boldsymbol{x}_{\boldsymbol{s}_n}=i\} = \int_{u=0+}^{\infty} \int_{v=0+}^{\infty} \Pr\{\boldsymbol{x}_{u+}=j \mid \boldsymbol{y}_n=i,\ \boldsymbol{s}_n=u,\ \boldsymbol{s}_{n+1}=v\}$$

$$\cdot\mathrm{d}_v \Pr\{\boldsymbol{s}_{n+1}<v \mid \boldsymbol{y}_n=i,\ \boldsymbol{s}_n=u\}\,\mathrm{d}_u \Pr\{\boldsymbol{y}_n=i,\ \boldsymbol{s}_n<u\}$$

$$= p_{ij}(0+)\Pr\{\boldsymbol{x}_{\boldsymbol{s}_n}=i\},$$

so that

$$\Pr\{\boldsymbol{x}_{\boldsymbol{s}_n+}=j \mid \boldsymbol{x}_{\boldsymbol{s}_n}=i\} = p_{ij}(0+), \qquad i,j\in S, \quad n=0,1,\ldots.$$

This shows that the matrix $p_{ij}(0+)$, $i,j\in S$, describes the probability distribution of the jumps of the process $\{\boldsymbol{x}_t,\ t\in[0,\infty)\}$ at renewal epochs. Let us define

$$r_{ij} \stackrel{\text{def}}{=} p_{ij}(0+), \qquad i,j\in S,$$

so that

$$\sum_{j\in S} r_{ij} = 1, \qquad i\in S.$$

DEFINITION. A process $\{\boldsymbol{x}_t,\ t\in[0,\infty)\}$ which is regenerative with respect to the renewal process $\{\boldsymbol{v}_t,\ t\in[0,\infty)\}$ is *Markovian* if a standard transition matrix $\pi_{ij}(t)$, $i, j\in S$, exists such that for all $i, j\in S$,

(*i*)
$$p_{ij}(t) = \sum_{k\in S} r_{ik}\,\pi_{kj}(t), \qquad t\geqq 0,$$

(*ii*) for $n=0, 1, \ldots;\ v>u\geqq 0$,

$$\Pr\{\boldsymbol{x}_{t_h}=j_h,\ u<t_h\leqq v,\ h=1,\ldots,m \mid \boldsymbol{s}_{n+1}=v,\ \boldsymbol{s}_n=u,\ \boldsymbol{x}_{\boldsymbol{s}_n}=i\}$$
$$= p_{ij_1}(t_1-u)\,\pi_{j_1j_2}(t_2-t_1)\ldots\pi_{j_{m-1}j_m}(t_m-t_{m-1}),$$

for $j_1,\ldots,j_m,\ i\in S$, with $u<t_1\leqq t_2\leqq\ldots\leqq t_m\leqq v$, and $m=2, 3, \ldots$.

Since $\pi_{ij}(t)$, $i, j\in S$, $t\geqq 0$, is a standard transition matrix the following limits exist (cf. section 3.2),

$$-q_i \overset{\text{def}}{=} q_{ii} \overset{\text{def}}{=} \lim_{t\downarrow 0}\frac{\pi_{ii}(t)-1}{t}, \qquad i\in S,$$

$$q_{ij} \overset{\text{def}}{=} \lim_{t\downarrow 0}\frac{\pi_{ij}(t)}{t}, \qquad i\neq j, \quad i, j\in S.$$

Henceforth it will be assumed that our regenerative process is Markovian and that

$$\sup_{i\in S} q_i < \infty. \tag{6.38}$$

Consider the discrete time parameter process $\{\boldsymbol{z}_n,\ n=0, 1, \ldots\}$, where

$$\boldsymbol{z}_n \overset{\text{def}}{=} \boldsymbol{x}_{\boldsymbol{s}_n+}, \qquad n=0, 1, \ldots; \quad \boldsymbol{z}_0 \overset{\text{def}}{=} \boldsymbol{x}_{0+}.$$

This is a discrete time parameter Markov chain with stationary transition probabilities. To show this we note that

$$\Pr\{\boldsymbol{z}_k=i_k,\ k=0,\ldots,n+1\} = \sum_{j\in S} r_{ji_{n+1}} \Pr\{\boldsymbol{y}_{n+1}=j,\ \boldsymbol{z}_k=i_k,\ k=0,\ldots,n\},$$

and

$$\Pr\{\boldsymbol{y}_{n+1}=j,\ \boldsymbol{z}_k=i_k,\ k=0,\ldots,n\}$$
$$= \sum_{h\in S} \Pr\{\boldsymbol{y}_{n+1}=j,\ \boldsymbol{z}_n=i_n \mid \boldsymbol{y}_n=h\}\Pr\{\boldsymbol{y}_n=h,\ \boldsymbol{z}_k=i_k,\ k=0,\ldots,n-1\}.$$

Further

$$\Pr\{\boldsymbol{y}_{n+1}=j,\ \boldsymbol{z}_n=i_n \mid \boldsymbol{y}_n=h\} = r_{hi_n}\int_0^\infty \pi_{i_nj}(\sigma)\,\mathrm{d}F(\sigma).$$

Hence

$\Pr\{z_k=i_k,\ k=0, \dots, n+1\}$

$$= \sum_{h\in S} \sum_{j\in S} r_{hi_n} \int_0^\infty \pi_{i_nj}(\sigma)\, r_{ji_{n+1}}\, \mathrm{d}F(\sigma) \Pr\{y_n=h,\ z_k=i_k,\ k=0, \dots, n-1\}$$

$$= \sum_{j\in S} \int_0^\infty \pi_{i_nj}(\sigma)\, r_{ji_{n+1}}\, \mathrm{d}F(\sigma) \Pr\{z_k=i_k,\ k=0, \dots, n\};$$

so that for $n=0, 1, \dots$,

$$\Pr\{z_{n+1}=i_{n+1} \mid z_k=i_k,\ k=0, \dots, n\} = \sum_{j\in S} \int_0^\infty \pi_{i_nj}(\sigma)\, r_{ji_{n+1}}\, \mathrm{d}F(\sigma).$$

Since in the last expression the right-hand side is independent of i_k, $k=0, \dots, n-1$, it follows that $\{z_n,\ n=0, 1, \dots\}$ is a discrete time parameter Markov chain with stationary transition probabilities given by

$$\Pr\{z_{n+1}=j \mid z_n=i\} = \sum_{k\in S} \int_0^\infty \pi_{ik}(\sigma)\, r_{kj}\, \mathrm{d}F(\sigma), \qquad i, j\in S.$$

If the Markov chain $\{y_n,\ n=0, 1, \dots\}$ is ergodic it follows from theorem 6.6 and

$$\Pr\{z_n=j\} = \sum_{i\in S} \Pr\{z_n=j \mid y_n=i\} \Pr\{y_n=i\},$$

that

$$v_j \overset{\text{def}}{=} \lim_{n\to\infty} \Pr\{z_n=j\} = \sum_{i\in S} u_i r_{ij}, \qquad j\in S, \tag{6.39}$$

so that the Markov chain $\{z_n,\ n=0, 1, \dots\}$ is ergodic also and the stationary distribution $\{v_j,\ j\in S\}$ satisfies (cf. (6.39)),

$$v_j = \sum_{i\in S} v_i \int_0^\infty \{\sum_{k\in S} \pi_{ik}(\sigma)\, r_{kj}\}\, \mathrm{d}F(\sigma), \qquad j\in S. \tag{6.40}$$

The transition matrix $\pi_{ij}(t)$, $t\geq 0$, $i, j\in S$, is a standard transition matrix, hence it follows from (6.38) and theorem 3.3 that $\pi_{ij}(t)$ satisfies the forward differential equations

$$\frac{\mathrm{d}}{\mathrm{d}t}\pi_{ij}(t) = -q_j\pi_{ij}(t) + \sum_{\substack{k\in S\\ k\neq j}} \pi_{ik}(t)\, q_{kj}, \qquad t>0, \quad i, j\in S, \tag{6.41}$$

with boundary conditions

$$\pi_{ij}(0+) = \delta_{ij}. \tag{6.42}$$

The solution of the set of equations (6.41) with boundary conditions (6.42) is given in theorem 3.2. We shall now transform this set of differential equations into a system of differential equations for a Markovian regenerative process $\{\boldsymbol{x}_t,\ t\in[0,\infty)\}$.

We multiply the expressions in (6.41) and (6.42) by r_{ki} and sum over all $i\in S$. Since it follows from theorem 3.2 that $(\mathrm{d}/\mathrm{d}t)\,\pi_{ij}(t)$ is uniformly bounded for $i\in S$ we obtain

$$\frac{\mathrm{d}}{\mathrm{d}t}p_{ij}(t) = -q_j p_{ij}(t) + \sum_{\substack{k\in S\\ k\neq j}} p_{ik}(t)\,q_{kj}, \qquad t>0,\quad i,j\in S, \tag{6.43}$$

with

$$p_{ij}(0+) = r_{ij}, \qquad i,j\in S. \tag{6.44}$$

From now on it will be assumed that the imbedded Markov chain $\{\boldsymbol{y}_n,\ n=0,1,\dots\}$ *is ergodic*, so that the limits in theorem 6.6 exist and are positive. From (6.37) it follows that the limit distribution

$$\lim_{t\to\infty} \Pr\{\boldsymbol{x}_t=j,\ \boldsymbol{\eta}_t<\eta\}, \qquad j\in S,\quad \eta\geqq 0,$$

has a density. Denoting this density function by $R_j(\eta)$, $j\in S$, then

$$R_j(\eta) = \begin{cases} Q_j(\eta)\{1-F(\eta)\}, & \eta>0,\\ 0, & \eta\leqq 0,\end{cases} \tag{6.45}$$

where

$$Q_j(t) \stackrel{\mathrm{def}}{=} \frac{1}{\mu}\sum_{i\in S} u_i p_{ij}(t), \qquad j\in S, \qquad 0<t<\infty. \tag{6.46}$$

Multiplying (6.43) by u_i/μ and summing over all $i\in S$ we obtain

$$\frac{\mathrm{d}}{\mathrm{d}t}Q_j(t) = -q_j Q_j(t) + \sum_{\substack{k\in S\\ k\neq j}} Q_k(t)\,q_{kj}, \qquad t>0,\quad j\in S. \tag{6.47}$$

In the derivation of this expression the interchange of the order of differentiation and summation is easily justified by starting from the explicit expression for $\pi_{ij}(t)$ given in theorem 3.2. Inserting (6.45) into (6.47) we get the system of differential equations

$$\frac{\mathrm{d}}{\mathrm{d}t}R_j(t) = -\left\{q_j + \frac{(\mathrm{d}/\mathrm{d}t)\,F(t)}{1-F(t)}\right\}R_j(t) + \sum_{\substack{k\in S\\ k\neq j}} R_k(t)\,q_{kj}, \qquad j\in S, \tag{6.48}$$

which hold for all values of $t>0$ for which the renewal distribution $F(t)$ has a derivative and for which $F(t)<1$.

To obtain boundary conditions for (6.47) and (6.48) we note that, since we assumed $F(0+)=0$,

$$R_j(0+) = Q_j(0+) = \frac{1}{\mu} \sum_{i \in S} u_i r_{ij}, \qquad j \in S,$$

so by (6.32) and the assumption that the imbedded Markov chain is ergodic, we have that

$$R_j(0+) = \frac{1}{\mu} \sum_{i \in S} \sum_{k \in S} u_k \int_0^\infty p_{ki}(\sigma)\, \mathrm{d}F(\sigma)\, r_{ij},$$

and hence that

$$Q_j(0+) = \sum_{k \in S} \int_0^\infty Q_k(\sigma)\, r_{kj}\, \mathrm{d}F(\sigma), \qquad j \in S, \tag{6.49}$$

$$R_j(0+) = \sum_{k \in S} \int_0^{T-} R_k(\sigma)\, r_{kj} \frac{\mathrm{d}F(\sigma)}{1-F(\sigma)}, \qquad j \in S,$$

where

$$T \stackrel{\text{def}}{=} \sup\{t:\ F(t)<1\}.$$

The first set of equations of (6.49) represents the boundary conditions for the differential equations (6.47), the second set those for the system (6.48).

It also follows from (6.32) that for $j \in S$,

$$R_j(0+) = \frac{1}{\mu} \sum_{i \in S} \sum_{k \in S} \sum_{h \in S} u_k \int_0^\infty r_{kh} \pi_{hi}(\sigma)\, \mathrm{d}F(\sigma)\, r_{ij},$$

so that

$$R_j(0+) = \sum_{h \in S} R_h(0+) \int_0^\infty \{\sum_{i \in S} \pi_{hi}(\sigma)\, r_{ij}\}\, \mathrm{d}F(\sigma). \tag{6.50}$$

Hence from (6.40) and

$$\sum_{i \in S} v_i = 1, \qquad \sum_{j \in S} R_j(0+) = \sum_{j \in S} Q_j(0+) = \frac{1}{\mu},$$

it follows that

$$v_i = \mu R_i(0+), \qquad i \in S,$$

since positive and bounded solutions of (6.50) only differ by a constant factor.

THEOREM 6.7. *If the Markovian regenerative process* $\{\boldsymbol{x}_t,\ t\in[0,\infty)\}$ *has a non-lattice renewal distribution* $F(t)$ *with* $F(0+)=0$, *and first moment* $\mu<\infty$, *if its transition matrix* $(\pi_{ij}(t))$ *satisfies condition* (6.38), *and if the imbedded Markov chain* $\{\boldsymbol{y}_n,\ n=0,1,\ldots\}$ *is ergodic then*

(*i*) $\Pr\{\boldsymbol{x}_t=j,\ \boldsymbol{\eta}_t<\eta\}$, $j\in S$, $t\geqq\eta\geqq 0$ *tends for* $t\to\infty$ *to a distribution which has a density with respect to* η,

$$\lim_{t\to\infty}\Pr\{\boldsymbol{x}_t=j,\ \boldsymbol{\eta}_t<\eta\}=\int_0^\eta R_j(\sigma)\,\mathrm{d}\sigma=\int_0^\eta Q_j(\sigma)\{1-F(\sigma)\}\,\mathrm{d}\sigma,\qquad \eta\geqq 0;$$

(*ii*) $R_j(t)$, $j\in S$, *is the unique non-negative solution, continuous from the left, satisfying the system of differential equations*

$$\frac{\mathrm{d}}{\mathrm{d}t}z_j(t)=-\left\{q_j+\frac{(\mathrm{d}/\mathrm{d}t)F(t)}{1-F(t)}\right\}z_j(t)+\sum_{\substack{k\in S\\k\neq j}}z_k(t)\,q_{kj},\qquad j\in S,\tag{6.51}$$

for all those values of $t\in(0,T)$ *for which* $F(t)$ *has a derivative, and the conditions*

$$z_j(0+)=\sum_{k\in S}\int_0^{T-}z_k(t)\,\frac{\mathrm{d}F(t)}{1-F(t)},\qquad j\in S;\tag{6.52}$$

$$\sum_{j\in S}z_j(t)=\frac{1}{\mu},\qquad t>0;$$

(*iii*) *the functions* $Q_j(t)$, $t>0$, $j\in S$ *are continuous and uniquely determined as the non-negative solution of*

$$\frac{\mathrm{d}}{\mathrm{d}t}y_j(t)=-q_jy_j(t)+\sum_{\substack{k\in S\\k\neq j}}y_k(t)\,q_{kj},\qquad j\in S,\quad t>0,\tag{6.53}$$

and

$$y_j(0+)=\sum_{k\in S}\int_0^\infty y_k(t)\,r_{kj}\,\mathrm{d}F(t),\qquad j\in S;\tag{6.54}$$

$$\sum_{j\in S}y_j(t)=\frac{1}{\mu},\qquad t>0.$$

Proof. The first statement follows from (6.37) and (6.45). To prove the second and third statements it is sufficient to consider the third only, since the systems (6.51), (6.52) and (6.53), (6.54) are equivalent, as we see by substituting

$$y_j(t)\{1-F(t)\}=z_j(t), \qquad j\in S, \quad t\geqq 0.$$

Since $\pi_{ij}(t)$ is a continuous function of t for $t\geqq 0$ (cf. theorem 3.2) it is seen from the definition that $p_{ij}(t)$ is continuous for $t\geqq 0$ since it is a uniformly convergent series of positive terms. Similarly it follows from (6.46) that $Q_j(t)$ is continuous for $t>0$. We now define ${}_n\pi_{ij}(t)$, $t>0$, $n=0, 1, \dots$, recursively, starting from the Q-matrix of $(\pi_{ij}(t))$, by means of the relations (3.17) and (3.19); so we rewrite (3.17) and (3.19) with ${}_np_{ij}(t)$ replaced by ${}_n\pi_{ij}(t)$. Let $y_j(t)$, $j\in S$, $t>0$ be a non-negative solution of (6.35) and (6.54), and define

$${}_ny_j(t) \overset{\text{def}}{=} \sum_{i\in S} y_i(0+)\,{}_n\pi_{ij}(t), \qquad j\in S, \quad t>0, \quad n=0, 1, \dots .$$

We then obtain from (3.17) and (3.19),

$${}_0y_j(t) = y_j(0+)\,\mathrm{e}^{-q_jt},$$

$${}_{n+1}y_j(t) = \sum_{i\in S} \sum_{\substack{k\in S\\ k\neq j}} \int_0^t y_i(0+)\,{}_n\pi_{ik}(\sigma)\, q_{kj}\, \mathrm{e}^{-q_j(t-\sigma)}\,\mathrm{d}\sigma,$$

for $j\in S$, $n=0, 1, \dots$. As in the proof of theorem 3.3 it now follows for $t\geqq 0, j\in S$,

$$\begin{aligned} y_j(t) \geqq \sum_{n=0}^{\infty} {}_ny_j(t) &= \sum_{i\in S} y_i(0+) \sum_{n=0}^{\infty} {}_n\pi_{ij}(t) \\ &= \sum_{i\in S} y_i(0+)\,\pi_{ij}(t). \end{aligned} \tag{6.55}$$

Summing over all $j\in S$ and using (6.54) we find that, for all $t>0$,

$$\frac{1}{\mu}\sum_{j\in S} y_j(t) \geqq \sum_{j\in S} y_j(0+) = \frac{1}{\mu}.$$

Consequently, there is no $t>0$ for which the inequality sign in (6.55) is valid, therefore

$$y_j(t) = \sum_{i\in S} y_i(0+)\,\pi_{ij}(t), \qquad t>0, \quad j\in S. \tag{6.56}$$

Inserting this expression into the first relation of (6.54) gives

$$y_j(0+) = \sum_{i\in S} y_i(0+) \int_0^\infty \{\sum_{k\in S} \pi_{ik}(t)\, r_{kj}\}\, \mathrm{d}F(t), \qquad j\in S. \tag{6.57}$$

Since the imbedded Markov chain $\{y_n, n=0, 1, ...\}$ is ergodic the Markov chain $\{z_n, n=0, 1, ...\}$ is ergodic also (see above). The stationary distribution $\{v_i, i\in S\}$ of the latter chain satisfies (6.40) so that from (6.40), (6.54) and (6.56),

$$y_i(0+) = \frac{1}{\mu} v_i, \qquad i\in S,$$

since positive solutions of (6.56) only differ by a factor which is independent of i. From (6.39), (6.46) and (6.56) we now obtain

$$\begin{aligned} y_j(t) &= \frac{1}{\mu}\sum_{i\in S} v_i\pi_{ij}(t) = \frac{1}{\mu}\sum_{k\in S}\sum_{i\in S} u_k r_{ki}\pi_{ij}(t) \\ &= Q_j(t), \qquad j\in S, \quad t>0. \end{aligned}$$

The proof is complete.

The above theorem provides us with a direct method of determining $R_j(\eta), j\in S, \eta\geqq 0$, from (6.51) and (6.52), a result which is often quite useful in queueing theory.

It should be noted that from (6.36), (6.45) and (6.46) we have, if $F(t)$ is not a lattice distribution,

$$\lim_{t\to\infty} \Pr\{x_t=j, \eta_t<\eta, \xi_t<\zeta\} = \int_0^\eta R_j(\sigma)\,\frac{F(\sigma+\zeta)-F(\sigma)}{1-F(\sigma)}\,\mathrm{d}\sigma, \tag{6.58}$$

for $\eta\in(0, T)$, $\zeta>0$. This relation has a direct probabilistic meaning, since for $t\to\infty$, $R_j(\sigma)\,\mathrm{d}\sigma$ is the probability that $x_t=j$ and the last renewal before t occurred between $t-\sigma-\mathrm{d}\sigma$ and $t-\sigma$, while

$$\frac{F(\sigma+\zeta)-F(\sigma)}{1-F(\sigma)}$$

is the conditional probability that whenever the past lifetime is σ, the residual lifetime is at most ζ.

The above theorem has been formulated for $F(t)$ not a lattice distribution. By starting from the definition of $Q_j(\sigma)$ (cf. (6.46)) it is easily verified, however, that the system of relations (6.53) and (6.54) also apply if $F(t)$ is a lattice distribution. Consequently, this system can also be used to determine $Q_j(\sigma)$, $j \in S$, $\sigma > 0$, if $F(t)$ is a lattice distribution, and may be used to find for instance $\Pr\{x_{\rho\tau+\sigma}=j\}$, $j \in S$, $0 \leqq \sigma < \tau$ for $\rho \to \infty$.

I.6.5. Some special renewal distributions

With a view to applications to queueing theory we shall briefly discuss here some special renewal processes, viz. renewal processes with renewal distribution (*i*) the negative exponential distribution, (*ii*) the Erlang distribution, (*iii*) the hyper-exponential distribution. The section will be concluded with some remarks on pooled renewal processes.

First we consider the renewal process for which the renewal time has the negative exponential distribution with parameter λ,

$$F(t) = \begin{cases} 0, & t \leqq 0, \\ 1 - e^{-\lambda t}, & t > 0, \quad \lambda > 0. \end{cases}$$

Hence

$$\mu = \int_0^\infty t \mathrm{d}F(t) = \frac{1}{\lambda}, \qquad \mu_2 = \int_0^\infty t^2 \,\mathrm{d}F(t) = \frac{2}{\lambda^2},$$

so that the coefficient of variation is given by

$$\left\{\frac{\mu_2 - \mu^2}{\mu^2}\right\}^{\frac{1}{2}} = 1.$$

Further, for $\mathrm{Re}\, s > -\lambda$ (cf. section 6.2),

$$f(s) = \int_0^\infty e^{-st}\,\mathrm{d}F(t) = \frac{\lambda}{\lambda + s}, \qquad \mu(s) = \int_0^\infty e^{-st}\,\mathrm{d}m(t) = \frac{\lambda}{s},$$

$$\mu_2(s) = \int_0^\infty e^{-st}\,\mathrm{d}m_2(t) = \frac{\lambda}{s} + 2\frac{\lambda^2}{s^2}.$$

Hence for $t \geqq 0$,

$$m(t) = \mathrm{E}\{\boldsymbol{v}_t\} = \lambda t, \qquad m_2(t) = \mathrm{E}\{\boldsymbol{v}_t^2\} = \lambda t + \lambda^2 t^2,$$

$$\mathrm{var}\,\{\boldsymbol{v}_t\} = \lambda t.$$

Evidently (cf. section 4.2) the renewal process $\{\boldsymbol{\nu}_t,\ t\in[0,\infty)\}$ is a Poisson process with parameter λ,

$$\Pr\{\boldsymbol{\nu}_t=k\} = \frac{(\lambda t)^k}{k!}\,\mathrm{e}^{-\lambda t}, \qquad t\geq 0,\quad k=0,1,\ldots .$$

Next, the renewal process with renewal distribution the *Erlang distribution* of type n and with parameter λ. This distribution function is given by

$$F(t) = \begin{cases} 0, & t\leq 0,\\ \displaystyle\int_0^t \frac{(n\lambda)^n\,\sigma^{n-1}}{(n-1)!}\,\mathrm{e}^{-n\lambda\sigma}\,\mathrm{d}\sigma, & t>0,\end{cases}$$

where $\lambda>0$ and n is a positive integer.

It is easily found that

$$f(s) = \frac{(n\lambda)^n}{(n\lambda+s)^n}, \qquad \mathrm{Re}\,s>-n\lambda.$$

From this relation we obtain

$$\mu = \frac{1}{\lambda}, \qquad \mu_2 = \frac{1}{n\lambda^2}+\frac{1}{\lambda^2},$$

so that the coefficient of variation is given by

$$\left\{\frac{\mu_2-\mu^2}{\mu^2}\right\}^{\frac{1}{2}} = \sqrt{\frac{1}{n}} < 1, \qquad n=2,3,\ldots .$$

For the Laplace-Stieltjes transform $\mu(s)$ of the renewal function $m(t)$ we obtain

$$\mu(s) = \frac{(n\lambda)^n}{(n\lambda+s)^n-(n\lambda)^n}, \qquad \mathrm{Re}\,s>-\lambda n.$$

It is not difficult to obtain $m(t)$ from $\mu(s)$ by applying the inversion theorem for the Laplace-Stieltjes transform since the poles of $\mu(s)$ can be easily found. The actual evaluation of $m(t)$ is, however, rather laborious and will be omitted. It is, however, easily found from the results of section 6.2 that for $t\to\infty$,

$$m(t)\approx\lambda t, \qquad \mathrm{var}\,\{\boldsymbol{\nu}_t\}\approx\frac{\lambda}{n}\,t.$$

To calculate the probability distribution of $\boldsymbol{\nu}_t$ one usually starts from (cf. (6.1)),

$$\Pr\{\boldsymbol{\nu}_t=k\}=F^{k*}(t)*\{U_1(t)-F(t)\}, \qquad k=0, 1, \dots;$$

in the present case, however, an easier derivation is possible. It is based on a property of the Erlang distribution which is frequently very helpful for solving queueing problems by theoretical methods and by simulation techniques (cf. section II.6.5).

From the Laplace-Stieltjes transform $f(s)$ of $F(t)$ it is seen that the Erlang distribution of type n, parameter λ, is the n-fold convolution of the negative exponential distribution with parameter $n\lambda$. Therefore, if the renewal time has the Erlang distribution of type n and parameter λ, the renewal time can be considered as the sum of n independent, identically distributed variables with negative exponential distribution function with parameter $n\lambda$. Hence, for the renewal process with Erlang renewal distribution of type n, parameter λ, we have $\boldsymbol{\nu}_t=k$ if and only if in the Poisson process with parameter $n\lambda$ during $[0, t)$ the number of renewals is kn or $kn+1$ or $\dots$ or $(k+1)\,n-1$. Therefore, for $t\geqq 0$,

$$\Pr\{\boldsymbol{\nu}_t=k\} = \sum_{h=kn}^{(k+1)n-1} \frac{(n\lambda t)^h}{h!}\,\mathrm{e}^{-n\lambda t}, \qquad k=0, 1, \dots .$$

It is noted that in the present case the renewal process $\{\boldsymbol{\nu}_t,\ t\in[0, \infty)\}$ is not a Markov process.

The Erlang distribution is unimodal and for increasing values of n the density function of the Erlang distribution is more and more concentrated about its mean $1/\lambda$; note that the coefficient of variation tends to zero. For $n\to\infty$ the Erlang distribution of type n, parameter λ tends to $U_1(t-1/\lambda)$, the probability distribution degenerated at $t=1/\lambda$. This is easily proved by starting from the Laplace-Stieltjes transform $f(s)$ and then applying Feller's convergence criterion for Laplace-Stieltjes transforms (cf. app. 5), since for $s>0$,

$$\lim_{n\to\infty} \frac{(n\lambda)^n}{(n\lambda+s)^n} = \mathrm{e}^{-s/\lambda} = \int_0^\infty \mathrm{e}^{-st}\,\mathrm{d}U_1\left(t-\frac{1}{\lambda}\right).$$

The *hyper-exponential distribution* is given by

$$F(t) = \begin{cases} 0, & t\leqq 0,\\ \sum_{k=1}^{n} a_k\{1-\mathrm{e}^{-\lambda_k t}\}, & t>0, \end{cases}$$

with $\lambda_k>0$, $a_k>0$, $k=1, \ldots, n$, and $a_1+\ldots+a_n=1$, where n is a positive integer.

It is easily seen that, for $\operatorname{Re} s>-\min_{1\leq k\leq n} \lambda_k$,

$$f(s) = \sum_{k=1}^{n} \frac{a_k\lambda_k}{\lambda_k+s},$$

and hence that

$$\mu = \sum_{k=1}^{n} \frac{a_k}{\lambda_k}, \qquad \mu_2 = 2\sum_{k=1}^{n} \frac{a_k}{\lambda_k^2}.$$

From Cauchy's inequality (cf. Hardy, Littlewood and Polya [1952]) it follows that always $\mu_2>2\mu^2$ for $n=2, 3, \ldots$, and hence that, for the hyper-exponential distribution, the coefficient of variation is always larger than one.

We consider the special case in which $n=2$, $a_1=\theta$, $a_2=1-\theta$. Then

$$f(s) = \theta\frac{\lambda_1}{\lambda_1+s} + (1-\theta)\frac{\lambda_2}{\lambda_2+s},$$

$$\mu = \frac{\theta}{\lambda_1} + \frac{1-\theta}{\lambda_2}, \qquad \mu_2 = 2\frac{\theta}{\lambda_1^2} + 2\frac{1-\theta}{\lambda_2^2},$$

$$\left\{\frac{\mu_2-\mu^2}{\mu^2}\right\}^{\frac{1}{2}} = \left\{1 + \frac{2\theta(1-\theta)(\lambda_1-\lambda_2)^2}{\{\theta\lambda_2+(1-\theta)\,\lambda_1\}^2}\right\}^{\frac{1}{2}} > 1,$$

$$\mu(s) = \frac{\lambda_1\lambda_2+s\{(1-\theta)\,\lambda_2+\theta\lambda_1\}}{s\{s+(1-\theta)\,\lambda_1+\theta\lambda_2\}}.$$

The form of the hyper-exponential distribution for $n=2$ shows that a realisation of a renewal process with this renewal distribution may be constructed by sampling the renewal times with probability θ from a negative exponential distribution with parameter λ_1, and with probability $1-\theta$ from a negative exponential distribution with parameter λ_2. A similar remark applies of course for higher values of n. This property of the hyper-exponential distribution is often useful not only in theoretical investigations but also in handling queueing problems by simulation techniques.

From what has been said above it is seen that the Erlang distribution has a coefficient of variation which is less than one, whereas for the hyper-exponential distribution this coefficient is larger than one. Since these distributions are unimodal, i.e. their densities have only one maximum, a simple calculation shows that a renewal process with hyper-exponential renewal distribution has more often bunches of short renewal times and also somewhat more long renewal times than the renewal process with negative exponential distribution and with the same mean of the renewal time. For the Erlang distribution as renewal time distribution it is just the other way. Here the renewal times show more regularity than those which are negative exponentially distributed.

In many situations occurring in practice it appears that a stationary process for which a renewal process is a good approximation often has negative exponentially distributed renewal times. For instance, this holds good with a high degree of accuracy for the occurrence of telephone calls originating from a large group of subscribers during the busy hour; also for pedestrians arriving at an intersection of two roads; for customers arriving at a ticket window; for ships arriving at a port. Here the originating of a telephone call, the arrival of a pedestrian or a customer is then considered as a renewal. PALM [1943] pointed out why so often these renewal times are negative exponentially distributed. We shall now discuss Palm's theorem, and mention the more general theorem due to Khintchine.

Suppose we have N stationary renewal processes $\{\boldsymbol{v}_t^{(i)},\ t\in[0,\infty)\}$, $i=1, 2, \ldots, N$, with renewal distributions $F_i(t)$, all having a finite first moment μ_i, and $F_i(0+)=0$, $i=1, \ldots, N$. The N processes being stochastically independent. Define

$$\boldsymbol{v}_t \overset{\text{def}}{=} \boldsymbol{v}_t^{(1)} + \ldots + \boldsymbol{v}_t^{(N)}, \qquad t \geqq 0,$$

so that the stochastic process $\{\boldsymbol{v}_t,\ t\in[0,\infty)\}$ represents the total number of renewals during $[0, t)$ of the N renewal processes together. In general $\{\boldsymbol{v}_t,\ t\in[0,\infty)\}$ will not be a renewal process. If $N\to\infty$, however, under rather weak conditions it will be a Poisson renewal process, i.e. a renewal process with negative exponentially distributed renewal times. Define

$$\frac{1}{\mu_N^{(N)}} \overset{\text{def}}{=} \frac{1}{\mu_1} + \ldots + \frac{1}{\mu_N},$$

and suppose that for $N\to\infty$,

(*i*) $\mu_i\to\infty$ uniformly in i for $i=1, 2, \ldots, N$, and $\mu_N^{(N)}\to\mu<\infty$;

(*ii*) $F_i(t)\to 0$ uniformly in i for every fixed $t>0$, and $i=1, \ldots, N$.

The first assumption implies that, for the individual renewal processes, the average number of renewals per unit of time tends to zero for increasing N but that the average total number of renewals per unit of time approaches a non-zero value. The second condition states that for $N\to\infty$ the probability of an accumulation of renewals of the same renewal process in the pooled output during any finite time interval tends to zero.

Palm's theorem (cf. PALM [1943]) now reads

$$\lim_{N\to\infty} \Pr\{\min_{1\leq i\leq N} \xi_t^{(i)} < \zeta\} = \begin{cases} 0, & \zeta\leq 0, \\ 1-e^{-\zeta/u}, & \zeta>0, \end{cases}$$

where $\xi_t^{(i)}$ is the residual lifetime at time t in the ith renewal process.

To prove this, we note that since the renewal processes are assumed to be stationary and independent (cf. section 6.3),

$$\Pr\{\min_{1\leq i\leq N} \xi_t^{(i)} > \zeta\} = \prod_{i=1}^{N} \frac{1}{\mu_i} \int_{\zeta}^{\infty} \{1-F_i(\sigma)\}\, d\sigma$$

$$= \prod_{i=1}^{N} \left\{1 - \frac{\zeta}{\mu_i} + \frac{1}{\mu_i} \int_0^{\zeta} F_i(\sigma)\, d\sigma\right\}, \qquad \zeta>0.$$

For any $\varepsilon>0$ and every $i=1, \ldots, N$ we have, by taking N sufficiently large,

$$\frac{1}{\mu_i} \int_0^{\zeta} F_i(\sigma)\, d\sigma < \frac{\varepsilon}{\mu_i}\, \theta_i \zeta \qquad \text{with} \quad 0 < \theta_i < 1,$$

and hence there exists a function $c(\zeta)>0$ depending only on ζ such that for N sufficiently large

$$\log\left\{\frac{1}{\mu_i} \int_{\zeta}^{\infty} \{1-F_i(\sigma)\}\, d\sigma + \frac{\zeta}{\mu_i}\right\} \leq \frac{\varepsilon}{\mu_i}\, c(\zeta), \qquad i=1, \ldots, N,$$

and consequently

$$\left|\log \Pr\{\min_{1\leq i\leq N} \xi_t^{(i)} > \zeta\} + \frac{\zeta}{\mu_N^{(N)}}\right| \leq \frac{\varepsilon c(\zeta)}{\mu_N^{(N)}},$$

from which the above statement follows.

Palm's theorem does not imply that $\{\boldsymbol{v}_t,\ t\in[0, \infty)\}$ tends for $N\to\infty$ to a renewal process. It is Khintchine's theorem which establishes this as well as that in the limit this process is a Poisson process. We shall not deal here with the proof of Khintchine's theorem but refer the reader to KHINTCHINE [1960].

I.6.6. Fluctuation theory

Many of the results recently obtained in fluctuation theory are important for queueing theory. In this section we shall, therefore, discuss a number of basic relations of fluctuation theory which will be interpreted later on in terms of queueing theory (cf. chapter II.5).

Let $\boldsymbol{x}_n$, $n=1, 2, \ldots$, be a sequence of independent, identically distributed stochastic variables with

$$G(x) \overset{\text{def}}{=} \Pr\{\boldsymbol{x}_n<x\}, \qquad -\infty<x<\infty, \tag{6.59}$$

and

$$g(\rho) \overset{\text{def}}{=} \int_{-\infty}^{\infty} \mathrm{e}^{-\rho x}\,\mathrm{d}G(x), \qquad \mathrm{Re}\,\rho=0. \tag{6.60}$$

Define

$$\boldsymbol{s}_0 \overset{\text{def}}{=} 0, \qquad \boldsymbol{s}_k \overset{\text{def}}{=} \boldsymbol{x}_1+\ldots+\boldsymbol{x}_k, \qquad k=1, 2, \ldots,$$

$$\boldsymbol{m}_k \overset{\text{def}}{=} \max(0, \boldsymbol{s}_1, \ldots, \boldsymbol{s}_{k-1}), \qquad \boldsymbol{n}_k \overset{\text{def}}{=} \min(0, \boldsymbol{s}_1, \ldots, \boldsymbol{s}_{k-1}), \qquad k=1, 2, \ldots . \tag{6.61}$$

The study of the variables $\boldsymbol{m}_k$ and $\boldsymbol{n}_k$ is one of the main objects of fluctuation theory. We also introduce the sequences $\boldsymbol{v}_k$, $k=1, 2, \ldots$, and $\boldsymbol{w}_k$, $k=1, 2, \ldots$, defined by

$$\boldsymbol{v}_{k+1} \overset{\text{def}}{=} \boldsymbol{s}_k-\boldsymbol{n}_{k+1}, \qquad \boldsymbol{w}_{k+1} \overset{\text{def}}{=} \boldsymbol{s}_k-\boldsymbol{m}_{k+1}, \qquad k=0, 1, \ldots . \tag{6.62}$$

Although these variables $\boldsymbol{v}_k$ and $\boldsymbol{w}_k$ are rather more meaningful in queueing theory than in fluctuation theory, the introduction of these variables is very helpful for the study of $\boldsymbol{m}_k$ and $\boldsymbol{n}_k$.

It follows from the definitions above that

$$\begin{aligned}
\boldsymbol{v}_1 &= 0, \\
\boldsymbol{v}_{k+1} &= \max(0, \boldsymbol{x}_k, \boldsymbol{x}_k+\boldsymbol{x}_{k-1}, \ldots, \boldsymbol{x}_k+\ldots+\boldsymbol{x}_1), \qquad k=1, 2, \ldots, \\
\boldsymbol{w}_1 &= 0, \\
\boldsymbol{w}_{k+1} &= \min(0, \boldsymbol{x}_k, \boldsymbol{x}_k+\boldsymbol{x}_{k-1}, \ldots, \boldsymbol{x}_k+\ldots+\boldsymbol{x}_1), \qquad k=1, 2, \ldots .
\end{aligned} \tag{6.63}$$

Using the notation

$$\boldsymbol{x} \sim \boldsymbol{y}$$

to indicate that the variables $\boldsymbol{x}$ and $\boldsymbol{y}$ have the same distribution, it is readily seen from (6.61) and (6.63) that for $k=1, 2, \dots,$

$$\boldsymbol{v}_k \sim \boldsymbol{m}_k, \qquad \boldsymbol{w}_k \sim \boldsymbol{n}_k, \qquad (\boldsymbol{v}_k, \boldsymbol{n}_k) \sim (\boldsymbol{m}_k, \boldsymbol{w}_k). \tag{6.64}$$

Applying the notation

$$[x]^+ = \max(0, x), \qquad [x]^- = \min(0, x), \qquad -\infty < x < \infty, \tag{6.65}$$

we see that it follows immediately from (6.62) and (6.63) for $k=1, 2, \dots,$

$$\begin{aligned} \boldsymbol{v}_{k+1} &= [\boldsymbol{x}_k + \boldsymbol{v}_k]^+, & \boldsymbol{n}_{k+1} - \boldsymbol{n}_k &= [\boldsymbol{x}_k + \boldsymbol{v}_k]^-, \\ \boldsymbol{v}_1 &= 0, & \boldsymbol{n}_1 &= 0, \\ \boldsymbol{w}_{k+1} &= [\boldsymbol{x}_k + \boldsymbol{w}_k]^-, & \boldsymbol{m}_{k+1} - \boldsymbol{m}_k &= [\boldsymbol{x}_k + \boldsymbol{w}_k]^+, \\ \boldsymbol{w}_1 &= 0, & \boldsymbol{m}_1 &= 0. \end{aligned} \tag{6.66}$$

The first two relations of (6.66) are of basic importance in queueing theory (cf. chapter II.5).

Obviously, the sequence $\{\boldsymbol{v}_k, k=1, 2, \dots\}$ is a discrete time parameter Markov process with stationary transition probabilities and with state space $[0, \infty)$. Also the sequence of vectors $\{(\boldsymbol{n}_k, \boldsymbol{v}_k), k=1, 2, \dots\}$ is a discrete time parameter vector Markov process with stationary transition probabilities and with state space $(-\infty, 0] \times [0, \infty)$. Similar remarks hold for the sequences $\{\boldsymbol{w}_k, k=1, 2, \dots\}$ and $\{(\boldsymbol{m}_k, \boldsymbol{w}_k), k=1, 2, \dots\}$.

To investigate the vector Markov process $\{(\boldsymbol{n}_k, \boldsymbol{v}_k), k=1, 2, \dots\}$ we start from the identity (cf. (6.65)),

$$\mathrm{e}^{-\rho_1[x]^+} + \mathrm{e}^{-\rho_2[x]^-} = \mathrm{e}^{-\rho_1[x]^+ - \rho_2[x]^-} + 1, \tag{6.67}$$

with x real and ρ_1, ρ_2 arbitrary complex numbers. From (6.67) and (6.66) we have for $k=1, 2, \dots,$

$$\begin{aligned} \exp(-\rho_1 \boldsymbol{v}_{k+1} - \rho_2 \boldsymbol{n}_{k+1}) &= \exp(-\rho_1[\boldsymbol{x}_k + \boldsymbol{v}_k]^+ - \rho_2[\boldsymbol{x}_k + \boldsymbol{v}_k]^- - \rho_2 \boldsymbol{n}_k) \\ &= \exp(-\rho_1[\boldsymbol{v}_k + \boldsymbol{x}_k]^+ - \rho_2 \boldsymbol{n}_k) + \exp(-\rho_2[\boldsymbol{x}_k + \boldsymbol{v}_k]^- - \rho_2 \boldsymbol{n}_k) - \exp(-\rho_2 \boldsymbol{n}_k) \\ &= \exp(-\rho_1 \boldsymbol{v}_k - \rho_2 \boldsymbol{n}_k - \rho_1 \boldsymbol{x}_k) \\ &\quad + \{\exp(-\rho_2[\boldsymbol{x}_k + \boldsymbol{v}_k]^-) - \exp(-\rho_1[\boldsymbol{x}_k + \boldsymbol{v}_k]^-)\} \exp(-\rho_2 \boldsymbol{n}_k). \end{aligned}$$

Define for $|r|<1$, $\operatorname{Re}\rho_1 \geqq 0$, $\operatorname{Re}\rho_2 \leqq 0$,

$$\varphi(r, \rho_1, \rho_2) \stackrel{\text{def}}{=} \sum_{k=1}^{\infty} r^k \, \mathrm{E}\{\exp(-\rho_1 \boldsymbol{v}_k - \rho_2 \boldsymbol{n}_k) \mid \boldsymbol{v}_1 = 0, \boldsymbol{n}_1 = 0\}. \tag{6.68}$$

Taking conditional expectations with respect to $\boldsymbol{v}_1=0$, $\boldsymbol{n}_1=0$, in the last but one relation, multiplying the result by r^k, adding for $k=1, 2, \dots,$ and noting that $\boldsymbol{x}_k$ is independent of $\boldsymbol{v}_k$ and of $\boldsymbol{n}_k$, we find that for $|r|<1$, $\text{Re}\,\rho_1=0$, $\text{Re}\,\rho_2 \leqq 0$,

$$\frac{1}{r}\varphi(r,\rho_1,\rho_2) = g(\rho_1)\,\varphi(r,\rho_1,\rho_2) + 1 + \sum_{k=1}^{\infty} r^k\,\text{E}\{(\exp(-\rho_2[\boldsymbol{x}_k+\boldsymbol{v}_k]^-)-\exp(-\rho_1[\boldsymbol{x}_k+\boldsymbol{v}_k]^-)) \cdot\exp(-\rho_2\boldsymbol{n}_k) \mid \boldsymbol{v}_1=0,\,\boldsymbol{n}_1=0\}. \tag{6.69}$$

From this relation we determine $\varphi(r,\rho_1,\rho_2)$. Define

$$K_+(r,\rho) \overset{\text{def}}{=} \exp\left\{-\sum_{n=1}^{\infty}\frac{r^n}{n}\,\text{E}\{\text{e}^{-\rho\boldsymbol{s}_n}\,(\boldsymbol{s}_n>0)\}\right\}, \qquad |r|<1, \quad \text{Re}\,\rho \geqq 0, \tag{6.70}$$

$$K_-(r,\rho) \overset{\text{def}}{=} \exp\left\{-\sum_{n=1}^{\infty}\frac{r^n}{n}\,\text{E}\{\text{e}^{-\rho\boldsymbol{s}_n}\,(\boldsymbol{s}_n\leqq 0)\}\right\}, \qquad |r|<1, \quad \text{Re}\,\rho \leqq 0,$$

using the notation $(\boldsymbol{x}<x)$ for the indicator function of the event $\{\boldsymbol{x}<x\}$, i.e.

$$(\boldsymbol{s}_n>0) = \begin{cases} 1 & \text{if } \boldsymbol{s}_n>0, \\ 0 & \text{if } \boldsymbol{s}_n \leqq 0, \end{cases} \qquad (\boldsymbol{s}_n \leqq 0) = \begin{cases} 1 & \text{if } \boldsymbol{s}_n \leqq 0, \\ 0 & \text{if } \boldsymbol{s}_n>0. \end{cases} \tag{6.71}$$

Evidently, for fixed $|r|<1$, the function $K_+(r,\rho)$ is analytic and has no zeros for $\text{Re}\,\rho \geqq 0$, while $K_-(r,\rho)$ is analytic and non-zero for $\text{Re}\,\rho \leqq 0$.

Since $\boldsymbol{x}_1, \boldsymbol{x}_2, \ldots,$ are independent and identically distributed it follows from (6.70) and (6.60) that for $|r|<1$, $\text{Re}\,\rho=0$,

$$K_+(r,\rho)\,K_-(r,\rho) = \exp\left\{-\sum_{n=1}^{\infty}\frac{r^n}{n}\,\text{E}\{\text{e}^{-\rho\boldsymbol{s}_n}\}\right\} = 1 - rg(\rho). \tag{6.72}$$

From (6.69) and (6.72) we obtain for $|r|<1$, $\text{Re}\,\rho_1=0$, $\text{Re}\,\rho_2 \leqq 0$,

$$\varphi(r,\rho_1,\rho_2)\,K_+(r,\rho_1) = r\{K_-(r,\rho_1)\}^{-1}\{1 + \sum_{k=1}^{\infty} r^k\,\text{E}\{(\exp(-\rho_2[\boldsymbol{x}_k+\boldsymbol{v}_k]^-) -\exp(-\rho_1[\boldsymbol{x}_k+\boldsymbol{v}_k]^-))\exp(-\rho_2\boldsymbol{n}_k) \mid \boldsymbol{v}_1=0,\,\boldsymbol{n}_1=0\}\}. \tag{6.73}$$

It is readily verified that, for fixed $|r|<1$, $\text{Re}\,\rho_2 \leqq 0$, the left-hand side of (6.73) is an analytic function of ρ_1 for $\text{Re}\,\rho_1>0$, whereas the right-hand side is an analytic function of ρ_1 for $\text{Re}\,\rho_1<0$. From their domain of analyticity these members approach their values at the axis $\text{Re}\,\rho_1=0$ continuously. Consequently, these members are each other's analytic continuations. It is easily verified that for fixed $|r|<1$, $\mathrm{Re}\,\rho_2 \leqq 0$, the left-hand side and the right-hand side are bounded for $\mathrm{Re}\,\rho_1 \geqq 0$ and $\mathrm{Re}\,\rho_1 \leqq 0$, respectively. Therefore, because of Liouville's theorem (cf. TITCHMARSH [1952]) both members of (6.73) are independent of ρ_1, i.e. a function $F(r, \rho_2)$ exists such that for $|r|<1$, $\mathrm{Re}\,\rho_1 \geqq 0$, $\mathrm{Re}\,\rho_2 \leqq 0$,

$$\varphi(r, \rho_1, \rho_2)\, K_+(r, \rho_1) = F(r, \rho_2). \tag{6.74}$$

To determine $F(r, \rho_2)$ take $\rho_1 = \rho_2 = \rho$ with $\mathrm{Re}\,\rho = 0$, then from (6.62), (6.68) and (6.74) for $|r|<1$, $\mathrm{Re}\,\rho = 0$,

$$\begin{aligned} F(r, \rho) &= K_+(r, \rho) \sum_{k=1}^{\infty} r^k\, \mathrm{E}\{\mathrm{e}^{-\rho \boldsymbol{s}_{k-1}} \mid \boldsymbol{s}_0 = 0\} \\ &= K_+(r, \rho)\, \frac{r}{1 - r g(\rho)} = r\{K_-(r, \rho)\}^{-1}. \end{aligned}$$

By analytic continuation it is seen that the latter relation also holds for $|r|<1$, $\mathrm{Re}\,\rho \leqq 0$. Hence, from (6.74), (6.68), (6.70) and (6.64) we have for $|r|<1$, $\mathrm{Re}\,\rho_1 \geqq 0$, $\mathrm{Re}\,\rho_2 \leqq 0$,

$$\begin{aligned} \varphi(r, \rho_1, \rho_2) &= \sum_{k=1}^{\infty} r^k\, \mathrm{E}\{\exp(-\rho_1 \boldsymbol{v}_k - \rho_2 \boldsymbol{n}_k) \mid \boldsymbol{v}_1 = 0, \boldsymbol{n}_1 = 0\} \\ &= \sum_{k=1}^{\infty} r^k\, \mathrm{E}\{\exp(-\rho_1 \boldsymbol{m}_k - \rho_2 \boldsymbol{w}_k) \mid \boldsymbol{m}_1 = 0, \boldsymbol{w}_1 = 0\} \\ &= r\{K_-(r, \rho_2)\, K_+(r, \rho_1)\}^{-1} \\ &= r \exp\left\{ \sum_{n=1}^{\infty} \frac{r^n}{n}\, \mathrm{E}\{\mathrm{e}^{-\rho_1 \boldsymbol{s}_n} (\boldsymbol{s}_n > 0) + (\boldsymbol{s}_n = 0) + \mathrm{e}^{-\rho_2 \boldsymbol{s}_n} (\boldsymbol{s}_n < 0)\} \right\}. \end{aligned} \tag{6.75}$$

This important relation is known as *Spitzer's identity*, and it is the starting point for the investigation of the Markov processes $\{\boldsymbol{v}_k, k=1, 2, \ldots\}$, $\{(\boldsymbol{n}_k, \boldsymbol{v}_k), k=1, 2, \ldots\}$ and $\{\boldsymbol{w}_k, k=1, 2, \ldots\}$. Obviously, for these processes the successive events $\{\boldsymbol{v}_k = 0\}$, $\{\boldsymbol{n}_k = 0, \boldsymbol{v}_k = 0\}$ and $\{\boldsymbol{w}_k = 0\}$, respectively (the so called zero states), constitute an imbedded renewal process. The renewal functions of these renewal processes are easily obtained from (6.75) by noting that for ρ real

$$\begin{aligned} \lim_{\rho \to \infty} \mathrm{E}\{\mathrm{e}^{-\rho \boldsymbol{v}_k} \mid \boldsymbol{v}_1 = 0\} &= \Pr\{\boldsymbol{v}_k = 0 \mid \boldsymbol{v}_1 = 0\}, \\ \lim_{\rho \to -\infty} \mathrm{E}\{\mathrm{e}^{-\rho \boldsymbol{n}_k} \mid \boldsymbol{n}_1 = 0\} &= \Pr\{\boldsymbol{n}_k = 0 \mid \boldsymbol{n}_1 = 0\}. \end{aligned}$$

Hence, we obtain from (6.75) for the renewal functions

$$\sum_{k=1}^{\infty} r^k \Pr\{\boldsymbol{v}_k=0 \mid \boldsymbol{v}_1=0\} = \sum_{k=1}^{\infty} r^k \Pr\{\boldsymbol{m}_k=0 \mid \boldsymbol{m}_1=0\}$$

$$= r \exp\left\{\sum_{n=1}^{\infty} \frac{r^n}{n} \Pr\{\boldsymbol{s}_n \leqq 0\}\right\}, \tag{6.76}$$

$$\sum_{k=1}^{\infty} r^k \Pr\{\boldsymbol{v}_k=0, \boldsymbol{n}_k=0 \mid \boldsymbol{v}_1=0, \boldsymbol{n}_1=0\}$$

$$= \sum_{k=1}^{\infty} r^k \Pr\{\boldsymbol{m}_k=0, \boldsymbol{w}_k=0 \mid \boldsymbol{m}_1=0, \boldsymbol{w}_1=0\}$$

$$= r \exp\left\{\sum_{n=1}^{\infty} \frac{r^n}{n} \Pr\{\boldsymbol{s}_n=0\}\right\},$$

$$\sum_{k=1}^{\infty} r^k \Pr\{\boldsymbol{w}_k=0 \mid \boldsymbol{w}_1=0\} = \sum_{k=1}^{\infty} r^k \Pr\{\boldsymbol{n}_k=0 \mid \boldsymbol{n}_1=0\}$$

$$= r \exp\left\{\sum_{n=1}^{\infty} \frac{r^n}{n} \Pr\{\boldsymbol{s}_n \geqq 0\}\right\},$$

with $|r|<1$.

The *return times* or *renewal times* of the imbedded renewal processes are defined by

$$\boldsymbol{h} \stackrel{\text{def}}{=} \min_{k=1,2,\dots} \{k: \boldsymbol{v}_{k+1}=0 \mid \boldsymbol{v}_1=0\}, \tag{6.77}$$

$$\boldsymbol{z} \stackrel{\text{def}}{=} \min_{k=1,2,\dots} \{k: \boldsymbol{v}_{k+1}=0, \boldsymbol{n}_{k+1}=0 \mid \boldsymbol{v}_1=0, \boldsymbol{n}_1=0\},$$

$$\boldsymbol{k} \stackrel{\text{def}}{=} \min_{k=1,2,\dots} \{k: \boldsymbol{w}_{k+1}=0 \mid \boldsymbol{w}_1=0\}.$$

It follows from renewal theory, that for $|r|<1$,

$$\sum_{n=1}^{\infty} \{\mathrm{E}\{r^{\boldsymbol{h}}\}\}^n = \sum_{k=1}^{\infty} r^k \Pr\{\boldsymbol{v}_{k+1}=0 \mid \boldsymbol{v}_1=0\},$$

and similar relations hold for $\boldsymbol{z}$ and $\boldsymbol{k}$. Consequently, for the return times we obtain from (6.76) for $|r| \leqq 1$,

$$\mathrm{E}\{r^{\boldsymbol{h}}\} = 1 - \exp\left\{-\sum_{n=1}^{\infty} \frac{r^n}{n} \Pr\{\boldsymbol{s}_n \leqq 0\}\right\}$$

$$= 1-(1-r) \exp\left\{\sum_{n=1}^{\infty} \frac{r^n}{n} \Pr\{\boldsymbol{s}_n>0\}\right\}, \tag{6.78}$$

$$\mathrm{E}\{r^{z}\} = 1 - \exp\left\{-\sum_{n=1}^{\infty} \frac{r^n}{n} \Pr\{s_n = 0\}\right\},$$

$$\mathrm{E}\{r^{k}\} = 1 - \exp\left\{-\sum_{n=1}^{\infty} \frac{r^n}{n} \Pr\{s_n \geqq 0\}\right\}$$

$$= 1 - (1-r)\exp\left\{\sum_{n=1}^{\infty} \frac{r^n}{n} \Pr\{s_n < 0\}\right\}.$$

From (6.61), (6.62) and (6.77) we have

$$\{z=1\} = \{s_1 = 0\},$$

$$\{z=k\} = \{s_k = 0,\ s_j > 0,\ j=1, \dots, k-1\}, \qquad k=2, 3, \dots .$$

From now we make the trivial assumption (cf. (6.59)),

$$\Pr\{x_n \neq 0\} > 0,$$

it follows immediately from the last but one relation that $\{z < \infty\}$ is not the sure event, hence from (6.78),

$$Q \overset{\text{def}}{=} \sum_{n=1}^{\infty} \frac{1}{n} \Pr\{s_n = 0\} < \infty. \tag{6.79}$$

With

$$P \overset{\text{def}}{=} \sum_{n=1}^{\infty} \frac{1}{n} \Pr\{s_n < 0\}, \qquad R \overset{\text{def}}{=} \sum_{n=1}^{\infty} \frac{1}{n} \Pr\{s_n > 0\},$$

so that

$$P + Q + R = \infty,$$

we have from (6.79),

$$P + R = \infty. \tag{6.80}$$

From (6.78) we obtain

$$\Pr\{h < \infty\} = 1 \quad \text{if and only if} \quad P = \infty, \tag{6.81}$$

$$\Pr\{k < \infty\} = 1 \quad \text{if and only if} \quad R = \infty,$$

$$\mathrm{E}\{h\} = \mathrm{e}^{R} < \infty \qquad \text{if} \quad R < \infty, \tag{6.82}$$

$$\mathrm{E}\{k\} = \mathrm{e}^{P} < \infty \qquad \text{if} \quad P < \infty,$$

$$\mathrm{E}\{h\} = \mathrm{E}\{k\} = \infty \qquad \text{if} \quad P = \infty,\ R = \infty.$$

From the above relations we obtain the interesting result that *for the Markov process* $\{v_k,\ k=1, 2, \dots\}$ *the zero state* $\{v_k = 0\}$ *is positive recurrent if* $P = \infty$,

$R<\infty$, is null recurrent if $P=\infty$, $R=\infty$, and is transient if $P<\infty$, $R=\infty$. The same conclusion holds for the zero state of the process $\{\boldsymbol{w}_k,\ k=1,2,\ldots\}$ with P and R interchanged. The zero state of the process $\{(\boldsymbol{n}_k,\boldsymbol{v}_k),\ k=1,2,\ldots\}$ is always transient. The results obtained above lead to interesting properties of the sequence $\boldsymbol{s}_0, \boldsymbol{s}_1, \boldsymbol{s}_2, \ldots$. For instance for $k=1,2,\ldots$,

$$\begin{aligned}
\{\boldsymbol{m}_k=0\}&=\{\boldsymbol{s}_{k-1}\leqq 0,\ \boldsymbol{s}_{k-2}\leqq 0,\ \ldots,\ \boldsymbol{s}_0\leqq 0\},\\
\{\boldsymbol{n}_k=0\}&=\{\boldsymbol{s}_{k-1}\geqq 0,\ \boldsymbol{s}_{k-2}\geqq 0,\ \ldots,\ \boldsymbol{s}_0\geqq 0\},\\
\{\boldsymbol{v}_k=0\}&=\{\boldsymbol{s}_{k-1}\leqq \boldsymbol{s}_j,\ j=0,\ldots,k-1\},\\
\{\boldsymbol{w}_k=0\}&=\{\boldsymbol{s}_{k-1}\geqq \boldsymbol{s}_j,\ j=0,\ldots,k-1\},
\end{aligned} \tag{6.83}$$

and the generating functions of the probabilities of these events are given by (6.76). Further, it is found from (6.63), that for $h=2,3,\ldots$,

$$\begin{aligned}
&\{\boldsymbol{v}_2=0\}=\{\boldsymbol{s}_1\leqq 0\},\\
&\{\boldsymbol{v}_{h+1}=0,\ \boldsymbol{v}_j>0,\ j=1,\ldots,h\}=\{\boldsymbol{s}_h\leqq 0,\ \boldsymbol{s}_j>0,\ j=1,\ldots,h-1\},\\
&\{\boldsymbol{w}_2=0\}=\{\boldsymbol{s}_1\geqq 0\},\\
&\{\boldsymbol{w}_{h+1}=0,\ \boldsymbol{w}_j>0,\ j=1,\ldots,h\}=\{\boldsymbol{s}_h\geqq 0,\ \boldsymbol{s}_j<0,\ j=1,\ldots,h-1\}.
\end{aligned} \tag{6.84}$$

Consequently, from (6.77) we have

$$\boldsymbol{h}=\min_{k=1,2,\ldots}\{k:\ \boldsymbol{s}_k-\boldsymbol{s}_0\leqq 0\},\qquad \boldsymbol{k}=\min_{k=1,2,\ldots}\{k:\ \boldsymbol{s}_k-\boldsymbol{s}_0\geqq 0\}. \tag{6.85}$$

We now introduce the concept of *ladder index*. An index h is called a (*weak*) *descending ladder index* of the sequence $\boldsymbol{s}_0, \boldsymbol{s}_1, \boldsymbol{s}_2, \ldots$, if

$$\boldsymbol{s}_h\leqq \boldsymbol{s}_j,\qquad j=0,1,\ldots,h-1, \tag{6.86}$$

and it is called a (*weak*) *ascending ladder index* of this sequence if

$$\boldsymbol{s}_h\geqq \boldsymbol{s}_j,\qquad j=0,1,\ldots,h-1. \tag{6.87}$$

Obviously, the descending, and similarly the ascending ladder indices form an ordered sequence which is finite or infinite. If in the Markov process $\{\boldsymbol{v}_k,\ k=1,2,\ldots\}$ a return to the zero state occurs at $k=h+1$, i.e. $\boldsymbol{v}_{h+1}=0$, then it follows immediately from (6.63) that h is a descending ladder index of the sequence $\boldsymbol{s}_0, \boldsymbol{s}_1, \boldsymbol{s}_2, \ldots$, and conversely. Similarly, the index h is an ascending ladder index if and only if $\boldsymbol{w}_{h+1}=0$. Consequently, the times between successive descending ladder indices are independent, identically distributed variables with the same distribution function as $\boldsymbol{h}$ (cf. (6.78)), and similarly for ascending ladder indices, here the distribution is that of $\boldsymbol{k}$.

Obviously, if h_1 and h_2 are two successive descending ladder indices, and k_1, k_2 two successive ascending ladder indices then

$$\begin{aligned} &\boldsymbol{s}_{h_2}-\boldsymbol{s}_{h_1}\leqq 0, \quad \boldsymbol{s}_h-\boldsymbol{s}_{h_1}>0, \quad h=h_1+1, h_1+2, \dots, h_2-1, \\ &\boldsymbol{s}_{k_2}-\boldsymbol{s}_{k_1}\geqq 0, \quad \boldsymbol{s}_k-\boldsymbol{s}_{k_1}<0, \quad k=k_1+1, k_1+2, \dots, k_2-1. \end{aligned} \tag{6.88}$$

To obtain the joint distribution of $\boldsymbol{h}$ and $\boldsymbol{s}_h$ we note that (cf. (6.62)) for $|r|<1$, $\mathrm{Re}\,\rho<0$,

$$\begin{aligned} \sum_{n=1}^{\infty} \{\mathrm{E}\{r^{\boldsymbol{h}} \exp(-\rho \boldsymbol{s}_h)\}\}^n &= \sum_{k=1}^{\infty} r^k\, \mathrm{E}\{\mathrm{e}^{-\rho \boldsymbol{s}_k}\, (\boldsymbol{v}_{k+1}=0) \mid \boldsymbol{v}_1=0\} \\ &= \sum_{k=1}^{\infty} r^k\, \mathrm{E}\{\mathrm{e}^{-\rho \boldsymbol{\eta}_{k+1}}\, (\boldsymbol{v}_{k+1}=0) \mid \boldsymbol{v}_1=0\}. \end{aligned}$$

Hence for $|r|<1$, $\mathrm{Re}\,\rho<0$, from (6.75),

$$[1-\mathrm{E}\{r^{\boldsymbol{h}} \exp(-\rho \boldsymbol{s}_h)\}]^{-1} = \exp\left\{\sum_{n=1}^{\infty} \frac{r^n}{n}\, \mathrm{E}\{\mathrm{e}^{-\rho \boldsymbol{s}_n}\, (\boldsymbol{s}_n\leqq 0)\}\right\}.$$

Consequently, we have for the joint distribution of $\boldsymbol{h}$ and $\boldsymbol{s}_h$, and similarly for that of $\boldsymbol{k}$ and $\boldsymbol{s}_k$,

$$\begin{aligned} \mathrm{E}\{r^{\boldsymbol{h}} \exp(-\rho \boldsymbol{s}_h)\} &= 1-\exp\left\{-\sum_{n=1}^{\infty} \frac{r^n}{n}\, \mathrm{E}\{\mathrm{e}^{-\rho \boldsymbol{s}_n}\, (\boldsymbol{s}_n\leqq 0)\}\right\}, \\ &\qquad\qquad |r|<1, \quad \mathrm{Re}\,\rho\leqq 0, \\ \mathrm{E}\{r^{\boldsymbol{k}} \exp(-\rho \boldsymbol{s}_k)\} &= 1-\exp\left\{-\sum_{n=1}^{\infty} \frac{r^n}{n}\, \mathrm{E}\{\mathrm{e}^{-\rho \boldsymbol{s}_n}\, (\boldsymbol{s}_n\geqq 0)\}\right\}, \\ &\qquad\qquad |r|<1, \quad \mathrm{Re}\,\rho\geqq 0. \end{aligned} \tag{6.89}$$

If in (6.86) the sign "$\leqq$" is replaced by "$<$" the index h is called a *strong descending ladder index*, and similarly, it is a *strong ascending ladder index* if in (6.87) the sign "$\geqq$" is replaced by "$>$". If in (6.88) the signs "$\leqq$, $>$, $\geqq$, $<$" are replaced by "$<$, $\geqq$, $>$, $\leqq$", respectively, then the resulting relations hold for strong ladder indices. The set of strong ladder indices is a subset of the set of weak ladder indices. Obviously, if the distribution function $G(x)$ of $\boldsymbol{x}_n$ is continuous, a weak ladder index is always a strong one. Evidently, the times between strong ascending ladder indices are independent, identically distributed variables, and similarly, for strong descending ladder indices. Define

$$\boldsymbol{h}_s \stackrel{\text{def}}{=} \min_{k=1,2,\dots} \{k: \boldsymbol{s}_k-\boldsymbol{s}_0<0\}; \qquad \boldsymbol{k}_s \stackrel{\text{def}}{=} \min_{k=1,2,\dots} \{k: \boldsymbol{s}_k-\boldsymbol{s}_0>0\}, \tag{6.90}$$

so that for the sequence $\{s_0, s_1, \dots\}$, h_s and k_s are the first strong descending and ascending ladder index, respectively. To obtain the distributions of h_s and k_s, note first that for $|r| \leqq 1$ (cf. (6.77)),

$$\mathrm{E}\{r^h\} = \mathrm{E}\{r^h(s_h<0)\} + \mathrm{E}\{r^h(s_h=0)\} = \mathrm{E}\{r^h(s_h<0)\} + \mathrm{E}\{r^z\},$$

hence from (6.89), for $|r| \leqq 1$,

$$\mathrm{E}\{r^h(s_h<0)\} = \exp\left\{-\sum_{n=1}^{\infty} \frac{r^n}{n} \Pr\{s_n=0\}\right\} - \exp\left\{-\sum_{n=1}^{\infty} \frac{r^n}{n} \Pr\{s_n \leqq 0\}\right\}. \tag{6.91}$$

Since the first strong descending ladder index is the sum of a number of return times of events $\{v_k=0, n_k=0\}$ and of an entrance time from such an event into $\{s_n<0, s_j>0, j=1, \dots, n-1\}$ and since all these times are independent, we have for $|r| \leqq 1$,

$$\mathrm{E}\{r^{h_s}\} = \sum_{m=0}^{\infty} \{\mathrm{E}\{r^z\}\}^m \, \mathrm{E}\{r^h(s_h<0)\}.$$

From the latter relation and from (6.91) we obtain the generating function of the distribution of h_s; that of k_s is derived in the same way. The result reads for $|r| \leqq 1$,

$$\mathrm{E}\{r^{h_s}\} = 1 - \exp\left\{-\sum_{n=1}^{\infty} \frac{r^n}{n} \Pr\{s_n<0\}\right\}, \tag{6.92}$$

$$\mathrm{E}\{r^{k_s}\} = 1 - \exp\left\{-\sum_{n=1}^{\infty} \frac{r^n}{n} \Pr\{s_n>0\}\right\}.$$

To obtain the joint distributions of h_s and s_{h_s} and of k_s and s_{k_s} we use a derivation similar to that of (6.89),

$$\mathrm{E}\{r^{h_s} \exp(-\rho s_{h_s})\} = 1 - \exp\left\{-\sum_{n=1}^{\infty} \frac{r^n}{n} \mathrm{E}\{\mathrm{e}^{-\rho s_n}(s_n<0)\}\right\},$$
$$|r| \leqq 1, \quad \mathrm{Re}\, \rho \leqq 0, \tag{6.93}$$

$$\mathrm{E}\{r^{k_s} \exp(-\rho s_{k_s})\} = 1 - \exp\left\{-\sum_{n=1}^{\infty} \frac{r^n}{n} \mathrm{E}\{\mathrm{e}^{-\rho s_n}(s_n>0)\}\right\},$$
$$|r| \leqq 1, \quad \mathrm{Re}\, \rho \geqq 0.$$

Strong descending ladder indices also constitute an imbedded renewal process of the Markov process $\{(n_k, v_k), k=1, 2, \dots\}$. Obviously, h is a strong descending ladder index if $v_{h+1}=0$, $s_h<0$, or equivalently, $v_{h+1}=0$, $n_{h+1}<0$.

From renewal theory we have for $|r|<1$,

$$\sum_{n=1}^{\infty} \{\mathrm{E}\{r^{\boldsymbol{h}_s}\}\}^n = \sum_{k=1}^{\infty} r^k \Pr\{\boldsymbol{v}_{k+1}=0, \boldsymbol{n}_{k+1}<0\}.$$

Consequently, from (6.92) and (6.64), and applying the symmetry property, we obtain for $|r|\leqq 1$,

$$\begin{aligned}\sum_{k=1}^{\infty} r^k \Pr\{\boldsymbol{v}_k=0, \boldsymbol{n}_k<0\} &= \sum_{k=1}^{\infty} r^k \Pr\{\boldsymbol{m}_k=0, \boldsymbol{w}_k<0\}\\ &= r\exp\left\{\sum_{n=1}^{\infty} \frac{r^n}{n} \Pr\{\boldsymbol{s}_n<0\}\right\}, \qquad (6.94)\\ \sum_{k=1}^{\infty} r^k \Pr\{\boldsymbol{w}_k=0, \boldsymbol{m}_k>0\} &= \sum_{k=1}^{\infty} r^k \Pr\{\boldsymbol{n}_k=0, \boldsymbol{v}_k>0\}\\ &= r\exp\left\{\sum_{n=1}^{\infty} \frac{r^n}{n} \Pr\{\boldsymbol{s}_n>0\}\right\}.\end{aligned}$$

It should be noted that for $h=1, 2, \ldots$,

$$\begin{aligned}\{\boldsymbol{v}_{h+1}=0, \boldsymbol{n}_{h+1}<0\} &=\{\boldsymbol{s}_h<0, \boldsymbol{s}_h\leqq \boldsymbol{s}_j, j=0, \ldots, h-1\}, \qquad (6.95)\\ \{\boldsymbol{m}_{h+1}=0, \boldsymbol{w}_{h+1}<0\}&=\{\boldsymbol{s}_h<0, \boldsymbol{s}_j\leqq 0, j=1, \ldots, h-1\},\\ \{\boldsymbol{w}_{h+1}=0, \boldsymbol{m}_{h+1}>0\}&=\{\boldsymbol{s}_h>0, \boldsymbol{s}_h\geqq \boldsymbol{s}_j, j=0, \ldots, h-1\},\\ \{\boldsymbol{n}_{h+1}=0, \boldsymbol{v}_{h+1}>0\} &=\{\boldsymbol{s}_h>0, \boldsymbol{s}_j\geqq 0, j=0, \ldots, h-1\},\end{aligned}$$

the generating functions of the probabilities of these events being given by (6.94).

From (6.61) it follows that for every real x,

$$\{\boldsymbol{m}_{k+1}<x\} \subset \{\boldsymbol{m}_k<x\}, \qquad \{\boldsymbol{n}_{k+1}<x\} \supset \{\boldsymbol{n}_k<x\}, \qquad k=0, 1, \ldots, \qquad (6.96)$$

hence $\boldsymbol{m}_k$ and $\boldsymbol{n}_k$ converge for $k\to\infty$. Define

$$\boldsymbol{m} \stackrel{\text{def}}{=} \lim_{k\to\infty} \boldsymbol{m}_k = \sup_{k=0,1,\ldots} \boldsymbol{s}_k, \qquad \boldsymbol{n} \stackrel{\text{def}}{=} \lim_{k\to\infty} \boldsymbol{n}_k = \inf_{k=0,1,\ldots} \boldsymbol{s}_k. \qquad (6.97)$$

We now formulate a result due to SPITZER [1956].

THEOREM 6.8. *For* $\Pr\{\boldsymbol{x}_n\neq 0\}>0$,
(*i*) *with probability one*

$$\begin{aligned}&\boldsymbol{n}= \inf_{k=0,1,\ldots} \boldsymbol{s}_k>-\infty, \boldsymbol{m}=\infty \text{ and } \lim_{n\to\infty} \boldsymbol{s}_n=\infty \qquad \text{if } P<\infty, \qquad (6.98)\\ &\boldsymbol{n}=-\infty, \lim_{n\to\infty} \boldsymbol{s}_n=\infty \text{ and } \boldsymbol{m}= \sup_{k=0,1,\ldots} \boldsymbol{s}_k<\infty \qquad \text{if } R<\infty,\\ &\boldsymbol{n}=-\infty, \liminf \boldsymbol{s}_n=-\infty, \boldsymbol{m}=\infty \text{ and } \limsup \boldsymbol{s}_n=\infty \qquad \text{if } P=\infty, \quad R=\infty;\end{aligned}$$

(*ii*) *if* $P<\infty$ *then for* $\text{Re}\,\rho \leqq 0$,

$$\lim_{k\to\infty} \text{E}\{\text{e}^{-\rho \boldsymbol{w}_k} \mid \boldsymbol{w}_1=0\} = \text{E}\{\text{e}^{-\rho \boldsymbol{n}}\}$$

$$= \exp\left\{-\sum_{n=1}^{\infty} \frac{1}{n} \text{E}\{(1-\text{e}^{-\rho \boldsymbol{s}_n})(\boldsymbol{s}_n<0)\}\right\}, \qquad (6.99)$$

if $R<\infty$ *then for* $\text{Re}\,\rho \geqq 0$,

$$\lim_{k\to\infty} \text{E}\{\text{e}^{-\rho \boldsymbol{v}_k} \mid \boldsymbol{v}_1=0\} = \text{E}\{\text{e}^{-\rho \boldsymbol{m}}\}$$

$$= \exp\left\{-\sum_{n=1}^{\infty} \frac{1}{n} \text{E}\{(1-\text{e}^{-\rho \boldsymbol{s}_n})(\boldsymbol{s}_n>0)\}\right\}.$$

Proof. From (6.81) and (6.85) it is seen that the sequence of ascending ladder indices is infinite and that of descending ladder indices is finite with probability one if $P<\infty$. Hence from (6.86) and (6.87) the first relation of (6.98) follows; the second is similarly proved. For $P=R=\infty$ both the ascending and descending sequence are infinite, so that from (6.86) and (6.87) the third relation of (6.98) results.

From (6.96) and (6.98) for $R<\infty$ it is seen that the distribution function of $\boldsymbol{m}_k$ converges for $k\to\infty$ to a proper probability distribution, viz. that of $\boldsymbol{m}$, at every point of continuity of the latter distribution. Note that this convergence is monotone. Hence from Helly-Bray's convergence theorem (cf. app. 5) we have for $R<\infty$, $\text{Re}\,\rho \geqq 0$,

$$\lim_{k\to\infty} \text{E}\{\text{e}^{-\rho \boldsymbol{m}_k}\} = \text{E}\{\text{e}^{-\rho \boldsymbol{m}}\}.$$

From (6.75) we have for $|r|<1$, $\text{Re}\,\rho \geqq 0$,

$$\sum_{k=1}^{\infty} r^k \text{E}\{\text{e}^{-\rho \boldsymbol{m}_k} \mid \boldsymbol{m}_1=0\} = \frac{r}{1-r} \exp\left\{-\sum_{n=1}^{\infty} \frac{r^n}{n} \text{E}\{(1-\text{e}^{-\rho \boldsymbol{s}_n})(\boldsymbol{s}_n>0)\}\right\}.$$

Hence by applying an Abelian theorem for power series (cf. app. 1) it follows from the latter relation for $R<\infty$, $\rho \geqq 0$, and by analytic continuation also for $\text{Re}\,\rho \geqq 0$, since $\boldsymbol{m} \geqq 0$, that the second relation of (6.99) is true; note that $\boldsymbol{m}_k$ and $\boldsymbol{v}_k$ have the same distribution. The proof of the first relation of (6.99) is similar.

From the above theorem it is seen that the existence of the limit distributions of $\boldsymbol{w}_k$ and $\boldsymbol{v}_k$ depends only on the values of P and R. It should be noted that if $\mathrm{E}\{|\boldsymbol{x}_n|\}<\infty$ then

$$P<\infty,\ R=\infty \Leftrightarrow \mathrm{E}\{\boldsymbol{x}_n\}>0, \tag{6.100}$$
$$P=\infty,\ R=\infty \Leftrightarrow \mathrm{E}\{\boldsymbol{x}_n\}=0,$$
$$P=\infty,\ R<\infty \Leftrightarrow \mathrm{E}\{\boldsymbol{x}_n\}<0.$$

The proof of this statement follows easily from the above theorem and theorem 1.1 of section II.1.3.

The properties of the variables $\max(0, \boldsymbol{s}_1, ..., \boldsymbol{s}_n)$ and $\min(0, \boldsymbol{s}_1, ..., \boldsymbol{s}_n)$ are completely described by the relation (6.75) and the theorem above. It should be noted that from (6.75) the characteristic function of the joint distribution of $\boldsymbol{s}_n$ and $\min(0, \boldsymbol{s}_1, ..., \boldsymbol{s}_n)$ can be easily obtained by taking $\mathrm{Re}\,\rho_1=0$, replacing $\boldsymbol{v}_k$ by $\boldsymbol{s}_{k-1}-\boldsymbol{n}_k$ (cf. (6.62)) and by writing ρ_3 for $\rho_2-\rho_1$ with $\mathrm{Re}\,\rho_3\leqq 0$. The result reads for $|r|\leqq 1$, $\mathrm{Re}\,\rho_1=0$, $\mathrm{Re}\,\rho_3\leqq 0$,

$$\sum_{k=1}^{\infty} r^k\, \mathrm{E}\{\exp(-\rho_1 \boldsymbol{s}_{k-1}-\rho_3 \boldsymbol{n}_k) \mid \boldsymbol{s}_0=0\}$$
$$= r\exp\left\{\sum_{n=1}^{\infty} \frac{r^n}{n}\, \mathrm{E}\{\mathrm{e}^{-\rho_1 \boldsymbol{s}_n}\, (\boldsymbol{s}_n\geqq 0) + \mathrm{e}^{-(\rho_3+\rho_1)\boldsymbol{s}_n}\, (\boldsymbol{s}_n<0)\}\right\}. \tag{6.101}$$

Important variables in fluctuation theory are the index $\boldsymbol{P}_n$ of the first maximum term and the number $\boldsymbol{Q}_n$ of positive terms of $\boldsymbol{s}_0, \boldsymbol{s}_1, ..., \boldsymbol{s}_n$. It is known (cf. FELLER [1966]) that $\boldsymbol{P}_n$ and $\boldsymbol{Q}_n$ have the same distribution, but explicit expressions for their distributions are not known. If $G(x)$ (cf. (6.59)) is a continuous and symmetric distribution then

$$\lim_{n\to\infty} \Pr\left\{\frac{1}{n}\boldsymbol{P}_n<x\right\} = \frac{2}{\pi}\arcsin\sqrt{x}, \qquad 0<x<1. \tag{6.102}$$

SPITZER [1956] has investigated the behaviour of $\boldsymbol{P}_n$ for $n\to\infty$ for non-symmetric $G(x)$.

Other variables important for queueing theory are

$$\boldsymbol{W}_+ \overset{\text{def}}{=} \max_{j=1,...,\boldsymbol{h}} \boldsymbol{v}_j, \qquad \boldsymbol{W}_- \overset{\text{def}}{=} \min_{j=1,...,\boldsymbol{k}} \boldsymbol{w}_j, \tag{6.103}$$
$$\boldsymbol{\alpha}_{0;0K} \overset{\text{def}}{=} \min_{k=1,2,...} \{\boldsymbol{k}: \boldsymbol{v}_{k+1}\geqq K,\ \boldsymbol{v}_1=0,\ \boldsymbol{v}_j>0,\ j=2, ..., k\},$$

where K is a positive constant and $\boldsymbol{h}$ and $\boldsymbol{k}$ are defined by (6.77). Obviously, $\boldsymbol{W}_+$ is the maximal value reached by the process $\{\boldsymbol{v}_k, k=1, 2, ...\}$ during its first return time to the zero state. $\boldsymbol{W}_-$ has a similar meaning for the process $\{\boldsymbol{w}_k, k=1, 2, ...\}$, and $\boldsymbol{\alpha}_{0;0K}$ is the first entrance time from the zero state into $\boldsymbol{v}_{k+1}\geqq K$ without a return to the zero state.

From (6.84) and (6.103) it is readily seen that

$$W_+ = \max_{j=0,\dots,h} s_j, \qquad W_- = \min_{j=0,\dots,k} s_j, \tag{6.104}$$

$$\alpha_{0;0K} = \min_{k=1,2,\dots} \{k: s_k \geqq K, s_0=0, s_j>0, j=1,\dots,k-1\},$$

so that W_+ is the maximal value of those s_j for which j is less than the first weak descending ladder index. Evidently, the variables described by (6.104) are also of interest for fluctuation theory. Explicit expressions for the distributions of the variables described by (6.104) are not known in the general case. The investigations of chapter III.4, however, will lead to explicit expressions for the distributions of W_+, of W_- and for the generating function of the distribution of $\alpha_{0;0K}$, for the case that every x_n is the difference of two positive stochastic variables τ_n and σ_n with finite first moments,

$$x_n = \tau_n - \sigma_n, \qquad n=1, 2, \dots,$$

with τ_n, $n=1, 2, \dots$, and similarly, σ_n, $n=1, 2, \dots$, sequences of independent, identically distributed variables, both sequences being independent families of variables and with τ_n or σ_n having a *negative exponential* distribution. These conditions are equivalent with the condition that $g(\rho)$ (cf. (6.60)) is a function of the type

$$(i) \quad g(\rho) = \frac{1}{1-\alpha\rho}\,\beta(\rho), \qquad \text{Re}\,\rho=0, \quad \alpha>0, \tag{6.105}$$

$$(ii) \quad g(\rho) = \frac{1}{1+\beta\rho}\,\alpha(-\rho), \quad \text{Re}\,\rho=0, \quad \beta>0,$$

with $\alpha(\rho)$ and $\beta(\rho)$ Laplace-Stieltjes transforms of distributions of positive stochastic variables with finite first moments.

Many of the results of chapter III.4 have an interpretation in fluctuation theory. For some of these results we shall give such an interpretation for those $g(\rho)$ described in (6.105).

We shall use the notation

$$\frac{1}{2\pi i}\int_{C_\zeta} \dots \mathrm{d}\zeta = \lim_{b\to\infty} \frac{1}{2\pi i}\int_{c-ib}^{c+ib} \dots \mathrm{d}\zeta, \qquad c=\text{Re}\,\zeta.$$

Define

$$\delta \stackrel{\text{def}}{=} \begin{cases} \text{the unique positive zero of } \beta(\rho)+\alpha\rho-1 & \text{if } \mathrm{E}\{\boldsymbol{x}_n\}>0, \\ 0 & \text{if } \mathrm{E}\{\boldsymbol{x}_n\}\leqq 0, \end{cases}$$

$$\varepsilon \stackrel{\text{def}}{=} \begin{cases} \text{the unique negative zero of } \alpha(-\rho)-\beta\rho-1 & \text{if } \mathrm{E}\{\boldsymbol{x}_n\}>0, \\ 0 & \text{if } \mathrm{E}\{\boldsymbol{x}_n\}\leqq 0. \end{cases}$$

For the case (i) of (6.105) with

$$\mathrm{E}\{\mathrm{e}^{-\rho\boldsymbol{\sigma}_n}\} = \frac{1}{1+\alpha\rho}, \qquad \mathrm{E}\{\mathrm{e}^{-\rho\boldsymbol{\tau}_n}\} = \beta(\rho), \qquad \mathrm{Re}\,\rho\geqq 0,$$

it follows from (6.103), (6.104) and (III.4.73) for $\delta<\mathrm{Re}\,\zeta<1/\alpha$, that

$$\Pr\{\boldsymbol{W}_+<w\}$$

$$= \begin{cases} \dfrac{1}{2\pi\mathrm{i}} \displaystyle\int_{C_\zeta} \frac{\mathrm{e}^{\zeta w}}{\beta(\zeta)+\alpha\zeta-1}\,\mathrm{d}\zeta \Big/ \frac{1}{2\pi\mathrm{i}} \int_{C_\zeta} \frac{\mathrm{e}^{\zeta w}}{1-\alpha\zeta}\,\frac{\mathrm{d}\zeta}{\beta(\zeta)+\alpha\zeta-1}, & w>0, \\ 0, & w\leqq 0. \end{cases} \tag{6.106}$$

For case (ii) with

$$\mathrm{E}\{\mathrm{e}^{-\rho\boldsymbol{\sigma}_n}\} = \alpha(\rho), \qquad \mathrm{E}\{\mathrm{e}^{-\rho\boldsymbol{\tau}_n}\} = \frac{1}{1+\beta\rho}, \qquad \mathrm{Re}\,\rho\geqq 0,$$

it follows from (6.103), (6.104) and (III.4.85) for $\varepsilon>\mathrm{Re}\,\zeta>-1/\beta$, that

$$\Pr\{\boldsymbol{W}_+<w\} = \begin{cases} 1+\left\{\dfrac{1}{2\pi\mathrm{i}} \displaystyle\int_{C_\zeta} \frac{\mathrm{e}^{-\zeta w}}{1+\beta\zeta}\,\frac{\beta\,\mathrm{d}\zeta}{1+\beta\zeta-\alpha(-\zeta)}\right\}^{-1}, & w>0, \\ 0, & w\leqq 0. \end{cases} \tag{6.107}$$

From (6.103) it follows that if we replace $\boldsymbol{x}_n$ by $-\boldsymbol{x}_n$ then $\boldsymbol{W}_+$ becomes $-\boldsymbol{W}_-$ and conversely. Replacing $\boldsymbol{x}_n$ by $-\boldsymbol{x}_n$ is equivalent to replacing $\boldsymbol{\tau}_n$ by $\boldsymbol{\sigma}_n$ and conversely. Hence we obtain from (6.106) and (6.107) with $w>0$, for case (i):

$$\Pr\{\boldsymbol{W}_->-w\}$$

$$= 1+\left\{\frac{1}{2\pi\mathrm{i}} \int_{C_\zeta} \frac{\mathrm{e}^{\zeta w}}{1-\zeta\alpha}\,\frac{\alpha\,\mathrm{d}\zeta}{\beta(\zeta)+\alpha\zeta-1}\right\}^{-1}, \qquad \frac{1}{\alpha}>\mathrm{Re}\,\zeta>\delta, \tag{6.108}$$

for case (ii):

$\Pr\{W_- > -w\}$

$$= \frac{1}{2\pi i}\int_{C_\zeta} \frac{e^{-\zeta w}}{1+\beta\zeta-\alpha(-\zeta)}\,d\zeta \Big/ \frac{1}{2\pi i}\int_{C_\zeta} \frac{e^{-\zeta w}}{1+\beta\zeta}\,\frac{d\zeta}{1+\beta\zeta-\alpha(-\zeta)},$$

$$\varepsilon > \operatorname{Re}\zeta > -\frac{1}{\beta}. \qquad (6.109)$$

For a discussion of the properties of the distributions (6.106), ..., (6.109) the reader is referred to sections III.7.4 and III.7.5.

Finally, it is noted that for case (*i*) the expression

$$E\{r^h \exp(-\rho s_h)(\max_{k=1,\dots,h} s_k < K) \mid s_0 = 0\}, \qquad |r| \leqq 1, \quad \operatorname{Re}\rho = 0,$$

of fluctuation theory is given by $\sum_{n=1}^{\infty} r^n f_{K;00}(\rho, \rho)$ according to (III.4.77) with $\rho = \rho_3 - \rho_4$, $\operatorname{Re}\rho_3 = 0$, $\operatorname{Re}\rho_4 = 0$. For case (*ii*) a similar remark holds for (III.4.85).

PART II

THE SINGLE SERVER QUEUE

II. 1. THE MODEL

II.1.1. Introduction

Nowadays queueing theory is a well developed branch of applied probability theory. A vast amount of literature, which is still growing rapidly, exists on this subject. The systematic study of queueing theory started in the beginning of this century. At that time the technological development of automatic telephone exchanges led to a class of problems which could be solved in a satisfactory way only by applying probabilistic methods. It is Erlang (cf. BROCKMEYER, HALSTRØM, JENSEN [1948]), a Danish mathematician, who may be considered as the founder of queueing theory. His studies, in the period 1909–1920, are now classical in queueing theory. Until about 1940 the development of queueing theory has mainly been directed by the needs encountered in the design of automatic telephone exchanges. After the Second World War, when applications of mathematical models and methods in technology and organization rose to a level hitherto unknown, it was soon recognized that queueing theory had a very broad field of application.

The best known queueing situation is that at a ticket window. Here people arrive and want to be served by the man at the window. To fix our terminology we shall speak of the *service station* or *service facility* where *customers* arrive to be served by the *server*. In this part of the book it is one of our main aims to confront the reader with the various mathematical techniques which are basic to the solving of queueing problems. These methods can be illustrated best by investigating the simplest queueing model, i.e. the single server queue, which will be the subject of this part.

In practical queueing situations wide variations in the behaviour of the customers and also in the organization of the service facility are encountered. It is therefore of great importance to describe the queueing situation in all its detail before constructing the mathematical model. To provide the reader from the outset with some insight into the variety of queueing situations, we shall discuss here some of the possibilities. The mathematical models relating to these ,however, will be discussed in part III.

The behaviour of customers has two characteristic features, viz. the arrival process and the waiting process. The arrival process is in general best described by a stochastic process. In the greater part of queueing situations the arrival process of customers at a service station can be considered as a renewal process, where the arrival of a single customer is a renewal. An important variant here is the simultaneous arrival of customers, the number of customers belonging to the same arrival group being constant or stochastic.

At his arrival the customer may find the server idle or busy. Usually an idle server will immediately provide service to an arriving customer. If, however, the server is busy, i.e. serving a customer, in general the arriving customer has to wait, and his waiting process is started. Here we have a large number of variants but only a few of them will be mentioned here: (*i*) the customer waits until he is served; (*ii*) he does not wait at all but disappears immediately and never returns; (*iii*) he returns after some time; (*iv*) the customer joins the queue but if after some time, which may be variable, he has still not been served, then he disappears.

Concerning the organization of the service station we should first mention the number of servers. In many queueing situations there is only one server available. However, frequently one server is not sufficient to serve the customers without the waiting times becoming too long. The management will then provide the service station with more servers. Again variants arise here, since there may be customers who prefer to be served by one particular server.

Usually waiting customers will be served in order of their arrival at the service station. Important variants here are *service in random order* and *last come first served.* Also priority service is important in applications. In that case customers of a special type obtain service before others. It may even happen that the service of a customer is interrupted to serve an arriving priority customer first.

When customers want to wait they form a queue. The management may restrict the number of customers in the queue. In such a case an arriving customer has to disappear if he arrives when the queue is at its maximum.

Usually a server serves only one customer at a time. There are situations, however, in which customers are served as a group by one server. The number of customers served simultaneously may be constant or stochastic. An important characteristic of queueing situations is the duration of the service of a customer: the so called *service time*. In general the service time is a stochastic variable. Sometimes the duration of the service time is determined by the customer himself, sometimes it depends on the management. For instance a customer who finds the server idle at his arrival may have a different service time than a customer who has had to wait.

In general an investigation of a queueing situation is initiated to obtain an insight into the effectiveness of the organization of the service station. Depending on the situation being considered various quantities may be used as a measure of this effectiveness, for example, the average delay time of a customer; the waiting time distribution of a customer; the probability that an arriving customer has to wait; the probability that all servers are idle; the distribution of the number of waiting customers or its average; the distribution of the busy period of a server. As far as the waiting time of a customer is concerned it is necessary to distinguish between the *actual waiting time* and the *virtual waiting time*. The actual waiting time of a customer is defined here as the time between the moment of his arrival and the moment at which his service is started. The virtual waiting time at time t is the waiting time of a customer for service if he had arrived at time t. In general the actual waiting time and the virtual waiting time are not identical. To illustrate the difference, consider a single server queue with customers arriving at regular time intervals, say the nth customer arrives at time $n\tau$, $n=1, 2, \dots$, where τ is a constant. Then the actual waiting time is always measured from a time point which is a multiple of τ, whereas the virtual waiting time is measured from an arbitrary time point t. So, if at the moment of arrival of the first customer the server is idle and if the service time is constant and equal to $\frac{1}{2}\tau$, then the actual waiting time is always zero, whereas the virtual waiting time at time t is positive for $m\tau<t<(m+\frac{1}{2})\tau$ and zero for $(m+\frac{1}{2})\tau<t<(m+1)\tau$, $m=1, 2, \dots$.

The state of the queueing system at a time t depends on t and on the state of the system at time $t=0$. Hence all quantities mentioned above depend on t and the state of the system at time $t=0$. An important topic in queueing theory is the investigation of the behaviour of the queueing system in the long run, i.e. for $t\to\infty$. As a rule the analysis of the system for $t\to\infty$ is much simpler than the study of the system for finite values of t, the so-called transient situation. It should be noted that often the mathematical description of the system for large values of t is of more importance in applications than that of the transient case.

II.1.2. Description of the model

As already stated in the introduction our main object here is to describe various mathematical methods which are available for solving queueing problems. These methods will all be demonstrated for the same model. We now describe this model for the single server queue.

To define the arrival process it will be assumed that customers always arrive individually. The arrival process is a stochastic process. Let $\boldsymbol{t}_n$ denote the moment of arrival after $t=0$ of the nth customer at the service station, $n=1, 2, \dots$. The times

$$\boldsymbol{\sigma}_n \stackrel{\text{def}}{=} \boldsymbol{t}_n - \boldsymbol{t}_{n-1}, \qquad n=1, 2, \dots,$$

with

$$\boldsymbol{t}_0 \stackrel{\text{def}}{=} 0,$$

will be called the *interarrival times*. These interarrival times are assumed to be independent and non-negative stochastic variables, identically distributed for $n=2, 3, \dots$, with

$$A_1(\sigma) \stackrel{\text{def}}{=} \Pr\{\boldsymbol{\sigma}_1 < \sigma\}, \qquad A(\sigma) \stackrel{\text{def}}{=} \Pr\{\boldsymbol{\sigma}_n < \sigma\}, \qquad n=2, 3, \dots,$$

and

$$A_1(\sigma)=0, \qquad A(\sigma)=0 \qquad \text{for } \sigma \leqq 0,$$

$$A_1(\sigma) \geqq 0, \qquad A(\sigma) \geqq 0 \qquad \text{for } \sigma > 0.$$

It is always assumed that the first moment of $A(.)$ is finite, and that

$$A(0+)=0.$$

We define

$$\alpha \stackrel{\text{def}}{=} \int_0^\infty t \, \mathrm{d}A(t) < \infty,$$

and the Laplace-Stieltjes transforms

$$\alpha_1(s) \stackrel{\text{def}}{=} \int_0^\infty \mathrm{e}^{-st} \, \mathrm{d}A_1(t), \qquad \alpha(s) \stackrel{\text{def}}{=} \int_0^\infty \mathrm{e}^{-st} \, \mathrm{d}A(t), \qquad \operatorname{Re} s \geqq 0.$$

From our definition of the arrival process it is evident that the number of customers arriving during $[0, t)$ at the service station is a general renewal process with $A(t)$ as the distribution of the renewal time. The renewal times are the interarrival times.

At the service station there is only one server. If at the moment of arrival of a customer the server is idle, the service of this customer is started immediately. An arriving customer who, at his arrival, finds the server busy, i.e. serving a customer, joins the queue and waits for service. Whenever waiting customers are present and the server has finished the service of a customer, he immediately starts the service of a waiting customer. Unless stated otherwise customers are served in order of their arrival. Denote by $\boldsymbol{\tau}_n$ the service time of the nth customer arriving after $t=0$; $n=1, 2, \dots$. The variables $\boldsymbol{\tau}_n$, $n=1, 2, \dots$, are assumed to be independent, identically distributed variables with

$$B(\tau) \stackrel{\text{def}}{=} \Pr\{\boldsymbol{\tau}_n < \tau\}, \qquad n=1, 2, \dots,$$

and

$$B(\tau)=0 \qquad \text{for } \tau \leqq 0; \qquad B(\tau) \geqq 0 \qquad \text{for } \tau > 0.$$

It is always assumed that the first moment of $B(.)$ is finite and that

$$B(0+)=0.$$

We define

$$\beta \stackrel{\text{def}}{=} \int_0^\infty t \, \mathrm{d}B(t) < \infty,$$

and the Laplace-Stieltjes transform

$$\beta(s) \stackrel{\text{def}}{=} \int_0^\infty \mathrm{e}^{-st} \, \mathrm{d}B(t), \qquad \operatorname{Re} s \geqq 0.$$

Above we have described the distribution of the service times of customers who have arrived after $t=0$. It is possible, however, that at $t=0$ the server is already serving a customer and that other customers are waiting. In this case the service times of the waiting customers will also be non-negative, independent, identically distributed variables with distribution function $B(\tau)$; moreover, it is always assumed that these service times are also independent of those of customers arriving after $t=0$. It remains to specify the properties of the residual service time of the customer who is being served at $t=0$. It is supposed that this residual service time is independent of all other service times. We shall not specify the distribution of the residual service time here but, if need be, refer to it by $B_1(\tau)$. We shall only assume that $B_1(\tau)$ has a finite first moment.

Finally, it is assumed that the arrival process, i.e. the family of interarrival times, and the service process, i.e. the family of service times, are independent stochastic processes.

The description of our model is complete. Next we shall discuss some general concepts.

Denote by $\boldsymbol{w}_n$, $n=1, 2, \ldots$, the actual waiting time of the nth customer, so that $\boldsymbol{w}_n=0$ if no customers are present just before $\boldsymbol{t}_n$. The nth arriving customer leaves the system at time $\boldsymbol{r}_n$,

$$\boldsymbol{r}_n \overset{\text{def}}{=} \boldsymbol{t}_n + \boldsymbol{w}_n + \boldsymbol{\tau}_n, \qquad n=1, 2, \ldots;$$

$\boldsymbol{r}_n$ is the departure time of the nth customer, and the time intervals $\boldsymbol{r}_{n+1}-\boldsymbol{r}_n$ are called the *interdeparture times.*

A *busy period* of the server is the time interval during which the server is continuously busy, i.e. if the nth arriving customer finds the server idle, a new busy period begins at time $\boldsymbol{t}_n$, and if $\boldsymbol{r}_m$ with $m \geqq n$ is the first moment after $\boldsymbol{t}_n$ at which all customers, who arrived during $[\boldsymbol{t}_n, \boldsymbol{r}_m)$, have departed, then this busy period ends at $\boldsymbol{r}_m$ and its duration is $\boldsymbol{r}_m-\boldsymbol{t}_n$.

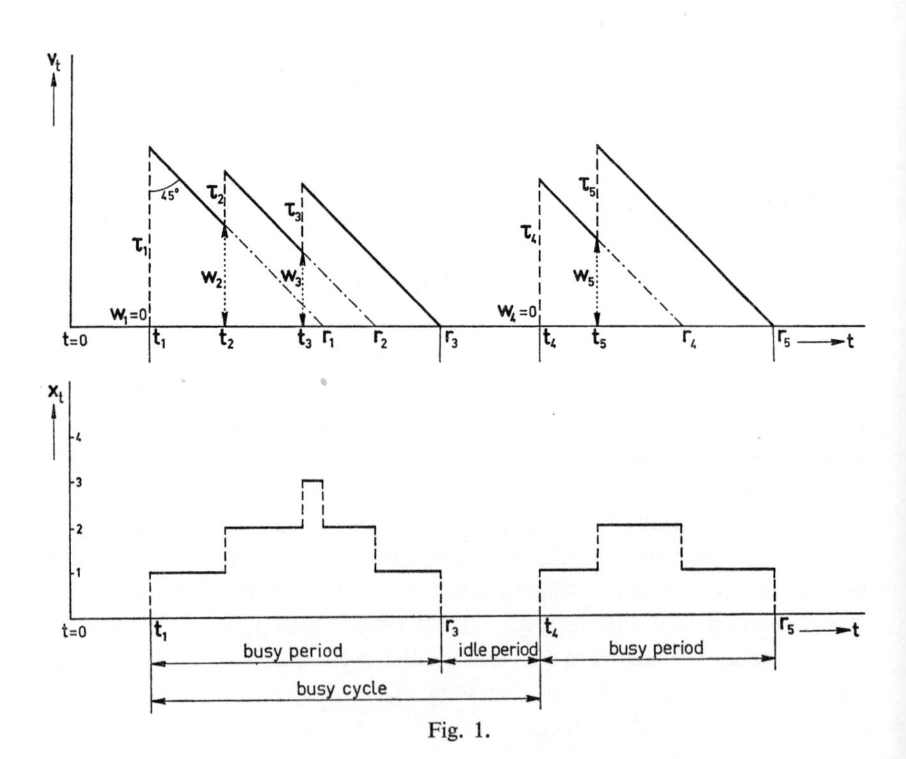

Fig. 1.

We shall denote *the number of customers present in the system at time t* by $\boldsymbol{x}_t$. Hence $\{\boldsymbol{x}_t=0\}$ stands for "the server is idle at time t"; $\{\boldsymbol{x}_t=1\}$ represents the event "at time t the server is busy and there are no customers waiting", while for $n=2, 3, \dots$, $\{\boldsymbol{x}_t=n\}$ is the event: "at time t the server is busy and $n-1$ waiting customers are present". In general the sample functions of the process $\{\boldsymbol{x}_t, t\in[0, \infty)\}$ are step-functions. These step-functions are defined as continuous from the left, so that $\boldsymbol{x}_{\boldsymbol{t}_n}$ represents the number of customers in the system just before the moment of arrival of the nth customer. With probability one the points $\boldsymbol{t}_n$ and $\boldsymbol{r}_n$ are discontinuity points of the $\boldsymbol{x}_t$ process if at least one of the distributions $A(t)$ and $B(t)$ is continuous.

Denote the *virtual waiting time at time t* by $\boldsymbol{v}_t$. If $\boldsymbol{x}_t=0$ then $\boldsymbol{v}_t=0$, and if $\boldsymbol{x}_t=1$ then $\boldsymbol{v}_t$ is equal to the residual service time at time t, whereas for $\boldsymbol{x}_t=n>1$ the virtual waiting time $\boldsymbol{v}_t$ is the sum of the residual service time at time t and the service times of the $n-1$ waiting customers. From the point of view of the server the virtual waiting time $\boldsymbol{v}_t$ represents the amount of work he has still to do at time t to complete the service of all customers present at that time. A graphical representation of $\boldsymbol{v}_t$ introduced by TAKÁCS [1962], see fig. 1, illustrates this. Here a sample function of the stochastic process $\{\boldsymbol{v}_t, t\in[0, \infty)\}$ and of the stochastic process $\{\boldsymbol{x}_t, t\in[0, \infty)\}$ are drawn. It is supposed that the server is idle at $t=0$. The sample function of $\boldsymbol{v}_t$ has upward jumps of magnitude $\boldsymbol{\tau}_n$ at time $\boldsymbol{t}_n$, $n=1, 2, \dots$; $\boldsymbol{v}_t$ decreases linearly with a slope of 45° between two successive jumps, or it becomes zero and then remains zero until the next jump. The sample function of $\boldsymbol{x}_t$ has upward jumps at $\boldsymbol{t}_n$ and downward jumps at $\boldsymbol{r}_n$; between the discontinuity points it is constant. The figure also shows the actual waiting times $\boldsymbol{w}_2$, $\boldsymbol{w}_3$ and $\boldsymbol{w}_5$, the other two ($\boldsymbol{w}_1$ and $\boldsymbol{w}_4$) are zero.

The stochastic processes $\{\boldsymbol{x}_t, t\in[0, \infty)\}$, $\{\boldsymbol{w}_n, n=1,2,\dots\}$ and $\{\boldsymbol{v}_t, t\in[0, \infty)\}$ are the most important processes in queueing theory. The techniques to be used to derive the mathematical description of these processes depend largely on the type of *interarrival time distribution* $A(.)$ and on the type of *service time distribution* $B(.)$.

KENDALL [1951] introduced a short-hand notation for describing which type of queueing situation is meant. The notation consists of three symbols, the first referring to the type of interarrival time distribution, the second to the service time distribution and the third symbol indicates the number of servers. In this notation, which is now in common use, the symbol M stands for the negative exponential distribution, E for the Erlang distribution, H for the hyperexponential distribution, D for the distribution degenerated at a positive value, K for a distribution of which the Laplace-Stieltjes transform is a rational function and G for a general distribution of a non-negative stochastic variable which is not further specified. For instance, $M/G/1$ is the single server queue with negative exponential interarrival time distribution and general service time distribution; $G/G/1$ is the most general single server queue for the model discussed above.

The negative exponential distribution plays an important role in queueing theory, the more so since it is a special case of both the Erlang distribution and the hyperexponential distribution (cf. section I.6.5). The reason is that *this distribution has no memory*. This property has already been discussed in part I (cf. sections I.4.2 and I.6.5). In queueing theory this property is utilized as follows.

LEMMA 1.1. *If the interarrival time distribution is negative exponential with parameter* α^{-1} *then for any time* $t>t_1$ *the conditional probability distribution of the residual interarrival time at time* t*, given the past interarrival time, is negative exponential with parameter* α^{-1} *and is independent of the past interarrival time.*

Here, the *residual* (*past*) *interarrival time* at t is the time between t and the moment of arrival of the next (last) customer after (before) t.

LEMMA 1.2. *If the service time distribution is negative exponential with parameter* β^{-1} *then if at a time* t *the server is busy the conditional probability distribution of the residual service time, given the past service time, is negative exponential with parameter* β^{-1} *and is independent of the past service time.*

The proof of the first lemma follows immediately from renewal theory or by the same reasoning as in the proof of the second lemma, which now follows. Generally, if at time t the server is busy, then the conditional probability that the service at hand will not be terminated before $t+\tau_1$, given that it was started at $t-\tau_2$, is given by

$$\frac{1-B(\tau_1+\tau_2)}{1-B(\tau_2)},$$

if $B(\tau_2)<1$. For $B(t)=1-\exp(-t/\beta)$, $t\geqq 0$, the conditional probability above becomes $\exp(-\tau_1/\beta)$ for all $\tau_1>0$, $\tau_2>0$. This proves the lemma.

II.1.3. Some general theorems and relations

In this section we shall discuss some general theorems to be used in the following chapters.

The actual waiting time $\boldsymbol{w}_{n+1}$ of the $(n+1)$th arriving customer is evidently zero if and only if he arrives after the departure of the nth arriving customer at time $\boldsymbol{r}_n$. Hence

$$\boldsymbol{w}_{n+1} = \begin{cases} 0 & \text{if} \quad \boldsymbol{r}_n - \boldsymbol{t}_{n+1} \leqq 0, \\ \boldsymbol{t}_n + \boldsymbol{w}_n + \boldsymbol{\tau}_n - \boldsymbol{t}_{n+1} & \text{if} \quad \boldsymbol{r}_n - \boldsymbol{t}_{n+1} > 0, \end{cases}$$

so that, using the notation described in (I.6.65),

$$\boldsymbol{w}_{n+1} = [\boldsymbol{w}_n + \boldsymbol{\tau}_n - \boldsymbol{\sigma}_{n+1}]^+, \qquad n=1, 2, \dots . \tag{1.1}$$

Putting

$$\boldsymbol{\rho}_n \overset{\text{def}}{=} \boldsymbol{\tau}_n - \boldsymbol{\sigma}_{n+1}, \qquad n=1, 2, \dots,$$

we see that $\boldsymbol{\rho}_n$, $n=1, 2, \dots$, is a sequence of independent, identically distributed variables with

$$\mathrm{E}\{\boldsymbol{\rho}_n\} = \beta - \alpha < \infty, \qquad n=1, 2, \dots .$$

From (1.1) it now follows that

$$\begin{aligned} \boldsymbol{w}_{n+1} &= \max(0, \boldsymbol{\rho}_n + \max(0, \boldsymbol{\rho}_{n-1} + \boldsymbol{w}_{n-1})) \\ &= \max(0, \boldsymbol{\rho}_n, \boldsymbol{\rho}_n + \boldsymbol{\rho}_{n-1} + \boldsymbol{w}_{n-1}), \end{aligned}$$

and generally for $n=1, 2, \dots$,

$$\boldsymbol{w}_{n+1} = \max(0, \boldsymbol{\rho}_n, \boldsymbol{\rho}_n + \boldsymbol{\rho}_{n-1}, \dots, \boldsymbol{\rho}_n + \dots + \boldsymbol{\rho}_2, \boldsymbol{\rho}_n + \dots + \boldsymbol{\rho}_1 + \boldsymbol{w}_1). \tag{1.2}$$

Defining

$$\boldsymbol{\rho}_0 \overset{\text{def}}{=} 0, \qquad \boldsymbol{s}_n \overset{\text{def}}{=} \sum_{i=0}^{n} \boldsymbol{\rho}_i, \qquad n=0, 1, \dots,$$

and

$$\boldsymbol{\omega}_{n+1} \overset{\text{def}}{=} \max(\boldsymbol{s}_0, \boldsymbol{s}_1, \dots, \boldsymbol{s}_{n-1}, \boldsymbol{s}_n + \boldsymbol{w}_1), \tag{1.3}$$

it is seen that the distribution functions of $\boldsymbol{w}_{n+1}$ and $\boldsymbol{\omega}_{n+1}$ are identical, since the variables $\boldsymbol{\rho}_i$, $i=1, 2, \dots$, are identically distributed. Further, define

$$a \overset{\text{def}}{=} \frac{\mathrm{E}\{\boldsymbol{\tau}_n\}}{\mathrm{E}\{\boldsymbol{\sigma}_n\}} = \frac{\beta}{\alpha}. \tag{1.4}$$

This quantity plays an important role in queueing theory and its value determines the behaviour of the process $\{x_t,\ t\in[0,\infty)\}$ for large values of t.

The distribution function of w_n for $n\to\infty$ is described by Lindley's theorem (cf. LINDLEY [1952]).

THEOREM 1.1. *For $n\to\infty$ the distribution function of w_n has a limit which is independent of w_1;*

$$W(\sigma)\overset{\text{def}}{=}\lim_{n\to\infty}\Pr\{w_n<\sigma\},\qquad -\infty<\sigma<\infty,$$

is a proper probability distribution if $a<1$; $W(\sigma)=0$ for every σ if $a>1$, or if $a=1$ and $\rho_n\neq 0$ with positive probability.

Proof. Suppose for the present

$$w_1\equiv 0,$$

so that

$$\omega_{n+1}=\max_{0\leq k\leq n} s_n,\qquad n=1,2,\ldots .$$

Since

$$\{\omega_{n+1}<\sigma\}\subset\{\omega_n<\sigma\},\qquad n=1,2,\ldots,$$

it follows for every σ that

$$\lim_{n\to\infty}\{\omega_n<\sigma\}=\bigcap_{n=1}^{\infty}\{\omega_n<\sigma\}=\{\sup_{k\geq 0} s_k<\sigma\};$$

so that (cf. LOÈVE [1960]) the following limit exists

$$\lim_{n\to\infty}\Pr\{\omega_n<\sigma\}=\Pr\{\sup_{k\geq 0} s_k<\sigma\}=\Pr\{s_k<\sigma,\ k=0,1,2,\ldots\}.$$

We now apply the strong law of large numbers to the sequence ρ_i, $i=0,1,\ldots$, so that with probability one

$$\lim_{n\to\infty}\frac{s_n}{n}=\mathrm{E}\{\rho_i\}\begin{cases}<0 & \text{if } a<1,\\ >0 & \text{if } a>1.\end{cases}\tag{1.5}$$

Hence, if $a>1$ there exists a positive number N such that with probability one $s_n\geq\frac{1}{2}n\,\mathrm{E}\{\rho_i\}$ for all $n>N$. Taking N sufficiently large we have with probability one $s_n\geq\sigma$ for every σ and hence $W(\sigma)=0$.

If $a<1$ we see from (1.5) that for any $x>0$ and any $\varepsilon>0$ a number N exists, with N independent of x, such that

$$\Pr\{s_n<x,\ n=N+1,N+2,\ldots\}\geq\Pr\{s_n<0,\ n=N+1,N+2,\ldots\}>1-\tfrac{1}{2}\varepsilon.$$

It is always possible to find a number $z>0$ such that for all $x>z$,

$$\Pr\{s_n<x,\ n=0,1,\ldots,N\}>1-\tfrac{1}{2}\varepsilon.$$

Since

$$\Pr\{\sup_{n\geq 0} s_n<x\}>\Pr\{s_n<x,\ n=0,\ldots,N\}+\Pr\{s_n<x,\ n=N+1,N+2,\ldots\}-1$$
$$>1-\varepsilon,$$

it follows by letting $x\to\infty$ that $\lim_{x\to\infty} W(x)=1$, so that since $W(x)$ is evidently non-decreasing and zero for $x<0$ it is seen that $W(x)$ is a proper probability distribution if $a<1$.

If $a=1$ then a theorem of Chung and Fuchs (1952) shows that for any x and any $\varepsilon>0$,

$$\Pr\{|s_n-x|<\varepsilon,\ \text{infinitely often}\}=1,$$

with the restriction that $\rho_i\neq 0$ with positive probability and that if ρ_i has a lattice distribution the relation only holds if x is a lattice point. It implies that $\limsup_{n\to\infty} s_n=\infty$ with probability one, since any value will be exceeded by s_n. Hence from (1.2) or (1.3), $W(\sigma)=0$ for every σ.
To show that $W(x)$ is independent of w_1 suppose $w_1=y>0$.
From

$$\Pr\{\omega_{n+1}<x \mid w_1=y\}=\Pr\{s_k<x,\ k=0,\ldots,n-1;\ s_n<x-y\}$$
$$\leqq\Pr\{s_k<x,\ k=0,\ldots,n-1\},$$

and

$$\Pr\{\omega_{n+1}<x \mid w_1=y\}\geqq\Pr\{s_k<x,\ k=0,\ldots,n-1\}-\Pr\{s_n\geqq x-y\},$$

we have since $\Pr\{s_n\geqq x-y\}\to 0$ for $n\to\infty$ if $a<1$,

$$\liminf_{n\to\infty}\Pr\{\omega_{n+1}<x \mid w_1=y\} \geqq W(x) \geqq \limsup_{n\to\infty}\Pr\{\omega_{n+1}<x \mid w_1=y\}.$$

The proof is complete.

The condition $a<1$ for w_n to have a proper limit distribution for $n\to\infty$ is intuitively clear. If $a<1$ the average amount of work measured in time units which is offered to the server per unit of time is less than one, so that in the long run the server can cope with the work brought in by the arriving customers; if $a>1$ this is not so. For this reason a is often referred to as the *load* of the queueing system.
Because of (1.2) and (I.6.63) Lindley's theorem may be interpreted as a theorem of fluctuation theory (cf. section I.6.6). It is readily seen from

the proof of the above theorem that this theorem is equivalent to theorem I.6.8 if $\mathrm{E}\{|\boldsymbol{x}_n|\}<\infty$.

Next, we consider the virtual waiting time $\boldsymbol{v}_t$ and some related quantities. Let $\boldsymbol{e}_t$ denote the *total idle time* of the server during $[0, t)$, and let $\boldsymbol{f}_t$ be the *total busy time* of the server in $[0, t)$, so that

$$\boldsymbol{e}_t+\boldsymbol{f}_t=t, \qquad t>0.$$

We then have

$$\boldsymbol{f}_t = t - \boldsymbol{e}_t = \int_0^t U_1(\boldsymbol{x}_u)\,\mathrm{d}u = \int_0^t U_1(\boldsymbol{v}_u)\,\mathrm{d}u \leqq t, \tag{1.6}$$

where $U_1(t)$ is the unit step-function defined to be continuous from the left,

$$U_1(x)=0 \quad \text{for } x\leqq 0, \qquad U_1(x)=1 \quad \text{for } x>0.$$

It follows (cf. Doob [1953]) that

$$\mathrm{E}\{\boldsymbol{e}_t\} = t - \int_0^t \mathrm{E}\{U_1(\boldsymbol{x}_u)\}\,\mathrm{d}u = \int_0^t \Pr\{\boldsymbol{x}_u=0\}\,\mathrm{d}u. \tag{1.7}$$

Let us further denote *the amount of work brought in by the customers arriving in* $[0, t)$ by $\boldsymbol{k}_t$.
Clearly we now have

$$\boldsymbol{v}_t=v+\boldsymbol{k}_t-\boldsymbol{f}_t=v+\boldsymbol{k}_t+\boldsymbol{e}_t-t, \qquad t>0, \tag{1.8}$$

where v is the amount of work present just before $t=0$. By eliminating $\boldsymbol{e}_t$ from (1.8) and (1.6) we arrive at an integral equation for $\boldsymbol{v}_t$. The solution of this integral equation is given in the following theorem, which is due to Reich [1958].

Theorem 1.2. *For* $t>0$,

(*i*) $\boldsymbol{e}_t=\max[0, \sup_{0<u<t} (u-\boldsymbol{k}_u-v)]$,

(*ii*) $\boldsymbol{v}_t=\max[v+\boldsymbol{k}_t-t, \sup_{0<u<t} (\boldsymbol{k}_t-t-(\boldsymbol{k}_u-u))]$.

Proof. The second statement follows immediately from (1.8) and the first statement. To prove the latter, define

$$\boldsymbol{z}_t \stackrel{\text{def}}{=} \begin{cases} \sup_{0<u<t} \{u\colon \boldsymbol{v}_u=0\} & \text{if } \boldsymbol{v}_u=0 \text{ for some } u\in[0, t), \\ t & \text{if } \boldsymbol{v}_u>0 \text{ for all } u\in[0, t). \end{cases}$$

Since

$$\boldsymbol{e}_{\boldsymbol{z}_t}=\boldsymbol{e}_t, \qquad \boldsymbol{e}_t \geqq \boldsymbol{e}_u, \qquad u \in (0, t),$$

and from (1.8),

$$\boldsymbol{e}_u = u - \boldsymbol{k}_u - v \qquad \text{if} \quad \boldsymbol{v}_u = 0, \tag{1.9}$$

it follows if $0<\boldsymbol{z}_t<t$ that

$$\boldsymbol{e}_{\boldsymbol{z}_t} = \boldsymbol{e}_t \geqq \sup_{0<u<t} \boldsymbol{e}_u = \sup_{u \in A_t} \boldsymbol{e}_u = \boldsymbol{e}_{\boldsymbol{z}_t},$$

where

$$A_t \stackrel{\text{def}}{=} (0, t) \cap \{u\colon \boldsymbol{v}_u = 0\}.$$

Hence

$$\boldsymbol{e}_t = \begin{cases} 0 & \text{if} \quad \boldsymbol{z}_t = t, \\ \sup\limits_{0<u<t} (u - \boldsymbol{k}_u - v) & \text{if} \quad \boldsymbol{z}_t < t, \end{cases}$$

which proves (*i*). The proof is complete.

If we denote *the overload of the server during* $[0, t)$ by $\boldsymbol{k}_t + v - t$ the second statement of theorem 1.2 may be written as

$$\boldsymbol{v}_t = \text{virtual delay at } t = \sup_{0<u<t} (\text{overload in } [u, t)).$$

Let us denote the *initial busy period* of the server by $\boldsymbol{\theta}_v$ if $\boldsymbol{v}_0 = v > 0$, so that (cf. (1.8)),

$$\boldsymbol{\theta}_v \stackrel{\text{def}}{=} \inf_{u>0} \{u\colon \boldsymbol{v}_u = 0\} = \inf_{u>0} \{u\colon v + \boldsymbol{k}_u - u \leqq 0\}. \tag{1.10}$$

Since

$$\Pr\{\boldsymbol{e}_t \geqq \sigma \mid \boldsymbol{v}_0 = v\} = \Pr\{v + \boldsymbol{k}_u + \sigma \leqq u \text{ for some } u \in [0, t)\},$$

$$\Pr\{\boldsymbol{\theta}_v < t \mid \boldsymbol{v}_0 = v + \sigma\} = \Pr\{v + \boldsymbol{k}_u + \sigma \leqq u \text{ for some } u \in [0, t)\},$$

it follows for $0<\sigma<t$, $v>0$, that

$$\Pr\{\boldsymbol{e}_t \geqq \sigma \mid \boldsymbol{v}_0 = v\} = \Pr\{\boldsymbol{\theta}_v < t \mid \boldsymbol{v}_0 = v + \sigma\}, \tag{1.11}$$

an interesting result due to TAKÁCS [1964[a]].

Define for $v \geqq 0$, $t > 0$,

$$\boldsymbol{p}_t(v) \stackrel{\text{def}}{=} \begin{cases} 1 & \text{if} \quad \boldsymbol{v}_t = 0 \quad \text{and} \quad \boldsymbol{v}_0 = v, \\ 0 & \text{if} \quad \boldsymbol{v}_t > 0 \quad \text{and} \quad \boldsymbol{v}_0 = v, \end{cases} \tag{1.12}$$

so that

$$\Pr\{\boldsymbol{p}_t(v) = 1\} = \Pr\{\boldsymbol{v}_t = 0 \mid \boldsymbol{v}_0 = v\}, \tag{1.13}$$

$$\Pr\{\boldsymbol{p}_t(v) = 0\} = \Pr\{\boldsymbol{v}_t > 0 \mid \boldsymbol{v}_0 = v\}.$$

Since

$$\mathrm{d}\boldsymbol{e}_t = \begin{cases} \mathrm{d}t & \text{if } \boldsymbol{v}_t = 0, \\ 0 & \text{if } \boldsymbol{v}_t > 0, \end{cases}$$

it follows from the identity

$$\exp(-\rho \boldsymbol{e}_t) = 1 - \rho \int_0^t \exp(-\rho \boldsymbol{e}_u)\, \mathrm{d}\boldsymbol{e}_u, \qquad -\infty < \rho < \infty,$$

that

$$\exp(-\rho \boldsymbol{e}_t) = 1 - \rho \int_0^t \exp(-\rho \boldsymbol{e}_u)\, \boldsymbol{p}_u(v)\, \mathrm{d}u.$$

Hence from (1.8) and (1.9),

$$\exp(-\rho \boldsymbol{v}_t) = \exp\{\rho(t-v) - \rho \boldsymbol{k}_t\} - \rho \int_0^t \exp\{-\rho(u-t) - \rho(\boldsymbol{k}_t - \boldsymbol{k}_u)\}\, \boldsymbol{p}_u(v)\, \mathrm{d}u,$$

so that for $v \geqq 0$, $t > 0$, $-\infty < \rho < \infty$,

$$\exp(-\rho \boldsymbol{v}_t) = \exp\{\rho(t-v) - \rho \boldsymbol{k}_t\}$$

$$- \rho U_1(t-v) \int_0^{t-v} \exp\{-\rho(u+v-t) - \rho(\boldsymbol{k}_t - \boldsymbol{k}_{u+v})\}\, \boldsymbol{p}_{u+v}(v)\, \mathrm{d}u. \tag{1.14}$$

This relation is due to Beneš [1963] (cf. also Kingman [1963]). It should be noted that in the proof of theorem 1.2 and in the derivations of the relations (1.11) and (1.14) the independence of interarrival times and of service times has not been used.

From theorem 1.2 and the above relations, especially (1.14), we see that important variables of the single server queue can be expressed as functions of the process $\{\boldsymbol{k}_t,\ t \in [0, \infty)\}$. Starting from the relations described above Beneš [1963] has obtained very interesting results for single server queues without introducing assumptions about the independence of interarrival times and service times. For this case see also Loynes [1962].

By definition of $\boldsymbol{k}_t$ we have

$$\boldsymbol{k}_t = \boldsymbol{\tau}_1 + \ldots + \boldsymbol{\tau}_{\boldsymbol{\nu}_t}, \qquad t > 0,$$

if $\boldsymbol{\nu}_t$ represents the number of customers arriving in $[0, t)$. The asymptotic distribution of $\boldsymbol{k}_t$ for $t \to \infty$ can be obtained from a theorem due to Dobrushin [1955]. Denote the second moments of the distribution functions $A(t)$ and $B(t)$ by α_2 and β_2, respectively, supposing they are finite. According to the central limit theorem $\boldsymbol{\tau}_1 + \ldots + \boldsymbol{\tau}_n$ is asymptotically normally distributed according to $N(n\beta, \{n(\beta_2-\beta^2)\}^{\frac{1}{2}})$ for $n\to\infty$. From renewal theory (cf. section I.6.2) it is known that $\boldsymbol{\nu}_t$ is asymptotically normally distributed according to $N(t/\alpha, \{(t/\alpha)(\alpha_2/\alpha^2-1)\}^{\frac{1}{2}})$. Now Dobrushin's theorem states that

$$\lim_{t\to\infty} \Pr\left\{\frac{\boldsymbol{k}_t - at}{\sqrt{\{a\beta t(\alpha_2/\alpha^2-1+\beta_2/\beta^2-1)\}}} \leqq x\right\} = \frac{1}{\sqrt{2\pi}} \int_{-\infty}^{x} e^{-\frac{1}{2}\xi^2}\, d\xi.$$

From the relation

$$\boldsymbol{f}_t = v + \boldsymbol{k}_t - \boldsymbol{v}_t,$$

it follows that $\boldsymbol{k}_t$ *and* $\boldsymbol{f}_t$ *have the same asymptotic distribution for* $t\to\infty$ *if* $\boldsymbol{v}_t$ *is finite with probability one for* $t\to\infty$, a result due to TAKÁCS [1964a]. In chapter 5 it will be proved that $\boldsymbol{v}_t$ is finite for $t\to\infty$ with probability one if and only if $a<1$.

In the above relations the time parameter is always t. Next we shall take as parameter the number of customers served and discuss some relations analogous to those above.

We take

$$\boldsymbol{x}_0 = k,$$

and denote by the stochastic variable $\boldsymbol{\nu}_n$, $n=1, 2, \ldots$, the number of customers arriving during the nth service time; further

$$\boldsymbol{n}_0 \overset{\text{def}}{=} 0, \qquad \boldsymbol{n}_n \overset{\text{def}}{=} \boldsymbol{\nu}_1 + \ldots + \boldsymbol{\nu}_n, \qquad n=1, 2, \ldots. \tag{1.15}$$

Denote the moment of the nth departure from the queue after $t=0$ by $\boldsymbol{r}'_n$, $n=1, 2, \ldots$; and

$$\boldsymbol{z}_n \overset{\text{def}}{=} \boldsymbol{x}_{\boldsymbol{r}'_n+}, \qquad n=1, 2, \ldots; \qquad \boldsymbol{z}_0 \overset{\text{def}}{=} k, \tag{1.16}$$

so that $\boldsymbol{z}_n$ is the number of customers left behind in the system by the nth departing customer.

Obviously, we have

$$\boldsymbol{z}_n = \boldsymbol{\nu}_n + [\boldsymbol{z}_{n-1} - 1]^+, \qquad n=1, 2, \ldots. \tag{1.17}$$

From this relation it will be shown that for $n=1, 2, \ldots$,

$$\boldsymbol{z}_n = \max(k+\boldsymbol{n}_n-n,\ \boldsymbol{n}_n-\boldsymbol{n}_r-(n-r-1) \text{ for } r=0, 1, \ldots, n-1). \tag{1.18}$$

Define for $n=1, 2, \ldots$,

$$\boldsymbol{\zeta}_n \overset{\text{def}}{=} \begin{cases} \max\limits_{0<i<n} \{i:\ \boldsymbol{z}_i=0\} & \text{if } \ \boldsymbol{z}_i=0 \text{ for a } \ i\in\{1, \ldots, n-1\}, \\ 0 & \text{if } \ \boldsymbol{z}_i\neq 0 \text{ for all } \ i\in\{1, \ldots, n-1\}, \end{cases} \tag{1.19}$$

then from (1.17),

$$z_n = \begin{cases} \boldsymbol{v}_{\zeta_n+1}+\ldots+\boldsymbol{v}_n-(n-1-\xi_n) & \text{if } z_i=0 \text{ for a } i\in\{1,\ldots,n-1\}, \\ \boldsymbol{v}_1+\ldots+\boldsymbol{v}_n+k-n & \text{if } z_i\neq 0 \text{ for all } i\in\{1,\ldots,n-1\}; \end{cases}$$

this proves (1.18).

Let $\boldsymbol{e}_n$ be the number of services among the first n that are preceded by an idle period, and $\boldsymbol{f}_n$ the number of services among the first n services that are not preceded by an idle period; an *idle period* is a time interval during which the server is continuously idle, and which is not a proper subset of such an interval.
Obviously for $n=1, 2, \ldots,$

$$\boldsymbol{e}_n=\max(0, r-\boldsymbol{n}_r-k+1 \text{ for } r=0, 1, \ldots, n-1), \tag{1.20}$$

$$\boldsymbol{f}_n=k+\boldsymbol{n}_n-z_n,$$

$$\boldsymbol{e}_n+\boldsymbol{f}_n=n.$$

Denoting the number of services during the initial busy period by $\boldsymbol{p}_k$ if $\boldsymbol{x}_0=k$ then

$$\boldsymbol{p}_k = \min_{i\geqq 1}\{i\colon k+\boldsymbol{n}_i=i\}, \qquad k\geqq 1. \tag{1.21}$$

Since for $n=1, 2, \ldots,$

$$\Pr\{\boldsymbol{e}_n>h \mid \boldsymbol{x}_0=k\}=\Pr\{k+\boldsymbol{n}_r+h\leqq r \text{ for some } r=0, 1, \ldots, n-1\},$$

and

$$\Pr\{\boldsymbol{p}_k<n \mid \boldsymbol{x}_0=k\}=\Pr\{k+\boldsymbol{n}_r\leqq r \text{ for some } r=0, 1, \ldots, n-1\},$$

it follows for $0\leqq h<n$ that

$$\Pr\{\boldsymbol{e}_n>h \mid \boldsymbol{x}_0=k\}=\Pr\{\boldsymbol{p}_k<n \mid \boldsymbol{x}_0=k+h\}. \tag{1.22}$$

The relations (1.18), ..., (1.22) are due to TAKÁCS [1964[a]]. For these relations the process $\{\boldsymbol{n}_n, n=1, 2, \ldots\}$ has a basic meaning similar to that of the process $\{\boldsymbol{k}_t, t\in[0, \infty)\}$ for the relations with continuous time parameter. Using combinatorial techniques for the investigation of the processes $\{\boldsymbol{k}_t, t\in[0, \infty)\}$ and $\{\boldsymbol{n}_n, n=1, 2, \ldots\}$ Takács obtained important results, see TAKÁCS [1967] and section 6.6.

II.2. THE QUEUEING SYSTEM $M/M/1$

II.2.1. The number of customers in the system

For the queueing process $M/M/1$ the service time distribution $B(t)$ is the negative exponential distribution with parameter β^{-1}. The interarrival time distribution is also negative exponential with parameter α^{-1}. It will always be assumed that $A_1(t) \equiv A(t)$ and $B_1(t) \equiv B(t)$. From the description of the model of the queueing system (cf. section 1.2) and from theorem I.4.3 it follows that the number of customers arriving during $[0, t)$ has a Poisson distribution with parameter t/α and that the arrival process is a Poisson process.

For the stochastic process $\{\boldsymbol{x}_t, t \in [0, \infty)\}$, where $\boldsymbol{x}_t$ denotes the number of customers present in the system at time t, it will be assumed, unless stated otherwise, that

$$\boldsymbol{x}_{0+} = k, \tag{2.1}$$

where k is a non-negative integer. We shall first show that the process $\{\boldsymbol{x}_t, t \in [0, \infty)\}$ is a birth and death process with stationary transition probabilities.

Transitions in the process $\{\boldsymbol{x}_t, t \in [0, \infty)\}$ are caused only by arrivals and departures of customers. Since the service and arrival processes are independent, while the latter is a stationary Poisson process, it follows that the number of transitions during $[t, t+T)$, $T>0$, of the process $\{\boldsymbol{x}_t, t \in [0, \infty)\}$ caused by arrivals is independent of t and of the states of the $\boldsymbol{x}_t$-process during $[0, t)$ for every $t>0$. The number of transitions in the $\boldsymbol{x}_t$-process during $[t, t+T)$ due to departures depends on the number of customers present at time t and on the number of arrivals in $[t, t+T)$. The service times are independent, identically distributed variables while the residual service time at time t is independent of the past service time (cf. lemma 1.2) and the preceding service times. Because of the independence of the service and arrival processes, the transitions in $[t, t+T)$ caused by the departures of customers present at time t are independent of the states of the system during $[0, t)$, and, *a fortiori*, this also holds for the transitions in $[t, t+T)$ caused by the departures of customers who arrive in $[t, t+T)$. Consequently, it follows from section I.3.1 that the process $\{\boldsymbol{x}_t,\ t\in[0, \infty)\}$ is a Markov process with continuous time parameter, with state space the set of non-negative integers and with stationary transition probabilities.

Since the arrival process is a Poisson process it follows that for every $t\geqq 0$ the elementary probability of i arriving customers in $(t, t+\Delta t)$ is given by

$$\frac{(\alpha^{-1}\Delta t)^i}{i!}\,\mathrm{e}^{-\Delta t/\alpha}=\begin{cases}\mathrm{o}(\Delta t) & \text{for } i=2, 3, \dots,\quad \Delta t\to 0,\\ \alpha^{-1}\Delta t+\mathrm{o}(\Delta t) & \text{for } i=1,\\ 1-\alpha^{-1}\Delta t+\mathrm{o}(\Delta t) & \text{for } i=0.\end{cases}$$

Because of lemma 1.2 we have for the elementary conditional probability that the service of a customer will be terminated during $(t, t+\Delta t)$, whenever he is being served at time t,

$$1-\mathrm{e}^{-\Delta t/\beta}=\frac{\Delta t}{\beta}+\mathrm{o}(\Delta t)\qquad \text{for } \Delta t\to 0.$$

Since service times are independent we may write for the elementary conditional probability that i customers leave the system during $(t, t+\Delta t)$ if at least i customers are present at time t,

$$\begin{aligned}&\mathrm{o}(\Delta t) && \text{for } i=2, 3, \dots,\quad \Delta t\to 0,\\ &\frac{\Delta t}{\beta}+\mathrm{o}(\Delta t) && \text{for } i=1,\\ &1-\frac{\Delta t}{\beta}+\mathrm{o}(\Delta t) && \text{for } i=0.\end{aligned}$$

The service and arrival processes are independent and hence the probability that one or more customers arrive during $(t, t+\Delta t)$ and are also completely served during this interval is of order $\mathrm{o}(\Delta t)$ for $\Delta t\to 0$. It follows from the relations for the probabilities discussed above that for any $t\geqq 0$ and $\Delta t\to 0$,

$$\Pr\{\boldsymbol{x}_{t+\Delta t}=j \mid \boldsymbol{x}_t=i\} = \begin{cases} \mathrm{o}(\Delta t) & \text{for } j\neq i,\ j\neq i\pm 1, \quad i=0, 1, \ldots; \\ 1-\dfrac{\Delta t}{\alpha}-\dfrac{\Delta t}{\beta}+\mathrm{o}(\Delta t), & j=i=1, 2, \ldots, \\ 1-\dfrac{\Delta t}{\alpha}+\mathrm{o}(\Delta t), & j=i=0; \\ \dfrac{\Delta t}{\alpha}+\mathrm{o}(\Delta t), \quad j=i+1, & i=0, 1, \ldots; \\ \dfrac{\Delta t}{\beta}+\mathrm{o}(\Delta t), \quad j=i-1, & i=1, 2, \ldots . \end{cases}$$

Defining

$$p_{ij}(t) \stackrel{\text{def}}{=} \Pr\{\boldsymbol{x}_{\tau+t}=j \mid \boldsymbol{x}_\tau=i\}, \qquad \tau\geqq 0, \quad t\geqq 0, \quad i,j=0,1, \ldots, \tag{2.3}$$

then it follows from the relations above that

$$q_{ij} \stackrel{\text{def}}{=} \lim_{t\downarrow 0} \frac{p_{ij}(t)-\delta_{ij}}{t}$$

exists for all $i,j=0, 1, \ldots$. Consequently, the Markov process $\{\boldsymbol{x}_t,\ t\in[0, \infty)\}$ has a standard transition matrix (cf. section I.3.2) and its Q-matrix (q_{ij}) is given by

$$q_{ij} = \begin{cases} 0 & \text{for } j\neq i,\ j\neq i\pm 1,\ i=0, 1, \ldots, \\ \alpha^{-1} & \text{for } j=i+1,\ i=0, 1, \ldots, \\ \beta^{-1} & \text{for } j=i-1,\ i=1, 2, \ldots, \end{cases} \tag{2.4}$$

$$-q_{ii} = q_i = \begin{cases} \alpha^{-1}+\beta^{-1} & \text{for } i=1, 2, \ldots, \\ \alpha^{-1} & \text{for } i=0. \end{cases}$$

Obviously, the q_i, $i=0, 1, \ldots$, are uniformly bounded in i. From section I.4.1 and the Q-matrix given above it is immediately seen that *the process* $\{\boldsymbol{x}_t,\ t\in[0, \infty)\}$ *is a birth and death process with constant birth rate and constant death rate and with state space the set of non-negative integers.* This proves our statement above concerning the process $\{\boldsymbol{x}_t,\ t\in[0, \infty)\}$.

From the Q-matrix of the $\boldsymbol{x}_t$-process and the Q-matrix of the birth and death process treated in section I.4.4 it is easily seen that these processes are identical if we take (cf. (2.1) and (I.4.9)),

$$\lambda=\alpha^{-1}, \qquad \mu=\beta^{-1}, \qquad a=\beta/\alpha, \qquad i=k. \tag{2.5}$$

From the results obtained in section I.4.4 it follows immediately that *the queueing process* $\{x_t,\ t\in[0,\infty)\}$ *is ergodic if* $a<1$, *null recurrent for* $a=1$ *and transient for* $a>1$. Since a is the product of the average service time β and the arrival rate α^{-1}, i.e. the average number of arriving customers per unit of time, the condition $a<1$ for ergodicity means that the average amount of work measured in units of time which is offered per unit of time to the server should be less than one for the process to be ergodic. Intuitively this condition is immediately clear, since for $a>1$ the server is unable to cope with the arriving customers, so that the queue length will not remain bounded. However, that $a=1$ does not lead to an ergodic system is more difficult to understand intuitively.

The number a is often referred to as the *load* offered to the server. In telephone engineering a is called the *traffic* offered, since here α^{-1} is the calling rate and β the average duration of a telephone call. Teletraffiic engineers denote the unit of traffic offered by one Erlang, in honour of the Danish mathematician Erlang.

The transition probabilities $p_{ij}(t)$ of the process $\{x_t,\ t\in[0,\infty)\}$ are now easily found from the results in section I.4.4. Three different expressions for $p_{ij}(t)$ are given but we shall mention only one here. From formula (I.4.33) we have for $k, j=0, 1, \ldots,$

$$p_{kj}(t) = (1-a)\,a^jU(1-a) + a^{\frac{1}{2}(j-k)}\,\mathrm{e}^{-(1+a)t/\beta}\,I_{j-k}\left(2\frac{t}{\beta}\sqrt{a}\right)$$

$$-\,a^{\frac{1}{2}(j-k)}\int_t^\infty \mathrm{e}^{-(1+a)\tau/\beta}\left\{I_{k+j}\left(2\frac{\tau}{\beta}\sqrt{a}\right) - 2a^{\frac{1}{2}}I_{k+j+1}\left(2\frac{\tau}{\beta}\sqrt{a}\right)\right.$$

$$\left. +\,aI_{k+j+2}\left(2\frac{\tau}{\beta}\sqrt{a}\right)\right\}\frac{\mathrm{d}\tau}{\beta},\qquad t\geqq 0. \tag{2.6}$$

Expressions for the generating function with respect to j and for the Laplace transform of $p_{kj}(t)$ with respect to t are given in section I.4.4.

To find an expression for the average number $\mathrm{E}\{x_t \mid x_0=k\}$ of customers present in the system at time t if initially k customers are present we start from the Laplace transform of the generating function of $p_{kj}(t)$ and derive for $k=0, 1, \ldots,$

$$\mathrm{E}\{x_t \mid x_0=k\} = \sum_{j=0}^{\infty} jp_{kj}(t) = k + \frac{t}{\alpha} - \frac{1}{\beta}\int_0^t \{1-p_{k0}(\tau)\}\,\mathrm{d}\tau. \tag{2.7}$$

Noting that t/α represents the average number of arrivals in $[0, t)$ it follows that

$$\frac{1}{\beta}\int_0^t \{1-p_{k0}(\tau)\}\,\mathrm{d}\tau$$

represents the average number of departures during $[0, t)$. In section 2.4 we shall give another derivation for the expression of the average number of departures in $[0, t)$.

If $a>1$ then all states are transient so that $\int_0^t p_{k0}(\tau)\,\mathrm{d}\tau$ has a finite limit for $t\to\infty$ (cf. theorem I.3.4). In fact we have from (I.4.25) and (I.4.26),

$$\int_0^\infty p_{k0}(t)\,\mathrm{d}t = \frac{1}{(a-1)\,a^k} \qquad \text{for} \quad a>1.$$

Hence if $a>1$ then $\mathrm{E}\{\boldsymbol{x}_t \mid \boldsymbol{x}_0=k\}$ increases linearly with t for the larger values of t.

To investigate $\mathrm{E}\{\boldsymbol{x}_t \mid \boldsymbol{x}_0=k\}$ for large values of t we apply (2.6) and obtain

$$\begin{aligned}\mathrm{E}\{\boldsymbol{x}_t \mid \boldsymbol{x}_0=k\} &= k + (a-1)\frac{t}{\beta} + (1-a)\frac{t}{\beta}U(1-a)\\ &\quad + a^{-\frac{1}{2}k}\int_0^t \mathrm{e}^{-(1+a)\sigma/\beta}\, I_k\Big(2\frac{\sigma}{\beta}\sqrt{a}\Big)\frac{\mathrm{d}\sigma}{\beta}\\ &\quad - a^{-\frac{1}{2}k}\int_0^t \frac{\mathrm{d}\tau}{\beta}\int_{\sigma=\tau}^\infty \mathrm{e}^{-(1+a)\sigma/\beta}\Big\{I_k\Big(2\frac{\sigma}{\beta}\sqrt{a}\Big)\\ &\quad - 2a^{\frac{1}{2}}I_{k+1}\Big(2\frac{\sigma}{\beta}\sqrt{a}\Big) + aI_{k+2}\Big(2\frac{\sigma}{\beta}\sqrt{a}\Big)\Big\}\frac{\mathrm{d}\sigma}{\beta}.\end{aligned}$$

Evaluation of the repeated integral yields for $a\neq 1$,

$$\begin{aligned}\mathrm{E}\{\boldsymbol{x}_t \mid \boldsymbol{x}_0=k\} &= k + (a-1)\frac{t}{\beta}U(a-1) + 2^k\frac{\{1+a+|1-a|\}^{-k}}{|1-a|}\\ &\quad - a^{-\frac{1}{2}k}\int_{t/\beta}^\infty \mathrm{e}^{-(1+a)\sigma}\, I_k(2\sigma\sqrt{a})\,\mathrm{d}\sigma -\end{aligned}$$

$$-\frac{2^k\{1-a+|1-a|\}}{|1-a|^3\{1+a+|1-a|\}^{k+2}}\{(1+a+k|1-a|)(1-a+|1-a|)-4a|1-a|\}$$

$$+a^{-\frac{1}{2}k}\int_{t/\beta}^{\infty}(\sigma-t)\,\mathrm{e}^{-(1+a)\sigma}\{I_k(2\sigma\sqrt{a})-2a^{\frac{1}{2}}I_{k+1}(2\sigma\sqrt{a})+aI_{k+2}(2\sigma\sqrt{a})\}\,\mathrm{d}\sigma. \tag{2.8}$$

From this relation we obtain, by applying the same asymptotic relations as in section I.4.4,

$$\mathrm{E}\{\boldsymbol{x}_t\,|\,\boldsymbol{x}_0=k\}=\begin{cases}\dfrac{a}{1-a}+\dfrac{a^{-\frac{1}{2}k}}{2\sqrt{\pi}\sqrt{a}}\,\dfrac{\exp\{-(1-\sqrt{a})^2\,t/\beta\}}{(1-\sqrt{a})^3\,(t/\beta)^{\frac{3}{2}}}\\ \qquad\cdot\left\{k-\dfrac{\sqrt{a}}{1-\sqrt{a}}+\mathrm{O}\left(\dfrac{1}{t}\right)\right\}, \qquad t\to\infty,\quad a<1,\\ k+(a-1)\dfrac{t}{\beta}+\dfrac{1}{(a-1)\,a^k}\\ \qquad+\dfrac{a^{-\frac{1}{2}k}}{2\sqrt{\pi}\sqrt{a}}\,\dfrac{\exp\{-(1-\sqrt{a})^2\,t/\beta\}}{(1-\sqrt{a})^3\,(t/\beta)^{\frac{3}{2}}}\\ \qquad\cdot\left\{k-\dfrac{\sqrt{a}}{1-\sqrt{a}}+\mathrm{O}\left(\dfrac{1}{t}\right)\right\}, \qquad t\to\infty,\quad a>1.\end{cases} \tag{2.9}$$

To obtain the asymptotic expression in the case $a=1$ we start from the Laplace transform of $p_{k0}(t)$ for which we have (cf. (I.4.26) and (2.1)),

$$\int_0^{\infty}\mathrm{e}^{-\rho t/\beta}\,\mathrm{d}_t\left\{\int_0^t p_{k0}(\sigma)\frac{\mathrm{d}\sigma}{\beta}\right\}=\frac{\rho^k}{2^{2k+1}}\left\{-1+2\frac{\sqrt{(1+\frac{1}{4}\rho)}}{\sqrt{\rho}}\right\}^{2k+1}$$

$$=\frac{1}{\sqrt{\rho}}\{1+\tfrac{1}{2}\rho-\sqrt{(\rho+\tfrac{1}{4}\rho^2)}\}^{k+\frac{1}{2}}, \qquad \mathrm{Re}\,\rho>0,\quad a=1.$$

Since this Laplace-Stieltjes transform behaves like $\rho^{-\frac{1}{2}}$ for $\rho\downarrow 0$ and $\int_0^t p_{k0}(\sigma)\,\mathrm{d}\sigma$ is non-decreasing we have (cf. app. 4),

$$\int_0^t p_{k0}(\sigma)\frac{\mathrm{d}\sigma}{\beta}\approx 2\sqrt{\frac{t}{\pi\beta}} \qquad \text{for}\quad t\to\infty,\quad a=1.$$

Using the first expression for the Laplace transform above it is possible to expand the Laplace transform of

$$\int_0^t p_{k0}(\sigma)\frac{d\sigma}{\beta} - 2\sqrt{\frac{t}{\pi\beta}},$$

into a convergent power series of $\sqrt{\rho}$ for $\rho \downarrow 0$. By using a Tauberian theorem (cf. Doetsch [1943] pp. 244, 277) we obtain for $a=1$,

$$\mathrm{E}\{\boldsymbol{x}_t \mid \boldsymbol{x}_0=k\}$$

$$= 2\sqrt{\frac{t}{\pi\beta}} - \tfrac{1}{2} + \left(\frac{\pi t}{\beta}\right)^{-\frac{1}{2}} \left\{\tfrac{1}{2}(k+\tfrac{1}{2})^2 + \mathrm{O}\left(\frac{1}{t}\right)\right\}, \qquad t\to\infty. \tag{2.10}$$

The asymptotic relations obtained above (cf. also (I.4.34)) provide us with a good insight into the behaviour of $\mathrm{E}\{\boldsymbol{x}_t \mid \boldsymbol{x}_0=k\}$ for $t\to\infty$. It is seen that for $a\neq 1$ the correction term decreases mainly exponentially, and hence we shall denote

$$\frac{\beta}{(1-\sqrt{a})^2} \tag{2.11}$$

as the *relaxation time* of the queueing system $M/M/1$. It gives an idea about the speed of approach to the stationary situation if $a<1$, and to the asymptote $(a-1)\,t/\beta$ if $a>1$ (cf. also section III.7.3).

As we have already said the queueing system $M/M/1$ is ergodic for $a<1$, and from (I.4.21) we find for *the stationary distribution*

$$u_j = \lim_{t\to\infty} \Pr\{\boldsymbol{x}_t=j \mid \boldsymbol{x}_0=k\} = \lim_{t\to\infty} \Pr\{\boldsymbol{x}_t=j\} = a^j(1-a), \qquad j=0, 1, \ldots, \tag{2.12}$$

thus

$$\lim_{t\to\infty} \mathrm{E}\{\boldsymbol{x}_t\} = \frac{a}{1-a}, \qquad \lim_{t\to\infty} \mathrm{var}\{\boldsymbol{x}_t\} = \frac{a^2}{(1-a)^2} + \frac{a}{1-a}. \tag{2.13}$$

The stochastic variable $\boldsymbol{x}_{t_n}$, where $\boldsymbol{t}_n$ is the moment of arrival of the nth customer after $t=0$, is the number of customers present in the system just before the arrival of the nth customer, since we have defined the sample functions to be continuous from the left. For the sake of simplicity we shall assume in the sequel that a customer arrives at $t=0$. Define

$$\boldsymbol{y}_n \stackrel{\text{def}}{=} \boldsymbol{x}_{t_n}, \qquad n=1, 2, \ldots; \tag{2.14}$$

$$\boldsymbol{y}_0 \stackrel{\text{def}}{=} \boldsymbol{x}_{0+} - 1, \qquad \boldsymbol{x}_{0+} = k, \qquad k=1, 2, \ldots .$$

Since the arrival and service processes are independent and all service times and interarrival times are independent and negative exponentially distributed it follows from lemmas 1.1 and 1.2 that the process $\{\boldsymbol{y}_n,\ n=0, 1, \dots\}$ is a discrete time parameter Markov chain. It is an imbedded Markov chain of the process $\{\boldsymbol{x}_t,\ t\in[0, \infty)\}$ and it is regenerative and Markovian (cf. section I.6.4); the regeneration points being the moments of arrival. Evidently, this Markov chain has stationary transition probabilities.

To deduce the one step transition matrix of the Markov chain $\{\boldsymbol{y}_n,\ n=0, 1, \dots\}$ note that between two successive arrivals the $\boldsymbol{x}_t$-process behaves like a pure death process with initial state $\boldsymbol{y}_n+1$ at $\boldsymbol{t}_n$, and with constant death rate β^{-1}. Hence, it follows from section I.4.3 for $i=0, 1, \dots,$

$$p_{ij}=\Pr\{\boldsymbol{y}_{n+1}=j\,|\,\boldsymbol{y}_n=i\}=\begin{cases}\displaystyle\int_0^\infty \frac{\{t/\beta\}^{i+1-j}}{(i+1-j)!}\,\mathrm{e}^{-t/\beta}\,\mathrm{e}^{-t/\alpha}\,\frac{\mathrm{d}t}{\alpha}, & j=1,2,\dots,i+1,\\[2ex] \displaystyle\int_0^\infty\left\{\int_0^t \frac{\{\tau/\beta\}^{i}}{i!}\,\mathrm{e}^{-\tau/\beta}\,\frac{\mathrm{d}\tau}{\beta}\right\}\mathrm{e}^{-t/\alpha}\,\frac{\mathrm{d}t}{\alpha}, & j=0,\\[2ex] 0, & j=i+2,i+3,\dots.\end{cases}$$

Therefore, for $i=0, 1, \dots,$

$$p_{ij}=\begin{cases}\dfrac{1}{(1+a)^{i+1}}, & j=0,\\[2ex] \dfrac{a}{(1+a)^{i-j+2}}, & j=1, 2, \dots, i+1,\\[2ex] 0, & j=i+2, i+3, \dots.\end{cases}\tag{2.15}$$

To investigate $p_{ij}^{(n)}$, the n-step transition probability of the Markov chain $\{\boldsymbol{y}_n,\ n=0, 1, \dots\}$, define

$$\pi_i^{(n)}(p)\stackrel{\text{def}}{=}\sum_{j=0}^{\infty} p_{ij}^{(n)}p^j,\qquad |p|\leq 1;\quad n=0, 1, \dots,\tag{2.16}$$

then from

$$\pi_i^{(n+1)}(p)=\sum_{j=0}^{\infty}\sum_{h=0}^{\infty} p_{ih}^{(n)}p_{hj}p^j,$$

and the expression (2.15) for p_{ij} it is found for $|p| \leqq 1$ that

$$\pi_i^{(n+1)}(p) = \frac{1-p}{(1+a)\{1-p(1+a)\}}\, \pi_i^{(n)} \frac{1}{1+a} - \frac{ap^2}{1-(1+a)\,p}\, \pi_i^{(n)}(p),$$

$$n=0, 1, \ldots . \qquad (2.17)$$

Introducing

$$\pi_i(p, r) \overset{\text{def}}{=} \sum_{n=0}^{\infty} \pi_i^{(n)}(p)\, r^n, \qquad |r|<1, \quad |p| \leqq 1, \qquad (2.18)$$

and noting that

$$\pi_i^{(0)}(p)=p^i, \qquad i=0, 1, \ldots,$$

it is found from (2.17) for $|p|\leqq 1$, $|r|<1$ that

$$\pi_i(p, r) = \frac{(1+a)\{1-p(1+a)\}\, p^i+(1-p)\, r\pi_i(1/(1+a), r)}{(1+a)\{1-(1+a)\, p+arp^2\}}. \qquad (2.19)$$

The zeros of the denominator of this expression are given by

$$z_2 \overset{\text{def}}{=} \frac{1}{2ar}\{1+a-\sqrt{[(1+a)^2-4ar]}\}, \quad z_1 \overset{\text{def}}{=} \frac{1}{2ar}\{1+a+\sqrt{[(1+a)^2-4ar]}\},$$

so that $|z_1|>1$, $|z_2|<1$ for $|r|<1$. By its definition $\pi_i(p, r)$ should be an analytic function of p for fixed r ($|r|<1$) inside the unit circle since $0 \leqq p_{ij} \leqq 1$; it follows that $p-z_2$ should be a divisor of the numerator of $\pi_i(p, r)$. This condition determines $\pi_i(1/(1+a), r)$, and from (2.19) for $|r|<1$,

$$\pi_i\left(\frac{1}{1+a}, r\right) = \frac{(1+a)\{(1+a)\, z_2-1\}\, z_2^i}{r(1-z_2)}, \qquad i=0, 1, \ldots .$$

It follows for $|p|\leqq 1$, $|r|<1$ that

$$\pi_i(p, r) = \frac{\{1-p(1+a)\}(1-z_2)\, p^i-\{1-(1+a)\, z_2\}(1-p)\, z_2^i}{ar(p-z_2)(p-z_1)(1-z_2)}$$

$$= \frac{(1-z_2)\, p^i-(1-p)\, z_2^i}{ar(p-z_2)(p-z_1)(1-z_2)} - (1+a)\, \frac{(1-z_2)\, p^{i+1}-(1-p)\, z_2^{i+1}}{ar(p-z_2)(p-z_1)(1-z_2)}.$$

$$(2.20)$$

(Compare this relation with (I.4.27).) By expanding the right-hand side in a power series of p we obtain

$$\sum_{n=0}^{\infty} p_{ij}^{(n)} r^n = (1-\lambda_{ij})(1+a)\frac{z_2^{-(j-i)}}{ar(z_1-z_2)} + (1+a)\lambda_{ij}\frac{(ar)^{j-i} z_2^{j-i}}{ar(z_1-z_2)}$$

$$-(1-\lambda_{i-1,j})\frac{z_2^{-(j-i+1)}}{ar(z_1-z_2)} - \lambda_{i-1,j}\frac{(ar)^{j-i+1} z_2^{j-i+1}}{ar(z_1-z_2)}$$

$$+\frac{(ar)^j z_2^{j+i+1}(1-az_2)^2}{(1-z_2)(z_1-z_2)}, \qquad i=1, 2, \ldots,$$

where

$$\lambda_{ij} = \begin{cases} 0, & j=0, \ldots, i, \\ 1, & j=i+1, i+2, \ldots . \end{cases}$$

To determine $p_{ij}^{(n)}$ from the above relation we derive the power series expansion of z_2 in p. Since (cf. ERDÉLYI [1954]),

$$z_2^j = \frac{j}{(ra)^{\frac{1}{2}j}} \int_0^{\infty} \exp\left\{-\frac{1+a}{\sqrt{r}}t\right\} t^{-1} I_j(2t\sqrt{a})\, dt, \tag{2.21}$$

it follows easily by using the series for the modified Bessel function

$$I_j(t) = \sum_{n=0}^{\infty} \frac{(\frac{1}{2}t)^{2n+j}}{n!(n+j)!}, \tag{2.22}$$

that

$$z_2^j = \sum_{n=0}^{\infty} \frac{j}{2n+j}\binom{2n+j}{n}\frac{a^n}{(1+a)^{2n+j}} r^n, \qquad j=1, 2, \ldots; \quad |r|<1. \tag{2.23}$$

The expression for $p_{ij}^{(n)}$ is rather complicated for $i>0$, therefore, we shall only consider the relation for $p_{0j}^{(n)}$. Since from (2.20),

$$\sum_{n=0}^{\infty} p_{0j}^{(n)} r^n = \frac{(arz_2)^{j+1}}{r(1-z_2)} = \frac{(arz_2)^j - (arz_2)^{j+1}}{1-r}, \qquad |r|<1, \tag{2.24}$$

it follows by expanding $(1-z_2)^{-1}$ in a geometrical series and using (2.23) that

$$p_{0j}^{(n)} = \begin{cases} 0, & n<j, \\ \sum_{k=0}^{\infty} \frac{j+1+k}{2n-j+1+k}\binom{2n-j+1+k}{n-j}\frac{a^{n+1}}{(1+a)^{2n-j+1+k}}, & n \geqq j. \end{cases} \tag{2.25}$$

From (2.24) we derive another representation for $p_{0j}^{(n)}$ which gives an insight into the behaviour for $n \to \infty$. It follows easily that, for $|r|<1$,

$$\frac{(arz_2)^j}{1-r} = \sum_{n=j}^{\infty} r^n \sum_{k=j}^{n} \frac{j}{2k-j} \binom{2k-j}{k} \frac{a^k}{(1+a)^{2k-j}}.$$

Since

$$\sum_{k=j}^{n} \frac{j}{2k-j} \binom{2k-j}{k} \frac{a^k}{(1+a)^{2k-j}}$$

$$= ja^{\frac{1}{2}j} \left\{ \int_0^\infty \exp\left\{-\frac{1+a}{\sqrt{a}}t\right\} t^{-1} I_j(2t)\, \mathrm{d}t \right.$$

$$\left. - \sum_{h=n-j+1}^{\infty} \int_0^\infty \exp\left\{-\frac{1+a}{\sqrt{a}}t\right\} \frac{t^{2h+j-1}}{h!(h+j)!}\, \mathrm{d}t \right\}$$

$$= \left\{\frac{2a}{1+a+|1-a|}\right\}^j - \sum_{k=n+1}^{\infty} \frac{j}{2k-j} \binom{2k-j}{k} \frac{a^k}{(1+a)^{2k-j}}, \qquad n \geqq j,$$

it follows from (2.24) for $j=0, 1, \ldots,$ and $n>j$ that

$$p_{0j}^{(n)} = (1-a)\, a^j U(1-a)$$

$$- \sum_{k=n+1}^{\infty} \left\{ \frac{j}{2k-j} \binom{2k-j}{k} - \frac{(1+a)(j+1)}{2k-j-1} \binom{2k-j-1}{k} \right\} \frac{a^k}{(1+a)^{2k-j}}. \tag{2.26}$$

Since

$$\binom{2k-j}{k} = 2^{2k-j} (\pi k)^{-\frac{1}{2}} \left\{1 + \mathrm{O}\left(\frac{1}{k}\right)\right\} \qquad \text{for} \quad k \to \infty,$$

we have

$$\left\{ \frac{j}{2k-j} \binom{2k-j}{k} - \frac{(1+a)(j+1)}{2k-j-1} \binom{2k-j-1}{k} \right\} \frac{a^k}{(1+a)^{2k-j}}$$

$$= \frac{1}{k\sqrt{\pi k}} \frac{2^{2k-j-2} a^k}{(1+a)^{2k-j}} \{(1-a)\, j-(1+a)\} \left\{1 + \mathrm{O}\left(\frac{1}{k}\right)\right\} \qquad \text{for} \quad k \to \infty,$$

so that from (2.26) for $n \to \infty$,

$$p_{0j}^{(n)} = (1-a)a^j U(1-a) + \{(1-a)j-(1+a)\} \sum_{k=n+1}^{\infty} \frac{2^{2k-j-2}a^k}{k(1+a)^{2k-j}\sqrt{\pi k}} \left\{1 + \mathrm{O}\left(\frac{1}{k}\right)\right\}. \tag{2.27}$$

By using Abel's formula for partial summation or the well-known relation from numerical analysis

$$\sum_{k=N}^{N+h} f(k) \approx \tfrac{1}{2}f(N) + \tfrac{1}{2}f(N+h) + \int_N^{N+h} f(x)\,\mathrm{d}x, \qquad h=1, 2, \ldots,$$

it follows easily from (2.27) that, for $n\to\infty$,

$$p_{0j}^{(n)} = \begin{cases} (1-a)a^j U(1-a) - \dfrac{(1-a)j-(1+a)}{8n\sqrt{\pi n}} \left\{1 + \dfrac{1}{\log(1+a)-\log(2\sqrt{a})}\right\} \\ \qquad \cdot \dfrac{2^{2n-j}a^n}{(1+a)^{2n-j}} \left\{1 + \mathrm{O}\left(\dfrac{1}{n}\right)\right\}, & a\neq 1, \\ \dfrac{1}{\sqrt{\pi n}} \left\{1 + \mathrm{O}\left(\dfrac{1}{n}\right)\right\}, & a=1. \end{cases} \tag{2.28}$$

From the generating function of $p_{ij}^{(n)}$ with respect to n it is easily verified that the asymptotic behaviour of $p_{ij}^{(n)}$, $i\neq 0$, for $n\to\infty$ is of the same type as that of $p_{0j}^{(n)}$, only the factors not depending on n are slightly different. Since

$$\frac{(1+a)^2}{4a} = 1 + \frac{1}{4a}(1-a)^2 \geqq 1,$$

and since the Markov chain $\{\boldsymbol{y}_n,\ n=0, 1, \ldots\}$ is irreducible and aperiodic, so that $p_{0j}^{(n)}$ and $p_{ij}^{(n)}$, $i\neq 0$, have the same limit for $n\to\infty$, it follows that

$$\lim_{n\to 0} p_{ij}^{(n)} = \begin{cases} (1-a)\,a^j, & a<1, \\ 0, & a\geqq 1. \end{cases} \tag{2.29}$$

Obviously

$$\sum_{n=0}^{\infty} p_{0j}^{(n)} \begin{cases} =\infty, & a\leqq 1, \\ <\infty, & a>1, \end{cases}$$

and hence for $a<1$ the chain is positive recurrent, for $a=1$ null recurrent, whereas for $a>1$ it is transient.

It is seen from (2.12) and (2.29) that the processes $\{\boldsymbol{y}_n,\ n=0, 1, \ldots\}$ and $\{\boldsymbol{x}_t,\ t\in[0, \infty)\}$ have the same stationary distribution.

II.2.2. The busy period

In section I.3.3 we introduced the entrance time α_{ij} from state E_i into state E_j; i.e. if at some time s the system is in state E_i then $s+\alpha_{ij}$ is the first time after s at which state E_j is reached from E_i. Hence for the queueing system $M/M/1$ this time $s+\alpha_{ij}$ denotes the first moment after s with j customers in the system when at time s the number of customers present is i. Consider α_{k0}, so if at time s the system contains k customers, $k-1$ waiting and one being served, then at time $s+\alpha_{k0}$ the first idle period after time s starts. Clearly the idle periods are independent, identically distributed variables having a negative exponential distribution with parameter α^{-1}. Since the residual service time has the same distribution as the service time, and the residual service time is independent of the past service time (cf. lemma 1.2) it follows that α_{10} and the busy period have the same distribution.

To determine $F_{k0}(t)$, the distribution function of α_{k0}, we start from (cf. (I.3.32)),

$$p_{k0}(t) = \int_0^t p_{00}(t-\tau)\,\mathrm{d}F_{k0}(\tau), \qquad k=1, 2, \ldots . \tag{2.30}$$

Using the expression for the Laplace transform of $p_{k0}(t)$ (cf. (I.4.26) and (2.1)) it is easily found from (2.30) that

$$\int_0^\infty \mathrm{e}^{-\rho t}\,\mathrm{d}F_{k0}(t) = \frac{\{1+a+\rho\beta-\sqrt{[(1+a+\rho\beta)^2-4a]}\}^k}{(2a)^k},$$

$$\operatorname{Re}\rho > -\frac{1}{\beta}(1-\sqrt{a})^2, \tag{2.31}$$

from which it follows for $k=1, 2, \ldots$, using a well-known relation of Bessel functions (cf. ERDÉLYI [1953] p. 79) that for $k=1, 2, \ldots$,

$$F_{k0}(t) = a^{-\frac{1}{2}k}\int_0^t \mathrm{e}^{-(1+a)\tau/\beta}\,\frac{k}{\tau}\,I_k\left(2\frac{\tau}{\beta}\sqrt{a}\right)\mathrm{d}\tau$$

$$= a^{-\frac{1}{2}(k-1)}\int_0^t \mathrm{e}^{-(1+a)\tau/\beta}\left\{I_{k-1}\left(2\frac{\tau}{\beta}\sqrt{a}\right)-I_{k+1}\left(2\frac{\tau}{\beta}\sqrt{a}\right)\right\}\frac{\mathrm{d}\tau}{\beta}, \qquad t>0,$$

$$F_{k0}(t) = 0, \qquad t\leqq 0. \tag{2.32}$$

Hence, *the distribution function* $D(t)$ *of the busy period* is given by (cf. also (I.4.35)),

$$D(t) = \begin{cases} \int_0^t e^{-(1+a)\tau/\beta} \left\{ I_0\left(2\frac{\tau}{\beta}\sqrt{a}\right) - I_2\left(2\frac{\tau}{\beta}\sqrt{a}\right) \right\} \frac{d\tau}{\beta}, & t>0, \\ 0, & t \leqq 0. \end{cases} \tag{2.33}$$

It is easily verified that

$$D(\infty) = \begin{cases} 1 & \text{if } a \leqq 1, \\ \frac{1}{a} & \text{if } a>1; \end{cases} \tag{2.34}$$

so that $D(t)$ is a true probability distribution if $a \leqq 1$. For $a<1$ the average length of the busy period is given by

$$\beta/(1-a). \tag{2.35}$$

Since (cf. (I.4.26)),

$$\frac{1}{\beta} \int_0^\infty e^{-\rho t/\beta} p_{k0}(t)\, dt = \frac{x_2^{k+1}}{1-x_2}, \qquad \operatorname{Re} \rho>0,$$

it follows from (2.31) and (2.32) that

$$\int_0^t p_{k0}(\tau) \frac{d\tau}{\beta} = \sum_{n=0}^\infty \{D(t)\}^{(k+1+n)*}, \qquad t \geqq 0.$$

Obviously we have

$$F_{k0}(t) = F_{10}(t) * F_{k-1,0}(t) = F_{10}^{k*}(t), \qquad k=1, 2, \dots .$$

This relation shows that the distribution of α_{k0} is the k-fold convolution of the busy period with itself. Its probabilistic meaning is easily explained and provides us with a property of busy periods which is also useful in deriving the distribution of the busy period for the $M/G/1$ queueing system. Suppose that at time s the system contains k customers. Let us remove from the system the $k-1$ waiting customers but new arrivals are handled as usual. If $s+\boldsymbol{s}_1$ is the time at which the server becomes idle for the first time after s then, by lemma 1.2, $\boldsymbol{s}_1$ has the distribution function of the busy period. At time $s+\boldsymbol{s}_1$ we replace one of the $k-1$ removed customers and a new busy period is started, let its duration be $\boldsymbol{s}_2$. At time $s+\boldsymbol{s}_1+\boldsymbol{s}_2$ again one of the removed customers is replaced in the system, and so on. Since service times are independent the variables $\boldsymbol{s}_1, \boldsymbol{s}_2, \dots, \boldsymbol{s}_k$ are also independent and moreover identically distributed with distribution function $F_{10}(t) \equiv D(t)$. Since clearly $\boldsymbol{\alpha}_{k0} = \boldsymbol{s}_1 + \dots + \boldsymbol{s}_k$ it follows immediately that $F_{k0}(t)$ is the k-fold convolution of $D(t)$.

Let $D_n(t)$ denote the probability that the busy period has a duration less than t and that it consists of n service times. Obviously

$$D_1(t) = \int_0^t \mathrm{e}^{-\tau/\alpha}\, \mathrm{e}^{-\tau/\beta} \frac{\mathrm{d}\tau}{\beta}, \qquad t \geqq 0. \tag{2.36}$$

To derive a relation for $D_n(t)$, $n>1$ suppose that the busy period starts at time s. If the busy period consists of more than one service time a second customer has to arrive during the service time of the customer who started the busy period. Say this second customer arrives at time $s+\boldsymbol{s}_0$. Remove him immediately from the system, but the server proceeds to serve the first customer and new arrivals, if any, until he becomes idle, say at time $s+\boldsymbol{s}_0+\boldsymbol{s}_1$. It follows from lemma 1.2 that $\boldsymbol{s}_1$ has the distribution of the busy period. Replacing the removed customer at time $s+\boldsymbol{s}_0+\boldsymbol{s}_1$ a new busy period starts, say of duration $\boldsymbol{s}_2$, so that the busy period started at s has a length $\boldsymbol{s}_0+\boldsymbol{s}_1+\boldsymbol{s}_2$. If $\boldsymbol{s}_1$ consists of $\boldsymbol{i}$ service times and $\boldsymbol{s}_2$ of $\boldsymbol{j}$ service times then $\boldsymbol{i}$ and $\boldsymbol{j}$ are independent variables since the arrival process is a Poisson process. Hence the busy period started at s consists of $\boldsymbol{i}+\boldsymbol{j}$ service times. Since (cf. lemma 1.2) $\boldsymbol{s}_0$, $\boldsymbol{s}_1$ and $\boldsymbol{s}_2$ are independent and

$$\Pr\{\boldsymbol{s}_0 < t\} = \int_0^t \mathrm{e}^{-(1-a)\tau/\beta} \frac{\mathrm{d}\tau}{\alpha}, \qquad t>0,$$

we have

$$D_n(t) = \sum_{j=1}^{n-1} \int_0^t \mathrm{e}^{-\tau/\beta - \tau/\alpha} D_j(t-\tau) * D_{n-j}(t-\tau) \frac{\mathrm{d}\tau}{\alpha}, \qquad n=2, 3, \dots. \tag{2.37}$$

Putting

$$\delta_n(\rho) \stackrel{\text{def}}{=} \int_0^\infty \mathrm{e}^{-\rho t}\, \mathrm{d}D_n(t), \qquad \operatorname{Re} \rho \geqq 0, \tag{2.38}$$

then from (2.36) and (2.37),

$$\delta_1(\rho) = \frac{1}{1+a+\beta\rho},$$

$$\delta_n(\rho) = \frac{a}{1+a+\beta\rho} \sum_{j=1}^{n-1} \delta_j(\rho)\, \delta_{n-j}(\rho), \qquad n=2, 3, \dots .$$

Introducing the generating function

$$\delta(p, \rho) \stackrel{\text{def}}{=} \sum_{n=1}^{\infty} p^n \delta_n(\rho), \qquad |p| \leqq 1, \quad \operatorname{Re} \rho \geqq 0, \tag{2.39}$$

then

$$a\delta^2(p, \rho) - (\beta\rho + a + 1)\, \delta(p, \rho) + p = 0.$$

This equation has two roots. Since the Laplace-Stieltjes transform of a distribution function of a non-negative stochastic variable has for $|\rho| \to \infty$, $\arg \rho = 0$, a limit which is non-negative and less than 1 it follows that we have to take

$$\delta(p, \rho) = \frac{1}{2a} \{1+a+\beta\rho - \sqrt{[(1+a+\beta\rho)^2 - 4ap]}\}, \qquad |p| \leqq 1, \quad \operatorname{Re} \rho \geqq 0. \tag{2.40}$$

Consequently

$$\sum_{n=1}^{\infty} p^n D_n(t) = \sqrt{\frac{p}{a}} \int_0^{t/\beta} \mathrm{e}^{-(1+a)\tau} \frac{I_1(2\tau\sqrt{ap})}{\tau} \, \mathrm{d}\tau$$

$$= \sum_{n=0}^{\infty} \frac{a^n p^{n+1}}{n!(n+1)!} \int_0^{t/\beta} \mathrm{e}^{-(1+a)\tau} \tau^{2n} \, \mathrm{d}\tau, \qquad |p| \leqq 1, \tag{2.41}$$

here we used the series for $I_1(2\tau\sqrt{ap})$ (cf. (2.22)).
It follows for *the joint distribution $D_n(t)$ of the busy period and the number of customers served during this period* that

$$D_n(t) = \frac{a^{n-1}}{(n-1)!n!} \int_0^{t/\beta} \mathrm{e}^{-(1+a)\tau} \tau^{2n-2} \, \mathrm{d}\tau, \qquad n=1, 2, \dots; \quad t \geqq 0. \tag{2.42}$$

For the probability that a busy period consists of exactly n service times we have

$$D_n(\infty) = \frac{1}{2n-1} \binom{2n-1}{n} \frac{a^{n-1}}{(1+a)^{2n-1}}, \qquad n=1, 2, \dots . \tag{2.43}$$

If $a<1$, the average number of customers served in a busy period is finite and it is easily found from (2.40) that it equals $(1-a)^{-1}$.

Next, we shall investigate *the distribution function of the maximum number of customers simultaneously present in the system during a busy period.* Denote by

$$p_{h;ij}(t)$$

the probability that, if at time s the queueing system is in state E_i, it is in E_j at time $s+t$ and has not entered state E_h during $(s, s+t)$. Since the queueing system $M/M/1$ is a birth and death process, so that a transition always leads to a neighbouring state, it follows that if the system starts from E_i and does not enter E_h it does not enter any state E_k with $k>h$ if $h \geqq i$, or with $k<h$ if $h \leqq i$. Hence, for the system $M/M/1$,

$$p_{h;ij}(t) = \begin{cases} \Pr\{\max\limits_{s<\tau<t+s} x_\tau < h, x_{t+s}=j \mid x_s=i\} & \text{if} \quad h>\max(i,j), \\ \Pr\{\min\limits_{s<\tau<t+s} x_\tau > h, x_{t+s}=j \mid x_s=i\} & \text{if} \quad h<\min(i,j). \end{cases} \tag{2.44}$$

Define for $h \neq j$, $h \neq i$,

$$\alpha_{h;ij} \stackrel{\text{def}}{=} \inf_{t>0} \{t:\ x_t=j,\ x_\tau \neq j,\ x_\tau \neq h,\ 0<\tau<t \mid x_0=i\},$$

so that $\alpha_{h;ij}$ is the first entrance time from E_i into E_j without having entered E_h. Define

$$F_{h;ij}(t) \stackrel{\text{def}}{=} \Pr\{\alpha_{h;ij}<t\}, \tag{2.45}$$

so that $F_{h;ij}(t)$ is the probability that the first entrance time from E_i to E_j is less than t and that during this entrance time the number of customers present in the system simultaneously is less (larger) than h if $h>\max(i,j)$ $(h<\min(i,j))$. In particular $F_{h;10}(\infty)$ for $h>1$ is the probability that the maximum number of customers present simultaneously in the system during a busy period is less than h. Obviously we have for $t \geqq 0$,

$$p_{ij}(t) = p_{h;ij}(t) + \int_0^t p_{hj}(t-s)\, \mathrm{d}F_{ih}(s), \qquad h \neq j, \tag{2.46}$$

$$p_{h;ij}(t) = \delta_{ij}\, \mathrm{e}^{-q_i t} + \int_0^t p_{h;jj}(t-s)\, \mathrm{d}F_{h;ij}(s), \qquad h \neq i, \quad h \neq j.$$

By writing

$$\pi_{ij}(\rho) \stackrel{\text{def}}{=} \int_0^\infty \mathrm{e}^{-\rho t}\, p_{ij}(t)\, \mathrm{d}t, \quad \pi_{h;ij}(\rho) \stackrel{\text{def}}{=} \int_0^\infty \mathrm{e}^{-\rho t}\, p_{h;ij}(t)\, \mathrm{d}t, \quad \operatorname{Re} \rho>0, \tag{2.47}$$

$$\varphi_{ij}(\rho) \stackrel{\text{def}}{=} \int_{0-}^\infty \mathrm{e}^{-\rho t}\, \mathrm{d}F_{ij}(t), \quad \varphi_{h;ij}(\rho) \stackrel{\text{def}}{=} \int_{0-}^\infty \mathrm{e}^{-\rho t}\, \mathrm{d}F_{h;ij}(t), \quad \operatorname{Re} \rho \geqq 0,$$

it follows from (2.46) for Re $\rho>0$ that

$$\pi_{h;ij}(\rho) = \pi_{ij}(\rho) - \pi_{hj}(\rho)\,\varphi_{ih}(\rho),$$

$$\varphi_{h;ij}(\rho) = \pi_{h;jj}^{-1}(\rho)\left\{\pi_{h;ij}(\rho) - \frac{\delta_{ij}}{\rho+q_i}\right\}.$$

Since (cf. (I.3.32)),

$$\varphi_{ih}(\rho) = \pi_{hh}^{-1}(\rho)\left\{\pi_{ih}(\rho) - \frac{\delta_{ih}}{\rho+q_i}\right\}, \qquad \text{Re}\,\rho>0,$$

we have for Re $\rho>0$,

$$\pi_{h;ij}(\rho) = \pi_{hh}^{-1}(\rho)\left\{\pi_{ij}(\rho)\,\pi_{hh}(\rho) - \pi_{ih}(\rho)\,\pi_{hj}(\rho) + \frac{\delta_{ih}}{\rho+q_i}\,\pi_{hj}(\rho)\right\}.$$

From the relations above and the Laplace transform of $p_{ij}(t)$ given in (I.4.28) the transforms $\pi_{h;ij}(\rho)$ and $\varphi_{h;ij}(\rho)$ can be calculated. It is found that

$$\frac{1}{\beta}\,\pi_{h;00}(\rho/\beta) = \frac{x_2(1-a^h x_2^{2h})}{1-x_2+(1-ax_2)\,a^h x_2^{2h+1}}, \qquad h=1, 2, \dots, \tag{2.48}$$

$$\frac{1}{\beta}\,\pi_{h;10}(\rho/\beta) = \frac{(1-a^{h-1}x_2^{2h-2})\,x_2^2}{1-x_2+(1-ax_2)\,a^h x_2^{2h+1}}, \qquad h=2, 3, \dots,$$

$$\varphi_{h;10}(\rho/\beta) = \frac{1-a^{h-1}x_2^{2h-2}}{1-a^h x_2^{2h}}\,x_2, \qquad h=2, 3, \dots,$$

where x_2 and x_1 are given by (I.4.25).

The inversion of the Laplace-Stieltjes transform of $F_{h;10}(t)$ may be performed by expanding $(1-a^h x_2^{2h})^{-1}$ into a geometrical series (cf. the derivation of (I.4.32)). Here, however, we are more interested in the probability $F_{h;10}(\infty)$ which can be directly obtained from $\varphi_{h;10}(\rho)$,

$$F_{h;10}(\infty) = \lim_{\rho\downarrow 0}\varphi_{h;10}(\rho). \tag{2.49}$$

It is then found that

$$F_{h;10}(\infty) = \begin{cases} \dfrac{1-a^{h-1}}{1-a^h} & \text{for } a\neq 1, \quad h=2, 3, \dots, \\[2ex] 1-\dfrac{1}{h} & \text{for } a=1, \quad h=2, 3, \dots, \end{cases}$$

where $F_{h;10}(\infty)$ is the probability that the maximum number of customers present simultaneously in the system during a busy period is less than h. In the same way it is found that

$$F_{h;00}(\infty)=\begin{cases}\dfrac{1-a^{h-1}}{1-a^h}, & a\neq 1,\quad h=1,2,\dots,\\[2ex] 1-\dfrac{1}{h} & a=1,\quad h=1,2,\dots;\end{cases}\tag{2.50}$$

$$F_{h;k0}(\infty)=\begin{cases}\dfrac{1-a^{h-k}}{1-a^h}, & a\neq 1,\quad h=k+1,k+2,\dots;\quad k=1,2,\dots,\\[2ex] 1-\dfrac{k}{h}, & a=1,\quad h=k+1,k+2,\dots;\quad k=1,2,\dots,\end{cases}$$

where $F_{h;00}(\infty)$ is the probability that the maximum number of customers present in the system at some moment during the first return time of the empty state is less than h.

It should be noted that $F_{h;k0}(\infty)$, $k=0, 1, \dots$, are proper probability distributions for $a\leqq 1$; however, for $a>1$ they are not. Hence for $a\leqq 1$ the maximum number of customers present in the system during a busy period is finite with probability one; for $a>1$ it is finite with probability a^{-1}, and infinite with probability $1-a^{-1}$. If $a<1$ the distributions in (2.50) have a finite first moment. For further properties of these distributions see sections III.7.4 and III.7.5.

II.2.3. The waiting time

As stated in section 1.2 service will be given in order of arrival. For this type of queueing we shall consider the virtual waiting time $\boldsymbol{v}_t$ at time t and the actual waiting time $\boldsymbol{w}_n$ for the nth arriving customer after $t=0$.

Obviously we have

$$\boldsymbol{v}_t=\begin{cases}0 & \text{if and only if}\quad \boldsymbol{x}_t=0,\\ \boldsymbol{\tau}_0+\boldsymbol{\tau}_1+\dots+\boldsymbol{\tau}_{j-1} & \text{if}\quad \boldsymbol{x}_t=j\geqq 1,\end{cases}$$

where $\boldsymbol{\tau}_0$ is the residual service time of the customer served at time t and $\boldsymbol{\tau}_1, \dots, \boldsymbol{\tau}_{j-1}$ are the service times of the $j-1$ waiting customers at time t. Since all these service times are independent and identically distributed (cf.

lemma 1.2) it follows that

$$\Pr\{\boldsymbol{v}_t < s \mid \boldsymbol{x}_{0+} = k\} = \begin{cases} 0, & s \leqq 0, \\ \sum_{j=0}^{\infty} p_{kj}(t)\{1-\mathrm{e}^{-s/\beta}\}^{j*}, & s > 0. \end{cases} \tag{2.51}$$

Consequently

$$\mathrm{E}\{\boldsymbol{v}_t \mid \boldsymbol{x}_{0+} = k\} = \beta \mathrm{E}\{\boldsymbol{x}_t \mid \boldsymbol{x}_{0+} = k\}; \tag{2.52}$$

the right-hand side of this expression has been discussed in section 2.1 for $t \to \infty$ (cf. (2.9) and (2.10)).

For $a \geqq 1$ it is easily verified by using the asymptotic expression of $p_{kj}(t)$ for $t \to \infty$ (cf. (I.4.34)) that $\Pr\{\boldsymbol{v}_t < s \mid \boldsymbol{x}_{0+} = k\}$ tends to zero for $t \to \infty$ and every finite s.

For $a < 1$ it follows immediately using the limits of $p_{kj}(t)$ for $t \to \infty$ (cf. (I.4.34)) that the distribution of the virtual waiting time has a limit for $t \to \infty$, which is independent of k and given by

$$\lim_{t \to \infty} \Pr\{\boldsymbol{v}_t < s \mid \boldsymbol{x}_{0+} = k\} = \begin{cases} 1 - a\,\mathrm{e}^{-(1-a)s/\beta}, & s > 0, \\ 0, & s \leqq 0, \end{cases} \tag{2.53}$$

and

$$\lim_{t \to \infty} \mathrm{E}\{\boldsymbol{v}_t \mid \boldsymbol{x}_{0+} = k\} = \frac{a\beta}{1-a}, \tag{2.54}$$

$$\lim_{t \to \infty} \operatorname{var}\{\boldsymbol{v}_t \mid \boldsymbol{x}_{0+} = k\} = \frac{(2a - a^2)\,\beta^2}{(1-a)^2}.$$

To find the expression for the distribution function of the actual waiting time $\boldsymbol{w}_n$ of the nth arriving customer after $t=0$ it is noted that $p_{k-1,j}^{(n)}$, $k = 1, 2, \ldots$, is the conditional probability that the nth arriving customer after $t=0$ finds j customers present in the system if $\boldsymbol{x}_{0+} = k$, $\boldsymbol{x}_0 = k-1$, $k = 1, 2, \ldots$. Hence for $k = 1, 2, \ldots$,

$$\Pr\{\boldsymbol{w}_n < t \mid \boldsymbol{x}_{0+} = k\} = \begin{cases} 0, & t \leqq 0, \\ \sum_{j=0}^{\infty} p_{k-1,j}^{(n)}\{1-\mathrm{e}^{-t/\beta}\}^{j*}, & t > 0. \end{cases} \tag{2.55}$$

This expression cannot be used for $k=0$; this is due to our definition of the Markov chain $\{\boldsymbol{y}_n, n = 0, 1, \ldots\}$. It is easily seen, however, that

$$\Pr\{\boldsymbol{w}_n < t \mid \boldsymbol{x}_{0+} = 1\} = \Pr\{\boldsymbol{w}_{n+1} < t \mid \boldsymbol{x}_{0+} = 0\}.$$

From the behaviour of $p_{kj}^{(n)}$ for $n\to\infty$ (cf. (2.29)) it is now easily found that the distribution function above has a limit for $n\to\infty$ which is independent of the initial state. For $k=0, 1, \dots; t>0$,

$$\lim_{n\to\infty} \Pr\{\boldsymbol{w}_n<t \mid \boldsymbol{x}_{0+}=k\} = \begin{cases} 1-a\,\mathrm{e}^{-(1-a)t/\beta}, & a<1, \\ 0, & a\geqq 1. \end{cases} \tag{2.56}$$

It should be noted that the limit distributions of $\boldsymbol{w}_n$ for $n\to\infty$ and of $\boldsymbol{v}_t$ for $t\to\infty$ are identical, and hence for $k=0, 1, \dots$, and $a<1$,

$$\lim_{n\to\infty} \mathrm{E}\{\boldsymbol{w}_n \mid \boldsymbol{x}_{0+}=k\} = \frac{a\beta}{1-a}, \qquad \lim_{n\to\infty} \operatorname{var}\{\boldsymbol{w}_n \mid \boldsymbol{x}_{0+}=k\} = \frac{(2a-a^2)\,\beta^2}{(1-a)^2}.$$

By using the generating function of $p_{kj}^{(n)}$ with respect to j (cf. (2.16)) the Laplace-Stieltjes transform of $\Pr\{\boldsymbol{w}_n<t \mid \boldsymbol{x}_{0+}=k\}$ with respect to t is easily obtained. Moreover, an asymptotic expression for $\Pr\{\boldsymbol{w}_n<t \mid \boldsymbol{x}_{0+}=1\}$ and for $\mathrm{E}\{\boldsymbol{w}_n \mid \boldsymbol{x}_{0+}=1\}$ for $n\to\infty$ can be derived from (2.55) and (2.28).

II.2.4. The departure process

A customer departs from the system when his service is finished. Let $\boldsymbol{h}_t$ denote the number of departures from the system during $[0, t)$ with $\boldsymbol{h}_0 \overset{\text{def}}{=} 0$; so that $\boldsymbol{h}_{t+s}-\boldsymbol{h}_t$ is the number of departures in $[t, t+s)$, $t\geqq 0$, $s>0$. The process $\{\boldsymbol{x}_t, t\in[0, \infty)\}$ is a continuous time parameter Markov chain, and consequently, the number of departures in $[t, t+s)$ depends on $\boldsymbol{x}_t$ but not on $\{\boldsymbol{x}_\tau, 0\leqq\tau<t\}$. Hence, since $\boldsymbol{h}_t$ depends only on $\{\boldsymbol{x}_\tau, 0\leqq\tau<t\}$ it follows from the defining property of a Markov chain that for all $i, j, h\in\{0, 1, \dots\}$,

$$\Pr\{\boldsymbol{h}_{t+s}-\boldsymbol{h}_t=i \mid \boldsymbol{x}_t=j, \boldsymbol{h}_t=h\}=\Pr\{\boldsymbol{h}_{t+s}-\boldsymbol{h}_t=i \mid \boldsymbol{x}_t=j\}, \quad t\geqq 0, \ s>0. \tag{2.57}$$

Moreover, it follows that the process $\{(\boldsymbol{x}_t, \boldsymbol{h}_t), t\in[0, \infty)\}$ is a vector Markov process with state space the set of ordered pairs $\{(i, j), i, j=0, 1, \dots\}$. Since the interarrival times and service times are negative exponentially distributed it follows from lemmas 1.1 and 1.2 that for $\Delta t\to 0$,

$$\Pr\{\boldsymbol{h}_{t+\Delta t}-\boldsymbol{h}_t=j,\ \boldsymbol{x}_{t+\Delta t}-\boldsymbol{x}_t=i\mid \boldsymbol{x}_t=n\}$$

$$=\begin{cases} 1-\dfrac{\Delta t}{\alpha}+\mathrm{o}(\Delta t), & i=0,\ j=0,\quad n=0;\\[2mm] \dfrac{\Delta t}{\alpha}+\mathrm{o}(\Delta t), & i=1,\ j=0,\quad n=0;\\[2mm] \dfrac{\Delta t}{\beta}+\mathrm{o}(\Delta t), & i=-1,\ j=1,\ n=1,2,\ldots;\\[2mm] 1-\dfrac{\Delta t}{\alpha}-\dfrac{\Delta t}{\beta}+\mathrm{o}(\Delta t), & i=0,\ j=0,\quad n=1,2,\ldots;\\[2mm] \dfrac{\Delta t}{\alpha}+\mathrm{o}(\Delta t), & i=1,\ j=0,\quad n=1,2,\ldots;\\[2mm] \mathrm{o}(\Delta t), & \text{otherwise.} \end{cases} \tag{2.58}$$

Consequently, the stochastic process $\{(\boldsymbol{h}_t, \boldsymbol{x}_t),\ t\in[0,\infty)\}$ is a vector birth and death process. For this process we now derive the set of forward differential equations. From (2.57) we have for $t\geqq 0$, $\Delta t\geqq 0$, $i, j\in\{0, 1, \ldots\}$,

$$\Pr\{\boldsymbol{h}_{t+\Delta t}=j,\ \boldsymbol{x}_{t+\Delta t}=i\mid \boldsymbol{x}_{0+}=k,\ \boldsymbol{h}_0=0\}$$

$$=\sum_{n=0}^{\infty}\sum_{m=0}^{j}\Pr\{\boldsymbol{h}_{t+\Delta t}=j,\ \boldsymbol{x}_{t+\Delta t}=i\mid \boldsymbol{h}_t=m,\ \boldsymbol{x}_t=n\}$$

$$\cdot\Pr\{\boldsymbol{h}_t=m,\ \boldsymbol{x}_t=n\mid \boldsymbol{x}_{0+}=k,\ \boldsymbol{h}_0=0\}$$

$$=\sum_{n=0}^{\infty}\sum_{m=0}^{j}\Pr\{\boldsymbol{h}_{t+\Delta t}-\boldsymbol{h}_t=j-m,\ \boldsymbol{x}_{t+\Delta t}-\boldsymbol{x}_t=i-n\mid \boldsymbol{x}_t=n\}$$

$$\cdot\Pr\{\boldsymbol{h}_t=m,\ \boldsymbol{x}_t=n\mid \boldsymbol{x}_{0+}=k,\ \boldsymbol{h}_0=0\}. \tag{2.59}$$

Defining

$$H_{ij}(t)\overset{\text{def}}{=}\Pr\{\boldsymbol{x}_t=i,\boldsymbol{h}_t=j\mid \boldsymbol{x}_{0+}=k,\boldsymbol{h}_0=0\},\qquad t\geqq 0,\quad i,j,k\in\{0,1,\ldots\}, \tag{2.60}$$

it follows from (2.58), …, (2.60) for $\Delta t\to 0$ that

$$H_{ij}(t+\Delta t)=\left\{1-\frac{\Delta t}{\beta}-\frac{\Delta t}{\alpha}\right\}H_{ij}(t)+\frac{\Delta t}{\alpha}H_{i-1,j}(t)$$

$$+\frac{\Delta t}{\beta}H_{i+1,j-1}(t)+\mathrm{o}(\Delta t),\qquad i,j\in\{1,2,\ldots\},\quad t\geqq 0. \tag{2.61}$$

This relation shows that $H_{ij}(t)$ is continuous from the right for $t \geqq 0$, and that it has a right-hand derivative. Replacing $t+\Delta t$ by s and t by $s-\Delta s$ in the above relations it is seen that $H_{ij}(t)$ is continuous from the left and has a left-hand derivative for $t>0$; moreover, it is seen that $H_{ij}(t)$ is continuous for $t>0$ and that its right-hand and left-hand derivatives are equal. Hence, it follows for $t>0$,

$$\beta \frac{\mathrm{d}}{\mathrm{d}t} H_{ij}(t) = -(1+a) H_{ij}(t) + aH_{i-1,j}(t) + H_{i+1,j-1}(t), \qquad i, j \in \{1, 2, \ldots\}. \tag{2.62}$$

Similarly, we obtain

$$\beta \frac{\mathrm{d}}{\mathrm{d}t} H_{0j}(t) = -aH_{0j}(t) + H_{1,j-1}(t), \qquad j=1, 2, \ldots, \tag{2.63}$$

$$\beta \frac{\mathrm{d}}{\mathrm{d}t} H_{i0}(t) = -(1+a) H_{i0}(t) + aH_{i-1,0}(t), \qquad i=1, 2, \ldots,$$

$$\beta \frac{\mathrm{d}}{\mathrm{d}t} H_{00}(t) = -aH_{00}(t).$$

The boundary conditions for this set of forward differential equations read

$$H_{ij}(0+) = \begin{cases} 1, & i=k, \quad j=0, \\ 0, & \text{otherwise.} \end{cases} \tag{2.64}$$

It is not difficult to see that the $H_{ij}(t)$, $i, j \in \{0, 1, \ldots\}$, can be calculated recursively from the set of differential equations and boundary conditions. Hence the functions $H_{ij}(t)$ are determined uniquely.

Defining for $|p| \leqq 1$, $|q| \leqq 1$, $\operatorname{Re} \rho > 0$,

$$h(\rho, p, q) \stackrel{\text{def}}{=} \sum_{i=0}^{\infty} \sum_{j=0}^{\infty} p^i q^j \int_0^{\infty} \mathrm{e}^{-\rho\tau} H_{ij}(\tau\beta) \, \mathrm{d}\tau, \tag{2.65}$$

$$h_{ij}(\rho) \stackrel{\text{def}}{=} \int_0^{\infty} \mathrm{e}^{-\rho\tau} H_{ij}(\tau\beta) \, \mathrm{d}\tau,$$

it is found from (2.62), ..., (2.64),

$$h(\rho, p, q) = \frac{(q-p) \sum_{j=0}^{\infty} h_{0j}(\rho) \, q^j - p^{k+1}}{ap^2 - (1+a+\rho) p + q}, \qquad |p| \leqq 1, \; |q| \leqq 1, \; \operatorname{Re} \rho > 0. \tag{2.66}$$

Denoting the zeros of the denominator of the expression above by $x_1(q)$, $x_2(q)$, so that

$$x_1(q) \stackrel{\text{def}}{=} \frac{1+a+\rho+\sqrt{\{(1+a+\rho)^2-4aq\}}}{2a}, \tag{2.67}$$

$$x_2(q) \stackrel{\text{def}}{=} \frac{1+a+\rho-\sqrt{\{(1+a+\rho)^2-4aq\}}}{2a},$$

then it is easily proved by Rouché's theorem (cf. app. 6) that for Re $\rho>0$, $|q|\leqq 1$,

$$|x_1(q)|>1, \qquad |x_2(q)|<1.$$

Since $h(\rho, p, q)$ should be an analytic function of p inside the unit circle for fixed ρ, q with Re $\rho>0$, $|q|\leqq 1$, it follows that x_2 should be also a zero of the numerator of the expression for $h(\rho, p, q)$.
Hence for $|q|\leqq 1$, Re $\rho>0$,

$$\sum_{j=0}^{\infty} h_{0j}(\rho)\, q^j = \frac{x_2^{k+1}(q)}{q-x_2(q)}, \tag{2.68}$$

$$h(\rho, p, q) = \frac{(q-p)\, x_2^{k+1}(q)-(q-x_2(q))\, p^{k+1}}{(q-x_2(q))\{ap^2-(1+a+\rho)\, p+q\}}.$$

To determine $H_{ij}(t)$ from the expression for $h(\rho, p, q)$ one should first develop the right-hand side above into a power series of p. This series contains terms of the type $x_2^n(q)$. Since (cf. (2.22)) for $|q|\leqq 1$, Re $\rho>0$,

$$\begin{aligned} x_2^n(q) &= \int_0^{\infty} \mathrm{e}^{-(1+a+\rho)\tau} \frac{nq^n}{\tau(aq)^{\frac{1}{2}n}} I_n(2\tau\sqrt{aq})\, \mathrm{d}\tau \\ &= \sum_{m=0}^{\infty} q^{n+m} \int_0^{\infty} \mathrm{e}^{-(1+a+\rho)\tau} \frac{na^m}{m!(m+n)!} \tau^{2m+n-1}\, \mathrm{d}\tau, \qquad n=1, 2, \ldots, \end{aligned} \tag{2.69}$$

it is now easy to determine $H_{ij}(t)$ and to show that the set of functions $H_{ij}(t)$ thus found is the solution of the forward differential equations and boundary conditions. For instance it follows from (2.68) and

$$\frac{x_2^{k+1}(q)}{q-x_2(q)} = \sum_{n=0}^{\infty} \frac{x_2^{k+1+n}(q)}{q^{n+1}},$$

for ρ sufficiently large, so that $|x_2(q)|<|q|$, that

$$\Pr\{\boldsymbol{h}_t=j,\ \boldsymbol{x}_t=0\,|\,\boldsymbol{x}_{0+}=k\}$$

$$=\begin{cases} H_{0j}(t)=0, & j=0,1,\dots,k-1;\quad k=1,2,\dots,\\ \displaystyle\sum_{n=0}^{\infty}\frac{(k+1+n)\,a^{j-k}}{(j-k)!(j+n+1)!}\left(\frac{t}{\beta}\right)^{2j-k+n}\mathrm{e}^{-(1+a)t/\beta}, & \\ & j=k,k+1,\dots;\quad k=0,1,\dots\,. \end{cases}\tag{2.70}$$

Taking $p=1$ then

$$h(\rho,1,q)=\sum_{j=0}^{\infty}q^j\int_0^{\infty}\mathrm{e}^{-\rho\tau}\Pr\{\boldsymbol{h}_{\tau\beta}=j\,|\,\boldsymbol{x}_0=k\}\,\mathrm{d}\tau$$

$$=\frac{(q-1)\,x_2^{k+1}(q)-\{q-x_2(q)\}}{\{q-x_2(q)\}(q-1-\rho)},\qquad \operatorname{Re}\rho>0,\quad |q|<1.\tag{2.71}$$

From which it follows for $\operatorname{Re}\rho>0;\ k=0,1,\dots,$ that

$$\int_0^{\infty}\mathrm{e}^{-\rho\tau}\,\mathrm{E}\{\boldsymbol{h}_{\tau\beta}\,|\,\boldsymbol{x}_{0+}=k\}\,\mathrm{d}\tau=\frac{1}{\rho^2}-\frac{1}{\rho}\frac{x_2^{k+1}}{1-x_2},\tag{2.72}$$

with $x_2\equiv x_2(1)$. From the Laplace transform $\pi_0(\rho)$ of $p_{k0}(t)$ (cf. section I.4.4) it now follows that, if $\boldsymbol{x}_0=k$, the average number of departures during $[0,t)$ is

$$\mathrm{E}\{\boldsymbol{h}_t\,|\,\boldsymbol{x}_{0+}=k\}=\mathrm{E}\{\boldsymbol{h}_{t+s}-\boldsymbol{h}_s\,|\,\boldsymbol{x}_s=k\}=\frac{1}{\beta}\int_0^t\{1-p_{k0}(\tau)\}\,\mathrm{d}\tau,\qquad t\geqq 0,\quad s\geqq 0,$$

a result which has also been obtained in section 2.1 (cf. (2.7)).
We now prove Burke's theorem (cf. BURKE [1956]).

THEOREM 2.1. *If the queueing process $M/M/1$ is stationary then the departure process $\{\boldsymbol{h}_t,\ t\in[0,\infty)\}$ is a homogeneous Poisson process with parameter α^{-1}.*

Proof. The process $\{\boldsymbol{x}_t,\ t\in[0,\infty)\}$ is stationary if $a<1$ and

$$\Pr\{\boldsymbol{x}_{0+}=k\}=(1-a)\,a^k,\qquad k=0,1,\dots,$$

since $\{(1-a)\,a^k,\ k=0,1,\dots\}$ is the stationary distribution of this Markov process (cf. section I.3.1, theorem I.3.5 and (2.12)). Hence, from (2.71) for $|q|\leqq 1$, Re $\rho>0$,

$$\sum_{j=0}^{\infty} q^j \int_0^{\infty} \mathrm{e}^{-\rho\tau} \sum_{k=0}^{\infty} \Pr\{\boldsymbol{h}_{\tau\beta}=j \mid \boldsymbol{x}_{0+}=k\} \Pr\{\boldsymbol{x}_{0+}=k\}\, \mathrm{d}\tau$$
$$= \sum_{k=0}^{\infty} h(\rho, 1, q)(1-a)\, a^k = \frac{1}{\rho+a(1-q)}.$$

Consequently, for $|q|\leqq 1$,

$$\sum_{j=0}^{\infty} q^j \Pr\{\boldsymbol{h}_t=j \mid \boldsymbol{h}_0=0\} = \mathrm{e}^{-a(1-q)t/\beta},$$

i.e.

$$\Pr\{\boldsymbol{h}_t=j \mid \boldsymbol{h}_0=0\} = \frac{(t/\alpha)^j}{j!}\mathrm{e}^{-t/\alpha}, \qquad j=0, 1, \ldots; \quad t\geqq 0. \tag{2.73}$$

From (2.57) and (2.73) it follows that if the $\boldsymbol{x}_t$-process is stationary then for $t>0, s>0$, arbitrary integer $N>0$, for all $t_1, \ldots, t_N$ with $0\leqq t_1<t_2<\ldots<t_N=s$, and for all $j, h_n \in \{0, 1, \ldots\}$,

$$\Pr\{\boldsymbol{h}_{t+s}-\boldsymbol{h}_s=j \mid \boldsymbol{h}_{t_n}=h_n,\ n=1, \ldots, N\}$$
$$= \sum_{i=0}^{\infty} \Pr\{\boldsymbol{h}_{t+s}-\boldsymbol{h}_s=j \mid \boldsymbol{x}_s=i,\ \boldsymbol{h}_{t_n}=h_n,\ n=1, \ldots, N\} \Pr\{\boldsymbol{x}_s=i\}$$
$$= \sum_{i=0}^{\infty} \Pr\{\boldsymbol{h}_{t+s}-\boldsymbol{h}_s=j \mid \boldsymbol{x}_s=i,\ \boldsymbol{h}_s=h_N\} \Pr\{\boldsymbol{x}_s=i\}$$
$$= \sum_{i=0}^{\infty} \Pr\{\boldsymbol{h}_{t+s}-\boldsymbol{h}_s=j \mid \boldsymbol{x}_s=i\} \Pr\{\boldsymbol{x}_s=i\}$$
$$= \sum_{i=0}^{\infty} \Pr\{\boldsymbol{h}_t=j \mid \boldsymbol{h}_0=0,\ \boldsymbol{x}_{0+}=i\} \Pr\{\boldsymbol{x}_{0+}=i\} = \frac{(t/\alpha)^j}{j!}\mathrm{e}^{-t/\alpha}.$$

From the above relation (cf. section I.3.1) it follows that the $\boldsymbol{h}_t$-process is a Markov process with stationary transition probabilities if the $\boldsymbol{x}_t$-process is stationary. Since

$$\Pr\{\boldsymbol{h}_{t+\Delta t}-\boldsymbol{h}_t=i \mid \boldsymbol{h}_t=j\} = \frac{(\Delta t/\alpha)^i}{i!}\mathrm{e}^{-\Delta t/\alpha},$$

it is easily seen that the $\boldsymbol{h}_t$-process is a birth process with constant birth rate and hence (cf. section I.4.2) it is a Poisson process. The proof is complete.

It should be noted that the theorem above implies that, if the queueing process $M/M/1$ is stationary, then the interdeparture times, i.e. the times between successive departures, are independent and identically distributed, their distribution function being negative exponential with parameter α^{-1}. Next, we prove

THEOREM 2.2. *For the queueing process $M/M/1$ with arbitrary initial distribution the number of departures in $[s, s+t)$, $t>0$ converges for $s\to\infty$ to a Poisson distribution with parameter t/α if $a<1$ and parameter t/β if $a\geqq 1$.*

Proof. Since the process $\{\boldsymbol{x}_t,\ t\in[0,\infty)\}$ has stationary transition probabilities we have

$$\begin{aligned}\Pr\{\boldsymbol{h}_{t+s}-\boldsymbol{h}_s=j\mid \boldsymbol{x}_{0+}=n\} &= \sum_{k=0}^{\infty}\Pr\{\boldsymbol{h}_{t+s}-\boldsymbol{h}_s=j,\ \boldsymbol{x}_s=k\mid \boldsymbol{x}_{0+}=n\}\\ &= \sum_{k=0}^{\infty}\Pr\{\boldsymbol{h}_{t+s}-\boldsymbol{h}_s=j\mid \boldsymbol{x}_s=k\}\,p_{nk}(s)\\ &= \sum_{k=0}^{\infty}\Pr\{\boldsymbol{h}_t=j\mid \boldsymbol{x}_{0+}=k\}\,p_{nk}(s),\\ &\qquad t\geqq 0,\quad s\geqq 0,\quad j\in\{0,1,\ldots\}.\end{aligned}$$

By using (I.4.31), noting that $x_2(q)<1$ and that the integrand below is non-negative and bounded, it follows for $|q|\leqq 1$, $\operatorname{Re}\rho>0$, that

$$\begin{aligned}&\sum_{j=0}^{\infty} q^j\int_0^{\infty} \mathrm{e}^{-\rho\tau}\lim_{s\to\infty}\sum_{k=0}^{\infty}\Pr\{\boldsymbol{h}_{\tau\beta}=j\mid \boldsymbol{x}_{0+}=k\}\,p_{nk}(s)\,\mathrm{d}\tau\\ &\qquad=\lim_{s\to\infty}\sum_{k=0}^{\infty}\sum_{j=0}^{\infty} q^j\int_0^{\infty}\mathrm{e}^{-\rho\tau}\Pr\{\boldsymbol{h}_{\tau\beta}=j\mid \boldsymbol{x}_{0+}=k\}\,p_{nk}(s)\,\mathrm{d}\tau\\ &\qquad=\frac{1}{\rho+1-q}+\lim_{s\to\infty}\sum_{k=0}^{\infty}\frac{q-1}{q-1-\rho}\,\frac{x_2^{k+1}(q)}{q-x_2(q)}\,p_{nk}(s)\\ &\qquad=\begin{cases}\dfrac{1}{\rho+a(1-q)} & \text{if } a<1,\\[2ex] \dfrac{1}{\rho+1-q} & \text{if } a\geqq 1.\end{cases}\end{aligned}$$

Therefore, for $j=0, 1, \ldots; t \geqq 0, s \geqq 0$, and every $n=0, 1, \ldots,$

$$\lim_{s\to\infty} \Pr\{\boldsymbol{h}_{t+s}-\boldsymbol{h}_s=j \mid \boldsymbol{x}_0=n\} = \begin{cases} \dfrac{(t/\alpha)^j}{j!}\,\mathrm{e}^{-t/\alpha} & \text{if } a<1, \\ \dfrac{(t/\beta)^j}{j!}\,\mathrm{e}^{-t/\beta} & \text{if } a\geqq 1. \end{cases}$$

The proof is complete.

II.3. THE QUEUEING SYSTEM $G/M/1$

II.3.1. Introduction

In this chapter the queueing system $G/M/1$ will be discussed; i.e. for the model described in section 1.2 the interarrival time distribution $A(t)$ will not be further specified, while the service time distribution is the negative exponential distribution with average service time β. Generally, we shall assume that $A_1(t) \equiv A(t)$, $-\infty < t < \infty$, so that the distribution of $\boldsymbol{t}_1$, the arrival time of the first arriving customer after $t=0$, is given by $A(t)$.

The process $\{\boldsymbol{x}_t,\ t \in [0, \infty)\}$, where $\boldsymbol{x}_t$ is the number of customers in the system at time t, is for the queueing process $G/M/1$ a regenerative process which is Markovian. To show this note that the interarrival times $\boldsymbol{\sigma}_n$, $n=1, 2, \ldots$,

$$\boldsymbol{\sigma}_1 = \boldsymbol{t}_1, \qquad \boldsymbol{\sigma}_n = \boldsymbol{t}_n - \boldsymbol{t}_{n-1}, \qquad n=2, 3, \ldots,$$

are independent and identically distributed variables. They determine the renewal process $\{\boldsymbol{\nu}_t,\ t \in [0, \infty)\}$ with

$$\boldsymbol{\nu}_t \overset{\text{def}}{=} \max\{n\colon \boldsymbol{t}_n < t\}, \qquad t > 0, \quad \boldsymbol{\nu}_0 \overset{\text{def}}{=} 0,$$

so that $\boldsymbol{\nu}_t$ represents the number of arrivals in $(0, t)$. Since the service times are independent and negative exponentially distributed with the same parameter it follows from lemma 1.2 and from the definition of section I.6.4 that *the $\boldsymbol{x}_t$-process is regenerative with respect to the renewal process* $\{\boldsymbol{\nu}_t,\ t \in [0, \infty)\}$ generated by the arrivals of the customers.

Between two successive arrivals or during $(0, \boldsymbol{t}_1)$ no increase of $\boldsymbol{x}_t$ is possible, and $\boldsymbol{x}_t$ will decrease only if one or more customers leave the system. Again from lemma 1.2 it follows that the $\boldsymbol{x}_t$-process or the departure process between two successive arrivals $(\boldsymbol{t}_n, \boldsymbol{t}_{n+1})$ is a continuous time parameter Markov process with initial state $\boldsymbol{x}_{t_n+}=\boldsymbol{x}_{t_n}+1$, $n=1, 2, \ldots$, since $B(t)$ is continuous and $A(0+)=0$. The same argument as in section 2.1 easily leads to the conclusion that this Markov process during $(\boldsymbol{t}_n, \boldsymbol{t}_{n+1})$ is a pure death process with constant death rate β^{-1} (cf. section I.4.3). Consequently, *the regenerative process* $\{\boldsymbol{x}_t, t\in[0, \infty)\}$ *is Markovian* (cf. section I.6.4).

The techniques described in section I.6.4 will be applied here to investigate the queueing process $G/M/1$. Obviously the arrival moments play an important role. Since, for every arrival moment, the future development of the $\boldsymbol{x}_t$-process is independent of its past, these arrival moments are called the *regeneration* points of the $\boldsymbol{x}_t$-process.

For the initial state of the $\boldsymbol{x}_t$-process we shall take

$$\boldsymbol{x}_0=k, \qquad \boldsymbol{x}_{0+}=k+1, \qquad k=0, 1, \ldots, \tag{3.1}$$

so that it is assumed that a customer arrives at $t=0$; this agrees with our assumption $A_1(t)\equiv A(t)$, which is not essential but has the advantage of simplifying notations and formulas.

II.3.2. Description of the process for $t\to\infty$

The imbedded process $\{\boldsymbol{y}_n, n=0, 1, \ldots\}$ with

$$\boldsymbol{y}_n \overset{\text{def}}{=} \boldsymbol{x}_{t_n}, \qquad n=1, 2, \ldots, \qquad \boldsymbol{y}_0 \overset{\text{def}}{=} \boldsymbol{x}_0,$$

is a discrete time parameter Markov chain with stationary transition probabilities (cf. section I.6.4). Clearly, $\boldsymbol{y}_n$ represents the number of customers in the system just before the arrival of the nth customer after $t=0$, $n=1, 2, \ldots$, and $\boldsymbol{y}_0=k$ (cf. (3.1)). Since between two successive arrivals the $\boldsymbol{x}_t$-process behaves as a pure death process with constant death rate β^{-1} it follows from sections I.6.4 and I.4.3 that for $n=0, 1, \ldots; i=0, 1, \ldots,$

$$p_{ij} \overset{\text{def}}{=} \Pr\{\boldsymbol{y}_{n+1}=j \mid \boldsymbol{y}_n=i\} = \begin{cases} \displaystyle\int_0^\infty \left\{\int_0^t \frac{(\sigma/\beta)^i}{i!} \mathrm{e}^{-\sigma/\beta} \frac{\mathrm{d}\sigma}{\beta}\right\} \mathrm{d}A(t), & j=0, \\ \displaystyle\int_0^\infty \frac{(t/\beta)^{i+1-j}}{(i+1-j)!} \mathrm{e}^{-t/\beta}\, \mathrm{d}A(t), & j=1,2,\ldots,i+1, \\ 0, & j=i+2,i+3,\ldots . \end{cases} \tag{3.2}$$

From this expression for the one-step transition matrix of the Markov chain it follows immediately that any state can be reached from any other state with positive probability, and since $p_{ii} \neq 0$ for all $i=0, 1, \ldots$, this Markov chain is irreducible and aperiodic.

To investigate the types of the states of this Markov chain we first study the function (cf. section 1.2),

$$\alpha\{(1-p)/\beta\} = \int_0^\infty \exp\{-(1-p)\, t/\beta\}\, \mathrm{d}A(t), \qquad \mathrm{Re}\, p \leqq 1.$$

Denote by

$$\left\{\frac{1}{\alpha} \int_0^t \{1-A(\tau)\}\, \mathrm{d}\tau\right\}^{n*}, \qquad n=0, 1, \ldots,$$

the n-fold convolution of the distribution function

$$\frac{1}{\alpha}\, U_1(t) \int_0^t \{1-A(\tau)\}\, \mathrm{d}\tau$$

with itself; for $n=0$ it represents the unit step-function $U_1(t)$. It follows by using elementary properties of the Laplace-Stieltjes transform that with $a=\beta/\alpha$ and $|p|<1$,

$$\left[\frac{1-\alpha\{(1-p)/\beta\}}{1-p}\right]^n = a^{-n} \int_{0-}^\infty \exp\{-(1-p)\, t/\beta\}\, \mathrm{d}_t \left\{\frac{1}{\alpha} \int_0^t \{1-A(\tau)\}\, \mathrm{d}\tau\right\}^{n*}$$

$$= \sum_{m=0}^\infty p^m \int_{0-}^\infty \mathrm{e}^{-t/\beta}\, \frac{(t/\beta)^m}{m!}\, a^{-n}\, \mathrm{d}_t \left\{\frac{1}{\alpha} \int_0^t \{1-A(\tau)\}\, \mathrm{d}\tau\right\}^{n*},$$

$$n=0, 1, \ldots. \qquad (3.3)$$

Let the series $\{b_n,\ n=0, 1, \ldots\}$ be defined by

$$\sum_{n=0}^\infty b_n p^n \overset{\mathrm{def}}{=} \frac{1-\alpha\{(1-p)/\beta\}}{1-p}, \qquad |p|<1, \qquad (3.4)$$

and denote by $\{b_n^{h*},\ n=0, 1, \ldots\}$ the series which is the h-fold convolution of the series $\{b_n, n=0, 1, \ldots\}$ with itself, $h=0, 1, \ldots$, where

$$b_0^{0*} \stackrel{\text{def}}{=} 1, \qquad b_n^{0*} \stackrel{\text{def}}{=} 0, \qquad n=1, 2, \ldots,$$

then from (3.3) and (3.4) for $n=0, 1, \ldots; h=0, 1, \ldots$,

$$b_n^{h*} = a^{-h} \int_{0-}^{\infty} \mathrm{e}^{-t/\beta} \frac{(t/\beta)^n}{n!} \mathrm{d}_t \left\{ \int_0^t \{1-A(\tau)\} \frac{\mathrm{d}\tau}{\alpha} \right\}^{h*}. \tag{3.5}$$

All b_n^{h*} are non-negative and it follows from (3.3) that

$$\sum_{n=0}^{\infty} b_n^{h*} = a^{-h}, \qquad h=0, 1, \ldots, \tag{3.6}$$

$$\lim_{n\to\infty} \sum_{i=0}^{n} \sum_{h=0}^{\infty} b_i^{h*} \begin{cases} < \infty & \text{if } \; a>1, \\ = \infty & \text{if } \; a \leqq 1. \end{cases} \tag{3.7}$$

In this chapter we often use the zero $\lambda(0, 1)$ of the function

$$p - \alpha\{(1-p)/\beta\}, \tag{3.8}$$

which is smallest in absolute value.

The properties of this zero are described by Takács' lemma (cf. app. 6): $\lambda(0, 1)=1$ if $a \geqq 1$, whereas $0<\lambda(0, 1)<1$ for $a<1$. It follows for $0<p<\lambda(0, 1)$,

$$1-p > 1-\alpha\{(1-p)/\beta\},$$

so that

$$\begin{aligned} \frac{1}{\alpha\{(1-p)/\beta\}-p} &= \frac{1}{1-p} \, \frac{1}{1-(1-p)^{-1}\{1-\alpha\{(1-p)/\beta\}\}} \\ &= \sum_{n=0}^{\infty} p^n \sum_{i=0}^{n} \sum_{h=0}^{\infty} b_i^{h*} \\ &= \sum_{m=0}^{\infty} p^m \sum_{n=0}^{\infty} \int_0^{\infty} \mathrm{e}^{-t/\beta} \frac{(t/\beta)^m}{m!} a^{-n} \left\{ \frac{1}{\alpha} \int_0^t \{1-A(\sigma)\} \, \mathrm{d}\sigma \right\}^{n*} \frac{\mathrm{d}t}{\beta}, \end{aligned} \tag{3.9}$$

for $|p|<\lambda(0, 1)$.

Consider the set of equations

$$z_i = \sum_{j=0}^{\infty} p_{ij} z_j, \qquad i=1, 2, \ldots . \tag{3.10}$$

Defining

$$Z(p) = \sum_{i=1}^{\infty} z_i p^i, \qquad |p| < \lambda(0, 1), \tag{3.11}$$

it follows from (3.2) and (3.10), if $z_0 = 0$, that

$$Z(p) = \frac{p}{\alpha\{(1-p)/\beta\} - p} z_1 \alpha\{1/\beta\}$$

$$= z_1 \alpha\{1/\beta\} \sum_{n=1}^{\infty} p^n \sum_{i=0}^{n-1} \sum_{h=0}^{\infty} b_i^{h*}, \qquad |p| < \lambda(0, 1). \tag{3.12}$$

Applying Foster's criteria (*iv*) and (*v*) (see section I.2.4) it is seen from (3.7) and (3.12) that the Markov chain $\{y_n, n=0, 1, \ldots\}$ is transient if $a > 1$, and recurrent if $a \leqq 1$.

If $a > 1$ it is easily seen that the maximal solution of all solutions of (3.10) bounded by one is given by (3.12) with z_1 satisfying

$$z_1 \alpha\{1/\beta\} \sum_{i=0}^{\infty} \sum_{h=0}^{\infty} b_i^{h*} = 1,$$

so that (cf. (3.6) and (3.7)) for $a > 1$,

$$z_1 \alpha\{1/\beta\} = 1 - a^{-1}.$$

Consequently, from theorem I.2.6,

$$z_j = \{1 - a^{-1}\} \sum_{i=0}^{j-1} \sum_{h=0}^{\infty} b_i^{h*}, \qquad j = 1, 2, \ldots; \quad a > 1, \tag{3.13}$$

is the probability that, whenever an arriving customer meets j customers in the system, the system will never return to its empty state.

Next, we consider the set of equations

$$z_j = \sum_{i=0}^{\infty} z_i p_{ij}, \qquad j = 0, 1, \ldots . \tag{3.14}$$

With the expression (3.2) for p_{ij} it is easily verified by substitution that $z_i = \lambda^i$, $i = 0, 1, \ldots$, is a solution of the set of equations (3.14) if λ satisfies

$$\lambda = \alpha\{(1-\lambda)/\beta\}.$$

Consequently, by applying Takács' lemma (cf. app. 6) it follows for $a < 1$, that $\lambda(0, 1)<1$, so that $z_i=\lambda^i(0, 1)$, $i=0, 1, \ldots$, is a solution of (3.14) satisfying Foster's first criterion. Hence the Markov chain $\{y_n, n=0, 1, \ldots\}$ is ergodic if $a<1$. Conversely, if λ is a non-negative zero of $z-\alpha\{(1-z)/\beta\}$ then $z_i=\lambda^i$, $i=0, 1, \ldots$, is a non-negative solution of the inequalities

$$z_j \geqq \sum_{i=0}^{\infty} z_i p_{ij}, \qquad i=0, 1, \ldots,$$

so that by Foster's first criterion

$$\sum_{i=0}^{\infty} |z_i| < \infty, \quad \text{i.e.} \quad 0<\lambda<1.$$

Therefore if the Markov chain $\{y_n, n=0, 1, \ldots\}$ is positive recurrent then $a<1$ as is seen from Takács' lemma.

The results obtained above show that *the imbedded Markov chain* $\{y_n, n=0, 1, \ldots\}$ *is ergodic if* $a<1$, *null recurrent if* $a=1$ *and transient if* $a>1$.

From theorem I.2.5 it now follows immediately that if $a<1$ then for every $k=0, 1, \ldots$,

$$u_i = \lim_{n\to\infty} \Pr\{y_n=i \mid y_0=k\} = \lim_{n\to\infty} p_{ki}^{(n)} = \{1-\lambda_0\}\, \lambda_0^i, \qquad i=0, 1, \ldots, \tag{3.15}$$

where

$$\lambda_0 \stackrel{\text{def}}{=} \lambda(0, 1), \tag{3.16}$$

is the root inside the unit circle of

$$z - \alpha\{(1-z)/\beta\} = 0, \qquad a<1.$$

Another important result can be obtained by applying theorem I.6.6 to the regenerative process $\{x_t, t\in[0, \infty)\}$.

From this theorem it follows that

$$\lim_{t\to\infty} \Pr\{x_t=j \mid y_0=k\} = 0, \qquad j=0, 1, \ldots, \qquad \text{if} \quad a\geqq 1, \tag{3.17}$$

and for $a<1$, if $A(t)$ is not a lattice distribution, that

$$\lim_{t\to\infty} \Pr\{x_t=j \mid y_0=k\}$$

$$= \sum_{i=0}^{\infty} \frac{(1-\lambda_0)\, \lambda_0^i}{\alpha} \int_0^{\infty} p_{ij}(\sigma)\{1-A(\sigma)\}\, \mathrm{d}\sigma, \qquad j=0, 1, \ldots, \tag{3.18}$$

whereas for $a<1$ and $A(t)$ a lattice distribution with period τ, that

$$\lim_{m\to\infty} \Pr\{x_{m\tau+\sigma}=j \mid y_0=k\} = \sum_{i=0}^{\infty} \frac{(1-\lambda_0)\,\lambda_0^i}{\alpha}\,\tau \sum_{n=0}^{\infty} p_{ij}(n\tau+\sigma)\{1-A(n\tau+)\}, \tag{3.19}$$

for $j=0, 1, \dots;\ 0\leqq\sigma<\tau$.

Since for $i=0, 1, \dots$ (cf. sections I.4.3 and I.6.4)

$$p_{ij}(\sigma) = \Pr\{x_t=j \mid x_u=i,\ t_n=u,\ t_{n+1}=v\}, \qquad 0\leqq u\leqq t\leqq v, \quad \sigma=t-u,$$

$$= \begin{cases} \displaystyle\int_0^{\sigma} \frac{(x/\beta)^i}{i!}\,\mathrm{e}^{-x/\beta}\,\frac{\mathrm{d}x}{\beta}, & j=0, \\ \displaystyle\frac{(\sigma/\beta)^{i+1-j}}{(i+1-j)!}\,\mathrm{e}^{-\sigma/\beta}, & j=1, \dots, i+1, \\ 0, & j=i+2, i+3, \dots, \end{cases} \tag{3.20}$$

it follows from (3.18) and (3.19) that if $a<1$,

$$\lim_{t\to\infty} \Pr\{x_t=j \mid y_0=k\} = \begin{cases} 1-a, & j=0, \\ a(1-\lambda_0)\,\lambda_0^{j-1}, & j=1, 2, \dots, \end{cases} \tag{3.21}$$

if $A(t)$ is not a lattice distribution, and

$$\lim_{m\to\infty} \Pr\{x_{m\tau+\sigma}=j \mid y_0=k\} = \begin{cases} 1-a(\sigma), & j=0, \\ a(\sigma)(1-\lambda_0)\,\lambda_0^{j-1}, & j=1, 2, \dots, \end{cases} \tag{3.22}$$

if $A(t)$ is a lattice distribution with period τ, where

$$a(\sigma) \stackrel{\text{def}}{=} a\exp(-\sigma(1-\lambda_0)/\beta)\,\frac{\tau(1-\lambda_0)/\beta}{1-\exp(-\tau(1-\lambda_0)/\beta)}, \qquad 0\leqq\sigma<\tau. \tag{3.23}$$

It should be noted that

$$\int_0^{\tau} a(\sigma)\,\frac{\mathrm{d}\sigma}{\tau} = a,$$

and consequently, if we take $\boldsymbol{\sigma}$ to be uniformly distributed over $[0, \tau)$ then

$$\lim_{m\to\infty} \Pr\{x_{m\tau+\boldsymbol{\sigma}}=j\} = \begin{cases} 1-a, & j=0, \\ a(1-\lambda_0)\,\lambda_0^{j-1}, & j=1, 2, \dots . \end{cases} \tag{3.24}$$

The initial situation of the x_t-process is $x_0=k$, $x_{0+}=k+1$, i.e. a customer arrives at $t=0$ and meets k customers in the system. It is seen from the above results and from renewal theory that the limit distributions obtained are independent of the initial state if $a<1$.

From the expressions given it is now easily found that for every $k=0, 1, \ldots,$

$$\lim_{n\to\infty} \mathrm{E}\{y_n \mid x_0=k\} = \frac{\lambda_0}{1-\lambda_0}, \tag{3.25}$$

$$\lim_{n\to\infty} \mathrm{var}\{y_n \mid x_0=k\} = \frac{\lambda_0}{(1-\lambda_0)^2},$$

$$\lim_{t\to\infty} \mathrm{E}\{x_t \mid x_0=k\} = \frac{a}{1-\lambda_0}, \tag{3.26}$$

$$\lim_{t\to\infty} \mathrm{var}\{x_t \mid x_0=k\} = \frac{a-a^2+a\lambda_0}{(1-\lambda_0)^2},$$

if $A(t)$ is not a lattice distribution.

Denoting the time between t and the moment of arrival of the next customer by ξ_t and the time between t and the moment of arrival of the last customer before t $(t>0)$ by η_t we see that it follows from (3.15) and the formulas (I.6.37) and (I.6.38) for $A(t)$ not a lattice distribution, $a<1$ and $\eta\geqq 0$,

$$\lim_{t\to\infty} \Pr\{x_t=j, \eta_t<\eta\} = \sum_{i=0}^{\infty} (1-\lambda_0)\,\lambda_0^i \int_0^{\eta} p_{ij}(\sigma)\{1-A(\sigma)\}\,\frac{\mathrm{d}\sigma}{\alpha}$$

$$= \begin{cases} \displaystyle\int_0^{\eta} \{1-\exp\{-(1-\lambda_0)\sigma/\beta\}\}\,\{1-A(\sigma)\}\,\frac{\mathrm{d}\sigma}{\alpha}, & j=0; \\[2ex] \displaystyle(1-\lambda_0)\,\lambda_0^{j-1} \int_0^{\eta} \exp\{-(1-\lambda_0)\sigma/\beta\}\,\{1-A(\sigma)\}\,\frac{\mathrm{d}\sigma}{\alpha}, & j=1, 2, \ldots; \end{cases} \tag{3.27}$$

and for $\xi\geqq 0$,

$$\lim_{t\to\infty} \Pr\{x_t=j, \xi_t<\zeta\} = \sum_{i=0}^{\infty} (1-\lambda_0)\lambda_0^i \int_0^\infty p_{ij}(\sigma)\{A(\sigma+\zeta)-A(\sigma)\} \frac{d\sigma}{\alpha}$$

$$= \begin{cases} \displaystyle\int_0^\infty \{1-\exp\{-(1-\lambda_0)\sigma/\beta\}\}\{A(\sigma+\zeta)-A(\sigma)\}\frac{d\sigma}{\alpha}, \\ \qquad\qquad j=0; \\ \displaystyle(1-\lambda_0)\lambda_0^{j-1}\int_0^\infty \exp\{-(1-\lambda_0)\sigma/\beta\}\{A(\sigma+\zeta)-A(\sigma)\}\frac{d\sigma}{\alpha}, \\ \qquad\qquad j=1,2,\dots. \end{cases} \tag{3.28}$$

II.3.3. The entrance and return time distributions

In this section we shall investigate the distribution functions of $\boldsymbol{\sigma}_{ij}(t)$ and $\boldsymbol{\tau}_{ij}$ which have been defined in section I.6.4. From that section it is seen that for $t>0, j=0, 1, \dots; k=0, 1, \dots; n=2, 3, \dots,$

$$F_{kj}^{(1)}(t)=\Pr\{\boldsymbol{\sigma}_{kj}(t)=1\}=\Pr\{\boldsymbol{y}_1=j, \boldsymbol{t}_1<t \mid \boldsymbol{y}_0=k\},$$

$$F_{kj}^{(n)}(t)=\Pr\{\boldsymbol{\sigma}_{kj}(t)=n\}=\Pr\{\boldsymbol{y}_n=j, \boldsymbol{t}_n<t, \boldsymbol{y}_h\neq j, h=1,\dots,n-1 \mid \boldsymbol{y}_0=k\},$$

and

$$F_{kj}(t) = \Pr\{\boldsymbol{\tau}_{kj}<t\} = \sum_{n=1}^{\infty} F_{kj}^{(n)}(t), \qquad t>0.$$

We investigate $F_{kj}^{(n)}(t)$ by starting from the integral relations (cf. section I.6.4),

$$F_{kj}^{(1)}(t) = \int_0^t p_{kj}(\sigma)\, dA(\sigma), \tag{3.29}$$

$$F_{kj}^{(n)}(t) = \sum_{\substack{i=0\\ i\neq j}}^{\infty} \int_0^t p_{ki}(u)\, F_{ij}^{(n-1)}(t-u)\, dA(u), \qquad n=2,3,\dots,$$

with $p_{kj}(\sigma)$ given by (3.20). Clearly, the functions $F_{kj}^{(n)}(t)$ are uniquely determined by the relations (3.29) and it follows easily that for $i, j=0, 1, \dots;$ $n=1, 2, \dots,$

$$1 \geqq F_{ij}^{(n)}(t) \geqq 0, \qquad \sum_{j=0}^{\infty} F_{ij}^{(n)}(t) \leqq 1, \qquad t>0. \tag{3.30}$$

Defining

$$F_j^{(n)}(t,p) \overset{\text{def}}{=} \sum_{i=0}^{\infty} p^i F_{ij}^{(n)}(t), \qquad |p|<1, \quad t \geqq 0, \tag{3.31}$$

we have

$$F_0^{(1)}(t,p) = \frac{1}{1-p} \int_0^t \{1-\exp\{-(1-p)u/\beta\}\} \, \mathrm{d}A(u),$$

$$F_0^{(n)}(t,p) = \frac{1}{p} \int_0^t \exp\{-(1-p)u/\beta\} \{F_0^{(n-1)}(t-u,p) - F_{00}^{(n-1)}(t-u)\} \, \mathrm{d}A(u),$$
$$n=2, 3, \ldots,$$

and for $j=1, 2, \ldots,$

$$F_j^{(1)}(t,p) = p^{j-1} \int_0^t \exp\{-(1-p)u/\beta\} \, \mathrm{d}A(u),$$

$$\begin{aligned} F_j^{(n)}(t,p) &= \frac{1}{p} \int_0^t \exp\{-(1-p)u/\beta\} \, F_j^{(n-1)}(t-u,p) \, \mathrm{d}A(u) \\ &\quad - p^{j-1} \int_0^t \exp\{-(1-p)u/\beta\} \, F_{jj}^{(n-1)}(t-u) \, \mathrm{d}A(u) \\ &\quad + \frac{1}{p(1-p)} \int_0^t \{p - \exp\{-(1-p)u/\beta\}\} \, F_{0j}^{(n-1)}(t-u) \, \mathrm{d}A(u), \end{aligned}$$
$$n=2, 3, \ldots .$$

Defining for $n=1, 2, \ldots;$ $\operatorname{Re} \rho \geqq 0,$

$$\varphi_{ij}^{(n)}(\rho) \overset{\text{def}}{=} \int_0^{\infty} \mathrm{e}^{-\rho t} \, \mathrm{d}F_{ij}^{(n)}(t), \tag{3.32}$$

$$\varphi_j^{(n)}(p,\rho) \overset{\text{def}}{=} \int_0^{\infty} \mathrm{e}^{-\rho t} \, \mathrm{d}F_j^{(n)}(t,p), \qquad |p|<1,$$

we see that the relations above lead to

$$\varphi_0^{(1)}(p, \rho) = \frac{1}{1-p} \{\alpha(\rho) - \alpha\{\rho + (1-p)/\beta\}\},$$

$$\varphi_0^{(n)}(p, \rho) = \frac{1}{p} \alpha\{\rho + (1-p)/\beta\} \{\varphi_0^{(n-1)}(p, \rho) - \varphi_{00}^{(n-1)}(\rho)\}, \qquad n = 2, 3, \ldots,$$

and for $j = 1, 2, \ldots$, to

$$\varphi_j^{(1)}(p, \rho) = p^{j-1}\alpha\{\rho + (1-p)/\beta\},$$

$$\begin{aligned} \varphi_j^{(n)}(p, \rho) = & \frac{1}{p} \varphi_j^{(n-1)}(p, \rho)\, \alpha\{\rho + (1-p)/\beta\} \\ & - p^{j-1}\varphi_{jj}^{(n-1)}(\rho)\, \alpha\{\rho + (1-p)/\beta\} \\ & + \frac{1}{p(1-p)} \{p\alpha(p) - \alpha\{\rho + (1-p)/\beta\}\}\, \varphi_{0j}^{(n-1)}(\rho), \\ & \qquad\qquad n = 2, 3, \ldots . \end{aligned}$$

Defining for $|r| \leqq 1$, $\text{Re}\, \rho \geqq 0$,

$$\varphi_{ij}(\rho, r) \stackrel{\text{def}}{=} \sum_{n=1}^{\infty} r^n \varphi_{ij}^{(n)}(\rho), \tag{3.33}$$

$$\varphi_j(p, \rho, r) \stackrel{\text{def}}{=} \sum_{n=1}^{\infty} r^n \varphi_j^{(n)}(p, \rho), \qquad |p| < 1,$$

we deduce from the relations above that for $|p| < 1$, $|r| \leqq 1$, $\text{Re}\, \rho \geqq 0$, $j = 1, 2, \ldots$,

$$\begin{aligned} \left[1 - \frac{r}{p}\, \alpha\{\rho + (1-p)/\beta\}\right] \varphi_0(p, \rho, r) = & \frac{r}{1-p} \{\alpha(\rho) - \alpha\{\rho + (1-p)/\beta\}\} \\ & - \frac{r}{p} \{\alpha\{\rho + (1-p)/\beta\}\}\, \varphi_{00}(\rho, r), \end{aligned} \tag{3.34}$$

$$\begin{aligned} \left[1 - \frac{r}{p}\, \alpha\{\rho + (1-p)/\beta\}\right] \varphi_j(p, \rho, r) = & rp^{j-1}\alpha\{\rho + (1-p)/\beta\}\{1 - \varphi_{jj}(\rho, r)\} \\ & + \frac{r}{p(1-p)} \{p\alpha(\rho) - \alpha\{\rho + (1-p)/\beta\}\}\, \varphi_{0j}(\rho, r). \end{aligned}$$

We now apply Takács' lemma (cf. app. 6) to the function $f(z)$,

$$f(z) \equiv z - r\alpha\{\rho + (1-z)/\beta\}, \qquad \text{Re}\, \rho \geqq 0, \quad |r| \leqq 1.$$

According to this lemma the zero $\lambda(\rho, r)$ of $f(z)$ which has the smallest absolute value is unique and $|\lambda(\rho, r)| \leqq 1$; while $|\lambda(\rho, r)| < 1$ for all $\operatorname{Re} \rho \geqq 0$, $|r| \leqq 1$, except $a \geqq 1$, $r=1$, $\rho=0$ if $A(t)$ is not a lattice distribution or $a \geqq 1$, $r=1$, $\rho=2\pi i n/\tau$, $n=0, 1, \dots$, if $A(t)$ is a lattice distribution with period τ. From the definition of $\varphi_j(p, \rho, r)$, $j=0, 1, \dots$, it follows that for fixed ρ and r with $\operatorname{Re} \rho \geqq 0$, $|r| \leqq 1$ the function $\varphi_j(p, \rho, r)$ of p is analytic for $|p| < 1$ for every $j=0, 1, \dots$. Hence $p=\lambda(\rho, r)$ must be a zero of the right-hand sides of (3.34) for $\operatorname{Re} \rho \geqq 0$, $|r| \leqq 1$. Consequently, for $\operatorname{Re} \rho \geqq 0$, $|r| \leqq 1$,

$$\varphi_{00}(\rho, r) = \frac{r\alpha(\rho) - \lambda(\rho, r)}{1-\lambda(\rho, r)} = 1 - \frac{1-r\alpha(\rho)}{1-\lambda(\rho, r)}, \tag{3.35}$$

and

$$\varphi_{0j}(\rho, r) = \frac{1-\lambda(\rho, r)}{1-r\alpha(\rho)} \lambda^j(\rho, r)\{1-\varphi_{jj}(\rho, r)\}, \qquad j=1, 2, \dots . \tag{3.36}$$

Hence from (3.33), ..., (3.36) for $|p| < 1$, $\operatorname{Re} \rho \geqq 0$, $|r| \leqq 1$,

$$\varphi_0(p, \rho, r) = \left[\frac{1}{1-p} \frac{1-\lambda(\rho, r)}{1-r\alpha(\rho)} - 1 + \frac{\lambda(\rho, r) - p}{(1-p)\{1-(r/p)\,\alpha\{\rho+(1-p)/\beta\}\}}\right] \cdot\{1-\varphi_{00}(\rho, r)\}, \tag{3.37}$$

and for $j=1, 2, \dots$,

$$\varphi_j(p, \rho, r) = \left[\frac{1-\lambda(\rho, r)}{1-r\alpha(\rho)} \frac{\lambda^j(\rho, r)}{1-p} - p^j + \frac{(1-p)\,p^j - (1-\lambda(\rho, r))\,\lambda^j(\rho, r)}{(1-p)\{1-(r/p)\,\alpha\{\rho+(1-p)/\beta\}\}}\right] \cdot\{1-\varphi_{jj}(\rho, r)\}. \tag{3.38}$$

From (3.38) it is now possible to determine $\varphi_{jj}(\rho, r)$, $j=1, 2, \dots$, by expanding the right- and left-hand side of (3.38) into a power series of p. In the same way we can calculate $\varphi_{0j}(\rho, r)$ and $\varphi_{j-1,j}(\rho, r)$; in fact for $\operatorname{Re} \rho \geqq 0$, $|r| \leqq 1$,

$$1 - \varphi_{jj}(\rho, r) = \left[\frac{1-\lambda(\rho, r)}{1-r\alpha(\rho)} \lambda^j(\rho, r) \cdot\left\{1 - \frac{1}{(j-1)!} \frac{\mathrm{d}^{j-1}}{\mathrm{d}p^{j-1}} \frac{1-r\alpha(\rho)}{(1-p)\{p-r\alpha\{\rho+(1-p)/\beta\}\}}\right\}\right]^{-1}\Bigg|_{p=0}, \quad j=1, 2, \dots, \tag{3.39}$$

$$\varphi_{0j}(\rho, r) = \left[1 - \frac{1}{(j-1)!}\,\frac{\mathrm{d}^{j-1}}{\mathrm{d}p^{j-1}}\,\frac{1-r\alpha(\rho)}{(1-p)\{p-r\alpha\{\rho+(1-p)/\beta\}\}}\right]^{-1}\Bigg|_{p=0}, \qquad j=1, 2, \ldots,$$

$$\varphi_{j-1,j}(\rho, r) = \frac{\varphi_{0j}(\rho, r)}{\varphi_{0,j-1}(\rho, r)}, \qquad j=2, 3, \ldots .$$

In the next part of this section we study some of the properties of the return and entrance time distributions $F_{ij}(t)$. The means of these distributions are given by

$$\mu_{ij} \overset{\text{def}}{=} \int_0^\infty t\,\mathrm{d}F_{ij}(t). \tag{3.40}$$

The first return time distribution $F_{00}(t)$ of the event: "an arriving customer finds the system empty" is given by

$$F_{00}(t) = \sum_{n=1}^{\infty} F_{00}^{(n)}(t) \leqq 1, \qquad t \geqq 0.$$

Hence

$$\varphi_{ij}(\rho) \overset{\text{def}}{=} \int_0^\infty \mathrm{e}^{-\rho t}\,\mathrm{d}F_{ij}(t) = \int_0^\infty \mathrm{e}^{-\rho t}\,\mathrm{d}\{\lim_{r\uparrow 1} \sum_{n=1}^{\infty} r^n F_{ij}^{(n)}(t)\}, \quad \operatorname{Re}\rho \geqq 0. \tag{3.41}$$

Writing the last integral in (3.41) as a sum of integrals over $[0, T]$ and (T, ∞), applying to the integral over $[0, T]$ Helly-Bray's theorem as given by WIDDER [1946] and choosing T sufficiently large it follows easily that

$$\varphi_{ij}(\rho) = \lim_{r\uparrow 1} \varphi_{ij}(\rho, r).$$

From Takács' lemma it follows that

$$\lim_{r\uparrow 1} \lambda(\rho, r) = \lambda(\rho, 1), \qquad \operatorname{Re}\rho \geqq 0,$$

hence from (3.35),

$$\varphi_{00}(\rho) = \frac{\alpha(\rho) - \lambda(\rho, 1)}{1 - \lambda(\rho, 1)}, \qquad \operatorname{Re}\rho \geqq 0, \tag{3.42}$$

where $\lambda(\rho, 1)$ is the root of

$$z - \alpha\{\rho + (1-z)/\beta\} = 0, \qquad \operatorname{Re}\rho \geqq 0, \tag{3.43}$$

which has the smallest absolute value.

Since $F_{00}^{(n)}(\infty)$, $n=1, 2, \dots$, represents for the imbedded Markov chain $\{y_n, n=0, 1, \dots\}$ the return time distribution of the event: "an arriving customer finds no customer present" it follows immediately from (3.33) that

$$\varphi_{00}(0, r) = \sum_{n=1}^{\infty} r^n F_{00}^{(n)}(\infty) = \lim_{\rho \downarrow 0} \varphi_{00}(\rho, r), \qquad |r| \leqq 1.$$

Since again by Takács' lemma

$$\lim_{\rho \downarrow 0} \lambda(\rho, r) = \lambda(0, r), \qquad |r| \leqq 1,$$

it follows from (3.35) that

$$\varphi_{00}(0, r) = \frac{r - \lambda(0, r)}{1 - \lambda(0, r)}, \qquad |r| \leqq 1, \tag{3.44}$$

where $\lambda(0, r)$ is the root of

$$z - r\alpha\{(1-z)/\beta\} = 0, \qquad |r| \leqq 1,$$

which has the smallest absolute value.
Since from (3.42) and (3.44),

$$\varphi_{00}(\rho) = 1 - (1 - \alpha(\rho)) \sum_{n=0}^{\infty} \lambda^n(\rho, 1), \qquad \operatorname{Re} \rho > 0, \tag{3.45}$$

$$\varphi_{00}(0, r) = 1 - (1 - r) \sum_{n=0}^{\infty} \lambda^n(0, r), \qquad |r| < 1,$$

explicit expressions for $F_{00}(t)$ and $F_{00}^{(n)}(\infty)$, $n=1, 2, \dots$, can easily be obtained by using the relations (cf. app. 6),

$$\lambda^j(\rho, 1) = \sum_{n=j}^{\infty} \frac{j}{n} \int_0^{\infty} \mathrm{e}^{-(\rho + 1/\beta)t} \frac{(t/\beta)^{n-j}}{(n-j)!} \, \mathrm{d}A^{n*}(t),$$

$$j=1, 2, \dots; \quad \operatorname{Re} \rho > 0,$$

$$\lambda^j(0, r) = \sum_{n=j}^{\infty} \frac{j}{n} r^n \int_0^{\infty} \mathrm{e}^{-t/\beta} \frac{(t/\beta)^{n-j}}{(n-j)!} \, \mathrm{d}A^{n*}(t), \qquad j=1, 2, \dots; \qquad |r| < 1.$$

If $a < 1$ then

$$\lambda_0 = \lim_{\rho \downarrow 0} \lambda(\rho, 1) < 1 \quad \text{and} \quad \lambda_0 = \lim_{r \uparrow 1} \lambda(0, r) < 1,$$

so that from (3.45),

$$F_{00}(\infty) = \varphi_{00}(0) = 1, \qquad \sum_{n=1}^{\infty} F_{00}^{(n)}(\infty) = \varphi_{00}(0, 1) = 1, \tag{3.46}$$

a result which agrees with that in the preceding section.

If $a \geqq 1$ then

$$\lim_{\rho \downarrow 0} \lambda(\rho, 1) = 1, \qquad \lim_{r \uparrow 1} \lambda(0, r) = 1,$$

so that (cf. app. 6),

$$\left\{\lim_{\rho \downarrow 0} \frac{\mathrm{d}}{\mathrm{d}\rho} \lambda(\rho, 1)\right\}^{-1} = -\frac{1-a^{-1}}{\alpha}, \qquad \left\{\lim_{r \uparrow 1} \frac{\mathrm{d}}{\mathrm{d}r} \lambda(0, r)\right\}^{-1} = 1 - a^{-1},$$
$$a \geqq 1. \tag{3.47}$$

Hence from (3.42) and (3.44),

$$F_{00}(\infty) = a^{-1}, \qquad \sum_{n=1}^{\infty} F_{00}^{(n)}(\infty) = a^{-1} \qquad \text{for } a \geqq 1, \tag{3.48}$$

and

$$\mu_{00} = \int_0^\infty t \,\mathrm{d}F_{00}(t) = \begin{cases} \dfrac{\alpha}{1-\lambda_0}, \\ \infty, \end{cases} \qquad \sum_{n=1}^{\infty} nF_{00}^{(n)}(\infty) = \begin{cases} \dfrac{1}{1-\lambda_0}, & a<1, \\ \infty, & a \geqq 1. \end{cases} \tag{3.49}$$

Since $\varphi_{0j}(0, 1)=1$ it is easily found from (3.36) that for $a<1$,

$$\mu_{jj} = \int_0^\infty t \,\mathrm{d}F_{jj}(t) = \frac{\alpha}{1-\lambda_0} \lambda_0^{-j}, \qquad \sum_{n=1}^{\infty} nF_{jj}^{(n)}(\infty) = \frac{1}{1-\lambda_0} \lambda_0^{-j},$$
$$j=1, 2, \ldots,$$

and from (3.39),

$$\mu_{0j} = -\alpha \frac{1}{(j-1)!} \frac{\mathrm{d}^{j-1}}{\mathrm{d}p^{j-1}} \left. \frac{1}{(1-p)\{p-\alpha\{(1-p)/\beta\}\}} \right|_{p=0},$$
$$j=1, 2, \ldots, \tag{3.50}$$

so that by using (3.9) an explicit expression for μ_{0j} may be obtained. From (3.39) and (3.41) it follows that

$$\varphi_{0j}(\rho) = \prod_{i=1}^{j} \varphi_{i-1,i}(\rho), \qquad \mathrm{Re}\, \rho \geqq 0,$$

so that

$$F_{0j}(t) = F_{01}(t) * F_{12}(t) * \dots * F_{j-1,j}(t), \qquad t \geq 0. \tag{3.51}$$

Since $F_{i-1,i}(t)$ is the distribution of the first entrance time $\tau_{i-1,i}$, $i=1, 2, \dots$, (cf. section I.6.4) from the state where an arriving customer finds $i-1$ customers present to the state where an arriving customer meets i customers in the system, the relation above is obvious since these entrance times are evidently independent and

$$\tau_{0j} = \sum_{i=1}^{j} \tau_{i-1,i}, \qquad j=1, 2, \dots .$$

It follows from

$$\mu_{0j} = \sum_{i=1}^{j} \mu_{i-1,i}, \qquad j=1, 2, \dots,$$

and (3.50) that

$$\mu_{j,j+1} = -\left\{\frac{1}{j!}\frac{\mathrm{d}^j}{\mathrm{d}p^j} - \frac{1}{(j-1)!}\frac{\mathrm{d}^{j-1}}{\mathrm{d}p^{j-1}}\right\}\frac{\alpha}{(1-p)\{p-\alpha\{(1-p)/\beta\}\}}\bigg|_{p=0},$$

hence

$$\mu_{j,j+1} = -\alpha\frac{1}{j!}\left[\frac{\mathrm{d}^j}{\mathrm{d}p^j}\frac{1}{p-\alpha\{(1-p)/\beta\}}\right]\bigg|_{p=0}, \qquad j=0, 1, \dots . \tag{3.52}$$

Consequently, from (3.9) we have

$$\mu_{j,j+1} = \sum_{n=0}^{\infty}\int_0^{\infty} \mathrm{e}^{-t/\beta}\frac{(t/\beta)^j}{j!} a^{-n}\left\{\int_0^t \{1-A(\tau)\}\,\mathrm{d}\tau\right\}^{n*}\frac{\mathrm{d}t}{\beta},$$
$$j=0, 1, \dots, \tag{3.53}$$

and

$$\mu(p) \overset{\text{def}}{=} \sum_{j=0}^{\infty} \mu_{j,j+1} p^j = \frac{-\alpha}{p-\alpha\{(1-p)/\beta\}}, \qquad |p| < \lambda(0, 1). \tag{3.54}$$

Moreover, it follows from (3.53) and (3.9) that

$$\lim_{j\to\infty} \mu_{j,j+1} = \begin{cases} \dfrac{\alpha}{1-a^{-1}} & \text{if } a>1, \\ \infty & \text{if } a \leq 1. \end{cases} \tag{3.55}$$

From (3.39) we have

$$\sum_{j=1}^{\infty}\left\{\frac{1}{\varphi_{0j}(\rho)}-1\right\}p^j=-\frac{p}{1-p}\,\frac{1-\alpha(\rho)}{p-\alpha\{\rho+(1-p)/\beta\}},$$

$$|p|<\lambda(\rho,1),\quad \operatorname{Re}\rho\geqq 0. \tag{3.56}$$

Denote by D_p a circle in the p-plane with centre at $p=0$ and radius $|p|$. It follows for $\operatorname{Re}\rho\geqq 0$, $j=1,2,\ldots$; and $|p|<|\lambda(\rho,1)|$ that

$$\{\varphi_{0j}(\rho)\}^{-1}=1-\frac{1}{2\pi i}\int_{D_p}\frac{p}{1-p}\,\frac{1-\alpha(\rho)}{p-\alpha\{\rho+(1-p)/\beta\}}\,\frac{dp}{p^{j+1}}, \tag{3.57}$$

the integration along D_p is taken in the counterclockwise direction. Suppose for the present $|\lambda(\rho,1)|<1$ and define for $j=1,2,\ldots,$

$$R(\rho)\overset{\text{def}}{=}\left[1-\int_0^{\infty}\frac{t}{\beta}\exp[-\{\rho+(1-\lambda(\rho,1))/\beta\}\,t]\,dA(t)\right]^{-1}, \tag{3.58}$$

$$S_j(\rho)\overset{\text{def}}{=}\frac{1}{2\pi i}\int_{D_q}\frac{1}{1-q}\,\frac{1}{q-\alpha\{\rho+(1-q)/\beta\}}\,\frac{\lambda^j(\rho,1)}{q^j}\,dq,$$

$$1>|q|>|\lambda(\rho,1)|. \tag{3.59}$$

Hence from (3.57) and (3.58) by contour integration

$$\{\varphi_{0j}(\rho)\}^{-1}=1+\lambda^{-j}(\rho,1)\frac{1-\alpha(\rho)}{1-\lambda(\rho,1)}\{R(\rho)-(1-\lambda(\rho,1))\,S_j(\rho)\}. \tag{3.60}$$

On D_q we have by definition of $\lambda(\rho,1)$,

$$|(1-q)[q-\alpha\{\rho+(1-q)/\beta\}]|^{-1}<\infty,\qquad \lim_{j\to\infty}|q^{-1}\lambda(\rho,1)|^j=0, \tag{3.61}$$

so that

$$\lim_{j\to\infty}|S_j(\rho)|=0\quad\text{if}\quad|\lambda(\rho,1)|<1. \tag{3.62}$$

From (3.51) we have for $\operatorname{Re}\rho\geqq 0$,

$$\varphi_{j,j+1}(\rho)=\frac{\varphi_{0,j+1}(\rho)}{\varphi_{0j}(\rho)}. \tag{3.63}$$

Hence from (3.59), ..., (3.62),

$$\lim_{j\to\infty}\varphi_{j,j+1}(\rho)=\lambda(\rho,1)\quad\text{if}\quad|\lambda(\rho,1)|<1. \tag{3.64}$$

Since for $\rho>0$ always $0<\lambda(\rho, 1)<1$ the limit in (3.64) holds for all $\rho>0$. Consequently, by Feller's convergence theorem (cf. app. 5),

$$\lim_{j\to\infty} F_{j,j+1}(t) = \sum_{n=1}^{\infty} \int_0^t \mathrm{e}^{-\tau/\beta} \frac{(\tau/\beta)^{n-1}}{n!} \mathrm{d}A^{n*}(\tau), \qquad t\geqq 0, \tag{3.65}$$

and the limit distribution is a proper probability distribution if and only if $a\geqq 1$. The relation (3.65) is due to HEATHCOTE [1965].

II.3.4. The transition probabilities

From section I.6.4 it is seen that the transition probabilities

$$P_{kj}^{(n)}(t) \stackrel{\text{def}}{=} \Pr\{\mathbf{y}_n=j,\ \mathbf{t}_n<t \mid \mathbf{y}_0=k\}, \qquad n=1, 2, \ldots, \tag{3.66}$$

satisfy

$$P_{kj}^{(1)}(t) = F_{kj}^{(1)}(t), \tag{3.67}$$

$$P_{kj}^{(n)}(t) = F_{kj}^{(n)}(t) + \sum_{m=1}^{n-1} \int_0^t F_{jj}^{(n-m)}(t-u)\, \mathrm{d}P_{kj}^{(m)}(u), \qquad n=2, 3, \ldots .$$

Obviously, the functions $P_{kj}^{(n)}(t)$ are uniquely determined by this set of equations. Writing

$$P_{kj}(t, r) \stackrel{\text{def}}{=} \sum_{n=1}^{\infty} r^n P_{kj}^{(n)}(t), \qquad |r|<1, \tag{3.68}$$

$$\pi_{kj}(\rho, r) \stackrel{\text{def}}{=} \int_0^{\infty} \mathrm{e}^{-\rho t}\, \mathrm{d}P_{kj}(t, r), \qquad \mathrm{Re}\,\rho\geqq 0,$$

it follows from (3.67), (3.35) and (3.36) for Re $\rho\geqq 0$, $|r|<1$ that

$$\pi_{kj}(\rho, r) = \frac{\varphi_{kj}(\rho, r)}{1-\varphi_{jj}(\rho, r)}$$

$$= \begin{cases} \dfrac{r\alpha(\rho)-\lambda(\rho, r)}{1-r\alpha(\rho)}, & k=0,\quad j=0, \\[2ex] \dfrac{1-\lambda(\rho, r)}{1-r\alpha(\rho)}\, \lambda^j(\rho, r), & k=0,\quad j=1, 2, \ldots, \end{cases} \tag{3.69}$$

and with

$$\pi_j(p, \rho, r) \overset{\text{def}}{=} \sum_{k=0}^{\infty} p^k \pi_{kj}(\rho, r), \qquad |p|<1, \quad |r|<1, \quad \operatorname{Re} \rho \geqq 0, \tag{3.70}$$

for $j=1, 2, \ldots$; $\operatorname{Re} \rho \geqq 0$, $|r|<1$, $|p|<1$, we have from (3.37) and (3.38),

$$\begin{aligned} \pi_j(p, \rho, r) &= \frac{1-\lambda(\rho, r)}{1-r\alpha(\rho)} \, \frac{\lambda^j(\rho, r)}{1-p} - p^j \\ &\quad + \frac{(1-p)\, p^j - \{1-\lambda(\rho, r)\}\, \lambda^j(\rho, r)}{(1-p)\{1-(r/p)\, \alpha\{\rho+(1-p)/\beta\}\}}, \end{aligned} \tag{3.71}$$

$$\begin{aligned} \pi_0(p, \rho, r) &= \frac{1-\lambda(\rho, r)}{1-r\alpha(\rho)} \, \frac{1}{1-p} - 1 \\ &\quad + \frac{\lambda(\rho, r)-p}{(1-p)\{1-(r/p)\, \alpha\{\rho+(1-p)/\beta\}\}}. \end{aligned}$$

Next we determine the generating functions for the n-step transition probabilities $p_{ij}^{(n)}$ of the imbedded Markov chain $\{\boldsymbol{y}_n, n=0, 1, \ldots\}$. These functions can be obtained by starting from the relation

$$p_{ij}^{(n+1)} = \sum_{h=0}^{\infty} p_{ih} p_{hj}^{(n)}, \qquad n=1, 2, \ldots,$$

with p_{ih} given by (3.2). From (3.66) we have

$$p_{kj}^{(n)} = P_{kj}^{(n)}(\infty),$$

and hence from (3.69) for $|r|<1$,

$$\sum_{n=1}^{\infty} p_{00}^{(n)} r^n = \frac{r-\lambda(0, r)}{1-r}, \tag{3.72}$$

$$\sum_{n=1}^{\infty} p_{0j}^{(n)} r^n = \frac{1-\lambda(0, r)}{1-r} \, \lambda^j(0, r), \qquad j=1, 2, \ldots;$$

$$\sum_{n=1}^{\infty} \sum_{k=0}^{\infty} p_{kj}^{(n)} p^k r^n = \pi_j(p, 0, r), \qquad j=0, 1, \ldots .$$

Note that the summation with respect to n starts with $n=1$.

In section I.6.4 we have introduced the renewal functions

$$m_{kj}(t) = \sum_{n=1}^{\infty} \Pr\{\boldsymbol{y}_n=j, \boldsymbol{t}_n<t \mid \boldsymbol{y}_0=k\} = \sum_{n=1}^{\infty} P_{kj}^{(n)}(t),$$

$m_{kj}(t)$ being the average number of times in the interval $[0, t)$ that an arriving customer meets j customers in the system. From (3.69) and (3.70) it is now easily found that for $\operatorname{Re} \rho > 0$,

$$\int_0^\infty e^{-\rho t}\, dm_{0j}(t) = \begin{cases} \dfrac{\alpha(\rho) - \lambda(\rho, 1)}{1 - \alpha(\rho)}, & j=0, \\[2ex] \dfrac{1 - \lambda(\rho, 1)}{1 - \alpha(\rho)} \lambda^j(\rho, 1), & j=1, 2, \ldots, \end{cases} \tag{3.73}$$

$$\sum_{k=0}^{\infty} p^k \int_0^\infty e^{-\rho t}\, dm_{kj}(t) = \pi_j(p, \rho, 1), \qquad j=0, 1, \ldots .$$

To obtain an expression for $\Pr\{\boldsymbol{x}_t = j \mid \boldsymbol{y}_0 = k\}$, it is noted that

$$\begin{aligned} \Pr\{\boldsymbol{x}_t = j \mid \boldsymbol{y}_0 = k\} &= p_{kj}(t)\{1 - A(t)\} \\ &\quad + \sum_{h=0}^{\infty} \int_0^t p_{hj}(t-u)\{1 - A(t-u)\}\, d_u \sum_{n=1}^{\infty} \Pr\{\boldsymbol{y}_n = h, \boldsymbol{t}_n < u \mid \boldsymbol{y}_0 = k\} \\ &= p_{kj}(t)\{1 - A(t)\} + \sum_{h=0}^{\infty} \int_0^t p_{hj}(t-u)\{1 - A(t-u)\}\, dm_{kh}(u). \end{aligned} \tag{3.74}$$

From (3.20) and (3.73) it now follows easily, for $|p| < 1$, $\operatorname{Re} \rho > 0$, that

$$\int_0^\infty e^{-\rho t} \Pr\{\boldsymbol{x}_t = j \mid \boldsymbol{y}_0 = 0\}\, dt = \begin{cases} \dfrac{1}{\rho} - \dfrac{1 - \lambda(\rho, 1)}{\{1 - \alpha(\rho)\}\{\rho + (1 - \lambda(\rho, 1))/\beta\}}, \\ \qquad\qquad\qquad j=0, \\[2ex] \dfrac{\{1 - \lambda(\rho, 1)\}^2 \lambda^{j-1}(\rho, 1)}{\{1 - \alpha(\rho)\}\{\rho + (1 - \lambda(\rho, 1))/\beta\}}, \\ \qquad\qquad\qquad j=1, 2, \ldots, \end{cases} \tag{3.75}$$

$$\begin{aligned} &\sum_{k=0}^{\infty} p^k \int_0^\infty e^{-\rho t} \Pr\{\boldsymbol{x}_t = j \mid \boldsymbol{y}_0 = k\}\, dt \\ &= \frac{1}{1-p} \frac{1}{\rho} - \frac{1}{1-p} \frac{1 - \alpha\{\rho + (1-p)/\beta\}}{1 - p^{-1}\alpha\{\rho + (1-p)/\beta\}} \frac{1}{\rho + (1-p)/\beta} \\ &\quad - \frac{1 - \lambda(\rho, 1)}{(1-p)\{\rho + (1 - \lambda(\rho, 1))/\beta\}} \left\{ \frac{1}{1 - \alpha(\rho)} - \frac{1}{1 - p^{-1}\alpha\{\rho + (1-p)/\beta\}} \right\}, \\ &\qquad\qquad\qquad \text{for } j=0; \end{aligned}$$

$$= \frac{[1-\alpha\{\rho+(1-p)/\beta\}]\, p^{j-1}}{[1-p^{-1}\alpha\{\rho+(1-p)/\beta\}][\rho+(1-p)/\beta\}}$$

$$+ \frac{\{1-\lambda(\rho,1)\}^2\, \lambda^{j-1}(\rho,1)}{(1-p)\{\rho+(1-\lambda(\rho,1))/\beta\}} \left\{\frac{1}{1-\alpha(\rho)} - \frac{1}{1-p^{-1}\alpha\{\rho+(1-p)/\beta\}}\right\},$$

for $j=1, 2, \dots$.

From (3.72) and (3.75) we obtain

$$\sum_{n=1}^{\infty} r^n \mathrm{E}\{\mathbf{y}_n \mid \mathbf{y}_0=0\} = \frac{1}{1-r}\,\frac{\lambda(0,r)}{1-\lambda(0,r)}, \qquad |r|<1, \quad (3.76)$$

$$\int_0^\infty \mathrm{e}^{-\rho t}\, \mathrm{E}\{\mathbf{x}_t \mid \mathbf{y}_0=0\}\, \mathrm{d}t = \frac{1}{\{1-\alpha(\rho)\}\{\rho+(1-\lambda(\rho,1))/\beta\}}, \qquad \mathrm{Re}\, \rho>0.$$

The relation (3.75) may be used as the starting point for the study of the asymptotic behaviour of $\mathrm{E}\{\mathbf{x}_t \mid \mathbf{y}_0=0\}$ as $t\to\infty$. For $a<1$ the first relation of (3.26) is easily deduced from (3.76). Since $\mathrm{E}\{\mathbf{x}_t \mid \mathbf{y}_0=0\}$ is non-negative it follows from a well-known Tauberian theorem (cf. app. 4) applied to (3.76) (cf. (3.74)),

$$\lim_{t\to\infty} \frac{1}{t^2} \int_0^t \mathrm{E}\{\mathbf{x}_\tau \mid \mathbf{y}_0=0\}\, \mathrm{d}\tau = \frac{a-1}{2\beta}, \qquad a>1. \quad (3.77)$$

In the derivations of this section we have always calculated probabilities with the condition $\mathbf{x}_0=k$, $\mathbf{x}_{0+}=k+1$, $k=0, 1, \dots$, i.e. at $t=0$ a customer arrives and finds k customers in the system. To show how other initial conditions may be handled we consider the case that $\mathbf{x}_{0+}=0$ and that the time after $t=0$ at which the first customer arrives has the distribution $A_1(t)$, the other interarrival times all having the same distribution $A(t)$ (cf. section 1.2). From (3.74) it now follows easily that

$$\Pr\{\mathbf{x}_t=j \mid \mathbf{x}_0=0\} = \begin{cases} 1 - A_1(t) + \int_0^t \Pr\{\mathbf{x}_{t-u}=0 \mid \mathbf{y}_0=0\}\, \mathrm{d}A_1(u), & j=0, \\ \int_0^t \Pr\{\mathbf{x}_{t-u}=j \mid \mathbf{y}_0=0\}\, \mathrm{d}A_1(u), & j=1, 2, 3, \dots . \end{cases} \quad (3.78)$$

Finally we consider *the maximum of* $\mathbf{x}_\tau$ for $\tau\in[0, t)$. Since the probability that all customers arriving in $[0, t)$ find less than j customers in the system is represented by $1-F_{0j}(t)$, it follows that

$$\Pr\{\max_{0\leq\tau<t} \boldsymbol{x}_\tau \leq j \mid \boldsymbol{y}_0=0\}=1-F_{0j}(t), \qquad t>0, \quad j=1, 2, \ldots,$$

hence

$$\Pr\{\max_{0\leq\tau<t} \boldsymbol{x}_\tau > j \mid \boldsymbol{y}_0=0\} = \begin{cases} 1, & j=0, \\ F_{0j}(t), & j=1, 2, \ldots, \end{cases} \tag{3.79}$$

and

$$\mathrm{E}\{\max_{0\leq\tau<t} \boldsymbol{x}_\tau \mid \boldsymbol{y}_0=0\} = 1 + \sum_{j=1}^{\infty} F_{0j}(t). \tag{3.80}$$

From (3.63) it is easily found that if $|\lambda(\rho, 1)|<1$,

$$\sum_{j=1}^{\infty} \varphi_{0j}(\rho)$$

$$= \frac{1-\lambda(\rho, 1)}{1-\alpha(\rho)} \sum_{j=1}^{\infty} \frac{\lambda^j(\rho, 1)}{\{1-\lambda(\rho, 1)\}\, \lambda^j(\rho, 1)/\{1-\alpha(\rho)\}+R(\rho)-\{1-\lambda(\rho, 1)\}\, S_j(\rho)}. \tag{3.81}$$

If $a>1$ then $\lambda(\rho, 1)\to 1$ for $\rho\to 0$, and hence from (3.58), ..., (3.62),

$$\sum_{j=1}^{\infty} \varphi_{0j}(\rho) \approx \frac{a-1}{\beta\rho}, \qquad \rho\downarrow 0.$$

Consequently, since the right-hand side of (3.80) is non-decreasing in t, it follows from a well-known Tauberian theorem (cf. app. 4) that

$$\lim_{t\to\infty} \frac{1}{t}\, \mathrm{E}\{\max_{0\leq\tau<t} \boldsymbol{x}_\tau \mid \boldsymbol{y}_0=0\} = \frac{a-1}{\beta}, \qquad a>1, \tag{3.82}$$

a result which is due to HEATHCOTE [1965] who in fact obtained the stronger asymptotic relation

$$\lim_{t\to\infty} \left\{ \mathrm{E}\{\max_{0\leq\tau<t} \boldsymbol{x}_\tau \mid \boldsymbol{y}_0=0\} - \frac{a-1}{\beta}t \right\} = \frac{\alpha_2(1+a^2)}{2\alpha^2 a(a-1)} - (1-a^{-1}), \qquad a>1, \tag{3.83}$$

if the second moment α_2 of $A(t)$ is finite.

II.3.5. The busy period

We shall denote by $D_{jn}(t), j=0, 1, \dots; n=1, 2, \dots$, the conditional probability that, if an arriving customer finds j customers present, the server will be idle for the first time after $n+j$ departures after this arrival and that the total time of these $n+j$ services is less than t, $t>0$. From this definition, from (3.20) and from section I.6.4 it follows immediately that for $j=0, 1, \dots$,

$$D_{j1}(t) = \int_0^t \left\{\frac{\mathrm{d}}{\mathrm{d}\tau} p_{j0}(\tau)\right\}\{1-A(\tau)\}\,\mathrm{d}\tau, \tag{3.84}$$

$$D_{j,n+1}(t) = \sum_{i=1}^{j+1} \int_0^t p_{ji}(u)\, D_{in}(t-u)\,\mathrm{d}A(u), \qquad n=1, 2, \dots .$$

Defining for $|p|<1$, $\operatorname{Re}\rho \geqq 0$,

$$D_n(p, t) \stackrel{\text{def}}{=} \sum_{j=0}^{\infty} D_{jn}(t)\, p^j, \tag{3.85}$$

$$\delta_n(p, \rho) \stackrel{\text{def}}{=} \int_0^\infty \mathrm{e}^{-\rho t}\,\mathrm{d}D_n(p, t),$$

$$\delta_{jn}(\rho) \stackrel{\text{def}}{=} \int_0^\infty \mathrm{e}^{-\rho t}\,\mathrm{d}D_{jn}(t),$$

it follows from (3.20) that the integral equations (3.84) are transformed into

$$\delta_1(p, \rho) = \frac{1}{\beta}\,\frac{1-\alpha\{\rho+(1-p)/\beta\}}{\rho+(1-p)/\beta}, \tag{3.86}$$

$$\delta_{n+1}(p, \rho) = \frac{1}{p}\,\{\delta_n(p, \rho)-\delta_{0n}(\rho)\}\,\alpha\,\{\rho+(1-p)/\beta\}, \qquad n=1, 2, \dots .$$

Putting for $|r|\leqq 1$,

$$\delta(p, \rho, r) \stackrel{\text{def}}{=} \sum_{n=1}^{\infty} \delta_n(p, \rho)\, r^n, \tag{3.87}$$

$$\delta_0(\rho, r) \stackrel{\text{def}}{=} \sum_{n=1}^{\infty} \delta_{0n}(\rho)\, r^n,$$

it follows from (3.85) that for $|p|<1$, $|r|\leqq 1$, $\operatorname{Re}\rho\geqq 0$,

$$\delta(p, \rho, r)$$

$$= \frac{r\beta^{-1}\{1-\alpha\{\rho+(1-p)/\beta\}\}\{\rho+(1-p)/\beta\}^{-1}-rp^{-1}\alpha\{\rho+(1-p)/\beta\}\,\delta_0(\rho, r)}{1-rp^{-1}\alpha\{\rho+(1-p)/\beta\}}. \tag{3.88}$$

Since $\delta(p, \rho, r)$ is an analytic function of p inside the unit circle for fixed ρ and r with $\mathrm{Re}\,\rho \geqq 0$, $|r| \leqq 1$, it follows from Takács' lemma (cf. app. 6) that

$$\delta_0(\rho, r) = \frac{r-\lambda(\rho, r)}{\beta\rho+1-\lambda(\rho, r)}, \qquad |r| \leqq 1, \quad \mathrm{Re}\,\rho \geqq 0, \tag{3.89}$$

where $\lambda(\rho, r)$ is the root with smallest absolute value of

$$z-r\alpha\{\rho+(1-z)/\beta\}=0, \qquad |r| \leqq 1, \quad \mathrm{Re}\,\rho \geqq 0.$$

Denoting the distribution function of the busy period by $D(t)$ it follows from (3.88), from Takács' lemma and the definition of $D_{jn}(t)$ since

$$D(t) = \sum_{n=1}^{\infty} D_{0n}(t),$$

that

$$\delta_0(\rho, 1) = \int_0^{\infty} \mathrm{e}^{-\rho t}\,\mathrm{d}D(t) = \frac{1-\lambda(\rho, 1)}{\beta\rho+1-\lambda(\rho, 1)}, \qquad \mathrm{Re}\,\rho \geqq 0. \tag{3.90}$$

Applying the expansion theorem of Lagrange (cf. app. 6) and taking for the present $\beta=1$ we have

$$\frac{1-\lambda(\rho, 1)}{\rho+1-\lambda(\rho, 1)} = \frac{1}{\rho+1} + \sum_{n=1}^{\infty} \frac{1}{n!}\frac{\mathrm{d}^{n-1}}{\mathrm{d}x^{n-1}}\left[\frac{-\rho}{(\rho+1-x)^2}\,\alpha^n\{\rho+1-x\}\right]\Bigg|_{x=0}.$$

Hence from (3.90),

$$D(t) = \sum_{n=1}^{\infty} \mathrm{e}^{-t/\beta}\frac{(t/\beta)^{n-1}}{n!}\int_0^t \{1-A^{n*}(\sigma)\}\,\frac{\mathrm{d}\sigma}{\beta}, \qquad t>0. \tag{3.91}$$

From (3.90) it follows easily that

$$D(\infty) = \begin{cases} 1 & \text{if } a<1, \\ a^{-1} & \text{if } a \geqq 1. \end{cases}$$

Moreover, it is found from (3.90) that

$$\int_0^\infty t\,\mathrm{d}D(t) = \begin{cases} \dfrac{\beta}{1-\lambda_0} & \text{if} \quad a<1, \\ \infty & \text{if} \quad a\geqq 1. \end{cases}$$

$D_{0n}(\infty)$ represents the probability that a busy period consists of n service times; its generating function is given by $\delta_0(0, r)$ and it is seen that this result agrees with the one given by (3.44).

Next, we investigate the distribution of *the maximum number of customers simultaneously present in the system during a busy period*, i.e. the maximum of $\boldsymbol{x}_t$ during a busy period. For the imbedded Markov chain $\{\boldsymbol{y}_n,\ n=0, 1, \ldots\}$ the n-step transition probabilities are denoted by $p_{ij}^{(n)}$ with $p_{ij}^{(1)}$ given by (3.2). Let $p_{h;ij}^{(n)}$ for this chain denote the n-step transition probability of going from E_i to E_j without passing E_h, where E_i denotes the state of the queueing system with i customers present at the moment of arrival of a customer. Further $\{f_{ij}^{(n)},\ n=1, 2, \ldots\}$ denotes for the chain $\{\boldsymbol{y}_n,\ n=0, 1, \ldots\}$ the first entrance distribution from E_i to E_j, and similarly $f_{h;ij}^{(n)}$, $n=1, 2, \ldots$, denotes the first entrance distribution from E_i into E_j without passing E_h. Hence for the $\boldsymbol{x}_t$-process

$$\sum_{n=1}^{\infty} f_{h-1;00}^{(n)}, \qquad h=2, 3, \ldots,$$

represents the probability that the maximum number of customers present simultaneously during a busy period is less than h.

Clearly we have for $h=1, 2, \ldots$,

$$p_{00}^{(1)} = p_{h;00}^{(1)},$$

$$p_{00}^{(n)} = p_{h;00}^{(n)} + \sum_{m=1}^{n-1} p_{h0}^{(n-m)} f_{0h}^{(m)}, \qquad n=2, 3, \ldots,$$

$$p_{h;00}^{(n)} = f_{h;00}^{(n)} + \sum_{m=1}^{n-1} p_{h;00}^{(n-m)} f_{h;00}^{(m)}, \qquad n=2, 3, \ldots .$$

Since by definition of $F_{0h}^{(n)}(t)$ (see sections I.6.4 and 3.3),

$$f_{0h}^{(n)} = F_{0h}^{(n)}(\infty), \qquad n=1, 2, \ldots; \quad h=1, 2, \ldots,$$

it follows from (3.32), (3.33), (3.36), (3.69) and (3.72) and from the above relations that for $h=1, 2, \dots; |r|<1$,

$$\sum_{n=1}^{\infty} p_{h;00}^{(n)} = -1 + \frac{1-\lambda(0, r)}{1-r}\left\{1 - \frac{\lambda^h(0, r)\,\pi_{h0}(0, r)}{1+\pi_{hh}(0, r)}\right\}, \tag{3.92}$$

$$\sum_{n=1}^{\infty} f_{h;00}^{(n)} r^n = \frac{\sum_{n=1}^{\infty} r^n p_{h;00}^{(n)}}{1 + \sum_{n=1}^{\infty} r^n p_{h;00}^{(n)}}.$$

Since the probabilities $f_{h;00}^{(n)}$, $n=1, 2, \dots$, are non-negative and their sum over n converges we have

$$\sum_{n=1}^{\infty} f_{h;00}^{(n)} = \lim_{r \uparrow 1} \sum_{n=1}^{\infty} f_{h;00}^{(n)} r^n.$$

This limit is easily calculated from (3.39), (3.69), (3.71) and (3.92). It is found by using (3.52) that

$$\sum_{n=1}^{\infty} f_{h;00}^{(n)} = 1 - \frac{\alpha}{\mu_{h-1,h}}, \qquad h=1, 2, \dots .$$

Consequently, the probability that the maximum number of customers present during a busy period is at least $h+1$ is given by

$$1 - \sum_{n=1}^{\infty} f_{h;00}^{(n)} = \frac{\alpha}{\mu_{h-1,h}}, \qquad h=1, 2, \dots . \tag{3.93}$$

Here $\mu_{h-1,h}$ represents the average of the first entrance time from E_{h-1} into E_h. An explicit expression for $\mu_{h-1,h}$ is given by (3.53). From (3.55) and (3.93) it follows by letting $h\to\infty$ that if $a\leq 1$ then the maximum number of customers present simultaneously during a busy period is finite with probability one, whereas if $a>1$ it is finite with probability a^{-1}, and infinite with probability $1-a^{-1}$. Moreover, it is not difficult to prove by starting from (3.54) that $\mu_{h-1,h}$ behaves as λ_0^{-h} for $h\to\infty$ if $a<1$, so that it is seen from (3.93) that the maximum number of customers present simultaneously during a busy period has a finite mean if $a<1$. For further properties of this distribution see section III.7.5.

II.3.6. The waiting time

Since the service time has the negative exponential distribution with mean β it follows immediately from lemma 1.2 that, for service in order of arrival, the distribution of the virtual waiting time $\boldsymbol{v}_t$ at time t is given by

$$\Pr\{\boldsymbol{v}_t<s \mid \boldsymbol{y}_0=k\} = \begin{cases} \sum_{j=0}^{\infty} \{1-\mathrm{e}^{-s/\beta}\}^{j*} \Pr\{\boldsymbol{x}_t=j \mid \boldsymbol{y}_0=k\}, & s>0, \\ 0, & s\leqq 0, \end{cases}$$

where $\Pr\{\boldsymbol{x}_t=j \mid \boldsymbol{y}_0=k\}$ is given by (3.75). From section 3.2 it is seen that $\Pr\{\boldsymbol{x}_t=j \mid \boldsymbol{y}_0=k\}$ has a limit for $t\to\infty$ if $A(t)$ is not a lattice distribution. From the above expression it is now easily seen that for $A(t)$ not a lattice distribution $\Pr\{\boldsymbol{v}_t<s \mid \boldsymbol{y}_0=k\}$ has a limit for $t\to\infty$, $k=0, 1, \ldots$; this limit distribution is given by

$$\lim_{t\to\infty} \Pr\{\boldsymbol{v}_t<s \mid \boldsymbol{y}_0=k\} = \begin{cases} 0, & \text{if } a\geqq 1, \\ 0, & s\leqq 0, \quad a<1, \\ 1-a\,\mathrm{e}^{-(1-\lambda_0)s/\beta}, & s>0, \quad a<1. \end{cases} \tag{3.94}$$

If $A(t)$ is a lattice distribution with period τ, we find for $a<1$, $0\leqq\sigma<\tau$, $k=0, 1, 2, \ldots$, (cf. section 3.2) that

$$\lim_{m\to\infty} \Pr\{\boldsymbol{v}_{m\tau+\sigma}<s \mid \boldsymbol{y}_0=k\} = \begin{cases} 0, & s\leqq 0, \\ 1-a(\sigma)\,\mathrm{e}^{-(1-\lambda_0)s/\beta}, & s>0. \end{cases} \tag{3.95}$$

For $A(t)$ not a lattice distribution and $a<1$, it is easily found from (3.94) that

$$\lim_{t\to\infty} \mathrm{E}\{\boldsymbol{v}_t \mid \boldsymbol{y}_0=k\} = \frac{a\beta}{1-\lambda_0}, \tag{3.96}$$

$$\lim_{t\to\infty} \operatorname{var}\{\boldsymbol{v}_t \mid \boldsymbol{y}_0=k\} = \frac{(2a-a^2)\,\beta^2}{\{1-\lambda_0\}^2}.$$

As above we now find for the distribution of the waiting time $\boldsymbol{w}_n$ of the nth arriving customer after $t=0$ that

$$\Pr\{\boldsymbol{w}_n<s \mid \boldsymbol{y}_0=k\} = \begin{cases} \sum_{j=0}^{\infty} \{1-\mathrm{e}^{-s/\beta}\}^{j*} \Pr\{\boldsymbol{y}_n=j \mid \boldsymbol{y}_0=k\}, & s>0, \\ 0, & s\leqq 0. \end{cases} \tag{3.97}$$

Since $\Pr\{\boldsymbol{y}_n=j \mid \boldsymbol{y}_0=k\}$ has always a limit for $n\to\infty$ it is easily found from (3.15) and the formula above that for every $k=0, 1, \ldots,$

$$\lim_{n\to\infty} \Pr\{\boldsymbol{w}_n<s \mid \boldsymbol{y}_0=k\} = \begin{cases} 0, & \text{if} \quad a\geqq 1, \\ 0, & s\leqq 0, \quad a<1, \\ 1-\lambda_0\,\mathrm{e}^{-(1-\lambda_0)s/\beta}, & s>0, \quad a<1. \end{cases} \tag{3.98}$$

Further

$$\lim_{n\to\infty} \mathrm{E}\{\boldsymbol{w}_n \mid \boldsymbol{y}_0=k\} = \frac{\lambda_0\beta}{1-\lambda_0}, \tag{3.99}$$

$$\lim_{n\to\infty} \mathrm{var}\{\boldsymbol{w}_n \mid \boldsymbol{y}_0=k\} = \frac{2\lambda_0-\lambda_0^2}{(1-\lambda_0)^2}\,\beta^2.$$

It should be noted that the limit distributions of $\boldsymbol{w}_n$ and $\boldsymbol{v}_t$ are not, in general, identical.

To obtain an expression for $\Pr\{\boldsymbol{w}_n<s \mid \boldsymbol{y}_0=0\}$ we note that from (3.97) and (3.72),

$$\sum_{n=1}^{\infty} r^n \Pr\{\boldsymbol{w}_n<s \mid \boldsymbol{y}_0=0\} = -1+\frac{1}{1-r}\{1-\lambda(0,r)\,\mathrm{e}^{-(1-\lambda(0,r))s/\beta}\}, \quad s>0, \quad |r|<1. \tag{3.100}$$

By applying Lagrange's expansion theorem (cf. app. 6) to the function

$$\lambda(0,r)\,\mathrm{e}^{-(1-\lambda(0,r))s/\beta},$$

we find for $s>0$ that

$$\Pr\{\boldsymbol{w}_n<s \mid \boldsymbol{y}_0=0\} = \begin{cases} 1-\mathrm{e}^{-s/\beta}\int_0^\infty \mathrm{e}^{-t/\beta}\,\mathrm{d}A(t), & n=1, \\ 1-\sum_{i=1}^{n}\mathrm{e}^{-s/\beta}\int_0^\infty \mathrm{e}^{-\sigma/\beta}\left(\frac{s+\sigma}{\beta}\right)^{i-1}\frac{\mathrm{d}A^{i*}(\sigma)}{i!} & \\ \quad -\sum_{i=2}^{n}\frac{s}{i\beta}\,\mathrm{e}^{-s/\beta}\int_0^\infty \mathrm{e}^{-\sigma/\beta}\left(\frac{s+\sigma}{\beta}\right)^{i-2}\frac{\mathrm{d}A^{i*}(\sigma)}{(i-2)!}, & n=2,3,\ldots. \end{cases} \tag{3.101}$$

II.3.7. The departure process

Let $\boldsymbol{h}_t$ denote the number of departures from the system during $[0, t)$ with $\boldsymbol{y}_0=k$, so that $\boldsymbol{h}_{t+\boldsymbol{t}_n}-\boldsymbol{h}_{\boldsymbol{t}_n}$ with $\boldsymbol{y}_n=j$ represents the number of departures during the time interval of length t after the arrival of the nth customer after $t=0$, if this customer finds j customers present at his arrival. Since the queueing process has stationary transition probabilities, $\boldsymbol{h}_t$ and $\boldsymbol{h}_{t+\boldsymbol{t}_n}-\boldsymbol{h}_{\boldsymbol{t}_n}$ have the same distribution if $\boldsymbol{y}_0=\boldsymbol{y}_n$.

We now have for $t>0$,

$$\Pr\{\boldsymbol{h}_t=h,\ t<\boldsymbol{t}_1 \mid \boldsymbol{y}_0=k\}=p_{k,k+1-h}(t)\{1-A(t)\}, \tag{3.102}$$

$$\Pr\{\boldsymbol{h}_t=h,\ \boldsymbol{t}_n<t\leqq\boldsymbol{t}_{n+1} \mid \boldsymbol{y}_0=k\}$$

$$=\sum_{i=0}^{\infty}\int_{u=0}^{t} p_{i,k+n+1-h}(t-u)\{1-A(t-u)\}\,\mathrm{d}_u\Pr\{\boldsymbol{y}_n=i,\ \boldsymbol{t}_n<u \mid \boldsymbol{y}_0=k\},$$

$$n=1, 2, \ldots .$$

Define

$$H_{kh}^{(n)}(t)=\begin{cases}\Pr\{\boldsymbol{h}_t=h,\ t<\boldsymbol{t}_1 \mid \boldsymbol{y}_0=k\}, & n=0,\\ \Pr\{\boldsymbol{h}_t=h,\ \boldsymbol{t}_n<t\leqq\boldsymbol{t}_{n+1} \mid \boldsymbol{y}_0=k\}, & n=1, 2, \ldots .\end{cases} \tag{3.103}$$

From these relations, from (3.15) and (3.66) it follows that

$$H_{kh}^{(0)}(t)=\begin{cases}0, & h>k+1,\\ \left\{\int_0^t \mathrm{e}^{-\tau/\beta}\dfrac{(\tau/\beta)^k}{k!}\dfrac{\mathrm{d}\tau}{\beta}\right\}(1-A(t)), & h=k+1,\\ \mathrm{e}^{-t/\beta}\dfrac{(t/\beta)^h}{h!}(1-A(t)), & 0\leqq h\leqq k,\end{cases} \tag{3.104}$$

and for $n=1, 2, \ldots$,

$$H_{kh}^{(n)}(t)=\begin{cases}0, & h>k+n+1,\\ \displaystyle\sum_{i=0}^{k+n}\int_0^t\left\{\int_0^{t-u}\mathrm{e}^{-\tau/\beta}\frac{(\tau/\beta)^i}{i!}\frac{\mathrm{d}\tau}{\beta}\right\}\{1-A(t-u)\}\,\mathrm{d}_u P_{ki}^{(n)}(u), & \\ & h=k+n+1,\\ \displaystyle\sum_{i=k+n-h}^{k+n}\int_0^t \mathrm{e}^{-(t-u)/\beta}\frac{((t-u)/\beta)^{i-(k+n-h)}}{\{i-(k+n-h)\}!}\{1-A(t-u)\}\,\mathrm{d}_u P_{ki}^{(n)}(u), & \\ & 0\leqq h<k+n+1.\end{cases} \tag{3.105}$$

Defining

$$h^{(n)}(p,\rho,q) \overset{\text{def}}{=} \sum_{k=0}^{\infty} p^k \sum_{h=0}^{\infty} q^h \int_0^{\infty} e^{-\rho t} H_{kh}^{(n)}(t)\,dt, \tag{3.106}$$

for $|p|<1$, $|q|\leqq 1$, $\text{Re}\,\rho>0$, $n=0, 1, \ldots$, it follows from (3.104) that

$$h^{(0)}(p,\rho,q) = \frac{q}{1-pq}\,\frac{1-\alpha(\rho)}{\rho} + \frac{1-q}{(1-p)(1-pq)}\,\frac{1-\alpha\{\rho+(1-pq)/\beta\}}{\rho+(1-pq)/\beta}.$$

From (3.105) and (3.106) we obtain, noting that

$$P_{ki}^{(n)}(t)=0 \qquad \text{for} \quad i>k+n,$$

for $|r|<1$, $|p|<1$, $|q|\leqq 1$, $\text{Re}\,\rho>0$,

$$\sum_{n=1}^{\infty} r^n h^{(n)}(p,\rho,q)$$

$$= q\sum_{i=0}^{\infty} \int_0^{\infty} e^{-\rho t}\left\{\int_0^t e^{-\tau/\beta}\,\frac{(\tau/\beta)^i}{i!}\,\frac{d\tau}{\beta}\right\}\{1-A(t)\}\,\pi_i(pq,\rho,rq)\,dt$$

$$+\sum_{m=0}^{\infty} q^{-m} \sum_{i=0}^{\infty} \int_0^{\infty} e^{-\rho t - t/\beta}\,\frac{(t/\beta)^i}{i!}\,\{1-A(t)\}\,\pi_{i+m}(pq,\rho,rq)\,dt. \tag{3.107}$$

Using the relations (3.71) it follows from (3.106) and (3.107) for $|p|<1$, $|q|\leqq 1$, $\text{Re}\,\rho>0$ that

$$\sum_{n=0}^{\infty} h^{(n)}(p,\rho,q) = \frac{q}{1-pq}\,\frac{1}{1-q\alpha(\rho)}\,\frac{1-\alpha(\rho)}{\rho}$$

$$+\frac{1-q}{(1-p)(1-pq)}\,\frac{1}{1-p^{-1}\alpha\{\rho+(1-pq)/\beta\}}\,\frac{1-\alpha\{\rho+(1-pq)/\beta\}}{\rho+(1-pq)/\beta}$$

$$-\frac{1-q}{1-pq}\,\frac{1}{\rho+(1-\lambda(\rho,q))/\beta}\left\{\frac{1}{1-p^{-1}\alpha\{\rho+(1-pq)/\beta\}}-\frac{1}{1-q\alpha(\rho)}\right\}. \tag{3.108}$$

From (3.108) we obtain immediately for $\text{Re}\,\rho>0$,

$$\int_0^{\infty} e^{-\rho t}\,E\{\boldsymbol{h}_t \mid \boldsymbol{y}_0=0\}\,dt = \frac{1}{\rho}\,\frac{1}{1-\alpha(\rho)} - \frac{1}{1-\alpha(\rho)}\,\frac{1}{\rho+(1-\lambda(\rho,1))/\beta}. \tag{3.109}$$

Denoting by $m(t)$ the renewal function of the renewal process with $A(t)$ as renewal distribution so that

$$\int_0^\infty e^{-\rho t}\, m(t)\, dt = \frac{1}{\rho} \frac{\alpha(\rho)}{1-\alpha(\rho)}, \qquad \operatorname{Re} \rho > 0,$$

it is seen that $1+m(t)$ represents the average number of arrivals in $[0, t)$, the arrival at $t=0$ being included. Hence the formulas (3.109) and (3.76) express the obvious relation that the average number of departures in $[0, t)$ is equal to the difference of the average number of arrivals in $[0, t)$ and the average number of customers present at time t.

Since the queueing process has stationary transition probabilities it follows that

$$\begin{aligned}\Pr\{\boldsymbol{h}_{t+t_n} - \boldsymbol{h}_{t_n} = h \mid \boldsymbol{y}_0 = k\} &= \sum_{i=0}^{\infty} \Pr\{\boldsymbol{h}_{t+t_n} - \boldsymbol{h}_{t_n} = h \mid \boldsymbol{y}_n = i\}\, p_{ki}^{(n)} \\ &= \sum_{i=0}^{\infty} \Pr\{\boldsymbol{h}_t = h \mid \boldsymbol{y}_0 = i\}\, p_{ki}^{(n)}. \qquad (3.110)\end{aligned}$$

Consequently, for $k=0$ from (3.72) and (3.106) for $\operatorname{Re} \rho \geqq 0$, $|r|<1$, we have

$$\sum_{n=1}^{\infty} r^n \sum_{h=0}^{\infty} q^h \int_0^\infty e^{-\rho t} \Pr\{\boldsymbol{h}_{t+t_n} - \boldsymbol{h}_{t_n} = h \mid \boldsymbol{y}_0 = 0\}\, dt = \sum_{m=0}^{\infty} h^{(m)}(\lambda(0, r), \rho, q) \frac{1-\lambda(0, r)}{1-r}.$$

It is readily verified that the right-hand side of (3.110) has a limit for $n \to \infty$, hence (cf. app. 1) we have

$$\begin{aligned}&\lim_{n\to\infty} \sum_{h=0}^{\infty} q^h \int_0^\infty e^{-\rho t} \Pr\{\boldsymbol{h}_{t+t_n} - \boldsymbol{h}_{t_n} = h \mid \boldsymbol{y}_0 = 0\}\, dt \\ &= \lim_{r\uparrow 1} (1-\lambda(0, r)) \sum_{m=0}^{\infty} h^{(m)}(\lambda(0, r), \rho, q), \qquad \operatorname{Re} \rho > 0, \quad |q| \leqq 1. \qquad (3.111)\end{aligned}$$

From (3.108) and (3.111) it is seen that if $a \geqq 1$ so that $\lambda(0, 1) = 1$ then for $\operatorname{Re} \rho > 0$, $|q| \leqq 1$,

$$\lim_{n\to\infty} \sum_{h=0}^{\infty} q^h \int_0^\infty e^{-\rho t} \Pr\{\boldsymbol{h}_{t+t_n} - \boldsymbol{h}_{t_n} = h \mid \boldsymbol{y}_0 = 0\}\, dt = \frac{1}{\rho + (1-q)/\beta},$$

and hence it follows easily that if $a \geqq 1$,

$$\lim_{n\to\infty} \Pr\{\boldsymbol{h}_{t+t_n} - \boldsymbol{h}_{t_n} = h \mid \boldsymbol{y}_0 = 0\} = \frac{(t/\beta)^h}{h!} e^{-t/\beta}, \qquad h = 0, 1, \ldots . \qquad (3.112)$$

II.4. THE QUEUEING SYSTEM $M/G/1$

II.4.1. Introduction

The model for the queueing system $M/G/1$ follows immediately from the general description given in section 1.2. In general we shall assume that the distribution function $A_1(t)$ of the moment of arrival of the first arriving customer after $t=0$ is the same as that of the interarrival times $\boldsymbol{\sigma}_n$, $n=2, 3, \dots$; so that (cf. section 1.2),

$$A_1(t) = A(t) = \begin{cases} 1 - \mathrm{e}^{-t/\alpha}, & t \geqq 0, \\ 0, & t < 0. \end{cases}$$

Consequently, the arrival process is a Poisson process.

The distribution function $B_1(t)$ of the residual service time of a customer, if any, being served at $t=0$ and the distribution $B(t)$ of the service times of customers whose services start after $t=0$ will not be specified except that they must satisfy the general conditions as described in section 1.2. To simplify our results it will in general be assumed that at $t=0$ a new service starts. This restriction, however, is not essential. The average service time is denoted by β and the second moment of the distribution $B(t)$ by

$$\beta_2 \stackrel{\text{def}}{=} \int_0^\infty t^2 \, \mathrm{d}B(t). \tag{4.1}$$

β_2 will be assumed to be finite, although this is not essential for a great number of the results to be derived.

We now introduce the departure time $\boldsymbol{r}_n'$, i.e. the moment of the nth departure from the queueing system after $t=0$; $n=1, 2, \dots$. For the process $\{\boldsymbol{x}_t,\ t \in [0, \infty)\}$, where $\boldsymbol{x}_t$ is the number of customers in the system at time t, the sequence $\{r'_n,\ n=1, 2, \ldots\}$ plays an important role. From the general description of our model it is clear that at a departure a new service is started immediately if the departing customer leaves at least one customer behind in the system, and, if not, the next service is started at the moment of arrival of the next customer. From lemma 1.1 and the independence of the service times it follows that at any time r'_n, $n=1, 2, \ldots$, the future development of the process $\{x_t,\ t\in[0, \infty)\}$ depends only on the state of the system at time r'_n, i.e. on $x_{r'_n+}$, and that this development is independent of the past before $t=r'_n$. Note that $A(t)$ is continuous and $B(0+)=0$.

Starting from these considerations it is not difficult to show that the discrete time parameter process $\{z_n,\ n=1, 2, \ldots\}$ defined by

$$z_n \stackrel{\text{def}}{=} x_{r'_n+}, \qquad n=1, 2, \ldots, \tag{4.2}$$

is a discrete time parameter Markov chain with state space the set of non-negative integers. (The proof of this statement is analogous to that for the sequence $\{y_n,\ n=0, 1, \ldots\}$ in section I.6.4.) Obviously z_n represents the number of customers left behind in the system by the customer departing at time r'_n.

If at $t=0$ a new service is started we can define

$$z_0 \stackrel{\text{def}}{=} x_{0+}, \qquad r'_0 \stackrel{\text{def}}{=} 0, \tag{4.3}$$

and whenever in the following z_0 is supposed to be defined this will always mean that at $t=0$ a new service is started with service-time distribution $B(t)$ if $z_0>0$. Clearly the process $\{z_n,\ n=0, 1, \ldots\}$ is also a discrete time parameter Markov chain with stationary transition probabilities.

The sequence $\{r'_n,\ n=0, 1, \ldots\}$ does not define a renewal process since the variables $r'_{n+1}-r'_n$, $n=0, 1, \ldots$, are not indentically distributed. The distribution of $r'_{n+1}-r'_n$ depends on the number of customers in the system at $t=r'_n$. Obviously for $n=0, 1, \ldots$,

$$\Pr\{r'_{n+1}-r'_n<t \mid z_n=i\} = \begin{cases} 0, & t\leqq 0,\\ B(t) \quad \text{if } i>0, & t>0,\\ \int_0^t \{1-\mathrm{e}^{-(t-\tau)/\alpha}\}\,\mathrm{d}B(\tau) & \text{if } i=0,\ t>0. \end{cases} \tag{4.4}$$

Consequently, the x_t-process is not regenerative with respect to the sequence $\{r'_n,\ n=0, 1, \ldots\}$ in the sense as defined in section I.6.4. Nevertheless the method discussed in that section is also useful for the investigation of the queueing system $M/G/1$.

II.4.2. The imbedded Markov chain z_n for $n\to\infty$

The Markov chain $\{z_n, n=0, 1, \ldots\}$ defined in the preceding section is an imbedded process of the x_t-process. To determine the one-step transition matrix (p_{ij}) of this Markov chain, which has stationary transition probabilities, we see that between two successive departures customers can only arrive. Hence during such a time interval the x_t-process behaves as a pure birth process with birth rate α.

Consequently, from section I.4.2,

$$p_{ij} \stackrel{\text{def}}{=} \Pr\{z_{n+1}=j \mid z_n=i\}, \qquad n=0, 1, \ldots,$$

$$= \begin{cases} \displaystyle\int_0^\infty \frac{(t/\alpha)^j}{j!}\, \mathrm{e}^{-t/\alpha}\, \mathrm{d}B(t), & j=0, 1, \ldots;\quad i=0, \\ 0, & j=0, 1, \ldots, i-2;\quad i=2, 3, \ldots, \\ \displaystyle\int_0^\infty \frac{(t/\alpha)^{j+1-i}}{(j+1-i)!}\, \mathrm{e}^{-t/\alpha}\, \mathrm{d}B(t), & j=i-1, i, \ldots;\quad i=1, 2, \ldots . \end{cases} \tag{4.5}$$

From this transition matrix it follows immediately that any two states of the Markov chain $\{z_n, n=0, 1, \ldots\}$ can be reached from each other with positive probability, so that the chain is irreducible. Moreover, since all $p_{ii}>0$, all states are aperiodic. Define

$$a \stackrel{\text{def}}{=} \beta/\alpha. \tag{4.6}$$

To investigate the type of the states of the chain $\{z_n, n=0, 1, \ldots\}$ we apply Foster's criteria (cf. section I.2.4). From (4.5) and (4.6) it is found that

$$\sum_{j=0}^{\infty} j p_{0j} = a, \tag{4.7}$$

$$\sum_{i=0}^{\infty} j p_{ij} = i + a - 1, \qquad i=1, 2, \ldots .$$

If $a<1$ choose ε such that $0<\varepsilon\leqq 1-a$, and take $z_j=j, j=0, 1, \ldots$; applying Foster's second criterion it is seen that all states of the chain are ergodic if $a<1$. Moreover, it follows in the same way from Foster's criterion (v) that all states are recurrent if $a\leqq 1$.

Denote the average first entrance time from state E_j into E_0 by ν_{j0}. From the structure of p_{ij} it follows that

$$\nu_{j,j-1}=\nu_{10}, \qquad j=1, 2, \ldots, \tag{4.8}$$

and since every passage from E_j to E_0 leads *via* E_{j-1} we have

$$\nu_{j0}=\nu_{10}+\nu_{j-1,0}, \qquad j=2, 3, \ldots,$$

hence

$$\nu_{j0}=j\nu_{10}, \qquad j=1, 2, \ldots . \tag{4.9}$$

From (4.7) and (4.9) we have

$$\sum_{j=1}^{\infty} p_{0j}\nu_{j0} = a\nu_{10},$$

$$\sum_{j=1}^{\infty} p_{ij}\nu_{j0} = \nu_{i0} - (1-a)\,\nu_{10}, \qquad i=1, 2, \ldots .$$

Applying Foster's third criterion it follows that if the chain is ergodic then

$$0<\nu_{10}<\infty, \qquad \nu_{i0}-(1-a)\,\nu_{10}=\nu_{i0}-1, \qquad i=1, 2, \ldots,$$

so that

$$\nu_{j0} = \frac{j}{1-a}, \qquad j=1, 2, \ldots, \tag{4.10}$$

and hence $a<1$.

From (4.5) it follows easily that

$$\sum_{j=0}^{\infty} p_{ij}p^j = p^{i-1}\beta\{(1-p)/\alpha\}, \qquad i=1, 2, \ldots; \quad |p|\leqq 1.$$

Consequently,

$$z_i = A\mu_0^i, \qquad i=0, 1, \ldots, \tag{4.11}$$

is a solution of

$$z_i = \sum_{j=0}^{\infty} p_{ij}z_j, \qquad i=1, 2, \ldots, \tag{4.12}$$

if μ_0 is a root of

$$p = \beta\{(1-p)/\alpha\}. \tag{4.13}$$

From Takács' lemma (cf. app. 6) it is seen that this equation has one root inside the unit circle if $a>1$ and no root inside the unit circle if $a\leqq 1$. Consequently, by Foster's criterion (*iv*) the Markov chain $\{z_n, n=0, 1, \ldots\}$ is transient if $a>1$.

From the results found above it follows that *the imbedded Markov chain* $\{z_n, n=0, 1, ...\}$ *is ergodic if* $a<1$, *is null recurrent if* $a=1$ *and transient if* $a>1$.

If we compare the set of equations (4.12) with the set of equations (3.14) we see that apart from the constants these sets are analogous. If the Markov chain $\{y_n, n=0, 1, ...\}$ of section 3.2 is ergodic then its stationary distribution is the only bounded non-constant solution with $\sum_{i=0}^{\infty} z_i=1$ of (3.14) (cf. theorem I.2.5). Therefore, it follows that if $a>1$ and μ_0 is the root, inside the unit circle, of (4.13), then (4.11) represents the only bounded non-constant solution of (4.12). Hence, from theorem I.2.6 it is seen that if $a>1$ then z_i, as given by (4.11) with $A=1$, is the probability for the chain $\{z_n, n=0, 1, ...\}$ that when starting in E_i the system reaches E_0.

For $a<1$ we have from theorem I.2.5,

$$v_j \stackrel{\text{def}}{=} \lim_{n\to\infty} p_{ij}^{(n)} = \lim_{n\to\infty} \Pr\{z_n=j \mid z_0=i\}, \qquad j=0, 1, ..., \tag{4.14}$$

and v_j, $j=0, 1, ...$, satisfy and are uniquely determined by

$$v_j = \sum_{i=0}^{\infty} v_i p_{ij}, \qquad j=0, 1, ..., \tag{4.15}$$

$$\sum_{j=0}^{\infty} v_j = 1.$$

Putting

$$v(p) \stackrel{\text{def}}{=} \sum_{j=0}^{\infty} v_j p^j, \qquad |p|\leqq 1, \tag{4.16}$$

it follows easily from (4.15) and (4.5) that

$$v(p) = \frac{(1-p)\,\beta\{(1-p)/\alpha\}}{\beta\{(1-p)/\alpha\}-p}\, v_0.$$

Since Takács' lemma shows that for $a<1$ the zero with the smallest absolute value of the denominator of this expression is $p=1$ and this is a simple zero, it follows from the norming condition in (4.15) that

$$v_0 = 1 - a, \tag{4.17}$$

$$v(p) = (1-a)(1-p)\,\frac{\beta\{(1-p)/\alpha\}}{\beta\{(1-p)/\alpha\}-p}, \qquad |p|\leqq 1.$$

II.4.3. Transition probabilities and first passage times

For $n=1, 2, \ldots; i, j=0, 1, \ldots; t>0$, let

$$P_{ij}^{(n)}(t) \stackrel{\text{def}}{=} \Pr\{z_n=j, r_n'<t \mid z_0=i\}, \tag{4.18}$$

denote the conditional probability that the nth customer departing after $t=0$ leaves the system at time $r_n'<t$ and that j customers stay behind in the system if i customers are present in the system at $t=0$. A new service is then started if $i>0$.

Since between two successive departures the x_t-process behaves like a Poisson process with parameter α^{-1} it follows from (4.18) that for $t>0$,

$$P_{0j}^{(1)}(t) = \int_0^t \{1-e^{-(t-\tau)/\alpha}\} e^{-\tau/\alpha} \frac{(\tau/\alpha)^j}{j!} dB(\tau), \qquad j=0, 1, \ldots, \tag{4.19}$$

$$P_{ij}^{(1)}(t) = \begin{cases} \displaystyle\int_0^t e^{-\tau/\alpha} \frac{(\tau/\alpha)^{j+1-i}}{(j+1-i)!} dB(\tau), & j=i-1, i, \ldots; \quad i=1, 2, \ldots, \\ 0, & j=0, 1, \ldots, i-2; \quad i=2, 3, \ldots . \end{cases}$$

Obviously we have for $t>0$; $i, j=0, 1, \ldots,$

$$P_{ij}^{(n)}(t) = \sum_{h=0}^{\infty} \int_{u=0}^{t} P_{ih}^{(n-1)}(t-u)\, d_u P_{hj}^{(1)}(u), \qquad n=2, 3, \ldots . \tag{4.20}$$

It is seen immediately that the functions $P_{ij}^{(n)}(t)$ are uniquely determined by the relations (4.19) and (4.20).

Define for $\operatorname{Re} \rho \geqq 0$, $|p| \leqq 1$, $|r|<1$,

$$\pi_{ij}^{(n)}(\rho) \stackrel{\text{def}}{=} \int_0^{\infty} e^{-\rho t} dP_{ij}^{(n)}(t), \tag{4.21}$$

$$\pi_{ij}(\rho, r) \stackrel{\text{def}}{=} \sum_{n=1}^{\infty} r^n \pi_{ij}^{(n)}(\rho),$$

$$\pi_i^{(n)}(p, \rho) \stackrel{\text{def}}{=} \sum_{j=0}^{\infty} p^j \pi_{ij}^{(n)}(\rho),$$

$$\pi_i(p, \rho, r) \stackrel{\text{def}}{=} \sum_{n=1}^{\infty} r^n \pi_i^{(n)}(p, \rho).$$

From (4.19), ..., (4.21) it is easily verified that for $\operatorname{Re}\rho \geqq 0$, $|p| \leqq 1$, $|r|<1$,

$$\pi_0(p,\rho,r)\{1-rp^{-1}\beta\{\rho+(1-p)/\alpha\}\} = \frac{r\beta\{\rho+(1-p)/\alpha\}}{p(1+\alpha\rho)}\{(p-1-\alpha\rho)\,\pi_{00}(\rho,r)+p\},$$

$$\pi_i(p,\rho,r)\{1-rp^{-1}\beta\{\rho+(1-p)/\alpha\}\} = \frac{r\beta\{\rho+(1-p)/\alpha\}}{p(1+\alpha\rho)}\{(p-1-\alpha\rho)\,\pi_{i0}(\rho,r)+(1+\alpha\rho)\,p^i\},$$
$$i=1,2,\ldots .$$

By applying Takács' lemma to the equation in z,

$$z = r\beta\{\rho+(1-z)/\alpha\}, \tag{4.22}$$

it follows that the root $\mu(\rho, r)$ with the smallest absolute value is less than 1 if $\operatorname{Re}\rho \geqq 0$, $|r|<1$. Since for $\operatorname{Re}\rho \geqq 0$, $|r|<1$, the functions $\pi_i(p,\rho,r)$, $i=0, 1, \ldots$, should be analytic functions of p for $|p| \leqq 1$ (cf. (4.21)), it follows that $\mu(\rho, r)$ is a zero of the right-hand sides of the expressions for $\pi_0(p,\rho,r)$ and $\pi_i(p,\rho,r)$ given above. Hence for $|p| \leqq 1$, $\operatorname{Re}\rho \geqq 0$, $|r|<1$, we have

$$\pi_{00}(\rho,r) = \frac{\mu(\rho,r)}{1+\alpha\rho-\mu(\rho,r)}, \tag{4.23}$$

$$\pi_{i0}(\rho,r) = \frac{(1+\alpha\rho)\,\mu^i(\rho,r)}{1+\alpha\rho-\mu(\rho,r)}, \qquad i=1,2,\ldots,$$

$$\pi_0(p,\rho,r) = \frac{r\beta\{\rho+(1-p)/\alpha\}}{p-r\beta\{\rho+(1-p)/\alpha\}}\,\frac{p-\mu(\rho,r)}{1+\alpha\rho-\mu(\rho,r)},$$

$$\pi_i(p,\rho,r) = \frac{r\beta\{\rho+(1-p)/\alpha\}}{p-r\beta\{\rho+(1-p)/\alpha\}}\left\{p^i - \frac{1+\alpha\rho-p}{1+\alpha\rho-\mu(\rho,r)}\,\mu^i(\rho,r)\right\},$$
$$i=1,2,\ldots .$$

From (4.18) it follows that

$$p_{ij}^{(n)} = \lim_{t\to\infty} P_{ij}^{(n)}(t), \qquad n=1,2,\ldots,$$

where $(p_{ij}^{(n)})$ is the n-step transition matrix of the Markov chain $\{z_n, n=0,1,\ldots\}$ (cf. (4.5)). Hence from (4.23) and the continuity properties of $\mu(\rho, r)$ for $\rho\to 0$

(cf. Takács' lemma) with $|p|\leqq 1$, $|r|<1$, we obtain the equations

$$\sum_{n=1}^{\infty}\sum_{j=0}^{\infty} r^n p^j p_{ij}^{(n)} = \begin{cases} \dfrac{r\beta\{(1-p)/\alpha\}}{p-r\beta\{(1-p)/\alpha\}}\,\dfrac{p-\mu(0,r)}{1-\mu(0,r)}, & i=0, \\ \dfrac{r\beta\{(1-p)/\alpha\}}{p-r\beta\{(1-p)/\alpha\}}\,\dfrac{p^i(1-\mu(0,r))-(1-p)\,\mu^i(0,r)}{1-\mu(0,r)}, & \\ & i=1,2,\dots . \end{cases} \tag{4.24}$$

Next, we introduce for $t>0$; $i,j=0,1,\ldots$, the conditional probability

$$F_{ij}^{(1)}(t) \stackrel{\text{def}}{=} P_{ij}^{(1)}(t), \tag{4.25}$$

$$F_{ij}^{(n)}(t) \stackrel{\text{def}}{=} \Pr\{z_n=j, r_n'<t, z_h\neq j,\ h=1,\ldots,n-1 \mid z_0=i\}, \qquad n=2,3,\ldots,$$

that the nth customer departing after $t=0$ leaves the system at time $r_n'<t$ and that at this departure for the first time after $t=0$, j customers stay behind in the system, if i customers are in the system at $t=0$ and a new service is started in case $i>0$.

Obviously,

$$F_{ij}(t) \stackrel{\text{def}}{=} \sum_{n=1}^{\infty} F_{ij}^{(n)}(t), \tag{4.26}$$

represents the distribution function of the first entrance time from the state with i customers present at a departure to the state with j customers present at a departure.

From the above definitions it follows for $t>0$ that

$$P_{ij}^{(n)}(t) = F_{ij}^{(n)}(t) + \sum_{m=1}^{n-1}\int_0^t F_{jj}^{(n-m)}(t-u)\,\mathrm{d}P_{ij}^{(m)}(u), \qquad n=2,3,\ldots . \tag{4.27}$$

Defining for $0<t<\infty$,

$$m_{ij}(t) \stackrel{\text{def}}{=} \sum_{n=1}^{\infty} P_{ij}^{(n)}(t), \tag{4.28}$$

then $m_{ij}(t)$ represents the renewal function of the general renewal process with renewal distribution $F_{jj}(t)$ and with $F_{ij}(t)$ as the distribution of the first renewal time (cf. chapter I.6).

Since $m_{ij}(t)$ is non-negative and finite for finite values of t it follows easily that

$$\sum_{j=0}^{\infty} p^j \int_0^{\infty} \mathrm{e}^{-\rho t}\,\mathrm{d}m_{ij}(t) = \lim_{r\uparrow 1} \pi_i(p,\rho,r), \qquad |p|<1, \quad \mathrm{Re}\,\rho>0, \tag{4.29}$$

and by Takács' lemma

$$\mu(\rho, 1) = \lim_{r\uparrow 1} \mu(\rho, r), \qquad \operatorname{Re} \rho \geqq 0. \tag{4.30}$$

From (4.26), (4.27) and (4.28) it follows that

$$m_{ij}(t) = F_{ij}(t) + \int_0^t F_{jj}(t-u)\,\mathrm{d}m_{ij}(u), \qquad t>0. \tag{4.31}$$

Define

$$\varphi_{ij}(\rho, r) \stackrel{\text{def}}{=} \sum_{n=1}^{\infty} r^n \int_0^{\infty} \mathrm{e}^{-\rho t}\,\mathrm{d}F_{ij}^{(n)}(t), \qquad \operatorname{Re}\rho\geqq 0, \quad |r|\leqq 1, \tag{4.32}$$

$$\varphi_{ij}(\rho) \stackrel{\text{def}}{=} \int_0^{\infty} \mathrm{e}^{-\rho t}\,\mathrm{d}F_{ij}(t), \qquad \operatorname{Re}\rho\geqq 0.$$

It follows from (4.27) that

$$\varphi_{ij}(\rho, r) = \frac{\pi_{ij}(\rho, r)}{1+\pi_{jj}(\rho, r)}, \qquad \operatorname{Re}\rho\geqq 0, \quad |r|<1. \tag{4.33}$$

Since the relation (4.23) shows that

$$1 + \pi_{jj}(\rho, r)$$

$$= \frac{\mu^j(\rho, r)}{1+\alpha\rho-\mu(\rho, r)}\,\frac{1}{j!}\,\frac{\mathrm{d}^j}{\mathrm{d}p^j}\,\frac{(1+\alpha\rho-p)\,r\beta\{\rho+(1-p)/\alpha\}}{r\beta\{\rho+(1-p)/\alpha\}-p}\Bigg|_{p=0}, \tag{4.34}$$

we have for $\operatorname{Re}\rho\geqq 0$, $|r|\leqq 1$, $j=0, 1, \ldots$,

$$\{1-\varphi_{jj}(\rho, r)\}^{-1}$$

$$= \frac{\mu^j(\rho, r)}{1+\alpha\rho-\mu(\rho, r)}\,\frac{1}{j!}\,\frac{\mathrm{d}^j}{\mathrm{d}p^j}\,\frac{(1+\alpha\rho-p)\,r\beta\{\rho+(1-p)/\alpha\}}{r\beta\{\rho+(1-p)/\alpha\}-p}\Bigg|_{p=0}, \tag{4.35}$$

$$\varphi_{0j}(\rho, r) = \mu^{-j}(\rho, r)$$

$$\cdot\left\{1-(1+\alpha\rho-\mu(\rho, r))\,\frac{\left.\dfrac{\mathrm{d}^j}{\mathrm{d}p^j}\,\dfrac{r\beta\{\rho+(1-p)/\alpha\}}{r\beta\{\rho+(1-p)/\alpha\}-p}\right|_{p=0}}{\left.\dfrac{\mathrm{d}^j}{\mathrm{d}p^j}\,\dfrac{(1+\alpha\rho-p)\,r\beta\{\rho+(1-p)/\alpha\}}{r\beta\{\rho+(1-p)/\alpha\}-p}\right|_{p=0}}\right\}.$$

Hence for $\operatorname{Re}\rho\geqq 0$, $|r|\leqq 1$,

$$\varphi_{00}(\rho, r) = \frac{\mu(\rho, r)}{1+\alpha\rho}, \qquad \varphi_{00}(\rho) = \frac{\mu(\rho, 1)}{1+\alpha\rho}. \tag{4.36}$$

In the same way we obtain for Re $\rho \geqq 0$, $|r| \leqq 1$,

$$\varphi_{i0}(\rho, r) = \mu^i(\rho, r), \qquad \varphi_{i,i-1}(\rho, r) = \mu(\rho, r), \qquad i=1, 2, \ldots, \tag{4.37}$$

with

$$\mu^i(\rho, r) = \sum_{n=i}^{\infty} \frac{i}{n} r^n \int_0^{\infty} \mathrm{e}^{-(\rho+1/\alpha)t} \frac{(t/\alpha)^{n-i}}{(n-i)!} \, \mathrm{d}B^{n*}(t).$$

From these relations it is seen that for $i=2, 3, \ldots$,

$$F_{i0}^{(n)}(t) = \sum_{m=1}^{n-1} \int_0^t F_{i,i-1}^{(n-m)}(t-u) \, \mathrm{d}F_{i-1,0}^{(m)}(u), \qquad n=2, 3, \ldots, \tag{4.38}$$

$$F_{i0}(t) = \int_0^t F_{i,i-1}(t-u) \, \mathrm{d}F_{i-1,0}(u).$$

Both relations are obvious. For instance the latter follows from the fact that the entrance time from a state with i customers left at a departure to the state with zero customers left is the sum of the entrance time from i to $i-1$ and the entrance time from $i-1$ to zero, these entrance times being independent.

Since Takács' lemma applied to the equation (4.22) shows that (cf. also (4.13)),

$$\mu_0 \stackrel{\text{def}}{=} \mu(0, 1) = \lim_{r \uparrow 1} \mu(0, r) = \lim_{\rho \downarrow 0} \mu(\rho, 1),$$

and

$$\mu_0 \begin{cases} =1 & \text{if } a \leqq 1, \\ <1 & \text{if } a > 1, \end{cases}$$

it follows from (4.36) and (4.37) that for all i, j,

$$F_{ij}(\infty) \begin{cases} =1 & \text{if } a \leqq 1, \\ <1 & \text{if } a > 1, \end{cases} \tag{4.39}$$

thus if $a \leqq 1$ the entrance time distributions are proper probability distributions.

From (4.22) we obtain

$$\lim_{\rho \downarrow 0} \frac{\mathrm{d}}{\mathrm{d}\rho} \mu(\rho, 1) = \begin{cases} -\dfrac{\beta}{1-a}, \\ \infty, \end{cases} \quad \lim_{r \uparrow 1} \frac{\mathrm{d}}{\mathrm{d}r} \mu(0, r) = \begin{cases} \dfrac{1}{1-a} & \text{if } a<1, \\ \infty & \text{if } a=1. \end{cases} \tag{4.40}$$

Putting

$$\mu_{ij} \stackrel{\text{def}}{=} \int_0^\infty t \, \mathrm{d}F_{ij}(t), \qquad \nu_{ij} \stackrel{\text{def}}{=} \sum_{n=1}^\infty nF_{ij}^{(n)}(\infty), \tag{4.41}$$

then it is easily found that these moments are infinite if $a=1$, while for $a<1$ they are finite. From (4.35), (4.36) and (4.37) it now follows using (4.14) and (4.17) that for $a<1$,

$$\mu_{00} = \frac{\alpha}{1-a}, \qquad \nu_{00} = \frac{1}{1-a}, \tag{4.42}$$

$$\mu_{j0} = \frac{j\beta}{1-a}, \qquad \nu_{j0} = \frac{j}{1-a}, \qquad j=1, 2, \ldots,$$

$$\mu_{jj} = \alpha\nu_{jj} = \frac{\alpha}{v_j}, \qquad j=1, 2, \ldots,$$

$$\mu_{0j} = -\frac{j\beta}{1-a} + \frac{\alpha}{1-a} \frac{v_0+\ldots+v_j}{v_j}, \qquad j=0, 1, \ldots,$$

$$\nu_{0j} = -\frac{j}{1-a} + \frac{1}{1-a} \frac{v_0+\ldots+v_j}{v_j}, \qquad j=0, 1, \ldots,$$

where

$$v_j = (1-a)\frac{1}{j!} \frac{\mathrm{d}^j}{\mathrm{d}p^j} \left. \frac{(1-p)\,\beta\{(1-p)/\alpha\}}{\beta\{(1-p)/\alpha\}-p} \right|_{p=0}, \qquad j=0, 1, \ldots. \tag{4.43}$$

The distribution function $F_{jj}(t)$ is a lattice distribution if and only if its Laplace-Stieltjes transform $\varphi_{jj}(\rho)=1$ for a value of ρ with $\mathrm{Re}\,\rho=0, \rho\neq 0$. From (4.35) it is seen that $\varphi_{jj}(\rho)=1$ is equivalent to

$$1+\alpha\rho=\mu(\rho, 1).$$

This relation and the fact that $\mu(\rho, 1)$ is a root of equation (4.22) implies that

$$1+\alpha\rho=\mu(\rho, 1)=1;$$

but this can never be true for $\mathrm{Re}\,\rho=0$, $\rho\neq 0$. Hence $F_{jj}(t)$ is not a lattice distribution.

We shall now investigate the distribution function of the number $\boldsymbol{x}_t$ of customers present in the system at time t. Let $\boldsymbol{v}_t$ denote the number of customers arriving in a time interval of length t, and define

$$p_j(t) \stackrel{\text{def}}{=} \Pr\{\boldsymbol{v}_t=j\} = \mathrm{e}^{-t/\alpha}\frac{(t/\alpha)^j}{j!}, \qquad j=0, 1, \ldots; \quad t>0. \tag{4.44}$$

We now have for $j, k=0, 1, \ldots; t>0$,

$$\begin{aligned}
\Pr\{\boldsymbol{x}_t=j \mid \boldsymbol{z}_0=k\} &= \Pr\{\boldsymbol{x}_t=j, \boldsymbol{r}_1'\geqq t \mid \boldsymbol{z}_0=k\} + \sum_{n=1}^{\infty} \Pr\{\boldsymbol{x}_t=j, \boldsymbol{r}_n'<t\leqq \boldsymbol{r}_{n+1}' \mid \boldsymbol{z}_0=k\} \\
&= \Pr\{\boldsymbol{x}_t=j, \boldsymbol{r}_1'\geqq t \mid \boldsymbol{z}_0=k\} \\
&\quad + \sum_{n=1}^{\infty} \sum_{h=0}^{j} \Pr\{\boldsymbol{z}_n=h, \boldsymbol{v}_{t-\boldsymbol{r}'_n}=j-h, \boldsymbol{r}_n'<t\leqq \boldsymbol{r}_{n+1}' \mid \boldsymbol{z}_0=k\}. \qquad (4.45)
\end{aligned}$$

For the probabilities in the expression (4.45) we may write (cf. (4.44) and (4.18)),

$$\Pr\{\boldsymbol{x}_t=j, \boldsymbol{r}_1'\geqq t \mid \boldsymbol{z}_0=k\} = \begin{cases} \mathrm{e}^{-t/\alpha}, & j=0, \quad k=0, \\ \displaystyle\int_0^t \frac{1}{\alpha}\, \mathrm{e}^{-\tau/\alpha}\, p_{j-1}(t-\tau)\{1-B(t-\tau)\}\, \mathrm{d}\tau, & j=1, 2, \ldots; \quad k=0, \\ p_{j-k}(t)\{1-B(t)\}, & j=k, k+1, \ldots; \quad k=1, 2, \ldots, \\ 0, & j=0, \ldots, k-1; \quad k=1, \ldots, \end{cases} \qquad (4.46)$$

and for $t>0$; $n=1, 2, \ldots$; $k=0, 1, \ldots$,

$$\Pr\{\boldsymbol{z}_n=h, \boldsymbol{v}_{t-\boldsymbol{r}'_n}=j-h, \boldsymbol{r}_n'<t\leqq \boldsymbol{r}_{n+1}' \mid \boldsymbol{z}_0=k\}$$

$$= \begin{cases} \displaystyle\int_0^t \mathrm{e}^{-(t-u)/\alpha}\, \mathrm{d}P_{k0}^{(n)}(u), & j=0, \quad h=0, \\ \displaystyle\int_0^t \left\{\int_0^{t-u} \frac{\mathrm{d}\tau}{\alpha}\, \mathrm{e}^{-\tau/\alpha}\, p_{j-1}(t-u-\tau)\{1-B(t-u-\tau)\}\right\} \mathrm{d}_u P_{k0}^{(n)}(u), & j=1, 2, \ldots; \quad h=0, \\ \displaystyle\int_0^t p_{j-h}(t-u)\{1-B(t-u)\}\, \mathrm{d}_u P_{kh}^{(n)}(u), & j=1, 2, \ldots; \quad h=1, 2, \ldots, j, \\ 0, \quad \text{otherwise}. \end{cases} \qquad (4.47)$$

From (4.28), (4.45), (4.46) and (4.47) it follows that

$$\Pr\{x_t=j \mid z_0=0\} = \begin{cases} e^{-t/\alpha} + \int_0^t e^{-(t-u)/\alpha}\, dm_{00}(u), & j=0, \\ \int_0^t \frac{1}{\alpha} e^{-\tau/\alpha} p_{j-1}(t-\tau)\{1-B(t-\tau)\}\, d\tau \\ \quad + \sum_{h=1}^{j} \int_0^t p_{j-h}(t-u)\{1-B(t-u)\}\, dm_{0h}(u) \\ \quad + \int_0^t \left\{ \int_0^{t-u} \frac{1}{\alpha} e^{-\tau/\alpha} p_{j-1}(t-u-\tau) \right. \\ \quad \left. \cdot\{1-B(t-u-\tau)\}\, d\tau \right\} dm_{00}(u), & j=1, 2, \dots . \end{cases} \tag{4.48}$$

In the same way the relation for $\Pr\{x_t=j, z_0=k\}$, $k=1, 2, \dots$, can be obtained, and it follows for $\operatorname{Re} \rho>0$, $|p|\leqq 1$, $k=0, 1, \dots$, that

$$\int_0^\infty e^{-\rho t} \Pr\{x_t=0 \mid z_0=k\}\, dt = \frac{\alpha\mu^k(\rho, 1)}{1+\alpha\rho-\mu(\rho, 1)}, \tag{4.49}$$

$$\sum_{j=1}^\infty p^j \int_0^\infty e^{-\rho t} \Pr\{x_t=j \mid z_0=k\}\, dt$$

$$= \frac{p}{p-\beta\{\rho+(1-p)/\alpha\}} \left\{ p^k - \frac{(1+\alpha\rho-p)\, \mu^k(\rho, 1)}{1+\alpha\rho-\mu(\rho, 1)} \right\} \frac{1-\beta\{\rho+(1-p)/\alpha\}}{\rho+(1-p)/\alpha}.$$

By expanding the right-hand side of the first relation into a geometric series of $\mu(\rho, 1)/(1+\alpha\rho)$ it is not difficult to obtain an explicit formula for $\Pr\{x_t=0 \mid z_0=k\}$ by using the relation for $\mu^i(\rho, r)$ (cf. equation after (4.37)). It is now possible to investigate $\Pr\{x_t=j \mid z_0=k\}$ for $t\to\infty$. Since $F_{jj}(t)$ is not a lattice distribution, it follows easily from theorem I.6.2 and (4.48) that $\Pr\{x_t=j \mid z_0=0\}$ has a limit for $t\to\infty$. Moreover, since $\mu_{jj}=\infty$ if and only if $a\geqq 1$ it follows from the same theorem that the value of the limit is zero if $a\geqq 1$. The same conclusion holds for $\Pr\{x_t=j \mid z_0=k\}$, $k=1, 2, \dots$. To calculate the limit in the case $a<1$ we apply Feller's convergence theorem for generating functions (cf. app. 5) and Abel's theorem for the Laplace transform (cf. app. 2) to the expression in (4.49). Since in the second relation of (4.49) summation and integration may be reversed we find for $|p|\leqq 1$,

$k=0, 1, \ldots,$ that

$$\sum_{j=0}^{\infty} p^j \lim_{t\to\infty} \Pr\{\boldsymbol{x}_t=j \mid z_0=k\} = \lim_{\rho\downarrow 0} \rho \sum_{j=0}^{\infty} p^j \int_0^\infty \mathrm{e}^{-\rho t} \Pr\{\boldsymbol{x}_t=j \mid z_0=k\}\, \mathrm{d}t$$

$$= (1-a)\frac{(1-p)\,\beta\{(1-p)/\alpha\}}{\beta\{(1-p)/\alpha\}-p}, \qquad a<1. \tag{4.50}$$

Hence from (4.14) and (4.17),

$$\lim_{t\to\infty} \Pr\{\boldsymbol{x}_t=j \mid z_0=k\} = \lim_{n\to\infty} \Pr\{z_n=j \mid z_0=k\} = \begin{cases} 0, & a\geqq 1, \\ v_j, & a<1. \end{cases} \tag{4.51}$$

The relation (4.51) shows that *the limit distribution for the imbedded Markov chain* $\{z_n,\ n=0, 1, \ldots\}$ *and that for the* $\boldsymbol{x}_t$*-process are identical and independent of the initial state* z_0. It is easily verified that this result is also true even if it is not assumed that a new service starts at $t=0$.

From (4.24) and (4.49) it is easily shown that

$$\sum_{n=1}^{\infty} r^n \mathrm{E}\{z_n \mid z_0=0\} = \frac{r}{(1-r)^2}\left\{\frac{1-r}{1-\mu(0, r)} - (1-a)\right\}, \qquad |r|<1, \tag{4.52}$$

$$\int_0^\infty \mathrm{e}^{-\rho t}\, \mathrm{E}\{\boldsymbol{x}_t \mid z_0=0\}\, \mathrm{d}t = \frac{1}{\alpha\rho^2} - \frac{1-\mu(\rho, 1)}{\rho\{1+\alpha\rho-\mu(\rho, 1)\}}\, \frac{\beta(\rho)}{1-\beta(\rho)}, \qquad \mathrm{Re}\, \rho>0.$$

From (4.48) and by applying theorem I.6.2 it can be proved that if $a<1$ and $\beta_2<\infty$ (cf. (4.1)) then $\mathrm{E}\{\boldsymbol{x}_t \mid z_0=0\}$ has a limit for $t\to\infty$. This limit is again calculated by applying Abel's theorem for the Laplace transform. We obtain (cf. app. 2),

$$\lim_{t\to\infty} \mathrm{E}\{\boldsymbol{x}_t \mid z_0=0\} = \lim_{\rho\downarrow 0} \rho \int_0^\infty \mathrm{e}^{-\rho t}\, \mathrm{E}\{\boldsymbol{x}_t \mid z_0=0\}\, \mathrm{d}t$$

$$= a + \frac{a^2}{2(1-a)}\, \frac{\beta_2}{\beta^2}, \qquad a<1. \tag{4.53}$$

From (4.23), (4.29) and (4.52) we have the relation

$$\frac{t}{\alpha} - \mathrm{E}\{\boldsymbol{x}_t \mid z_0=0\} = \sum_{j=0}^{\infty} m_{0j}(t), \qquad t>0. \tag{4.54}$$

The right-hand side of (4.54) is non-decreasing in t, and it follows easily from (4.23) and app. 4 that it behaves as t/β for $t\to\infty$ if $a>1$. Hence

$$\lim_{t\to\infty}\frac{1}{t}\,\mathrm{E}\{\boldsymbol{x}_t \mid \boldsymbol{z}_0=0\} = \frac{a-1}{\beta} \qquad \text{if} \quad a>1. \tag{4.55}$$

Next, we define for $t>0$; $h=0, 1, \dots; j=0, 1, \dots,$

$$\begin{aligned} P^{(n)}_{h;0j}(t) &\overset{\text{def}}{=} \Pr\{\boldsymbol{z}_n=j,\ \boldsymbol{r}'_n<t,\ \boldsymbol{z}_k\neq h,\ k=1,\dots,n-1 \mid \boldsymbol{z}_0=0\}, \\ &\qquad\qquad\qquad\qquad\qquad\qquad n=2,3,\dots, \\ P^{(1)}_{h;0j}(t) &\overset{\text{def}}{=} P^{(1)}_{0j}(t). \end{aligned} \tag{4.56}$$

Clearly we have

$$P^{(n)}_{0j}(t) = P^{(n)}_{h;0j}(t) + \sum_{m=1}^{n-1}\int_0^t P^{(n-m)}_{hj}(t-u)\,\mathrm{d}F^{(m)}_{0h}(u), \qquad n=2,3,\dots. \tag{4.57}$$

Defining for $|p|\leqq 1$, $|r|<1$, $\operatorname{Re}\rho\geqq 0$,

$$\pi_{h;0}(p,\rho,r) = \sum_{n=1}^{\infty}\sum_{j=0}^{\infty} r^n p^j \int_0^\infty \mathrm{e}^{-\rho t}\,\mathrm{d}P^{(n)}_{h;0j}(t),$$

it follows immediately for $|p|<1$, $|r|<1$, $\operatorname{Re}\rho\geqq 0$ that

$$\pi_{h;0}(p,\rho,r)=\pi_0(p,\rho,r)-\pi_h(p,\rho,r)\,\varphi_{0h}(\rho,r). \tag{4.58}$$

From the relations (4.56), ..., (4.58) we shall deduce a relation for the distribution of $\max \boldsymbol{x}_\tau$ during a period between two (not necessarily successive) idle states of the server.

It follows for $h=0, 1, \dots; t>0$ that

$$\begin{aligned} &\Pr\{\boldsymbol{x}_t=0,\ \max_{0<\tau<t}\boldsymbol{x}_\tau\leqq h+1 \mid \boldsymbol{z}_0=0\} \\ &\qquad\qquad = \mathrm{e}^{-t/\alpha} + \sum_{n=1}^{\infty}\int_0^t \mathrm{e}^{-(t-u)/\alpha}\,\mathrm{d}_u P^{(n)}_{h;00}(u). \end{aligned} \tag{4.59}$$

From (4.59), (4.58), (4.23) and (4.35) it is found for $\operatorname{Re}\rho>0$, $h=0, 1, \dots,$ that

$$\begin{aligned} &\int_0^\infty \mathrm{e}^{-\rho t}\Pr\{\boldsymbol{x}_t=0,\ \max_{0<\tau<t}\boldsymbol{x}_\tau\leqq h+1 \mid \boldsymbol{z}_0=0\}\,\mathrm{d}t \\ &\qquad = \alpha\,\frac{\left.\dfrac{\mathrm{d}^h}{\mathrm{d}p^h}\,\dfrac{\beta\{\rho+(1-p)/\alpha\}}{\beta\{\rho+(1-p)/\alpha\}-p}\right|_{p=0}}{\left.\dfrac{\mathrm{d}^h}{\mathrm{d}p^h}\,\dfrac{(1+\alpha\rho-p)\beta\{\rho+(1-p)/\alpha\}}{\beta\{\rho+(1-p)/\alpha\}-p}\right|_{p=0}}. \end{aligned} \tag{4.60}$$

Further we note that the definition of $F_{j,j-1}(t)$ together with (4.37) implies that for $t>0$,

$$\Pr\{\min_{0<\tau\leqq t} x_\tau \geqq j \mid z_0=j\}=1-F_{j,j-1}(t)=1-F_{10}(t), \qquad j=1, 2, \ldots . \tag{4.61}$$

Finally, a remark about the initial conditions. Till now it has always been assumed that if, at $t=0$, customers are present then a new service starts at $t=0$. If we drop this assumption and assume instead that i customers are present at $t=0$ with $i=1, 2, \ldots$, while the residual service time of the customer being served at $t=0$ is given by $B_1(t)$ then for $t>0$, $i=1, 2, \ldots; j=i-1, i, \ldots$,

$$\Pr\{z_1=j, r_1'<t \mid x_0=i\} = \int_0^t \mathrm{e}^{-\tau/\alpha} \frac{(\tau/\alpha)^{j+1-i}}{(j+1-i)!} \, \mathrm{d}B_1(t).$$

All the derivations given above proceed in the same way if in (4.19), $P_{ij}^{(1)}(t)$, $i=1, 2, \ldots$, is replaced by the expression for $\Pr\{z_1=j, r_1'<t \mid x_0=i\}$ given above.

II.4.4. The busy period

We shall denote the duration of a busy period by $\boldsymbol{p}$ and the number of customers served during $\boldsymbol{p}$ by $\boldsymbol{n}$. The joint distribution $D_n(t)$ of $\boldsymbol{p}$ and $\boldsymbol{n}$ may be immediately obtained from the results of section 4.3. The following direct derivation which is due to Takács, is worth mentioning (cf. also section 2.2).

Let $\boldsymbol{\tau}$ denote the duration of the first service time of the busy period and let $\boldsymbol{\nu}$ represent the number of customers arriving during $\boldsymbol{\tau}$. If $\boldsymbol{\nu}=0$ then the busy period consists of one service. If $\boldsymbol{\nu}\geqq 1$ remove $\boldsymbol{\nu}-1$ customers from the system at the end of the first service time and let the server start serving the remaining customer and subsequent new arrivals until the busy period started by this customer ends. Let $\boldsymbol{n}_1$ be the number of service times and $\boldsymbol{p}_{n_1}$ the duration of this busy period. At the end of $\boldsymbol{p}_{n_1}$ replace one of the customers removed, and suppose he starts a new busy period say with $\boldsymbol{n}_2$ service times and duration $\boldsymbol{p}_{n_2}$. This procedure is repeated until all $\boldsymbol{\nu}-1$ customers removed have been replaced and served. Since the order of service of arriving customers is irrelevant for the length of the busy period we have

$$\boldsymbol{p}=\boldsymbol{\tau}, \qquad \boldsymbol{n}=1 \quad \text{if} \quad \boldsymbol{\nu}=0,$$

$$\boldsymbol{p}=\boldsymbol{\tau}+\boldsymbol{p}_{n_1}+\ldots+\boldsymbol{p}_{n_\nu}, \qquad \boldsymbol{n}=1+\boldsymbol{n}_1+\ldots+\boldsymbol{n}_\nu \qquad \text{if} \quad \boldsymbol{\nu}>0.$$

Obviously, the joint distribution of $\boldsymbol{p}$ and $\boldsymbol{n}$ is the same as the joint distribution of $\boldsymbol{p}_{n_j}$ and $\boldsymbol{n}_j$. Since $\boldsymbol{p}_{n_j}$, $j=1, \ldots, \boldsymbol{\nu}$, and $\boldsymbol{n}_j$, $j=1, \ldots, \boldsymbol{\nu}$ are independent variables we have for $|r|\leqq 1$, $\operatorname{Re}\rho\geqq 0$,

$$\begin{aligned}
\sum_{n=1}^{\infty} r^n \int_0^\infty \mathrm{e}^{-\rho t}\,\mathrm{d}D_n(t) &= \mathrm{E}\{r^{\boldsymbol{n}}\,\mathrm{e}^{-\rho\boldsymbol{p}}\} \\
&= \mathrm{E}\{\exp[-\rho(\boldsymbol{\tau}+\boldsymbol{p}_1+\ldots+\boldsymbol{p}_{n_\nu})]\, r^{1+\boldsymbol{n}_1+\ldots+\boldsymbol{n}_\nu}\} \\
&= r\sum_{n=0}^{\infty} \mathrm{E}\{\mathrm{e}^{-\rho\boldsymbol{\tau}}\,(\boldsymbol{\nu}=n)\} \prod_{i=1}^{n} \mathrm{E}\{r^{\boldsymbol{n}_i}\,\mathrm{e}^{-\rho\boldsymbol{p}_i}\} \\
&= r\sum_{n=0}^{\infty} \int_0^\infty \mathrm{e}^{-\rho t}\,\mathrm{e}^{-t/\alpha}\,\frac{(t/\alpha)^n}{n!}\,[\mathrm{E}\{r^{\boldsymbol{n}}\,\mathrm{e}^{-\rho\boldsymbol{p}}\}]^n\,\mathrm{d}B(t) \\
&= r\beta\{\rho+(1-\mathrm{E}\{r^{\boldsymbol{n}}\,\mathrm{e}^{-\rho\boldsymbol{p}}\})/\alpha\}.
\end{aligned}$$

Here $(\boldsymbol{\nu}=n)$ is the indicator function of the event $\{\boldsymbol{\nu}=n\}$.
Consequently, for $|r|\leqq 1$, $\operatorname{Re}\rho\geqq 0$, $\mathrm{E}\{r^{\boldsymbol{n}}\,\mathrm{e}^{-\rho\boldsymbol{p}}\}$ is a zero of the function $z-r\beta\{\rho+(1-z)/\alpha\}$. Since

$$|\mathrm{E}\{r^{\boldsymbol{n}}\,\mathrm{e}^{-\rho\boldsymbol{p}}\}|\leqq 1, \qquad |r|\leqq 1, \qquad \operatorname{Re}\rho\geqq 0,$$

it follows from Takács' lemma (cf. app. 6) that

$$\mu(\rho, r) = \mathrm{E}\{r^{\boldsymbol{n}}\,\mathrm{e}^{-\rho\boldsymbol{p}}\} = \sum_{n=1}^{\infty} \int_0^\infty r^n\,\mathrm{e}^{-\rho t}\,\mathrm{e}^{-t/\alpha}\,\frac{(t/\alpha)^{n-1}}{n!}\,\mathrm{d}B^{n*}(t). \tag{4.62}$$

Hence for $n=1, 2, \ldots;\ t>0$,

$$D_n(t) = \Pr\{\boldsymbol{n}=n, \boldsymbol{p}<t\} = \int_0^t \frac{(u/\alpha)^{n-1}}{n!}\,\mathrm{e}^{-u/\alpha}\,\mathrm{d}B^{n*}(u), \tag{4.63}$$

$$D(t) = \sum_{n=1}^{\infty} D_n(t),$$

and (cf. (4.25) and (4.37)) for $n=1, 2, \ldots;\ t>0$,

$$D_n(t)=F_{10}^{(n)}(t). \tag{4.64}$$

It is easily found that

$$\Pr\{\boldsymbol{n}<\infty\} = \Pr\{\boldsymbol{p}<\infty\}\begin{cases} =1 & \text{if } a\leqq 1, \\ <1 & \text{if } a>1, \end{cases} \tag{4.65}$$

and for $a<1$,

$$\mathrm{E}\{\boldsymbol{n}\} = (1-a)^{-1}, \qquad \mathrm{E}\{\boldsymbol{p}\} = (1-a)^{-1}\beta, \tag{4.66}$$

$$\mathrm{E}\{\boldsymbol{p}^2\} = \frac{\beta_2}{(1-a)^3}, \qquad \mathrm{E}\{\boldsymbol{p}^3\} = \frac{\beta_3}{(1-a)^4} + \frac{3}{(1-a)^5}\,\frac{\beta_2^2}{\alpha},$$

with β_2 and β_3 the second and third moments of $B(t)$, respectively.
Next, we shall determine *the distribution of the maximum number of customers present simultaneously during a busy period*, i.e. the maximum of $\boldsymbol{x}_t$ during a busy period. Whenever, in a busy period, the maximum of $\boldsymbol{x}_t$ equals $j+1$, then during this period the maximum number of customers left behind at a departure is j, i.e. max $\boldsymbol{z}_n=j$ during this busy period, and conversely. We define for the imbedded Markov chain $\{\boldsymbol{z}_n,\ n=0, 1, \ldots\}$ for $h=1, 2, \ldots$,

$$p_{h;00}^{(1)} \stackrel{\text{def}}{=} p_{00}, \qquad f_{h;00}^{(1)} \stackrel{\text{def}}{=} p_{00}, \tag{4.67}$$

$$p_{h;00}^{(n)} \stackrel{\text{def}}{=} \Pr\{\boldsymbol{z}_n=0,\ \boldsymbol{z}_k\neq h,\ k=1,\ldots,n-1 \mid \boldsymbol{z}_0=0\}, \qquad n=2,3,\ldots,$$

$$f_{h;00}^{(n)} \stackrel{\text{def}}{=} \Pr\{\boldsymbol{z}_n=0,\ \boldsymbol{z}_k\neq h,\ \boldsymbol{z}_k\neq 0,\ k=1,\ldots,n-1 \mid \boldsymbol{z}_0=0\}, \quad n=2,3,\ldots.$$

Hence $f_{h;00}^{(n)}$ denotes, for the Markov chain $\{\boldsymbol{z}_n,\ n=0, 1, \ldots\}$, the probability of a first return from state E_0 to E_0 in exactly n steps without passing state E_h. Consequently,

$$P_h \stackrel{\text{def}}{=} \sum_{n=1}^{\infty} f_{h;00}^{(n)}, \qquad h=1, 2, \ldots,$$

is the probability that during a busy period the maximum number of customers present simultaneously is, at most, h.
It follows from (4.67) for $h=1, 2, \ldots$, that

$$p_{00}^{(n)} = p_{h;00}^{(n)} + \sum_{m=1}^{n-1} p_{h0}^{(n-m)} f_{0h}^{(m)}, \qquad n=2, 3, \ldots, \tag{4.68}$$

$$p_{h;00}^{(n)} = f_{h;00}^{(n)} + \sum_{m=1}^{n-1} p_{h;00}^{(n-m)} f_{h;00}^{(m)}, \qquad n=2, 3, \ldots,$$

where (cf. (4.25)),

$$f_{0h}^{(m)} = F_{0h}^{(m)}(\infty), \qquad m=1, 2, \ldots. \tag{4.69}$$

From (4.68), (4.23), (4.24) and from (4.33) for $\rho=0$ it is not difficult to calculate

$$\sum_{n=1}^{\infty} r^n p_{h;00}^{(n)} \quad \text{and} \quad \sum_{n=1}^{\infty} r^n f_{h;00}^{(n)}.$$

By letting $r\uparrow 1$ and applying Abel's theorem it is found that for $h=1, 2, \ldots$, (cf. (4.17) and (4.42))

$$P_h = \begin{cases} \dfrac{\dfrac{1}{(h-1)!}\dfrac{\mathrm{d}^{h-1}}{\mathrm{d}p^{h-1}}\dfrac{\beta\{(1-p)/\alpha\}}{\beta\{(1-p)/\alpha\}-p}\Big|_{p=0}}{\dfrac{1}{h!}\dfrac{\mathrm{d}^{h}}{\mathrm{d}p^{h}}\dfrac{\beta\{(1-p)/\alpha\}}{\beta\{(1-p)/\alpha\}-p}\Big|_{p=0}}, & a<1 \quad \text{or} \quad a\geqq 1, \\[2ex] \dfrac{v_0+\ldots+v_{h-1}}{v_0+\ldots+v_h} = 1-\dfrac{v_h}{v_0+\ldots+v_h} = 1-\dfrac{\mu_{00}}{\mu_{0h}+\mu_{h0}} & \text{if} \quad a<1. \end{cases} \tag{4.70}$$

If $a>1$ then $\mu_0<1$ (cf. (4.13)) and it is easily shown that for $h=1, 2, \ldots$,

$$\frac{1}{h!}\frac{\mathrm{d}^h}{\mathrm{d}p^h}\frac{\beta\{(1-p)/\alpha\}}{\beta\{(1-p)/\alpha\}-p}\Big|_{p=0} = \frac{1}{h!}\frac{\mathrm{d}^h}{\mathrm{d}p^h}\frac{p}{\beta\{(1-p)/\alpha\}-p}\Big|_{p=0} \approx A\mu_0^{-h}, \qquad h\to\infty, \tag{4.71}$$

where $-A$ is the residue of $[\beta\{(1-p)/\alpha\}-p]^{-1}$ at $p=\mu_0$.
Hence

$$\lim_{h\to\infty} P_h = \mu_0 \qquad \text{if} \quad a>1, \tag{4.72}$$

so that if $a>1$ the maximum of $\boldsymbol{x}_t$ during a busy period is finite with probability μ_0 and infinite with probability $1-\mu_0$.
If $a<1$ then $v_h\to 0$ for $h\to\infty$ and by using the properties of the generating function (3.54) it is easily verified that the maximum of $\boldsymbol{x}_t$ during a busy period is finite with probability one if $a\leqq 1$. Moreover, it follows from (4.70) that this maximum has a finite mean if $a<1$. Clearly this mean is given by

$$\sum_{h=1}^{\infty} \frac{v_{h-1}}{v_0+v_1+\ldots+v_{h-1}} < (1-a)^{-1}, \qquad a<1.$$

Finally, we note that (4.42) implies

$$\frac{v_0+\ldots+v_j}{v_j}=j+(1-a)\,\nu_{0j},\qquad j=1,2,\ldots,\qquad \text{if}\quad a<1,$$

and hence

$$P_j=1-\frac{1}{(j+1-a)\,\nu_{0j}}>1-\frac{1}{j+1-a}>1-\frac{1}{j}\qquad \text{for}\quad a<1. \tag{4.73}$$

For other derivations of the expression for P_h see (6.77) and (III.6.51).

II.4.5. The waiting time

Denoting the actual waiting time of the nth arriving customer after $t=0$ by $\boldsymbol{w}_n$, we have

$$\Pr\{\boldsymbol{w}_n=0\mid \boldsymbol{z}_0=i\}=p_{i0}^{(n+i-1)},\qquad i=0,1,\ldots;\quad n=1,2,\ldots. \tag{4.74}$$

From (1.1),

$$\boldsymbol{w}_{n+1}=[\boldsymbol{w}_n+\boldsymbol{\tau}_n-\boldsymbol{\sigma}_{n+1}]^+,\qquad n=1,2,\ldots. \tag{4.75}$$

Since $\boldsymbol{w}_n$, $\boldsymbol{\tau}_n$ and $\boldsymbol{\sigma}_{n+1}$ are independent it follows for Re $\rho\geqq 0$, $x\geqq 0$, $n=1,2,\ldots$, that

$$\mathrm{E}\{\exp(-\rho[\boldsymbol{w}_n+\boldsymbol{\tau}_n-\boldsymbol{\sigma}_{n+1}]^+)\,|\,\boldsymbol{w}_n+\boldsymbol{\tau}_n=x\}=\frac{1}{1-\alpha\rho}\{\mathrm{e}^{-\rho x}-\alpha\rho\,\mathrm{e}^{-x/\alpha}\},\qquad \rho\neq\frac{1}{\alpha},$$

$$\Pr\{[\boldsymbol{w}_n+\boldsymbol{\tau}_n-\boldsymbol{\sigma}_{n+1}]^+=0\mid \boldsymbol{w}_n+\boldsymbol{\tau}_n=x\}=\mathrm{e}^{-x/\alpha},$$

hence

$$\mathrm{E}\{\mathrm{e}^{-\rho\boldsymbol{w}_{n+1}}\}=\int_{x=0}^{\infty}\mathrm{E}\{\mathrm{e}^{-\rho\boldsymbol{w}_{n+1}}\mid \boldsymbol{w}_n+\boldsymbol{\tau}_n=x\}\,\mathrm{d}\Pr\{\boldsymbol{w}_n+\boldsymbol{\tau}_n<x\}$$

$$=\frac{1}{1-\alpha\rho}\left[\int_{x=0}^{\infty}\mathrm{e}^{-\rho x}\,\mathrm{d}\Pr\{\boldsymbol{w}_n+\boldsymbol{\tau}_n<x\}-\alpha\rho\Pr\{\boldsymbol{w}_{n+1}=0\}\right],$$

so that for Re $\rho\geqq 0$, $n=1,2,\ldots$,

$$(1-\alpha\rho)\,\omega_k^{(n+1)}(\rho)=\omega_k^{(n)}(\rho)\,\beta(\rho)-\alpha\rho\Pr\{\boldsymbol{w}_{n+1}=0\mid \boldsymbol{z}_0=k\}, \tag{4.76}$$

where

$$\omega_k^{(n)}(\rho) \stackrel{\text{def}}{=} \mathrm{E}\{e^{-\rho w_n} \mid z_0 = k\}, \qquad \mathrm{Re}\,\rho \geqq 0.$$

Noting that

$$p_{k0}^{(n)} = 0 \qquad \text{for} \quad n = 0, 1, \ldots, k-1; \quad k = 1, 2, \ldots,$$

and defining for $|r| < 1$, $\mathrm{Re}\,\rho \geqq 0$, $k = 0, 1, \ldots,$

$$\omega_k(\rho, r) \stackrel{\text{def}}{=} \sum_{n=1}^{\infty} r^n \omega_k^{(n)}(\rho),$$

we obtain from (4.24), (4.74) and (4.76) for $|r| < 1$, $\mathrm{Re}\,\rho \geqq 0$, $k = 1, 2, \ldots,$

$$\begin{aligned} \omega_k(\rho, r)(1 - \alpha\rho - r\beta(\rho)) &= r(1-\alpha\rho)\,\omega_k^{(1)}(\rho) - \alpha\rho\{\sum_{j=1}^{\infty} r^{j-k-1} p_{k0}^{(j)} - r p_{k0}^{(k)}\}, \\ \omega_0(\rho, r) &= \frac{r}{1-\mu(0,r)}\,\frac{1-\alpha\rho-\mu(0,r)}{1-\alpha\rho-r\beta(\rho)}. \end{aligned} \tag{4.77}$$

With $z = 1 - \alpha\rho$ we have

$$1 - \alpha\rho - r\beta(\rho) = z - r\beta\{(1-z)/\alpha\}.$$

From Rouché's theorem (cf. app. 6) it is seen that

$$1 - \alpha\rho - r\beta(\rho), \qquad |r| < 1,$$

has only one zero in the right semi-plane $\mathrm{Re}\,\rho > 0$.
Takács' lemma yields that for $|r| < 1$, $\rho = (1 - \mu(0, r))/\alpha$ is the zero of $1 - \alpha\rho - r\beta(\rho)$ in the semi-plane $\mathrm{Re}\,\rho > 0$. Since $\omega_k(\rho, r)$ should be analytic for $\mathrm{Re}\,\rho > 0$ and fixed r with $|r| < 1$, it follows from (4.77) that for $|r| < 1$,

$$\begin{aligned} &\omega_k^{(1)}\{(1-\mu(0,r))/\alpha\} \\ &\quad = \frac{1-\mu(0,r)}{\mu(0,r)}\{\sum_{j=1}^{\infty} r^{j-k} p_{k0}^{(j)} - p_{k0}^{(k)}\}, \qquad k = 1, 2, \ldots, \end{aligned} \tag{4.78}$$

so that from (4.24), (4.77) and (4.78), $\omega_k^{(1)}(\rho)$ and $\omega_k(\rho, r)$ can be determined. It is readily seen that $\omega_0(\rho, r)$ with $|r| < 1$ is analytic for $\mathrm{Re}\,\rho > 0$. It follows from the second relation of (4.77) by differentiating with respect to ρ and taking $\rho = 0$ that for $|r| < 1$,

$$\sum_{n=1}^{\infty} r^n \mathrm{E}\{w_n \mid z_0 = 0\} = (a-1)\frac{\alpha r^2}{(1-r)^2} + \frac{\alpha r \mu(0,r)}{(1-r)(1-\mu(0,r))}. \tag{4.79}$$

Hence for $n=2, 3, \ldots$, (cf. (4.24))

$$\mathrm{E}\{\boldsymbol{w}_n \mid \boldsymbol{z}_0=0\} = (n-1)(a-1)\,\alpha + \alpha \sum_{i=1}^{n-1} p_{00}^{(i)},$$

therefore, if $a>1$,

$$\lim_{n\to\infty} \frac{\mathrm{E}\{\boldsymbol{w}_n \mid \boldsymbol{z}_0=0\}}{n-1} = (a-1)\,\alpha, \tag{4.80}$$

$$\lim_{n\to\infty} [\mathrm{E}\{\boldsymbol{w}_n \mid \boldsymbol{z}_0=0\} - (n-1)(a-1)\,\alpha] = \frac{\alpha\mu_0}{1-\mu_0}.$$

From Lindley's theorem (cf. section 1.3) it is known that, for $a<1$, as $n\to\infty$ the distribution of $\boldsymbol{w}_n$ tends to a limit distribution $W(t)$ at every continuity point of $W(t)$. This distribution is a proper probability distribution and independent of the initial state. If $a\geqq 1$ the distribution of $\boldsymbol{w}_n$ tends to zero for all finite t. Applying Helly-Bray's theorem (cf. app. 5) it follows for $\mathrm{Re}\,\rho\geqq 0$, $a<1$ that

$$\omega(\rho) = \int_{0-}^{\infty} \mathrm{e}^{-\rho t}\,\mathrm{d}W(t) = \int_{0-}^{\infty} \mathrm{e}^{-\rho t}\,\mathrm{d}\{\lim_{n\to\infty} \Pr\{\boldsymbol{w}_n<t \mid \boldsymbol{z}_0=k\}\}$$
$$= \lim_{n\to\infty} \omega_k^{(n)}(\rho).$$

Hence from (4.14), (4.17), (4.74) and (4.76) or by multiplying $\omega_0(\rho, r)$ by $1-r$ and taking the limit for $r\uparrow 1$ (cf. app. 1) we have for $a<1$,

$$\begin{aligned} \omega(\rho) &= (1-a)\frac{\alpha\rho}{\beta(\rho)-1+\alpha\rho}, \qquad \mathrm{Re}\,\rho\geqq 0, \\ W(0+) &= \lim_{n\to\infty} \Pr\{\boldsymbol{w}_n=0 \mid \boldsymbol{z}_0=k\} = 1-a. \end{aligned} \tag{4.81}$$

Since for $a<1$, $\rho>0$,

$$1 > \frac{1-\beta(\rho)}{\alpha\rho} = a\int_0^{\infty} \mathrm{e}^{-\rho t}\,\mathrm{d}_t\left\{\frac{1}{\beta}\int_0^t \{1-B(\tau)\}\,\mathrm{d}\tau\right\},$$

it is easily found from (4.81) that for $a<1$,

$$W(t) = \begin{cases} 0, & t<0, \\ (1-a)\displaystyle\sum_{n=0}^{\infty} a^n \left[\frac{1}{\beta}\int_0^t \{1-B(\tau)\}\,\mathrm{d}\tau\right]^{n*}, & t>0. \end{cases} \tag{4.82}$$

Obviously, $W(t)$ has a discontinuity at $t=0$ and is absolutely continuous for $t>0$. From (4.81) it is easily found that

$$\int_0^\infty t\,\mathrm{d}W(t) = \frac{1}{2}\,\frac{a\beta}{1-a}\,\frac{\beta_2}{\beta^2} \qquad \text{if } \beta_2<\infty,$$

$$\int_0^\infty t^2\,\mathrm{d}W(t) = \frac{a}{3(1-a)}\,\frac{\beta_3}{\beta} + \frac{a^2}{2(1-a)^2}\,\frac{\beta_2^2}{\beta^2} \qquad \text{if } \beta_3<\infty,$$

$$\int_0^\infty t^3\,\mathrm{d}W(t) = \frac{a}{4(1-a)}\,\frac{\beta_4}{\beta} + \frac{a^2}{(1-a)^2}\,\frac{\beta_2\beta_3}{\beta^2} + \frac{3a^3}{4(1-a)^3}\,\frac{\beta_2^3}{\beta^3} \quad \text{if } \beta_4<\infty,$$

with β_n the nth moment of $B(t)$. The relation for the first moment of $W(t)$ is known as the Pollaczek-Khintchine formula.

A more direct calculation of the Laplace-Stieltjes transform of $W(t)$ for $a<1$ follows from the fact that the number of customers left behind by a departing customer is equal to the number of customers that arrived during the time interval which is the sum of the waiting time $\boldsymbol{w}$ and the service time of this departing customer, since customers are served in order of arrival, so that (cf. (4.16)),

$$v_j = \int_{0-}^\infty \mathrm{e}^{-\tau/\alpha}\,\frac{(\tau/\alpha)^j}{j!}\,\mathrm{d}_\tau\{W(\tau)*B(\tau)\}, \qquad j=0, 1, \dots .$$

Consequently for $a<1$,

$$v(p) = \omega\,\{(1-p)/\alpha\}\;\beta\{(1-p)/\alpha\}, \qquad |p|\leqq 1,$$

and from (4.17) the expression (4.81) for $\omega(\rho)$ is now easily obtained.

To investigate the virtual waiting time $\boldsymbol{v}_t$ at time t it is noted that

$$\Pr\{\boldsymbol{v}_t<\sigma\,|\,\boldsymbol{z}_0=k\} = \Pr\{\boldsymbol{v}_t=0\,|\,\boldsymbol{z}_0=k\} + \Pr\{0<\boldsymbol{v}_t<\sigma\,|\,\boldsymbol{z}_0=k\}, \qquad \sigma>0. \quad (4.83)$$

Since

$$\Pr\{\boldsymbol{v}_t=0\,|\,\boldsymbol{z}_0=k\} = \Pr\{\boldsymbol{x}_t=0\,|\,\boldsymbol{z}_0=k\},$$

it follows from (4.51) that for $k=0, 1, \dots,$

$$\lim_{t\to\infty} \Pr\{\boldsymbol{v}_t=0 \mid \boldsymbol{z}_0=k\} = \begin{cases} 0 & \text{if } a\geqq 1, \\ 1-a & \text{if } a<1. \end{cases} \tag{4.84}$$

To determine the last term in the expression (4.83), let $\boldsymbol{\xi}_t$ denote the residual service time at time t supposing the server is busy at t. We have (cf. (4.44)) for $k=1, 2, \dots; j=1, 2, \dots; t>0, \zeta>0,$

$$\begin{aligned}
&\Pr\{\boldsymbol{x}_t=j,\ \boldsymbol{\xi}_t<\zeta \mid \boldsymbol{z}_0=k\} \\
&\quad= \Pr\{\boldsymbol{x}_t=j,\ \boldsymbol{\xi}_t<\zeta,\ \boldsymbol{r}_1'\geqq t \mid \boldsymbol{z}_0=k\} \\
&\qquad+ \sum_{n=1}^{j} \Pr\{\boldsymbol{x}_t=j,\ \boldsymbol{\xi}_t<\zeta,\ \boldsymbol{r}_n'<t\leqq \boldsymbol{r}_{n+1}' \mid \boldsymbol{z}_0=k\} \\
&\quad= p_{j-k}(t)\{B(t+\zeta)-B(t)\} \\
&\qquad+ \sum_{h=1}^{j} \int_{u=0}^{t} p_{j-h}(t-u)\{B(t+\zeta-u)-B(t-u)\}\,\mathrm{d}_u m_{kh}(u) \\
&\qquad+ \int_{u=0}^{t} \Bigg[\int_{\tau=0}^{t-u} \frac{\mathrm{d}\tau}{\alpha}\,\mathrm{e}^{-\tau/\alpha}\, p_{j-1}(t-u-\tau) \\
&\qquad\qquad\cdot\{B(t+\zeta-u-\tau)-B(t-u-\tau)\}\Bigg]\,\mathrm{d}m_{k0}(u),
\end{aligned} \tag{4.85}$$

whereas for $k=0$ the first term in the right-hand side of (4.85) should be replaced by

$$\int_0^t \frac{\mathrm{d}\tau}{\alpha}\,\mathrm{e}^{-\tau/\alpha}\, p_{j-1}(t-\tau)\{B(t+\zeta-\tau)-B(t-\tau)\}.$$

As in section 4.3 (cf. (4.51)) it can be shown by an application of the key-renewal theorem I.6.2 that $\Pr\{\boldsymbol{x}_t=j,\ \boldsymbol{\xi}_t<\zeta \mid \boldsymbol{z}_0=k\}$ has a limit for $t\to\infty$; it is found that for $k=0, 1, \dots; \zeta>0; j=1, 2, \dots; |p|\leqq 1,$

$$\begin{aligned}
&\sum_{j=1}^{\infty} p^j \lim_{t\to\infty} \Pr\{\boldsymbol{x}_t=j,\ \boldsymbol{\xi}_t<\zeta \mid \boldsymbol{z}_0=k\} \\
&= \begin{cases} 0, & a\geqq 1, \\ \dfrac{(1-a)\,p(1-p)}{\beta\{(1-p)/\alpha\}-p} \displaystyle\int_{\eta=0}^{\infty} \mathrm{e}^{-\eta(1-p)/\alpha}\,[B(\eta+\zeta)-B(\eta)]\,\frac{\mathrm{d}\eta}{\alpha}, & a<1. \end{cases}
\end{aligned} \tag{4.86}$$

The evaluation of this limit is performed by first determining the generating function with respect to j and the Laplace transform with respect to t of $\Pr\{\boldsymbol{x}_t=j,\ \boldsymbol{\xi}_t<\zeta \mid \boldsymbol{z}_0=k\}$ and then applying Abel's theorem for the Laplace transform (cf. the evaluation of the limit in (4.51)).

Since for $\operatorname{Re}\rho>0$, $\operatorname{Re}\delta\geqq 0$, $\delta\neq\rho$,

$$\int_0^\infty e^{-\delta\zeta}\,\mathrm{d}_\zeta \int_0^\infty e^{-\rho t}\{B(t+\zeta)-B(t)\}\,\mathrm{d}t = \frac{\beta(\rho)-\beta(\delta)}{\delta-\rho}, \tag{4.87}$$

it follows from (4.86) by using Helly-Bray's theorem (cf. WIDDER [1946]) that for $\operatorname{Re}\rho>0$, $|p|\leqq 1$, $a<1$,

$$\lim_{t\to\infty}\sum_{j=1}^{\infty} p^j \mathrm{E}\{e^{-\rho\boldsymbol{\xi}_t}(\boldsymbol{x}_t=j)\mid \boldsymbol{z}_0=k\} = \frac{(1-a)\,p(1-p)}{\beta\{(1-p)/\alpha\}-p}\;\frac{\beta\{(1-p)/\alpha\}-\beta(\rho)}{\alpha\rho-(1-p)}. \tag{4.88}$$

The virtual waiting time $\boldsymbol{v}_t$ is the sum of the residual service time $\boldsymbol{\xi}_t$ and the total service time of the waiting customers at t, hence for $\sigma>0$, $t>0$, $k=0, 1, \ldots,$

$$\Pr\{0<\boldsymbol{v}_t<\sigma \mid \boldsymbol{z}_0=k\}$$

$$= \sum_{j=1}^{\infty}\int_{\tau=0}^{\sigma}\Pr\{\boldsymbol{x}_t=j,\ \boldsymbol{\xi}_t<\sigma-\tau \mid \boldsymbol{z}_0=k\}\,\mathrm{d}_\tau B^{(j-1)*}(\tau). \tag{4.89}$$

Hence from (4.88) and (4.89) for $\operatorname{Re}\rho>0$,

$$\lim_{t\to\infty}\mathrm{E}\{e^{-\rho\boldsymbol{v}_t}(\boldsymbol{v}_t>0)\mid \boldsymbol{z}_0=k\} = \lim_{t\to\infty}\sum_{j=1}^{\infty}\beta^{j-1}(\rho)\,\mathrm{E}\{e^{-\rho\boldsymbol{\xi}_t}(\boldsymbol{x}_t=j)\mid \boldsymbol{z}_0=k\}$$

$$= (1-a)\frac{1-\beta(\rho)}{\alpha\rho-1+\beta(\rho)}.$$

Since for $a<1$ (cf. (4.84)),

$$\lim_{t\to\infty}\Pr\{\boldsymbol{v}_t=0\mid \boldsymbol{z}_0=k\} = 1-a,$$

we have for $\operatorname{Re}\rho>0$, if $a<1$,

$$\lim_{t\to\infty}\mathrm{E}\{e^{-\rho\boldsymbol{v}_t}\mid \boldsymbol{z}_0=k\} = (1-a)\,\frac{\alpha\rho}{\beta(\rho)+\alpha\rho-1}.$$

Consequently, from Feller's convergence theorem, it follows that, for $a<1$, the distribution of $\boldsymbol{v}_t$ converges for $t\to\infty$ to a proper distribution $V(\tau)$ and

$$\int_{0-}^{\infty} \mathrm{e}^{-\rho t}\,\mathrm{d}V(t) = (1-a)\frac{\alpha\rho}{\beta(\rho)+\alpha\rho-1}, \qquad \operatorname{Re}\rho\geqq 0, \quad a<1. \tag{4.90}$$

Obviously, the distributions of $\boldsymbol{w}_n$ for $n\to\infty$ and that of $\boldsymbol{v}_t$ for $t\to\infty$ converge to the same limit (cf. (4.81)).

From (4.23), (4.49), (4.83), (4.85), (4.87) and (4.89) it follows for $\operatorname{Re}\rho>0$, $\operatorname{Re}\delta\geqq 0$, $k=0, 1, 2, \ldots$, that

$$\int_{0-}^{\infty} \mathrm{e}^{-\delta\sigma}\,\mathrm{d}_\sigma \int_{0}^{\infty} \mathrm{e}^{-\rho t}\,\Pr\{\boldsymbol{v}_t<\sigma\mid \boldsymbol{z}_0=k\}\,\mathrm{d}t$$
$$= \left\{\beta^k(\delta) - \frac{\alpha\delta\mu^k(\rho, 1)}{1+\alpha\rho-\mu(\rho, 1)}\right\}\frac{\alpha}{1+\alpha\rho-\alpha\delta-\beta(\delta)}.$$

Inversion of the Laplace transform with respect to ρ yields, using (4.49), for $\operatorname{Re}\delta\geqq 0$, $k=0, 1, \ldots;\ t>0$,

$$\int_{0-}^{\infty} \mathrm{e}^{-\delta\sigma}\,\mathrm{d}_\sigma\,\Pr\{\boldsymbol{v}_t<\sigma \mid \boldsymbol{z}_0=k\}$$
$$= \mathrm{e}^{[\delta-\{1-\beta(\delta)\}/\alpha]t}\,\beta^k(\delta) - \delta\int_0^t \mathrm{e}^{\{\delta-(1-\beta(\delta))/\alpha\}(t-u)}\,\Pr\{\boldsymbol{x}_u=0 \mid \boldsymbol{z}_0=k\}\,\mathrm{d}u. \tag{4.91}$$

This relation may also be derived by taking the Laplace-Stieltjes transform with respect to σ of the expression given by (4.89) and (4.83).

In the above relations we have taken as the initial condition that k customers are present in the system at $t=0$ and a new service is started if $k=1, 2, \ldots$. We shall now consider another initial condition.

In section 1.2 it was pointed out that the virtual waiting time $\boldsymbol{v}_t$ can be interpreted as the amount of work the server has still to do at time t if after t no new customers arrive. We shall now consider the situation that $\boldsymbol{v}_0=v$, i.e. at time $t=0$ there is already present for the server an amount of work equal to v.

If $\boldsymbol{v}_0=v$ then at time $t=v$ the last one of the customers present at $t=0+$ will leave the system, so that if k customers have arrived in $[0, v)$ then at time $t=v+$ a new service is started with k customers in the system, hence at $t=v$ the queueing process is started with k customers in the system. Therefore we have for the process after $t=v$ the same initial condition as we considered before. Since the probability that in a time interval of length v

exactly k customers arrive is given by

$$\mathrm{e}^{-v/\alpha}\frac{(v/\alpha)^k}{k!}, \qquad k=0, 1, \ldots,$$

it follows from (4.49) for Re $\rho>0$, $v\geqq 0$ that

$$\int_0^\infty \mathrm{e}^{-\rho t}\,\Pr\{\boldsymbol{x}_t=0 \mid \boldsymbol{v}_0=v\}\,\mathrm{d}t$$
$$= \int_v^\infty \mathrm{e}^{-\rho t}\sum_{k=0}^\infty \mathrm{e}^{-v/\alpha}\frac{(v/\alpha)^k}{k!}\Pr\{\boldsymbol{x}_{t-v}=0 \mid \boldsymbol{z}_0=k\}\,\mathrm{d}t$$
$$= \int_0^\infty \mathrm{e}^{-\rho t}\Pr\{\boldsymbol{v}_t=0 \mid \boldsymbol{v}_0=v\}\,\mathrm{d}t = \frac{\mathrm{e}^{-\{\rho+(1-\mu(\rho,1))/\alpha\}v}}{\rho+\{1-\mu(\rho,1)\}/\alpha}. \tag{4.92}$$

This relation is due to Beneš [1957]. By applying Lagrange's expansion theorem (see app. 6) we can invert the Laplace transform in (4.92) and obtain

$$\Pr\{\boldsymbol{v}_t=0 \mid \boldsymbol{v}_0=v\} = \begin{cases} \displaystyle\sum_{n=0}^\infty \mathrm{e}^{-t/\alpha}\frac{(t/\alpha)^n}{n!}\int_0^{t-v}\frac{t-u}{t}\,\mathrm{d}B^{n*}(u), & t>v,\\ 0, & t<v. \end{cases} \tag{4.93}$$

The relation (4.92) or (4.93) has some simple direct probabilistic interpretations. We shall illustrate one of them here, and another in section 6.6 (cf. (6.81)). Consider a general renewal process with renewal function $m_{v,0}(t)$, with $F_{00}(t)$ as renewal distribution and $G_v(t)$ as distribution of the first renewal time. $F_{00}(t)$ is the distribution function of the sum of a busy period and an idle period of the server, where the idle period is defined as the time during which the server is continuously idle.

For $G_v(t)$ we shall take the distribution function of the *initial busy period* $\boldsymbol{\theta}_v$ of the queueing process with $\boldsymbol{v}_0=v>0$ (cf. section 1.3).

Hence, by the same argument as above for $t>0$,

$$G_v(t) = \Pr\{\boldsymbol{\theta}_v<t \mid \boldsymbol{v}_0=v\}$$
$$= \mathrm{e}^{-v/\alpha}U_1(t-v) + \sum_{n=1}^\infty \mathrm{e}^{-v/\alpha}\frac{(v/\alpha)^n}{n!}F_{n0}(t-v).$$

It follows that

$$\int_0^\infty \mathrm{e}^{-\rho t}\,\mathrm{d}G_v(t) = \mathrm{e}^{-\{\rho+(1-\mu(\rho,1))/\alpha\}v}, \qquad \mathrm{Re}\,\rho \geqq 0, \tag{4.94}$$

so that

$$G_v(t) = \begin{cases} \sum_{n=0}^{\infty} \int_v^t \mathrm{e}^{-\tau/\alpha} \frac{(\tau/\alpha)^n}{n!} \frac{v}{\tau}\,\mathrm{d}_\tau B^{n*}(\tau - v), & t \geqq v, \\ 0, & t < v. \end{cases} \tag{4.95}$$

Clearly,

$$\Pr\{\boldsymbol{x}_t = 0 \mid \boldsymbol{v}_0 = v\} = \int_0^t \mathrm{e}^{-(t-u)/\alpha}\,\mathrm{d}m_{v,0}(u), \qquad t > v, \tag{4.96}$$

with

$$m_{v,0}(t) = G_v(t) + \sum_{n=1}^{\infty} G_v(t) * F_{00}^{n*}(t), \qquad t > 0.$$

The expression (4.92) for the Laplace transform of $\Pr\{\boldsymbol{x}_t = 0 \mid \boldsymbol{v}_0 = v\}$ follows immediately from the last four relations.

From the relation (4.95) and (1.11) it is now possible to determine the probability distribution of *the total idle time* $\boldsymbol{e}_t$ of the server during $[0, t)$. It follows (cf. (4.100)) that

$$\begin{aligned} \Pr\{\boldsymbol{e}_t \geqq \sigma - v \mid \boldsymbol{v}_0 = v\} &= \sum_{n=1}^{\infty} \mathrm{e}^{-\sigma/\alpha} \frac{(\sigma/\alpha)^n}{n!} F_{n0}(t-\sigma) + \mathrm{e}^{-\sigma/\alpha} U_1(t-\sigma) \\ &= \sum_{n=0}^{\infty} \int_\sigma^t \mathrm{e}^{-\tau/\alpha} \frac{(\tau/\alpha)^n}{n!} \frac{\sigma}{\tau}\,\mathrm{d}_\tau B^{n*}(\tau-\sigma) \\ &= \begin{cases} \int_\sigma^t \frac{\sigma}{\tau}\,\mathrm{d}_\tau \Pr\{\boldsymbol{k}_\tau < \tau - \sigma\}, & v \leqq \sigma < t, \\ 0, & \sigma \geqq t. \end{cases} \end{aligned} \tag{4.97}$$

Next, we shall determine the distribution function of $\boldsymbol{v}_t$ given $\boldsymbol{v}_0$. Since

$$\Pr\{\boldsymbol{v}_t < \sigma \mid \boldsymbol{v}_0 = v\} = \begin{cases} 0, & \sigma < v - t,\ t < v, \\ \sum_{n=0}^{\infty} \mathrm{e}^{-t/\alpha} \frac{(t/\alpha)^n}{n!} B^{n*}(\sigma - (v-t)), & \sigma \geqq v - t,\ t < v, \\ \sum_{n=0}^{\infty} \mathrm{e}^{-v/\alpha} \frac{(v/\alpha)^n}{n!} \Pr\{\boldsymbol{v}_{t-v} < \sigma \mid \boldsymbol{z}_0 = n\}, & t > v, \end{cases} \tag{4.98}$$

it follows from (4.91) and (4.92) for Re $\delta \geqq 0$, $t \geqq 0$ that

$$\int_{0-}^{\infty} e^{-\delta\sigma} \, d_\sigma \Pr\{\boldsymbol{v}_t < \sigma \mid \boldsymbol{v}_0 = v\} = e^{\delta(t-v) - t\{1-\beta(\delta)\}/\alpha} - \delta U_1(t-v) \int_0^{t-v} e^{\{\delta - (1-\beta(\delta))/\alpha\}(t-v-u)} \Pr\{\boldsymbol{v}_{u+v} = 0 \mid \boldsymbol{v}_0 = v\} \, du, \quad (4.99)$$

where $U_1(t)$ is the unit step-function. Denoting the amount of work brought in by arriving customers during an interval of length $[0, t)$ by $\boldsymbol{k}_t$ (cf. section 1.3), we have

$$K(\sigma, t) \stackrel{\text{def}}{=} \Pr\{\boldsymbol{k}_t < \sigma\} = \sum_{n=0}^{\infty} e^{-t/\alpha} \frac{(t/\alpha)^n}{n!} B^{n*}(\sigma), \qquad \sigma > 0, \quad t > 0, \quad (4.100)$$

so that

$$\int_{0-}^{\infty} e^{-\delta\sigma} \, d\Pr\{\boldsymbol{k}_t < \sigma\} = e^{-t\{1-\beta(\delta)\}/\alpha}, \qquad \text{Re } \delta \geqq 0, \quad t > 0. \quad (4.101)$$

The relation (4.99) can be obtained immediately from formulas (1.13) and (1.14) by noting that in the case studied here the arrival process is a Poisson process so that $\boldsymbol{k}_t - \boldsymbol{k}_{u+v}$ is independent of $\boldsymbol{p}_{u+v}(v)$, and that $K(\sigma, t-u-v)$ is the distribution function of $\boldsymbol{k}_t - \boldsymbol{k}_{u+v}$.

Using (4.98) and (4.101) the inversion of the relation (4.99) can be easily performed and it is found for $t > 0$, $\sigma > 0$, $v \geqq 0$ that

$$\Pr\{\boldsymbol{v}_t < \sigma \mid \boldsymbol{v}_0 = v\} = \begin{cases} K(t+\sigma-v, t), & \sigma > v-t > 0, \\ K(t+\sigma-v, t) - \dfrac{\partial}{\partial\sigma} \displaystyle\int_{u=0}^{t-v} K(\sigma+u, u) \Pr\{\boldsymbol{v}_{t-u} = 0 \mid \boldsymbol{v}_0 = v\} \, du, & t > v. \end{cases} \quad (4.102)$$

From (4.89), (4.93) and (4.98) it is easily verified that $\Pr\{\boldsymbol{v}_t < \sigma \mid \boldsymbol{v}_0 = v\}$ and $\Pr\{\boldsymbol{v}_t < \sigma \mid \boldsymbol{z}_0 = k\}$ have the same limit for $t \to \infty$. We can also find from the above relations the average virtual waiting time at t given $\boldsymbol{v}_0 = v$. The following argument, however, is somewhat simpler.

Starting from the relations (1.7) and (1.8) we immediately obtain, since

$$E\{\boldsymbol{k}_t\} = at,$$

that

$$E\{\boldsymbol{v}_t \mid \boldsymbol{v}_0 = v\} = v + \int_0^t \Pr\{\boldsymbol{v}_u = 0 \mid \boldsymbol{v}_0 = v\} \, du - (1-a)\, t, \qquad t > 0. \quad (4.103)$$

From this relation and (4.92) it is easily shown that for $a>1$ (cf. app. 2),

$$\lim_{t\to\infty} \mathrm{E}\left\{\frac{1}{t}\, \boldsymbol{v}_t-(a-1) \mid \boldsymbol{v}_0=v\right\} = 0, \tag{4.104}$$

$$\lim_{t\to\infty} [\mathrm{E}\{\boldsymbol{v}_t \mid \boldsymbol{v}_0=v\}-(a-1)\,t] = v + \frac{\alpha\,\mathrm{e}^{-v(1-\mu_0)/\alpha}}{1-\mu_0}.$$

From (1.8) we further have

$$\boldsymbol{v}_{t+\Delta t}=\boldsymbol{v}_t+\boldsymbol{k}_{t+\Delta t}-\boldsymbol{k}_t-\theta\Delta t,$$

with $0\leqq\theta\leqq 1$, and $\theta=1$ if $\boldsymbol{v}_t>\Delta t$. In the interval $t\div t+\Delta t$ a new customer may arrive with probability $\alpha^{-1}\Delta t+\mathrm{o}(\Delta t)$ and not arrive with probability $1-\alpha^{-1}\Delta t+\mathrm{o}(\Delta t)$. Since $\boldsymbol{v}_t$ and $\boldsymbol{k}_{t+\Delta t}-\boldsymbol{k}_t$ are independent variables it follows with

$$V_t(\sigma, v) \stackrel{\text{def}}{=} \Pr\{\boldsymbol{v}_t<\sigma \mid \boldsymbol{v}_0=v\}, \qquad \sigma\geqq 0, \quad v\geqq 0, \tag{4.105}$$

that for $\sigma>0$, $t>0$,

$$1 - V_{t+\Delta t}(\sigma, v) = (1-\alpha^{-1}\Delta t)(1-V_t(\sigma+\Delta t, v))$$
$$+ \alpha^{-1}\Delta t\{1 - \int_0^\sigma B(\sigma-y)\, \mathrm{d}_y V_t(y+\theta\Delta t, v)\} + \mathrm{o}(\Delta t).$$

Assuming that we may write for $\Delta t\to 0$,

$$V_{t+\Delta t}(\sigma, v) = V_t(\sigma, v) + \Delta t \frac{\partial}{\partial t} V_t(\sigma, v) + \mathrm{o}(\Delta t), \tag{4.106}$$

$$V_t(\sigma+\Delta t, v) = V_t(\sigma, v) + \Delta t \frac{\partial}{\partial \sigma} V_t(\sigma, v) + \mathrm{o}(\Delta t),$$

we obtain for $\sigma>0$, $t>0$, $v\geqq 0$,

$$\frac{\partial}{\partial t} V_t(\sigma, v) = \frac{\partial}{\partial \sigma} V_t(\sigma, v) - \alpha^{-1} V_t(\sigma, v)$$
$$+ \alpha^{-1} \int_0^\sigma B(\sigma-y)\, \mathrm{d}_y\, V_t(y, v). \tag{4.107}$$

This relation is known as *the integro-differential equation of Takács*. This equation can be solved by determining the Laplace-Stieltjes transform of (4.107) with respect to σ. It is then not difficult to obtain the relation (4.99) and to show that the solution is given by (4.102). Concerning the assumptions (4.106) we note that if $B(t)$ is absolutely continuous and has a bounded derivative for all $t>0$ then the validity of (4.106) follows from (4.85) and (4.89) (cf. HASOFER [1963]).

II.4.6. The departure process

Denote the number of departures from the system in the time interval $(0, t]$ by $\boldsymbol{h}_t$. It follows that

$$\Pr\{\boldsymbol{h}_t=0 \mid \boldsymbol{z}_0=i\} = \begin{cases} \mathrm{e}^{-t/\alpha} + \int_0^t \frac{1}{\alpha}\, \mathrm{e}^{-\tau/\alpha}\{1-B(t-\tau)\}\,\mathrm{d}\tau, & i=0, \\ 1-B(t), & i=1, 2, \dots; \end{cases}$$

and from (4.18) for $t>0$; $i=0, 1, \dots$; $n=1, 2, \dots$, that

$$\Pr\{\boldsymbol{h}_t \geqq n \mid \boldsymbol{z}_0=i\} = \Pr\{\boldsymbol{r}_n' < t \mid \boldsymbol{z}_0=i\} = \sum_{j=0}^{\infty} P_{ij}^{(n)}(t). \tag{4.108}$$

Consequently from (4.23) for $\mathrm{Re}\, \rho>0$, $|r|\leqq 1$,

$$\int_0^\infty \mathrm{e}^{-\rho t}\, \mathrm{d}_t \sum_{n=1}^{\infty} r^n \Pr\{\boldsymbol{h}_t \geqq n \mid \boldsymbol{z}_0=i\}$$

$$= \frac{r\beta(\rho)}{1-r\beta(\rho)} \left\{1 - \frac{\alpha\rho}{1+\alpha\rho-\mu(\rho, r)}\, \mu^i(\rho, r)\right\}, \qquad i=0, 1, \dots . \tag{4.109}$$

Since

$$\mathrm{E}\{\boldsymbol{h}_t \mid \boldsymbol{z}_0=i\} = \sum_{n=1}^{\infty} \sum_{j=0}^{\infty} P_{ij}^{(n)}(t),$$

we obtain from (4.109) and (4.40) for $\mathrm{Re}\, \rho>0$,

$$\int_0^\infty \mathrm{e}^{-\rho t}\, \mathrm{d}_t\, \mathrm{E}\{\boldsymbol{h}_t \mid \boldsymbol{z}_0=i\} = \frac{\beta(\rho)}{1-\beta(\rho)} \left\{1 - \frac{\alpha\rho}{1+\alpha\rho-\mu(\rho, 1)}\, \mu^i(\rho, 1)\right\}$$

$$\approx \begin{cases} \dfrac{1}{\beta\rho} & \text{if } a>1, \quad \rho\downarrow 0, \\ \dfrac{1}{\alpha\rho} & \text{if } a<1, \quad \rho\downarrow 0. \end{cases} \tag{4.110}$$

Since $\mathrm{E}\{\boldsymbol{h}_t \mid \boldsymbol{z}_0=i\}$ is a non-decreasing function of t (cf. (4.28)) it follows that for $t\to\infty$ (cf. app. 4),

$$\frac{1}{t}\,\mathrm{E}\{\boldsymbol{h}_t \mid z_0=i\} \to \begin{cases} \dfrac{1}{\beta} & \text{if } a>1, \\ \dfrac{1}{\alpha} & \text{if } a<1. \end{cases} \tag{4.111}$$

Note that (cf. 4.54),

$$\mathrm{E}\{\boldsymbol{h}_t \mid z_0=i\}+\mathrm{E}\{\boldsymbol{x}_t \mid z_0=i\}=i+t/\alpha, \qquad t>0.$$

Obviously

$$\boldsymbol{h}_{\tau+\boldsymbol{r}'_m}-\boldsymbol{h}_{\boldsymbol{r}'_m}$$

represents the number of departing customers during $(\boldsymbol{r}'_m, \boldsymbol{r}'_m+\tau]$, i.e. during the time interval of length τ immediately after the departure of the mth departing customer after $t=0$, and for $m=1, 2, \dots; \tau>0$,

$$\begin{aligned}\Pr\{\boldsymbol{h}_{\tau+\boldsymbol{r}'_m}-\boldsymbol{h}_{\boldsymbol{r}'_m}=0 \mid z_0=k\} &= p_{k0}^{(m)}\left\{\mathrm{e}^{-\tau/\alpha}+\int_0^\tau \frac{1}{\alpha}\{1-B(\tau-\sigma)\}\,\mathrm{e}^{-\sigma/\alpha}\mathrm{d}\sigma\right\} \\ &\quad +\{1-p_{k0}^{(m)}\}\{1-B(\tau)\},\end{aligned} \tag{4.112}$$

$$\Pr\{\boldsymbol{h}_{\tau+\boldsymbol{r}'_m}-\boldsymbol{h}_{\boldsymbol{r}'_m}\geqq n \mid z_0=k\} = \sum_{i=0}^{\infty} p_{ki}^{(m)} \sum_{j=0}^{\infty} P_{ij}^{(n)}(\tau), \qquad n=1, 2, \dots .$$

We now prove for $\tau>0$,

$$\lim_{m\to\infty} \Pr\{\boldsymbol{h}_{\tau+\boldsymbol{r}'_m}-\boldsymbol{h}_{\boldsymbol{r}'_m}=0 \mid z_0=k\}$$

$$= \begin{cases} 1-B(\tau) & \text{if } a\geqq 1, \\ (1-a)\left\{\mathrm{e}^{-\tau/\alpha}+\dfrac{1}{\alpha}\displaystyle\int_0^\tau \{1-B(\tau-\sigma)\}\,\mathrm{e}^{-\sigma/\alpha}\,\mathrm{d}\sigma\right\}+a\{1-B(\tau)\} & \\ & \text{if } a<1, \end{cases} \tag{4.113}$$

and for $\operatorname{Re}\rho\geqq 0$, $|r|<1$,

$$\int_0^\infty \mathrm{e}^{-\rho\tau}\,\mathrm{d}_\tau \sum_{n=1}^{\infty} r^n \lim_{m\to\infty} \Pr\{\boldsymbol{h}_{\tau+\boldsymbol{r}'_m}-\boldsymbol{h}_{\boldsymbol{r}'_m}\geqq n \mid z_0=k\}$$

$$= \begin{cases} \dfrac{r\beta(\rho)}{1-r\beta(\rho)} & \text{if } a\geqq 1, \\ \dfrac{r\beta(\rho)}{1-r\beta(\rho)}\left\{1-\dfrac{\alpha\rho(1-a)(1-\mu(\rho,r))}{1+\alpha\rho-\mu(\rho,r)}\,\dfrac{\beta\{(1-\mu(\rho,r))/\alpha\}}{\beta\{(1-\mu(\rho,r))/\alpha\}-\mu(\rho,r)}\right\} & \\ & \text{if } a<1. \end{cases} \tag{4.114}$$

The relation (4.113) follows directly from (4.112) and (4.17). It is easily verified (cf. (4.18), (4.21), (4.23)) that the right-hand side of (4.114) is the limit for $m\to\infty$ of the Laplace-Stieltjes transform with respect to τ of the function

$$\sum_{n=1}^{\infty} r^n \sum_{i=0}^{\infty} p_{ki}^{(m)} \sum_{j=0}^{\infty} P_{ij}^{(n)}(\tau), \qquad |r|<1. \tag{4.115}$$

If this function has for every $|r|<1$ a limit for $m\to\infty$ then the right-hand side of (4.112) has a limit for $m\to\infty$ and conversely. The function given by (4.115) is a probability distribution in τ for every r with $|r|<1$, when it is multiplied by $r^{-1}(1-r)$. Its Laplace-Stieltjes transform has a limit for $m\to\infty$ if $\operatorname{Re}\rho\geqq 0$, $|r|<1$; using Feller's convergence theorem for Laplace-Stieltjes transforms (cf. app. 5) it is seen that (4.114) is true.

It should be noted that if $a<1$ and the initial distribution is the stationary distribution of the process, i.e. $\Pr\{z_0=k\}=v_k$, $k=0, 1, \ldots$, then $\Pr\{\boldsymbol{h}_{\tau+r'_m}-\boldsymbol{h}_{r'_m}\geqq n\}$, $n=0, 1, \ldots$, is independent of m, and the Laplace-Stieltjes transform of

$$\sum_{n=1}^{\infty} r^n \Pr\{\boldsymbol{h}_{\tau+r'_m}-\boldsymbol{h}_{r'_m}\geqq n\}, \qquad |r|<1,$$

is then given by the right-hand side of (4.114) for $a<1$. FINCH [1959b] proved that the interdeparture times are in general not independent, even if the process $\{z_n, n=0, 1, \ldots\}$ is stationary (cf. section 2.4).

II.5. THE QUEUEING SYSTEM $G/G/1$

II.5.1. Introduction

In this chapter we shall discuss the queueing system $G/G/1$ as described in section 1.2. For general distribution functions $A(t)$ and $B(t)$ of interarrival time and service time, respectively, neither the process $\{\boldsymbol{x}_{\boldsymbol{t}_n}, n=1,2,\ldots\}$ nor the process $\{\boldsymbol{x}_{\boldsymbol{r}_n}, n=1,2,\ldots\}$ are in general imbedded Markov chains of the $\boldsymbol{x}_t$-process. Here, $\boldsymbol{x}_t$ denotes the number of customers in the system at time t and $\boldsymbol{t}_n$ and $\boldsymbol{r}_n$ are the moments of arrival and departure, respectively, of the nth arriving customer after $t=0$. If we consider the process $\{\boldsymbol{v}_t, t\in[0,\infty)\}$, however, where $\boldsymbol{v}_t$ is the virtual waiting time at time t, it is easily seen that the process $\{\boldsymbol{w}_n, n=1,2,\ldots\}$ with $\boldsymbol{w}_n$ $(=\boldsymbol{v}_{\boldsymbol{t}_n})$ the waiting time of the nth arriving customer is a discrete time parameter Markov process with stationary transition probabilities and state space the set of real numbers (cf. (1.1)); this process is an imbedded Markov process of the $\boldsymbol{v}_t$-process. Extending the concept of a regenerative process with a denumerable state space in an obvious manner to the one with a non-denumerable state space, it is easily seen that the $\boldsymbol{v}_t$-process is regenerative with respect to the renewal process generated by the arrivals of customers.

We shall denote *the total idle time* of the server during the time interval $[0,\boldsymbol{t}_n)$ by $\boldsymbol{d}_n$, $n=1,2,\ldots$, so that $\boldsymbol{d}_n=\boldsymbol{e}_{\boldsymbol{t}_n}$ (cf. section 1.3). As initial conditions we shall take

$$\boldsymbol{w}_1=w \quad\text{and}\quad \boldsymbol{d}_1=d, \qquad w\geqq 0,\quad d\geqq 0. \tag{5.1}$$

We have already defined (cf. section 1.3),

$$\boldsymbol{\rho}_n \overset{\text{def}}{=} \boldsymbol{\tau}_n-\boldsymbol{\sigma}_{n+1}, \qquad \boldsymbol{s}_n \overset{\text{def}}{=} \sum_{i=1}^{n}\boldsymbol{\rho}_i, \qquad n=1,2,\ldots .$$

We further define

$$\boldsymbol{a}_n \stackrel{\text{def}}{=} \sum_{i=1}^{n} \boldsymbol{\sigma}_{i+1}, \qquad \boldsymbol{b}_n \stackrel{\text{def}}{=} \sum_{i=1}^{n} \boldsymbol{\tau}_i, \qquad n=1, 2, \ldots,$$

$$\boldsymbol{a}_0 \stackrel{\text{def}}{=} 0, \qquad \boldsymbol{b}_0 \stackrel{\text{def}}{=} 0.$$

From the definition of $\boldsymbol{d}_n$ we have for $n=1, 2, \ldots,$

$$\begin{aligned} &\boldsymbol{d}_{n+1}=\boldsymbol{d}_n && \text{if} \quad \boldsymbol{w}_n+\boldsymbol{\tau}_n-\boldsymbol{\sigma}_{n+1}>0, \quad \text{i.e.} \quad \boldsymbol{w}_{n+1}>0, \\ &\boldsymbol{d}_{n+1}=\boldsymbol{d}_n-(\boldsymbol{w}_n+\boldsymbol{\tau}_n-\boldsymbol{\sigma}_{n+1}) && \text{if} \quad \boldsymbol{w}_n+\boldsymbol{\tau}_n-\boldsymbol{\sigma}_{n+1} \leqq 0, \quad \text{i.e.} \quad \boldsymbol{w}_{n+1}=0; \end{aligned}$$

therefore,

$$\boldsymbol{d}_{n+1}-\boldsymbol{d}_n=-\min(0, \boldsymbol{w}_n+\boldsymbol{\tau}_n-\boldsymbol{\sigma}_{n+1}), \qquad n=1, 2, \ldots . \tag{5.2}$$

Moreover,

$$\boldsymbol{d}_{n+1}-\boldsymbol{d}_n-\boldsymbol{\sigma}_{n+1}=\boldsymbol{w}_{n+1}-\boldsymbol{w}_n-\boldsymbol{\tau}_n, \qquad n=1, 2, \ldots, \tag{5.3}$$

so that

$$\boldsymbol{d}_{n+1}-\boldsymbol{d}_1-\boldsymbol{a}_n=\boldsymbol{w}_{n+1}-\boldsymbol{w}_1-\boldsymbol{b}_n, \qquad n=1, 2, \ldots . \tag{5.4}$$

From the last relation and from (1.2) it follows that

$$\boldsymbol{d}_{n+1}-\boldsymbol{d}_1=-\boldsymbol{w}_1-\min(\boldsymbol{s}_n, \boldsymbol{s}_{n-1}, \ldots, \boldsymbol{s}_1, -\boldsymbol{w}_1), \qquad n=1, 2, \ldots . \tag{5.5}$$

Consequently, for $x \geqq 0$, $n=1, 2, \ldots$, (cf. section 1.3), we have

$$\begin{aligned} &\Pr\{\boldsymbol{w}_{n+1}<x \mid \boldsymbol{w}_1=0\}=\Pr\{\max(0, \boldsymbol{s}_1, \ldots, \boldsymbol{s}_n)<x\}, \\ &\Pr\{\boldsymbol{d}_{n+1}<x \mid \boldsymbol{w}_1=0, \boldsymbol{d}_1=0\}=\Pr\{\max(0, -\boldsymbol{s}_1, \ldots, -\boldsymbol{s}_n)<x\}. \end{aligned} \tag{5.6}$$

For the interdeparture time $\boldsymbol{r}_{n+1}-\boldsymbol{r}_n$ (cf. section 1.2) we now have

$$\begin{aligned} \boldsymbol{r}_{n+1}-\boldsymbol{r}_n&=\boldsymbol{\tau}_{n+1}+\boldsymbol{d}_{n+1}-\boldsymbol{d}_n \\ &=\boldsymbol{\tau}_{n+1}-\min(0, \boldsymbol{w}_n+\boldsymbol{\tau}_n-\boldsymbol{\sigma}_{n+1}). \end{aligned} \tag{5.7}$$

We shall first study the joint distribution of $\boldsymbol{w}_n$, $\boldsymbol{d}_n$, $\boldsymbol{a}_{n-1}$ and $\boldsymbol{b}_{n-1}$. We first derive an integral relation for the Laplace-Stieltjes transform of this distribution.

Denoting the normalized unit step-function by $U(x)$ then the Dirichlet integral representation of $U(x)$ is given by (cf. WIDDER [1946]),

$$U(x) = \frac{1}{2\pi \mathrm{i}} \lim_{\delta \to \infty} \int_{\varepsilon-\mathrm{i}\delta}^{\varepsilon+\mathrm{i}\delta} \mathrm{e}^{x\xi} \frac{\mathrm{d}\xi}{\xi} = \begin{cases} 0, & x<0, \\ \frac{1}{2}, & x=0, \\ 1, & x>0, \end{cases}$$

for $\varepsilon>0$. In the sequel we shall use the notation

$$\int_{C_\xi} \dots \mathrm{d}\xi \equiv \int_{\xi=\varepsilon-\mathrm{i}\infty}^{\varepsilon+\mathrm{i}\infty} \dots \mathrm{d}\xi \equiv \lim_{\delta\to\infty} \int_{\xi=\varepsilon-\mathrm{i}\delta}^{\varepsilon+\mathrm{i}\delta} \dots \mathrm{d}\xi, \qquad \varepsilon=\operatorname{Re}\xi.$$

Since

$$\mathrm{e}^{-\rho x}\,U(x)=\frac{1}{2\pi\mathrm{i}}\int_{C_\xi}\mathrm{e}^{-x\xi}\frac{\mathrm{d}\xi}{\rho-\xi}, \qquad \operatorname{Re}\rho>\operatorname{Re}\xi,\quad x\in(-\infty,\infty), \tag{5.8}$$

and

$$\exp\{-\rho_1[x]^+-\rho_2[x]^-\} = \mathrm{e}^{-\rho_1 x}\,U(x) + \mathrm{e}^{-\rho_2 x}\,U(-x),$$

with

$$[x]^+ \overset{\text{def}}{=} \max(0,x), \qquad [x]^- \overset{\text{def}}{=} \min(0,x), \qquad x \ \text{ real},$$

it follows for real x and complex numbers ρ_1, ρ_2 that

$$\exp\{-\rho_1[x]^+-\rho_2[x]^-\} = \frac{1}{2\pi\mathrm{i}}\int_{C_\xi}\mathrm{e}^{-x\xi}\left\{\frac{1}{\rho_1-\xi}+\frac{1}{\xi-\rho_2}\right\}\mathrm{d}\xi,$$
$$\operatorname{Re}\rho_2<\operatorname{Re}\xi<\operatorname{Re}\rho_1. \tag{5.9}$$

Since (cf. (1.1)),

$$\boldsymbol{w}_{n+1}=[\boldsymbol{w}_n+\boldsymbol{\tau}_n-\boldsymbol{\sigma}_{n+1}]^+, \qquad n=1,2,\dots,$$

it follows from (5.2) and (5.9) for $n=1,2,\dots$; $\operatorname{Re}\rho_1>0$, $\operatorname{Re}\rho_2<0$, $\operatorname{Re}\rho_3>0$, $\operatorname{Re}\rho_4>0$ that

$$\begin{aligned}
&\mathrm{E}\{\exp(-\rho_1\boldsymbol{w}_{n+1}+\rho_2\boldsymbol{d}_{n+1}-\rho_3\boldsymbol{t}_{n+1}-\rho_4\boldsymbol{b}_n)\mid \boldsymbol{w}_1=w,\,\boldsymbol{d}_1=d\}\\
&\quad=\mathrm{E}\{\exp(-\rho_1[\boldsymbol{w}_n+\boldsymbol{\tau}_n-\boldsymbol{\sigma}_{n+1}]^+-\rho_2[\boldsymbol{w}_n+\boldsymbol{\tau}_n-\boldsymbol{\sigma}_{n+1}]^-\\
&\qquad+\rho_2\boldsymbol{d}_n-\rho_3\boldsymbol{t}_{n+1}-\rho_4\boldsymbol{b}_n)\mid \boldsymbol{w}_1=w,\,\boldsymbol{d}_1=d\}\\
&\quad=\frac{1}{2\pi\mathrm{i}}\int_{C_\xi}\mathrm{d}\xi\left\{\frac{1}{\rho_1-\xi}+\frac{1}{\xi-\rho_2}\right\}\\
&\qquad\cdot\mathrm{E}\{\exp(-\xi\boldsymbol{w}_n+\rho_2\boldsymbol{d}_n-\rho_3\boldsymbol{t}_n-\rho_4\boldsymbol{b}_{n-1}-(\rho_3-\xi)\,\boldsymbol{\sigma}_{n+1}\\
&\qquad-(\rho_4+\xi)\,\boldsymbol{\tau}_n)\mid \boldsymbol{w}_1=w,\,\boldsymbol{d}_1=d\}\\
&\quad=\frac{1}{2\pi\mathrm{i}}\int_{C_\xi}\mathrm{d}\xi\left\{\frac{1}{\rho_1-\xi}+\frac{1}{\xi-\rho_2}\right\}\\
&\qquad\cdot\mathrm{E}\{\exp(-\xi\boldsymbol{w}_n+\rho_2\boldsymbol{d}_n-\rho_3\boldsymbol{t}_n-\rho_4\boldsymbol{b}_{n-1})\mid \boldsymbol{w}_1=w,\,\boldsymbol{d}_1=d\}\\
&\qquad\cdot\beta(\rho_4+\xi)\,\alpha(\rho_3-\xi),
\end{aligned} \tag{5.10}$$

with $0 < \mathrm{Re}\,\xi < \min(\mathrm{Re}\,\rho_1, \mathrm{Re}\,\rho_3)$; the last equality in (5.10) is based on the fact that $\boldsymbol{\tau}_n$ and $\boldsymbol{\sigma}_{n+1}$ are independent and also independent of $\boldsymbol{w}_n$, $\boldsymbol{d}_n$, $\boldsymbol{t}_n$ and $\boldsymbol{b}_{n-1}$. The permutation in (5.10) of the operators E and $\int_{C_\xi}$ is easily justified since in the domain of integration $\beta(\rho_4+\xi)\,\alpha(\rho_3-\xi)$ and the conditional expectation E{... | ...} are bounded whereas

$$\int_{C_\xi} \left| \frac{1}{\rho_1-\xi} + \frac{1}{\xi-\rho_2} \right| |\mathrm{d}\xi| < \infty.$$

Substituting in (5.10)

$$\boldsymbol{t}_{n+1} = \boldsymbol{t}_1 + \boldsymbol{a}_n, \qquad n=0, 1, \ldots,$$

and defining for $\mathrm{Re}\,\rho_1 \geqq 0$, $\mathrm{Re}\,\rho_2 \leqq 0$, $\mathrm{Re}\,\rho_3 \geqq 0$, $\mathrm{Re}\,\rho_4 \geqq 0$,

$$\omega^{(n)}(\rho_1, \rho_2, \rho_3, \rho_4, w, d) \stackrel{\text{def}}{=} \mathrm{E}\{\exp(-\rho_1 \boldsymbol{w}_n + \rho_2 \boldsymbol{d}_n - \rho_3 \boldsymbol{a}_{n-1} - \rho_4 \boldsymbol{b}_{n-1}) |\, \boldsymbol{w}_1 = w, \boldsymbol{d}_1 = d\}, \tag{5.11}$$

we have for $n=2, 3, \ldots$; $\mathrm{Re}\,\rho_1 > 0$, $\mathrm{Re}\,\rho_2 < 0$, $\mathrm{Re}\,\rho_3 > 0$, $\mathrm{Re}\,\rho_4 > 0$,

$$\omega^{(n)}(\rho_1, \ldots) = \frac{1}{2\pi \mathrm{i}} \int_{C_\xi} \mathrm{d}\xi \left\{ \frac{1}{\rho_1-\xi} + \frac{1}{\xi-\rho_2} \right\} \cdot \beta(\rho_4+\xi)\,\alpha(\rho_3-\xi)\,\omega^{(n-1)}(\xi, \rho_2, \ldots), \tag{5.12}$$

$$\omega^{(1)}(\rho_1, \ldots) = \exp(-\rho_1 w + \rho_2 d),$$

with $0 < \mathrm{Re}\,\xi < \min(\mathrm{Re}\,\rho_1, \mathrm{Re}\,\rho_3)$.

Introducing the generating function

$$\omega(r, \rho_1, \rho_2, \rho_3, \rho_4, w, d) \stackrel{\text{def}}{=} \sum_{n=1}^{\infty} r^n \omega^{(n)}(\rho_1, \ldots), \qquad |r| < 1, \tag{5.13}$$

we obtain for $|r| < 1$ and for the same conditions as in (5.12),

$$\omega(r, \rho_1, \rho_2, \ldots) = r \exp(-\rho_1 w + \rho_2 d) + \frac{r}{2\pi \mathrm{i}} \int_{C_\xi} \mathrm{d}\xi \left\{ \frac{1}{\rho_1-\xi} + \frac{1}{\xi-\rho_2} \right\} \beta(\rho_4+\xi)\,\alpha(\rho_3-\xi)\,\omega(r, \xi, \rho_2, \ldots). \tag{5.14}$$

In the following section we shall study this integral equation. Integral equations of this type were studied for the first time by POLLACZEK [1957] in his research of the $G/G/1$ queueing system.

II.5.2. Pollaczek's integral equation

Let $\Psi(\rho)$ be a bounded analytic function for $\operatorname{Re}\rho>\delta$ with $\delta<0$. For $|r|<1$, $\operatorname{Re}\rho_1>0$, $\operatorname{Re}\rho_2<0$, $\operatorname{Re}\rho_3>0$, $\operatorname{Re}\rho_4>0$, we define

$$w(\rho_1,\rho_2)\overset{\text{def}}{=}\exp\left\{-\frac{1}{2\pi\mathrm{i}}\int\limits_{C_\xi}\mathrm{d}\xi\left\{\frac{1}{\rho_1-\xi}+\frac{1}{\xi-\rho_2}\right\}\cdot\log\{1-r\beta(\rho_4+\xi)\,\alpha(\rho_3-\xi)\}\right\}, \tag{5.15}$$

where $0<\operatorname{Re}\xi<\min(\operatorname{Re}\rho_1,\operatorname{Re}\rho_3)$, and

$$v(\rho_1,\rho_2)\overset{\text{def}}{=}\frac{1}{2\pi\mathrm{i}}\int\limits_{C_\eta}\Psi(\eta)\,w(\rho_1,\eta)\left\{\frac{1}{\rho_1-\eta}+\frac{1}{\eta-\rho_2}\right\}\mathrm{d}\eta, \tag{5.16}$$

with $\max(\delta,\operatorname{Re}\rho_2)<\operatorname{Re}\eta<0$. We shall show that for $|r|<1$, $\operatorname{Re}\rho_1>0$, $\operatorname{Re}\rho_2<0$, $\operatorname{Re}\rho_3>0$, $\operatorname{Re}\rho_4>0$, the function $v(\rho_1,\rho_2)$ is a solution of Pollaczek's integral equation

$$\omega(\rho_1,\rho_2)=\frac{r}{2\pi\mathrm{i}}\int\limits_{C_\xi}\mathrm{d}\xi\left\{\frac{1}{\rho_1-\xi}+\frac{1}{\xi-\rho_2}\right\}\beta(\rho_4+\xi)\,\alpha(\rho_3-\xi)\,\omega(\xi,\rho_2)+\Psi(\rho_1), \tag{5.17}$$

with $0<\operatorname{Re}\xi<\min(\operatorname{Re}\rho_1,\operatorname{Re}\rho_3)$ and

$$w(\rho_1,\rho_2)=\exp\left\{\sum_{n=1}^{\infty}\frac{r^n}{n}\,\mathrm{E}\{\exp(-\rho_4\boldsymbol{b}_n-\rho_3\boldsymbol{a}_n)(\mathrm{e}^{-\rho_1\boldsymbol{s}_n}(\boldsymbol{s}_n>0)+\mathrm{e}^{-\rho_2\boldsymbol{s}_n}(\boldsymbol{s}_n\leqq 0))\}\right\}, \tag{5.18}$$

where (A) is the indicator function of the event A, i.e.

$$(\boldsymbol{s}_n>0)=\begin{cases}1 & \text{if } \boldsymbol{s}_n>0,\\ 0 & \text{if } \boldsymbol{s}_n\leqq 0,\end{cases}\qquad (\boldsymbol{s}_n\leqq 0)=\begin{cases}1 & \text{if } \boldsymbol{s}_n\leqq 0,\\ 0 & \text{if } \boldsymbol{s}_n>0.\end{cases}$$

To prove this statement note that for $0<\operatorname{Re}\xi<\operatorname{Re}\rho_3$,

$$\beta(\rho_4+\xi)\,\alpha(\rho_3-\xi)=\mathrm{E}\{\exp(-(\rho_4+\xi)\,\boldsymbol{\tau}_1-(\rho_3-\xi)\,\boldsymbol{\sigma}_1)\},$$

so that by definition of $\boldsymbol{a}_n$, $\boldsymbol{b}_n$ and $\boldsymbol{s}_n$ (see preceding section) and the independence of $\boldsymbol{\tau}_1,\ldots,\boldsymbol{\tau}_n,\boldsymbol{\sigma}_1,\ldots,\boldsymbol{\sigma}_n$ for $n=1,2,\ldots$,

$$\begin{aligned}\beta^n(\rho_4+\xi)\,\alpha^n(\rho_3-\xi)&=\mathrm{E}\{\exp(-\rho_4\boldsymbol{b}_n-\rho_3\boldsymbol{a}_n-\xi\boldsymbol{s}_n)\}\\&=\mathrm{E}\{\exp(-\rho_4\boldsymbol{b}_n-\rho_3\boldsymbol{a}_n)(\mathrm{e}^{-\xi\boldsymbol{s}_n}(\boldsymbol{s}_n>0)+\mathrm{e}^{-\xi\boldsymbol{s}_n}(\boldsymbol{s}_n\leqq 0))\},\end{aligned}$$

and hence by Cauchy's theorem for $0<\text{Re}\,\xi<\min(\text{Re}\,\rho_1,\,\text{Re}\,\rho_3)$,

$$\frac{1}{2\pi i}\int_{C_\xi}\beta^n(\rho_4+\xi)\,\alpha^n(\rho_3-\xi)\left\{\frac{1}{\rho_1-\xi}+\frac{1}{\xi-\rho_2}\right\}\mathrm{d}\xi$$

$$=\mathrm{E}\{\exp(-\rho_4\boldsymbol{b}_n-\rho_3\boldsymbol{a}_n)(\mathrm{e}^{-\rho_1\boldsymbol{s}_n}(\boldsymbol{s}_n>0)+\mathrm{e}^{-\rho_2\boldsymbol{s}_n}(\boldsymbol{s}_n\leqq 0))\},$$

the permutation of the operators E and $\int_{C_\xi}$ being justified as above (cf. (5.10)).

From (5.15) and the last relation the expression (5.18) follows since for $|r|<1$,

$$w(\rho_1,\rho_2)=\exp\left\{\sum_{n=1}^{\infty}\frac{r^n}{n}\frac{1}{2\pi i}\int_{C_\xi}\left\{\frac{1}{\rho_1-\xi}+\frac{1}{\xi-\rho_2}\right\}\beta^n(\rho_4+\xi)\,\alpha^n(\rho_3-\xi)\,\mathrm{d}\xi\right\}$$

$$=\exp\left\{\sum_{n=1}^{\infty}\frac{r^n}{n}\mathrm{E}\{\exp(-\rho_4\boldsymbol{b}_n-\rho_3\boldsymbol{a}_n)(\mathrm{e}^{-\rho_1\boldsymbol{s}_n}(\boldsymbol{s}_n>0)+\mathrm{e}^{-\rho_2\boldsymbol{s}_n}(\boldsymbol{s}_n\leqq 0))\}\right\},\quad (5.19)$$

since the expectation in (5.19) is bounded by one and $|\beta(\rho_4+\xi)\,\alpha(\rho_3-\xi)|<1$. We now have from (5.19) for $|r|<1$,

$$\{1-r\beta(\rho_4+\xi)\,\alpha(\rho_3-\xi)\}\,w(\xi,\rho_2)$$

$$=w(\xi,\rho_2)\exp\left\{\sum_{n=1}^{\infty}-\frac{r^n}{n}\mathrm{E}\{\exp(-\rho_4\boldsymbol{b}_n-\rho_3\boldsymbol{a}_n-\xi\boldsymbol{s}_n)\}\right\}$$

$$=\exp\left\{\sum_{n=1}^{\infty}\frac{r^n}{n}\mathrm{E}\{\exp(-\rho_4\boldsymbol{b}_n-\rho_3\boldsymbol{a}_n)(\mathrm{e}^{-\rho_2\boldsymbol{s}_n}-\mathrm{e}^{-\xi\boldsymbol{s}_n})(\boldsymbol{s}_n\leqq 0)\}\right\},\quad (5.20)$$

the right-hand side being an analytic and bounded function of ξ for $\text{Re}\,\xi<0$. Hence by Cauchy's theorem for $|r|<1$, $0<\text{Re}\,\xi<\min(\text{Re}\,\rho_1,\,\text{Re}\,\rho_3)$,

$$\frac{1}{2\pi i}\int_{C_\xi}\{1-r\beta(\rho_4+\xi)\,\alpha(\rho_3-\xi)\}\,w(\xi,\rho_2)\left\{\frac{1}{\rho_1-\xi}+\frac{1}{\xi-\rho_2}\right\}\mathrm{d}\xi=1.\quad (5.21)$$

The right-hand side of (5.19) shows that $w(\rho_1,\rho_2)$ is also analytic and bounded for $\text{Re}\,\rho_1\geqq 0$, so that for $0<\text{Re}\,\xi<\min(\text{Re}\,\rho_1,\,\text{Re}\,\rho_3)$,

$$w(\rho_1,\rho_2)=\frac{1}{2\pi i}\int_{C_\xi}\left\{\frac{1}{\rho_1-\xi}+\frac{1}{\xi-\rho_2}\right\}w(\xi,\rho_2)\,\mathrm{d}\xi.\quad (5.22)$$

Summation of the relations (5.21) and (5.22) shows that $w(\rho_1,\rho_2)$ as given by (5.15) is a solution of the integral equation (5.17) if $\Psi(\rho_1)\equiv 1$.

In (5.17) replace ρ_2 by η, $\omega(.,.)$ by $w(.,.)$ and $\Psi(\rho_1)$ by 1 and multiply the relation by

$$\frac{1}{2\pi i}\,\Psi(\eta)\left\{\frac{1}{\rho_1-\eta}+\frac{1}{\eta-\rho_2}\right\},$$

integrate the result with respect to η along C_η with $\max(\operatorname{Re}\rho_2, \delta)<\operatorname{Re}\eta<0$, then the first term gives $v(\rho_1, \rho_2)$ according to (5.16), and the second term in the right-hand side becomes $\Psi(\rho_1)\,U(\rho_1)$. For the first term in the right-hand side we have for $0<\operatorname{Re}\xi<\min(\operatorname{Re}\rho_1, \operatorname{Re}\rho_3)$,

$$\frac{r}{(2\pi i)^2}\int_{C_\eta}\Psi(\eta)\left\{\frac{1}{\rho_1-\eta}+\frac{1}{\eta-\rho_2}\right\}$$

$$\cdot\left\{\int_{C_\xi} w(\xi,\eta)\left\{\frac{1}{\rho_1-\xi}+\frac{1}{\xi-\eta}\right\}\beta(\rho_4+\xi)\,\alpha(\rho_3-\xi)\,d\xi\right\}d\eta$$

$$=\frac{r}{(2\pi i)^2}\int_{C_\xi}\beta(\rho_4+\xi)\,\alpha(\rho_3-\xi)\,\frac{\rho_1-\rho_2}{(\rho_1-\xi)(\xi-\rho_2)}$$

$$\cdot\left\{\int_{C_\eta} w(\xi,\eta)\,\Psi(\eta)\,\frac{\xi-\rho_2}{(\eta-\rho_2)(\xi-\eta)}\,d\eta\right\}d\xi$$

$$=\frac{r}{2\pi i}\int_{C_\xi}\frac{\rho_1-\rho_2}{(\rho_1-\xi)(\rho_2-\xi)}\,\beta(\rho_4+\xi)\,\alpha(\rho_3-\xi)\,v(\xi,\rho_2)\,d\xi.$$

The integrations can be interchanged since in the domain of integration

$$|\Psi(\eta)\,\beta(\rho_4+\xi)\,\alpha(\rho_3-\xi)\,w(\xi,\eta)|<\infty,\qquad \int_{C_\eta}\int_{C_\xi}\left|\frac{d\eta\,d\xi}{(\rho_1-\xi)(\xi-\eta)(\eta-\rho_2)}\right|<\infty.$$

Consequently, $v(\rho_1, \rho_2)$ is a solution of (5.17) and the proof is complete.

Above it has been shown that $w(\rho_1, \rho_2)$ satisfies the relations (cf. (5.21) and (5.22)),

$$u(\rho_1,\rho_2)=\frac{1}{2\pi i}\int_{C_\xi}\left\{\frac{1}{\rho_1-\xi}+\frac{1}{\xi-\rho_2}\right\}u(\xi,\rho_2)\,d\xi, \tag{5.23}$$

$$u(\rho_1,\rho_2)-\frac{1}{2\pi i}\int_{C_\xi}\left\{\frac{1}{\rho_1-\xi}+\frac{1}{\xi-\rho_2}\right\}u(\xi,\rho_2)\,\varphi(\xi)\,d\xi=1, \tag{5.24}$$

for $0<\mathrm{Re}\,\xi<\mathrm{Re}\,\rho_1$ with $\varphi(\xi)$ bounded and analytic in this region. By applying the theory of linear operators it is easy to prove that the solution of the relations above is given by (5.16). We shall give a sketch of such a proof, for a rigorous proof the reader is referred to WENDEL [1958, 1960] and KINGMAN [1966]. Define the linear operator A by

$$A(.) = \frac{1}{2\pi i}\int_{C_\xi}\left\{\frac{1}{\rho_1-\xi}+\frac{1}{\xi-\rho_2}\right\}(.)\,d\xi,$$

the domain of A being the set of functions $u(.)$ for which the integral in (5.23) exists. Assuming that (5.23) holds then it follows that e^u also satisfies (5.23). Hence $v=\log u$ is a solution of (5.23), i.e. $A\log u=\log u$. Since from (5.23), $A1=1$ we have from (5.24),

$$0 = A\log u(1-\varphi) = A\log u + A\log(1-\varphi) = \log u + A\log(1-\varphi),$$

so that

$$u = \exp\{-A\log(1-\varphi)\},$$

which corresponds with (5.15).

Next we consider $w(\rho_1, \rho_2)$ as a function of r. As the expectation in (5.18) is bounded, the function $w(\rho_1, \rho_2)$ possesses a series expansion in r convergent for $|r|<1$. So we may define for $\mathrm{Re}\,\rho_1>0$, $\mathrm{Re}\,\rho_2<0$, $\mathrm{Re}\,\rho_3>0$, $\mathrm{Re}\,\rho_4>0$,

$$\sum_{n=0}^{\infty} r^n a_n(\rho_1, \rho_2) \stackrel{\text{def}}{=} w(\rho_1, \rho_2), \qquad |r|<1, \tag{5.25}$$

moreover, since $w(\rho_1, \rho_2)$ is bounded a constant c exists such that

$$|w(\rho_1, \rho_2)|<c.$$

Hence

$$|a_n(\rho_1, \rho_2)| = \left|\frac{1}{2\pi i}\int_{D_{r_0}}\frac{w(\rho_1, \rho_2)}{r^{n+1}}\,dr\right| < c r_0^{-n}, \qquad 0<r_0<1,$$

where D_{r_0} is a circular contour in the r-plane with radius r_0 and centre at $r=0$. Substitution of (5.25) in (5.21) and integrating term by term along D_{r_0} yields by equating terms with equal powers of r,

$$\begin{aligned} a_n(\rho_1, \rho_2) &= \frac{1}{2\pi i}\int_{C_\xi} a_{n-1}(\xi, \rho_2)\,\beta(\rho_4+\xi)\,\alpha(\rho_3-\xi)\left\{\frac{1}{\rho_1-\xi}+\frac{1}{\xi-\rho_2}\right\}d\xi, \\ a_0(\rho_1, \rho_2) &= 1, \end{aligned} \tag{5.26}$$

for $n=1, 2, \dots; 0<\mathrm{Re}\,\xi<\min(\mathrm{Re}\,\rho_1, \mathrm{Re}\,\rho_3)$, $\mathrm{Re}\,\rho_1>0$, $\mathrm{Re}\,\rho_2<0$, $\mathrm{Re}\,\rho_3>0$, $\mathrm{Re}\,\rho_4>0$.

The sequence $a_n(\rho_1, \rho_2)$ is uniquely determined by the recurrence relations (5.26), so that the generating function (5.25) of this sequence is the solution of the integral equation (5.17) with $\Psi(\rho)\equiv 1$.

Defining for $\max(\mathrm{Re}\,\rho_2, \delta)<\mathrm{Re}\,\eta<0$,

$$A_n(\rho_1, \rho_2) \stackrel{\text{def}}{=} \frac{1}{2\pi \mathrm{i}} \int_{C_\eta} a_n(\rho_1, \eta)\, \Psi(\eta) \left\{ \frac{1}{\rho_1-\eta} + \frac{1}{\eta-\rho_2} \right\} \mathrm{d}\eta, \qquad n=0, 1, \dots, \tag{5.27}$$

then it follows since $a_n(\rho_1, \rho_2)$ is analytic in ρ_1 for $\mathrm{Re}\,\rho_1>0$, from (5.26) for $n=1, 2, \dots;\ 0<\mathrm{Re}\,\xi<\min(\mathrm{Re}\,\rho_1, \mathrm{Re}\,\rho_3)$ that

$$\begin{aligned} A_n(\rho_1, \rho_2) &= \frac{1}{2\pi \mathrm{i}} \int_{C_\xi} A_{n-1}(\xi, \rho_2)\, \beta(\rho_4+\xi)\, \alpha(\rho_3-\xi) \left\{ \frac{1}{\rho_1-\xi} + \frac{1}{\xi-\rho_2} \right\} \mathrm{d}\xi, \\ A_0(\rho_1, \rho_2) &= \Psi(\rho_1). \end{aligned} \tag{5.28}$$

Again this set of recurrence relations determines the functions $A_n(\rho_1, \rho_2)$, $n=0, 1, 2, \dots$, uniquely, and it is easily verified that its generating function is the solution (5.16) of the integral equation (5.17).

Extensions of the type of integral equation (5.17) have been discussed by LE GALL [1962].

II.5.3. The joint distribution of $\boldsymbol{w}_n$, $\boldsymbol{d}_n$, $\boldsymbol{a}_{n-1}$, $\boldsymbol{b}_{n-1}$

We combine the results of the preceding sections. From the integral equations (5.14) and (5.17), the definitions (5.11), (5.13) and (5.25) and from the similarity between the set of recurrence relations (5.12) and those given by (5.26) it follows that the Laplace-Stieltjes transform of the joint distribution of the waiting time $\boldsymbol{w}_n$ of the nth arriving customer after $t=0$, the total idle time $\boldsymbol{d}_n$ of the server during $[0, \boldsymbol{t}_n)$, the sum $\boldsymbol{a}_{n-1}=\boldsymbol{\sigma}_2+\dots+\boldsymbol{\sigma}_n$ of the first $n-1$ interarrival times, and the sum $\boldsymbol{b}_{n-1}=\boldsymbol{\tau}_1+\dots+\boldsymbol{\tau}_{n-1}$ of the first $n-1$ service times is given by (cf. (5.16) and (5.19)),

$$\omega(r, \rho_1, \rho_2, \rho_3, \rho_4, 0, 0)$$

$$\stackrel{\text{def}}{=} \sum_{n=1}^{\infty} r^n\, \mathrm{E}\{\exp(-\rho_1 \boldsymbol{w}_n+\rho_2 \boldsymbol{d}_n-\rho_3 \boldsymbol{a}_{n-1}-\rho_4 \boldsymbol{b}_{n-1}) \mid \boldsymbol{w}_1=0, \boldsymbol{d}_1=0\} =$$

$$=r\exp\left\{\sum_{n=1}^{\infty}\frac{r^n}{n}\,\mathrm{E}\{\exp(-\rho_4\boldsymbol{b}_n-\rho_3\boldsymbol{a}_n)(\mathrm{e}^{-\rho_1\boldsymbol{s}_n}(\boldsymbol{s}_n>0)+(\boldsymbol{s}_n=0)+\mathrm{e}^{-\rho_2\boldsymbol{s}_n}(\boldsymbol{s}_n<0))\}\right\}$$

$$=r\exp\left\{-\frac{1}{2\pi\mathrm{i}}\int_{C_\xi}\left\{\frac{1}{\rho_1-\xi}+\frac{1}{\xi-\rho_2}\right\}\log\{1-r\beta(\rho_4+\xi)\,\alpha(\rho_3-\xi)\}\,\mathrm{d}\xi\right\},$$

$$\omega(r,\rho_1,\rho_2,\rho_3,\rho_4,w,d) \tag{5.29}$$

$$\stackrel{\text{def}}{=}\sum_{n=1}^{\infty}r^n\,\mathrm{E}\{\exp(-\rho_1\boldsymbol{w}_n+\rho_2\boldsymbol{d}_n-\rho_3\boldsymbol{a}_{n-1}-\rho_4\boldsymbol{b}_{n-1})\mid\boldsymbol{w}_1=w,\,\boldsymbol{d}_1=d\}$$

$$=\frac{1}{2\pi\mathrm{i}}\int_{C_\eta}\exp(-\eta w+\rho_2 d)\left\{\frac{1}{\rho_1-\eta}+\frac{1}{\eta-\rho_2}\right\}\omega(r,\rho_1,\eta,\rho_3,\rho_4,0,0)\,\mathrm{d}\eta, \tag{5.30}$$

for

$$0<\mathrm{Re}\,\xi<\min(\mathrm{Re}\,\rho_1,\mathrm{Re}\,\rho_3),\quad \mathrm{Re}\,\rho_2<\mathrm{Re}\,\eta<0,$$

$$|r|<1,\quad \mathrm{Re}\,\rho_1>0,\quad \mathrm{Re}\,\rho_2<0,\quad \mathrm{Re}\,\rho_3>0,\quad \mathrm{Re}\,\rho_4>0.$$

Obviously, for the path of integration in (5.29) we may also choose $\max(-\mathrm{Re}\,\rho_4, \mathrm{Re}\,\rho_2)<\mathrm{Re}\,\xi<\min(\mathrm{Re}\,\rho_1, \mathrm{Re}\,\rho_3)$ and for that in (5.30) $\mathrm{Re}\,\rho_2<\mathrm{Re}\,\eta\leqq 0$. Since by its definition $\omega(r, \rho_1, \rho_2, \rho_3, \rho_4, w, d)$ is in each of its variables ρ_1, ρ_2, ρ_3, ρ_4 analytic for $\mathrm{Re}\,\rho_1\geqq 0$, $\mathrm{Re}\,\rho_2\leqq 0$, $\mathrm{Re}\,\rho_3\geqq 0$, $\mathrm{Re}\,\rho_4\geqq 0$, it follows by analytic continuation that the first equality sign in (5.29) holds for $\mathrm{Re}\,\rho_1\geqq 0$, $\mathrm{Re}\,\rho_2\leqq 0$, $\mathrm{Re}\,\rho_3\geqq 0$, $\mathrm{Re}\,\rho_4\geqq 0$. Similarly, each of the functions expressed by the integrals in (5.29) and (5.30) has an analytic continuation which represents the left-hand side of (5.29) and (5.30), respectively, for $\mathrm{Re}\,\rho_1\geqq 0$, $\mathrm{Re}\,\rho_2\leqq 0$, $\mathrm{Re}\,\rho_3\geqq 0$, $\mathrm{Re}\,\rho_4\geqq 0$. Hence, for $|r|<1$, $\mathrm{Re}\,\rho_1\geqq 0$, $\mathrm{Re}\,\rho_2\leqq 0$, $\mathrm{Re}\,\rho_3\geqq 0$, $\mathrm{Re}\,\rho_4\geqq 0$,

$$\omega(r,\rho_1,\rho_2,\rho_3,\rho_4,0,0) \tag{5.31}$$

$$=r\exp\left[\sum_{n=1}^{\infty}\frac{r^n}{n}\,\mathrm{E}\{\exp(-\rho_3\boldsymbol{a}_n-\rho_4\boldsymbol{b}_n)(\mathrm{e}^{-\rho_1\boldsymbol{s}_n}(\boldsymbol{s}_n>0)+\mathrm{e}^{-\rho_2\boldsymbol{s}_n}(\boldsymbol{s}_n\leqq 0))\}\right],$$

$$\omega(r,\rho_1,\rho_2,\rho_3,\rho_4,w,d)=\exp(-\rho_2 w+\rho_2 d)\,\omega(r,\rho_1,\rho_2,\rho_3,\rho_4,0,0)$$

$$+\frac{1}{2\pi\mathrm{i}}\int_{C_\eta}\exp(-\eta w+\rho_2 d)\left\{\frac{1}{\rho_1-\eta}+\frac{1}{\eta-\rho_2}\right\}\omega(r,\rho_1,\eta,\rho_3,\rho_4,0,0)\,\mathrm{d}\eta,$$

with $\mathrm{Re}\,\eta<\mathrm{Re}\,\rho_2$.

It should be noted that in the derivation of the above results no use has been made of the assumption that interarrival times and service times have finite means.

The queueing system $G/G/1$ with interarrival time distribution $A(t)$ and service time distribution $B(t)$, and the system $G/G/1$ with interarrival time distribution $B(t)$ and service time distribution $A(t)$ are called dual systems. From (5.29) it is seen that if we replace $\boldsymbol{a}_n$ by $\boldsymbol{b}_n$ and conversely, interchange ρ_3 and ρ_4, and also ρ_1 and ρ_2 and replace ξ by $-\xi$, then we have to interchange $\boldsymbol{w}_n$ and $\boldsymbol{d}_n$ in the dual system. For this reason $\boldsymbol{w}_n$ and $\boldsymbol{d}_n$ are called dual variables. For the dual system $\boldsymbol{w}_n(\boldsymbol{d}_n)$ has the same distribution as $\boldsymbol{d}_n(\boldsymbol{w}_n)$ for the original system. This conclusion may also be obtained from (5.6).

The relation between the queueing process $G/G/1$ and fluctuation theory follows from the relations (1.1), (5.2) and (I.6.66). It is readily seen from these relations that by identifying $\boldsymbol{w}_n$ with $\boldsymbol{v}_n$, $\boldsymbol{d}_n$ with $-\boldsymbol{n}_n$ and $\boldsymbol{\tau}_n-\boldsymbol{\sigma}_{n+1}$ with $\boldsymbol{x}_n$ that the stochastic processes $\{(\boldsymbol{w}_n, \boldsymbol{d}_n), n=1, 2, \dots\}$ with $\boldsymbol{w}_1=0, \boldsymbol{d}_1=0$, and $\{(\boldsymbol{v}_n, -\boldsymbol{n}_n), n=1, 2, \dots\}$ with $\boldsymbol{v}_1=\boldsymbol{n}_1=0$ are identical. Consequently, the relation (5.29) with $\rho_3=\rho_4=0$ may be immediately obtained from (I.6.75). The method used in fluctuation theory for the derivation of (I.6.75) can also be used to derive directly (i.e. without using an integral equation) the relation (5.29) for $\rho_3\neq0$, $\rho_4\neq0$. In the next section the results obtained for the process $\{(\boldsymbol{v}_n, \boldsymbol{n}_n), n=1, 2, \dots\}$ in section I.6.6 will be used for the description of the properties of the process $\{(\boldsymbol{w}_n, \boldsymbol{d}_n), n=1, 2, \dots\}$ with $\boldsymbol{w}_1=\boldsymbol{d}_1=0$.

II.5.4. The actual waiting time

As stated before (cf. section 1.3) the actual waiting time $\boldsymbol{w}_n$ of the nth arriving customer after $t=0$ satisfies

$$\boldsymbol{w}_{n+1}=[\boldsymbol{w}_n+\boldsymbol{\tau}_n-\boldsymbol{\sigma}_{n+1}]^+, \qquad n=1, 2, \dots, \tag{5.32}$$

$$\boldsymbol{w}_1=w.$$

This relation implies that the process $\{\boldsymbol{w}_n, n=0, 1, \dots\}$ is a discrete time parameter Markov process with state space the set of real numbers and with stationary transition probabilities.

Define for $n=1, 2, \dots,$

$$P^{(n)}(w, v, t) \stackrel{\text{def}}{=} \Pr\{\boldsymbol{w}_n<v, \boldsymbol{t}_n<t \mid \boldsymbol{w}_1=w\}, \qquad w\geq0, \quad v\geq0, \quad t\geq0, \tag{5.33}$$

so that for $n=1, 2, \dots$; $w \geqq 0$, $v \geqq 0$,

$$P^{(n)}(w, v) \stackrel{\text{def}}{=} \Pr\{\boldsymbol{w}_n < v \mid \boldsymbol{w}_1 = w\} = \lim_{t\to\infty} P^{(n)}(w, v, t), \tag{5.34}$$

represents the n-step transition probability of the Markov process $\{\boldsymbol{w}_n, n=1, 2, \dots\}$.
Since

$$(\boldsymbol{s}_n \leqq 0) = 1 - (\boldsymbol{s}_n > 0),$$

it follows immediately from the results of the preceding section by choosing $\rho_2=0$, $\rho_4=0$ that for $|r|<1$, $\operatorname{Re}\rho_1 \geqq 0$, $\operatorname{Re}\rho_3 \geqq 0$,

$$\sum_{n=1}^{\infty} r^n \operatorname{E}\{\exp(-\rho_1 \boldsymbol{w}_n - \rho_3 \boldsymbol{t}_n) \mid \boldsymbol{t}_1 = 0, \boldsymbol{w}_1 = 0\}$$

$$= \sum_{n=1}^{\infty} r^n \int_{v=0-}^{\infty} \int_{t=0}^{\infty} \exp(-\rho_1 v - \rho_3 t)\, \mathrm{d}_v\, \mathrm{d}_t\, P^{(n)}(0, v, t)$$

$$= \frac{r}{1 - r\alpha(\rho_3)} \exp\left\{-\sum_{n=1}^{\infty} \frac{r^n}{n} \operatorname{E}\{\exp(-\rho_3 \boldsymbol{a}_n)(1 - \exp(-\rho_1 \boldsymbol{s}_n))(\boldsymbol{s}_n > 0)\}\right\}$$

$$= r \exp\left\{-\frac{1}{2\pi \mathrm{i}} \int_{C_\xi} \left\{\frac{1}{\xi} + \frac{1}{\rho_1 - \xi}\right\} \log\{1 - r\beta(\xi)\,\alpha(\rho_3 - \xi)\}\, \mathrm{d}\xi\right\},$$

$$0 < \operatorname{Re}\xi < \min(\operatorname{Re}\rho_1, \operatorname{Re}\rho_3). \tag{5.35}$$

For the sake of simplicity we have chosen the time origin just before the arrival of the first customer.
Further, by taking $\rho_3 = 0$ we have for $|r|<1$, $\operatorname{Re}\rho \geqq 0$,

$$\sum_{n=1}^{\infty} r^n \operatorname{E}\{\mathrm{e}^{-\rho \boldsymbol{w}_n} \mid \boldsymbol{w}_1 = 0\}$$

$$= \frac{r}{1-r} \exp\left\{-\sum_{n=1}^{\infty} \frac{r^n}{n} \int_0^{\infty} \{1 - \mathrm{e}^{-\rho t}\}\, \mathrm{d}_t \Pr\{\boldsymbol{s}_n < t\}\right\}, \tag{5.36}$$

whence by taking ρ real and $\rho \to \infty$,

$$\sum_{n=1}^{\infty} r^n \Pr\{\boldsymbol{w}_n = 0 \mid \boldsymbol{w}_1 = 0\} = \frac{r}{1-r} \exp\left\{-\sum_{n=1}^{\infty} \frac{r^n}{n} \Pr\{\boldsymbol{s}_n > 0\}\right\}$$

$$= r \exp\left\{\sum_{n=1}^{\infty} \frac{r^n}{n} \Pr\{\boldsymbol{s}_n \leqq 0\}\right\}, \qquad |r|<1. \tag{5.37}$$

From here onwards we shall exclude the trivial case that $\boldsymbol{\rho}_n=0$ with probability 1, i.e. *we assume*

$$\Pr\{\boldsymbol{\rho}_n\neq 0\}>0, \qquad n=1, 2, \ldots . \tag{5.38}$$

Since interarrival times and service times are independent variables, it follows that the event: "an arriving customer finds the server idle" generates a discrete renewal process. Denote the renewal variable of this renewal process by $\boldsymbol{n}$ i.e.

$$\boldsymbol{n} \stackrel{\text{def}}{=} \min\{n\colon \boldsymbol{w}_{n+1}=0,\ n=1, 2, \ldots\} \qquad \text{for} \quad \boldsymbol{w}_1=0. \tag{5.39}$$

From renewal theory (cf. (I.6.78)), it follows that

$$\sum_{n=1}^{\infty} r^n \Pr\{\boldsymbol{w}_{n+1}=0 \mid \boldsymbol{w}_1=0\} = \frac{\mathrm{E}\{r^{\boldsymbol{n}}\}}{1-\mathrm{E}\{r^{\boldsymbol{n}}\}}, \qquad |r|<1, \tag{5.40}$$

so that from (5.37) for $|r|<1$,

$$\begin{aligned}\mathrm{E}\{r^{\boldsymbol{n}}\} &= 1 - \exp\left\{-\sum_{n=1}^{\infty} \frac{r^n}{n} \Pr\{\boldsymbol{s}_n\leqq 0\}\right\} \\ &= 1 - (1-r)\exp\left\{\sum_{n=1}^{\infty} \frac{r^n}{n} \Pr\{\boldsymbol{s}_n>0\}\right\}.\end{aligned} \tag{5.41}$$

Define

$$A \stackrel{\text{def}}{=} \sum_{n=1}^{\infty} \frac{1}{n} \Pr\{\boldsymbol{s}_n<0\}, \quad B \stackrel{\text{def}}{=} \sum_{n=1}^{\infty} \frac{1}{n} \Pr\{\boldsymbol{s}_n>0\}, \quad C \stackrel{\text{def}}{=} \sum_{n=1}^{\infty} \frac{1}{n} \Pr\{\boldsymbol{s}_n=0\}. \tag{5.42}$$

Obviously,

$$A+B+C=\infty. \tag{5.43}$$

However, it follows from (5.38), (I.6.80) and the remark at the end of the preceding section that

$$C<\infty, \qquad A+B=\infty. \tag{5.44}$$

It follows from (5.31) for $\rho_3=\rho_4=0$, $\boldsymbol{w}_1=0$, $\boldsymbol{d}_1=0$, that

$$\sum_{n=1}^{\infty} r^n \Pr\{\boldsymbol{w}_n=0, \boldsymbol{d}_n=0 \mid \boldsymbol{w}_1=0, \boldsymbol{d}_1=0\} = r\exp\left\{\sum_{n=1}^{\infty} \frac{r^n}{n} \Pr\{\boldsymbol{s}_n=0\}\right\}, \qquad |r|<1. \tag{5.45}$$

From (5.6) we have the relation

$$\Pr\{\boldsymbol{w}_{n+1}=0 \mid \boldsymbol{w}_1=0\}=\Pr\{\max(0, \boldsymbol{s}_1, \ldots, \boldsymbol{s}_n)=0\}; \tag{5.46}$$

hence the sequence $\Pr\{w_n=0 \mid w_1=0\}$, $n=1, 2, \ldots$, is non-increasing and its limit exists. Therefore, by applying the last theorem of app. 1 to (5.37) we obtain

$$\lim_{n\to\infty} \Pr\{w_n=0 \mid w_1=0\} = \begin{cases} e^{-B} & \text{if } B<\infty, \\ 0 & \text{if } B=\infty. \end{cases} \tag{5.47}$$

From renewal theory and (5.41), (5.44) and (5.47) it now follows that

$$\Pr\{n<\infty\} \begin{cases} =1, & \mathrm{E}\{n\} \begin{cases} <\infty & \text{if } B<\infty, \quad A=\infty, \\ =\infty & \text{if } B=\infty, \quad A=\infty, \end{cases} \\ =1, & \\ <1, & \qquad\qquad \text{if } B=\infty, \quad A<\infty. \end{cases} \tag{5.48}$$

Hence the event: "an arriving customer finds the server idle" is positive recurrent if $B<\infty$, null recurrent if $B=\infty$, $A=\infty$ and transient if $A<\infty$.

From (1.3) we have $\omega_{n+1} \geqq \omega_n$ so that, since ω_n and w_n have the same distribution $\Pr\{w_n<x \mid w_1=0\}$, $n=1, 2, \ldots$, and $\mathrm{E}\{\exp(-\rho w_n) \mid w_1=0\}$, $n=1, 2, \ldots$, with $\rho>0$, are both bounded and non-increasing sequences and, therefore, have limits for $n\to\infty$. Since for $\rho>0$, $B<\infty$,

$$\lim_{r\uparrow 1} \sum_{n=1}^{\infty} \frac{r^n}{n} \int_0^\infty (1-e^{-\rho t})\, \mathrm{d}\Pr\{s_n<t\} < \infty,$$

it follows from (5.36) and app. 1 for $\rho>0$, $B<\infty$ that

$$\lim_{n\to\infty} \mathrm{E}\{e^{-\rho w_n} \mid w_1=0\} = \exp\left\{-\sum_{n=1}^{\infty} \frac{1}{n} \int_0^\infty (1-e^{-\rho t})\, \mathrm{d}\Pr\{s_n<t\}\right\}. \tag{5.49}$$

By applying Feller's convergence theorem (cf. app. 5) it is seen that if and only if $B<\infty$ the distribution of w_n converges for $n\to\infty$ to a proper probability distribution $W(t)$ for every t which is a continuity point of $W(t)$, and for Re $\rho \geqq 0$, $B<\infty$,

$$\int_{0-}^\infty e^{-\rho t}\, \mathrm{d}W(t) = \exp\left[-\sum_{n=1}^{\infty} \frac{1}{n} \int_0^\infty (1-e^{-\rho t})\, \mathrm{d}\Pr\{s_n<t\}\right]. \tag{5.50}$$

The relation (5.49) also holds if $w_1 \neq 0$, as may be shown by using an argument similar to the one used in the proof of theorem 1.1. Obviously, (5.49) and (5.50) also follow from theorem I.6.8. From this theorem we further have that if $B=\infty$ then for every finite x,

$$\lim_{n\to\infty} \Pr\{\boldsymbol{w}_n<x\}=0.$$

Next we shall investigate, for $B<\infty$, the limit of $\Pr\{\boldsymbol{w}_n<x\}$ for $n\to\infty$ for those x which are discontinuity points of $W(x)$.

For $B<\infty$, $|r|\leqq 1$, define

$$B_0(r) \overset{\text{def}}{=} \sum_{n=1}^{\infty} \frac{r^n}{n} \Pr\{\boldsymbol{s}_n>0\}, \tag{5.51}$$

$$G(r,t) \overset{\text{def}}{=} \begin{cases} B_0^{-1}(r) \sum\limits_{n=1}^{\infty} \dfrac{r^n}{n} \Pr\{0<\boldsymbol{s}_n<t\}, & t>0, \\ 0, & t\leqq 0, \end{cases}$$

$$G(t) \overset{\text{def}}{=} \lim_{r\uparrow 1} G(r,t) = G(1,t).$$

Obviously, $G(r, t)$ is a proper probability distribution in t for every fixed r with $r\in(0, 1]$, and the same holds for $G^{n*}(r, t)$, the n-fold convolution of $G(r, t)$ with respect to t. From (5.36), (5.50) and (5.51) it follows for $|r|<1$, $\operatorname{Re}\rho\geqq 0$ that

$$\sum_{n=1}^{\infty} r^n \operatorname{E}\{e^{-\rho \boldsymbol{w}_n} \mid \boldsymbol{w}_1=0\} = \frac{r}{1-r} e^{-B_0(r)} \sum_{n=0}^{\infty} \frac{B_0^n(r)}{n!} \int_{0-}^{\infty} e^{-\rho t}\, d_t G^{n*}(r,t),$$

$$\int_{0-}^{\infty} e^{-\rho t}\, dW(t) = e^{-B} \sum_{n=0}^{\infty} \frac{B^n}{n!} \int_{0-}^{\infty} e^{-\rho t}\, d_t G^{n*}(t).$$

Write

$$W_n(t) \overset{\text{def}}{=} \Pr\{\boldsymbol{w}_n<t \mid \boldsymbol{w}_1=0\}, \qquad n=1,2,\dots .$$

Since we take distributions to be continuous from the left we have, from the inversion formula for the Laplace-Stieltjes transform, that

$$\sum_{n=1}^{\infty} r^n W_n(t) = \begin{cases} \dfrac{r}{1-r} e^{-B_0(r)} \sum\limits_{n=0}^{\infty} \dfrac{B_0^n(r)}{n!} G^{n*}(r,t), & |r|<1,\quad t>0, \\ 0, & t\leqq 0, \end{cases} \tag{5.52}$$

$$W(t) = \begin{cases} e^{-B} \sum\limits_{n=0}^{\infty} \dfrac{B^n}{n!} G^{n*}(t), & t>0, \\ 0, & t\leqq 0. \end{cases} \tag{5.53}$$

Denote by T_n the set of discontinuity points of $\boldsymbol{s}_n$, $n=1, 2, \ldots$, and write

$$T \stackrel{\text{def}}{=} \bigcup_{n=1}^{\infty} T_n.$$

Obviously

$$S \stackrel{\text{def}}{=} \{t=0\} \bigcup T, \tag{5.54}$$

is the set of discontinuity points of the right-hand side of (5.52) and also of (5.53). Consequently, $W_n(t)$ converges for $n \to \infty$ to $W(t)$ for every $t \notin S$. Since we take distributions to be continuous from the left it follows that $W_n(t)$ converges to $W(t)$ for every t. Note, that for every t, $W_n(t)$ is non-increasing with n.

Clearly, the set T is empty if $A(t)$ or $B(t)$ or both are continuous for all t. Moreover, it is seen that if $A(t)$ and $B(t)$ are both lattice distributions with commensurable periods ω_1 and ω_2, respectively, then $W(t)$ is a lattice distribution with period the least common multiple of ω_1 and ω_2.

The distribution $W(t)$ is infinitely divisible. This property of $W(t)$ follows readily from (5.50) and the general representation of Laplace-Stieltjes transforms of infinitely divisible distributions (cf. LUKACS [1960]).

Writing

$$C(t) \stackrel{\text{def}}{=} \Pr\{\boldsymbol{s}_1 < t\} = \int_0^{\infty} B(t+\tau)\, \mathrm{d}A(\tau), \qquad -\infty < t < \infty, \tag{5.55}$$

we have

$$C^{n*}(t) = \Pr\{\boldsymbol{s}_n < t\}, \qquad n=1, 2, \ldots,$$

and

$$\int_0^{\infty} t\, \mathrm{d}W(t) = \sum_{n=1}^{\infty} \frac{1}{n} \int_0^{\infty} t\, \mathrm{d}C^{n*}(t),$$

if the right-hand side is finite.

It should be noted that in the above derivations the assumption of section 1.2 that the service times and interarrival times have finite first moments has not been used, only the trivial assumption (5.38) is needed. Consequently, it has been shown that the waiting time distribution for the nth arriving customer may have a non-degenerate limit distribution for $n \to \infty$ even if α and β are infinite. In case α and β are finite it follows immediately from Lindley's theorem (cf. section 1.3) that

$$\begin{aligned} a&<1 \quad \text{if} \quad B<\infty, \quad A=\infty, \\ a&=1 \quad \text{if} \quad B=\infty, \quad A=\infty, \\ a&>1 \quad \text{if} \quad B=\infty, \quad A<\infty. \end{aligned} \tag{5.56}$$

II.5.5. The busy period, the idle period and the busy cycle

The duration of the busy period will be denoted by $\boldsymbol{p}$ and that of the busy cycle by $\boldsymbol{c}$; the busy cycle is the time between two successive moments at which an arriving customer has a zero delay. From the definition of $\boldsymbol{n}$ (cf. (5.39)) it follows that $\boldsymbol{n}$ is the number of customers served during a busy period, and hence also during a busy cycle. For the idle period $\boldsymbol{i}$ of a busy cycle we have

$$\boldsymbol{i}=\boldsymbol{c}-\boldsymbol{p}. \tag{5.57}$$

Since successive busy cycles are independent we have from renewal theory that for $|r|<1$, $\mathrm{Re}\,\rho_2 \leqq 0$, $\mathrm{Re}\,\rho_3 \geqq 0$, $\mathrm{Re}\,\rho_4 \geqq 0$,

$$\begin{aligned}\sum_{n=1}^{\infty} r^n\, \mathrm{E}\{(\boldsymbol{w}_{n+1}=0)\exp(\rho_2 \boldsymbol{d}_{n+1}-\rho_3 \boldsymbol{a}_n-\rho_4 \boldsymbol{b}_n) \mid \boldsymbol{w}_1=0, \boldsymbol{d}_1=0\}\\ = \sum_{m=1}^{\infty} [\mathrm{E}\{r^{\boldsymbol{n}} \exp(\rho_2 \boldsymbol{i}-\rho_3 \boldsymbol{a}_{\boldsymbol{n}}-\rho_4 \boldsymbol{b}_{\boldsymbol{n}})\}]^m\\ = \sum_{m=1}^{\infty} [\mathrm{E}\{r^{\boldsymbol{n}} \exp(\rho_2 \boldsymbol{i}-\rho_3 \boldsymbol{c}-\rho_4 \boldsymbol{p})\}]^m,\end{aligned} \tag{5.58}$$

because $\boldsymbol{c}=\boldsymbol{a}_{\boldsymbol{n}}$ and $\boldsymbol{p}=\boldsymbol{b}_{\boldsymbol{n}}$.
Moreover, from (5.29), we have

$$\begin{aligned}\sum_{n=1}^{\infty} r^n\, \mathrm{E}\{(\boldsymbol{w}_n=0)\exp(\rho_2 \boldsymbol{d}_n-\rho_3 \boldsymbol{a}_{n-1}-\rho_4 \boldsymbol{b}_{n-1}) \mid \boldsymbol{w}_1=0, \boldsymbol{d}_1=0\}\\ = r\exp\left[\sum_{n=1}^{\infty} \frac{r^n}{n}\, \mathrm{E}\{(\boldsymbol{s}_n \leqq 0)\exp(-\rho_3 \boldsymbol{a}_n-\rho_4 \boldsymbol{b}_n-\rho_2 \boldsymbol{s}_n)\}\right].\end{aligned} \tag{5.59}$$

Since for $|r|\leqq 1$, $\mathrm{Re}\,\rho_2 \leqq 0$, $\mathrm{Re}\,\rho_3 \geqq 0$, $\mathrm{Re}\,\rho_4 \geqq 0$, the geometric series in the right-hand side of (5.58) converges, it follows easily for *the joint distribution of* $\boldsymbol{n}$, $\boldsymbol{i}$, $\boldsymbol{c}$ *and* $\boldsymbol{p}$ that for $|r|\leqq 1$, $\mathrm{Re}\,\rho_2 \leqq 0$, $\mathrm{Re}\,\rho_3 \geqq 0$, $\mathrm{Re}\,\rho_4 \geqq 0$,

$$\begin{aligned}\mathrm{E}\{r^{\boldsymbol{n}} \exp(\rho_2 \boldsymbol{i}-\rho_3 \boldsymbol{c}-\rho_4 \boldsymbol{p})\}\\ = 1-\exp\left\{-\sum_{n=1}^{\infty} \frac{r^n}{n}\, \mathrm{E}\{(\boldsymbol{s}_n \leqq 0)\exp(-\rho_3 \boldsymbol{a}_n-\rho_4 \boldsymbol{b}_n-\rho_2 \boldsymbol{s}_n)\}\right\}.\end{aligned} \tag{5.60}$$

To obtain the Laplace-Stieltjes transforms of the marginal distributions of $\boldsymbol{i}$, $\boldsymbol{c}$ and $\boldsymbol{p}$, that of $\boldsymbol{n}$ is already given by (5.41), note that

$$\Pr\{\boldsymbol{a}_n<t,\, \boldsymbol{b}_n \leqq \boldsymbol{a}_n\} = \int_0^t B^{n*}(u+)\, \mathrm{d}A^{n*}(u), \qquad t\geqq 0.$$

It follows for Re $\rho \geqq 0$ that

$$\mathrm{E}\{\mathrm{e}^{-\rho \boldsymbol{i}}\} = 1 - \exp\left[-\sum_{n=1}^{\infty} \frac{1}{n} \mathrm{E}\{\mathrm{e}^{\rho \boldsymbol{s}_n} (\boldsymbol{s}_n \leqq 0)\}\right]$$

$$= 1 - \exp\left[-\sum_{n=1}^{\infty} \frac{1}{n} \int_{-\infty}^{0+} \mathrm{e}^{\rho t}\, \mathrm{d}_t \Pr\{\boldsymbol{s}_n < t\}\right], \tag{5.61}$$

$$\mathrm{E}\{\mathrm{e}^{-\rho \boldsymbol{p}}\} = 1 - \exp\left[-\sum_{n=1}^{\infty} \frac{1}{n} \mathrm{E}\{\mathrm{e}^{-\rho \boldsymbol{b}_n} (\boldsymbol{s}_n \leqq 0)\}\right]$$

$$= 1 - \exp\left[-\sum_{n=1}^{\infty} \frac{1}{n} \int_{0}^{\infty} \mathrm{e}^{-\rho t} \{1 - A^{n*}(t)\}\, \mathrm{d}B^{n*}(t)\right], \tag{5.62}$$

$$\mathrm{E}\{\mathrm{e}^{-\rho \boldsymbol{c}}\} = 1 - \exp\left[-\sum_{n=1}^{\infty} \frac{1}{n} \mathrm{E}\{\mathrm{e}^{-\rho \boldsymbol{a}_n} (\boldsymbol{s}_n \leqq 0)\}\right]$$

$$= 1 - \exp\left[-\sum_{n=1}^{\infty} \frac{1}{n} \int_{0}^{\infty} \mathrm{e}^{-\rho t} B^{n*}(t+)\, \mathrm{d}A^{n*}(t)\right], \tag{5.63}$$

$$\mathrm{E}\{r^{\boldsymbol{n}}\, \mathrm{e}^{-\rho \boldsymbol{i}}\} = 1 - \exp\left[-\sum_{n=1}^{\infty} \frac{r^n}{n} \int_{-\infty}^{0+} \mathrm{e}^{\rho t}\, \mathrm{d}\Pr\{\boldsymbol{s}_n < t\}\right], \qquad |r| \leqq 1. \tag{5.64}$$

From (5.61) it is seen that

$$\Pr\{\boldsymbol{i} = 0\} = 1 - \mathrm{e}^{-C}. \tag{5.65}$$

Consequently, if $\Pr\{\boldsymbol{\rho}_1 = 0\} > 0$, which is possible if $A(t)$ and $B(t)$ are lattice distributions with commensurable periods, then $C > 0$ (cf. (5.42)), so that there is a positive probability that a busy period is followed by an idle period of duration zero. $C = 0$ if at least one of these distributions is continuous or both are lattice distributions with incommensurable periods. An idle period of duration zero occurs if at the moment the last customer leaves the system a new customer arrives. This possibility makes it desirable to sharpen the definition of busy period. By the *strong busy period* we shall from now on indicate the time between the end of an idle period and the beginning of the next idle period, the latter idle period having a non-zero duration. We shall speak of a *weak busy period* if the latter idle period has a zero or positive duration. Clearly, $\boldsymbol{n}$ and $\boldsymbol{p}$ defined above refer to the number of customers served during a weak busy period and the length of a weak busy period, respectively. By $\boldsymbol{p}_s$ we shall denote the length of a strong busy period and by $\boldsymbol{n}_s$ the number of customers served during a strong busy period.

From (5.64) it follows that

$$\sum_{n=1}^{\infty} r^n \Pr\{\boldsymbol{i}=0, \boldsymbol{n}=n\} = 1 - \exp\left[-\sum_{n=1}^{\infty} \frac{r^n}{n} \Pr\{\boldsymbol{s}_n=0\}\right], \qquad |r|\leqq 1, \tag{5.66}$$

so that for $|r|\leqq 1$, $\operatorname{Re} \rho \geqq 0$, we have from (5.64) that

$$\begin{aligned}&\sum_{n=1}^{\infty} r^n \int_{0+}^{\infty} e^{-\rho t} \, d\Pr\{\boldsymbol{i}<t, \boldsymbol{n}=n\} \\ &= \exp\left[-\sum_{n=1}^{\infty} \frac{r^n}{n} \Pr\{\boldsymbol{s}_n=0\}\right]\left[1 - \exp\left[-\sum_{n=1}^{\infty} \frac{r^n}{n} \int_{-\infty}^{0-} e^{\rho t} \, d\Pr\{\boldsymbol{s}_n<t\}\right]\right].\end{aligned} \tag{5.67}$$

Since a strong busy period can be the sum of any number of weak busy periods, all interrupted by idle periods of duration zero, and one busy period followed by a positive idle period, since all these components are independent it follows that for $|r|\leqq 1$, $\operatorname{Re} \rho \geqq 0$,

$$\begin{aligned}&\sum_{n=1}^{\infty} r^n \int_{0+}^{\infty} e^{-\rho t} \, d\Pr\{\boldsymbol{n}_s=n, \boldsymbol{i}<t\} \\ &\qquad = \sum_{m=0}^{\infty} \Big\{\sum_{n=1}^{\infty} r^n \Pr\{\boldsymbol{n}=n, \boldsymbol{i}=0\}\Big\}^m \sum_{n=1}^{\infty} r^n \int_{0+}^{\infty} e^{-\rho t} \, d\Pr\{\boldsymbol{i}<t, \boldsymbol{n}=n\}.\end{aligned}$$

Hence from (5.67) for $|r|\leqq 1$, $\operatorname{Re} \rho \geqq 0$, we have

$$\mathrm{E}\{r^{\boldsymbol{n}_s} e^{-\rho \boldsymbol{i}} \mid \boldsymbol{i}>0\} = 1 - \exp\left[-\sum_{n=1}^{\infty} \frac{r^n}{n} \int_{-\infty}^{0-} e^{\rho t} \, d\Pr\{\boldsymbol{s}_n<t\}\right], \tag{5.68}$$

$$\mathrm{E}\{r^{\boldsymbol{n}_s}\} = 1 - \exp\left[-\sum_{n=1}^{\infty} \frac{r^n}{n} \Pr\{\boldsymbol{s}_n<0\}\right]. \tag{5.69}$$

In exactly the same way we obtain from (5.60) for $|r|\leqq 1$, $\operatorname{Re} \rho_2 \leqq 0$, $\operatorname{Re} \rho_3 \geqq 0$, $\operatorname{Re} \rho_4 \geqq 0$,

$$\begin{aligned}&\mathrm{E}\{r^{\boldsymbol{n}_s} \exp(\rho_2 \boldsymbol{i} - \rho_3 \boldsymbol{c}_s - \rho_4 \boldsymbol{p}_s) \mid \boldsymbol{i}>0\} \\ &\qquad = 1 - \exp\left\{-\sum_{n=1}^{\infty} \frac{r^n}{n} \mathrm{E}\{(\boldsymbol{s}_n<0) \exp(-\rho_3 \boldsymbol{a}_n - \rho_4 \boldsymbol{b}_n - \rho_2 \boldsymbol{s}_n)\}\right\},\end{aligned} \tag{5.70}$$

where $\boldsymbol{c}_s$ is the duration of a *strong busy cycle*, i.e. a busy cycle containing a positive idle period. It follows for Re $\rho \geqq 0$ that

$$\mathrm{E}\{\mathrm{e}^{-\rho \boldsymbol{p}_s}\} = 1 - \exp\left[-\sum_{n=1}^{\infty} \frac{1}{n} \mathrm{E}\{\mathrm{e}^{-\rho \boldsymbol{b}_n}(\boldsymbol{s}_n < 0)\}\right]$$

$$= 1 - \exp\left[-\sum_{n=1}^{\infty} \frac{1}{n} \int_0^{\infty} \mathrm{e}^{-\rho t}\{1 - A^{n*}(t+)\}\,\mathrm{d}B^{n*}(t)\right], \tag{5.71}$$

$$\mathrm{E}\{\mathrm{e}^{-\rho \boldsymbol{c}_s}\} = 1 - \exp\left[-\sum_{n=1}^{\infty} \frac{1}{n} \mathrm{E}\{\mathrm{e}^{-\rho \boldsymbol{a}_n}(\boldsymbol{s}_n < 0)\}\right]$$

$$= 1 - \exp\left[-\sum_{n=1}^{\infty} \frac{1}{n} \int_0^{\infty} \mathrm{e}^{-\rho t} B^{n*}(t)\,\mathrm{d}A^{n*}(t)\right]. \tag{5.72}$$

From the relations obtained above it is easily verified that all variables, $\boldsymbol{n}$, $\boldsymbol{n}_s$, $\boldsymbol{i}$, $\boldsymbol{c}$, $\boldsymbol{c}_s$, $\boldsymbol{p}$ and $\boldsymbol{p}_s$ are finite with probability 1 if and only if $A = \infty$, i.e. if $B < \infty$, or $B = \infty$, $A = \infty$.

Further from (5.41) and (5.69),

$$\mathrm{E}\{\boldsymbol{n}\} = \mathrm{e}^B, \qquad \mathrm{E}\{\boldsymbol{n}_s\} = \mathrm{e}^{B+C} \qquad \text{if} \quad B < \infty. \tag{5.73}$$

If $B = \infty$, $A = \infty$, then $\boldsymbol{n}$ and $\boldsymbol{n}_s$ have no finite first moments, and hence $\boldsymbol{i}$, $\boldsymbol{c}$, $\boldsymbol{c}_s$, $\boldsymbol{p}$ and $\boldsymbol{p}_s$ cannot have finite first moments, since,

$$\mathrm{E}\{\boldsymbol{c}\} = \mathrm{E}\{\boldsymbol{\sigma}_2 + \ldots + \boldsymbol{\sigma}_{n+1}\}, \tag{5.74}$$

$$\mathrm{E}\{\boldsymbol{p}\} = \mathrm{E}\{\boldsymbol{\tau}_1 + \ldots + \boldsymbol{\tau}_n\}.$$

If α and β are finite and $B < \infty$, so that $a < 1$ (cf. (5.56)), then by using Wald's theorem (cf. app. 7) we see that

$$\mathrm{E}\{\boldsymbol{c}\} = \alpha\,\mathrm{e}^B, \quad \mathrm{E}\{\boldsymbol{c}_s\} = \alpha\,\mathrm{e}^{B+C}, \quad \mathrm{E}\{\boldsymbol{p}\} = \beta\,\mathrm{e}^B, \quad \mathrm{E}\{\boldsymbol{p}_s\} = \beta\,\mathrm{e}^{B+C}, \tag{5.75}$$

$$\mathrm{E}\{\boldsymbol{i}\} = \mathrm{E}\{\boldsymbol{c}\} - \mathrm{E}\{\boldsymbol{p}\} = (\alpha - \beta)\,\mathrm{e}^B, \quad \mathrm{E}\{\boldsymbol{i} \mid \boldsymbol{i} > 0\} = (\alpha - \beta)\,\mathrm{e}^{B+C}.$$

Let us denote the number of arriving customers during $(0, t)$ whose services start immediately after their arrival by $\boldsymbol{\alpha}_t$, $t > 0$, supposing that at $t = 0$ a new service starts with only one customer in the system, i.e. $\boldsymbol{t}_1 = 0$. Obviously $\boldsymbol{\alpha}_t$ is the number of renewals in $(0, t)$ of the renewal process with renewal distribution the distribution function of the weak busy cycle. Denoting the distribution function of the weak busy cycle by $P(t)$, it follows immediately from renewal theory that

$$\Pr\{\boldsymbol{\alpha}_t \geqq n\} = P^{n*}(t), \qquad n=0, 1, \ldots,$$

and hence

$$\sum_{n=0}^{\infty} r^n \int_0^{\infty} \mathrm{e}^{-\rho t}\, \mathrm{d}_t \Pr\{\boldsymbol{\alpha}_t \geqq n\} = \frac{1}{1-r\, \mathrm{E}\{\mathrm{e}^{-\rho \boldsymbol{c}}\}}, \qquad |r|<1, \quad \mathrm{Re}\, \rho \geqq 0. \tag{5.76}$$

In (5.76) $\boldsymbol{c}$ should be replaced by $\boldsymbol{c}_s$ if we want to have the distribution of the number of arriving customers during $(0, t)$ who find the server idle, the preceding idle period at an arrival being positive.

Let $\boldsymbol{\alpha}_n$ be the number of customers among the first n arriving customers who do not have to wait, supposing that at $\boldsymbol{t}_1$ a new service starts with only one customer in the system, $n=0, 1, \ldots,$ (note that $\boldsymbol{\alpha}_{n+1} = \boldsymbol{e}_n$ if in (1.20) $k=1$).
Hence $\boldsymbol{\alpha}_n$ is the number of renewals of the discrete renewal process with renewal distribution the distribution of $\boldsymbol{n}$, the number of customers served during a weak busy cycle.
It is easily found that

$$\sum_{k=0}^{\infty} \sum_{n=1}^{\infty} p^k r^n \Pr\{\boldsymbol{\alpha}_n \geqq k\} = \frac{1}{1-r} \frac{1}{1-p\, \mathrm{E}\{r^{\boldsymbol{n}}\}}, \qquad |r|<1, \quad |p|<1. \tag{5.77}$$

So far we have only considered the distribution function of the busy period, the busy cycle and so on, under the condition that the busy period was started with only one customer in the system. Next, we shall investigate the *residual busy period*, i.e. the distribution function of the time between $\boldsymbol{t}_1$ and the first moment thereafter at which the server becomes idle assuming that the arriving customer at $\boldsymbol{t}_1$ has a waiting time $\boldsymbol{w}_1 = w > 0$; we shall also consider the residual busy cycle, and the number of customers served during a residual busy period.
For the derivation of the above distribution we need the limit for $\rho_1 \to \infty$, $\arg \rho_1 = 0$ of the expression (5.31). For the sake of simplicity we shall calculate this limit for $\rho_3 = \rho_4 = 0$.
For $|r|<1$, $\mathrm{Re}\, \rho_1 \geqq 0$, $\mathrm{Re}\, \xi < \mathrm{Re}\, \rho_2 \leqq 0$ we have for the last term in (5.31),

$$\frac{1}{2\pi i}\int_{C_\xi}\left\{\frac{1}{\rho_1-\xi}+\frac{1}{\xi-\rho_2}\right\}$$

$$\cdot\sum_{n=1}^{\infty} r^n\, \mathrm{E}\{\exp(-\rho_1 \boldsymbol{w}_n-\xi(w-\boldsymbol{d}_n)) \mid \boldsymbol{w}_1=0,\, \boldsymbol{d}_1=0\}\, \mathrm{d}\xi$$

$$=\frac{1}{2\pi i}\int_{C_\xi}\left\{\frac{1}{\rho_1-\xi}+\frac{1}{\xi-\rho_2}\right\}$$

$$\cdot\sum_{n=1}^{\infty} r^n\, \mathrm{E}\{((w\leqq \boldsymbol{d}_n)+(w>\boldsymbol{d}_n))\exp(-\rho_1 \boldsymbol{w}_n-\xi(w-\boldsymbol{d}_n)) \mid \boldsymbol{w}_1=0,\, \boldsymbol{d}_1=0\}\, \mathrm{d}\xi$$

$$=\sum_{n=1}^{\infty} r^n\, \mathrm{E}\{(w>\boldsymbol{d}_n)\exp(-\rho_1 \boldsymbol{w}_n-\rho_1(w-\boldsymbol{d}_n)) \mid \boldsymbol{w}_1=0,\, \boldsymbol{d}_1=0\}$$

$$-\sum_{n=1}^{\infty} r^n\, \mathrm{E}\{(w>\boldsymbol{d}_n)\exp(-\rho_1 \boldsymbol{w}_n-\rho_2(w-\boldsymbol{d}_n)) \mid \boldsymbol{w}_1=0,\, \boldsymbol{d}_1=0\}. \tag{5.78}$$

In the same way we obtain for $|r|<1$, $\mathrm{Re}\,\xi<\mathrm{Re}\,\rho_2\leqq 0$,

$$\frac{1}{2\pi i}\int_{C_\xi}\frac{\mathrm{d}\xi}{\xi-\rho_2}\sum_{n=1}^{\infty} r^n\, \mathrm{E}\{(\boldsymbol{w}_n=0)\exp(-\xi(w-\boldsymbol{d}_n)) \mid \boldsymbol{w}_1=0,\, \boldsymbol{d}_1=0\}$$

$$=-\sum_{n=1}^{\infty} r^n\, \mathrm{E}\{(\boldsymbol{w}_n=0)(w>\boldsymbol{d}_n)\exp(-\rho_2(w-\boldsymbol{d}_n)) \mid \boldsymbol{w}_1=0,\, \boldsymbol{d}_1=0\}$$

$$-\tfrac{1}{2}\sum_{n=1}^{\infty} r^n\, \Pr\{\boldsymbol{w}_n=0,\, \boldsymbol{d}_n=w \mid \boldsymbol{w}_1=0,\, \boldsymbol{d}_1=0\}. \tag{5.79}$$

Hence for $|r|<1$, $\mathrm{Re}\,\xi<\mathrm{Re}\,\rho_2\leqq 0$, $\arg\rho_1=0$,

$$\lim_{\rho_1\to\infty}\frac{1}{2\pi i}\int_{C_\xi}\left\{\frac{1}{\rho_1-\xi}+\frac{1}{\xi-\rho_2}\right\}$$

$$\cdot\sum_{n=1}^{\infty} r^n\, \mathrm{E}\{\exp(-\rho_1 \boldsymbol{w}_n-\xi(w-\boldsymbol{d}_n)) \mid \boldsymbol{w}_1=0,\, \boldsymbol{d}_1=0\}\, \mathrm{d}\xi$$

$$=\frac{1}{2\pi i}\int_{C_\xi}\frac{\mathrm{d}\xi}{\xi-\rho_2}\sum_{n=1}^{\infty} r^n\, \mathrm{E}\{(\boldsymbol{w}_n=0)\exp(-\xi(w-\boldsymbol{d}_n)) \mid \boldsymbol{w}_1=0,\, \boldsymbol{d}_1=0\}$$

$$+\tfrac{1}{2}\sum_{n=1}^{\infty} r^n\, \Pr\{\boldsymbol{w}_n=0,\, \boldsymbol{d}_n=w \mid \boldsymbol{w}_1=0,\, \boldsymbol{d}_1=0\}. \tag{5.80}$$

Obviously, $\Pr\{\boldsymbol{d}_n = w \mid \boldsymbol{w}_1 = 0, \boldsymbol{d}_1 = 0\} = 0$ if the distribution of $\boldsymbol{d}_n$ is continuous for $w > 0$, i.e. if $A(t)$ or $B(t)$ has no discontinuity points. In the following derivations this will be assumed.
From (5.30) and from the above relations we have for $|r| < 1$, $\operatorname{Re} \rho_2 \leqq 0$, $\operatorname{Re} \rho_3 \geqq 0$, $\operatorname{Re} \rho_4 \geqq 0$, $w > 0$, $\operatorname{Re} \xi > \operatorname{Re} \rho_2$,

$$\sum_{n=1}^{\infty} r^n \mathrm{E}\{(\boldsymbol{w}_n = 0) \exp(\rho_2 \boldsymbol{d}_n - \rho_3 \boldsymbol{a}_{n-1} - \rho_4 \boldsymbol{b}_{n-1}) \mid \boldsymbol{w}_1 = w, \boldsymbol{d}_1 = 0\} \tag{5.81}$$

$$= \frac{1}{2\pi \mathrm{i}} \int_{C_\xi} \frac{r \, \mathrm{d}\xi}{\xi - \rho_2} \mathrm{e}^{-\xi w} \exp\left[\sum_{n=1}^{\infty} \frac{r^n}{n} \mathrm{E}\{(\boldsymbol{s}_n < 0) \exp(-\rho_3 \boldsymbol{a}_n - \rho_4 \boldsymbol{b}_n - \xi \boldsymbol{s}_n)\} \right].$$

Denoting the duration of the residual busy cycle by $\boldsymbol{c}_1$, the residual busy period by $\boldsymbol{p}_1$, the idle time during the residual busy cycle by $\boldsymbol{i}_1$, and the number of customers served during the residual busy cycle by $\boldsymbol{n}_1$ and assuming that the busy cycle starts with an initial waiting time $\boldsymbol{w}_1 = w > 0$, it follows from renewal theory and (5.60) that for $|r| < 1$, $\operatorname{Re} \rho_2 \leqq 0$, $\operatorname{Re} \rho_3 \geqq 0$, $\operatorname{Re} \rho_4 \geqq 0$ (cf. (5.58)),

$$\frac{1}{r} \sum_{n=1}^{\infty} r^n \mathrm{E}\{(\boldsymbol{w}_n = 0) \exp(\rho_2 \boldsymbol{d}_n - \rho_3 \boldsymbol{a}_{n-1} - \rho_4 \boldsymbol{b}_{n-1}) \mid \boldsymbol{w}_1 = w, \boldsymbol{d}_1 = 0\}$$

$$= \sum_{n=1}^{\infty} r^n \mathrm{E}\{(\boldsymbol{w}_{n+1} = 0) \exp(\rho_2 \boldsymbol{d}_{n+1} - \rho_3 \boldsymbol{a}_n - \rho_4 \boldsymbol{b}_n) \mid \boldsymbol{w}_1 = w, \boldsymbol{d}_1 = 0\}$$

$$= \frac{\mathrm{E}\{r^{\boldsymbol{n}_1} \exp(\rho_2 \boldsymbol{i}_1 - \rho_3 \boldsymbol{c}_1 - \rho_4 \boldsymbol{p}_1)\}}{1 - \mathrm{E}\{r^{\boldsymbol{n}} \exp(\rho_2 \boldsymbol{i} - \rho_3 \boldsymbol{c} - \rho_4 \boldsymbol{p})\}}$$

$$= \mathrm{E}\{r^{\boldsymbol{n}_1} \exp(\rho_2 \boldsymbol{i}_1 - \rho_3 \boldsymbol{c}_1 - \rho_4 \boldsymbol{p}_1)\}$$

$$\cdot \exp\left[\sum_{n=1}^{\infty} \frac{r^n}{n} \mathrm{E}\{(\boldsymbol{s}_n < 0) \exp(-\rho_3 \boldsymbol{a}_n - \rho_4 \boldsymbol{b}_n - \rho_2 \boldsymbol{s}_n)\} \right]. \tag{5.82}$$

Hence, from (5.81) and (5.82) we obtain for $|r| \leqq 1$, $\operatorname{Re} \rho_2 \leqq 0$, $\operatorname{Re} \rho_3 \geqq 0$, $\operatorname{Re} \rho_4 \geqq 0$, $w > 0$, assuming that $A(t)$ or $B(t)$ has no discontinuity points,

$$\mathrm{E}\{r^{\boldsymbol{n}_1} \exp(\rho_2 \boldsymbol{i}_1 - \rho_3 \boldsymbol{c}_1 - \rho_4 \boldsymbol{p}_1)\} = \frac{1}{2\pi \mathrm{i}} \int_{C_\xi} \frac{\mathrm{d}\xi}{\xi - \rho_2} \mathrm{e}^{-\xi w}$$

$$\cdot \exp\left[\sum_{n=1}^{\infty} \frac{r^n}{n} \mathrm{E}\{\exp(-\rho_3 \boldsymbol{a}_n - \rho_4 \boldsymbol{b}_n)(\mathrm{e}^{-\xi \boldsymbol{s}_n} - \mathrm{e}^{-\rho_2 \boldsymbol{s}_n})(\boldsymbol{s}_n < 0)\} \right], \tag{5.83}$$

with $\operatorname{Re} \xi > \operatorname{Re} \rho_2$.

II.5.6. The virtual waiting time and the idle time

As usual the virtual waiting time at time t is denoted by $\boldsymbol{v}_t$. Denoting by $\boldsymbol{\nu}_t$ the number of renewals* in $[0, t)$ of the renewal process with renewal distribution $A(t)$ it is easily seen that with probability one

$$\boldsymbol{v}_t > \boldsymbol{w}_{\boldsymbol{\nu}_t+1}, \qquad t \geqq 0. \tag{5.84}$$

For $a \geqq 1$ Lindley's theorem 1.1 shows that

$$\lim_{n\to\infty} \Pr\{\boldsymbol{w}_n < \infty\} = 0,$$

moreover, $\boldsymbol{\nu}_t \to \infty$ as $t \to \infty$ with probability one since $\alpha < \infty$. Consequently, the above relation shows that

$$\lim_{t\to\infty} \Pr\{\boldsymbol{v}_t < \infty\} = 0 \qquad \text{if} \quad a \geqq 1. \tag{5.85}$$

For the present *it will be assumed that* $a<1$. Let us denote by $D(t)$ and $P(t)$ the distribution functions of the busy period and the busy cycle respectively (cf. (5.62) and (5.63)). Assuming for the sake of simplicity that $\boldsymbol{w}_1 = 0$, $\boldsymbol{t}_1 = 0$, it follows that

$$\Pr\{\boldsymbol{v}_t > 0\} = \int_0^t \{1 - D(t-u)\} \,\mathrm{d}_u \sum_{n=0}^{\infty} P^{n*}(u), \qquad t > 0. \tag{5.86}$$

From (5.71) and (5.72) it is seen that for $\mathrm{Re}\, \rho \geqq 0$,

$$\int_0^\infty \mathrm{e}^{-\rho t}\,\mathrm{d}P(t) = 1 - (1-\alpha(\rho)) \exp\left[\sum_{n=1}^{\infty} \frac{1}{n} \mathrm{E}\{\mathrm{e}^{-\rho \boldsymbol{a}_n} (\boldsymbol{s}_n > 0)\}\right], \tag{5.87}$$

$$\int_0^\infty \mathrm{e}^{-\rho t}\,\mathrm{d}D(t) = 1 - (1-\beta(\rho)) \exp\left[\sum_{n=1}^{\infty} \frac{1}{n} \mathrm{E}\{\mathrm{e}^{-\rho \boldsymbol{b}_n} (\boldsymbol{s}_n > 0)\}\right];$$

so that $P(t)$ is a lattice distribution if and only if $A(t)$ is a lattice distribution; the same relation exists between $B(t)$ and $D(t)$. Suppose that $A(t)$ is not a lattice distribution. Applying the key renewal theorem of Smith (cf. theorem I.6.2) it follows from (5.86) that $\Pr\{\boldsymbol{v}_t > 0\}$ has a limit for $t \to \infty$

*) In the case when $\boldsymbol{t}_1 = 0$, the renewal at $t=0$ is supposed to be included in $\boldsymbol{\nu}_t$, so that then e.g. $\boldsymbol{\nu}_t = 1$ for $\boldsymbol{t}_1 = 0 < t < \boldsymbol{t}_2$.

since $a<1$ implies $\mathrm{E}\{\boldsymbol{c}\}<\infty$, and (cf. (5.75)),

$$\lim_{t\to\infty} \Pr\{\boldsymbol{v}_t>0\} = \frac{1}{\mathrm{E}\{\boldsymbol{c}\}} \int_0^\infty \{1-D(\tau)\}\,\mathrm{d}\tau = \frac{\mathrm{E}\{\boldsymbol{p}\}}{\mathrm{E}\{\boldsymbol{c}\}} = \frac{\beta}{\alpha} = a, \tag{5.88}$$

$$\lim_{t\to\infty} \Pr\{\boldsymbol{v}_t=0\} = 1-a.$$

If $A(t)$ is a lattice distribution with period τ then we obtain from (5.86) for $0\leqq\sigma<\tau$, using the method described in section I.6.2,

$$a(\sigma) \stackrel{\text{def}}{=} \lim_{n\to\infty} \Pr\{\boldsymbol{v}_{n\tau+\sigma}>0\} = \frac{\tau}{\alpha\,\mathrm{e}^B} \sum_{k=0}^{\infty} \{1-D(k\tau+\sigma)\}. \tag{5.89}$$

Note that if we consider for the present case $\boldsymbol{\sigma}$ to be a stochastic variable uniformly distributed on $[0, \tau)$ then

$$\lim_{n\to\infty} \Pr\{\boldsymbol{v}_{n\tau+\sigma}=0\} = 1-a.$$

We shall now investigate the distribution function of $\boldsymbol{v}_t$, and show first that *this distribution has for $t\to\infty$ a limit which is a proper probability distribution if $a<1$ and if $A(t)$ is not a lattice distribution.*

It is easily verified that

$$\boldsymbol{v}_t=\max(0, \boldsymbol{w}_{\nu_t}+\boldsymbol{\tau}_{\nu_t}-(t-\boldsymbol{t}_{\nu_t})), \qquad t>0. \tag{5.90}$$

Obviously, $\boldsymbol{v}_t$ will have a limit distribution for $t\to\infty$ if the distribution of $\boldsymbol{w}_{\nu_t}-(t-\boldsymbol{t}_{\nu_t})$ has a limit for $t\to\infty$, since the service time $\boldsymbol{\tau}_{\nu_t}$ is independent of $\boldsymbol{v}_t$ and of $\boldsymbol{w}_{\nu_t}$. Taking (again for the sake of simplicity) $\boldsymbol{w}_1=0$, $\boldsymbol{d}_1=0$, $\boldsymbol{t}_1=0$ (for the general case the proof is not essentially different) it follows from (5.31) that for $|r|<1$, $\mathrm{Re}\,\rho_1\geqq 0$, $\mathrm{Re}\,\rho_3\geqq 0$,

$$\sum_{n=1}^{\infty} r^n \int_{x=0-}^{\infty} \int_{u=0}^{\infty} \mathrm{e}^{-\rho_1 x-\rho_3 u}\,\mathrm{d}_x\,\mathrm{d}_u \Pr\{\boldsymbol{w}_n<x, \boldsymbol{a}_{n-1}<u \mid \boldsymbol{w}_1=0\}$$

$$= \frac{r}{1-r\alpha(\rho_3)} \exp\left\{\sum_{n=1}^{\infty} \frac{r^n}{n} \mathrm{E}\{\mathrm{e}^{-\rho_3 \boldsymbol{a}_n}(\mathrm{e}^{-\rho_1 \boldsymbol{s}_n}-1)(\boldsymbol{s}_n>0)\}\right\}. \tag{5.91}$$

For $a<1$ we have $B<\infty$ so that if $\mathrm{Re}\,\rho_3>0$ then (5.91) is also true for $r=1$.

Since for $x \geqq 0$, $t>0$,

$$\Pr\{\boldsymbol{w}_{\boldsymbol{v}_t}<x \mid \boldsymbol{w}_1=0\} = \sum_{n=1}^{\infty} \Pr\{\boldsymbol{w}_n<x,\ \boldsymbol{v}_t=n \mid \boldsymbol{w}_1=0\}$$

$$= \int_0^t \{1-A(t-u)\}\, \mathrm{d}_u \sum_{n=1}^{\infty} \Pr\{\boldsymbol{w}_n<x,\ \boldsymbol{a}_{n-1}<u \mid \boldsymbol{w}_1=0\}, \tag{5.92}$$

we obtain from (5.91) and (5.92) for $\operatorname{Re} \rho_1 \geqq 0$, $\operatorname{Re} \rho_3>0$ by taking $r=1$,

$$\int_{x=0-}^{\infty} \int_{t=0}^{\infty} \mathrm{e}^{-\rho_1 x-\rho_3 t}\, \mathrm{d}_x\, \mathrm{d}_t \Pr\{\boldsymbol{w}_{\boldsymbol{v}_t}<x \mid \boldsymbol{w}_1=0\}$$

$$= \exp\left\{\sum_{n=1}^{\infty} \frac{1}{n} \mathrm{E}\{\mathrm{e}^{-\rho_3 \boldsymbol{a}_n}(\mathrm{e}^{-\rho_1 \boldsymbol{s}_n}-1)(\boldsymbol{s}_n>0)\}\right\}. \tag{5.93}$$

Since $\mathrm{E}\{\exp(-\rho \boldsymbol{w}_{\boldsymbol{v}_t}) \mid \boldsymbol{w}_1=0\}$ is a non-increasing and bounded function of t for $\operatorname{Re} \rho \geqq 0$ (cf. the derivation of (5.49)) it follows from (5.93) and a well-known Tauberian theorem (cf. app. 4) that for $\operatorname{Re} \rho \geqq 0$,

$$\lim_{t\to\infty} \int_{0-}^{\infty} \mathrm{e}^{-\rho_1 x}\, \mathrm{d}_x \Pr\{\boldsymbol{w}_{\boldsymbol{v}_t}<x \mid \boldsymbol{w}_1=0\}$$

$$= \exp\left[-\sum_{n=1}^{\infty} \frac{1}{n} \int_0^{\infty} \{1-\mathrm{e}^{-\rho_1 t}\}\, \mathrm{d}\Pr\{\boldsymbol{s}_n<t\}\right]. \tag{5.94}$$

Using Feller's convergence theorem (cf. app. 5) it is seen that the distribution of $\boldsymbol{w}_{\boldsymbol{v}_t}$ converges for $t\to\infty$ to a limit distribution. This limit distribution is that of $\boldsymbol{w}_n$ for $n\to\infty$.

Denoting the renewal function of the renewal process $\boldsymbol{v}_t$ by $n(t)$ with $\boldsymbol{v}_{0+}=1$, so that

$$1+m(t) \stackrel{\text{def}}{=} n(t) = \sum_{n=0}^{\infty} A^{n*}(t) = \mathrm{E}\{\boldsymbol{v}_t\}, \tag{5.95}$$

then it follows from (5.93) and (5.91) for $r=1$, $\operatorname{Re} \rho_3>0$ that for $t>0$,

$$\sum_{n=1}^{\infty} \Pr\{\boldsymbol{w}_n<x,\ \boldsymbol{a}_{n-1}<t \mid \boldsymbol{w}_1=0\} = \int_{0-}^{t} \Pr\{\boldsymbol{w}_{\boldsymbol{v}_{t-u}}<x \mid \boldsymbol{w}_1=0\}\, \mathrm{d}n(u), \tag{5.96}$$

or

$$L(x, t) = H(x, t) * n(t),$$

if we define for the present

$$L(x,t) \overset{\text{def}}{=} \begin{cases} \sum_{n=1}^{\infty} \Pr\{\boldsymbol{w}_n < x,\ \boldsymbol{a}_{n-1} < t \mid \boldsymbol{w}_1 = 0\}, & t>0,\quad x\geqq 0,\\ 0, & \text{otherwise,} \end{cases} \tag{5.97}$$

$$H(x,t) \overset{\text{def}}{=} \begin{cases} \sum_{n=1}^{\infty} \Pr\{\boldsymbol{w}_{\boldsymbol{v}_t} < x \mid \boldsymbol{w}_1 = 0\}, & t>0,\quad x\geqq 0,\\ 0, & \text{otherwise.} \end{cases}$$

Since $\boldsymbol{w}_n$ and $\boldsymbol{\sigma}_{n+1}$ are independent we have for $t>0$, $x\geqq 0$, $y\geqq 0$,

$$\Pr\{\boldsymbol{w}_{\boldsymbol{v}_t} < x,\ t-\boldsymbol{t}_{\boldsymbol{v}_t} < y \mid \boldsymbol{w}_1 = 0\} = \int_{t-y}^{t} \{1-A(t-u)\}\,\mathrm{d}_u L(x,u). \tag{5.98}$$

Putting (cf. derivation of (I.6.22)) for fixed y,

$$g(u) \overset{\text{def}}{=} \{1-A(u)\}\{1-U(u-y)\}, \qquad u\geqq 0,$$

then from (5.95) and (5.98) for $t>0$, $x\geqq 0$, $y\geqq 0$ we have

$$\Pr\{\boldsymbol{w}_{\boldsymbol{v}_t} < x,\ t-\boldsymbol{t}_{\boldsymbol{v}_t} < y \mid \boldsymbol{w}_1 = 0\} = g(t) * H(x,t) * n(t). \tag{5.99}$$

It is now easily verified that the function $g(t)*H(x,t)$ satisfies the conditions of the key renewal theorem of Smith (theorem I.6.2). Consequently, if $A(t)$ is not a lattice distribution and $a<1$, then the left-hand side (cf. (5.99)) has a limit for $t\to\infty$ and, since (5.94) shows that

$$\lim_{t\to\infty} H(x,t) = W(x),$$

it follows from (cf. (I.6.22)),

$$\lim_{t\to\infty} g(t) * H(x,t) * n(t) = \frac{1}{\alpha} \int_0^{\infty} \left\{ \int_0^t g(t-u)\,\mathrm{d}_u H(x,u) \right\} \mathrm{d}t,$$

that

$$\lim_{t\to\infty} \Pr\{\boldsymbol{w}_{\boldsymbol{v}_t} < x,\ t-\boldsymbol{t}_{\boldsymbol{v}_t} < y\} = \begin{cases} \dfrac{1}{\alpha} W(x) \displaystyle\int_0^y \{1-A(v)\}\,\mathrm{d}v, & x\geqq 0,\quad y\geqq 0,\\ 0 \quad \text{for} \quad x<0 \quad \text{or} \quad y<0. \end{cases} \tag{5.100}$$

Obviously, the right-hand side of (5.100) represents a proper two dimensional probability distribution if $a<1$. If $a<1$ and $A(t)$ a lattice distribution with period τ then by applying a similar argument as that used in section I.6.3 (cf. the derivation of (I.6.24)) it is seen that the left-hand side of (5.99) has a limit for $n\to\infty$ if $t=n\tau+\sigma$ with $0\leqq\sigma<\tau$.

Since, for $a<1$ and $A(t)$ not a lattice distribution, the result obtained above implies that the distribution of $\boldsymbol{w}_{\boldsymbol{v}_t}-(t-\boldsymbol{t}_{\boldsymbol{v}_t})$ has for $t\to\infty$ a limit distribution, it is seen from (5.90) that in this case the distribution of $\boldsymbol{v}_t$ converges for $t\to\infty$ to a proper probability distribution; for $A(t)$ a lattice distribution the distribution of $\boldsymbol{v}_{n\tau+\sigma}$ tends for $n\to\infty$ to a proper probability distribution.

Next, we shall derive the Laplace-Stieltjes transform of the distribution of $\boldsymbol{v}_t$. Since $\boldsymbol{e}_t$ represents the total idle time of the server in $[0, t)$ (cf. section 1.3) we have for $\boldsymbol{d}_1=0$, $\boldsymbol{w}_1=0$, $\boldsymbol{t}_1=0$,

$$\boldsymbol{e}_t=\boldsymbol{d}_{\boldsymbol{v}_t}-\min(0,\, \boldsymbol{w}_{\boldsymbol{v}_t}+\boldsymbol{\tau}_{\boldsymbol{v}_t}-(t-\boldsymbol{t}_{\boldsymbol{v}_t})), \qquad t>0. \tag{5.101}$$

From (5.90), (5.101) and (5.9) we obtain for $\operatorname{Re}\rho_1\geqq 0$, $\operatorname{Re}\rho_2\leqq 0$ with $\operatorname{Re}\rho_2<\operatorname{Re}\xi<\operatorname{Re}\rho_1$,

$$\exp(-\rho_1\boldsymbol{v}_t+\rho_2\boldsymbol{e}_t-\rho_4\boldsymbol{k}_t)$$
$$=\frac{1}{2\pi \mathrm{i}}\int\limits_{C_\xi}\left\{\frac{1}{\rho_1-\xi}+\frac{1}{\xi-\rho_2}\right\}\exp(-\xi\{\boldsymbol{w}_{\boldsymbol{v}_t}+\boldsymbol{\tau}_{\boldsymbol{v}_t}-(t-\boldsymbol{t}_{\boldsymbol{v}_t})\}+\rho_2\boldsymbol{d}_{\boldsymbol{v}_t}-\rho_4\boldsymbol{k}_t)\,\mathrm{d}\xi, \tag{5.102}$$

where (cf. section 1.3),

$$\boldsymbol{k}_t=\boldsymbol{b}_{\boldsymbol{v}_t}, \qquad t>0. \tag{5.103}$$

From (5.102), by taking the expectation and the Laplace transform with respect to t, we obtain

$$\int\limits_0^\infty \mathrm{e}^{-\rho t}\,\mathrm{E}\{\exp(-\rho_1\boldsymbol{v}_t+\rho_2\boldsymbol{e}_t-\rho_4\boldsymbol{k}_t)\mid \boldsymbol{w}_1=0,\, \boldsymbol{d}_1=0,\, \boldsymbol{t}_1=0\}\,\mathrm{d}t$$

$$=\frac{1}{2\pi \mathrm{i}}\int\limits_{C_\xi}\left\{\frac{\mathrm{d}\xi}{\rho_1-\xi}+\frac{\mathrm{d}\xi}{\xi-\rho_2}\right\}\int\limits_0^\infty \mathrm{e}^{-\rho t}\sum_{n=1}^\infty \mathrm{E}\{\exp(-\xi\boldsymbol{w}_n+\rho_2\boldsymbol{d}_n-\rho_4\boldsymbol{b}_{n-1})$$

$$\cdot\exp(-(\rho_4+\xi)\,\boldsymbol{\tau}_n+\xi(t-\boldsymbol{t}_n))(\boldsymbol{v}_t=n)\mid \boldsymbol{w}_1=0,\, \boldsymbol{d}_1=0\}\,\mathrm{d}t$$

$$=\frac{1}{2\pi \mathrm{i}}\int\limits_{C_\xi}\left\{\frac{\mathrm{d}\xi}{\rho_1-\xi}+\frac{\mathrm{d}\xi}{\xi-\rho_2}\right\}\int\limits_0^\infty \mathrm{e}^{-\rho t}\,\mathrm{d}t\int\limits_{u=0-}^{t}\mathrm{e}^{\xi(t-u)}\,\{1-A(t-u)\}\,\beta(\rho_4+\xi)$$

$$\cdot\mathrm{d}_u\sum_{n=1}^\infty \mathrm{E}\{(\boldsymbol{a}_{n-1}<u)\exp(-\xi\boldsymbol{w}_n+\rho_2\boldsymbol{d}_n-\rho_4\boldsymbol{b}_{n-1})\mid \boldsymbol{w}_1=0,\, \boldsymbol{d}_1=0\}=$$

$$= \frac{1}{2\pi i} \int_{C_\xi} \left\{ \frac{d\xi}{\rho_1 - \xi} + \frac{d\xi}{\xi - \rho_2} \right\} \beta(\rho_4 + \xi) \frac{1}{\rho - \xi} \{1 - \alpha(\rho - \xi)\}$$

$$\cdot \sum_{n=1}^{\infty} \mathrm{E}\{\exp(-\xi \boldsymbol{w}_n + \rho_2 \boldsymbol{d}_n - \rho \boldsymbol{a}_{n-1} - \rho_4 \boldsymbol{b}_{n-1}) \mid \boldsymbol{w}_1 = 0, \boldsymbol{d}_1 = 0\}, \tag{5.104}$$

for $0 < \mathrm{Re}\, \xi < \min(\mathrm{Re}\, \rho_1, \mathrm{Re}\, \rho)$, $\mathrm{Re}\, \rho_1 \geqq 0$, $\mathrm{Re}\, \rho_2 \leqq 0$, $\mathrm{Re}\, \rho > 0$, $\mathrm{Re}\, \rho_4 \geqq 0$; the interchanging of the various integrations in the relation above is easily justified (cf. section 5.2). It follows from (5.31) that for $\mathrm{Re}\, \rho > 0$,

$$\omega(1, \rho_1, \rho_2, \rho, \rho_4, 0, 0)$$

$$= \sum_{n=1}^{\infty} \mathrm{E}\{\exp(-\rho_1 \boldsymbol{w}_n + \rho_2 \boldsymbol{d}_n - \rho \boldsymbol{a}_{n-1} - \rho_4 \boldsymbol{b}_{n-1}) \mid \boldsymbol{w}_1 = 0, \boldsymbol{d}_1 = 0\}. \tag{5.105}$$

Further

$$\beta(\rho_4 + \xi) \frac{1}{\rho - \xi} \{1 - \alpha(\rho - \xi)\} = -\frac{1 - \beta(\rho_4 + \xi)}{\rho - \xi} + \frac{1 - \alpha(\rho - \xi)\, \beta(\rho_4 + \xi)}{\rho - \xi}. \tag{5.106}$$

Since (cf. (5.20)) for the conditions of (5.104),

$$\frac{1}{2\pi i} \int_{C_\xi} \left\{ \frac{d\xi}{\rho_1 - \xi} + \frac{d\xi}{\xi - \rho_2} \right\} \frac{1 - \alpha(\rho - \xi)\, \beta(\rho_4 + \xi)}{\rho - \xi} \omega(1, \xi, \rho_2, \rho, \rho_4, 0, 0) = \frac{1}{\rho - \rho_2},$$

and

$$\frac{-1}{2\pi i} \int_{C_\xi} \left\{ \frac{d\xi}{\rho_1 - \xi} + \frac{d\xi}{\xi - \rho_2} \right\} \frac{1 - \beta(\rho_4 + \xi)}{\rho - \xi} \omega(1, \xi, \rho_2, \rho, \rho_4, 0, 0)$$

$$= -\frac{1 - \beta(\rho_4 + \rho_1)}{\rho - \rho_1} \omega(1, \rho_1, \rho_2, \rho, \rho_4, 0, 0)$$

$$- \frac{1 - \beta(\rho_4 + \rho)}{\rho_1 - \rho} \omega(1, \rho, \rho_2, \rho, \rho_4, 0, 0)$$

$$- \frac{1 - \beta(\rho_4 + \rho)}{\rho - \rho_2} \omega(1, \rho, \rho_2, \rho, \rho_4, 0, 0),$$

it follows from the above relations that

$$\int_0^\infty \mathrm{e}^{-\rho t}\, \mathrm{E}\{\exp(-\rho_1 \boldsymbol{v}_t + \rho_2 \boldsymbol{e}_t - \rho_4 \boldsymbol{k}_t) \mid \boldsymbol{w}_1 = 0, \boldsymbol{d}_1 = 0, \boldsymbol{t}_1 = 0\}\, \mathrm{d}t$$

$$= \frac{1-\beta(\rho_4+\rho_1)}{\rho_1 - \rho}\, \omega(1, \rho_1, \rho_2, \rho, \rho_4, 0, 0) + \frac{1}{\rho - \rho_2}$$

$$+ \frac{(\rho_2-\rho_1)\{1-\beta(\rho_4+\rho)\}}{(\rho_1-\rho)(\rho-\rho_2)}\, \omega(1, \rho, \rho_2, \rho, \rho_4, 0, 0), \qquad (5.107)$$

for $\mathrm{Re}\, \rho_1 \geqq 0$, $\mathrm{Re}\, \rho_2 \leqq 0$, $\mathrm{Re}\, \rho > 0$, $\mathrm{Re}\, \rho_4 \geqq 0$; the function $\omega(1, \ldots)$ being given by (5.31).

The latter relation is of special interest in connection with the formula (1.8) for $v = 0$.

Taking $\rho_2 = 0$, $\rho_4 = 0$ we obtain by using the relation (5.31) for $\mathrm{Re}\, \rho > 0$, $\mathrm{Re}\, \rho_1 \geqq 0$,

$$\int_0^\infty \mathrm{e}^{-\rho t}\, \mathrm{E}\{\mathrm{e}^{-\rho_1 \boldsymbol{v}_t} \mid \boldsymbol{w}_1 = 0, \boldsymbol{d}_1 = 0, \boldsymbol{t}_1 = 0\}\, \mathrm{d}t$$

$$= \frac{1}{1-\alpha(\rho)}\, \frac{1-\beta(\rho_1)}{\rho_1 - \rho} \exp\left[-\sum_{n=1}^{\infty} \frac{1}{n}\, \mathrm{E}\{\mathrm{e}^{-\rho \boldsymbol{a}_n}(1-\mathrm{e}^{-\rho_1 \boldsymbol{s}_n})(\boldsymbol{s}_n > 0)\}\right] + 1/\rho$$

$$+ \frac{\rho_1}{\rho - \rho_1}\, \frac{1-\beta(\rho)}{\{1-\alpha(\rho)\}\, \rho} \exp\left[-\sum_{n=1}^{\infty} \frac{1}{n}\, \mathrm{E}\{\mathrm{e}^{-\rho \boldsymbol{a}_n}(1-\mathrm{e}^{-\rho \boldsymbol{s}_n})(\boldsymbol{s}_n > 0)\}\right]. \qquad (5.108)$$

We have shown above that if $A(t)$ is not a lattice distribution and if $a < 1$ then the distribution function of the virtual waiting time $\boldsymbol{v}_t$ has a limit for $t \to \infty$. Denoting this limit distribution by $V(\tau)$ it follows from a well-known Abelian theorem (cf. app. 3) by taking ρ real, multiplying by ρ and then $\rho \downarrow 0$ in (5.108) that (cf. (5.50)),

$$\int_0^\infty \mathrm{e}^{-\rho_1 \tau}\, \mathrm{d}V(\tau) = \lim_{t\to\infty} \mathrm{E}\{\mathrm{e}^{-\rho_1 \boldsymbol{v}_t} \mid \boldsymbol{w}_1 = 0, \boldsymbol{d}_1 = 0, \boldsymbol{t}_1 = 0\}$$

$$= 1 - a + a\, \frac{1-\beta(\rho_1)}{\rho_1 \beta} \exp\left[-\sum_{n=1}^{\infty} \frac{1}{n} \int_0^\infty \{1-\mathrm{e}^{-\rho_1 x}\}\, \mathrm{d}\Pr\{\boldsymbol{s}_n < x\}\right]$$

$$= 1 - a + a\, \frac{1-\beta(\rho_1)}{\rho_1 \beta} \int_{0-}^\infty \mathrm{e}^{-\rho_1 x}\, \mathrm{d}W(x), \qquad \mathrm{Re}\, \rho_1 \geqq 0. \qquad (5.109)$$

Consequently,

$$V(t)=\begin{cases}1-a+a\displaystyle\int_0^t W(t-u)\{1-B(u)\}\frac{du}{\beta}, & t>0,\\ 0, & t\leqq 0.\end{cases}\tag{5.110}$$

From (5.107) and from (5.108) we have for Re $\rho>0$, Re $\rho_2\leqq 0$,

$$\int_0^\infty e^{-\rho t}\,\mathrm{E}\{e^{\rho_2 \boldsymbol{e}_t}\mid \boldsymbol{w}_1=0,\boldsymbol{d}_1=0,\boldsymbol{t}_1=0\}\,dt$$
$$=\frac{1}{\rho-\rho_2}+\frac{\rho_2}{(\rho_2-\rho)\,\rho}\exp\left[-\sum_{n=1}^{\infty}\frac{1}{n}\mathrm{E}\{e^{-\rho \boldsymbol{b}_n}(1-e^{-(\rho_2-\rho)\boldsymbol{s}_n})(\boldsymbol{s}_n<0)\}\right],\tag{5.111}$$

$$\int_0^\infty e^{-\rho t}\Pr\{\boldsymbol{v}_t=0\mid \boldsymbol{w}_1=0,\boldsymbol{t}_1=0\}\,dt=\frac{1}{\rho}\left\{1-\frac{1-\mathrm{E}\{e^{-\rho\boldsymbol{p}}\}}{1-\mathrm{E}\{e^{-\rho\boldsymbol{c}}\}}\right\}.$$

From (5.108) we obtain for Re $\rho>0$,

$$\int_0^\infty e^{-\rho t}\,d_t\,\mathrm{E}\{\boldsymbol{v}_t\mid \boldsymbol{w}_1=0,\boldsymbol{t}_1=0\}$$
$$=\frac{\beta}{1-\alpha(\rho)}-\frac{1}{\rho}\,\frac{1-\beta(\rho)}{1-\alpha(\rho)}\exp\left[-\sum_{n=1}^{\infty}\frac{1}{n}\mathrm{E}\{e^{-\rho \boldsymbol{a}_n}(1-e^{-\rho\boldsymbol{s}_n})(\boldsymbol{s}_n>0)\}\right]$$
$$=\frac{\beta}{1-\alpha(\rho)}-\frac{1}{\rho}\int_{0-}^\infty e^{-\rho t}\,d\Pr\{\boldsymbol{v}_t>0\mid \boldsymbol{w}_1=0,\boldsymbol{t}_1=0\},\tag{5.112}$$

and if $a<1$ and $A(t)$ not a lattice distribution then (cf. (5.55)),

$$\lim_{t\to\infty}\mathrm{E}\{\boldsymbol{v}_t\}=\int_0^\infty t\,dV(t)=a\left\{\frac{\beta_2}{2\beta}+\sum_{n=1}^{\infty}\frac{1}{n}\int_0^\infty t\,dC^{n*}(t)\right\}$$
$$=a\left\{\frac{\beta_2}{2\beta}+\int_0^\infty t\,dW(t)\right\},\tag{5.113}$$

if the second moment β_2 of the service time distribution is finite.

If $A(t)$ is a lattice distribution with period τ, then (cf. (5.89)) a similar argument leads to

$$\lim_{n\to\infty} \Pr\{\boldsymbol{v}_{n\tau+\sigma}<x\} = 1 - a(\sigma) + a(\sigma)\int_0^x W(x-u)\{1-B(u)\}\frac{\mathrm{d}u}{\beta}, \quad x\geqq 0,\ a<1,$$

for $0\leqq\sigma<\tau$; a result which can be established by noting that the right-hand side of (5.108) is a power series in $\mathrm{e}^{-\rho}$.

II.5.7. The queue length

The number of customers in the system at time t is denoted by $\boldsymbol{x}_t$. For the sake of simplicity it will be assumed here that $\boldsymbol{w}_1=0$, $\boldsymbol{t}_1=0$. The sample functions of the $\boldsymbol{x}_t$-process are taken to be continuous from the left, so that if $\boldsymbol{r}_n$ is the moment of departure of the nth arriving customer after $t=0$, it is seen that $\boldsymbol{x}_{\boldsymbol{r}_n}$ and $\boldsymbol{x}_{\boldsymbol{r}_n+}$ represent the number of customers present in the system immediately before and immediately after the nth departure, respectively. Service is in order of arrival so that the waiting customers at a departure have arrived during the waiting time and service time of the departing customer. It follows for $n=1, 2, \ldots$, that

$$\Pr\{\boldsymbol{x}_{\boldsymbol{r}_n}=1 \mid \boldsymbol{w}_1=0\}=\Pr\{\boldsymbol{w}_n+\boldsymbol{\tau}_n-\boldsymbol{\sigma}_{n+1}\leqq 0 \mid \boldsymbol{w}_1=0\}, \tag{5.114}$$

$$\Pr\{\boldsymbol{x}_{\boldsymbol{r}_n+}=0 \mid \boldsymbol{w}_1=0\}=\Pr\{\boldsymbol{w}_n+\boldsymbol{\tau}_n-\boldsymbol{\sigma}_{n+1}<0 \mid \boldsymbol{w}_1=0\},$$

$$\Pr\{\boldsymbol{x}_{\boldsymbol{r}_n}\geqq j+1 \mid \boldsymbol{w}_1=0\}=\Pr\{\boldsymbol{w}_n+\boldsymbol{\tau}_n>\boldsymbol{\sigma}_{n+1}+\ldots+\boldsymbol{\sigma}_{n+j} \mid \boldsymbol{w}_1=0\},\quad j=1, 2, \ldots,$$

$$\Pr\{\boldsymbol{x}_{\boldsymbol{r}_n+}\geqq j \mid \boldsymbol{w}_1=0\}=\Pr\{\boldsymbol{w}_n+\boldsymbol{\tau}_n\geqq\boldsymbol{\sigma}_{n+1}+\ldots+\boldsymbol{\sigma}_{n+j} \mid \boldsymbol{w}_1=0\},\quad j=1, 2, \ldots .$$

In general the events $\{\boldsymbol{x}_{\boldsymbol{r}_n}\geqq j+1\}$ and $\{\boldsymbol{x}_{\boldsymbol{r}_n+}\geqq j\}$ do not have the same probability, since the probability of an arrival and a departure at the same moment is not necessarily zero.

In section 5.4 it has been shown that the distribution of $\boldsymbol{w}_n$ converges to $W(t)$ for every t if $B<\infty$. Hence for $B<\infty$ it follows that for $n\to\infty$ the distributions of $\boldsymbol{x}_{\boldsymbol{r}_n}$ and of $\boldsymbol{x}_{\boldsymbol{r}_n+}$ converge for $n\to\infty$ to proper probability distributions, which are identical if $A(t)$ or $B(t)$ has no discontinuity points. Assuming from now on that the latter condition holds we have, since $\boldsymbol{\sigma}_n$ is positive with probability one (cf. (5.32)),

$$\Pr\{\boldsymbol{x}_{r_n+} \geqq 1 \mid \boldsymbol{w}_1 = 0\} = \Pr\{\boldsymbol{w}_{n+1} > 0 \mid \boldsymbol{w}_1 = 0\}, \tag{5.115}$$

$$\Pr\{\boldsymbol{x}_{r_n+} \geqq j \mid \boldsymbol{w}_1 = 0\} = \Pr\{\boldsymbol{w}_{n+1} > \boldsymbol{\sigma}_{n+2} + \ldots + \boldsymbol{\sigma}_{n+j} \mid \boldsymbol{w}_1 = 0\}, \quad j = 2, 3, \ldots .$$

From (5.114) and (5.115) we obtain for $n=1, 2, \ldots,$

$$\Pr\{\boldsymbol{x}_{r_n+} = j \mid \boldsymbol{w}_1 = 0\}$$

$$= \int_{u=0-}^{\infty} \{A^{j*}(u) - A^{(j+1)*}(u)\} \, \mathrm{d}_u \Pr\{\boldsymbol{w}_n + \boldsymbol{\tau}_n < u \mid \boldsymbol{w}_1 = 0\}, \qquad j = 0, 1, \ldots, \tag{5.116}$$

$$= \begin{cases} \Pr\{\boldsymbol{w}_{n+1} = 0 \mid \boldsymbol{w}_1 = 0\}, & j = 0, \\ \int_{u=0-}^{\infty} \{A^{(j-1)*}(u) - A^{j*}(u)\} \, \mathrm{d}_u \Pr\{\boldsymbol{w}_{n+1} < u \mid \boldsymbol{w}_1 = 0\}, & j = 1, 2, \ldots . \end{cases}$$

Hence if $B < \infty$, since $W_n(t)$ converges to $W(t)$ for every t, we have (cf. (5.47)),

$$v_j \stackrel{\text{def}}{=} \lim_{n \to \infty} \Pr\{\boldsymbol{x}_{r_n+} = j \mid \boldsymbol{w}_1 = 0\}$$

$$= \begin{cases} \mathrm{e}^{-B}, & j = 0, \\ \int_{u=0-}^{\infty} \{A^{j*}(u) - A^{(j+1)*}(u)\} \, \mathrm{d}\{W(u) * B(u)\}, & j = 1, 2, \ldots, \\ \int_{u=0-}^{\infty} \{A^{(j-1)*}(u) - A^{j*}(u)\} \, \mathrm{d}W(u), & j = 1, 2, \ldots . \end{cases} \tag{5.117}$$

With $m(t)$ the renewal function of the renewal process with $A(t)$ as the renewal distribution

$$m(t) = \sum_{n=1}^{\infty} A^{n*}(t), \tag{5.118}$$

we have

$$\mathrm{E}\{\boldsymbol{x}_{r_n+} \mid \boldsymbol{w}_1 = 0\} = \int_{u=0}^{\infty} m(u) \, \mathrm{d}_u \Pr\{\boldsymbol{w}_n + \boldsymbol{\tau}_n < u \mid \boldsymbol{w}_1 = 0\}, \tag{5.119}$$

so that

$$\sum_{j=0}^{\infty} j v_j = \int_{u=0}^{\infty} m(u) \, \mathrm{d}_u\{W(u) * B(u)\}, \tag{5.120}$$

$$= 1 + \int_{u=0}^{\infty} m(u) \, \mathrm{d}W(u),$$

the integrals being finite if $W(t)$ has a finite first moment since $m(t)$ behaves as $\alpha^{-1}t$ for $t \to \infty$; the integral in (5.119) is always finite since $\mathrm{E}\{\boldsymbol{w}_n\}$ is finite.

Next we shall investigate the distribution of *the number of customers in the system just before the arrival of a customer*. Obviously,

$$\{\boldsymbol{r}_{n-1} < \boldsymbol{t}_{n+j} \leq \boldsymbol{r}_n\}, \quad n=1, 2, \ldots; \qquad \boldsymbol{r}_0 \stackrel{\text{def}}{=} 0; \quad j=0, 1, \ldots,$$

is the event that the $(n+j)$th arriving customer finds j customers in the system on his arrival. Since for $\boldsymbol{w}_1=0$, $\boldsymbol{t}_1=0$,

$$\{\boldsymbol{r}_n < \boldsymbol{t}_{n+j}\} \cup \{\boldsymbol{r}_{n-1} < \boldsymbol{t}_{n+j} \leq \boldsymbol{r}_n\} = \{\boldsymbol{r}_{n-1} < \boldsymbol{t}_{n+j}\}, \tag{5.121}$$

and because the events in the left-hand side of (5.121) are disjoint we have for $n=2, 3, \ldots; j=0, 1, \ldots,$

$$\begin{aligned}\Pr\{\boldsymbol{x}_{t_{n+j}}=j \mid \boldsymbol{w}_1=0\} &= \Pr\{\boldsymbol{r}_{n-1} < \boldsymbol{t}_{n+j} \leq \boldsymbol{r}_n \mid \boldsymbol{w}_1=0\} \\ &= \Pr\{\boldsymbol{r}_{n-1} < \boldsymbol{t}_{n+j} \mid \boldsymbol{w}_1=0\} - \Pr\{\boldsymbol{r}_n < \boldsymbol{t}_{n+j} \mid \boldsymbol{w}_1=0\}.\end{aligned}$$

Since for $j=1, 2, \ldots,$

$$\Pr\{\boldsymbol{r}_n < \boldsymbol{t}_{n+j} \mid \boldsymbol{w}_1=0\} = \int_0^\infty \{1-A^{j*}(u)\}\, \mathrm{d}_u \Pr\{\boldsymbol{w}_n+\boldsymbol{\tau}_n < u \mid \boldsymbol{w}_1=0\}, \tag{5.122}$$

and

$$\Pr\{\boldsymbol{r}_0 < \boldsymbol{t}_{1+j} < \boldsymbol{r}_1 \mid \boldsymbol{w}_1=0\} = \begin{cases} 0, & j=0, \\ \int_0^\infty \{1-B(u)\}\, \mathrm{d}A^{j*}(u), & j=1, 2, \ldots, \end{cases}$$

we obtain

$$\Pr\{\boldsymbol{x}_{t_{n+j}}=j \mid \boldsymbol{w}_1=0\}$$

$$= \begin{cases} 0, & n=1, \quad j=0, \\ \int_0^\infty \{1-B(u)\}\, \mathrm{d}A^{j*}(u), & n=1, \quad j=1, 2, \ldots, \\ \int_0^\infty \{1-A^{(j+1)*}(u)\}\, \mathrm{d}_u \Pr\{\boldsymbol{w}_{n-1}+\boldsymbol{\tau}_{n-1} < u \mid \boldsymbol{w}_1=0\} & \\ \quad - \int_0^\infty \{1-A^{j*}(u)\}\, \mathrm{d}_u \Pr\{\boldsymbol{w}_n+\boldsymbol{\tau}_n < u \mid \boldsymbol{w}_1=0\}, & \\ & n=2, 3, \ldots; \quad j=0, 1, \ldots. \end{cases} \tag{5.123}$$

Moreover, for $n=1, 2, \ldots; j=0, 1, \ldots,$

$$\begin{aligned}\Pr\{\boldsymbol{x}_{t_{n+j}} \geq j \mid \boldsymbol{w}_1=0\} &= \Pr\{\boldsymbol{t}_{n+j} \leq \boldsymbol{r}_n \mid \boldsymbol{w}_1=0\} \\ &= \int_0^\infty A^{j*}(u)\, \mathrm{d}_u \Pr\{\boldsymbol{w}_n+\boldsymbol{\tau}_n < u \mid \boldsymbol{w}_1=0\}.\end{aligned} \tag{5.124}$$

As above (cf. the derivation of (5.117)) it is now seen that if $B<\infty$ then

$$u_j \overset{\text{def}}{=} \lim_{n\to\infty} \Pr\{\boldsymbol{x}_{t_{n+j}}=j \mid \boldsymbol{w}_1=0\}$$

$$=\begin{cases} W(0+) = \mathrm{e}^{-B}, & j=0, \\ \int_0^\infty \{A^{j*}(u) - A^{(j+1)*}(u)\}\, \mathrm{d}\{W(u) * B(u)\}, & j=1, 2, \dots . \end{cases} \tag{5.125}$$

From (5.117) and (5.125) it is seen that *for the ergodic situation of the queueing system the distribution function of the number of customers present just before the arrival of a customer is the same as the distribution function of the number of customers in the system just after a departure*. This statement is also true if both $A(t)$ and $B(t)$ have discontinuity points.

As before (cf. (5.117)) it follows from (5.123) that

$$\Pr\{\boldsymbol{x}_{t_n}=0 \mid \boldsymbol{w}_1=0\}=\Pr\{\boldsymbol{w}_n=0 \mid \boldsymbol{w}_1=0\}.$$

To find the distribution function of $\boldsymbol{x}_t$, i.e. of the number of customers in the system at time t, suppose that the nth arriving customer is being served at time t and that $\boldsymbol{x}_t=i$, so that $i-1$ customers have arrived during $(\boldsymbol{t}_n, t)$; hence

$$\boldsymbol{w}_n<t-\boldsymbol{t}_n\leqq \boldsymbol{w}_n+\boldsymbol{\tau}_n, \qquad \boldsymbol{t}_{n+i}<t\leqq \boldsymbol{t}_{n+i+1}.$$

Generally, we have with $\boldsymbol{w}_1=0$, $\boldsymbol{t}_1=0$,

$$\{\boldsymbol{x}_t=i+1\}=\{\boldsymbol{w}_{\nu_t-i}<t-\boldsymbol{t}_{\nu_t-i}\leqq \boldsymbol{w}_{\nu_t-i}+\boldsymbol{\tau}_{\nu_t-i}\}, \qquad i=0, 1, \dots; \quad t>0. \tag{5.126}$$

Clearly

$$\Pr\{\boldsymbol{x}_t=0\}=\Pr\{\boldsymbol{v}_t=0\}, \tag{5.127}$$

whereas we obtain from (5.126) for $i=0, 1, \dots; t>0$,

$$\Pr\{\boldsymbol{x}_t=i+1 \mid \boldsymbol{w}_1=0, \boldsymbol{t}_1=0\}$$

$$= \int_{u=0}^{t} \{A^{i*}(t-u)-A^{(i+1)*}(t-u)\}$$

$$\cdot \sum_{n=i}^{\infty} \Pr\{\boldsymbol{w}_{n-i+1}<t-u\leqq \boldsymbol{w}_{n-i+1}+\boldsymbol{\tau}_{n-i+1} \mid \boldsymbol{a}_{n-i}=u, \boldsymbol{w}_1=0\}\, \mathrm{d}_u \Pr\{\boldsymbol{a}_{n-i}<u\}$$

$$= \int_{v=0}^{\infty} \int_{u=t-y}^{t} \{A^{i*}(t-u)-A^{(i+1)*}(t-u)\}\{U_1(t-u-v)-B(t-u-v)\}$$

$$\cdot \mathrm{d}_u \sum_{n=i}^{\infty} \mathrm{d}_v \Pr\{\boldsymbol{w}_{n-i+1}<v, \boldsymbol{a}_{n-1}<u \mid \boldsymbol{w}_1=0\}. \tag{5.128}$$

The second integral in the right-hand side is a uniformly bounded function of v for $-\infty<v<\infty$. Applying theorem I.6.2 it follows easily (cf. the derivation of (5.100)) that if $A(t)$ is not a lattice distribution then the distribution of $\boldsymbol{x}_t$ converges for $t\to\infty$. The limit is zero for $a\geqq 1$, and is a proper probability distribution if $a<1$.

It follows for $a<1$ and $A(t)$ not a lattice distribution (cf. (5.96) and (5.100) for $y\to\infty$) that

$$w_i \overset{\text{def}}{=} \lim_{t\to\infty} \Pr\{\boldsymbol{x}_t=i \mid \boldsymbol{w}_1=0, \boldsymbol{t}_1=0\}$$

$$= \int_{v=0}^{\infty}\int_{0}^{\infty} \{A^{(i-1)*}(y)-A^{i*}(y)\}\{U_1(y-v)-B(y-v)\}\,\frac{\mathrm{d}y}{\alpha}\,\mathrm{d}W(v)$$

$$= a\int_0^{\infty} \{A^{(i-1)*}(y)-A^{i*}(y)\}\,\mathrm{d}_y\Big\{W(y)*\frac{1}{\beta}\int_0^y \{1-B(u)\}\,\mathrm{d}u\Big\}, \qquad i=1,2,\dots, \tag{5.129}$$

$$w_0 = \lim_{t\to\infty} \Pr\{\boldsymbol{x}_t=0 \mid \boldsymbol{w}_1=0, \boldsymbol{t}_1=0\} = 1-a.$$

Further, we obtain from (5.129) (cf. (5.95)),

$$\sum_{i=0}^{\infty} iw_i = a + a\int_0^{\infty} m(t)\,\mathrm{d}_t\Big\{W(t)*\frac{1}{\beta}\int_0^t \{1-B(u)\}\,\mathrm{d}u\Big\}, \tag{5.130}$$

the integral being finite if the second moment of $B(t)$ and the first moment of $W(t)$ are finite since $m(t)$ behaves as $\alpha^{-1}t$ for $t\to\infty$. It is interesting to compare the latter relation with (5.120).

In section 5.9 (cf. (5.181)) we shall prove that

$$\sum_{i=0}^{\infty} iw_i = \frac{1}{\alpha}\Big\{\beta + \int_0^{\infty} t\,\mathrm{d}W(t)\Big\}, \tag{5.131}$$

for $a<1$ and $A(t)$ not a lattice distribution. This is an important relation since, for the ergodic situation, it relates the average actual waiting time and the average number of customers in the system at some moment.

II.5.8. Integral representations

In this section we shall derive integral representations for the transforms of the distributions of the variables which have been studied in the preceding sections. These integral representations are often very useful as a starting point for the inversion of the transforms in case the distribution functions $A(t)$ and $B(t)$ are known. To obtain the integral representations we apply Hewitt's inversion theorem (cf. HEWITT [1953]). Let $F(t)$ be a distribution function and $G(t)$ a function of bounded variation which is absolutely integrable on $(-\infty, \infty)$; define

$$f(\xi) \stackrel{\text{def}}{=} \int_{-\infty}^{\infty} \mathrm{e}^{-\xi t}\, \mathrm{d}F(t), \qquad g(\xi) \stackrel{\text{def}}{=} \int_{-\infty}^{\infty} \mathrm{e}^{-\xi t}\, G(t)\, \mathrm{d}t, \qquad \mathrm{Re}\, \xi = 0, \tag{5.132}$$

then Hewitt's inversion formula reads

$$\lim_{T\to\infty} \frac{1}{2\pi\mathrm{i}} \int_{-\mathrm{i}T}^{\mathrm{i}T} f(\xi)\, g(-\xi)\, \mathrm{d}\xi = \tfrac{1}{2} \int_{-\infty}^{\infty} \{G(u+) + G(u-)\}\, \mathrm{d}F(u); \tag{5.133}$$

in the following we shall again use the notation

$$\lim_{T\to\infty} \int_{-\mathrm{i}T}^{\mathrm{i}T} \dots \equiv \int_{-\mathrm{i}\infty}^{\mathrm{i}\infty} \dots,$$

i.e. we have to take for the latter integral its principal value.
From (5.133) we have for $\mathrm{Re}\, \rho \geqq 0$,

$$\tfrac{1}{2}\mathrm{E}\{\mathrm{e}^{-\rho \boldsymbol{b}_n} \{(\boldsymbol{s}_n < 0) + (\boldsymbol{s}_n \leqq 0)\}\} = \mathrm{E}\{\mathrm{e}^{-\rho \boldsymbol{b}_n} \{(\boldsymbol{s}_n < 0) + \tfrac{1}{2}(\boldsymbol{s}_n = 0)\}\}$$

$$= \int_0^{\infty} \mathrm{e}^{-\rho t} \{1 - \tfrac{1}{2} A^{n*}(t+) - \tfrac{1}{2} A^{n*}(t-)\}\, \mathrm{d}B^{n*}(t)$$

$$= \frac{1}{2\pi\mathrm{i}} \int_{-\mathrm{i}\infty}^{\mathrm{i}\infty} \beta^n(\xi)\, \frac{1 - \alpha^n(\rho - \xi)}{\rho - \xi}\, \mathrm{d}\xi.$$

Since for $|r| < 1$, $\mathrm{Re}\, \rho > 0$,

$$\sum_{n=1}^{\infty} \frac{r^n}{n} \frac{1}{2\pi i} \int_{-i\infty}^{i\infty} \beta^n(\xi) \frac{1-\alpha^n(\rho-\xi)}{\rho-\xi} d\xi$$

$$= - \frac{1}{2\pi i} \int_{-i\infty}^{i\infty} \frac{d\xi}{\rho-\xi} \log \frac{1-r\beta(\xi)}{1-r\beta(\xi)\,\alpha(\rho-\xi)},$$

an integral representation for $E\{r^{\boldsymbol{n}} e^{-\rho \boldsymbol{p}}\}$ is easily found from (5.60). Similarly, integral representations may be found for (5.61), (5.63) and (5.64). Assuming that $A(t)$ or $B(t)$ has no discontinuity points, so that $\Pr\{\boldsymbol{a}_n = \boldsymbol{b}_n\} = 0$, $n = 1, 2, \ldots$, we obtain with $|r| < 1$, $\mathrm{Re}\, \rho > 0$, for the joint distribution of the busy period $\boldsymbol{p}$, and number $\boldsymbol{n}$ of customers served during the busy period,

$$E\{r^{\boldsymbol{n}} e^{-\rho \boldsymbol{p}}\} = 1 - \exp\left\{\frac{1}{2\pi i} \int_{-i\infty}^{i\infty} \frac{d\xi}{\rho-\xi} \log \frac{1-r\beta(\xi)}{1-r\beta(\xi)\,\alpha(\rho-\xi)}\right\}; \tag{5.134}$$

for the joint distribution of $\boldsymbol{n}$ and the busy cycle $\boldsymbol{c}$,

$$E\{r^{\boldsymbol{n}} e^{-\rho \boldsymbol{c}}\} = 1 - \exp\left\{\frac{1}{2\pi i} \int_{-i\infty}^{i\infty} \frac{d\xi}{\rho-\xi} \log\{1-r\beta(\rho-\xi)\,\alpha(\xi)\}\right\}; \tag{5.135}$$

and for that of $\boldsymbol{n}$ and the idle period $\boldsymbol{i}$,

$$E\{r^{\boldsymbol{n}} e^{-\rho \boldsymbol{i}}\} = 1 - \exp\left\{\frac{1}{2\pi i} \int_{-i\infty}^{i\infty} \frac{d\xi}{\rho+\xi} \log\{1-r\beta(\xi)\,\alpha(-\xi)\}\right\}. \tag{5.136}$$

If $A(t)$ and $B(t)$ are lattice distributions both with period τ we write

$$a_m \stackrel{\text{def}}{=} \Pr\{\boldsymbol{\sigma}_n = m\tau\}, \qquad b_m \stackrel{\text{def}}{=} \Pr\{\boldsymbol{\tau}_n = m\tau\}, \qquad m = 1, 2, \ldots,$$

$$a(q) \stackrel{\text{def}}{=} \sum_{m=1}^{\infty} a_m q^m, \qquad b(q) \stackrel{\text{def}}{=} \sum_{m=1}^{\infty} b_m q^m, \qquad |q| \leqq 1.$$

To obtain an integral representation for the present case we apply the relation

$$\frac{1}{2\pi i} \int_{D_\zeta} \frac{d\zeta}{\zeta} a(q/\zeta)\, b(\zeta) = \sum_{m=1}^{\infty} a_m b_m q^m, \qquad |q| \leqq 1,$$

with D_ζ a circle with centre at $\zeta=0$ and radius $|\zeta|=1$. The direction of integration is counterclockwise.
Taking

$$q=e^{-\rho\tau}, \qquad \text{Re}\,\rho\geqq 0,$$

we have

$$E\{e^{-\rho b_1}(s_1\leqq 0)\} = \sum_{m=1}^{\infty}\{1-\sum_{k=1}^{m-1}a_k\}\,b_m q^m$$

$$= \frac{1}{2\pi i}\int_{D_\zeta}\left\{\frac{b(\zeta)}{\zeta-q}-\frac{q}{(\zeta-q)\,\zeta}\,a(q/\zeta)\,b(\zeta)\right\}d\zeta, \qquad |q|<1.$$

Hence for $|r|<1$, $|q|<1$,

$$\sum_{n=1}^{\infty}\frac{r^n}{n}\,E\{e^{-\rho b_n}(s_n\leqq 0)\} = \sum_{n=1}^{\infty}\frac{r^n}{n}\,\frac{1}{2\pi i}\int_{D_\zeta}\frac{b^n(\zeta)}{\zeta-q}\left\{1-\left(\frac{q}{\zeta}\right)^n a^n(q/\zeta)\right\}d\zeta.$$

Consequently, if both $A(t)$ and $B(t)$ are lattice distributions with period τ then for $|r|<1$, $|q|<1$,

$$E\{r^n q^{p/\tau}\} = 1-\exp\left\{\frac{1}{2\pi i}\int_{D_\zeta}\frac{d\zeta}{\zeta-q}\log\frac{1-rb(\zeta)}{1-(r\,q/\zeta)\,a(q/\zeta)\,b(\zeta)}\right\},$$

similar relations may be obtained for $E\{r^n q^{i/\tau}\}$ and $E\{r^n q^{c/\tau}\}$.
Next we consider the distribution of w_n, the actual waiting time of the nth arriving customer (cf. (5.36)). For $\text{Re}\,\rho>0$, $|r|<1$, and arbitrary $\varepsilon>0$ we have

$$-\sum_{n=1}^{\infty}r^{n-1}\int_0^{\infty}\{e^{-\varepsilon t}-e^{-\rho t}\}\,d_t\Pr\{s_n<t\}$$

$$= \frac{1}{2\pi i}\int_{-i\infty}^{i\infty}\frac{\beta(\xi)\,\alpha(-\xi)}{1-r\beta(\xi)\,\alpha(-\xi)}\left\{\frac{1}{\rho-\xi}+\frac{1}{\varepsilon-\xi}\right\}d\xi$$

$$= \frac{1}{2\pi i}\int_{-i\infty}^{i\infty}\left\{\frac{1}{\rho-\xi}+\frac{1}{\varepsilon-\xi}\right\}\frac{\beta(\xi)\,\alpha(-\xi)-1}{(1-r)(1-r\beta(\xi)\,\alpha(-\xi))}\,d\xi,$$

since the integral of the difference between the integrands of the second and third member is zero as it is seen by contour integration. From

$$1-\beta(\xi)\,\alpha(-\xi)=(\beta-\alpha)\,\xi+o(\xi), \qquad \mathrm{Re}\,\xi=0, \quad |\xi|\to 0,$$

it is easily verified that for $\varepsilon\to 0$ we obtain

$$\sum_{n=1}^{\infty} r^{n-1}\int_0^\infty \{1-e^{-\rho t}\}\,d_t\,\Pr\{s_n<t\}$$
$$=\frac{-1}{2\pi i}\int_{-i\infty}^{i\infty}\frac{\rho}{\xi(\rho-\xi)}\,\frac{\beta(\xi)\,\alpha(-\xi)-1}{1-r\beta(\xi)\,\alpha(-\xi)}\,\frac{d\xi}{1-r},$$

for $|r|<1$, $\mathrm{Re}\,\rho\geqq 0$. Integration of this relation with respect to r yields (cf. (5.36)) for $|r|<1$, $\mathrm{Re}\,\rho>0$,

$$\sum_{n=1}^{\infty} r^n\,E\{e^{-\rho w_n}\mid w_1=0\}$$
$$=\frac{r}{1-r}\exp\left\{\frac{1}{2\pi i}\int_0^r dx\int_{-i\infty}^{i\infty}\frac{\rho}{\xi(\rho-\xi)}\,\frac{\beta(\xi)\,\alpha(-\xi)-1}{1-x\beta(\xi)\,\alpha(-\xi)}\,\frac{d\xi}{1-x}\right\}. \tag{5.137}$$

Hence, if $a<1$ then for $\mathrm{Re}\,\rho>0$,

$$\int_{0-}^{\infty} e^{-\rho t}\,dW(t)=\exp\left\{\frac{1}{2\pi i}\int_0^{1-} dx\int_{-i\infty}^{i\infty}\frac{\rho}{\xi(\rho-\xi)}\,\frac{\beta(\xi)\,\alpha(-\xi)-1}{1-x\beta(\xi)\,\alpha(-\xi)}\,\frac{d\xi}{1-x}\right\}. \tag{5.138}$$

If $B<\infty$ and if $A(t)$ or $B(t)$ has no discontinuity points then the distribution $\{v_j, j=0, 1, \ldots\}$ represents for the ergodic situation of the queueing process the distribution of the number of customers in the system just before an arrival, and also just after a departure (cf. (5.125) and (5.117)). Since $\alpha<\infty$, it follows that $1-A(t)$ is integrable over $(0,\infty)$, and hence we obtain from (5.117) or (5.125) by applying Hewitt's inversion formula

$$v_j=u_j=\frac{1}{2\pi i}\int_{-i\infty}^{i\infty}E\{e^{-\xi w}\}\,\beta(\xi)\,\frac{\alpha^j(-\xi)-\alpha^{j+1}(-\xi)}{\xi}\,d\xi, \qquad j=0, 1, \ldots . \tag{5.139}$$

Here w denotes a stochastic variable with distribution function $W(t)$.

In the same way we obtain, if $a<1$ and $A(t)$ not a lattice distribution, for the probability w_j that j customers are present in the system at some moment for the ergodic situation (cf. (5.129)),

$$w_0 = 1 - a, \tag{5.140}$$

$$w_j = \frac{a}{2\pi i}\int_{-i\infty}^{i\infty} \mathrm{E}\{e^{-\xi w}\}\,\frac{1-\beta(\xi)}{\beta\xi}\,\frac{\alpha^{j-1}(-\xi)-\alpha^{j}(-\xi)}{-\xi}\,d\xi, \qquad j=1,2,\ldots;$$

as in (5.139), the integrand is continuous at $\xi=0$, since

$$\beta(\xi)=1-\beta\xi+o(\xi), \qquad \alpha(\xi)=1-\alpha\xi+o(\xi), \qquad \mathrm{Re}\,\xi=0, \quad |\xi|\to 0.$$

The relation (5.140) is true only if $A(t)$ is not a lattice distribution. If $A(t)$ is a lattice distribution, then (5.140) represents for the ergodic situation the distribution of the number of customers in the system at some moment if this moment is chosen according to a uniform distribution over the period of $A(t)$ (cf. the discussion following (5.89)).

To obtain an expression for the transform of the interdeparture time $\boldsymbol{r}_{n+1}-\boldsymbol{r}_n$ (cf. section 1.2) we use the relations (5.7), (5.9) and formula (1.1). It is found that for $\mathrm{Re}\,\rho_1>0$, $\mathrm{Re}\,\rho_2>0$,

$$\begin{aligned}
&\mathrm{E}\{\exp(-\rho_1\boldsymbol{w}_{n+1}-\rho_2(\boldsymbol{r}_{n+1}-\boldsymbol{r}_n))\mid \boldsymbol{w}_1=0\}\\
&=\mathrm{E}\{\exp(-\rho_1\boldsymbol{w}_{n+1}-\rho_2(\boldsymbol{d}_{n+1}-\boldsymbol{d}_n)-\rho_2\boldsymbol{\tau}_{n+1})\mid \boldsymbol{w}_1=0\}\\
&=\frac{1}{2\pi i}\,\beta(\rho_2)\int_{-i\infty}^{i\infty}\left\{\frac{1}{\rho_1-\xi}+\frac{1}{\rho_2+\xi}\right\}\beta(\xi)\,\alpha(-\xi)\,\mathrm{E}\{e^{-\xi\boldsymbol{w}_n}\mid \boldsymbol{w}_1=0\}\,d\xi,\\
&\qquad\qquad n=1,2,\ldots.
\end{aligned} \tag{5.141}$$

II.5.9. Truncated service time distribution, the system $G/G_c/1$

In general the Laplace-Stieltjes transform $\beta(\rho)$ of the service time distribution $B(t)$ is not an analytic function of ρ in a strip with $\mathrm{Re}\,\rho<0$. If we truncate the distribution function $B(t)$ then the Laplace-Stieltjes transform of the truncated distribution is analytic for $\mathrm{Re}\,\rho<0$, $|\rho|<\infty$.

Define

$$B_c(t) \overset{\text{def}}{=} \begin{cases} B(t) & \text{for } t\leq c,\\ 1 & \text{for } t>c,\end{cases}$$

where c is some finite positive number. For the Laplace-Stieltjes transform of $B_c(t)$ we have

$$\beta_c(\rho) \stackrel{\text{def}}{=} \int_0^{\infty} e^{-\rho t}\, dB_c(t) = \int_0^{c-} e^{-\rho t}\, dB(t) + e^{-\rho c}\{1-B(c)\}; \qquad (5.142)$$

obviously, $|\beta_c(\rho)| < \infty$ for all finite $|\rho|$. It is easily seen from (5.142) that

$$|\beta_c(\rho)| = O(e^{c\rho}) \qquad \text{for} \quad |\rho| \to \infty. \qquad (5.143)$$

Hence $\beta_c(\rho)$ is an entire function of order 1 and in fact of exponential type (cf. TITCHMARSH [1952] pp. 246 and 276). By Hadamard's factorization theorem every entire function $f(\rho)$ of order r can be represented as

$$f(\rho) = e^{Q(\rho)}\, P(\rho),$$

where $Q(\rho)$ is a polynomial of degree not greater than r and $P(\rho)$ is the canonical product formed with the zeros of $f(\rho)$. If $f(\rho)$ has no zeros then $P(\rho) \equiv 1$. Clearly, for Re $\rho \geqq 0$,

$$f(\rho) \stackrel{\text{def}}{=} \frac{1-\beta_c(\rho)}{\rho\beta_c} = \int_{-\infty}^{\infty} e^{-\rho t}\, d_t \left\{U_1(t) \int_0^t \{1-B_c(v)\} \frac{dv}{\beta_c}\right\},$$

is also an entire function of order 1 and of exponential type; here β_c is the first moment of the truncated distribution. Hence, if $f(\rho)$ has no zeros then a positive constant k_0 exists such that

$$f(\rho) = e^{-k_0 \rho}, \qquad \text{Re}\, \rho \geqq 0,$$

since $f(\rho)$ is the Laplace-Stieltjes transform of a distribution function. However, this distribution function is continuous so that we have a contradiction. Therefore $f(\rho)$ has zeros and since it is not a polynomial, because $|f(\rho)| \leqq 1$ for Re $\rho \geqq 0$, it has infinitely many zeros. Denote these zeros by $\eta_1, \eta_2, \ldots$, with $|\eta_1| \leqq |\eta_2| \leqq \ldots$, then from Hadamard's factorization theorem

$$f(\rho) = \frac{1-\beta_c(\rho)}{\rho\beta_c} = e^{-k_1 \rho} \prod_{i=1}^{\infty} \left(1 - \frac{\rho}{\eta_i}\right) e^{\rho/\eta_i}, \qquad |\rho| < \infty, \qquad (5.144)$$

where k_1 is a constant.

As $f(\rho)$ is of exponential type and of order 1 we have (cf. TITCHMARSH [1952] p. 286, ex. 17) that

$$\sum_{i=1}^{\infty} \frac{1}{|\eta_i|^{1+\varepsilon}}, \qquad (5.145)$$

converges for every $\varepsilon>0$, and diverges for $\varepsilon=0$; the zeros occur in (5.145) according to their multiplicity. To determine k_1 note that

$$\log \prod_{i=1}^{\infty}\left(1-\frac{\rho}{\eta_i}\right)\mathrm{e}^{\rho/\eta_i} = \sum_{i=1}^{\infty}\left\{\frac{\rho}{\eta_i}+\log\left(1-\frac{\rho}{\eta_i}\right)\right\}$$

$$= -\sum_{i=1}^{\infty}\sum_{n=2}^{\infty}\frac{1}{n}\left(\frac{\rho}{\eta_i}\right)^n \quad \text{for} \quad \left|\frac{\rho}{\eta_1}\right|<1;$$

so that

$$\frac{\mathrm{d}}{\mathrm{d}\rho}\log \prod_{i=1}^{\infty}\left(1-\frac{\rho}{\eta_i}\right)\mathrm{e}^{\rho/\eta_i} = -\sum_{i=1}^{\infty}\sum_{n=2}^{\infty}\frac{\rho^{n-1}}{\eta_i^n} \quad \text{for} \quad \left|\frac{\rho}{\eta_1}\right|<1,$$

since the latter series converges uniformly for $|\rho|<|\eta_1|$ on behalf of (5.145). From

$$\frac{\beta_{2c}}{2\beta_c} = -\lim_{\substack{\rho\to 0\\ \arg\rho=0}}\frac{\mathrm{d}}{\mathrm{d}\rho}f(\rho),$$

where β_{2c} is the second moment of $B_c(t)$, we obtain from (5.144) that

$$k_1=\beta_{2c}/2\beta_c,$$

hence

$$f(\rho)=\frac{1-\beta_c(\rho)}{\rho\beta_c} = \exp\left\{-\rho\frac{\beta_{2c}}{2\beta_c}\right\}\prod_{i=1}^{\infty}\left(1-\frac{\rho}{\eta_i}\right)\mathrm{e}^{\rho/\eta_i}, \qquad |\rho|<\infty. \tag{5.146}$$

In the following we shall often use $B_c(t)$ instead of $B(t)$ as the service time distribution. The queueing system with interarrival time distribution $A(t)$ and service time distribution $B_c(t)$ is indicated by $G/G_c/1$. *Quantities referring to this queueing system will be indicated by an additional index* c. Obviously, for every continuity point t of $B(t)$,

$$\lim_{c\to\infty} B_c(t)=B(t),$$

so that by Helly-Bray's theorem (cf. app. 5),

$$\lim_{c\to\infty}\beta_c(\rho)=\beta(\rho) \qquad \text{for} \quad \operatorname{Re}\rho\geqq 0. \tag{5.147}$$

We shall always take c to be a continuity point of $B(t)$. Since

$$\beta=\int_0^{\infty} t\,\mathrm{d}B(t) \geqq \int_0^{c} t\,\mathrm{d}B(t) + c\{1-B(c)\} = \int_0^{\infty} t\,\mathrm{d}B_c(t)=\beta_c,$$

it follows that if $a<1$ then $a_c<1$. In the following it will always be assumed that $a<1$, unless stated otherwise. Moreover, it will always be assumed that (cf. (5.47)),

$$W(0+)<1 \quad \text{and} \quad W_c(0+)<1, \tag{5.148}$$

i.e. we exclude the trivial case that with probability one the interarrival time is not less than the service time, and similarly for the truncated service time.

Since $\beta_c(\xi)$ is analytic for $\text{Re}\,\xi<0$ it follows for $|\text{Re}\,\xi|$ and $\text{Re}\,\rho_3$ both sufficiently small that

$$|\beta_c(\xi)\,\alpha(\rho_3-\xi)|\leqq\beta_c(\text{Re}\,\xi)\,\alpha(\text{Re}(\rho_3-\xi))$$
$$=1-(a_c-1)\,\alpha\,\text{Re}\,\xi-\alpha\,\text{Re}\,\rho_3+\text{o}(\text{Re}\,\xi)+\text{o}(\text{Re}\,\rho_3).$$

Hence for $\text{Re}\,\rho_3\geqq0$ and sufficiently small and $|\text{Re}\,\xi|$ sufficiently small

$$|\beta_c(\xi)\,\alpha(\rho_3-\xi)|<1 \qquad \text{for} \quad \text{Re}\,\xi<0. \tag{5.149}$$

From (5.35) it now follows by contour integration for $\text{Re}\,\rho_1\geqq0$, $|r|\leqq1$ and $\text{Re}\,\rho_3>0$, but sufficiently small that

$$\sum_{n=1}^{\infty} r^n\,\text{E}\{\exp(-\rho_1\boldsymbol{w}_{n,c}-\rho_3\boldsymbol{t}_n)\mid \boldsymbol{t}_1=0,\ \boldsymbol{w}_1=0\}$$
$$=\frac{r}{1-r\alpha(\rho_3)}\exp\left\{-\frac{1}{2\pi\text{i}}\int_{-\text{i}\infty-0}^{\text{i}\infty-0}\left\{\frac{1}{\rho_1-\xi}+\frac{1}{\xi}\right\}\cdot\log\{1-r\beta_c(\xi)\,\alpha(\rho_3-\xi)\}\,\text{d}\xi\right\}.$$

By letting $|\rho_3|\to0$ and then $n\to\infty$ we obtain in the same way as in section 5.4 (cf. also (5.50)) for $\text{Re}\,\rho\geqq0$,

$$\int_{0-}^{\infty}\text{e}^{-\rho t}\,\text{d}W_c(t)=\lim_{n\to\infty}\text{E}\{\text{e}^{-\rho\boldsymbol{w}_{n,c}}\mid\boldsymbol{w}_1=0\}$$
$$=\exp\left\{-\frac{1}{2\pi\text{i}}\int_{-\text{i}\infty-0}^{\text{i}\infty-0}\left\{\frac{1}{\rho-\xi}+\frac{1}{\xi}\right\}\log\{1-\beta_c(\xi)\,\alpha(-\xi)\}\,\text{d}\xi\right\}$$
$$=\exp\left\{-\sum_{n=1}^{\infty}\frac{1}{n}\int_0^{\infty}\{1-\text{e}^{-\rho t}\}\,\text{d}\Pr\{\boldsymbol{s}_{n,c}<t\}\right\}. \tag{5.150}$$

We now prove that

$$\lim_{c\to\infty} W_c(t) = W(t), \tag{5.151}$$

at every continuity point t of $W(t)$. To prove this it suffices to show by Feller's convergence theorem (cf. app. 5) that for $\rho>0$ the right-hand side of (5.150) converges to that of (5.50) for $c\to\infty$. From (5.147) we have for $\rho>0$,

$$\lim_{c\to\infty} \mathrm{E}\{\mathrm{e}^{-\rho s_{n,c}}\} = \lim_{c\to\infty} \beta_c^n(\rho)\, \alpha^n(-\rho) = \beta^n(\rho)\, \alpha^n(-\rho) = \mathrm{E}\{\mathrm{e}^{-\rho s_n}\}, \tag{5.152}$$

so that $s_{n,c}$ converges in distribution to s_n. Hence

$$\lim_{c\to\infty} \int_0^\infty \{1-\mathrm{e}^{-\rho t}\}\, \mathrm{d}_t \Pr\{s_{n,c}<t\} = \int_0^\infty \{1-\mathrm{e}^{-\rho t}\}\, \mathrm{d}_t \Pr\{s_n<t\},$$

by Helly-Bray's theorem. Since

$$\Pr\{s_{n,c_i}>0\} \leqq \Pr\{s_{n,c_j}>0\} \qquad \text{for} \quad c_i<c_j,$$

we have for Re $\rho\geqq 0$ and an arbitrary positive integer N,

$$\left| \sum_{n=N}^{\infty} \frac{1}{n} \int_0^\infty \{1-\mathrm{e}^{-\rho t}\}\, \mathrm{d}_t \Pr\{s_{n,c}<t\} \right| \leqq 2 \sum_{n=N}^{\infty} \frac{1}{n} \Pr\{s_n>0\};$$

the last term can be made arbitrarily small since $a<1$ implies $B<\infty$ (cf. (5.56)). The relation (5.151) now follows immediately. Next we shall show that if $W(t)$ has a finite first moment then

$$\lim_{c\to\infty} \int_0^\infty t\, \mathrm{d}W_c(t) = \int_0^\infty t\, \mathrm{d}W(t). \tag{5.153}$$

From (5.6), (5.49) and (5.50) it is seen that

$$W(t) = \Pr\{\sup_{n\geq 1}(0, s_n)<t\}, \qquad -\infty<t<\infty. \tag{5.154}$$

We now truncate for the present the stochastic variables $\tau_1, \tau_2, \ldots$, i.e. the service times, and define for $i=1, 2, \ldots;\ c>0$,

$$\tau_{i,c} \overset{\text{def}}{=} \begin{cases} \tau_i & \text{if} \quad \tau_i<c, \\ c & \text{if} \quad \tau_i\geqq c, \end{cases}$$

so that $\tau_{i,c}$ has distribution function $B_c(t)$. Consider now the realisations of the usual queueing process with service time distribution $B(t)$ and those of the queueing process with same initial conditions, with the same realisations of interarrival times and service times but the latter all truncated at c. For these realisations we have for $n=1, 2, \dots,$

$$\boldsymbol{s}_{n,c}=\boldsymbol{\tau}_{1,c}+\dots+\boldsymbol{\tau}_{n,c}-\boldsymbol{a}_n \leqq \boldsymbol{b}_n-\boldsymbol{a}_n=\boldsymbol{s}_n,$$

with probability one. Therefore with probability one

$$\sup_{n\geqq 1}(0, \boldsymbol{s}_n) \geqq \sup_{n\geqq 1}(0, \boldsymbol{s}_{n,c}),$$

so that by (5.154),

$$W_c(t)=\Pr\{\sup_{n\geqq 1}(0, \boldsymbol{s}_{n,c})<t\} \geqq \Pr\{\sup_{n\geqq 1}(0, \boldsymbol{s}_n)<t\}=W(t). \tag{5.155}$$

Consequently, if $W(t)$ has a finite first moment

$$\int_0^\infty t\,\mathrm{d}W_c(t) = \int_0^\infty \{1-W_c(t)\}\,\mathrm{d}t \leqq \int_0^\infty \{1-W(t)\}\,\mathrm{d}t = \int_0^\infty t\,\mathrm{d}W(t). \tag{5.156}$$

Since $W_c(t)$ converges for $c\to\infty$ from above to $W(t)$ for almost every t (cf. (5.151)), it is seen from the dominated convergence theorem (cf. LOÈVE [1960]) that (5.153) is true.

We shall now derive an explicit expression for the Laplace-Stieltjes transform of the distribution $W_c(t)$ of the waiting time for the stationary state of the queueing system $G/G_c/1$. The method is based on a Wiener-Hopf decomposition. As in (5.139) we define two stochastic variables $\boldsymbol{w}$ and $\boldsymbol{w}_c$ with distribution functions $W(t)$ and $W_c(t)$, respectively.
Define

$$\begin{aligned} Z_c(\xi) &\overset{\text{def}}{=} \exp\left\{\sum_{n=1}^\infty \frac{1}{n}\,\mathrm{E}\{(1-\mathrm{e}^{-\xi \boldsymbol{s}_{n,c}})(\boldsymbol{s}_{n,c}>0)\}\right\} \\ &= \{\mathrm{E}\{\mathrm{e}^{-\xi \boldsymbol{w}_c}\}\}^{-1}, \qquad \operatorname{Re}\xi \geqq 0. \end{aligned} \tag{5.157}$$

Since for $a<1$ and hence $a_c<1$, the function $\mathrm{E}\{\exp(-\xi \boldsymbol{w}_c)\}$ is analytic for $\operatorname{Re}\xi\geqq 0$ it is seen that $Z_c(\xi)$ has no zeros for $\operatorname{Re}\xi\geqq 0$. Moreover, the distribution function $W_c(t)$ of $\boldsymbol{w}_c$ is infinitely divisible (cf. (5.53)) and hence its Laplace-Stieltjes transform has no zeros in the interior of its domain of convergence (cf. LUKACS [1960] p. 187); therefore $Z_c(\xi)$ is analytic for $\operatorname{Re}\xi>0$.
We further define

$$\begin{aligned} Z_c(\xi) &\overset{\text{def}}{=} \{1-\beta_c(\xi)\,\alpha(-\xi)\}\exp\left\{B_c + \sum_{n=1}^\infty \frac{1}{n}\,\mathrm{E}\{\mathrm{e}^{-\xi \boldsymbol{s}_{n,c}}(\boldsymbol{s}_{n,c}\leqq 0)\}\right\} \\ &= \{1-\beta_c(\xi)\,\alpha(-\xi)\}\,\mathrm{e}^{B_c}\{1-\mathrm{E}\{\mathrm{e}^{\xi \boldsymbol{i}_c}\}\}^{-1}, \qquad \operatorname{Re}\xi<0, \end{aligned} \tag{5.158}$$

where $\boldsymbol{i}_c$ is the idle period (cf. (5.61)). Since

$$|\mathrm{E}\{\mathrm{e}^{-\rho \boldsymbol{i}_c}\}|<1 \qquad \text{for} \quad \mathrm{Re}\,\rho>0,$$

and $\beta_c(\xi)$ and $\alpha(-\xi)$ are analytic for $\mathrm{Re}\,\xi<0$, $|\xi|<\infty$, it is seen that $Z_c(\xi)$ is analytic for $\mathrm{Re}\,\xi<0$, $|\xi|<\infty$.

From the identity

$$1-\beta_c(\xi)\,\alpha(-\xi)$$
$$=\exp\left\{-\sum_{n=1}^{\infty}\frac{1}{n}\,\mathrm{E}\{\mathrm{e}^{-\xi \boldsymbol{s}_{n,c}}\,(\boldsymbol{s}_{n,c}\leqq 0)+\mathrm{e}^{-\xi \boldsymbol{s}_{n,c}}\,(\boldsymbol{s}_{n,c}>0)\}\right\}, \tag{5.159}$$

for $\mathrm{Re}\,\xi=0$, it is seen that for every zero ξ_0 with $\mathrm{Re}\,\xi_0=0$ of $1-\beta_c(\xi)\,\alpha(-\xi)$,

$$\lim_{\substack{\xi\to\xi_0\\ \mathrm{Re}\,\xi=0}}\ \{1-\beta_c(\xi)\,\alpha(-\xi)\}\exp\left\{\sum_{n=1}^{\infty}\frac{1}{n}\,\mathrm{E}\{\mathrm{e}^{-\xi \boldsymbol{s}_{n,c}}\,(\boldsymbol{s}_{n,c}\leqq 0)\}\right\}$$
$$=\exp\left\{-\sum_{n=1}^{\infty}\frac{1}{n}\,\mathrm{E}\{\mathrm{e}^{-\xi_0 \boldsymbol{s}_{n,c}}\,(\boldsymbol{s}_{n,c}>0)\}\right\}, \tag{5.160}$$

the right-hand side having a finite norm. Consequently, the right-hand sides of (5.157) and (5.158) are equal for $\mathrm{Re}\,\xi=0$. Since these functions approach their values at $\mathrm{Re}\,\xi=0$ continuously, it follows that they are analytic continuations of each other; in fact it is a Wiener-Hopf decomposition. Consequently, $Z_c(\xi)$ as defined by (5.157) and (5.158) is an analytic function of ξ in the whole ξ plane except for a singularity at $|\xi|=\infty$, $\mathrm{Re}\,\xi<0$ (cf. (5.143)); *hence $Z_c(\xi)$ is an entire function.*

From (5.47) it is seen that

$$Z_c(\xi)\to \mathrm{e}^{B_c} \qquad \text{for} \quad |\xi|\to\infty, \quad \arg\xi=0. \tag{5.161}$$

Now $Z_c(\xi)$ cannot be a constant for all ξ, since this would imply that $W_c(t)$ is the distribution function degenerated at zero. This, however, contradicts our assumption (5.148). The relation (5.161) also implies, since $Z_c(\xi)$ is an entire function, that $Z_c(\xi)$ is not a polynomial. We now prove that $Z_c(\xi)$ *has infinitely many zeros* $\xi_1, \xi_2, \ldots$.

If the entire function $Z_c(\xi)$ has no zeros and is not a constant then it is of the form (cf. Hadamard's factorization theorem)

$$Z_c(\xi)=\mathrm{e}^{g(\xi)}, \tag{5.162}$$

where $g(\xi)$ is a polynomial in ξ, so that

$$\mathrm{E}\{\mathrm{e}^{-\xi \boldsymbol{w}_c}\}=\mathrm{e}^{-g(\xi)}, \qquad \mathrm{Re}\,\xi\geqq 0. \tag{5.163}$$

Since

$$|Z_c(\xi)|<\infty, \quad \text{Re}\,\xi=0,$$
$$|Z_c(\xi)|\approx e^{B_c+C_c}\,|1-\beta_c(\xi)\,\alpha(-\xi)|, \qquad |\xi|\to\infty, \quad \tfrac{1}{2}\pi<\arg\xi<1\tfrac{1}{2}\pi,$$
$$|\alpha(-\xi)|\to 0, \qquad |\xi|\to\infty, \quad \tfrac{1}{2}\pi<\arg\xi<1\tfrac{1}{2}\pi,$$

and $\beta_c(\xi)$ is an entire function of order 1, it follows that the entire function $Z_c(\xi)$ is at most of order 1, so that $g(\xi)$ can be only a polynomial of at most the first degree. (This result follows also from the fact that in (5.163) $g(\xi)$ can be at most of degree 2 (cf. Marcinkiewicz' theorem, LUKACS [1960] p. 147), but if $g(\xi)$ is of degree 2 then $W_c(t)$ is a normal distribution, which is impossible since $\boldsymbol{w}_c$ is a non-negative variable.) If $g(\xi)$ is of degree 1 then $W_c(t)$ is a degenerate distribution, but since $a_c<1$ and $W_c(0+)<1$ (cf. (5.148)), this is also impossible. Consequently, $Z_c(\xi)$ has at least one zero with finite norm, so that, since $Z_c(\xi)$ is not a polynomial, it has infinitely many zeros. This proves our statement.

It has been shown above that $Z_c(\xi)$ has no zeros with $\text{Re}\,\xi\geqq 0$; since

$$|\text{E}\{e^{\xi \boldsymbol{l}_c}\}|<1 \qquad \text{for} \quad \text{Re}\,\xi<0,$$

it follows that the zeros of $1-\beta_c(\xi)\,\alpha(-\xi)$ with $\text{Re}\,\xi<0$ are the zeros of $Z_c(\xi)$;

$$1-\beta_c(\xi_i)\,\alpha(-\xi_i)=0, \qquad i=1,2,\ldots; \quad \text{Re}\,\xi_i<0; \quad |\xi_1|\leqq|\xi_2|\leqq,\ldots; \tag{5.164}$$

the complex zeros occur in conjugate pairs since $\beta_c(\xi)$ and $\alpha(-\xi)$ are real for real ξ, we shall take $\xi_i=\bar{\xi}_{i+1}$, $i=2,4,6,\ldots$.

We now prove that $Z_c(\xi)$ is of exponential type. Obviously we need only to investigate $Z_c(\xi)$ for $\text{Re}\,\xi<0$, $|\xi|\to\infty$. From (5.158) we see that the behaviour of $Z_c(\xi)$ for $|\xi|\to\infty$, $\text{Re}\,\xi<0$ is determined by that of $\alpha(-\xi)\,\beta_c(\xi)$. On a semicircle with $\text{Re}\,\xi<0$ and centre at $\xi=0$ the function $\alpha(-\xi)\,\beta_c(\xi)$ takes its maximum value at $\arg\xi=\pi$, since $\beta_c(\xi)$ and $\alpha(-\xi)$ are both real for real ξ. From (5.143) it follows that

$$\beta_c(\xi)\,\alpha(-\xi) = \text{O}(e^{c|\xi|}\int_0^\infty e^{-|\xi|t}\,\text{d}A(t)), \qquad |\xi|\to\infty, \quad \arg\xi=\pi.$$

For real $\xi>0$ we have

$$e^{\xi c}\int_0^\infty e^{-\xi t}\,\text{d}A(t) = \int_0^{c+} e^{-\xi(t-c)}\,\text{d}A(t) + \int_{c+}^\infty e^{-\xi(t-c)}\,\text{d}A(t);$$

the second integral tends to zero for $\xi\to\infty$, while by the first mean value theorem a constant k exists such that

$$\int_0^{c+} e^{-\xi(t-c)}\,dA(t) = e^{\xi k}(A(c+) - A(0+)) = e^{\xi k} A(c), \qquad c \geqq k \geqq 0.$$

Since our assumption (5.148) implies that an interval $(c-\delta, c)$, $\delta>0$, exists where $A(t)$ is non-zero, it follows that $k>0$. Consequently,

$$\beta_c(\xi)\,\alpha(-\xi) = O(e^{k|\xi|}) \qquad \text{for} \quad |\xi|\to\infty, \quad \arg\xi=\pi; \tag{5.165}$$

hence $Z_c(\xi)$ is of exponential type.

This result leads to the conclusion (cf. TITCHMARSH [1952] p. 286, ex. 17) that

$$\sum_{i=1}^{\infty} |\xi_i|^{-(1+\varepsilon)}, \tag{5.166}$$

converges for any $\varepsilon>0$, and diverges for $\varepsilon=0$.
From Hadamard's factorization theorem we now conclude that the entire function $Z_c(\xi)$ has the representation

$$Z_c(\xi) = e^{k_2+k_3\xi} \prod_{i=1}^{\infty} \left\{1 - \frac{\xi}{\xi_i}\right\} e^{\xi/\xi_i}, \tag{5.167}$$

here k_2 and k_3 are constants. Hence, from (5.157),

$$E\{e^{-\xi w_c}\} = e^{-(k_2+k_3\xi)} \prod_{i=1}^{\infty} \left\{1 - \frac{\xi}{\xi_i}\right\}^{-1} e^{-\xi/\xi_i}, \qquad \operatorname{Re}\xi > \max_{i\geqq 1}(\operatorname{Re}\xi_i). \tag{5.168}$$

Since, for a Laplace-Stieltjes transform of a probability distribution the singularity nearest to the imaginary axis is always real (cf. WIDDER [1946]), it follows from (5.164) that

$$\xi_1 < 0, \qquad \xi_1 > \max_{i\geqq 2}(\operatorname{Re}\xi_i),$$

so that $E\{\exp(-\xi w_c)\}$ exists for a ξ with $\operatorname{Re}\xi<0$; this implies that all moments of w_c are finite.
To determine k_2 and k_3 note that the left-hand side of (5.167) is 1 for $\xi=0$, hence $k_2=0$. The value of k_3 is found by the same argument which yielded the constant k_1 in (5.144) (cf. (5.146)). It is found that

$$k_3 = E\{w_c\} = \int_0^{\infty} t\,dW_c(t) = -\frac{1}{2\pi i}\int_{-i\infty-0}^{i\infty-0} \frac{d\xi}{\xi^2}\log\{1-\beta_c(\xi)\,\alpha(-\xi)\}, \tag{5.169}$$

$k_3<\infty$ since $E\{w_c\}<\infty$.

We summarize the results obtained above.

If the service time distribution is truncated at c, if $a_c<1$ *and if* (5.148) *holds then*

$$Z_c(\xi) = e^{\xi E\{w_c\}} \prod_{i=1}^{\infty} \left\{1 - \frac{\xi}{\xi_i}\right\} e^{\xi/\xi_i}, \tag{5.170}$$

$$E\{e^{-\xi w_c}\} = \int_{0-}^{\infty} e^{-\xi t}\, dW_c(t) = e^{-\xi E\{w_c\}} \prod_{i=1}^{\infty} \left\{1 - \frac{\xi}{\xi_i}\right\}^{-1} e^{-\xi/\xi_i}, \qquad \operatorname{Re} \xi > \xi_1, \tag{5.171}$$

where ξ_i, $i=1, 2, \dots$, *are determined by* (5.164) *and satisfy* (5.166)*;* $Z_c(\xi)$ *is an entire function of exponential type, and* $E\{\exp(-\xi w_c)\}$ *is a meromorphic function with poles at* ξ_i, $i=1, 2, \dots$, *and analytic for* $\operatorname{Re} \xi > \xi_1$, *all moments of* w_c *are finite.*

It is now easily shown that from (5.158) and from (5.169) and (5.171) we have for the Laplace-Stieltjes transform of the distribution function of the idle period i_c,

$$E\{e^{-\xi i_c}\} = 1 - \{1-\beta_c(-\xi)\,\alpha(\xi)\} \exp[B+\xi E\{w_c\}] \prod_{i=1}^{\infty} \left\{1 + \frac{\xi}{\xi_i}\right\}^{-1} e^{\xi/\xi_i}, \qquad \operatorname{Re} \xi \geqq 0. \tag{5.172}$$

The latter result can be also obtained by starting from the integral representation (5.136) if $A(t)$ has no discontinuity points. From (5.149) it is seen that we can replace the path of integration in (5.136) by a line parallel to the imaginary axis, and just left of it. We may now take $r=1$ and obtain

$$E\{e^{-\rho i_c}\} = 1 - \exp\left\{\frac{1}{2\pi i} \int_{-i\infty-0}^{i\infty-0} \log\{1-\alpha(-\xi)\,\beta_c(\xi)\} \frac{d\xi}{\rho+\xi}\right\}, \qquad \operatorname{Re} \rho > 0. \tag{5.173}$$

To evaluate the integral in (5.173) it is observed that for arbitrary constants P and Q,

$$\frac{1}{2\pi i} \int_{-i\infty-0}^{i\infty-0} \log\{1-\alpha(-\xi)\,\beta_c(\xi)\} \left\{\frac{P}{\xi} + \frac{Q}{\xi^2}\right\} d\xi$$

$$= \frac{-1}{2\pi i} \int_{-i\infty-0}^{i\infty-0} \sum_{n=1}^{\infty} \frac{1}{n} \Big[E\{e^{-\xi s_{n,c}}(s_{n,c} \leqq 0)\} +$$

$$+\ \mathrm{E}\{\mathrm{e}^{-\xi s_{n,c}}(s_{n,c}>0)\}\Big]\Big]\left\{\frac{P}{\xi}+\frac{Q}{\xi^2}\right\}\mathrm{d}\xi$$

$$=\frac{-1}{2\pi\mathrm{i}}\int_{-\mathrm{i}\infty-0}^{\mathrm{i}\infty-0}\sum_{n=1}^{\infty}\frac{1}{n}\mathrm{E}\{\mathrm{e}^{-\xi s_{n,c}}(s_{n,c}>0)\}\left\{\frac{P}{\xi}+\frac{Q}{\xi^2}\right\}\mathrm{d}\xi$$

$$=PB_c-Q\mathrm{E}\{w_c\}. \tag{5.174}$$

Since

$$\frac{1}{\rho+\xi}-\frac{1}{\xi}+\frac{\rho}{\xi^2}=\frac{\rho^2}{\xi^2(\rho+\xi)}=\mathrm{O}(|\xi|^{-3})\qquad\text{for}\quad|\xi|\to\infty,$$

it follows (cf. (5.165)) that

$$\left|\log\{1-\alpha(-\xi)\,\beta_c(\xi)\}\left\{\frac{1}{\rho+\xi}-\frac{1}{\xi}+\frac{\rho}{\xi^2}\right\}\right|$$
$$=\mathrm{O}(|\xi|^{-2}),\quad|\xi|\to\infty,\qquad\mathrm{Re}\,\xi<0.$$

Therefore, if $A(t)$ has no discontinuity points,

$$\left|\frac{1}{2\pi\mathrm{i}}\int_{\mathrm{G}_n}\log\{1-\alpha(-\xi)\,\beta_c(\xi)\}\left\{\frac{1}{\rho+\xi}-\frac{1}{\xi}+\frac{\rho}{\xi^2}\right\}\mathrm{d}\xi\right|\to 0\qquad\text{for}\quad n\to\infty,$$

here G_n is a semicircle with radius r_n, centre at $\xi=0$ and $\mathrm{Re}\,\xi\leqq 0$ on G_n, $r_n\to\infty$ for $n\to\infty$. Denoting by N_n that value of i for which

$$|\xi_i|<r_n<|\xi_{i+1}|,$$

then N_n always exists, since the sequence $|\xi_i|$, $i=1,2,\ldots$, has no finite limit point (cf. Titchmarsh), moreover

$$N_n\to\infty\qquad\text{for}\quad n\to\infty.$$

We now have

$$\lim_{n\to\infty}\frac{1}{2\pi\mathrm{i}}\int_{-\mathrm{i}r_n-0}^{\mathrm{i}r_n-0}\log\{1-\alpha(-\xi)\,\beta_c(\xi)\}\left\{\frac{1}{\rho+\xi}-\frac{1}{\xi}+\frac{\rho}{\xi^2}\right\}\mathrm{d}\xi$$

$$=\lim_{n\to\infty}\frac{1}{2\pi\mathrm{i}}\int_{-\mathrm{i}r_n-0}^{\mathrm{i}r_n-0}\log\{1-\alpha(-\xi)\,\beta_c(\xi)\}\,\mathrm{d}_\xi\left\{\log\left(\frac{\rho+\xi}{\xi}\right)-\frac{\rho}{\xi}\right\}=$$

$$= \lim_{n\to\infty} \frac{1}{2\pi i}\left[\log\{1-\alpha(-\xi)\,\beta_c(\xi)\}\left\{\log\left(\frac{\rho+\xi}{\xi}\right)-\frac{\rho}{\xi}\right\}\right]\Bigg|_{\xi=-ir_n-0}^{\xi=ir_n-0}$$

$$-\lim_{n\to\infty}\frac{1}{2\pi i}\int_{-ir_n-0}^{ir_n-0}\frac{(d/d\xi)\{1-\alpha(-\xi)\,\beta_c(\xi)\}}{1-\alpha(-\xi)\,\beta_c(\xi)}\left\{\log\left(\frac{\rho+\xi}{\xi}\right)-\frac{\rho}{\xi}\right\}d\xi$$

$$=\log\{1-\alpha(\rho)\,\beta_c(-\rho)\}-\lim_{n\to\infty}\sum_{i=1}^{N_n}\log\left\{\left(1+\frac{\rho}{\xi_i}\right)e^{-\rho/\xi_i}\right\}$$

$$+\lim_{n\to\infty}\frac{1}{2\pi i}\int_{G_n}\frac{(d/d\xi)\{1-\alpha(-\xi)\,\beta_c(\xi)\}}{1-\alpha(-\xi)\,\beta_c(\xi)}\left\{\log\left(\frac{\rho+\xi}{\xi}\right)-\frac{\rho}{\xi}\right\}d\xi.$$

Since

$$\log\left(\frac{\rho+\xi}{\xi}\right)-\frac{\rho}{\xi}=\log\left(1+\frac{\rho}{\xi}\right)-\frac{\rho}{\xi}=O(|\xi|^{-2})\qquad\text{for}\quad|\xi|\to\infty,$$

and $(d/d\xi)\,\beta_c(\xi)$ is also an entire function of the same order as $\beta_c(\xi)$ (cf. TITCHMARSH [1952]) it is easily verified that the last integral in the above expression tends to zero for $n\to\infty$. Moreover, from (5.166) the limit of the sum in the last expression exists, therefore

$$\frac{1}{2\pi i}\int_{-i\infty-0}^{i\infty-0}\log\{1-\alpha(-\xi)\,\beta_c(\xi)\}\left\{\frac{1}{\rho+\xi}-\frac{1}{\xi}+\frac{\rho}{\xi^2}\right\}d\xi$$
$$=\log\left[\{1-\alpha(\rho)\,\beta_c(-\rho)\}\prod_{i=1}^{\infty}\left(1+\frac{\rho}{\xi_i}\right)^{-1}e^{\rho/\xi_i}\right].$$

If $A(t)$ has no discontinuity points then

$$B_c=-\frac{1}{2\pi i}\int_{-i\infty-0}^{i\infty-0}\frac{d\xi}{\xi}\log\{1-\beta_c(\xi)\,\alpha(-\xi)\}.$$

Hence, from (5.169), (5.173) and (5.174) with $P=-1$, $Q=\rho$ we obtain

$$E\{e^{-\rho i_c}\}=1-\{1-\alpha(\rho)\,\beta_c(-\rho)\}\exp(B_c+\rho E\{w_c\})\prod_{i=1}^{\infty}\left(1+\frac{\rho}{\xi_i}\right)^{-1}e^{\rho/\xi_i},$$
$$\operatorname{Re}\rho>0,$$

which is the expression (5.172).

The same method as described above can be used to evaluate the expression (5.136). Denote by $\xi_i(r)$, $i=1, 2, \ldots$; $|r|\leq 1$, the zeros in the left half plane of

$$1-r\beta_c(\xi)\,\alpha(-\xi),$$

then these zeros have the same properties as ξ_i, $i=1, 2, \ldots$, (cf. (5.164) and (5.166)); this is easily proved from the theory of entire functions. By applying the same technique of integration as above we obtain from (5.136) for the Laplace-Stieltjes transform of the joint distribution of $\boldsymbol{i}_c$ and $\boldsymbol{n}_c$, the number of customers served during a busy period, if $A(t)$ has no discontinuity points,

$$\mathrm{E}\{r^{\boldsymbol{n}_c}\mathrm{e}^{-\rho \boldsymbol{i}_c}\}$$

$$= 1 - \{1-r\alpha(\rho)\,\beta_c(-\rho)\}\exp[B_{0,c}(r)+\rho k(r)]\prod_{i=1}^{\infty}\left\{1+\frac{\rho}{\xi_i(r)}\right\}^{-1}\mathrm{e}^{\rho/\xi_i(r)}, \tag{5.175}$$

for $|r|\leqq 1$, Re $\rho\geqq 0$, where $B_{0,c}(r)$ is defined analogous to (5.51) and $k(r)$ by (note: $k(1)=\mathrm{E}\{\boldsymbol{w}_c\}$),

$$k(r) \overset{\text{def}}{=} \sum_{n=1}^{\infty}\frac{r^n}{n}\int_0^{\infty} t\,\mathrm{d}C_c^{n*}(t), \qquad |r|\leqq 1, \qquad C_c(t) = \int_0^{\infty} B_c(t+\tau)\,\mathrm{d}A(\tau). \tag{5.176}$$

The integral expressions (5.134) and (5.135) can now be handled in exactly the same way as (5.136). However, the resulting expressions are rather intricate.

Next, we consider the distribution $\{w_{j,c},\ j=0, 1, \ldots\}$ (cf. (5.129)) which is the distribution of the number of customers in the system at some moment for the ergodic situation if $A(t)$ is not a lattice distribution. From (5.129) and (5.140) we have

$$w_{0,c} = 1 - a_c,$$

$$w_{j,c} = \begin{cases} a_c \displaystyle\int_0^{\infty} \{A^{(j-1)*}(u)-A^{j*}(u)\}\,\mathrm{d}\left\{W_c(u) * \frac{1}{\beta_c}\int_0^{u}\{1-B_c(v)\}\,\mathrm{d}v\right\}, & j=1, 2, \ldots, \\[2ex] \dfrac{a_c}{2\pi \mathrm{i}}\displaystyle\int_{-\mathrm{i}\infty-0}^{\mathrm{i}\infty-0} \mathrm{E}\{\mathrm{e}^{-\xi \boldsymbol{w}_c}\}\,\frac{1-\beta_c(\xi)}{\xi\beta_c}\,\frac{\alpha^{j-1}(-\xi)-\alpha^j(-\xi)}{-\xi}\,\mathrm{d}\xi, & j=1, 2, \ldots . \end{cases} \tag{5.177}$$

The integrals can be evaluated by using the expression (5.171). Using (5.147) and (5.151) it follows easily from (5.129) that

$$w_j = \lim_{c\to\infty} w_{j,c}, \qquad j=0, 1, \dots, \tag{5.178}$$

and if $\mathrm{E}\{\boldsymbol{w}\} < \infty$ then

$$\lim_{c\to\infty} \sum_{j=1}^{\infty} j w_{j,c} = \sum_{j=1}^{\infty} j w_j,$$

here w_j refers to the queueing system with $B(t)$ as service time distribution. From (5.177) it is easily found that

$$\sum_{j=0}^{\infty} j w_{j,c} = \frac{-a_c}{2\pi\mathrm{i}} \int_{-\mathrm{i}\infty-0}^{\mathrm{i}\infty-0} \mathrm{E}\{\mathrm{e}^{-\xi \boldsymbol{w}_c}\} \frac{1-\beta_c(\xi)}{\xi^2 \beta_c} \frac{\mathrm{d}\xi}{1-\alpha(-\xi)}. \tag{5.179}$$

This relation may be simplified. We first note that

$$\frac{1}{1-\alpha(-\xi)} + \frac{\beta_c(\xi)}{1-\beta_c(\xi)} = \frac{1-\alpha(-\xi)\,\beta_c(\xi)}{\{1-\alpha(-\xi)\}\{1-\beta_c(\xi)\}}.$$

Hence, from (5.157), (5.158) and (5.168) we have

$$\frac{1}{2\pi\mathrm{i}} \int_{-\mathrm{i}\infty-0}^{\mathrm{i}\infty-0} \mathrm{E}\{\mathrm{e}^{-\xi \boldsymbol{w}_c}\} \frac{1-\beta_c(\xi)}{\xi^2 \beta_c} \left\{ \frac{1}{1-\alpha(-\xi)} + \frac{\beta_c(\xi)}{1-\beta_c(\xi)} \right\} \mathrm{d}\xi$$

$$= \frac{1}{2\pi\mathrm{i}} \int_{-\mathrm{i}\infty-0}^{\mathrm{i}\infty-0} \exp\left\{ -B_c - \sum_{n=1}^{\infty} \frac{1}{n} \mathrm{E}\{\mathrm{e}^{-\xi \boldsymbol{s}_{n,c}} (\boldsymbol{s}_{n,c} \leqq 0)\} \right\}$$

$$\cdot \frac{\mathrm{d}\xi}{\xi^2 (1-\alpha(-\xi))\, \beta_c} = 0,$$

since the latter integrand is analytic for $\mathrm{Re}\,\xi < 0$ and of order $\mathrm{O}(|\xi|^{-2})$ for $|\xi| \to \infty$, $\frac{1}{2}\pi < \arg \xi < 1\frac{1}{2}\pi$.
Therefore from (5.179),

$$\sum_{j=0}^{\infty} j w_{j,c} = \frac{a_c}{2\pi\mathrm{i}} \int_{-\mathrm{i}\infty-0}^{\mathrm{i}\infty-0} \mathrm{E}\{\mathrm{e}^{-\xi \boldsymbol{w}_c}\}\, \beta_c(\xi) \frac{\mathrm{d}\xi}{\xi^2 \beta_c}.$$

The latter integrand is analytic for $\mathrm{Re}\,\xi > \xi_1$ except for a double pole at

$\xi=0$; the residue at this pole is

$$-\frac{1}{\beta_c}\{\mathrm{E}\{\boldsymbol{w}_c\}+\beta_c\}.$$

Consequently,

$$\sum_{j=0}^{\infty} j w_{j,c} = \frac{1}{\alpha}\{\mathrm{E}\{\boldsymbol{w}_c\}+\beta_c\}, \tag{5.180}$$

since $\boldsymbol{w}_c$ has a finite first moment. Since β_c and $\mathrm{E}\{\boldsymbol{w}_c\}$ converge from below to β and $\mathrm{E}\{\boldsymbol{w}\}$, respectively, for $c\to\infty$ (cf. (5.153)), it follows from (5.178) and (5.180) that if $A(t)$ is not a lattice distribution and $\mathrm{E}\{\boldsymbol{w}\}$ is finite then

$$\sum_{j=0}^{\infty} j w_j = \frac{1}{\alpha}\{\mathrm{E}\{\boldsymbol{w}\}+\beta\}; \tag{5.181}$$

this proves the relation (5.131). It is easily verified that this result is independent of the assumption (5.148).

Finally, a remark about the Laplace-Stieltjes transform of the distribution function $V(t)$ of the virtual delay time for the case that $A(t)$ is not a lattice distribution. From (5.109), (5.146) and (5.171) it is easily found that

$$\int_{0-}^{\infty} \mathrm{e}^{-\xi t}\,\mathrm{d}V_c(t)$$

$$= 1 - a_c + a_c \exp\left[-\xi\left\{\mathrm{E}\{\boldsymbol{w}_c\} + \frac{\beta_{2c}}{2\beta_c}\right\}\right]\prod_{i=1}^{\infty}\frac{(1-\xi/\eta_i)\,\mathrm{e}^{\xi/\eta_i}}{(1-\xi/\xi_i)\,\mathrm{e}^{\xi/\xi_i}}, \qquad \mathrm{Re}\,\xi>\xi_1. \tag{5.182}$$

The discussions in this section were mainly concentrated on the ergodic situation of the queueing system. However, it is possible to study also the distribution function of $\boldsymbol{w}_{n,c}$, the actual waiting time of the nth arriving customer when the service time distribution is truncated at c. The discussion of this case is completely analogous. We have to consider instead of the Wiener-Hopf decomposition of $1-\beta_c(\xi)\,\alpha(-\xi)$ (cf. (5.160)) that of $1-r\beta_c(\xi)\,\alpha(-\xi)$.

II.5.10. The queueing system $G/K_n/1$

For the queueing system $G/K_n/1$ the Laplace-Stieltjes transform $\beta(\xi)$ of the service time distribution is a rational function with n the degree of the denominator. We shall write

$$\beta(\xi)=\beta_1(\xi)/\beta_2(\xi), \tag{5.183}$$

with $\beta_2(\xi)$ a polynomial in ξ of degree n, and $\beta_1(\xi)$ a polynomial of degree $n-1$ at most, since we always assume $B(0+)=0$ (cf. section 1.2) and

$$B(0+) = \lim_{\xi\to\infty} \beta(\xi), \qquad \xi \text{ real}.$$

Moreover, we shall take the coefficient of ξ^n in $\beta_2(\xi)$ equal to 1. The function $\beta_2(\xi)$ has no zeros with $\operatorname{Re}\xi \geqq 0$. Denote its zeros by ζ_i, $i=1, 2, \ldots, n$; the zero nearest to the imaginary axis is real. Let ζ_1 be this zero. Taking in (5.29) $\rho_3=0$, $\rho_4=0$, then we have for $\operatorname{Re}\rho_2<0$, $\operatorname{Re}\rho_1\geqq 0$, $|r|<1$,

$$\sum_{n=1}^{\infty} r^n \operatorname{E}\{\exp(-\rho_1 \boldsymbol{w}_n+\rho_2 \boldsymbol{d}_n) \mid \boldsymbol{w}_1=0, \boldsymbol{d}_1=0\}$$

$$= r\exp\left\{-\frac{1}{2\pi\mathrm{i}}\int_{-\mathrm{i}\infty-0}^{\mathrm{i}\infty-0}\left\{\frac{1}{\rho_1-\xi}+\frac{1}{\xi-\rho_2}\right\}\log\{1-r\beta(\xi)\,\alpha(-\xi)\}\,\mathrm{d}\xi\right\}.$$

It follows that for $\operatorname{Re}\rho_1\geqq 0$, $\operatorname{Re}\rho_2=0$, $|r|<1$,

$$\sum_{n=1}^{\infty} r^n \operatorname{E}\{\exp(-\rho_1 \boldsymbol{w}_n+\rho_2 \boldsymbol{d}_n) \mid \boldsymbol{w}_1=0, \boldsymbol{d}_1=0\} = \frac{r}{1-r\beta(\rho_2)\,\alpha(-\rho_2)}$$

$$\cdot\exp\left\{-\frac{1}{2\pi\mathrm{i}}\int_{-\mathrm{i}\infty-0}^{\mathrm{i}\infty-0}\left\{\frac{1}{\rho_1-\xi}+\frac{1}{\xi-\rho_2}\right\}\log\{1-r\beta(\xi)\,\alpha(-\xi)\}\,\mathrm{d}\xi\right\}. \tag{5.184}$$

To evaluate the integral consider the function

$$1-r\beta(\xi)\,\alpha(-\xi)\equiv\frac{1}{\beta_2(\xi)}\{\beta_2(\xi)-r\beta_1(\xi)\,\alpha(-\xi)\}. \tag{5.185}$$

Since

$$|\beta(\xi)\,\alpha(-\xi)|\leqq\beta(\operatorname{Re}\xi)\,\alpha(-\operatorname{Re}\xi)=1-(a-1)\alpha\operatorname{Re}\xi+o(\operatorname{Re}\xi), \qquad a=\beta/\alpha,$$

for $\operatorname{Re}\xi<0$ and $|\operatorname{Re}\xi|$ sufficiently small, we have

$$|\beta(\xi)|^{-1}\geqq|r||\alpha(-\xi)|, \tag{5.186}$$

for $|r|<1$ or for $|r|=1$, $a<1$. ($\operatorname{Re}\xi$ can always be taken such that $\beta_1(\xi)\neq 0$.) Since $|\alpha(-\xi)|<1$ for $\operatorname{Re}\xi<0$ and the degree of $\beta_2(\xi)$ is higher than that of

$\beta_1(\xi)$ it follows that (5.186) is also true for $\text{Re}\,\xi<0$ and $|\xi|$ sufficiently large. Hence, by Rouché's theorem (cf. app. 6) the function of (5.185) has exactly n zeros $\xi_i(r)$, $i=1,\ldots,n$, in the left half-plane if $|r|<1$ or if $|r|=1$, $a<1$; it is easily seen that these zeros are continuous in r for $|r|\leqq 1$. It follows from (5.184) that for $\text{Re}\,\rho_1\geqq 0$, $\text{Re}\,\rho_2=0$, $|r|<1$,

$$\frac{1}{2\pi \mathrm{i}}\int_{-\mathrm{i}\infty-0}^{\mathrm{i}\infty-0}\left\{\frac{1}{\rho_1-\xi}+\frac{1}{\xi-\rho_2}\right\}\log\{1-r\beta(\xi)\,\alpha(-\xi)\}\,\mathrm{d}\xi$$

$$=\frac{1}{2\pi \mathrm{i}}\int_{-\mathrm{i}\infty-0}^{\mathrm{i}\infty-0}\log\{1-r\beta(\xi)\,\alpha(-\xi)\}\,\mathrm{d}_\xi\log\frac{\xi-\rho_2}{\rho_1-\xi}$$

$$=\lim_{\delta\to\infty}\frac{1}{2\pi \mathrm{i}}\left[\log\{1-r\beta(\xi)\,\alpha(-\xi)\}\log\frac{\xi-\rho_2}{\rho_1-\xi}\right]\Bigg|_{\xi=-0-\mathrm{i}\delta}^{\xi=-0+\mathrm{i}\delta}$$

$$-\frac{1}{2\pi \mathrm{i}}\int_{-\mathrm{i}\infty-0}^{\mathrm{i}\infty-0}\left\{\frac{\mathrm{d}}{\mathrm{d}\xi}\log\{1-r\beta(\xi)\,\alpha(-\xi)\}\right\}\log\frac{\xi-\rho_2}{\rho_1-\xi}\,\mathrm{d}\xi$$

$$=-\frac{1}{2\pi \mathrm{i}}\int_{-\mathrm{i}\infty-0}^{\mathrm{i}\infty-0}\frac{(\mathrm{d}/\mathrm{d}\xi)\{1-r\beta(\xi)\,\alpha(-\xi)\}}{1-r\beta(\xi)\,\alpha(-\xi)}\log\frac{\xi-\rho_2}{\rho_1-\xi}\,\mathrm{d}\xi$$

$$=-\sum_{i=1}^{n}\log\left\{\frac{\rho_1-\zeta_i}{\rho_2-\zeta_i}\,\frac{\rho_2-\xi_i(r)}{\rho_1-\xi_i(r)}\right\},\tag{5.187}$$

since the integrand is of order $o(|\xi|^{-1})$ for $|\xi|\to\infty$, $\text{Re}\,\xi<0$. Therefore, by analytic continuation with respect to ρ_2 we obtain from (5.184) and (5.187) for $\text{Re}\,\rho_1\geqq 0$, $\text{Re}\,\rho_2\leqq 0$, $|r|<1$,

$$\sum_{m=1}^{\infty} r^m\,\mathrm{E}\{\exp(-\rho_1\boldsymbol{w}_m+\rho_2\boldsymbol{d}_m)\mid \boldsymbol{w}_1=0,\,\boldsymbol{d}_1=0\}$$

$$=\frac{r}{1-r\beta(\rho_2)\,\alpha(-\rho_2)}\prod_{i=1}^{n}\left\{\frac{\rho_1-\zeta_i}{\rho_2-\zeta_i}\,\frac{\rho_2-\xi_i(r)}{\rho_1-\xi_i(r)}\right\}$$

$$=\frac{r\beta_2(\rho_1)}{\beta_2(\rho_2)-r\beta_1(\rho_2)\,\alpha(-\rho_2)}\prod_{i=1}^{n}\frac{\rho_2-\xi_i(r)}{\rho_1-\xi_i(r)}.\tag{5.188}$$

Hence, for $\text{Re}\,\rho\geqq 0$, $|r|<1$,

$$\sum_{m=1}^{\infty} r^m \mathrm{E}\{\mathrm{e}^{-\rho w_m} \mid w_1=0\} = \frac{r}{1-r}\,\frac{\beta_2(\rho)}{\beta_2(0)} \prod_{i=1}^{n} \frac{\xi_i(r)}{\xi_i(r)-\rho}, \tag{5.189}$$

$$\sum_{m=1}^{\infty} r^m \mathrm{E}\{\mathrm{e}^{-\rho d_m} \mid d_1=0\} = \frac{r\beta_2(0)}{\beta_2(-\rho)-r\beta_1(-\rho)\,\alpha(\rho)} \prod_{i=1}^{n} \frac{\rho+\xi_i(r)}{\xi_i(r)},$$

note that always $\beta_2(0)\neq 0$ since $\beta(0)=1$. Further,

$$\lim_{m\to\infty} \mathrm{E}\{\mathrm{e}^{-\rho w_m} \mid w_1=0\} = \frac{\beta_2(\rho)}{\beta_2(0)} \prod_{i=1}^{n} \frac{\xi_i(1)}{\xi_i(1)-\rho}, \qquad \operatorname{Re}\rho \geqq 0, \quad a<1, \tag{5.190}$$

and

$$\lim_{m\to\infty} \mathrm{E}\{w_m \mid w_1=0\} = -\frac{\beta_2'(0)}{\beta_2(0)} - \sum_{i=1}^{n} \frac{1}{\xi_i(1)}, \qquad a<1, \tag{5.191}$$

$$\lim_{m\to\infty} \Pr\{w_m=0 \mid w_1=0\} = \frac{(-1)^n}{\beta_2(0)} \prod_{i=1}^{n} \xi_i(1), \qquad a<1;$$

here $\beta_2'(0)$ is the derivative of $\beta_2(\xi)$ at $\xi=0$. From (5.190) it is seen that for the stationary state of the queueing system the Laplace-Stieltjes transform of the distribution of the actual waiting time is a rational function of the same degree as the transform of the service time distribution. This statement is the content of Smith's theorem (SMITH [1953]). From (5.109) and (5.190) it is easily seen that for the stationary situation the distribution function of the virtual delay time has a rational Laplace-Stieltjes transform of degree n. We shall prove that the converse statement is also true.

For $a<1$ and under the usual conditions for $A(t)$ and $B(t)$ the Laplace-Stieltjes transform $\beta(\xi)$ of $B(t)$ is a rational function if the Laplace-Stieltjes transform of $W(t)$ is a rational function; the denominators of these rational functions have the same degree.

To prove this statement let $w_1(\xi)$ and $w_2(\xi)$ denote two polynomials such that

$$\lim_{n\to\infty} \mathrm{E}\{\mathrm{e}^{-\xi w_n} \mid w_1=0\} = \int_{0-}^{\infty} \mathrm{e}^{-\xi t}\,\mathrm{d}W(t) = \frac{w_1(\xi)}{w_2(\xi)}, \qquad \operatorname{Re}\xi \geqq 0.$$

Since

$$0 < W(0+) = \lim_{\xi\to\infty} \frac{w_1(\xi)}{w_2(\xi)}, \qquad \xi \text{ real},$$

$w_1(\xi)$ and $w_2(\xi)$ have the same degree, say n, i.e. $w_1(\xi)$ and $w_2(\xi)$ are nth degree polynomials without common zeros. Analogous to (5.157) and (5.158)

we have

$$\{\int_{0-}^{\infty} \mathrm{e}^{-\xi t}\,\mathrm{d}W(t)\}^{-1} = \{1-\beta(\xi)\,\alpha(-\xi)\}\,\mathrm{e}^{B}\,\{1-\mathrm{E}\{\mathrm{e}^{\xi i}\}\}^{-1}, \quad \mathrm{Re}\,\xi=0, \quad (5.192)$$

hence

$$w_2(\xi)=w_1(\xi)\{1-\beta(\xi)\,\alpha(-\xi)\}\,\mathrm{e}^{B}\,\{1-\mathrm{E}\{\mathrm{e}^{\xi i}\}\}^{-1}, \qquad \mathrm{Re}\,\xi=0. \qquad (5.193)$$

The functions $w_2(\xi)$ and $w_1(\xi)$ are analytic in the whole plane, and $\alpha(-\xi)$, $\mathrm{E}\{\mathrm{e}^{\xi i}\}$ are analytic for $\mathrm{Re}\,\xi<0$, both being Laplace-Stieltjes transforms of non-negative variables. Hence, we can determine $\beta(\xi)$ from (5.193) for $\mathrm{Re}\,\xi<0$ uniquely, except at the zeros of $w_1(\xi)$. Since $w_2(\xi)$ and $w_1(\xi)$ have no common zeros it follows that $\beta(\xi)$ must have poles at the zeros of $w_1(\xi)$. Since $|\beta(\xi)|\leqq 1$ for $\mathrm{Re}\,\xi=0$, all poles of $\beta(\xi)$ have negative real parts. For $\mathrm{Re}\,\xi>0$ the function $\beta(\xi)$ is analytic, hence from the definition of $\beta(\xi)$ for $\mathrm{Re}\,\xi<0$ it is seen that $\beta(\xi)$ as defined above is a single-valued function for all ξ, it approaches its values on the imaginary axis continuously from the left half-plane and from the right half-plane. Since a single-valued function with no singularities, except poles, is a rational function (cf. WHITTAKER and WATSON [1946] p. 105) it follows that $\beta(\xi)$ is a rational function. The number of poles of $\beta(\xi)$ counted according to their multiplicity is n, the degree of $w_1(\xi)$, i.e. of $w_2(\xi)$. This proves our statement.

An immediate consequence is that the distribution $W(t)$ contains only one exponential term if and only if $B(t)$ is a negative exponential distribution.

From (5.190) and (5.192) we find that, for $a<1$,

$$1-\mathrm{E}\{\mathrm{e}^{\rho i}\} = \{1-\beta(\rho)\,\alpha(-\rho)\}\,\mathrm{e}^{B}\,\frac{\beta_2(\rho)}{\beta_2(0)}\prod_{i=1}^{n}\frac{\xi_i(1)}{\xi_i(1)-\rho}, \qquad \mathrm{Re}\,\rho\leqq 0,$$

and hence by using (5.191) and (5.47) that

$$\mathrm{E}\{\mathrm{e}^{-\rho i}\} = 1-\{\beta_2(-\rho)-\beta_1(-\rho)\,\alpha(\rho)\}(-1)^n\prod_{i=1}^{n}\{\rho+\xi_i(1)\}^{-1},$$

$$\mathrm{Re}\,\rho\geqq 0, \quad a<1. \qquad (5.194)$$

Note, that in general the distribution function of the idle time $\boldsymbol{i}$ will not have a rational Laplace-Stieltjes transform.

We consider the transform of the joint distribution of $\boldsymbol{n}$ the number of customers served during a busy period and of the busy period $\boldsymbol{p}$. From (5.134) we have for $\mathrm{Re}\,\rho\geqq 0$, $|r|\leqq 1$, since $B(t)$ is continuous,

$$\mathrm{E}\{r^n \mathrm{e}^{-\rho p}\} = 1 - \exp\left\{\frac{1}{2\pi \mathrm{i}} \int_{-\mathrm{i}\infty-0}^{\mathrm{i}\infty-0} \frac{\mathrm{d}\xi}{\rho-\xi} \log \frac{1-r\beta(\xi)}{1-r\beta(\xi)\,\alpha(\rho-\xi)}\right\}$$

$$= 1 - \exp\left\{\frac{1}{2\pi \mathrm{i}} \int_{-\mathrm{i}\infty-0}^{\mathrm{i}\infty-0} \frac{\mathrm{d}\xi}{\rho-\xi} \log \frac{\beta_2(\xi)-r\beta_1(\xi)}{\beta_2(\xi)-r\beta_1(\xi)\,\alpha(\rho-\xi)}\right\}. \tag{5.195}$$

Noting that

$$|\beta_2(\xi)| \geqq |r|\,|\beta_1(\xi)|, \qquad \mathrm{Re}\,\xi = 0,$$

it is easily verified by Rouché's theorem that

$$\beta_2(\xi) - r\beta_1(\xi),$$

has exactly n zeros $\eta_i(r)$, $i=1, \dots, n$, with $\mathrm{Re}\,\eta_i(r)<0$ for $|r|<1$. Again by Rouché's theorem it can be shown that

$$\beta_2(\xi) - r\beta_1(\xi)\,\alpha(\rho-\xi),$$

has exactly n zeros $\xi_i(r, \rho)$, $i=1, \dots, n$, with $\mathrm{Re}\,\xi_i(r,\rho)<0$ for $|r|\leqq 1$, $\mathrm{Re}\,\rho>0$ or $|r|<1$, $\mathrm{Re}\,\rho\geqq 0$, or $|r|\leqq 1$, $\mathrm{Re}\,\rho \geqq 0$, $a<1$. We have

$$\frac{1}{2\pi \mathrm{i}} \int_{-\mathrm{i}\infty-0}^{\mathrm{i}\infty-0} \frac{\mathrm{d}\xi}{\rho-\xi} \log \frac{\beta_2(\xi)-r\beta_1(\xi)}{\beta_2(\xi)-r\beta_1(\xi)\,\alpha(\rho-\xi)}$$

$$= \frac{1}{2\pi \mathrm{i}} \int_{-\mathrm{i}\infty-0}^{\mathrm{i}\infty-0} \left\{\frac{(\mathrm{d}/\mathrm{d}\xi)\{\beta_2(\xi)-r\beta_1(\xi)\}}{\beta_2(\xi)-r\beta_1(\xi)} - \frac{(\mathrm{d}/\mathrm{d}\xi)\{\beta_2(\xi)-r\beta_1(\xi)\,\alpha(\rho-\xi)\}}{\beta_2(\xi)-r\beta_1(\xi)\,\alpha(\rho-\xi)}\right\} \cdot \log(\rho-\xi)\,\mathrm{d}\xi.$$

The expression between the brackets in the last integral behaves as $\mathrm{O}(|\xi|^{-2})$ for $|\xi|\to\infty$, $\frac{1}{2}\pi<\arg\xi<1\frac{1}{2}\pi$, except the term with $(\mathrm{d}/\mathrm{d}\xi)\,\alpha(\rho-\xi)$. However, for $|\xi|\to\infty$, $\frac{1}{2}\pi<\arg\xi<1\frac{1}{2}\pi$,

$$\left\{\frac{\mathrm{d}}{\mathrm{d}\xi}\,\alpha(\rho-\xi)\right\} \log(\rho-\xi) = \int_0^\infty t\,\mathrm{e}^{-t(\rho-\xi)} \log(\rho-\xi)\,\mathrm{d}A(t) \to 0.$$

It follows immediately from (5.195) for $|r|\leqq 1$, $\mathrm{Re}\,\rho>0$, or $|r|<1$, $\mathrm{Re}\,\rho\geqq 0$ that

$$\mathrm{E}\{r^n \mathrm{e}^{-\rho p}\} = 1 - \prod_{i=1}^{n} \frac{\rho-\eta_i(r)}{\rho-\xi_i(r,\rho)}$$

$$= 1 - \{\beta_2(\rho)-r\beta_1(\rho)\} \prod_{i=1}^{n} \{\rho-\xi_i(r,\rho)\}^{-1}. \tag{5.196}$$

Hence, since $\beta_2(0)=\beta_1(0)$,

$$\mathrm{E}\{r^{\boldsymbol{n}}\} = 1 - \beta_2(0)(1-r) \prod_{i=1}^{n} \{-\xi_i(r, 0)\}^{-1}, \qquad |r|<1, \tag{5.197}$$

$$\mathrm{E}\{\mathrm{e}^{-\rho \boldsymbol{p}}\} = 1 - \{\beta_2(\rho)-\beta_1(\rho)\} \prod_{i=1}^{n} \{\rho-\xi_i(1, \rho)\}^{-1}, \qquad \mathrm{Re}\, \rho>0.$$

The distributions of $\boldsymbol{n}$ and $\boldsymbol{p}$ are proper probability distributions if and only if $a \leqq 1$ (see below (5.72)). This follows also from (5.197) by taking the limits for $r \uparrow 1$ and $\rho \downarrow 0$, respectively.

From (5.135) we have for the joint distribution of $\boldsymbol{n}$ and of the busy cycle $\boldsymbol{c}$ with $\mathrm{Re}\, \rho > \mathrm{Re}\, \xi$,

$$\begin{aligned}
\mathrm{E}\{r^{\boldsymbol{n}}\, \mathrm{e}^{-\rho \boldsymbol{c}}\} &= 1 - \exp\left\{\frac{1}{2\pi \mathrm{i}} \int_{-\mathrm{i}\infty+0}^{\mathrm{i}\infty+0} \log\{1-r\beta(\rho-\xi)\, \alpha(\xi)\} \frac{\mathrm{d}\xi}{\rho-\xi}\right\} \\
&= 1 - \exp\left\{\frac{1}{2\pi \mathrm{i}} \int_{-\mathrm{i}\infty+\rho-0}^{\mathrm{i}\infty+\rho-0} \log\{1-r\beta(\eta)\, \alpha(\rho-\eta)\} \frac{\mathrm{d}\eta}{\eta}\right\} \\
&= 1 - (1-r\alpha(\rho)) \exp\left\{\frac{1}{2\pi \mathrm{i}} \int_{-\mathrm{i}\infty-0}^{\mathrm{i}\infty-0} \log\{1-r\beta(\eta)\, \alpha(\rho-\eta)\} \frac{\mathrm{d}\eta}{\eta}\right\} \\
&= 1 - \beta_2(0)\{1-r\alpha(\rho)\} \prod_{i=1}^{n} \{-\xi_i(r, \rho)\}^{-1}, \quad |r| \leqq 1,\ \mathrm{Re}\, \rho>0.
\end{aligned} \tag{5.198}$$

To obtain expressions for the distribution functions of both types of queue-length we have to substitute the relation (5.190) in the expressions (5.139) and (5.140). Explicit expressions may then be easily obtained by evaluating the integrals by contour integration.

Finally, we consider the distribution function of the interdeparture time $\boldsymbol{r}_{n+1}-\boldsymbol{r}_n$. From (5.141) we obtain for $\mathrm{Re}\, \rho_1 \geqq 0$, $\mathrm{Re}\, \rho_2>0$, $n=1, 2, \ldots$,

$$\begin{aligned}
&\mathrm{E}\{\exp\{-\rho_1 \boldsymbol{w}_{n+1}-\rho_2(\boldsymbol{r}_{n+1}-\boldsymbol{r}_n)\} \mid \boldsymbol{w}_1=0\} \\
&= \frac{1}{2\pi \mathrm{i}}\, \beta(\rho_2) \int_{-\mathrm{i}\infty-0}^{\mathrm{i}\infty-0} \left\{\frac{1}{\rho_1-\xi} + \frac{1}{\rho_2+\xi}\right\} \beta(\xi)\, \alpha(-\xi)\, \mathrm{E}\{\mathrm{e}^{-\xi \boldsymbol{w}_n} \mid \boldsymbol{w}_1=0\}\, \mathrm{d}\xi;
\end{aligned}$$

hence for Re $\rho \geqq 0$; $n=1, 2, \ldots,$

$$\mathrm{E}\{\exp\{-\rho(\boldsymbol{r}_{n+1}-\boldsymbol{r}_n)\} \mid \boldsymbol{w}_1=0\}$$

$$= \frac{1}{2\pi\mathrm{i}} \beta(\rho) \int_{-\mathrm{i}\infty-0}^{\mathrm{i}\infty-0} \left\{\frac{1}{\rho+\xi} - \frac{1}{\xi}\right\} \beta(\xi)\,\alpha(-\xi)\,\mathrm{E}\{\mathrm{e}^{-\xi \boldsymbol{w}_n} \mid \boldsymbol{w}_1=0\}\,\mathrm{d}\xi.$$

Since the Laplace-Stieltjes transform of the distribution of $\boldsymbol{w}_n$ has a limit for $n\to\infty$ if $a<1$, it is easily shown that for $a<1$, Re $\rho>0$,

$$\lim_{n\to\infty} \mathrm{E}\{\exp\{-\rho(\boldsymbol{r}_{n+1}-\boldsymbol{r}_n)\} \mid \boldsymbol{w}_1=0\}$$

$$= \frac{1}{2\pi\mathrm{i}} \beta(\rho) \int_{-\mathrm{i}\infty-0}^{\mathrm{i}\infty-0} \left\{\frac{1}{\rho+\xi} - \frac{1}{\xi}\right\} \beta(\xi)\,\alpha(-\xi) \lim_{\to\infty} \mathrm{E}\{\mathrm{e}^{-\xi \boldsymbol{w}_n} \mid \boldsymbol{w}_1=0\}\,\mathrm{d}\xi.$$

Hence, from (5.190) for $a<1$, Re $\rho \geqq 0$,

$$\lim_{n\to\infty} \mathrm{E}\{\exp\{-\rho(\boldsymbol{r}_{n+1}-\boldsymbol{r}_n)\} \mid \boldsymbol{w}_1=0\}$$

$$= -\frac{\beta(\rho)}{2\pi\mathrm{i}} \int_{-\mathrm{i}\infty-0}^{\mathrm{i}\infty-0} \left\{\frac{1}{\rho+\xi} - \frac{1}{\xi}\right\}$$

$$\cdot \{1-\beta(\xi)\,\alpha(-\xi)\}\, W(0+)\,\beta_2(\xi) \prod_{i=1}^{n} (\xi-\xi_i(1))^{-1}\,\mathrm{d}\xi$$

$$+ \frac{\beta(\rho)}{2\pi\mathrm{i}} \int_{-\mathrm{i}\infty-0}^{\mathrm{i}\infty-0} \left\{\frac{1}{\rho+\xi} - \frac{1}{\xi}\right\} \frac{\beta_2(\xi)}{\beta_2(0)} \prod_{i=1}^{n} \frac{\xi_i(1)}{\xi_i(1)-\xi}\,\mathrm{d}\xi$$

$$= \beta(\rho)[1-\{1-\beta(-\rho)\,\alpha(\rho)\}\, W(0+)\,\beta_2(-\rho)(-1)^n \prod_{i=1}^{n} (\rho+\xi_i(1))^{-1}]$$

$$= W(0+)\,\beta(\rho)[\beta_1(-\rho)\,\alpha(\rho) \prod_{i=1}^{n} \{\rho+\xi_i(1)\}^{-1}$$

$$+ \sum_{i=1}^{n} \frac{\rho}{\rho+\xi_i(1)} \frac{\beta_2(\xi_i(1))}{\xi_i(1)} \prod_{\substack{j=1\\ j\neq i}}^{n} \{\xi_i(1)-\xi_j(1)\}^{-1}].$$

Consequently, for the queueing system $G/K_n/1$, the distribution function for the interdeparture time does not in general have a rational Laplace-Stieltjes transform. It has such a transform if $\alpha(\rho)$ is a rational function of ρ.

II.5.11. The queueing system $K_m/G/1$

For the queueing system $K_m/G/1$ the Laplace-Stieltjes transform $\alpha(\xi)$ of the interarrival time distribution is a rational function with m the degree of the denominator. We write

$$\alpha(\xi)=\alpha_1(\xi)/\alpha_2(\xi), \tag{5.199}$$

with $\alpha_2(\xi)$ a polynomial of degree m and $\alpha_1(\xi)$ a polynomial of degree $m-1$ at most, since $A(0+)=0$. It is further assumed that the coefficient of ξ^m in $\alpha_2(\xi)$ is one. We start again from (5.29) with $\rho_3=0$, $\rho_4=0$, $\operatorname{Re}\rho_1 > \operatorname{Re}\xi>0$, $\operatorname{Re}\rho_2\leqq 0$, and have for $|r|<1$,

$$\sum_{n=1}^{\infty} r^n \operatorname{E}\{\exp(-\rho_1 \boldsymbol{w}_n+\rho_2 \boldsymbol{d}_n) \mid \boldsymbol{w}_1=0, \boldsymbol{d}_1=0\}$$
$$= r\exp\left\{-\frac{1}{2\pi i}\int_{-i\infty+0}^{i\infty+0}\left\{\frac{1}{\rho_1-\xi}+\frac{1}{\xi-\rho_2}\right\}\log\{1-r\alpha(-\xi)\,\beta(\xi)\}\,d\xi\right\}.$$

By applying Rouché's theorem it is seen that

$$\alpha_2(-\xi)-r\beta(\xi)\,\alpha_1(-\xi),$$

has exactly m zeros $\delta_i(r)$, $i=1, \dots, m$, with $\operatorname{Re}\delta_i(r)>0$ if $|r|<1$. By evaluating the above integral it is found that for $|r|<1$, $\operatorname{Re}\rho_1>0$, $\operatorname{Re}\rho_2<0$,

$$\sum_{n=1}^{\infty} r^n \operatorname{E}\{\exp(-\rho_1 \boldsymbol{w}_n+\rho_2 \boldsymbol{d}_n) \mid \boldsymbol{w}_1=0, \boldsymbol{d}_1=0\}$$
$$=\frac{r\alpha_2(-\rho_2)}{\alpha_2(-\rho_1)-r\beta(\rho_1)\,\alpha_1(-\rho_1)}\prod_{i=1}^{m}\frac{\rho_1-\delta_i(r)}{\rho_2-\delta_i(r)}. \tag{5.200}$$

Hence, for $\operatorname{Re}\rho\geqq 0$, $|r|<1$,

$$\sum_{n=1}^{\infty} r^n \operatorname{E}\{e^{-\rho \boldsymbol{w}_n} \mid \boldsymbol{w}_1=0\} = \frac{r\alpha_2(0)}{\alpha_2(-\rho)-r\beta(\rho)\,\alpha_1(-\rho)}\prod_{i=1}^{m}\frac{\delta_i(r)-\rho}{\delta_i(r)}, \tag{5.201}$$

$$\sum_{n=1}^{\infty} r^n \operatorname{E}\{e^{-\rho \boldsymbol{d}_n} \mid \boldsymbol{d}_1=0\} = \frac{r\alpha_2(\rho)}{\alpha_2(0)(1-r)}\prod_{i=1}^{m}\frac{\delta_i(r)}{\rho+\delta_i(r)}. \tag{5.202}$$

It should be noted that the relations (5.200), ..., (5.202) can be obtained immediately from (5.189), ..., (5.191) by using the duality property (cf. end of section 5.3).

To obtain the limit distribution $W(t)$ in case $a<1$, it is noted that the zeros $\delta_i(r)$ are continuous functions of r for $|r|\leqq 1$ and that one of the zeros $\delta_i(r)$ tends to zero if $r\uparrow 1$. Denote this zero by $\delta_1(r)$. Then from

$$\alpha_2(-\xi)-r\alpha_1(-\xi)\beta(\xi)=\alpha_2(0)-\xi\alpha_2'(0)-r\{\alpha_1(0)-\xi\alpha_1'(0)\}(1-\beta\xi)+o(|\xi|),$$

for $|\xi|\to 0$, Re $\xi>0$, and

$$\alpha=\frac{\alpha_2'(0)-\alpha_1'(0)}{\alpha_2(0)},\qquad \alpha_2(0)=\alpha_1(0),$$

it follows that

$$\delta_1(r)=\frac{1-r}{r(\alpha-\beta)}+o(|1-r|)\qquad\text{for}\quad r\to 1. \tag{5.203}$$

Hence, we obtain from (5.201) that for Re $\rho\geqq 0$, $a<1$,

$$\lim_{n\to\infty}\mathrm{E}\{\mathrm{e}^{-\rho w_n}\mid w_1=0\}=\int_{0-}^{\infty}\mathrm{e}^{-\rho t}\,\mathrm{d}W(t)$$

$$=\frac{-\alpha_2(0)\,\alpha\rho(1-a)}{\alpha_2(-\rho)-\beta(\rho)\,\alpha_1(-\rho)}\prod_{i=2}^{m}\frac{\delta_i(1)-\rho}{\delta_i(1)}, \tag{5.204}$$

if $m=1$ the empty product should be replaced by 1. It follows for $a<1$ that

$$\lim_{n\to\infty}\mathrm{E}\{w_n\mid w_1=0\}=\frac{a}{2(1-a)\,\beta}\left\{\beta_2+\alpha_2+2\beta\frac{\alpha_1'(0)}{\alpha_1(0)}-2\alpha\frac{\alpha_2'(0)}{\alpha_2(0)}\right\}$$

$$+\sum_{i=2}^{m}\frac{1}{\delta_i(1)}, \tag{5.205}$$

$$\lim_{n\to\infty}\Pr\{w_n=0\mid w_1=0\}=(1-a)\,\alpha\alpha_2(0)\prod_{i=2}^{m}\frac{1}{\delta_i(1)},$$

here β_2 and α_2 are the second moments of the service time and interarrival time distributions. The relation (5.192) is also valid here. From (5.204) and (5.192) we obtain

$$\mathrm{E}\{\mathrm{e}^{-\rho i}\}=1-\frac{\rho}{\alpha_2(\rho)}\prod_{i=2}^{m}\{\delta_i(1)+\rho\},\qquad \text{Re}\,\rho\geqq 0,\quad a<1. \tag{5.206}$$

It follows that the distribution function of the idle period has a rational Laplace-Stieltjes transform if the interarrival time distribution has a rational transform; moreover they are of the same degree and have the same poles.

The converse statement is also true. This is proved by starting from (5.192), and the proof is similar to the proof of the analogous statement in section 5.10 (see below (5.193)).

To obtain the transform of the joint distribution of the busy period $\boldsymbol{p}$ and the number $\boldsymbol{n}$ of customers served during a busy period we start from (5.134) (note that $A(t)$ is continuous),

$$\mathrm{E}\{r^{\boldsymbol{n}}\,\mathrm{e}^{-\rho\boldsymbol{p}}\}$$

$$= 1 - \exp\left\{\frac{1}{2\pi\mathrm{i}} \int_{-\mathrm{i}\infty+0}^{\mathrm{i}\infty+0} \log \frac{1-r\beta(\xi)}{1-r\beta(\xi)\,\alpha(\rho-\xi)} \frac{\mathrm{d}\xi}{\rho-\xi}\right\}, \qquad |r|<1, \quad \mathrm{Re}\,\rho>0.$$

Denoting by ε_i, $i=1, \dots, m$, the zeros of $\alpha_2(\xi)$ and by $\delta_i(r, \rho)$ the zeros of

$$\alpha_2(\rho-\xi)-r\beta(\xi)\,\alpha_1(\rho-\xi), \qquad \mathrm{Re}\,\xi>0,$$

then it follows by Rouché's theorem that there are exactly m of these zeros if $|r|\leqq 1$, $\mathrm{Re}\,\rho>0$, or $|r|<1$, $\mathrm{Re}\,\rho\geqq 0$, or $|r|\leqq 1$, $\mathrm{Re}\,\rho\geqq 0$, $a<1$. The integral can be evaluated by contour integration and it follows for $|r|<1$, $\mathrm{Re}\,\rho>0$,

$$\mathrm{E}\{r^{\boldsymbol{n}}\,\mathrm{e}^{-\rho\boldsymbol{p}}\} = 1 - \prod_{i=1}^{m} \frac{\rho-\delta_i(r, \rho)}{\varepsilon_i} = 1 - \frac{(-1)^m}{\alpha_2(0)} \prod_{i=1}^{m} (\rho-\delta_i(r, \rho)). \tag{5.207}$$

For the joint distribution of $\boldsymbol{n}$ and of the busy cycle $\boldsymbol{c}$ it is found from (5.135) for $|r|<1$, $\mathrm{Re}\,\rho>0$,

$$\mathrm{E}\{r^{\boldsymbol{n}}\,\mathrm{e}^{-\rho\boldsymbol{c}}\} = 1 - \exp\left\{\frac{1}{2\pi\mathrm{i}} \int_{-\mathrm{i}\infty-0}^{\mathrm{i}\infty-0} \log\{1-r\beta(\rho-\xi)\,\alpha(\xi)\} \frac{\mathrm{d}\xi}{\rho-\xi}\right\}$$

$$= 1 - \frac{1}{\alpha_2(\rho)} \prod_{i=1}^{m} \delta_i(r, \rho). \tag{5.208}$$

II.6. SOME SPECIAL METHODS

II.6.1. Introduction

In the four preceding chapters we have discussed some basic techniques for investigating the single server queue as described in chapter 1. These methods play an important role in the investigation of queueing problems. In this chapter we shall discuss some approaches to queueing problems which are often very effective in dealing with special questions. For the more general questions, however, they are usually too restricted.

II.6.2. The method of the supplementary variable for the $M/G/1$ queue

In section I.6.4 regenerative processes have been discussed. It appeared that the imbedded process defined at the moments of renewals could be studied easily. To investigate the process at those time points, which are not regeneration points, it turned out to be advantageous to consider the pair of variables $\boldsymbol{x}_t$ and $\boldsymbol{\eta}_t$, i.e. the variable of the process itself and the past renewal time, particularly if the regenerative process is Markovian (cf. theorem I.6.7). Here we have a typical example of the method of inclusion of a supplementary variable. In general this method may be used if the process to be studied has an imbedded process which is a discrete time parameter Markov process. Once the behaviour of the imbedded process has been investigated the $\boldsymbol{x}_t$-process itself can often be successfully investigated by incorporating a second variable $\boldsymbol{\eta}_t$ which represents at time t the time between t and the preceding regeneration point. In many cases the process $\{(\boldsymbol{x}_t, \boldsymbol{\eta}_t), t \in [0, \infty)\}$ is then a vector valued Markov process.

Since the application of the present method to the queueing process $G/M/1$ is obvious because of the results of section I.6.4 we shall demonstrate the method for the queueing system $M/G/1$.

For the system $M/G/1$ we shall denote by $\boldsymbol{x}_t$ the number of customers in the system at time t, while $\boldsymbol{\eta}_t$ will be the past service time of the customer being served at time t if there is a customer in the system. Define for $t>0$,

$$R_0(t) \stackrel{\text{def}}{=} \Pr\{\boldsymbol{x}_t=0 \mid \boldsymbol{x}_0=0\}, \tag{6.1}$$

$$R_j(\eta, t)\,\mathrm{d}\eta \stackrel{\text{def}}{=} \Pr\{\boldsymbol{x}_t=j,\ \eta \leqq \boldsymbol{\eta}_t<\eta+\mathrm{d}\eta \mid \boldsymbol{x}_0=0\}, \quad \eta>0,\ j=1, 2, \ldots .$$

We derive a set of relations by considering the process $\{(\boldsymbol{x}_t, \boldsymbol{\eta}_t),\ t\in[0, \infty)\}$ during a small time interval $(t, t+\Delta t)$.

For $\Delta t>0$ and sufficiently small the event $\{\boldsymbol{x}_{t+\Delta t}=0\}$ is preceded by the event $\{\boldsymbol{x}_t=0\}$ or the event $\{\boldsymbol{x}_t=1\}$. The arrival process is a Poisson process and for $B(\tau)<1$,

$$\frac{B(\sigma+\tau)-B(\tau)}{1-B(\tau)}$$

is the conditional probability that a service time has a duration less than $\sigma+\tau$, whenever it has a duration of at least τ. It follows, for those t for which $B(t)<1$, with $0<\Delta\eta<\Delta t$, that

$$R_0(t+\Delta t) = \left\{1 - \frac{1}{\alpha}\,\Delta t + \mathrm{o}(\Delta t)\right\} R_0(t)$$
$$+ \int_{\eta=\Delta\eta}^{t} R_1(\eta-\Delta\eta, t)\,\mathrm{d}\eta\,\frac{B(\eta-\Delta\eta+\Delta t)-B(\eta-\Delta\eta)}{1-B(\eta-\Delta\eta)} + \mathrm{o}(\Delta t), \qquad t>0. \tag{6.2}$$

If $\boldsymbol{\eta}_t>0$, Δt can be chosen such that $\Delta t<\boldsymbol{\eta}_t$ and hence for $t>0$, $\eta>0$,

$$R_1(\eta, t)\,\mathrm{d}\eta$$
$$= R_1(\eta-\Delta t, t-\Delta t)\,\mathrm{d}\eta\left\{1 - \frac{1}{\alpha}\Delta t + \mathrm{o}(\Delta t)\right\}\frac{1-B(\eta)}{1-B(\eta-\Delta t)} + \mathrm{o}(\Delta t), \tag{6.3}$$

$$R_j(\eta, t)\,\mathrm{d}\eta = R_j(\eta-\Delta t, t-\Delta t)\,\mathrm{d}\eta\left\{1 - \frac{1}{\alpha}\Delta t + \mathrm{o}(\Delta t)\right\}\frac{1-B(\eta)}{1-B(\eta-\Delta t)}$$
$$+ \frac{1}{\alpha}\Delta t R_{j-1}(\eta-\Delta t, t-\Delta t)\,\mathrm{d}\eta + \mathrm{o}(\Delta t), \qquad j=2, 3, \ldots . \tag{6.4}$$

Consider now the situation in which $\boldsymbol{\eta}_t=0+$, i.e. just before t a new service has been started; it is then impossible to choose Δt so small that $\Delta t<\boldsymbol{\eta}_t$. Let θ be a number such that $0<\theta<1$.

Then for $t>0, j=2, 3, \ldots,$

$$R_1(\theta\Delta t, t)\,\Delta t = \int_{\eta=\Delta\eta}^{t-\Delta t} R_2(\eta-\Delta\eta, t-\Delta t)\,\mathrm{d}\eta\,\frac{B(\eta-\Delta\eta+\Delta t)-B(\eta-\Delta\eta)}{1-B(\eta-\Delta\eta)} + \frac{1}{\alpha}\,\Delta t R_0(t-\Delta t) + \mathrm{o}(\Delta t), \tag{6.5}$$

$$R_j(\theta\Delta t, t)\,\Delta t = \int_{\eta=\Delta\eta}^{t-\Delta t} R_{j+1}(\eta-\Delta\eta, t-\Delta t)\,\mathrm{d}\eta\,\frac{B(\eta-\Delta\eta+\Delta t)-B(\eta-\Delta\eta)}{1-B(\eta-\Delta\eta)}. \tag{6.6}$$

Next we divide the relations (6.2), ..., (6.6) by Δt and proceed to the limit $\Delta t \to 0$. We shall assume here that the relevant limits exist. Moreover, we should recall the implicit assumption made in the definition (6.1); we tacitly assumed that the joint distribution of $\boldsymbol{x}_t$ and $\boldsymbol{\eta}_t$ possesses a density with respect to the second variable. It may be shown that these assumptions are justified if $B(t)$ is absolutely continuous (cf. also the relations (I.6.48), ..., (I.6.50)). From now on we shall assume that $B(t)$ is absolutely continuous.

It follows from (6.2), ..., (6.6) that for $t>0$, $B(t)<1$,

$$\frac{\mathrm{d}}{\mathrm{d}t} R_0(t) + \alpha^{-1}R_0(t) = \int_{\eta=0}^{t} R_1(\eta, t)\frac{\mathrm{d}B(\eta)}{1-B(\eta)}, \tag{6.7}$$

$$\frac{\partial}{\partial\eta} R_1(\eta, t) + \frac{\partial}{\partial t} R_1(\eta, t) = -\left\{\alpha^{-1} + \frac{B'(\eta)}{1-B(\eta)}\right\} R_1(\eta, t), \qquad \eta>0, \tag{6.8}$$

$$\frac{\partial}{\partial\eta} R_j(\eta, t) + \frac{\partial}{\partial t} R_j(\eta, t) = -\left\{\alpha^{-1} + \frac{B'(\eta)}{1-B(\eta)}\right\} R_j(\eta, t) + \alpha^{-1}R_{j-1}(\eta, t), \qquad \eta>0,\quad j=2, 3, \ldots, \tag{6.9}$$

$$R_1(0+, t) = \int_0^t R_2(\eta, t)\frac{\mathrm{d}B(\eta)}{1-B(\eta)} + \alpha^{-1}R_0(t), \tag{6.10}$$

$$R_j(0+, t) = \int_0^t R_{j+1}(\eta, t)\frac{\mathrm{d}B(\eta)}{1-B(\eta)}, \qquad j=2, 3, \ldots. \tag{6.11}$$

Further from (6.1),

$$R_0(0+)=1, \qquad R_j(\eta, 0+)=0, \qquad \eta>0;\ j=1, 2, \ldots. \tag{6.12}$$

Defining for $t \geqq 0$, $\eta \geqq 0$,

$$Q_0(t) \stackrel{\text{def}}{=} R_0(t), \qquad Q_j(\eta, t) \stackrel{\text{def}}{=} \frac{1}{1-B(\eta)} R_j(\eta, t), \qquad j=1, 2, \ldots, \tag{6.13}$$

we obtain from (6.7), ..., (6.11) for $t>0$, $\eta>0$,

$$\frac{\mathrm{d}}{\mathrm{d}t} Q_0(t) + \alpha^{-1} Q_0(t) = \int_{\eta=0}^{t} Q_1(\eta, t)\,\mathrm{d}B(\eta), \tag{6.14}$$

$$\frac{\partial}{\partial\eta} Q_1(\eta, t) + \frac{\partial}{\partial t} Q_1(\eta, t) = -\alpha^{-1} Q_1(\eta, t), \tag{6.15}$$

$$\frac{\partial}{\partial\eta} Q_j(\eta, t) + \frac{\partial}{\partial t} Q_j(\eta, t) = -\alpha^{-1} Q_j(\eta, t) + \alpha^{-1} Q_{j-1}(\eta, t),$$

$$j=2, 3, \ldots, \tag{6.16}$$

$$Q_1(0+, t) = \int_0^t Q_2(\eta, t)\,\mathrm{d}B(\eta) + \alpha^{-1} Q_0(t), \tag{6.17}$$

$$Q_j(0+, t) = \int_0^t Q_{j+1}(\eta, t)\,\mathrm{d}B(\eta), \qquad j=2, 3, \ldots . \tag{6.18}$$

Introducing the generating function

$$G(p, \eta, t) \stackrel{\text{def}}{=} \sum_{j=1}^{\infty} Q_j(\eta, t)\, p^j, \qquad |p|<1, \quad \eta>0, \quad t>0, \tag{6.19}$$

we obtain from (6.15) and (6.16),

$$\frac{\partial}{\partial\eta} G(p, \eta, t) + \frac{\partial}{\partial t} G(p, \eta, t) = -\alpha^{-1}(1-p)\, G(p, \eta, t). \tag{6.20}$$

For this partial differential equation the characteristic equations are

$$\mathrm{d}\eta = \mathrm{d}t = -\frac{\alpha\, \mathrm{d}G(p, \eta, t)}{(1-p)\, G(p, \eta, t)}.$$

It follows that

$$G(p, \eta, t) = c_2 \exp\{-(1-p)\,\eta/\alpha\}, \qquad \eta = t + c_1,$$

c_1 and c_2 being independent of η and t. Hence, the general solution of (6.20) reads

$$G(p, \eta, t) = \exp\{-(1-p)\,\eta/\alpha\}\, F(t-\eta, p), \tag{6.21}$$

where $F(.,.)$ is a function to be determined by (6.12), (6.14), (6.17) and (6.18). From (6.12) it follows that

$$F(t-\eta, p)=0 \qquad \text{for} \quad \eta \geqq t.$$

From (6.17) and (6.18) we obtain

$$G(p, 0+, t) = \alpha^{-1} p Q_0(t) + \frac{1}{p} \int_0^t G(p, \eta, t) \, \mathrm{d}B(\eta) - \int_0^t Q_1(\eta, t) \, \mathrm{d}B(\eta),$$

so that by using (6.14) and (6.21) we have

$$F(t, p) = -\frac{\mathrm{d}}{\mathrm{d}t} Q_0(t) - \alpha^{-1}(1-p) \, Q_0(t) + \frac{1}{p} \int_0^t \exp\{-(1-p)\,\eta/\alpha\} \, F(t-\eta, p) \, \mathrm{d}B(\eta).$$

Noting that $Q_0(0+)=1$ we deduce the equation

$$\int_0^\infty \mathrm{e}^{-\rho t} F(t, p) \, \mathrm{d}t = \frac{1-\{\rho+(1-p)/\alpha\} \int_0^\infty \mathrm{e}^{-\rho t} Q_0(t) \, \mathrm{d}t}{1-\beta\{\rho+(1-p)/\alpha\}/p}, \qquad \operatorname{Re} \rho>0. \tag{6.22}$$

The left-hand side should be an analytic function of p for $\operatorname{Re} \rho>0$, $|p|<1$. From Takács' lemma (cf. app. 6) it follows that for $\operatorname{Re} \rho>0$,

$$p - \beta\{\rho + (1-p)/\alpha\}$$

has exactly one zero inside the unit circle $|p|=1$. Denoting this root by $\mu(\rho, 1)$ (cf. (4.22)) we deduce from (6.22) (cf. (4.49)), that

$$\int_0^\infty \mathrm{e}^{-\rho t} Q_0(t) \, \mathrm{d}t = \frac{\alpha}{1+\alpha\rho-\mu(\rho, 1)}, \qquad \operatorname{Re} \rho>0. \tag{6.23}$$

From (6.21), ..., (6.23) we now obtain

$$\int_0^\infty \mathrm{e}^{-\rho t} G(p, \eta, t) \, \mathrm{d}t = \frac{\{p-\mu(\rho, 1)\} \exp[-\eta\{\rho+(1-p)/\alpha\}]}{\{1-\beta\{\rho+(1-p)/\alpha\}/p\}\{1+\alpha\rho-\mu(\rho, 1)\}}, \qquad \operatorname{Re} \rho>0. \tag{6.24}$$

From this relation and from (6.13) it is not difficult to obtain the relations (4.49) for $k=0$. We may now proceed as in chapter 4 since we can determine from (6.24) the Laplace transform with respect to t and the generating function with respect to j of $R_j(\eta, t)$. Note that

$$\Pr\{\boldsymbol{x}_t=j \mid \boldsymbol{x}_0=0\} = \int_0^t R_j(\eta, t)\,\mathrm{d}\eta.$$

Further, we have (cf. (4.85)),

$$\Pr\{\boldsymbol{x}_t=j, \boldsymbol{\xi}_t<\zeta \mid \boldsymbol{x}_0=0\} = \int_0^t R_j(\eta, t)\,\frac{B(\zeta+\eta)-B(\eta)}{1-B(\eta)}\,\mathrm{d}\eta, \qquad j=1,2,\ldots;\ t>0,$$

and from this relation we find as in chapter 4 the distribution function of the virtual delay time $\boldsymbol{v}_t$.

We shall not continue the analysis since the results can be found in chapter 4 and our main object was to illustrate the method of supplementary variables.

II.6.3. Lindley's integral equation

LINDLEY [1952] derived for the queueing system $G/G/1$ an integral equation for the limit distribution of the actual waiting time for $a<1$. We shall shortly discuss this integral equation. An extensive study of this integral equation has been made by SPITZER [1956, 1957, 1960], see also LOYNES [1962].

From relation (1.1) it follows that

$$\begin{aligned}\Pr\{\boldsymbol{w}_{n+1}<t \mid \boldsymbol{w}_1=0\} &= \Pr\{\max(0, \boldsymbol{w}_n+\boldsymbol{\tau}_n-\boldsymbol{\sigma}_{n+1})<t \mid \boldsymbol{w}_1=0\} \\ &= \Pr\{\boldsymbol{w}_n+\boldsymbol{\tau}_n-\boldsymbol{\sigma}_{n+1}\leqq 0 \mid \boldsymbol{w}_1=0\} + \Pr\{0<\boldsymbol{w}_n+\boldsymbol{\tau}_n-\boldsymbol{\sigma}_{n+1}<t \mid \boldsymbol{w}_1=0\} \\ &= \int_{\tau=-\infty}^{t} \Pr\{\boldsymbol{w}_n<t-\tau \mid \boldsymbol{w}_1=0\}\,\mathrm{d}C(\tau), \qquad t>0;\quad n=1,2,\ldots, \end{aligned} \tag{6.25}$$

with

$$C(t) = \int_0^\infty B(t+\tau)\,\mathrm{d}A(\tau), \qquad -\infty<t<\infty.$$

From theorem 1.1 it is known that if $a<1$ then $\Pr\{\boldsymbol{w}_n<t \mid \boldsymbol{w}_1=w\}$ converges to a limit distribution $W(t)$ which is independent of w. Hence, if $a<1$ it follows from (6.25) that

$$W(t) = \begin{cases} \int_{\tau=-\infty}^{t} W(t-\tau)\,\mathrm{d}C(\tau), & t>0, \\ 0, & t<0; \end{cases} \tag{6.26}$$

this is Lindley's integral equation, which is in fact a Wiener-Hopf integral equation. Applying the usual Wiener-Hopf technique, we define

$$\text{for}\quad t>0: \qquad W_+(t) \overset{\text{def}}{=} W(t), \quad W_-(t) \overset{\text{def}}{=} 0, \tag{6.27}$$

$$\text{for}\quad t<0: \qquad W_+(t) \overset{\text{def}}{=} 0, \qquad W_-(t) \overset{\text{def}}{=} \int_{-\infty}^{t} W(t-\tau)\,\mathrm{d}C(\tau).$$

It now follows from (6.26) and (6.27) that

$$W_+(t) + W_-(t) = \int_{-\infty}^{\infty} W(t-\tau)\,\mathrm{d}C(\tau), \qquad -\infty<t<\infty. \tag{6.28}$$

Putting

$$\omega_+(\rho) \overset{\text{def}}{=} \int_{0-}^{\infty} \mathrm{e}^{-\rho t}\,\mathrm{d}W_+(t), \qquad \mathrm{Re}\,\rho \geqq 0, \tag{6.29}$$

$$\omega_-(\rho) \overset{\text{def}}{=} \int_{-\infty}^{0} \mathrm{e}^{-\rho t}\,\mathrm{d}W_-(t), \qquad \mathrm{Re}\,\rho \leqq 0,$$

it is easily verified that $\omega_+(\rho)$ and $\omega_-(\rho)$ exist for $\mathrm{Re}\,\rho=0$, and that $\omega_+(\rho)$ is bounded and analytic for $\mathrm{Re}\,\rho>0$, while $\omega_-(\rho)$ is bounded and analytic for $\mathrm{Re}\,\rho<0$. From (6.28) and (6.29) we obtain

$$\omega_+(\rho)+\omega_-(\rho)=\omega_+(\rho)\,\beta(\rho)\,\alpha(-\rho), \qquad \mathrm{Re}\,\rho=0,$$

i.e.

$$\omega_+(\rho)\{1-\beta(\rho)\,\alpha(-\rho)\} = -\omega_-(\rho), \qquad \mathrm{Re}\,\rho=0. \tag{6.30}$$

The problem has now been reduced to the construction of two functions $\omega_+(\rho)$ and $\omega_-(\rho)$ which should be analytic and bounded for $\mathrm{Re}\,\rho>0$ and $\mathrm{Re}\,\rho<0$ respectively, and should satisfy (6.30); moreover, we must have

$$\omega_+(0+)=1, \qquad \lim_{\substack{|\rho|\to\infty \\ \arg\rho=0}} \omega_+(\rho)=W(0+). \tag{6.31}$$

Hence, we have to consider the Wiener-Hopf decomposition of $1-\beta(\rho)\,\alpha(-\rho)$. In the preceding chapters we have performed this decomposition several times (cf. (I.6.73) and (5.58)). Since (cf. (5.56)),

$$B = \sum_{n=1}^{\infty} \frac{1}{n} \Pr\{s_n>0\} < \infty \qquad \text{if}\quad a<1,$$

we have for Re $\rho=0$,

$$\{1-\beta(\rho)\,\alpha(-\rho)\}\exp\left[B+\sum_{n=1}^{\infty}\frac{1}{n}\mathrm{E}\{\mathrm{e}^{-\rho s_n}(s_n\leqq 0)\}\right]$$
$$=\exp\left[\sum_{n=1}^{\infty}\frac{1}{n}\mathrm{E}\{(1-\mathrm{e}^{-\rho s_n})(s_n>0)\}\right], \tag{6.32}$$

so that we again obtain from (6.30), ..., (6.32) by applying Liouville's theorem (cf. (5.50)),

$$\omega_+(\rho)=\exp\left[-\sum_{n=1}^{\infty}\frac{1}{n}\mathrm{E}\{(1-\mathrm{e}^{-\rho s_n})(s_n>0)\}\right],\qquad \mathrm{Re}\,\rho\geqq 0. \tag{6.33}$$

As an example we take the queueing system $E_n/E_m/1$, where E_n is the Erlang distribution (cf. section I.6.5). We have

$$\alpha(\rho)=\frac{n^n}{(n+\alpha\rho)^n},\qquad \beta(\rho)=\frac{m^m}{(m+\beta\rho)^m},\qquad \mathrm{Re}\,\rho\geqq 0.$$

Since it is assumed that $\beta\alpha^{-1}=a<1$ it follows easily from Rouché's theorem that exactly m, say ξ_i for $i=1, ..., m$, of the zeros ξ_i, $i=1, ..., m+n$ of the function

$$\left(1-\frac{\alpha\rho}{n}\right)^n\left(1+\frac{\beta\rho}{m}\right)^m-1,$$

have Re $\xi_i<0$ (cf. SYSKI [1960] p. 310). From (6.30) we now have

$$\omega_+(\rho)(-1)^n\frac{\alpha^n}{n^n}\frac{\beta^m}{m^m}\frac{\prod_{i=1}^{m}(\rho-\xi_i)}{(1+\beta\rho/m)^m}=-\omega_-(\rho)\frac{(1-\alpha\rho/n)^n}{\prod_{i=m+1}^{m+n}(\rho-\xi_i)},\qquad \mathrm{Re}\,\rho=0.$$

The left-hand side of this relation is analytic for Re $\rho>0$, the right-hand side is analytic for Re $\rho<0$; they are the analytic continuations of each other so that these members are constant because of (6.29), (6.31) and Liouville's theorem; therefore

$$\omega_+(\rho)=c(-1)^n\frac{n^n}{\alpha^n}\frac{m^m}{\beta^m}\frac{(1+\beta\rho/m)^m}{\prod_{i=1}^{m}(\rho-\xi_i)},\qquad \mathrm{Re}\,\rho\geqq 0,$$

the constant c follows from (6.31). Hence

$$\omega_+(\rho) = \int_{0-}^{\infty} \mathrm{e}^{-\rho t}\, \mathrm{d}W(t) = \frac{\{1+\beta\rho/m\}^m}{\prod_{i=1}^{m} (1-\rho/\xi_i)}, \qquad \mathrm{Re}\,\rho \geqq 0, \tag{6.34}$$

$$W(0+) = (-1)^m \frac{\beta^m}{m^m} \prod_{i=1}^{m} \xi_i.$$

II.6.4. The method of collective marks

The method of collective marks has been introduced by van Dantzig and may be used to derive generating functions directly by a probabilistic argument. This method often yields very elegant derivations of important results in queueing theory. Till now, however, no results of the method of collective marks are known which cannot be derived by the usual techniques. A detailed account of the method of collective marks has been given by RUNNENBURG [1965] (cf. also the discussion of this paper by Van der Vaart). We shall illustrate the method here by applying it to two simple examples.

As a first example consider an arrival process with negative exponential interarrival time distribution and average α. Every arriving customer is marked with probability $1-x$ and not marked with probability x, $0<x<1$; moreover, customers are marked (or not marked) independently. We are interested in the probability $p(t, x)$ that all arriving customers in $(0, t)$ are unmarked. Clearly,

$$p(t, x) = \sum_{n=0}^{\infty} \frac{(t/\alpha)^n}{n!} \mathrm{e}^{-t/\alpha} x^n = \mathrm{e}^{-(1-x)t/\alpha}.$$

This result shows that $p(t, x)$ is the generating function of the distribution of the variable representing the number of customers who arrived in $(0, t)$.

As a second example we consider the queueing system $M/G/1$ with $a<1$ (cf. chapter 4). By the event {no C_X in $\boldsymbol{w}_n$} we shall denote the event that all customers arriving during the waiting time of the nth arriving customer are unmarked. It follows with $\boldsymbol{\tau}_n$ the service time of the nth arriving customer that

$$\sum_{k=0}^{\infty} \int_0^{\infty} \frac{(t/\alpha)^k}{k!} \mathrm{e}^{-t/\alpha} x^k \,\mathrm{d}_t \Pr\{\boldsymbol{w}_n+\boldsymbol{\tau}_n<t\}$$
$$= \mathrm{E}\{\exp(-(1-x)\boldsymbol{w}_n/\alpha)\}\, \beta\{(1-x)/\alpha\} =$$

$$= \Pr\{\text{no } C_X \text{ in } \boldsymbol{w}_n+\boldsymbol{\tau}_n\}$$

$$= \Pr\{\text{no } C_X \text{ in } \boldsymbol{w}_n+\boldsymbol{\tau}_n \text{ and } (n+1)\text{th arriving customer marked}\}$$

$$+ \Pr\{\text{no } C_X \text{ in } \boldsymbol{w}_n+\boldsymbol{\tau}_n \text{ and } (n+1)\text{th arriving customer unmarked}\}$$

$$= (1-x)\Pr\{\boldsymbol{w}_{n+1}=0\} + x\Pr\{\text{no } C_X \text{ in } \boldsymbol{w}_{n+1}\}$$

$$= (1-x)\,\mathrm{E}\{\exp(-\boldsymbol{w}_{n+1}/\alpha)\}\,\beta(1/\alpha) + x\mathrm{E}\{\exp(-(1-x)\boldsymbol{w}_{n+1}/\alpha)\}, \qquad n=1,2,\dots.$$

Since $a<1$ we have (cf. (4.81)),

$$\omega(\rho) = \lim_{n\to\infty} \mathrm{E}\{\mathrm{e}^{-\rho \boldsymbol{w}_n}\}, \qquad \mathrm{Re}\,\rho\geqq 0.$$

Hence we obtain from the above relation

$$\omega\,\{(1-x)/\alpha\} = (1-x)\,\frac{\omega(1/\alpha)\,\beta(1/\alpha)}{\beta\{(1-x)/\alpha\}-x}, \qquad |x|<1.$$

Since $\omega(0)=1$ it follows from the last relation that

$$1 = \frac{1}{1-a}\,\omega\,(1/\alpha)\,\beta\,(1/\alpha),$$

hence

$$\omega(\rho) = (1-a)\,\frac{\alpha\rho}{\beta(\rho)-1+\alpha\rho}, \qquad \mathrm{Re}\,\rho\geqq 0,$$

a result which agrees with formula (4.81).

For further examples of the method of collective marks we refer the reader to RUNNENBURG's paper [1965].

II.6.5. The phase method and its variants

In section I.6.5 we have discussed the Erlang distribution of type m with parameter α^{-1},

$$F(t) = \begin{cases} 0, & t\leqq 0, \\ \displaystyle\int_0^t \frac{\{m/\alpha\}^m\, u^{m-1}}{(m-1)!}\,\mathrm{e}^{-mu/\alpha}\,\mathrm{d}u, & t>0. \end{cases}$$

Clearly, it is the distribution function of the sum of m independent variables, all having the same negative exponential distribution with first moment α/m. This property enables us to describe the queueing process $E_m/M/1$ by a birth and death process. To show this we introduce for every $n=1, 2, \ldots$, the independent and negative exponentially distributed variables $\boldsymbol{\sigma}_n^{(1)}, \ldots, \boldsymbol{\sigma}_n^{(m)}$, such that

$$\boldsymbol{\sigma}_n = \boldsymbol{\sigma}_n^{(1)} + \ldots + \boldsymbol{\sigma}_n^{(m)}, \qquad n=1, 2, \ldots, \tag{6.35}$$

$$\Pr\{\boldsymbol{\sigma}_n^{(i)} < \sigma\} = \begin{cases} 1 - \mathrm{e}^{-m\sigma/\alpha}, & \sigma>0, \quad i=1, 2, \ldots, m; \quad n=1, 2, \ldots, \\ 0, & \sigma \leqq 0, \end{cases}$$

where $\boldsymbol{\sigma}_n$ denotes the nth interarrival time (cf. section 1.2, we take here $A_1(t) \equiv A(t)$). Further we introduce the stochastic variable $\boldsymbol{\xi}_t$ defined as follows: for $i=1, 2, \ldots, m$,

$$\{\boldsymbol{\xi}_t = i\} = \bigcup_{n=0}^{\infty} \{\boldsymbol{\sigma}_1 + \ldots + \boldsymbol{\sigma}_n + \boldsymbol{\sigma}_{n+1}^{(1)} + \ldots + \boldsymbol{\sigma}_{n+1}^{(i-1)} < t \leqq \boldsymbol{\sigma}_1 + \ldots + \boldsymbol{\sigma}_n + \boldsymbol{\sigma}_{n+1}^{(1)} + \ldots + \boldsymbol{\sigma}_{n+1}^{(i)}\}, \tag{6.36}$$

with $\boldsymbol{\sigma}_n^{(0)} \stackrel{\text{def}}{=} 0$, $n=0, 1, 2, \ldots$;
the arrival process is at time t in the ith phase if $\boldsymbol{\xi}_t = i$. Consider the stochastic process $\{(\boldsymbol{x}_t, \boldsymbol{\xi}_t), t \in [0, \infty)\}$, where $\boldsymbol{x}_t$ is the number of customers in the system at time t. Since the sojourn times of the phases are independent and negative exponentially distributed as well as the service times it follows as in section 2.1 that the process $\{(\boldsymbol{x}_t, \boldsymbol{\xi}_t), t \in [0, \infty)\}$ is a birth and death process. Consequently, we can use the birth and death technique to describe the queueing process $E_m/M/1$. This will be illustrated for the case $m=2$; the general case is similar. Indicating by E_{ij} the event that the arrival process is in phase i and that j customers are in the system it is easily verified that the birth and death coefficients or the elements of the Q-matrix are given below:

$$\begin{aligned} &E_{1j} \to E_{2j}: && 2\alpha^{-1}, && j=0, 1, \ldots, \\ &E_{1j} \to E_{1,j-1}: && \beta^{-1}, && j=1, 2, \ldots, \\ &E_{2j} \to E_{1,j+1}: && 2\alpha^{-1}, && j=0, 1, \ldots, \\ &E_{2j} \to E_{2,j-1}: && \beta^{-1}, && j=1, 2, \ldots . \end{aligned}$$

Taking for the sake of simplicity

$$\boldsymbol{x}_0 = 0, \qquad \boldsymbol{\xi}_0 = 1,$$

so that at $t=0$ there is no customer in the system and the arrival process is in phase 1 then, writing

$$P_{ji}(t) \stackrel{\text{def}}{=} \Pr\{\boldsymbol{x}_t=i,\ \boldsymbol{\xi}_t=j \mid \boldsymbol{x}_0=0,\ \boldsymbol{\xi}_0=1\}, \qquad i=0, 1, \dots;\ j=1, 2,$$

we see that the forward differential equations for the process $\{(\boldsymbol{x}_t, \boldsymbol{\xi}_t), t\in[0, \infty)\}$ read for $t>0$,

$$\frac{\mathrm{d}}{\mathrm{d}t} P_{10}(t) = -\frac{2}{\alpha} P_{10}(t) + \frac{1}{\beta} P_{11}(t),$$

$$\frac{\mathrm{d}}{\mathrm{d}t} P_{20}(t) = -\frac{2}{\alpha} P_{20}(t) + \frac{2}{\alpha} P_{10}(t) + \frac{1}{\beta} P_{21}(t),$$

$$\frac{\mathrm{d}}{\mathrm{d}t} P_{1i}(t) = -\left\{\frac{2}{\alpha} + \frac{1}{\beta}\right\} P_{1i}(t) + \frac{2}{\alpha} P_{2,i-1}(t) + \frac{1}{\beta} P_{1,i+1}(t), \qquad i=1, 2, \dots,$$

$$\frac{\mathrm{d}}{\mathrm{d}t} P_{2i}(t) = -\left\{\frac{2}{\alpha} + \frac{1}{\beta}\right\} P_{2i}(t) + \frac{2}{\alpha} P_{1i}(t) + \frac{1}{\beta} P_{2,i+1}(t), \quad i=1, 2, \dots,$$

with initial conditions

$$P_{1i}(0+) = \begin{cases} 1, & i=0; \\ 0, & i\neq 0; \end{cases} \qquad P_{2i}(0+) = 0, \qquad i=0, 1, \dots .$$

The above system of differential-difference relations can be solved by the technique used in section I.4.4, i.e. by introducing generating functions for the sequences $P_{1i}(t)$, $i=0, 1, \dots$, and $P_{2i}(t)$, $i=0, 1, \dots$, and using a Laplace transform with respect to t. Since the present queueing system is a special case of that studied in chapter 3 we shall not continue the analysis.

We next consider the queueing system $E_m/E_n/1$, so that both the inter-arrival time and the service time have an Erlang distribution. This process can also be described by a birth and death process. To see this we again introduce a stochastic variable $\boldsymbol{\xi}_t$ to describe the phase of the arrival process, and a variable $\boldsymbol{\eta}_t$ to specify the phase of the service process at time t. Since E_n denotes the distribution function of the service time, every service time can be considered as the sum of n independent and identically distributed variables, the distribution function of these variables being negative exponential with parameter β/n. Hence, every service time consists of n phases, with every phase having a negative exponential distributed sojourn time. We now define $\boldsymbol{\eta}_t=0$ if at time t the system is empty, and $\boldsymbol{\eta}_t=i$, $i=1, \dots, n$, if at time t a service is proceeding and the service time of the customer being served at t is in its ith phase. The process $\{(\boldsymbol{x}_t, \boldsymbol{\xi}_t, \boldsymbol{\eta}_t), t\in[0, \infty)\}$

is clearly a birth and death process. We shall not analyse the system $E_m/E_n/1$ since it is a special case of the system $K_m/K_n/1$ which can be handled by the more general techniques of sections 5.10 and 5.11.

The hyperexponential distribution H_m of type m has been defined in section I.6.5 as the distribution function

$$F(t) = \begin{cases} 0, & t \leqq 0, \\ \sum_{k=1}^{m} a_k(1-e^{-\lambda_k t}), & t>0, \end{cases}$$

with

$$\lambda_k > 0, \qquad a_k \geqq 0, \qquad k=1, \dots, m; \qquad \sum_{k=1}^{m} a_k = 1.$$

Define

$$F_k(t) \stackrel{\text{def}}{=} \begin{cases} 0, & t \leqq 0, \\ 1-e^{-\lambda_k t}, & t>0, \end{cases}$$

and consider the following arrival process. With probability a_k the distribution function of the nth interarrival time is the function $F_k(t)$ for every $n=1, 2, \dots$. This arrival process has an interarrival time distribution of type H_m, i.e. the distribution given by $F(t)$ above, if it is assumed that the events "the nth interarrival time distribution is $F_k(t)$", $k=1, 2, \dots, m$, are independent events which are also independent of n and of the state of the queueing system. Noting this construction of the hyperexponential distribution we shall show how to describe the queueing system $H_m/M/1$ by a birth and death process.

Introduce the stochastic variable $\xi_t^{(o)}$, and define $\xi_t^{(o)}=k$, $k=1, \dots, m$, if at time t the interarrival time between the last arriving customer before t and the next arriving customer after t has distribution function $F_k(t)$. The stochastic process $\{(\boldsymbol{x}_t, \xi_t^{(o)}), t \in [0, \infty)\}$ is clearly a birth and death process; here $\boldsymbol{x}_t$ is as usual the number of customers in the system at time t.

We shall consider the situation $H_2/M/1$ more closely; the general case is analogous. Define by $E_{k,n}$ the event with $\boldsymbol{x}_t=n$, $\xi_t^{(o)}=k$, $k=1, 2$; then the list of possible transitions in Δt and of transition densities is as follows:

$$\begin{array}{lll} E_{1,j} \to E_{1,j+1}: & \lambda_1 a_1, & j=0, 1, \dots, \\ E_{1,j} \to E_{2,j+1}: & \lambda_1 a_2, & j=0, 1, \dots, \\ E_{1,j} \to E_{1,j-1}: & \beta^{-1}, & j=1, 2, \dots; \\ E_{2,j} \to E_{1,j+1}: & \lambda_2 a_1, & j=0, 1, \dots, \\ E_{2,j} \to E_{2,j+1}: & \lambda_2 a_2, & j=0, 1, \dots, \\ E_{2,j} \to E_{2,j-1}: & \beta^{-1}, & j=1, 2, \dots . \end{array}$$

Defining

$$P_{1j}(t) \stackrel{\text{def}}{=} \Pr\{\boldsymbol{x}_t=j,\ \xi_t^{(o)}=1 \mid \boldsymbol{x}_0=0,\ \xi_0^{(o)}=1\},$$

$$P_{2j}(t) \stackrel{\text{def}}{=} \Pr\{\boldsymbol{x}_t=j,\ \xi_t^{(o)}=2 \mid \boldsymbol{x}_0=0,\ \xi_0^{(o)}=1\},$$

then the forward differential equations for the process $\{(\boldsymbol{x}_t, \xi_t^{(o)}),\ t\in[0, \infty)\}$ read

$$\frac{\mathrm{d}}{\mathrm{d}t} P_{10}(t) = -\lambda_1 P_{10}(t) + \frac{1}{\beta} P_{11}(t),$$

$$\frac{\mathrm{d}}{\mathrm{d}t} P_{20}(t) = -\lambda_2 P_{20}(t) + \frac{1}{\beta} P_{21}(t),$$

and for $j=1, 2, \ldots,$

$$\frac{\mathrm{d}}{\mathrm{d}t} P_{1j}(t) = -\left(\lambda_1 + \frac{1}{\beta}\right) P_{1j}(t) + \lambda_1 a_1 P_{1,j-1}(t) + \lambda_2 a_1 P_{2,j-1}(t) + \frac{1}{\beta} P_{1,j+1}(t),$$

$$\frac{\mathrm{d}}{\mathrm{d}t} P_{2j}(t) = -\left(\lambda_2 + \frac{1}{\beta}\right) P_{2j}(t) + \lambda_2 a_2 P_{2,j-1}(t) + \lambda_1 a_2 P_{1,j-1}(t) + \frac{1}{\beta} P_{2,j+1}(t),$$

with initial conditions

$$P_{1j}(0+) = \begin{cases} 1, & j=0; \\ 0, & j\neq 0; \end{cases} \qquad P_{2j}(0+) = 0, \qquad j=0, 1, \ldots .$$

The above system can be solved by introducing the generating functions of the sequences $P_{1i}(t)$, $i=0, 1, \ldots,$ and $P_{2i}(t)$, $i=0, 1, \ldots,$ and applying a Laplace transform with respect to t. We shall not continue the analysis.

Next we consider the queueing process $M/H_n/1$ with service time distribution given by

$$B(t) = \begin{cases} \sum_{h=0}^{n} b_h B_h(t), & t>0, \\ 0, & t\leqq 0, \end{cases}$$

$$b_h \geqq 0, \quad \sum_{h=0}^{n} b_h = 1, \quad B_h(t) \stackrel{\text{def}}{=} \begin{cases} 1-\mathrm{e}^{-\mu_h t}, & t>0, \quad \mu_h>0, \quad h=1, \ldots, n, \\ 0, & t\leqq 0. \end{cases}$$

It is easily seen that if we consider a single server queue with Poisson arrival process and a service process such that if a new service starts this service time has distribution function $B_h(t)$ with probability b_h, $h=1, ..., n$, then this process is the queueing process $M/H_n/1$, assuming that the events "a service time has distribution function $B_h(t)$", $h=1, ..., n$, are independent events and also independent of the state of the queueing system. If we define $\eta_t^{(0)}=0$ if the system is empty at time t, and $\eta_t^{(0)}=h$ if a customer is served at time t and his service time distribution is $B_h(.)$, it is easily shown that the process $\{(\boldsymbol{x}_t, \boldsymbol{\eta}_t^{(0)}),\ t\in[0, \infty)\}$ is a birth and death process.

From what has been said above it is now not difficult to see how the more general process $H_m/H_n/1$ can be described by a birth and death process $\{(\boldsymbol{x}_t, \boldsymbol{\xi}_t^{(0)}, \boldsymbol{\eta}_t^{(0)}),\ t\in[0, \infty)\}$; $\boldsymbol{\xi}_t^{(0)}$ is the stochastic variable specifying the parameter of the distribution of the interarrival time covering t, while $\boldsymbol{\eta}_t^{(0)}$ specifies the parameter of the service time distribution.

The same technique can be used to describe the queueing processes $E_m/H_n/1$ and $H_m/E_n/1$ by birth and death processes.

The approach followed above can be used even for the queueing system $K_m/K_n/1$. Here K_m denotes a distribution function of which the Laplace-Stieltjes transform is a rational function with denominator of degree m.

This technique has been used extensively by MORSE [1955]. It is a very powerful technique. Generally, however, it is very laborious, since to describe the process as a birth and death process we need two variables $\boldsymbol{x}_t$, $\boldsymbol{\xi}_t$ or sometimes three $\boldsymbol{x}_t$, $\boldsymbol{\xi}_t$, $\boldsymbol{\eta}_t$ to specify its states. This leads to a rather complicated system of forward differential equations. Although no essentially new difficulties are involved in solving this set of equations, the analysis is intricate. It should be mentioned that the method of phases can be very helpful and effective in simulation studies of queueing processes.

Next we discuss the queueing systems $E_m/G/1$, $H_m/G/1$, $G/E_n/1$ and $G/H_n/1$. Here the special construction of the Erlang distribution and the hyperexponential distribution may also be used to describe these processes by Markov chains.

Let us consider the system $G/E_m/1$. Denote the number of customers in the system at time t by $\boldsymbol{x}_t$, and the phase of the service time at time t if $\boldsymbol{x}_t>0$ by $\boldsymbol{\eta}_t$; i.e. $\boldsymbol{\eta}_t=0$ if the system is empty at time t, while $\boldsymbol{\eta}_t=k$ if, at time t, a service is proceeding and is in its kth phase, $k=1, ..., m$. Suppose, for the sake of simplicity, that $\boldsymbol{x}_0=0$, and that at $t=0$ a customer arrives. Defining

$$\boldsymbol{y}_n = \boldsymbol{x}_{t_n}, \qquad \tilde{\boldsymbol{\eta}}_n = \boldsymbol{\eta}_{t_n}, \qquad n=0, 1, ...; \qquad \boldsymbol{t}_0 \stackrel{\text{def}}{=} 0,$$

it is not difficult to show that the process $\{(y_n, \tilde{\eta}_n),\ n=0, 1, ...\}$ is a discrete time parameter Markov chain with stationary transition probabilities and with state space the product space of the set of non-negative integers and the set of numbers $\{0, 1, ..., m\}$. Clearly the process $\{(y_n, \eta_n),\ n=0, 1, ...\}$ is an imbedded Markov chain of the process $\{(x_t, \tilde{\eta}_t),\ t\in[0, \infty)\}$. The transition probabilities can be found easily and the investigation of this process is similar to that of chapter 3.

It is evident that for the queueing systems $E_m/G/1$, $H_m/G/1$ and $G/H_m/1$, it is also possible to define imbedded Markov chains for the description of the queueing process by introducing stochastic variables specifying the phase of the service time or of the interarrival time. In all these cases the imbedded Markov chain has a state space which is a product space of the set of non-negative integers and a finite set. The analysis is similar to that of chapters 3 and 4.

JACKSON and NICKOLS [1956] have indicated the possibility of describing the queueing process $E_m/M/1$ by a continuous time parameter Markov chain with state space the set of non-negative integers. With x_t the number of customers in the system at time t and ξ_t the phase of the arrival process at t (cf. (6.36)) they introduced the variable

$$\zeta_t \overset{\text{def}}{=} mx_t + \xi_t - 1, \qquad t \geqq 0,$$

so that

$$x_t = \left[\frac{\zeta_t}{m}\right], \qquad \xi_t - 1 = \zeta_t - m\left[\frac{\zeta_t}{m}\right];$$

here $[x]$ is the greatest integer not exceeding x. It is not difficult to prove that the process $\{\zeta_t,\ t\in[0, \infty)\}$ is a continuous time parameter Markov chain with state space the set of non-negative integers. The Q-matrix of this process is easily determined. Putting

$$P_n(t) \overset{\text{def}}{=} \Pr\{\zeta_t = n \mid x_0 = 0,\ \xi_0 = 1\}, \qquad n=1, 2, ...,$$

we see that the forward differential equations for this process are for $t>0$,

$$\frac{\mathrm{d}}{\mathrm{d}t} P_0(t) = -\frac{m}{\alpha} P_0(t) + \frac{1}{\beta} P_m(t),$$

$$\frac{\mathrm{d}}{\mathrm{d}t} P_n(t) = -\frac{m}{\alpha} P_n(t) + \frac{m}{\alpha} P_{n-1}(t) + \frac{1}{\beta} P_{m+n}(t), \qquad n=1, ..., m-1,$$

$$\frac{\mathrm{d}}{\mathrm{d}t} P_n(t) = -\left(\frac{m}{\alpha} + \frac{1}{\beta}\right) P_n(t) + \frac{m}{\alpha} P_{n-1}(t) + \frac{1}{\beta} P_{m+n}(t),$$

$$n=m, m+1, ...,$$

with initial conditions

$$P_n(0+) = \begin{cases} 1, & n=0, \\ 0, & n \neq 0. \end{cases}$$

This system of equations can be solved by introducing a generating function with respect to n, $n=0, 1, \ldots$, and a Laplace transform with respect to t. Note that the probability of k customers in the system at time t is given by

$$\sum_{n=mk}^{(m+1)k-1} P_n(t).$$

The idea of Jackson and Nickols can also be used for describing the queueing processes $E_m/G/1$ and $G/E_m/1$ by an imbedded Markov chain with state space the set of non-negative integers.

Consider first the queueing process $E_m/G/1$. Assume $x_{0+}=0$, denote the number of customers in the system just after the nth departure by z_n, and the phase of the arrival process at this moment by ξ_n. Putting

$$\tilde{\xi}_n \overset{\text{def}}{=} mz_n + \xi_n - 1, \qquad n=0, 1, \ldots,$$

then the process $\{\tilde{\xi}_n, n=1, 2, \ldots\}$ is a discrete time parameter Markov chain with stationary transition probabilities and state space the set of non-negative integers. Its one-step transition matrix is given by

$$P_{ij} \overset{\text{def}}{=} \Pr\{\tilde{\xi}_{n+1}=j \mid \tilde{\xi}_n=i\}, \qquad n=1, 2, \ldots,$$

$$= \begin{cases} \displaystyle\int_0^\infty \frac{(mt/\alpha)^{j-(i-m)}}{\{j-(i-m)\}!} \mathrm{e}^{-mt/\alpha} \, \mathrm{d}B(t), & j \geqq i-m; \quad i \geqq m, \\ \displaystyle\int_0^\infty \frac{(mt/\alpha)^j}{j!} \mathrm{e}^{-mt/\alpha} \, \mathrm{d}B(t), & j \geqq 0; \qquad i=0, 1, \ldots, m-1. \end{cases}$$

The stochastic variable $\tilde{\xi}_n$ has the following interpretation. Consider a single server queue with customers arriving according to a Poisson process with parameter m/α and service in order of arrival, the rth arriving customer has service time distribution $B(t)$ if r is a multiple of m, otherwise his service time is zero with probability 1 (supposing that at $t=0$ the system is empty). Then $\tilde{\xi}_n$ denotes the number of customers in the system just after the departure of a customer with a non-zero service time distribution.

Finally, consider the queueing system $G/E_m/1$. Denote the number of customers in the system just before the nth arrival by y_n, and the phase of the service time distribution just after this arrival by $\tilde{\eta}_n$. Define

$$\tilde{\xi}_n \stackrel{\text{def}}{=} m(y_n-1) + m - (\tilde{\eta}_n-1), \qquad n=1, 2, \ldots,$$

then $\tilde{\xi}_n$ denotes the total number of service time phases which, at the nth arrival, the server has still to handle before he becomes idle, if after the nth arrival no new customer arrives. The process $\{\tilde{\xi}_n, n=1, 2, \ldots\}$ is again a discrete time parameter Markov chain with state space the set of non-negative integers and with stationary transition probabilities. The one-step transition matrix of this chain is given by

$$P_{ij} \stackrel{\text{def}}{=} \Pr\{\tilde{\xi}_{n+1}=j \mid \tilde{\xi}_n=i\}, \qquad n=1, 2, \ldots,$$

$$= \begin{cases} 0, & j>i+m; \quad i=0, 1, \ldots, \\ \displaystyle\int_0^\infty \frac{(mt/\beta)^{i+m-j}}{(i+m-j)!} \mathrm{e}^{-mt/\beta} \,\mathrm{d}A(t), & j=1, \ldots, i+m; \quad i=0, 1, \ldots, \\ \displaystyle\int_0^\infty \left\{ \int_0^t \frac{(m\sigma/\beta)^{i+m-1}}{(i+m-1)!} \mathrm{e}^{-m\sigma/\beta} \frac{\beta}{m} \,\mathrm{d}\sigma \right\} \mathrm{d}A(t), & j=0; \quad i=0, 1, \ldots . \end{cases}$$

II.6.6. The combinatorial method

In a number of papers Takács has developed combinatorial theorems which are quite interesting for queueing theory. A good account of his beautiful results can be found in TAKÁCS [1965 and 1967]. The origin of these theorems is the classical ballot theorem, reading:

If in a ballot candidate A scores a votes and candidate B scores b votes, and $a \geqq \mu b$, μ being a non-negative integer, then the probability P that throughout the counting the number of votes registered for A is always greater than μ times the number of votes registered for B is given by

$$P = \frac{a-\mu b}{a+b}, \tag{6.37}$$

provided that all possible voting results are equally probable.

For $\mu=1$ this theorem has been found by J. Bertrand in 1887; for further notes on the history of the ballot theorem, see TAKÁCS [1965]. We shall first discuss some of Takács' combinatorial theorems and then apply these theorems to the queueing system $M/G/1$.

DEFINITION. The stochastic variables $\boldsymbol{v}_1, \ldots, \boldsymbol{v}_n$ are called *interchangeable* if for every $m=1, 2, \ldots, n$, and every permutation $j_1, \ldots, j_m$ of $1, \ldots, m$ the joint distribution of $\boldsymbol{v}_{j_1}, \ldots, \boldsymbol{v}_{j_m}$ is the same as that of $\boldsymbol{v}_1, \ldots, \boldsymbol{v}_m$. The stochastic variables $\boldsymbol{v}_n$, $n=1, 2, \ldots$, are interchangeable if every finite subset of this sequence is a set of interchangeable variables.

Clearly if $\boldsymbol{v}_1, \ldots, \boldsymbol{v}_n$ are independent, identically distributed variables then they are interchangeable.

A generalization of the classical ballot theorem is

THEOREM 6.1. *If $\boldsymbol{v}_1, \ldots, \boldsymbol{v}_n$ are interchangeable, non-negative, integer valued variables and*

$$\boldsymbol{n}_i \overset{\text{def}}{=} \boldsymbol{v}_1 + \ldots + \boldsymbol{v}_i, \qquad i=1, \ldots, n; \qquad \boldsymbol{n}_0 \overset{\text{def}}{=} 0, \tag{6.38}$$

then for $n=1, 2, \ldots$,

$$\Pr\{\boldsymbol{n}_i<i,\ i=1, \ldots, n \mid \boldsymbol{n}_n=k\} = \begin{cases} 1-\dfrac{k}{n}, & k=0, 1, \ldots, n, \\ 0, & \text{otherwise.} \end{cases} \tag{6.39}$$

Proof. We have

$$\sum_{i=1}^{n} \mathrm{E}\{\boldsymbol{v}_i \mid \boldsymbol{n}_n=k\} = \mathrm{E}\{\boldsymbol{n}_n \mid \boldsymbol{n}_n=k\} = k, \tag{6.40}$$

so that

$$\mathrm{E}\{\boldsymbol{v}_i \mid \boldsymbol{n}_n=k\} = \frac{k}{n}, \tag{6.41}$$

since every term in the left-hand side of (6.40) is independent of i due to the interchangeability of $\boldsymbol{v}_1, \ldots, \boldsymbol{v}_n$. Relation (6.39) is obviously true for $k=1$, $n=1$ and $k=0$, $n=1$; and also for $k=n$. We complete the proof by induction. Suppose (6.39) holds for all k with $0\leqq k<m$ and all $n=1, 2, \ldots, m-1$; $m\geqq 2$. Take $0\leqq j\leqq k$ then from the induction hypothesis, since $\boldsymbol{n}_k=j$, $\boldsymbol{n}_{j+i}<j+i<k$, implies $\boldsymbol{n}_k-\boldsymbol{n}_{j+i}>-i$,

$$\Pr\{\boldsymbol{n}_i<i,\ i=1, \ldots, k \mid \boldsymbol{n}_k=j,\ \boldsymbol{n}_m=k\} = \Pr\{\boldsymbol{n}_i<i,\ i=1, \ldots, j \mid \boldsymbol{n}_k=j\} = 1-\frac{j}{k}.$$

Therefore, using (6.41),

$$\Pr\{\boldsymbol{n}_i<i,\ i=1,\dots,k \mid \boldsymbol{n}_m=k\} = \sum_{j=0}^{k}\left(1-\frac{j}{k}\right)\Pr\{\boldsymbol{n}_k=j \mid \boldsymbol{n}_m=k\}$$

$$= 1-\frac{1}{k}\,\mathrm{E}\{\boldsymbol{n}_k \mid \boldsymbol{n}_m=k\} = 1-\mathrm{E}\{\boldsymbol{\nu}_i \mid \boldsymbol{n}_m=k\} = 1-\frac{k}{m}.$$

Hence, the theorem holds for $n=m$ and $0\leqq k\leqq m$ and the proof is complete.

Note that (6.39) implies for $n=1, 2, \dots$; $h=1, 2, \dots$,

$$\Pr\{\boldsymbol{n}_i<i,\ i=1,\dots,n;\ \boldsymbol{n}_n\leqq h\} = \sum_{j=0}^{h}\left\{1-\frac{j}{n}\right\}\Pr\{\boldsymbol{n}_n=j\}. \tag{6.42}$$

THEOREM 6.2. *Under the same conditions as in the previous theorem*

$$\Pr\{\max_{1\leqq r\leqq n}(\boldsymbol{n}_i-r)<k\} = \Pr\{\boldsymbol{n}_n<n+k\}$$

$$-\sum_{j=1}^{n-1}\sum_{i=0}^{n-j}\left(1-\frac{i}{n-j}\right)\Pr\{\boldsymbol{n}_j=j+k,\ \boldsymbol{n}_n=j+k+i\}, \qquad n=1, 2, \dots, \tag{6.43}$$

for $k=\dots, -1, 0, 1, \dots$; *if* k *is negative then both sides of* (6.43) *are zero.*

Proof. For $h=1, 2, \dots, n-1$; $k=\dots, -1, 0, 1, \dots$, relation (6.43) is a special case of relation (6.44) below (take $h=1$),

$$\Pr\{\boldsymbol{n}_r<r+k,\ r=1,\dots,n;\ \boldsymbol{n}_n\leqq n+k-h\} = \Pr\{\boldsymbol{n}_n\leqq n+k-h\}$$

$$-\sum_{j=1}^{n-h}\sum_{i=0}^{n-h-j}\left(1-\frac{i}{n-j}\right)\Pr\{\boldsymbol{n}_j=j+k,\ \boldsymbol{n}_n=j+k+i\}. \tag{6.44}$$

It clearly suffices to prove that the last term in (6.44) is equal to

$$\Pr\{\boldsymbol{n}_n\leqq n+k-h,\ \boldsymbol{n}_r\geqq r+k \text{ for some } r=1, 2, \dots, n\}$$

$$= \Pr\{\boldsymbol{n}_n\leqq n+k-h,\ \bigcup_{j=0}^{n-h}\{\boldsymbol{n}_j=j+k,\ \boldsymbol{n}_r<r+k,\ r=j+1,\dots,n\}\}. \tag{6.45}$$

We have for $j=1, 2, \dots, n$,

$$\{\boldsymbol{n}_j=j+k, \boldsymbol{n}_r<r+k,\ r=j+1,\dots,n\} \equiv \{\boldsymbol{n}_j=j+k, \boldsymbol{n}_r-\boldsymbol{n}_j<r-j, r=j+1,\dots,n\},$$

and by theorem 6.1 for $i=1, \dots, n-j$,

$$\Pr\{\boldsymbol{n}_i-\boldsymbol{n}_j<i-j,\ i=j+1,\dots,n \mid \boldsymbol{n}_j=j+k,\ \boldsymbol{n}_n=j+k+i\} = 1-\frac{i}{n-j}. \tag{6.46}$$

Noting that for varying j the events in the right-hand side of the last but one relation are disjoint, we see that the probability in (6.45) equals the sum over i and j with $1 \leqq j \leqq j+i \leqq n-h$ of the product of $1-i/(n-j)$ and the probability of the condition in the left-hand side of (6.46).
This proves relation (6.44) and hence the theorem.

Note that

$$\Pr\{\boldsymbol{n}_j=j+k,\, \boldsymbol{n}_n=j+k+i\} = \Pr\{\boldsymbol{n}_j=j+k\}\,\Pr\{\boldsymbol{n}_{n-j}=i\}, \tag{6.47}$$

if $\boldsymbol{v}_1, \ldots, \boldsymbol{v}_n$ are independent and identically distributed variables.

Define for $j=0, 1, \ldots,$

$$p_j \overset{\text{def}}{=} \Pr\{\boldsymbol{v}_i=j\}, \qquad \boldsymbol{m}_j \overset{\text{def}}{=} \inf_{i \geqq 1} \{i\colon \boldsymbol{n}_i \leqq i-j\}, \tag{6.48}$$

assuming that $p_0>0$.

THEOREM 6.3. *If $\boldsymbol{v}_1, \boldsymbol{v}_2, \ldots,$ are independent, identically distributed, non-negative, integer valued variables and $p_0>0$ then for $1 \leqq i \leqq k;\ k=1, 2, \ldots,$*

$$\Pr\{\sup_{1 \leqq r \leqq \boldsymbol{m}_i} (\boldsymbol{n}_r-r)<k-i\} = \frac{w_{k-i}}{w_k}, \tag{6.49}$$

where w_0 is an arbitrary positive number, and w_k, $k=1, 2, \ldots,$ is determined by

$$w_k = \sum_{j=0}^{k} p_j w_{k-j+1}, \qquad k=0, 1, \ldots . \tag{6.50}$$

Proof. Since for $j=0, 1, \ldots,$

$$\boldsymbol{n}_{\boldsymbol{m}_j} \leqq \boldsymbol{m}_j-j \quad \text{and} \quad \boldsymbol{n}_{\boldsymbol{m}_j-1} > \boldsymbol{m}_j-1-j,$$

it follows, if the "<" sign holds in the first inequality, that

$$\boldsymbol{n}_{\boldsymbol{m}_j} < \boldsymbol{m}_j-j < \boldsymbol{n}_{\boldsymbol{m}_j-1}+1,$$

so that

$$\boldsymbol{v}_{\boldsymbol{m}_j}=0, \qquad \boldsymbol{n}_{\boldsymbol{m}_j}=\boldsymbol{n}_{\boldsymbol{m}_j-1},$$

and

$$0<\boldsymbol{m}_j-j-\boldsymbol{n}_{\boldsymbol{m}_j}<1,$$

which is impossible. Hence

$$\boldsymbol{n}_{\boldsymbol{m}_j}=\boldsymbol{m}_j-j. \tag{6.51}$$

Consequently, for $j=0, 1, \ldots,$

$$\boldsymbol{\nu}_{\boldsymbol{m}_j}=\boldsymbol{n}_{\boldsymbol{m}_j}-\boldsymbol{n}_{\boldsymbol{m}_j-1}=\boldsymbol{m}_j-\boldsymbol{m}_{j-1}+1. \tag{6.52}$$

Therefore, since $\boldsymbol{\nu}_1, \boldsymbol{\nu}_2, \ldots,$ are independent, identically distributed variables, the variables $\boldsymbol{m}_{i-1}$ and $\boldsymbol{m}_i-\boldsymbol{m}_{i-1}$ are independent. Moreover, it follows that

$$\boldsymbol{m}_i=\boldsymbol{m}_1+(\boldsymbol{m}_2-\boldsymbol{m}_1)+(\boldsymbol{m}_3-\boldsymbol{m}_2)+\ldots+(\boldsymbol{m}_i-\boldsymbol{m}_{i-1}), \qquad i=2, 3, \ldots, \tag{6.53}$$

the terms of the sum being independent and identically distributed variables. Define for $k=1, 2, \ldots,$

$$w_{ki} \overset{\text{def}}{=} \Pr\{\boldsymbol{n}_j<j+k-i,\ j=1, \ldots, \boldsymbol{m}_i\}, \qquad i=0, 1, \ldots, k. \tag{6.54}$$

We now have

$$\begin{aligned}\bigcap_{j=1}^{\boldsymbol{m}_i} \{\boldsymbol{n}_j<j+k-i\} = \bigcap_{j=1}^{\boldsymbol{m}_1} \{\boldsymbol{n}_j<j+k-i\} \\ \times \bigcap_{j=\boldsymbol{m}_1+1}^{\boldsymbol{m}_2} \{\boldsymbol{n}_j-\boldsymbol{n}_{\boldsymbol{m}_1}<j-\boldsymbol{m}_1+k-i+1,\ \boldsymbol{n}_{\boldsymbol{m}_1}\leqq \boldsymbol{m}_1-1\} \\ \times \bigcap_{j=\boldsymbol{m}_2+1}^{\boldsymbol{m}_3} \{\boldsymbol{n}_j-\boldsymbol{n}_{\boldsymbol{m}_2}<j-\boldsymbol{m}_2+k-i+2,\ \boldsymbol{n}_{\boldsymbol{m}_2}\leqq \boldsymbol{m}_2-2\} \\ \times \cdots\cdots\cdots \\ \times \bigcap_{j=\boldsymbol{m}_{i-1}+1}^{\boldsymbol{m}_i} \{\boldsymbol{n}_j-\boldsymbol{n}_{\boldsymbol{m}_{i-1}}<j-\boldsymbol{m}_{i-1}+k-1, \boldsymbol{n}_{\boldsymbol{m}_{i-1}}\leqq \boldsymbol{m}_{i-1}-(i-1)\},\end{aligned}$$

so that from (6.53) and (6.54),

$$w_{ki} = \prod_{j=k-i+1}^{k} w_{j1}. \tag{6.55}$$

Since $p_0>0$ and hence $w_{ki}>0$ for $1\leqq i\leqq k$, we may define

$$W_k \overset{\text{def}}{=} \frac{W_0}{\prod_{i=1}^{k} w_{i1}}, \qquad k=1, 2, \ldots,$$

with $W_0>0$, but arbitrary. Hence from (6.55),

$$w_{ki} = \frac{W_{k-i}}{W_k}, \qquad i=1, 2, \ldots, k. \tag{6.56}$$

Since for $k=0, 1, \ldots; i=2, 3, \ldots,$

$$w_{k+i,i} = \Pr\{\boldsymbol{n}_r<r+k,\ r=1,\ldots,\boldsymbol{m}_i\}$$
$$= \sum_{j=0}^{k} \Pr\{\boldsymbol{n}_r-\boldsymbol{\nu}_1<r+k-j,\ r=2,\ldots,\inf_{h\geqq 2}\{h:\ \boldsymbol{n}_h-\boldsymbol{\nu}_1\leqq h-i-j\}\mid \boldsymbol{\nu}_1=j\}p_j$$
$$= \sum_{j=0}^{k} \Pr\{\boldsymbol{n}_{r-1}<r+k-j,\ r=2,\ldots,\inf_{h\geqq 2}\{h:\ \boldsymbol{n}_{h-1}\leqq h-i-j\}\}p_j$$
$$= \sum_{j=0}^{k} \Pr\{\boldsymbol{n}_r<r+k-j+1,\ r=1,\ldots,\inf_{h\geqq 1}\{h:\ \boldsymbol{n}_h\leqq h-i-j+1\}\}p_j$$
$$= \sum_{j=0}^{k} \Pr\{\boldsymbol{n}_r<r+k-j+1,\ r=1,\ldots,\boldsymbol{m}_{i+j-1}\}p_j$$
$$= \sum_{j=0}^{\infty} p_j w_{k+i,i+j-1}. \tag{6.57}$$

Hence from (6.56) we obtain (6.50). This proves the theorem.

Defining

$$w(p) \stackrel{\text{def}}{=} \sum_{k=0}^{\infty} w_k p^k, \qquad \pi(p) \stackrel{\text{def}}{=} \sum_{k=0}^{\infty} p_k p^k, \qquad |p|<1,$$

we obtain from (6.50),

$$w(p) = \frac{\pi(p)}{\pi(p)-p}\, w_0, \tag{6.58}$$

for p smaller in absolute value than the smallest positive zero of $\pi(p)-p$.

THEOREM 6.4. *Under the same conditions as in theorem 6.1 we have for* $k=1, 2, \ldots,$

$$\Pr\{\max_{1\leqq r\leqq n} (r-\boldsymbol{n}_r)<k\} = 1 - \sum_{j=k}^{n} \frac{k}{j} \Pr\{\boldsymbol{n}_j=j-k\}. \tag{6.59}$$

Proof. We determine the probability of the complementary event, i.e. of

$$\{\boldsymbol{n}_r\leqq r-k \text{ for some } r=1, \ldots, n\}.$$

This event is the union of the disjoint events: "the smallest r such that $\boldsymbol{n}_r\geqq r-k$ is $r=j$", $j=k, \ldots, n$. Hence from (6.51) the latter event is equivalent with

$$\{\boldsymbol{n}_j=j-k,\ \boldsymbol{n}_r>r-k,\ r=1, \ldots, j-1\},$$

so that

$$\Pr\{\boldsymbol{n}_r \leqq r-k \text{ for some } r=1, \dots, k\}$$

$$= \sum_{j=1}^{n} \Pr\{\boldsymbol{n}_j - \boldsymbol{n}_r < j-r,\ \boldsymbol{n}_j = j-k,\ r=1, \dots, j-1\}. \tag{6.60}$$

From theorem 6.1 we have

$$\Pr\{\boldsymbol{n}_j - \boldsymbol{n}_r < j-r,\ r=1, \dots, j-1 \mid \boldsymbol{n}_j = j-k\} = \frac{k}{j}, \qquad k=0, 1, \dots, j,$$

hence from (6.60),

$$\Pr\{\boldsymbol{n}_r \leqq r-k, \text{ for some } r=1, \dots, n\} = \sum_{j=k}^{n} \frac{k}{j} \Pr\{\boldsymbol{n}_j = j-k\};$$

this proves the theorem.

THEOREM 6.5. *Under the same conditions as in theorem* 6.3,

$$\Pr\{\sup_{r \geqq 1} (\boldsymbol{n}_r - r) < k\} = w_k, \qquad k=0, 1, \dots, \tag{6.61}$$

with

$$w_0 = \begin{cases} 1 - \mathrm{E}\{\boldsymbol{\nu}_r\} & \text{if } \mathrm{E}\{\boldsymbol{\nu}_r\} < 1, \\ 0 & \text{if } \mathrm{E}\{\boldsymbol{\nu}_r\} \geqq 1, \end{cases} \tag{6.62}$$

and w_k, $k=1, 2, \dots$, *being defined by* (6.50).

Proof. For $k=0, 1, \dots$, (cf. (6.48))

$$\Pr\{\boldsymbol{n}_r < r+k,\ r=1, \dots, n+1\} = \sum_{j=0}^{k} p_j \Pr\{\boldsymbol{n}_r < r+k+1-j,\ r=1, \dots, n\}.$$

Letting $n \to \infty$ we obtain (6.50) by using the continuity theorem of probability. From theorem 6.1 we have

$$w_0 = \lim_{n\to\infty} \Pr\{\boldsymbol{n}_r < r,\ r=1, \dots, n\} = \lim_{n\to\infty} \mathrm{E}\{\max(0, 1-\boldsymbol{n}_n/n)\}.$$

The strong law of large numbers yields

$$\Pr\left\{\lim_{n\to\infty} \frac{\boldsymbol{n}_n}{n}\right\} = \mathrm{E}\{\boldsymbol{\nu}_i\},$$

so that, since $\max(0, 1-\boldsymbol{n}_n/n)$ is bounded, (6.62) follows. This proves the theorem.

The theorems considered so far concern a discrete time parameter process $\{\boldsymbol{n}_r, r=0, 1, \ldots\}$. Below we mention analogous theorems for continuous time parameter processes. For details we refer to TAKÁCS [1967].

Consider a separable stochastic process $\{\boldsymbol{k}_t, t\in[0, \infty)\}$ with non-negative increments and with almost all its sample functions non-decreasing step-functions. Take $\boldsymbol{k}_0 \overset{\text{def}}{=} 0$, and for all finite t and $n=2, 3, \ldots$, it is assumed that the stochastic variables

$$\boldsymbol{k}_{t_i} - \boldsymbol{k}_{t_{i-1}}, \qquad i=1, \ldots, n, \qquad t_i \overset{\text{def}}{=} \frac{it}{n}, \tag{6.63}$$

are interchangeable stochastic variables. This process $\{\boldsymbol{k}_t, t\in[0, \infty)\}$ will be called *a process with non-negative, interchangeable stochastic increments.* In the special case in which the variables in (6.63) are independent and identically distributed the process $\{\boldsymbol{k}_t, t\in[0, \infty)\}$ is a process with stationary independent increments.

Consider the variables

$$\boldsymbol{h}_i \overset{\text{def}}{=} \frac{2^m}{t}\{\boldsymbol{k}_{t_i,s} - \boldsymbol{k}_{t_{i-1},s}\}, \qquad i=1, \ldots, 2^m, \quad m=0, 1, \ldots,$$

with

$$t_i = it/2^m,$$

and where $\boldsymbol{k}_{t_i,s}$ is the sum of the jumps not smaller than $s>0$ occurring during $[0, t_i)$ in the process $\{\boldsymbol{k}_t, t\in[0, \infty)\}$. Denoting the greatest integer not exceeding x by $[x]$ then for fixed m and fixed t the variables

$$\boldsymbol{\nu}_i \overset{\text{def}}{=} [\boldsymbol{h}_i], \qquad i=1, \ldots, 2^m,$$

form a sequence of interchangeable, non-negative, integer valued stochastic variables. The above theorems may be applied to this sequence, and, by performing appropriate limit operations, theorems analogous to those above may be obtained for the process $\{\boldsymbol{k}_t, t\in[0, \infty)\}$. For instance taking in theorem 6.1,

$$i = \left[\frac{2^m u}{t}\right], \qquad k = \left[\frac{2^m z}{t}\right], \qquad 0 \leqq z \leqq t,$$

and letting $m\to\infty$, $s\to 0$ then formally from (6.42),

$$\Pr\{\boldsymbol{k}_u \leqq u, 0 \leqq u \leqq t, \boldsymbol{k}_t \leqq z\} = \int_0^z \left\{1 - \frac{y}{t}\right\} \mathrm{d}_t \Pr\{\boldsymbol{k}_t \leqq y\}, \tag{6.64}$$

and hence

$$\Pr\{\boldsymbol{k}_u \leqq u,\ 0 \leqq u \leqq t \mid \boldsymbol{k}_t = y\} = \begin{cases} 1 - \dfrac{y}{t}, & 0 \leqq y \leqq t, \\ 0, & \text{otherwise}, \end{cases} \tag{6.65}$$

the conditional probability being defined up to an equivalence.

We now quote for the process $\{\boldsymbol{k}_t,\ t \in [0, \infty)\}$ generalizations of the theorems for the process $\{\boldsymbol{n}_r,\ r=0, 1, \ldots\}$. For the proofs we refer the reader to TAKÁCS [1965 and 1967]; they are similar to those of theorems 6.1, ..., 6.5.

THEOREM 6.6. *If* $\{\boldsymbol{k}_t,\ t \in [0, \infty)\}$ *has non-negative interchangeable increments then for finite* $t>0$ *the relation* (6.65) *holds.*

THEOREM 6.7. *Under the same conditions as in theorem 6.6,*

$$\Pr\{\sup_{0 \leqq u \leqq t} (\boldsymbol{k}_u - u) \leqq x\} = \Pr\{\boldsymbol{k}_t \leqq t + x\} - \iint\limits_{0 \leqq y \leqq z \leqq t} \frac{t-z}{t-y}\, \mathrm{d}_y\, \mathrm{d}_z \Pr\{\boldsymbol{k}_y \leqq y + x,\ \boldsymbol{k}_t \leqq z + x\}, \qquad \textit{for all } x,$$

and

$$\Pr\{\boldsymbol{k}_u \leqq u + x,\ 0 \leqq u \leqq t;\ \boldsymbol{k}_t \leqq t + x - c\}$$

$$= \Pr\{\boldsymbol{k}_t \leqq t + x - c\} - \iint\limits_{0 \leqq y \leqq z \leqq t-c} \frac{t-z}{t-y}\, \mathrm{d}_y\, \mathrm{d}_z \Pr\{\boldsymbol{k}_y \leqq y + x, \boldsymbol{k}_t \leqq z + x\}, \qquad c>0.$$

Whenever $\{\boldsymbol{k}_t,\ t \in [0, \infty)\}$ is a process with non-negative stationary increments then it is well-known that the distribution function of $\boldsymbol{k}_t$ for fixed t is infinitely divisible (cf. DOOB [1953]) and $t^{-1}\,\mathrm{E}\{\boldsymbol{k}_t\}$ is independent of t while the Laplace-Stieltjes transform of the distribution function of $\boldsymbol{k}_t$ is of the form

$$\mathrm{E}\{\mathrm{e}^{-\rho \boldsymbol{k}_t}\} = \mathrm{e}^{-t\Phi(\rho)}, \qquad \mathrm{Re}\, \rho \geqq 0, \tag{6.66}$$

$\Phi(\rho)$ being uniquely determined (cf. also section I.5.1). For finite $c>0$ let $\boldsymbol{\theta}_c$ be defined by

$$\boldsymbol{\theta}_c \stackrel{\text{def}}{=} \inf_{u>0} \{u:\ \boldsymbol{k}_u < u - c,\ u < \infty\}, \tag{6.67}$$

if such a u exists, otherwise

$$\boldsymbol{\theta}_c \stackrel{\text{def}}{=} \infty.$$

THEOREM 6.8. *If the process* $\{\boldsymbol{k}_t,\ t\in[0,\infty)\}$ *has non-negative, stationary independent increments then for* $c\leqq x$,

$$\Pr\{\boldsymbol{k}_u\leqq u+x-c,\ 0\leqq u\leqq\boldsymbol{\theta}_c\} = \frac{W(x-c)}{W(x)}, \tag{6.68}$$

where $W(0+)>0$, *but further arbitrary, and for* $\operatorname{Re}\rho$ *sufficiently large*

$$\int_0^\infty \mathrm{e}^{-\rho x}\,\mathrm{d}W(x) = \frac{\rho W(0)}{\rho-\Phi(\rho)}. \tag{6.69}$$

THEOREM 6.9. *Under the same conditions as in theorem* 6.6 *we have for* $0<x<t<\infty$,

$$\Pr\{\sup_{0\leqq u\leqq t}(u-\boldsymbol{k}_u)\leqq x\} = 1 - \int_x^t \frac{x}{y}\,\mathrm{d}_y \Pr\{\boldsymbol{k}_y\leqq y-x\}. \tag{6.70}$$

THEOREM 6.10. *Under the same conditions as in theorem* 6.8,

$$\Pr\{\sup_{u\geqq 0}(\boldsymbol{k}_u-u)\leqq x\}=W(x); \tag{6.71}$$

with

$$W(x)\equiv 0 \qquad \text{if} \quad t^{-1}\,\mathrm{E}\{\boldsymbol{k}_t\}\geqq 1,$$

and

$$W(0)=1-t^{-1}\,\mathrm{E}\{\boldsymbol{k}_t\} \qquad \text{if} \quad t^{-1}\,\mathrm{E}\{\boldsymbol{k}_t\}<1;$$

in the latter case

$$W(x) = \begin{cases} 1-(1-t^{-1}\,\mathrm{E}\{\boldsymbol{k}_t\})\int_0^\infty \mathrm{d}_y \Pr\{\boldsymbol{k}_t\leqq y+x\}, & x>0,\\ 0, & x<0, \end{cases} \tag{6.72}$$

$$\int_{0-}^\infty \mathrm{e}^{-\rho x}\,\mathrm{d}W(x) = \frac{\rho}{\rho-\Phi(\rho)}(1-t^{-1}\,\mathrm{E}\{\boldsymbol{k}_t\}), \tag{6.73}$$

for $\operatorname{Re}\rho$ *sufficiently large.*

We now apply the above theorems to the queueing system $M/G/1$. For $\boldsymbol{v}_i$ we take the number of customers arriving during the ith service time (cf.

(1.15)) then for $j=0, 1, \ldots$,

$$p_j = \int_0^\infty \frac{(t/\alpha)^j}{j!} e^{-t/\alpha} \mathrm{d}B(t), \qquad \Pr\{\boldsymbol{n}_i=j\} = \int_0^\infty \frac{(t/\alpha)^j}{j!} e^{-t/\alpha} \mathrm{d}B^{i*}(t), \quad i=1,2,\ldots, \tag{6.74}$$

$$\pi(p) = \sum_{j=0}^\infty p_j p^j = \beta\{(1-p)/\alpha\}, \qquad |p| \leqq 1.$$

Clearly, $\boldsymbol{\nu}_i$, $i=1, 2, \ldots$, is a sequence of independent, non-negative, integer valued stochastic variables. From (1.18) we have for the number $\boldsymbol{z}_n$ of customers left behind at the nth departure after $t=0$,

$$\boldsymbol{z}_n = \max\{k+\boldsymbol{n}_n-n,\ \boldsymbol{n}_n-\boldsymbol{n}_r-(n-r-1),\ r=0, 1, \ldots, n-1\}, \qquad n=1, 2, \ldots, \tag{6.75}$$

if $\boldsymbol{x}_0=k$. Obviously, $\boldsymbol{z}_n$ has the same distribution as the variable

$$\tilde{\boldsymbol{z}}_n = \max\{k+\boldsymbol{n}_n-n,\ \boldsymbol{n}_r-r+1,\ r=1, \ldots, n\}, \qquad n=1, 2, \ldots .$$

Hence, by applying (6.42), (6.44) and (6.47) we obtain for $n=1,2,\ldots$; $h=0, 1, 2, \ldots$,

$$\begin{aligned}
&\Pr\{\boldsymbol{z}_n \leqq h \mid \boldsymbol{x}_0=k\} \\
&= \Pr\{\boldsymbol{n}_n \leqq n+h-k\} - \sum_{j=1}^{n-k} \sum_{i=0}^{n-j-k} \left\{1 - \frac{i}{n-j}\right\} \Pr\{\boldsymbol{n}_j=j+h,\ \boldsymbol{n}_n=j+h+i\} \\
&= \Pr\{\boldsymbol{n}_n \leqq n+h-k\} - \sum_{j=1}^{n-k} \Pr\{\boldsymbol{z}_{n-j}=0 \mid \boldsymbol{x}_0=k\} \Pr\{\boldsymbol{n}_j=j+h\}.
\end{aligned}$$

This relation should be compared with the relations discussed in section 4.3.

From the definition of $\boldsymbol{e}_n$, the number of services among the first n that are preceded by an idle period, we have (cf. (1.19)),

$$\Pr\{\boldsymbol{e}_n > h-k \mid \boldsymbol{x}_0=k\} = 1 - \Pr\{r-\boldsymbol{n}_r < h,\ r=0, 1, \ldots, n-1\}, \qquad n=1, 2, \ldots .$$

Hence, from theorem 6.4 for $n=1, 2, \ldots$; $h=k, k+1, \ldots$,

$$\Pr\{\boldsymbol{e}_n > h-k \mid \boldsymbol{x}_0=k\} = \sum_{j=h}^{n-1} \frac{h}{j} \Pr\{\boldsymbol{n}_j=j-h\}.$$

For $\boldsymbol{f}_n$, the number of services among the first n that are not preceded by an idle period, we have (cf. (1.19)),

$$\boldsymbol{f}_n = k+\boldsymbol{n}_n-\boldsymbol{z}_n \quad \text{if} \quad \boldsymbol{x}_0=k. \tag{6.76}$$

To investigate the distribution of $\boldsymbol{f}_n$ for $n\to\infty$ we first consider that of $\boldsymbol{n}_n$ for $n\to\infty$.

If $\beta/\alpha=a<1$ then z_n remains bounded for $n\to\infty$ with probability one. Since

$$\mathrm{E}\{\boldsymbol{v}_i\}=a,\qquad \mathrm{var}\{\boldsymbol{v}_i\}=a+a^2\left(\frac{\beta_2}{\beta^2}-1\right),$$

if β_2, the second moment of $B(t)$, is finite we have by the strong law of large numbers

$$\Pr\{\lim_{n\to\infty}\boldsymbol{n}_n/n=a\}=1,$$

and by the central limit theorem it is seen that the distribution of

$$\frac{\boldsymbol{n}_n-na}{\sqrt{n\,\mathrm{var}\{\boldsymbol{v}_i\}}},$$

is asymptotically normally distributed with zero mean and unit variance for $n\to\infty$. From (6.76) and the remarks above it is seen that $\boldsymbol{n}$ -and $\boldsymbol{f}_n$ have for large n asymptotically the same distribution if $a<1$.

From (6.74) and (6.58) we obtain

$$w(p)=w_0\,\frac{\beta\{(1-p)/\alpha\}}{\beta\{(1-p)/\alpha\}-p},\qquad |p|\leq 1,$$

and from theorem 6.5,

$$w_0=1-a\qquad\text{if}\quad a<1;$$

so that (cf. (4.14) and (4.17)) if $a<1$,

$$\lim_{n\to\infty}\Pr\{z_n\leq h\mid \boldsymbol{x}_0=k\}=w_h=v_0+\ldots+v_h,\qquad h=0,1,\ldots,$$

independent of k. Further, by definition of $\boldsymbol{m}_j$ (cf. (6.48) and (6.75)) it is seen that the event

$$\{\sup_{1\leq r\leq \boldsymbol{m}_i}(\boldsymbol{n}_r-r)<k-i\}\equiv\{\boldsymbol{n}_r<r+k-i,\ r=1,\ldots,\boldsymbol{m}_i\},$$

represents the event that during the initial busy period starting with i customers the maximum queue length is at most k. Hence, from theorem 6.3

$$\Pr\{\boldsymbol{n}_r<r+k-i,\ r=1,\ldots,\boldsymbol{m}_i\}=\frac{w_{k-i}}{w_k},\qquad i=1,\ldots,k;\quad k=1,2,\ldots. \tag{6.77}$$

For $i=1$ this represents the probability that *during a busy period the maximum queue length is at most k* (cf. this result with (4.70) and (III.6.51)).

We next consider the virtual waiting time $\boldsymbol{v}_t$ for the queueing process $M/G/1$. In section 1.3 we derived a relation between $\boldsymbol{v}_t$ and $\boldsymbol{k}_t$, the total amount of work brought in by customers arriving during an interval of length t, viz.

$$\boldsymbol{v}_t = \max[v+\boldsymbol{k}_t-t, \sup_{0<u<t} (\boldsymbol{k}_t-t-(\boldsymbol{k}_u-u))], \qquad \boldsymbol{v}_0=v. \tag{6.78}$$

For the process $\{\boldsymbol{k}_t, t\in[0, \infty)\}$ we have (cf. (4.100)),

$$\Pr\{\boldsymbol{k}_t<\sigma\} = \sum_{n=0}^{\infty} \frac{(t/\alpha)^n}{n!} \mathrm{e}^{-t/\alpha} B^{n*}(\sigma), \qquad \sigma>0, \quad t>0, \tag{6.79}$$

$$\mathrm{E}\{\boldsymbol{k}_t\} = at, \qquad \mathrm{var}\{\boldsymbol{k}_t\} = a\frac{\beta_2}{\beta}\, t,$$

$$\mathrm{E}\{\mathrm{e}^{-\rho\boldsymbol{k}_t}\} = \exp\left[-\{1-\beta(\rho)\}t/\alpha\right], \qquad \mathrm{Re}\,\rho\geqq 0.$$

It is easily seen that the process $\{\boldsymbol{k}_t, t\in[0, \infty)\}$ is a process with non-negative, stationary independent increments. Since $\boldsymbol{k}_{t-u}$ and $\boldsymbol{k}_t-\boldsymbol{k}_u$ have the same distribution, it follows that $\boldsymbol{v}_t$ and

$$\tilde{\boldsymbol{v}}_t \stackrel{\text{def}}{=} \max[v+\boldsymbol{k}_t-t, \sup_{0<u<t} (\boldsymbol{k}_u-u)],$$

have the same distribution. Applying theorem 6.7 we obtain for all σ, and $v\geqq 0$,

$$\Pr\{\boldsymbol{v}_t<\sigma \mid \boldsymbol{v}_0=v\} = \Pr\{\boldsymbol{k}_t<t+\sigma-v\}$$

$$- \iint\limits_{\substack{u+w\leqq t-v\\ u\geqq 0, w\geqq 0}} \left\{1 - \frac{w}{t-u}\right\} \mathrm{d}_u\, \mathrm{d}_w \Pr\{\boldsymbol{k}_u<u+\sigma,\ \boldsymbol{k}_t-\boldsymbol{k}_u<w\}, \tag{6.80}$$

and in particular for $\sigma=0+$ (cf. (6.64) and theorem 1.2),

$$\Pr\{\boldsymbol{v}_t=0 \mid \boldsymbol{v}_0=v\} = \begin{cases} \displaystyle\int_0^{t-v} \left(1-\frac{y}{t}\right) \mathrm{d}_y \Pr\{\boldsymbol{k}_t<y\}, & t>v,\\ 0, & t<v. \end{cases} \tag{6.81}$$

Relation (6.80) should be compared with (4.91) and relation (6.81) with (4.93). Further, we deduce relation (4.102) from (6.80) by noting that

$$\Pr\{\boldsymbol{k}_t-\boldsymbol{k}_u<w,\ \boldsymbol{k}_u<u+\sigma\}=\Pr\{\boldsymbol{k}_u<u+\sigma\}\ \Pr\{\boldsymbol{k}_{t-u}<w\}.$$

For the empty time $\boldsymbol{e}_t$ we have (cf. theorem 1.2),

$$\boldsymbol{e}_t=\max[0,\ \sup_{0<u<t}\ (u-\boldsymbol{k}_u-v)],$$

hence by theorem 6.9 we obtain (4.97), i.e.

$$\Pr\{\boldsymbol{e}_t\geqq\sigma-v \mid \boldsymbol{v}_0=v\} = 1 - \Pr\{u-\boldsymbol{k}_u<\sigma,\ 0<u<t\}$$

$$= \begin{cases} \displaystyle\int_\sigma^t \frac{\sigma}{\tau}\, \mathrm{d}_\tau \Pr\{\boldsymbol{k}_\tau<\tau-\sigma\}, & v\leqq\sigma<t, \\ 0, & \sigma>t. \end{cases} \tag{6.82}$$

From relations (1.11) and (6.82) we immediately obtain for the distribution of the initial busy period $\boldsymbol{\theta}_v$ (cf. also (6.67)),

$$\Pr\{\boldsymbol{\theta}_v<t \mid \boldsymbol{v}_0=v\} = \begin{cases} \displaystyle\int_v^t \frac{v}{\tau}\, \mathrm{d}\Pr\{\boldsymbol{k}_\tau<\tau-v\}, & t>v, \\ 0, & t<v. \end{cases} \tag{6.83}$$

The distribution function $D(t)$ of the busy period can now be deduced from (6.83) by noting that if a busy period starts at $t=0$ then $\boldsymbol{v}_0$ has the distribution function $B(.)$. Hence,

$$D(t) = \int_0^\infty \Pr\{\boldsymbol{\theta}_u<t \mid \boldsymbol{v}_0=u\}\ \mathrm{d}B(u)$$

$$= \alpha \int_0^t \frac{1}{u}\, \mathrm{d}_u \Pr\{0<\boldsymbol{k}_u<u\}, \qquad t>0; \tag{6.84}$$

and from (6.79) it is easily verified that we obtain (4.63).

Next, we shall consider *the distribution of the supremum of the virtual waiting time during a busy period for the system M/G/1.*
With $\boldsymbol{\theta}_v$ defined by (6.67) it is seen that $\boldsymbol{\theta}_v$ is the length of the initial busy

period. Hence from theorem 6.8,

$$\Pr\{\sup_{0<t<\boldsymbol{\theta}_v} \boldsymbol{v}_t<\sigma \mid \boldsymbol{v}_0=v\} = \Pr\{\boldsymbol{v}_t<\sigma,\ 0<t<\boldsymbol{\theta}_v \mid \boldsymbol{v}_0=v\}$$

$$= \Pr\{\boldsymbol{k}_t<t+\sigma-v,\ 0<t<\boldsymbol{\theta}_v\} = \frac{W(\sigma-v)}{W(\sigma)}, \qquad \sigma>v, \qquad (6.85)$$

where $W(\sigma)/W(0)$ is given by (cf. (6.79)),

$$\int_0^\infty \mathrm{e}^{-\rho\sigma}\,\mathrm{d}W(\sigma) = W(0)\,\frac{\rho}{\rho-\{1-\beta(\rho)\}/\alpha},$$

for Re ρ sufficiently large, so that if $a<1$,

$$W(t) = W(0)\sum_{n=0}^{\infty} a^n \left[\frac{1}{\beta}\int_0^t \{1-B(\tau)\}\,\mathrm{d}\tau\right]^{n*}, \qquad t>0, \qquad (6.86)$$

with $W(0)>0$ but arbitrary.

To obtain the distribution of the supremum of $\boldsymbol{v}_t$ during a busy period $\boldsymbol{p}$ take $\boldsymbol{v}_0$ as having distribution function $B(.)$. Hence

$$\Pr\{\sup_{0<t<\boldsymbol{p}} \boldsymbol{v}_t<\sigma\} = \frac{W(0)}{W(\sigma)}\int_0^\sigma \frac{W(\sigma-v)}{W(0)}\,\mathrm{d}B(v), \qquad \sigma>0. \qquad (6.87)$$

It should be noted that if $a<1$ then $W(t)$ represents for $t\to\infty$ the limit distribution of the virtual and of the actual waiting time if we take $W(0)=1-a$ (cf. (4.81)), a result which also follows from theorem 6.10. In section III.5 (cf. (III.5.87)) another derivation of (6.87) is given. In section III.7 the properties of the distribution given by (6.87) are studied. In particular we have ($a<1$) (cf. (III.770)),

$$\mathrm{E}\{\sup_{0<t<\boldsymbol{p}} \boldsymbol{v}_t\} = \alpha\log\frac{1}{1-a}.$$

PART III

VARIANTS OF THE SINGLE SERVER QUEUE

III.1. INTRODUCTION

In part II all the discussions concerned the single server queue as described in section II.1.1, and no variants of this model have been considered. The number of variants of the single server model seems to be unlimited, and it is hardly possible to give a detailed account of all its variants which have been considered in literature up to now.

Many server queueing systems which are a generalisation rather than a variant of the single server model will not be discussed here. A vast amount of literature on many server systems is available but, unfortunately, relatively little is yet known about these systems as compared with the single server model. The interested reader is referred to SYSKI [1962], LE GALL [1962], SAATY [1961], POLLACZEK [1961, 1965], GNEDENKO and KOVALENKO [1968].

The most important features of the single server model are:
(*i*) customers arrive singly and are served individually; (*ii*) interarrival times are independent identically distributed variables, the same applies to the service times; moreover, both families are independent; (*iii*) service is given in order of arrival; (*iv*) the queue length and similarly the waiting time are not bounded; (*v*) customers remain in the queue until they are served.

Many simple variants of this model may be discussed by a straightforward application of the methods described in part II.

Important variants from a practical point of view are those where customers arrive in groups and/or are served in batches; such variants are often found in road traffic and transportation theory. In chapter 2 queueing models of this type will be investigated.

Although service in order of arrival is very common, other types especially priority disciplines frequently arise. Such disciplines may be used to influence the distributions of the queue length and the waiting time. Chapter 3 is devoted to priority disciplines.

Relatively little is yet known about systems for which interarrival times and/or service times are not independent or depend on the state of the system. Some studies, however, are available and the reader is referred to RUNNENBURG [1960], FABENS [1961], BENEŠ [1963], NEUTS [1966[a], 1966[b] and 1967], LAMBOTTE and TEGHEM [1966]. For systems with a finite number of sources see SYSKI [1962].

In modern data processing systems queueing situations are often found where the queue length or the waiting time is bounded. Models for such situations may often be represented as random walks with two boundaries. Problems of this type have been extensively studied by KEILSON [1965], who uses Green's functions. A more direct approach starting from the integral relations for these models is developed in chapters 4, 5 and 6. The investigations in these chapters lead to new results for the single server model of part II, they are discussed in chapter 7.

III.2. THE BULK QUEUE $G/G/1$

III.2.1. The model

In the queueing model described in chapter II.1 customers arrive singly, and are served individually. We shall consider the same model here with two major modifications: viz. firstly the arrival of a customer is replaced by the arrival of a group of customers, the number of customers belonging to the same arrival group being a stochastic variable and secondly, customers are served in batches, the number of customers belonging to the same service batch being a stochastic variable.

At time $\mathbf{t}_n$ the nth group of customers arrives. The interarrival times of successive groups are independent, identically distributed variables with distribution function $A(t)$. Let $\mathbf{g}_{1,n}$ denote the total number of customers belonging to the nth arrival group. It will be assumed that the variables $\mathbf{g}_{1,n}$, $n=1, 2, \ldots$, are independent, identically distributed variables with distribution given by

$$g_{1,i} = \Pr\{\mathbf{g}_{1,n}=i\}, \qquad i=1, 2, \ldots; \qquad \sum_{i=1}^{\infty} g_{1,i}=1, \tag{2.1}$$

$$\gamma_1 \overset{\text{def}}{=} \mathrm{E}\{\mathbf{g}_{1,n}\}, \qquad g_1(p) \overset{\text{def}}{=} \sum_{i=1}^{\infty} g_{1,i}p^i, \qquad |p| \leqq 1.$$

We further introduce the independent and identically distributed variables $\mathbf{g}_{2,n}$, $n=1, 2, \ldots$, with distribution

$$g_{2,i} \overset{\text{def}}{=} \Pr\{\mathbf{g}_{2,n}=i\}, \qquad i=1, 2, \ldots; \qquad \sum_{i=1}^{\infty} g_{2,i}=1, \tag{2.2}$$

$$\gamma_2 \overset{\text{def}}{=} \mathrm{E}\{\mathbf{g}_{2,n}\}, \qquad g_2(p) \overset{\text{def}}{=} \sum_{i=1}^{\infty} g_{2,i}p^i, \qquad |p| \leqq 1.$$

Obviously, these variables are positive and integer valued.

It will be assumed that the four families of stochastic variables, i.e. the interarrival times of the groups, the service times of the batches, the group sizes $\boldsymbol{g}_{1,n}$, $n=1, 2, \ldots$, and $\boldsymbol{g}_{2,n}$, $n=1, 2, \ldots$, are independent families.

The variable $\boldsymbol{g}_{2,n}$ represents the *capacity for service* of the nth batch served after $t=0$, i.e. $\boldsymbol{g}_{2,n}$ is the number of customers that can be served in the nth batch. Whether this nth batch will actually contain $\boldsymbol{g}_{2,n}$ customers depends on the number of customers present in the system at the moment that the server can start with the nth service after $t=0$. Denote the number of customers left behind in the system at the moment the batch with service capacity $\boldsymbol{g}_{2,n}$ departs by $\boldsymbol{z}_{n+1}$. The next batch will contain $\boldsymbol{g}_{2,n+1}$ customers if $\boldsymbol{z}_{n+1} \geqq \boldsymbol{g}_{2,n+1}$, and the service capacity of this batch is completely used. If, however, $\boldsymbol{g}_{2,n+1} > \boldsymbol{z}_{n+1}$ a number of variants for the behaviour of the server is possible. We shall describe the most important of these variants here.

(*i*) $\boldsymbol{g}_{2,n+1} > \boldsymbol{z}_{n+1} \geqq 0$, the server waits for arriving customers until the $(n+1)$th service capacity can be fully utilized (delayed service);

(*ii*) $\boldsymbol{g}_{2,n+1} > \boldsymbol{z}_{n+1} > 0$, the $(n+1)$th batch contains $\boldsymbol{z}_{n+1}$ customers and customers arriving during the service of this batch cannot join; the service capacity of the $(n+1)$th batch is partly used;

(*ii*)a $\boldsymbol{g}_{2,n+1} > \boldsymbol{z}_{n+1} = 0$, the server remains idle until the next group of customers arrives; the service capacity of the $(n+1)$th batch can be completely or partly used depending on the number of customers in the arriving group and on $\boldsymbol{g}_{2,n+1}$, cf. (*ii*) above;

(*ii*)b $\boldsymbol{g}_{2,n+1} > \boldsymbol{z}_{n+1} = 0$, the server starts a service with distribution function $B(t)$ but without serving customers and customers arriving during this service time have at least to wait until this is finished; the server handles an "empty batch" and $\boldsymbol{z}_{n+2}$ is equal to the number of arriving customers during this service time; $\min(\boldsymbol{z}_{n+2}, \boldsymbol{g}_{2,n+2})$ will be the number of customers served in the $(n+2)$th batch if $\boldsymbol{z}_{n+2} > 0$.

Other variants are obtained if arriving customers can join a batch which is being served but which does not utilize the full capacity for service (accessible batches).

It should be noted that the model (*ii*)b can be used for queueing at a bus stop or a transportation problem in general. Interesting and important models are obtained for all variants described above when group size and/or service capacity are constant.

III.2.2. Group arrivals, individual service for $G/G/1$

In this section we shall consider the model which is described in the preceding section but supposing that customers are served individually, i.e.

$$g_{2,i} = \begin{cases} 1 & \text{for } i=1, \\ 0 & \text{for } i=2, 3, \ldots, \end{cases} \tag{2.3}$$

and that the server is idle only when there are no customers present (cf. variant (*ii*)a of the preceding section).

In this case the analysis of the model can be reduced to that of the single server queue $G/G/1$ discussed in part II. To illustrate this, we note that an arriving group of customers may be considered as a single arriving *super customer* with service time distribution $B_g(t)$ defined by

$$B_g(t) \overset{\text{def}}{=} \sum_{i=1}^{\infty} g_{1,i} B^{i*}(t), \tag{2.4}$$

because $B_g(t)$ represents the distribution function of the total time required by the server to serve all customers belonging to the same arrival group. Hence, in the terminology of the $G/G/1$ queue we interpret $\boldsymbol{w}_n$ as the actual waiting time of the nth super customer, or $\boldsymbol{w}_n$ is the actual waiting time of that customer in the nth arrival group who is the first one of this group to be served, supposing that customers belonging to different arrival groups will be served in order of arrival.

Since

$$\int_0^{\infty} t \, \mathrm{d}B_g(t) = \gamma_1 \beta,$$

it follows from the results of section II.5.4 that the distribution function of the actual waiting time of the nth super customer has a limit distribution for $n \to \infty$ which is a proper probability distribution if and only if

$$a_g \overset{\text{def}}{=} \frac{\gamma_1 \beta}{\alpha} < 1. \tag{2.5}$$

A similar statement applies to the distribution function of the virtual delay time $\boldsymbol{v}_t$ at time t.

From the results of chapter II.5 we can now immediately obtain the expressions for the distributions of the busy period, the busy cycle, the idle period and so on if we replace the function $B(t)$ by $B_g(t)$ in the relevant formulas of chapter II.5.

For the determination of the distribution of the actual waiting time of the individual customers it is necessary to know by which rule the server chooses the next customer to be served. Usually, it is supposed that customers are served in order of arrival; it may be difficult, however, to fix the order of arrival for those customers who arrive simultaneously, i.e. belong to the same arrival group. Assuming that customers belonging to different arrival groups are served in order of arrival it follows for $a_g<1$ that $W(t)$ as given by (II.5.50) with $B(t)$ replaced by $B_g(t)$ is, for the stationary situation, the distribution function of the actual waiting time of the customer who is the first one of his group to be served.

Clearly

$$W(t) * \sum_{k=1}^{\infty} g_{1,k} B^{(k-1)*}(t),$$

is the distribution of the actual waiting time for the customer who is the last one of his group to be served. If the group consists of at least j customers

$$W(t) * B^{(i-1)*}(t), \qquad i=1, \ldots, j,$$

is the distribution of the actual waiting time of the customer who is the ith one of his group to be served.

Whenever the server selects at random the customers to be served out of the same arrival group the distribution of the actual waiting time of a customer is given by

$$W(t) * \sum_{k=1}^{\infty} \frac{g_{1,k}}{k} \sum_{i=1}^{k} B^{(i-1)*}(t).$$

By the same argument and by using the results of chapter II.5 it is also possible to determine the distribution of the number of groups or super customers in the system at an arrival moment of a group, at a departure moment of the last customer of a group, or at an arbitrary time t. However, it is much more difficult to obtain the distribution of the number of customers present at these moments. For instance if $a_g<1$ the limit distribution of the number of customers left behind in the system by a departing customer exists. For the stationary situation, denoting the probability that j customers are left behind by a departing customer by v_j, then for $j=0, 1, \ldots,$

$$\begin{aligned} v_j = {} & \sum_{i=1}^{\infty} g_{1,i+j} \int_0^{\infty} \{1-A(u)\} \, \mathrm{d}\{W(u) * B^{i*}(u)\} \\ & + \sum_{n=1}^{\infty} \sum_{i=1}^{n} g_{1,n} \sum_{k=1}^{\infty} g_{1,j-(n-i)}^{k*} \int_0^{\infty} \{A^{k*}(u) - A^{(k+1)*}(u)\} \, \mathrm{d}\{W(u) * B^{i*}(u)\}, \end{aligned} \tag{2.6}$$

where $g_{1,i}^{k*}$, $i=1, 2, \ldots$, is the k-fold convolution of the sequence $g_{1,i}$, $i=1, 2, \ldots$, with itself, and $g_{1,i}^{k*}=0$ for $i \leqq k-1$; $k=1, 2, \ldots$. The derivation of relation (2.6) is analogous to that of (II.5.117).

In the model discussed above all customers have the same service time distribution; we shall now consider a generalisation. Suppose that customers of each arrival group are numbered, so that the customers of the same arrival group are distinguishable by their number. Within each group they are numbered 1, 2, ..., and service of customers of the same arrival group is given in the order in which they are numbered. Moreover, in each arrival group customers with number 1 have service time distribution $B_1(t)$, customers with number 2 have service time distribution $B_2(t)$, and so on; all service times are supposed to be independent. Again considering an arrival group as a "super customer" with service time distribution

$$\sum_{k=1}^{\infty} g_{1,k} B_1(t) * B_2(t) * \ldots * B_k(t),$$

we see that the queueing system $G/G/1$ of part II applies to the super customers. The present model can be analysed along the lines indicated above.

It should be noted that the queueing system $G/E_n/1$ can be considered as a special case of the latter model, because the service time with a distribution function of the type E_n may be interpreted as the sum of n independent service times each having a negative exponential distribution. Therefore, every arrival may be considered as a group of n arriving customers, every customer having as service time distribution one of the components of E_n.

III.2.3. Group arrivals, individual service for $M/G/1$

With $\boldsymbol{x}_t$ the number of customers in the system at time t and $\boldsymbol{r}_n'$ the moment of the nth departure from the queue after $t=0$ (cf. section II.4.1) the process $\{\boldsymbol{z}_n, n=1, 2, \ldots\}$

$$\boldsymbol{z}_n \stackrel{\text{def}}{=} \boldsymbol{x}_{\boldsymbol{r}'_n+}, \qquad n=1, 2, \ldots,$$

is for the system $M/G/1$ with group arrivals and individual service an imbedded Markov chain of the process $\{\boldsymbol{x}_t, t \in [0, \infty)\}$ with stationary transition probabilities. Supposing that the server is busy when at least one customer is present, the one-step transition matrix (p_{ij}) is given by

$$p_{ij} = \begin{cases} \sum_{k=1}^{j+1} g_{1,k} \int_{t=0}^{\infty} \sum_{h=0}^{\infty} \frac{(s/\alpha)^h}{h!} \mathrm{e}^{-s/\alpha} g_{1,j-(k-1)}^{h*} \, \mathrm{d}B(s), \\ \qquad\qquad j=0, 1, \ldots; \quad i=0, \\ 0, \qquad j=0, 1, \ldots, i-2; \quad i=2, 3, \ldots, \\ \int_0^{\infty} \sum_{h=0}^{\infty} \frac{(s/\alpha)^h}{h!} \mathrm{e}^{-s/\alpha} g_{1,j-(i-1)}^{h*} \, \mathrm{d}B(s), \\ \qquad\qquad j=i-1, i, \ldots; \quad i=1, 2, \ldots, \end{cases} \tag{2.7}$$

where $g_{1,i}^{k*}$, $i=1, 2, \ldots$, is the k-fold convolution of the sequence $g_{1,i}$, $i=1, 2, \ldots$, with itself, $k=1, 2, \ldots$; while

$$g_{1,0}^{0*} \overset{\text{def}}{=} 1, \quad g_{1,i}^{0*} \overset{\text{def}}{=} 0, \quad i=1, 2, \ldots; \quad g_{1,0}^{k*} = 0, \quad k=1, 2, \ldots . \tag{2.8}$$

As in section II.4.1 it is clear that the chain $\{z_n, n=1, 2, \ldots\}$ is ergodic if and only if $a_g = \beta\gamma/\alpha$ is less than 1. With

$$v_j = \lim_{n\to\infty} p_{ij}^{(n)} = \lim_{n\to\infty} \Pr\{z_n = j\}, \qquad j=0, 1, \ldots, \tag{2.9}$$

it is found for $a_g < 1$, that

$$\begin{aligned} v_0 &= (1-a_g)/\gamma_1, \\ v(p) = \sum_{j=0}^{\infty} v_j p^j &= \frac{1-a_g}{\gamma_1} \frac{\{1-g_1(p)\}\, \beta\{(1-g_1(p))/\alpha\}}{\beta\{(1-g_1(p))/\alpha\} - p}, \qquad |p| \leq 1, \end{aligned} \tag{2.10}$$

for the generating function of the limiting distribution of the number of customers left behind in the system at a departure.

Note that v_0 differs from the probability $1-a_g$ that a departing super customer leaves the system empty. The reason for this is that the set of moments of departure of super customers is a subset of the set of moments of departure of customers, and hence probabilities defined on elements of these sets will in general be different.

With

$$z_0 \overset{\text{def}}{=} x_{0+},$$

we define

$$P_{ij}^{(n)}(t) \overset{\text{def}}{=} \Pr\{z_n = j, r_n' < t \mid z_0 = i\},$$

for $n=1, 2, \ldots$; $i, j=0, 1, \ldots$; $t>0$. It follows for $t>0$,

$$P_{ij}^{(1)}(t) = \begin{cases} \int_0^t \{1-e^{-(t-s)/\alpha}\} \sum_{k=1}^{j+1} g_{1,k} \sum_{h=0}^{\infty} e^{-s/\alpha} \frac{(s/\alpha)^h}{h!} g_{1,j-(k-1)}^{h*} \, dB(s), \\ \qquad j=0, 1, \ldots; \quad i=0, \\ \int_0^t \sum_{h=0}^{\infty} \frac{(s/\alpha)^h}{h!} e^{-s/\alpha} g_{1,j-(i-1)}^{h*} \, dB(s), \\ \qquad j=i-1, i, \ldots; \quad i=1, 2, \ldots, \\ 0, \qquad j=0, 1, \ldots, i-2; \quad i=2, 3, \ldots . \end{cases} \tag{2.11}$$

Hence for $\operatorname{Re} \rho > 0$, $|p| \leqq 1$,

$$\sum_{j=0}^{\infty} p^j \int_0^{\infty} e^{-\rho t} \, dP_{0j}(t) = \frac{1}{p} g_1(p) \frac{1}{1+\alpha\rho} \beta\{\rho + (1-g_1(p))/\alpha\}, \tag{2.12}$$

$$\sum_{j=0}^{\infty} p^j \int_0^{\infty} e^{-\rho t} \, dP_{ij}(t) = p^{i-1} \beta\{\rho + (1-g_1(p))/\alpha\}, \qquad i=1, 2, \ldots .$$

With the aid of the relations (II.4.20) we can now proceed as in section II.4.3. However, we propose to leave the details of the analysis to the reader. We quote one result, viz. if $a_g < 1$,

$$\lim_{t\to\infty} \Pr\{\boldsymbol{x}_t = 0 \mid \boldsymbol{z}_0 = i\} = 1 - a_g,$$

$$\sum_{j=0}^{\infty} p^j \lim_{t\to\infty} \Pr\{\boldsymbol{x}_t = j \mid \boldsymbol{z}_0 = i\} = (1-a_g) \frac{(1-p)\, \beta\{(1-g_1(p))/\alpha\}}{\beta\{(1-g_1(p))/\alpha\} - p}, \qquad |p| < 1. \tag{2.13}$$

This result can also be obtained by using the method of the supplementary variable as described in section II.6.2.

Other quantities of interest for the present queueing system may be obtained from chapter II.4 by first considering the super customers, as discussed in the previous section. As an example we consider the busy period. From formulas (II.4.22) and (II.4.62) for $D_n(t)$, the probability that the busy period has duration less than t and consists of n service times of super customers, we have

$$\sum_{n=1}^{\infty} r^n \int_0^{\infty} e^{-\rho t} \, d_t D_n(t) = \mu(\rho, r), \qquad \operatorname{Re} \rho \geqq 0, \quad |r| < 1,$$

where $\mu(\rho, r)$ is the smallest root, in absolute value, of the equation

$$z = rg_1\{\beta\{\rho + (1-z)/\alpha\}\}, \qquad \mathrm{Re}\,\rho \geqq 0, \quad |r|<1. \tag{2.14}$$

Consequently, for the probability $\tilde{D}_m(t)$ that the busy period has duration less than t and consists of m service times of customers

$$\tilde{D}_m(t) = \sum_{n=1}^{\infty} g_{1,m}^{n*} D_n(t).$$

Hence

$$\sum_{m=1}^{\infty} r^m \int_0^{\infty} \mathrm{e}^{-\rho t}\, \mathrm{d}\tilde{D}_m(t) = \mu(\rho, g_1(r)), \qquad \mathrm{Re}\,\rho \geqq 0, \quad |r|<1. \tag{2.15}$$

III.2.4. Group arrivals and batch service for $M/G/1$

In this section we shall consider for the bulk queue $M/G/1$ the model of section 2.1, the behaviour of the server for the situation $\boldsymbol{g}_{2,n+1} > \boldsymbol{z}_{n+1}$ being described by the variant (*ii*)a of that section. With $\boldsymbol{x}_t$ the number of customers in the system at time t and $\boldsymbol{r}_n'$ the moment of the nth batch departure from the system after $t=0$ it follows from section 2.1 (cf. also section II.4.1) that the process $\{\boldsymbol{z}_n, n=1, 2, \ldots\}$,

$$\boldsymbol{z}_n \stackrel{\text{def}}{=} \boldsymbol{x}_{\boldsymbol{r}'_n+}, \qquad n=1, 2, \ldots,$$

is an imbedded Markov chain of the process $\{\boldsymbol{x}_t, t \in [0, \infty)\}$. The analysis of the process $\{\boldsymbol{x}_t, t \in [0, \infty)\}$ may be carried out along the same lines as in chapter II.4; but we shall use a slightly different approach here.

We shall denote the number of customers arriving during the service of the nth batch by $\boldsymbol{g}_{3,n}$. Since interarrival times are negative exponentially distributed with average α we have for $n=1, 2, \ldots$, (cf. (2.8))

$$\Pr\{\boldsymbol{g}_{3,n}=j\} = \int_0^{\infty} \sum_{h=0}^{\infty} \frac{(t/\alpha)^h}{h!}\, \mathrm{e}^{-t/\alpha}\, g_{1,j}^{h*}\, \mathrm{d}B(t), \qquad j=0, 1, \ldots; \tag{2.16}$$

the variables $\boldsymbol{g}_{3,n}$, $n=1, 2, \ldots$, are mutually independent and also independent of $\boldsymbol{g}_{2,n}$, $n=1, 2, \ldots$. Obviously,

$$g_3(p) \stackrel{\text{def}}{=} \sum_{j=0}^{\infty} p^j \Pr\{\boldsymbol{g}_{3,n}=j\} = \beta\{(1-g_1(p))/\alpha\}, \qquad |p| \leqq 1. \tag{2.17}$$

Denoting the number of customers in the first arrival group after r_n' by $g_{0,n}$ if $z_n=0$, so that every $g_{0,n}$ is a member of the family of variables $\{g_{1,n}, n=1, 2, \dots\}$ it follows from the description of the model that for $n=1, 2, \dots$,

$$z_{n+1} = \begin{cases} [z_n - g_{2,n}]^+ + g_{3,n} & \text{if} \quad z_n>0, \\ [g_{0,n} - g_{2,n}]^+ + g_{3,n} & \text{if} \quad z_n=0. \end{cases} \tag{2.18}$$

We shall first discuss the general solution of the relations (2.18). Defining

$$\gamma_i(\rho) = g_i(\mathrm{e}^{-\rho}), \qquad i=1, 2, 3, \quad \mathrm{Re}\, \rho \geqq 0, \tag{2.19}$$

then for $n=1, 2, \dots;$ $\mathrm{Re}\,\rho\geqq 0$,

$$\begin{aligned}\frac{1}{\gamma_3(\rho)}\, &\mathrm{E}\{\exp(-\rho z_{n+1}) \mid z_1=i\} \\ &= \mathrm{E}\{(z_n>0)\exp(-\rho[z_n-g_{2,n}]^+) \mid z_1=i\} \\ &\quad + \mathrm{E}\{\exp(-\rho[g_{0,n}-g_{2,n}]^+)\}\, \Pr\{z_n=0 \mid z_1=i\} \\ &= \mathrm{E}\{\exp(-\rho[z_n-g_{2,n}]^+) \mid z_1=i\} \\ &\quad - [1-\mathrm{E}\{\exp(-\rho[g_{0,n}-g_{2,n}]^+)\}]\, \Pr\{z_n=0 \mid z_1=i\}. \end{aligned} \tag{2.20}$$

Introduce the generating functions

$$\zeta_j(\rho, r) \stackrel{\text{def}}{=} \sum_{n=2}^{\infty} r^n\, \mathrm{E}\{\mathrm{e}^{-\rho z_n} \mid z_1=j\}, \qquad |r|<1, \tag{2.21}$$

$$P_j(r) \stackrel{\text{def}}{=} \sum_{n=1}^{\infty} r^n \Pr\{z_n=0 \mid z_1=j\}, \qquad |r|<1,$$

it follows from (2.20) for $\mathrm{Re}\,\rho\geqq 0$, $|r|<1$ that

$$\begin{aligned}\frac{\zeta_j(\rho, r)}{r\,\gamma_3(\rho)} &= \sum_{n=2}^{\infty} r^n\, \mathrm{E}\{\exp(-\rho[z_n-g_{2,n}]^+) \mid z_1=j\} \\ &\quad - \{1-\mathrm{E}\{\exp(-\rho[g_1-g_2]^+)\}\}\, P_j(r) + r\, \mathrm{E}\{\exp(-\rho[j-g_2]^+)\},\end{aligned}$$

where g_1 stands for $g_{0,n}$ or $g_{1,n}$ and g_2 for $g_{2,n}$, $n=1, 2, \dots$.
For the sake of simplicity *it will be assumed that the series which defines $g_2(p)$ converges also for $|p|$ slightly larger than one*, and hence $\gamma_2(\rho)$ exists also for $\mathrm{Re}\,\rho$ slightly less than zero, say,

$$\gamma_2(\rho) \quad \text{analytic for} \quad \mathrm{Re}\,\rho>-\delta, \;\delta>0. \tag{2.22}$$

This assumption is not essential and can be removed (cf. chapter II.5). From formula (II.5.9) it follows that for $0<\operatorname{Re}\xi<\operatorname{Re}\rho$, $\operatorname{Re}\xi<\delta$,

$$\mathrm{E}\{\exp(-\rho[z_n-g_{2,n}]^+)\mid z_1=j\} = \frac{1}{2\pi i}\int_{C_\xi}\left\{\frac{1}{\rho-\xi}+\frac{1}{\xi}\right\}\frac{\mathrm{E}\{e^{-\xi z_n}\mid z_1=j\}}{\gamma_3(\xi)}\gamma_3(\xi)\,\gamma_2(-\xi)\,d\xi, \tag{2.23}$$

$$\Psi_j(\rho)\stackrel{\text{def}}{=}\mathrm{E}\{\exp(-\rho[j-g_2]^+)\} = \frac{1}{2\pi i}\int_{C_\xi}\left\{\frac{1}{\rho-\xi}+\frac{1}{\xi}\right\}\gamma_2(-\xi)\,e^{-\xi j}\,d\xi, \qquad j=0,1,\dots,$$

$$\Phi(\rho)\stackrel{\text{def}}{=}1-\mathrm{E}\{\exp(-\rho[g_1-g_2]^+)\} = 1-\frac{1}{2\pi i}\int_{C_\xi}\left\{\frac{1}{\rho-\xi}+\frac{1}{\xi}\right\}\gamma_1(\xi)\,\gamma_2(-\xi)\,d\xi.$$

Hence, we obtain from (2.21) ff. for $|r|<1$, $\operatorname{Re}\xi<\delta$, $0<\operatorname{Re}\xi<\operatorname{Re}\rho$,

$$\frac{\zeta_j(\rho,r)}{r\,\gamma_3(\rho)} = \frac{r}{2\pi i}\int_{C_\xi}\left\{\frac{1}{\rho-\xi}+\frac{1}{\xi}\right\}\frac{\zeta_j(\xi,r)}{r\,\gamma_3(\xi)}\gamma_3(\xi)\,\gamma_2(-\xi)\,d\xi - \Phi(\rho)\,P_j(r)+r\Psi_j(\rho). \tag{2.24}$$

From (2.21) it is clear that its left-hand side is an analytic function of ρ for $\operatorname{Re}\rho>0$. The integral equation above is a Pollaczek integral equation (cf. (II.5.17)). Its solution may be immediately obtained from the results obtained in section II.5.2. Defining

$$\boldsymbol{k}_n\stackrel{\text{def}}{=}\sum_{m=1}^{n}(g_{3,m}-g_{2,m}), \qquad n=1,2,\dots, \tag{2.25}$$

and for $|r|<1$, $\operatorname{Re}\rho_1\geqq0$, $\operatorname{Re}\rho_2\leqq0$ (cf. (II.5.19)),

$$\eta(r,\rho_1,\rho_2)\stackrel{\text{def}}{=}\exp\left\{\sum_{n=1}^{\infty}\frac{r^n}{n}\mathrm{E}\{e^{-\rho_1\boldsymbol{k}_n}(\boldsymbol{k}_n>0)+e^{-\rho_2\boldsymbol{k}_n}(\boldsymbol{k}_n\leqq0)\}\right\}$$

$$= \exp\left\{-\frac{1}{2\pi i}\int_{C_\xi}\left\{\frac{1}{\rho_1-\xi}+\frac{1}{\xi-\rho_2}\right\}\log\{1-r\,\gamma_3(\xi)\,\gamma_2(-\xi)\}\,d\xi\right\}, \tag{2.26}$$

with $0<\operatorname{Re}\xi<\min(\delta,\operatorname{Re}\rho_1)$; it follows for $|r|\leqq1$, $\operatorname{Re}\rho\geqq0$ (cf. (II.5.16)) that

$$\frac{\zeta_j(\rho,r)}{r\,\gamma_3(\rho)} = \frac{1}{2\pi i}\int_{C_\xi}\left\{\frac{1}{\rho-\xi}+\frac{1}{\xi}\right\}\eta(r,\rho,\xi)\,[-\Phi(\xi)\,P_j(r)+r\Psi_j(\xi)]\,d\xi, \tag{2.27}$$

for $0<\operatorname{Re}\xi<\delta$, $\operatorname{Re}\xi<\operatorname{Re}\rho$.

To determine $P_j(r)$ note that for $\operatorname{Im}\rho=0$, and $\rho\to\infty$ the relation (2.20) yields an inhomogeneous difference equation for $\Pr\{z_n=0 \mid z_1=i\}$, $n=1, 2, \dots$. This difference equation together with (2.27) determines $P_j(r)$. A more direct approach for the determination of $P_j(r)$ is obtained by noting that (cf. (2.21)),

$$P_j(r) - r\Pr\{z_1=0 \mid z_1=j\} = \lim_{\rho\to\infty} \zeta_j(\rho, r), \qquad \rho \text{ real}, \tag{2.28}$$

where the right-hand side is given by (2.27) for $\operatorname{Im}\rho=0$, $\rho\to\infty$. Along the same lines as in section II.5.2 it can now be proved that the solution of the recurrence relations (2.20) has been found.

It is noted that

$$\Phi(\rho)\equiv 0 \quad \text{if} \quad \Pr\{g_2\geqq g_1\}=1;$$

if this condition holds, $\zeta_j(\rho, r)$ is immediately given by (2.27). Obviously, this condition is fulfilled if customers do not arrive in groups but arrive singly.

For $z_1=j=0$ it follows for $\operatorname{Re}\rho>0$, $|r|<1$,

$$\frac{\zeta_0(\rho, r)}{r\,\gamma_3(\rho)} = r\,\eta(r, \rho, 0) - \frac{1}{2\pi i}\,P_0(r)\int_{-i\infty+0}^{i\infty+0}\left\{\frac{1}{\rho-\xi}+\frac{1}{\xi}\right\}\eta(r, \rho, \xi)\,\Phi(\xi)\,d\xi. \tag{2.29}$$

Since for any real x

$$e^{-\rho[x]^+}+e^{-\rho[x]^-}=e^{-\rho x}+1, \tag{2.30}$$

it follows that

$$\Phi(\xi)=-\gamma_1(\xi)\,\gamma_2(-\xi)+E\{\exp(-\xi[g_1-g_2]^-)\}, \qquad 0\leqq \operatorname{Re}\xi<\delta, \tag{2.31}$$

hence for $\operatorname{Re}\rho>0$,

$$\begin{aligned}
&\frac{1}{2\pi i}\int_{-i\infty+0}^{i\infty+0}\left\{\frac{1}{\rho-\xi}+\frac{1}{\xi}\right\}\eta(r, \rho, \xi)\,\Phi(\xi)\,d\xi \\
&= \frac{1}{2\pi i}\int_{-i\infty+0}^{i\infty+0}\left\{\frac{1}{\rho-\xi}+\frac{1}{\xi}\right\}\{1-\gamma_1(\xi)\,\gamma_2(-\xi)\}\,\eta(r, \rho, \xi)\,d\xi \\
&\quad - \frac{1}{2\pi i}\int_{-i\infty+0}^{i\infty+0}\left\{\frac{1}{\rho-\xi}+\frac{1}{\xi}\right\}\{1-E\{\exp(-\xi[g_1-g_2]^-)\}\}\,\eta(r, \rho, \xi)\,d\xi.
\end{aligned} \tag{2.32}$$

Noting that $\eta(r, \rho, \xi)$ and the second term in the right-hand side of (2.31) are both analytic for Re $\xi<0$, it follows that the last term in (2.32) is zero, since its residue at $\xi=0$ is zero.

Assuming for the sake of simplicity that $\gamma_1(\xi)$ *is analytic in a strip containing in its interior the axis* Re $\xi=0$ (this assumption is not essential), it follows that

$$\frac{1}{2\pi i}\int_{-i\infty+0}^{i\infty+0}\left\{\frac{1}{\rho-\xi}+\frac{1}{\xi}\right\}\eta(r,\rho,\xi)\,\Phi(\xi)\,d\xi$$

$$=\frac{1}{2\pi i}\int_{-i\infty-0}^{i\infty-0}\left\{\frac{1}{\rho-\xi}+\frac{1}{\xi}\right\}\{1-\gamma_1(\xi)\,\gamma_2(-\xi)\}\,\eta(r,\rho,\xi)\,d\xi,$$

$$\text{Re}\,\rho>0,\quad |r|<1.$$

Note that the latter integral tends to zero for $\rho\to 0$.
Hence from (2.29) for Re $\rho\geqq 0$, $|r|<1$, we have

$$\frac{\zeta_0(\rho,r)}{r\,\gamma_3(\rho)}=r\,\eta(r,\rho,0)$$

$$-\frac{1}{2\pi i}P_0(r)\int_{-i\infty-0}^{i\infty-0}\left\{\frac{1}{\rho-\xi}+\frac{1}{\xi}\right\}\{1-\gamma_1(\xi)\,\gamma_2(-\xi)\}\,\eta(r,\rho,\xi)\,d\xi, \tag{2.33}$$

$$P_0(r)=r+\lim_{\rho\to\infty}\zeta_0(\rho,r),\qquad \rho \text{ real.}$$

Assuming further that $\beta(\xi)$ is analytic in a strip containing in its interior the axis Re $\xi=0$ it follows for Re $\rho>0$, $|r|<1$ that

$$\frac{1}{2\pi i}\int_{-i\infty-0}^{i\infty-0}\left\{\frac{1}{\rho-\xi}+\frac{1}{\xi}\right\}\{1-\gamma_1(\xi)\,\gamma_2(-\xi)\}\,\eta(r,\rho,\xi)\,d\xi$$

$$=\exp\left\{-\sum_{n=1}^{\infty}\frac{r^n}{n}\,\mathrm{E}\{(1-e^{-\rho\boldsymbol{k}_n})(\boldsymbol{k}_n>0)\}\right\}$$

$$\cdot\frac{1}{2\pi i}\int_{-i\infty-0}^{i\infty-0}\left\{\frac{1}{\rho-\xi}+\frac{1}{\xi}\right\}\frac{1-\gamma_1(\xi)\,\gamma_2(-\xi)}{1-r\,\gamma_3(\xi)\,\gamma_2(-\xi)}$$

$$\cdot\exp\left[\sum_{n=1}^{\infty}\frac{r^n}{n}\,\mathrm{E}\{(1-e^{-\xi\boldsymbol{k}_n})(\boldsymbol{k}_n>0)\}\right]d\xi. \tag{2.34}$$

Hence from (2.33) with

$$B(r) \overset{\text{def}}{=} \sum_{n=1}^{\infty} \frac{r^n}{n} \Pr\{\boldsymbol{k}_n > 0\}, \qquad |r| \leqq 1, \tag{2.35}$$

we obtain for $|r|<1$ (cf. (2.17)),

$$\begin{aligned} P_0(r) \Big\{1 + \frac{r\beta(1/\alpha)}{2\pi \mathrm{i}} \mathrm{e}^{-B(r)} \int_{-\mathrm{i}\infty-0}^{\mathrm{i}\infty-0} \frac{1}{\xi} \frac{1-\gamma_1(\xi)\,\gamma_2(-\xi)}{1-r\,\gamma_3(\xi)\,\gamma_2(-\xi)} \\ \cdot \exp\Big[\sum_{n=1}^{\infty} \frac{r^n}{n} \mathrm{E}\{(1-\mathrm{e}^{-\xi \boldsymbol{k}_n})(\boldsymbol{k}_n>0)\}\Big] \mathrm{d}\xi\Big\} \\ = \frac{r^2\beta(1/\alpha)}{1-r} \mathrm{e}^{-B(r)} + r, \end{aligned} \tag{2.36}$$

since it is easily verified that the limit with respect to ρ and the integration with respect to ξ may be interchanged.

Excluding the trivial case $\boldsymbol{g}_{3,n} - \boldsymbol{g}_{2,n} = 0$ with probability one for $n=1, 2, \ldots$, it follows, as in section II.5.4 by an argument based on renewal theory, that $\Pr\{\boldsymbol{z}_{n+1}=0 \mid \boldsymbol{z}_1=0\}$ has a non-zero limit if and only if

$$B \overset{\text{def}}{=} B(1) = \sum_{n=1}^{\infty} \frac{1}{n} \Pr\{\boldsymbol{k}_n>0\} < \infty.$$

The results of section II.5.4 show that this condition is equivalent to

$$a_g \overset{\text{def}}{=} \frac{\beta}{\alpha} \frac{\gamma_1}{\gamma_2} < 1. \tag{2.37}$$

Since the Markov chain $\{\boldsymbol{z}_n, n=1, 2, \ldots\}$ is irreducible and aperiodic the condition (2.37) is necessary and sufficient for the chain to be ergodic.

Defining for $a_g<1$,

$$v_j = \lim_{n\to\infty} \Pr\{\boldsymbol{z}_n = j \mid \boldsymbol{z}_1 = 0\}, \qquad j=0, 1, \ldots,$$

we obtain from (2.34) and (2.36) for the stationary distribution $\{v_j, j=0, 1, \ldots\}$ with $\mathrm{Re}\,\rho>0$,

$$\begin{aligned} v_0 = \beta(1/\alpha)\,\mathrm{e}^{-B}\Big[1 + \frac{\beta(1/\alpha)}{2\pi\mathrm{i}} \mathrm{e}^{-B} \int_{-\mathrm{i}\infty-0}^{\mathrm{i}\infty-0} \frac{\mathrm{d}\xi}{\xi} \frac{1-\gamma_1(\xi)\,\gamma_2(-\xi)}{1-\gamma_3(\xi)\,\gamma_2(-\xi)} \\ \cdot \exp\Big[\sum_{n=1}^{\infty} \frac{1}{n} \mathrm{E}\{(1-\mathrm{e}^{-\xi\boldsymbol{k}_n})(\boldsymbol{k}_n>0)\}\Big]\Big]^{-1}, \end{aligned} \tag{2.38}$$

$$v(\mathrm{e}^{-\rho}) \stackrel{\text{def}}{=} \sum_{j=0}^{\infty} \mathrm{e}^{-\rho j} v_j = \gamma_3(\rho) \exp\left\{-\sum_{n=1}^{\infty} \frac{1}{n} \mathrm{E}\{(1-\mathrm{e}^{-\rho \boldsymbol{k}_n})(\boldsymbol{k}_n>0)\}\right\}$$

$$\cdot\left\{1 - \frac{v_0}{2\pi\mathrm{i}} \int_{-\mathrm{i}\infty-0}^{\mathrm{i}\infty-0} \left\{\frac{1}{\rho-\xi} + \frac{1}{\xi}\right\} \frac{1-\gamma_1(\xi)\,\gamma_2(-\xi)}{1-\gamma_3(\xi)\,\gamma_2(-\xi)}\right.$$

$$\left.\cdot \exp\left[\sum_{n=1}^{\infty} \frac{1}{n} \mathrm{E}\{(1-\mathrm{e}^{-\xi \boldsymbol{k}_n})(\boldsymbol{k}_n>0)\}\right] \mathrm{d}\xi\right\}.$$

The above results are based on the assumptions that $\gamma_1(\xi)$, $\gamma_2(\xi)$ and $\beta(\xi)$ are all analytic in a strip containing the axis $\mathrm{Re}\,\xi=0$ in its interior. These assumptions are not essential; this can be shown by applying a more elaborate analysis along the same lines as in chapter II.5.

Denote the number of service times contained in a busy period by $\boldsymbol{n}$,

$$\boldsymbol{n} \stackrel{\text{def}}{=} \min\{n: \ z_{n+1}=0, \ n=1, 2, \dots\} \qquad \text{if} \quad z_1=0. \tag{2.39}$$

Therefore (cf. II.5.40) from renewal theory it follows that

$$\frac{1}{r}\,\zeta_0(\infty, r) \stackrel{\text{def}}{=} \lim_{\rho\to\infty} \frac{1}{r}\,\zeta_0(\rho, r) = \frac{\mathrm{E}\{r^{\boldsymbol{n}}\}}{1-\mathrm{E}\{r^{\boldsymbol{n}}\}}, \qquad |r|<1, \quad \rho \text{ real},$$

so that

$$\mathrm{E}\{r^{\boldsymbol{n}}\} = \frac{\zeta_0(\infty, r)}{r+\zeta_0(\infty, r)}, \qquad |r|<1. \tag{2.40}$$

With $D(t)$ the distribution function of the busy period we have, by applying Wald's theorem (cf. app. 7),

$$\int_0^{\infty} t \,\mathrm{d}D(t) = \beta\, \mathrm{E}\{\boldsymbol{n}\}. \tag{2.41}$$

From the latter relations we shall determine the probability that the system is empty at time t. Since the interarrival times of groups have the negative exponential distribution, the distribution function of the busy cycle is the convolution of the negative exponential distribution and the distribution of the busy period; the busy cycle being the time between the starts of two successive busy periods. Hence, with $U_1(t)$ the unit step-function,

$$\Pr\{\boldsymbol{x}_t=0 \mid \boldsymbol{x}_0=0\} = \int_0^t \mathrm{e}^{-(t-u)/\alpha}\,\mathrm{d}_u \sum_{n=0}^{\infty} \{(1-\mathrm{e}^{-u/\alpha})\, U_1(u) * D(u)\}^{n*}, \qquad t>0. \tag{2.42}$$

If $B<\infty$, i.e. $a_g<1$, it is obvious from (2.42) and the relations above by applying the key renewal theorem I.6.2 that

$$R_0 \overset{\text{def}}{=} \lim_{t\to\infty} \Pr\{\boldsymbol{x}_t=0\} = \frac{1}{1+\beta\, \mathrm{E}\{\boldsymbol{n}\}/\alpha}. \tag{2.43}$$

Clearly, the latter probability is also the probability that in the stationary situation an arriving group of customers finds the server idle.

Next we shall consider the waiting time distribution of a customer. The waiting time of a customer is the time between the moment of his arrival and that at which his service starts. When the server finishes a batch service and his service capacity for the next batch is less than the number of waiting customers he has to decide which of the waiting customers will join the next batch. The waiting time of customers depends on the server's decision. We suppose that the server acts as follows. Assume that the customers of each arrival group are numbered so that of a given arrival group we can distinguish customer number one, customer number two and so on. The rule for service is now first come first served, such that if in a batch not all remaining customers of the same arrival group can be served those with the lower numbers have priority. Of course there are a number of variants. The analysis for the determination of the waiting time distribution, however, is not essentially different from the one just described.

By applying the usual methods it can be shown that if $a_g<1$ the distribution function of the waiting time of the first customer of the nth arrival group has a limit for $n\to\infty$, similarly for the last customer of the nth arrival group. Here we shall only consider the limit distributions of the waiting times for $n\to\infty$.

For the stationary situation, denote the probability that at some moment j customers are waiting, $j=0, 1, \dots$, and that the past service time of the batch being served at that moment is $\sigma \div \sigma+\mathrm{d}\sigma$ by $R_j(\sigma)\,\mathrm{d}\sigma$. Denoting the number of customers arriving during an interval of duration t by $\boldsymbol{g}_t$ it follows that for $\sigma>0, j=0, 1, \dots$,

$$R_j(\sigma)\,\mathrm{d}\sigma = \{1-B(\sigma)\}\frac{\mathrm{d}\sigma}{\beta}\Big[\Pr\{[\boldsymbol{z}-\boldsymbol{g}_2]^+ +\boldsymbol{g}_\sigma=j,\ \boldsymbol{z}>0\} + \Pr\{[\boldsymbol{g}_1-\boldsymbol{g}_2]^+ +\boldsymbol{g}_\sigma=j,\ \boldsymbol{z}=0\}\Big], \tag{2.44}$$

since $\{1-B(\sigma)\}/\beta$, $\sigma>0$ is the density of the distribution function of the past service time for the stationary situation; here $\boldsymbol{z}$ denotes a variable independent of $\boldsymbol{g}_1$, $\boldsymbol{g}_2$ and $\boldsymbol{g}_\sigma$ and whose distribution is the stationary distribution of the Markov chain $\{z_n,\ n=1, 2, ...\}$. Since

$$\mathrm{E}\{p^{g_\sigma}\} = \sum_{k=0}^{\infty} p^k \sum_{h=0}^{\infty} \mathrm{e}^{-\sigma/\alpha} \frac{(\sigma/\alpha)^h}{h!} g_{1,k}^{h*} = \exp\{-(1-g_1(p))\sigma/\alpha\}, \qquad |p| \leqq 1,$$

and from (2.20) for $a_g<1$,

$$\lim_{n\to\infty} \sum_{k=0}^{\infty} p^k[\Pr\{[z_n-g_2]^+=k,\ z_n>0\} + \Pr\{[g_1-g_2]^+=k,\ z_n=0\}] = \frac{v(p)}{g_3(p)},$$

we obtain from (2.44),

$$\sum_{j=0}^{\infty} p^j R_j(\sigma)\,\mathrm{d}\sigma = \{1-B(\sigma)\} \frac{\mathrm{d}\sigma}{\beta} \frac{v(p)}{g_3(p)} \exp\{-(1-g_1(p))\sigma/\alpha\},$$
$$\sigma>0, \quad |p|\leqq 1. \qquad (2.45)$$

If an arriving group of customers finds the server busy and j waiting customers are present, every customer of this group has to wait at least until the service at hand is finished. Supposing that the arrival group contains at least m customers the mth customer will be taken for service as soon as the total service capacity of the batches served after the arrival exceeds $m+j-1$. Hence

$$\sum_{k=0}^{m+j-1} \{g_{2,k}^{n*} - g_{2,k}^{(n+1)*}\}$$
$$= \Pr\{g_{2,1}+\ldots+g_{2,n} \leqq m+j-1,\ g_{2,1}+\ldots+g_{2,n+1} \geqq m+j\},$$

is the probability that the mth customer will be served in the $(n+1)$th batch after the server has finished the service at hand at the moment of arrival of the group.

Denoting, for the stationary situation ($a_g<1$), the distribution of the waiting time of the mth customer of an arrival group by $W_m(t)$ (supposing the group contains at least m customers) it follows for $m=1, 2, ...;\ t>0$ that

$$W_m(t) = R_0 \sum_{n=0}^{\infty} B^{n*}(t) \sum_{k=0}^{m-1} \{g_{2,k}^{n*} - g_{2,k}^{(n+1)*}\}$$
$$+ (1-R_0) \sum_{j=0}^{\infty} \left\{ \int_{\sigma=0}^{\infty} R_j(\sigma)\,\mathrm{d}\sigma \frac{B(t+\sigma)-B(\sigma)}{1-B(\sigma)} \right\}$$
$$* \{ \sum_{n=0}^{\infty} B^{n*}(t) \sum_{k=0}^{m+j-1} \{g_{2,k}^{n*} - g_{2,k}^{(n+1)*}\}. \qquad (2.46)$$

For two discrete probability distributions $\{h_j, j=0, 1, \dots\}$, $\{k_j, j=0, 1, \dots\}$ with generating functions

$$h(p) \overset{\text{def}}{=} \sum_{j=0}^{\infty} h_j p^j, \qquad k(p) \overset{\text{def}}{=} \sum_{j=0}^{\infty} k_j p^j, \qquad h(1)=k(1)=1, \quad |p| \leqq 1,$$

we have

$$\frac{1}{2\pi \mathrm{i}} \int_{D_\xi} \frac{p}{\xi - p} h(1/\xi)\, k(\xi)\, \mathrm{d}\xi = \sum_{m=1}^{\infty} p^m \sum_{j=0}^{\infty} h_j k_{m+j-1}, \qquad 1 = |\xi| > |p|,$$

where D_ξ is the unit circle in the complex ξ domain and the integration is counterclockwise. Applying this relation it follows from (2.46) that for $\operatorname{Re} \rho > 0$, $1 = |\xi| > |p|$,

$$\sum_{m=1}^{\infty} p^m \int_{0-}^{\infty} \mathrm{e}^{-\rho t}\, \mathrm{d}W_m(t) = R_0 \frac{p}{1-p} \frac{1-g_2(p)}{1-\beta(\rho) g_2(p)}$$

$$+ (1-R_0) \frac{1}{2\pi \mathrm{i}} \int_{D_\xi} \frac{p\, \mathrm{d}\xi}{\xi - p} \frac{1-g_2(\xi)}{1-\beta(\rho) g_2(\xi)} \frac{v(1/\xi)}{(1-\xi) g_3(1/\xi)}$$

$$\cdot \int_{\sigma=0}^{\infty} \exp\left(-(1-g_1(1/\xi))\sigma/\alpha\right) \frac{\mathrm{d}\sigma}{\beta} \{1 - B(\sigma)\} \int_{0}^{\infty} \mathrm{e}^{-\rho t} \frac{\mathrm{d}_t B(t+\sigma)}{1 - B(\sigma)}$$

$$= R_0 \frac{p}{1-p} \frac{1-g_2(p)}{1-\beta(\rho) g_2(p)}$$

$$+ (1-R_0) \frac{1}{2\pi \mathrm{i}} \int_{D_\xi} \frac{p\, \mathrm{d}\xi}{(\xi - p)(1-\xi)} \frac{1-g_2(\xi)}{1-\beta(\rho) g_2(\xi)} \frac{v(1/\xi)}{g_3(1/\xi)}$$

$$\cdot \frac{\beta(\rho) - g_3(1/\xi)}{(1-g_1(1/\xi))\beta/\alpha - \beta\rho}. \tag{2.47}$$

It should be noted that the zero's of

$$1 - g_1(1/\xi) - \alpha\rho, \qquad \operatorname{Re} \rho > 0,$$

are also zero's of

$$\beta(\rho) - g_3(1/\xi);$$

moreover, $v(1/\xi)/g_3(1/\xi)$ is analytic for $|\xi|>1$, so that the poles of the integral above outside the unit circle are the zero's of

$$1-\beta(\rho)\,g_2(\xi), \qquad \text{Re}\,\rho>0.$$

Hence, the evaluation of the above integral should be performed by contour integration along a contour outside the unit circle.

If it is assumed that the generating functions with argument ξ^{-1} in the relation (2.47) have a radius of convergence which is slightly larger than 1, then the contour D_ξ may be replaced by D'_ξ, i.e. the circle with center at $\xi=0$ and radius $|\xi|=1-$. In this case we may derive an expression for the average waiting time of the mth customer of an arrival group by differentiating (2.47) with respect to ρ and putting $\rho=0$. It is found for $|p|<|\xi|=1-$, that

$$\sum_{m=1}^{\infty} p^m \int_0^\infty t\,\mathrm{d}W_m(t) = \beta R_0 \frac{p}{1-p}\,\frac{g_2(p)}{1-g_2(p)}$$
$$+\frac{1-R_0}{2\pi\mathrm{i}} \int_{D'_\xi} \frac{p\alpha}{(\xi-p)(1-\xi)}\,\frac{v(1/\xi)}{g_3(1/\xi)}\,\frac{1-g_3(1/\xi)}{1-g_1(1/\xi)}$$
$$\cdot\left\{\frac{g_2(\xi)}{1-g_2(\xi)}+\frac{(1-g_1(1/\xi))\beta/\alpha-(1-g_3(1/\xi))}{(1-g_1(1/\xi))\beta/\alpha}\right\}\mathrm{d}\xi. \tag{2.48}$$

Note that the integrand has a double pole at $\xi=1$.

Next, we shall consider some special cases.

(*i*) $g_2(p)=p^{\gamma_2}$, $g_1(p)=p^{\gamma_1}$, $1\leqq\gamma_1<\gamma_2$, i.e. the service capacity is constant and equal to γ_2, and the number of customers of every arrival group is constant and equal to γ_1. For the sake of simplicity it will be assumed that γ_1 and γ_2 have no common factors larger than one. From (2.23) it follows immediately that

$$\Phi(\rho)\equiv 0,$$

so that from (2.26) and (2.27) for Re $\rho>0$,

$$\zeta_0(\rho,r) = r^2\gamma_3(\rho)\exp\left\{-\frac{1}{2\pi\mathrm{i}}\int_{-\mathrm{i}\infty+0}^{\mathrm{i}\infty+0}\left\{\frac{1}{\rho-\xi}+\frac{1}{\xi}\right\}\log\{1-r\gamma_3(\xi)\,\mathrm{e}^{\xi\gamma_2}\}\,\mathrm{d}\xi\right\}. \tag{2.49}$$

By applying Rouché's theorem it is seen that the function

$$p^{\gamma_2}-rg_3(p), \tag{2.50}$$

has exactly γ_2 zeros $\mu_j(r)$, $j=1, \dots, \gamma_2$, inside the unit circle if $|r|<1$ or $|r|=1$ and $a_g>1$, while for $|r|=1$ and $a_g<1$ it has γ_2-1 zeros $\mu_j(1)$, $j=2, \dots, \gamma_2$, inside the unit circle. The zeros $\mu_j(r)$ are distinct and continuous functions of r, and $\mu_1(r)$ tends to one for $r\to 1$. These statements may be proved by starting from

$$p=\{rg_3(p)\}^{1/\gamma_2}\, \mathrm{e}^{2k\pi\mathrm{i}/\gamma_2}, \qquad k=1, \dots, \gamma_2.$$

Substituting $p=\exp(-\rho)$ in (2.49) we may write for $|r|<1$,

$$\log\{1-r\,\mathrm{e}^{\xi\gamma_2}\,\gamma_3(\xi)\} = \sum_{j=1}^{\gamma_2} \log\{1-\mathrm{e}^{\xi}\,\mu_j(r)\} + \log \frac{1-r\,\mathrm{e}^{\xi\gamma_2}\,\gamma_3(\xi)}{\prod_{i=1}^{\gamma_2}(1-\mathrm{e}^{\xi}\,\mu_i(r))}. \tag{2.51}$$

With

$$0 < \operatorname{Re}\rho < \operatorname{Re}\eta < \min_{1\leq j\leq \gamma_2} \{-\operatorname{Re}\log \mu_j(r)\},$$

we have

$$\frac{1}{2\pi\mathrm{i}} \int_{-\mathrm{i}\infty+0}^{\mathrm{i}\infty+0} \left\{\frac{1}{\rho-\xi}+\frac{1}{\xi}\right\} \log\{1-r\,\mathrm{e}^{\xi\gamma_2}\,\gamma_3(\xi)\}\,\mathrm{d}\xi$$

$$= \log\{1-r\,\mathrm{e}^{\rho\gamma_2}\,\gamma_3(\rho)\} + \frac{1}{2\pi\mathrm{i}} \int_{C_\eta} \left\{\frac{1}{\rho-\eta}+\frac{1}{\eta}\right\} \log\{1-r\,\mathrm{e}^{\eta\gamma_2}\,\gamma_3(\eta)\}\,\mathrm{d}\eta. \tag{2.52}$$

Using (2.51) it follows that

$$\frac{1}{2\pi\mathrm{i}} \int_{C_\eta} \left\{\frac{1}{\rho-\eta}+\frac{1}{\eta}\right\} \log\{1-r\,\mathrm{e}^{\eta\gamma_2}\,\gamma_3(\eta)\}\,\mathrm{d}\eta$$

$$= \sum_{j=1}^{\gamma_2} \frac{1}{2\pi\mathrm{i}} \int_{C_\eta} \left\{\frac{1}{\rho-\eta}+\frac{1}{\eta}\right\} \log\{1-\mathrm{e}^{\eta}\,\mu_j(r)\}\,\mathrm{d}\eta = \sum_{j=1}^{\gamma_2} \log \frac{1-\mu_j(r)}{1-\mathrm{e}^{\rho}\,\mu_j(r)}.$$

Consequently, by analytic continuation (cf. (2.50)) we obtain from (2.49), ..., (2.52),

$$\zeta_0(\rho, r) = \frac{r^2\gamma_3(\rho)}{\mathrm{e}^{-\rho\gamma_2}-r\gamma_3(\rho)} \prod_{j=1}^{\gamma_2} \frac{\mathrm{e}^{-\rho}-\mu_j(r)}{1-\mu_j(r)}, \qquad |r|<1, \quad \operatorname{Re}\rho\geq 0. \tag{2.53}$$

Since

$$\mu_1(r)\to 1 \quad \text{for} \quad r\to 1 \quad \text{if} \quad a_g<1,$$

it follows from

$$p^{\gamma_2} - rg_3(p) = 1 - (1-p)\gamma_2 - r\{1 - (1-p)\gamma_1\, \beta/\alpha\} + \mathrm{o}(1-p) \qquad \text{for} \quad 1-p \to 0,$$

that if $a_g < 1$ then

$$1 - \mu_1(r) = \frac{1}{\gamma_2}\, \frac{1-r}{1-a_g} + \mathrm{o}(1-\mu_1(r)) \qquad \text{for} \quad r \to 1.$$

Since for $a_g < 1$,

$$v(\mathrm{e}^{-\rho}) = \lim_{r \uparrow 1} (1-r)\, \zeta_0(\rho, r),$$

we have from (2.53) that

$$\begin{aligned} v(p) &= \lim_{r\uparrow 1} (1-r) \frac{r^2 g_3(p)}{p^{\gamma_2} - rg_3(p)} \prod_{j=1}^{\gamma_2} \frac{p-\mu_j(r)}{1-\mu_j(r)} \\ &= (1-a_g)\,\gamma_2 \frac{(1-p)\, g_3(p)}{g_3(p) - p^{\gamma_2}} \prod_{j=2}^{\gamma_2} \frac{p-\mu_j(1)}{1-\mu_j(1)}, \qquad |p| \leqq 1. \end{aligned} \tag{2.54}$$

It is now not difficult by starting from (2.54) to obtain the expressions for the Laplace-Stieltjes transforms of the waiting time distributions by applying formula (2.47). The resulting expressions are rather complicated and will be omitted.

An important simplification is obtained if

(*ii*) $\quad g_1(p) = p, \qquad g_2(p) = p^{\gamma_2}, \qquad B(t) = 1 - \mathrm{e}^{-t/\beta}, \qquad t > 0,$

i.e. customers arrive singly, the batch size is constant and the service time distribution is negative exponential. It follows

$$p^{\gamma_2} - rg_3(p) = \frac{-\beta/\alpha}{1+(1-p)\beta/\alpha} \prod_{j=0}^{\gamma_2} (p - \mu_j(r)), \qquad |r| \leqq 1,$$

where $\mu_0(r)$ is the zero of the left-hand side outside the unit circle. It is now easily found that (cf. (2.40), (2.43) and (2.53)),

$$\mathrm{E}\{r^{\mathbf{n}}\} = r\, \frac{1-\mu_0(r)}{r-\mu_0(r)}, \qquad |r| < 1,$$

$$\zeta_0(\rho, r) = \frac{r^2}{1-r}\, \frac{1-\mu_0(r)}{\mathrm{e}^{-\rho} - \mu_0(r)}, \qquad |r| < 1, \quad \operatorname{Re} \rho \geqq 0,$$

$$v(p) = \frac{1-\mu_0}{p-\mu_0}, \qquad R_0 = \frac{\mu_0 - 1}{(1+\beta/\alpha)\,\mu_0 - 1} \qquad \text{if } a_g<1,$$

with

$$\mu_0 \stackrel{\text{def}}{=} \mu(1).$$

Since customers arrive singly only $W_1(t)$ is relevant; we obtain from (2.47) for $a_g<1$,

$$\int_{0-}^{\infty} e^{-\rho t}\, dW_1(t) = R_0 + \frac{1-R_0}{2\pi i} \int_{D_\xi} \frac{d\xi}{\xi(1-\xi)} \frac{1-\xi^{\gamma_2}}{1-\beta(\rho)\,\xi^{\gamma_2}}$$

$$\cdot \frac{1-\mu_0}{1-\mu_0\xi} \left\{\xi + \frac{\beta}{\alpha}(\xi-1)\right\} \frac{\beta(\rho)-g_3(1/\xi)}{(1-\xi^{-1})\beta/\alpha-\beta\rho}.$$

The only pole of the integrand inside the unit circle is $\xi=\mu_0^{-1}$, and it follows that

$$\int_{0-}^{\infty} e^{-\rho t}\, dW_1(t) = \frac{1}{(1+\beta/\alpha)\,\mu_0-1}\left\{\mu_0 - 1 + \frac{\beta}{\alpha}\mu_0 \frac{\mu_0^{\gamma_2}-1}{(1+\beta\rho)\,\mu_0^{\gamma_2}-1}\right\}$$

$$\operatorname{Re}\rho \geqq 0,$$

$$\int_{0}^{\infty} t\, dW_1(t) = \frac{\beta\mu_0/\alpha}{(1+\beta/\alpha)\,\mu_0-1}\, \frac{\beta}{1-\mu_0^{-\gamma_2}},$$

$$W_1(t) = 1 - \frac{\beta\mu_0/\alpha}{(1+\beta/\alpha)\,\mu_0-1} \exp\left\{-\frac{t}{\beta}(1-\mu_0^{-\gamma_2})\right\}, \qquad t\geqq 0.$$

As a third example we consider

(*iii*) $$g_2(p) = \frac{(1-d)\,p}{1-dp}, \qquad g_1(p) = \frac{(1-c)\,p}{1-cp}, \qquad |p|\leqq 1,$$

with c and d given by

$$0<d=1-\gamma_2^{-1}<1, \qquad 0<c=1-\gamma_1^{-1}<1. \tag{2.55}$$

By applying Rouché's theorem it can be shown that the function

$$p-d-r(1-d)\,g_3(p),$$

has exactly one zero $\mu(r)$ inside the unit circle if $|r|<1$ or if $|r|\leqq 1$ and $a_g>1$; if $a_g<1$ and $|r|=1$ there is no zero inside the unit circle. For $|r|<1$

the zero $\mu(r)$ is a continuous function of r (apply Lagrange's expansion theorem, cf. app. 6) and it tends to 1 if $r\to 1$ and $a_g<1$.

To determine the function $\eta(r, \rho_1, \rho_2)$ (cf. (2.26)) note that (cf. (2.51) and (2.52)),

$$\frac{1}{2\pi\mathrm{i}}\int_{-\mathrm{i}\infty+0}^{\mathrm{i}\infty+0}\left\{\frac{1}{\rho_1-\xi}+\frac{1}{\xi-\rho_2}\right\}\log\left\{1-r\frac{(1-d)\,\mathrm{e}^{\xi}}{1-d\,\mathrm{e}^{\xi}}\gamma_3(\xi)\right\}\mathrm{d}\xi$$

$$=\log\left\{1-r\frac{(1-d)\,\mathrm{e}^{\rho_1}}{1-d\,\mathrm{e}^{\rho_1}}\gamma_3(\rho_1)\right\}+\frac{1}{2\pi\mathrm{i}}\int_{C_\eta}\left\{\frac{1}{\rho_1-\eta}+\frac{1}{\eta-\rho_2}\right\}\log\frac{1-\mathrm{e}^{\eta}\,\mu(r)}{1-d\,\mathrm{e}^{\eta}}\,\mathrm{d}\eta$$

$$=-\log\left\{\frac{1-d\,\mathrm{e}^{\rho_2}}{1-d\,\mathrm{e}^{\rho_1}-r(1-d)\,\mathrm{e}^{\rho_1}\gamma_3(\rho_1)}\,\frac{1-\mu(r)\,\mathrm{e}^{\rho_1}}{1-\mu(r)\,\mathrm{e}^{\rho_2}}\right\},$$

$$|r|<1,\quad \mathrm{Re}\,\rho_1>0,\quad \mathrm{Re}\,\rho_2\leqq 0,$$

with $\mathrm{Re}\,\rho_1<\mathrm{Re}\,\eta<\min\{\mathrm{Re}(-\log\mu(r)),\,-\log d\}$. Hence for $|r|<1$, $\mathrm{Re}\,\rho_1>0$,

$$\eta(r,\rho_1,\rho_2)=\frac{1-d\,\mathrm{e}^{\rho_2}}{1-d\,\mathrm{e}^{\rho_1}-r(1-d)\,\mathrm{e}^{\rho_1}\gamma_3(\rho_1)}\,\frac{1-\mu(r)\,\mathrm{e}^{\rho_1}}{1-\mu(r)\,\mathrm{e}^{\rho_2}},$$

$$\mathrm{Re}\,\rho_2<\mathrm{Re}(-\log\mu(r)).\tag{2.56}$$

From (2.23) we have with $\mathrm{Re}\,\rho>\mathrm{Re}\,\xi>0$,

$$\Phi(\xi)=1-\frac{1}{2\pi\mathrm{i}}\int_{-\mathrm{i}\infty+0}^{\mathrm{i}\infty+0}\left\{\frac{1}{\rho-\xi}+\frac{1}{\xi}\right\}\frac{(1-c)(1-d)}{(1-c\,\mathrm{e}^{-\xi})(1-d\,\mathrm{e}^{\xi})}\,\mathrm{d}\xi$$

$$=\frac{(1-d)\,c}{1-dc}\,\frac{1-\mathrm{e}^{-\rho}}{1-c\,\mathrm{e}^{-\rho}},\tag{2.57}$$

since

$$\frac{1}{1-c\,\mathrm{e}^{-\xi}}\,\frac{1}{1-d\,\mathrm{e}^{\xi}}=\frac{1}{1-cd}\left\{\frac{1}{1-d\,\mathrm{e}^{\xi}}+\frac{c}{\mathrm{e}^{\xi}-c}\right\}.\tag{2.58}$$

Further from (2.56) and (2.58) for $\mathrm{Re}\,\rho>\mathrm{Re}\,\xi>0$, we have

$$\frac{1}{2\pi\mathrm{i}}\int_{-\mathrm{i}\infty+0}^{\mathrm{i}\infty+0}\left\{\frac{1}{\rho-\xi}+\frac{1}{\xi}\right\}\eta(r,\rho,\xi)\,\Phi(\xi)\,\mathrm{d}\xi$$

$$=\frac{(1-d)\,c}{1-dc}\,\frac{1-\mu(r)\,\mathrm{e}^{\rho}}{1-d\,\mathrm{e}^{\rho}-r(1-d)\,\mathrm{e}^{\rho}\gamma_3(\rho)}$$

$$\cdot \frac{1}{2\pi i} \int_{-i\infty+0}^{i\infty+0} \frac{1-e^{-\xi}}{1-c\,e^{-\xi}} \frac{1-d\,e^{\xi}}{1-\mu(r)\,e^{\xi}} \left\{ \frac{1}{\rho-\xi} + \frac{1}{\xi} \right\} d\xi$$

$$= \frac{(1-d)\,c}{1-c\mu(r)} \frac{1-e^{-\rho}}{1-c\,e^{-\rho}} \frac{1-\mu(r)\,e^{\rho}}{1-d\,e^{\rho}-r(1-d)\,e^{\rho}\,\gamma_3(\rho)}. \tag{2.59}$$

Now from (2.29) for $|r|<1$, $\mathrm{Re}\,\rho \geqq 0$, it follows that

$$\frac{\zeta_0(\rho, r)}{r\gamma_3(\rho)} = \frac{(1-d)(e^{-\rho}-\mu(r))}{e^{-\rho}-d-r(1-d)\,\gamma_3(\rho)} \cdot \left\{ \frac{r}{1-\mu(r)} - P_0(r) \frac{c}{1-c\mu(r)} \frac{1-e^{-\rho}}{1-c\,e^{-\rho}} \right\}, \tag{2.60}$$

with $P_0(r)$ given by

$$P_0(r) = \left\{ r + \frac{r^2\mu(r)}{1-\mu(r)} \frac{(1-d)\,\beta(1/\alpha)}{d+r(1-d)\,\beta(1/\alpha)} \right\} \cdot \left\{ 1 + \frac{(1-d)\,\mu(r)}{d+r(1-d)\,\beta(1/\alpha)} \frac{cr\beta(1/\alpha)}{1-c\mu(r)} \right\}^{-1}, \qquad |r|<1. \tag{2.61}$$

Since

$$\lim_{r\uparrow 1} \frac{1-\mu(r)}{1-r} = \frac{1}{(1-a_g)\gamma_2} \qquad \text{if} \quad a_g<1,$$

we get for $a_g<1$, $|p|\leqq 1$,

$$v_0 = \frac{\gamma_2\beta(1/\alpha)(1-a_g)}{\gamma_2-(1-\beta(1/\alpha))} \left\{ 1 + \frac{c\gamma_1\beta(1/\alpha)}{\gamma_2-(1-\beta(1/\alpha))} \right\}^{-1}, \tag{2.62}$$

$$v(p) = \frac{(1-p)\,g_3(p)}{g_3(p)+\gamma_2-1-p\gamma_2} \left\{ (1-a_g)\gamma_2 - v_0 c\gamma_1 \frac{1-p}{1-cp} \right\}.$$

It is now possible to determine from (2.40) and (2.43) the probability R_0. Moreover, from (2.47) for $|p|<1$, $\mathrm{Re}\,\rho>0$, it follows that

$$\sum_{m=1}^{\infty} p^m \int_{0-}^{\infty} e^{-\rho t}\, dW_m(t) = R_0 \frac{p}{1-p\{d+(1-d)\,\beta(\rho)\}}$$

$$+ (1-R_0) \frac{1}{2\pi i} \int_{D_\xi} \frac{p\, d\xi}{\xi-p} \frac{v(1/\xi)}{g_3(1/\xi)} \frac{(c-\xi)\alpha/\beta}{1-\xi\{d+(1-d)\,\beta(\rho)\}} \frac{\beta(\rho)-g_3(1/\xi)}{1-\alpha c\rho-\xi(1-\alpha\rho)}. \tag{2.63}$$

The integrand is analytic outside the unit circle except for the pole

$$\xi=\{d+(1-d)\,\beta(\rho)\}^{-1};$$

it is easily verified that the value of the integral is equal to its residue at this pole. We shall not evaluate the integral, but shall consider the special case $c=0$, i.e.

$$(iv)\qquad g_1(p)=p,\qquad g_2(p)=\frac{(1-d)\,p}{1-dp},\qquad |p|<1.$$

In this situation the customers arrive singly. We apply the formulas above with $c=0$, and obtain

$$\mathrm{E}\{\boldsymbol{n}\}=\frac{\beta(1/\alpha)+\gamma_2-1}{\beta(1/\alpha)}\,\frac{\mathrm{d}}{\mathrm{d}r}\,\mu(r)\Big|_{r=1}.$$

Since $\mu(r)$ is a zero of

$$p-d-r(1-d)\,g_3(p),$$

it is found that

$$\frac{\mathrm{d}}{\mathrm{d}r}\,\mu(r)\Big|_{r=1}=\frac{1}{1-a_g}\qquad\text{if}\quad a_g<1.$$

Hence

$$R_0=\left[1+\frac{1}{1-a_g}\,\frac{\beta}{\alpha}\,\frac{\beta(1/\alpha)+\gamma_2-1}{\beta(1/\alpha)}\right]^{-1}.$$

Further we obtain from (2.63),

$$\int_{0-}^{\infty}\mathrm{e}^{-\rho t}\,\mathrm{d}W_1(t)=R_0+(1-R_0)\,\frac{1-a_g}{a_g}\,\frac{1-\beta(\rho)}{\alpha\rho\gamma_2-1+\beta(\rho)},\qquad \mathrm{Re}\,\rho\geqq 0.$$

III.2.5. The transportation problem

The queueing model considered in this section is the $M/G/1$ bulk queue with the behaviour of the server described by the variant (*ii*)b of section 2.1. This queueing problem is known as the transportation problem. In many transportation situations service is given at constant time intervals; this corresponds with the $M/D/1$ bulk queue, i.e. the bulk queue (variant (*ii*)b) with constant service time and group arrivals according to a Poisson process.

We shall use the same notation as in the preceding section, so that z_n, $n=1, 2, \ldots$, represents the number of customers left behind in the system at the moment of the nth batch departure. Clearly, we now have for $n=1, 2, \ldots$,

$$z_{n+1}=[z_n-g_{2,n}]^+ +g_{3,n}; \tag{2.64}$$

so that for $n=1, 2, \ldots$; $\operatorname{Re} \rho \geqq 0$ (cf. (2.20)),

$$\frac{1}{\gamma_3(\rho)} \mathrm{E}\{\exp(-\rho z_{n+1}) \mid z_1=j\} = \mathrm{E}\{\exp(-\rho[z_n-g_{2,n}]^+) \mid z_1=j\}. \tag{2.65}$$

This set of recurrence relations is a special case of the relations (2.20). In fact the problem is identical with that of the preceding section by taking there

$$\Phi(\rho)\equiv 0,$$

hence the results of the preceding section may be applied.

Define for $i=2, 3$,

$$\boldsymbol{k}_{i,n} \stackrel{\text{def}}{=} \begin{cases} \sum_{m=1}^{n} g_{i,m}, & n=1, 2, \ldots, \\ 0, & n=0, \end{cases} \tag{2.66}$$

so that $\boldsymbol{k}_{2,n}$ is the total service capacity offered by the server to the customers for the first n service periods. We shall further denote by $\boldsymbol{d}_{n+1}$, $n=1, 2, \ldots$, with $\boldsymbol{d}_1 \stackrel{\text{def}}{=} 0$, the total unused service capacity during the first n service periods. Clearly

$$\boldsymbol{d}_{n+1} - \boldsymbol{d}_n = -[z_n-g_{2,n}]^-, \qquad n=1, 2, \ldots; \qquad \boldsymbol{d}_1 \stackrel{\text{def}}{=} 0. \tag{2.67}$$

It follows for $n=1, 2, \ldots$; $\operatorname{Re} \rho_1 \geqq 0$, $\operatorname{Re} \rho_2 \leqq 0$, $\operatorname{Re} \rho_3 \geqq 0$ that

$$\begin{aligned}
&\mathrm{E}\{\exp(-\rho_1 z_{n+1}+\rho_2 \boldsymbol{d}_{n+1}-\rho_3 \boldsymbol{k}_{2,n}) \mid z_1=j\} \\
&= \mathrm{E}\{\exp(-\rho_1[z_n-g_{2,n}]^+ -\rho_2[z_n-g_{2,n}]^- -\rho_3 \boldsymbol{k}_{2,n}-\rho_1 g_{3,n}+\rho_2 \boldsymbol{d}_n) \mid z_1=j\} \\
&= \gamma_3(\rho_1)\, \mathrm{E}\{\exp(-\rho_1[z_n-g_{2,n}]^+ -\rho_2[z_n-g_{2,n}]^- -\rho_3 \boldsymbol{k}_{2,n}+\rho_2 \boldsymbol{d}_n) \mid z_1=j\} \\
&= \gamma_3(\rho_1) \frac{1}{2\pi \mathrm{i}} \int_{C_\xi} \left\{\frac{1}{\rho_1-\xi}+\frac{1}{\xi-\rho_2}\right\} \\
&\quad \cdot \mathrm{E}\{\exp(-\xi z_n+\rho_2 \boldsymbol{d}_n-\rho_3 \boldsymbol{k}_{2,n-1}) \mid z_1=j\}\, \gamma_2(\rho_3-\xi)\, \mathrm{d}\xi,
\end{aligned}$$

or

$$\frac{1}{\gamma_3(\rho_1)} \mathrm{E}\{\exp(-\rho_1 \boldsymbol{z}_{n+1} + \rho_2 \boldsymbol{d}_{n+1} - \rho_3 \boldsymbol{k}_{2,n}) \mid \boldsymbol{z}_1 = j\}$$

$$= \frac{1}{2\pi \mathrm{i}} \int_{C_\xi} \left\{ \frac{1}{\rho_1 - \xi} + \frac{1}{\xi - \rho_2} \right\} \frac{1}{\gamma_3(\xi)}$$

$$\cdot \mathrm{E}\{\exp(-\xi \boldsymbol{z}_n + \rho_2 \boldsymbol{d}_n - \rho_3 \boldsymbol{k}_{2,n-1}) \mid \boldsymbol{z}_1 = j\} \, \gamma_3(\xi) \, \gamma_2(\rho_3 - \xi) \, \mathrm{d}\xi, \tag{2.68}$$

with $\operatorname{Re} \rho_2 < \operatorname{Re} \xi < \operatorname{Re} \rho_1$, $\operatorname{Re} \rho_3 > \operatorname{Re} \xi > 0$. Introducing the generating function

$$\zeta_j(r, \rho_1, \rho_2, \rho_3) \stackrel{\text{def}}{=} \sum_{n=2}^{\infty} r^n \, \mathrm{E}\{\exp(-\rho_1 \boldsymbol{z}_n + \rho_2 \boldsymbol{d}_n - \rho_3 \boldsymbol{k}_{2,n-1}) \mid \boldsymbol{z}_1 = j\}, \tag{2.69}$$

with $|r| < 1$, $\operatorname{Re} \rho_1 \geqq 0$, $\operatorname{Re} \rho_2 \leqq 0$, $\operatorname{Re} \rho_3 \geqq 0$, it follows from (2.68) and (2.66) since $\boldsymbol{d}_1 = 0$ that

$$\frac{1}{r\gamma_3(\rho_1)} \zeta_j(r, \rho_1, \rho_2, \rho_3)$$

$$= \frac{r}{2\pi \mathrm{i}} \int_{C_\xi} \left\{ \frac{1}{\rho_1 - \xi} + \frac{1}{\xi - \rho_2} \right\} \frac{1}{r\gamma_3(\xi)} \zeta_j(r, \xi, \rho_2, \rho_3) \, \gamma_3(\xi) \, \gamma_2(\rho_3 - \xi) \, \mathrm{d}\xi$$

$$+ \frac{r}{2\pi \mathrm{i}} \int_{C_\xi} \left\{ \frac{1}{\rho_1 - \xi} + \frac{1}{\xi - \rho_2} \right\} \mathrm{e}^{-j\xi} \, \gamma_2(\rho_3 - \xi) \, \mathrm{d}\xi, \tag{2.70}$$

for $|r| < 1$, $\operatorname{Re} \rho_2 \leqq 0 < \operatorname{Re} \xi < \min(\operatorname{Re} \rho_1, \operatorname{Re} \rho_3)$. Relation (2.70) represents Pollackzek's integral equation for the function $r^{-1} \gamma_3^{-1}(\rho_1) \, \zeta_j(r, \rho_1, \rho_2, \rho_3)$ (cf. (II.5.17)). The solution is immediately obtained from the results of sections II.5.2 and II.5.3. In fact, we have for $j = 0$, $\operatorname{Re} \rho_1 \geqq 0$, $\operatorname{Re} \rho_2 \leqq 0$, $\operatorname{Re} \rho_3 \geqq 0$, $|r| < 1$ (cf. (II.5.15), ..., (II.5.18)),

$$\zeta_0(r, \rho_1, \rho_2, \rho_3) = r^2 \gamma_3(\rho_1) \, \gamma_2(\rho_3 - \rho_2) \, \eta(r, \rho_1, \rho_2, \rho_3), \tag{2.71}$$

with

$$\eta(r, \rho_1, \rho_2, \rho_3)$$

$$\stackrel{\text{def}}{=} \exp\left\{ \sum_{n=1}^{\infty} \frac{r^n}{n} \, \mathrm{E}\{\exp(-\rho_3 \boldsymbol{k}_{2,n})(\mathrm{e}^{-\rho_1 \boldsymbol{k}_n}(\boldsymbol{k}_n > 0) + \mathrm{e}^{-\rho_2 \boldsymbol{k}_n}(\boldsymbol{k}_n \leqq 0))\} \right\}$$

$$= \exp\left\{ \frac{-1}{2\pi \mathrm{i}} \int_{C_\xi} \left\{ \frac{1}{\rho_1 - \xi} + \frac{1}{\xi - \rho_2} \right\} \log\{1 - r\gamma_3(\xi) \, \gamma_2(\rho_3 - \xi)\} \, \mathrm{d}\xi \right\}, \tag{2.72}$$

for $\operatorname{Re} \rho_2 < \operatorname{Re} \xi < \operatorname{Re} \rho_1$, $0 < \operatorname{Re} \xi < \operatorname{Re} \rho_3$.

We obtain for $|r|<1$, $\operatorname{Re}\rho\geqq 0$ (cf. (II.5.36)),

$$\sum_{n=2}^{\infty} r^n \operatorname{E}\{\mathrm{e}^{-\rho z_n} \mid z_1=0\}$$
$$=\frac{r^2}{1-r}\gamma_3(\rho)\exp\left\{-\sum_{n=1}^{\infty}\frac{r^n}{n}\int_0^{\infty}\{1-\mathrm{e}^{-\rho t}\}\,\mathrm{d}_t \Pr\{\boldsymbol{k}_n<t\}\right\}, \tag{2.73}$$

or for $|r_1|<1$, $|r_2|\leqq 1$,

$$\sum_{n=2}^{\infty} r_1^n \operatorname{E}\{r_2^{z_n} \mid z_1=0\}$$
$$=\frac{r_1^2}{1-r_1} g_3(r_2)\exp\left\{-\sum_{n=1}^{\infty}\frac{r_1^n}{n}\sum_{j=1}^{\infty}\{1-r_2^j\}\Pr\{\boldsymbol{k}_n=j\}\right\}. \tag{2.74}$$

Excluding the trivial case $\boldsymbol{g}_{3,n}-\boldsymbol{g}_{2,n}=0$ with probability one, it follows by the same argument as in section II.5.4 that the Markov chain $\{\boldsymbol{z}_n, n=1,2,\ldots\}$ is ergodic if and only if

$$B=\sum_{n=1}^{\infty}\frac{1}{n}\Pr\{\boldsymbol{k}_n>0\}<\infty. \tag{2.75}$$

This condition is equivalent with

$$a_g<1. \tag{2.76}$$

Putting for $a_g<1$,

$$v_j \stackrel{\text{def}}{=} \lim_{n\to\infty}\Pr\{\boldsymbol{z}_n=j \mid \boldsymbol{z}_1=0\}, \qquad j=0,1,\ldots, \tag{2.77}$$

then for $|p|\leqq 1$, $a_g<1$,

$$v(p)\stackrel{\text{def}}{=}\sum_{j=0}^{\infty}p^j v_j = g_3(p)\exp\left\{-\sum_{n=1}^{\infty}\frac{1}{n}\sum_{i=1}^{\infty}(1-p^i)\Pr\{\boldsymbol{k}_n=i\}\right\}. \tag{2.78}$$

Note that $v(p)/g_3(p)$ represents the generating function of a discrete, infinitely divisible, probability distribution.

Next, we shall consider the return time $\boldsymbol{u}$ of the event: "a departing batch leaves the system empty". This return time $\boldsymbol{u}$ consists of an integral number $\boldsymbol{n}$ of service times. Define for $\boldsymbol{z}_1=0$,

$$\boldsymbol{g}\stackrel{\text{def}}{=}\boldsymbol{k}_{2,\boldsymbol{n}}, \qquad \boldsymbol{d}\stackrel{\text{def}}{=}\boldsymbol{d}_{\boldsymbol{n}+1}-\boldsymbol{d}_1, \qquad \boldsymbol{m}\stackrel{\text{def}}{=}\boldsymbol{g}-\boldsymbol{d},$$

so that $\boldsymbol{g}$ represents the total capacity for service during the return time $\boldsymbol{u}$, while $\boldsymbol{d}$ is the unused capacity and $\boldsymbol{m}$ the total number of customers served during the return time $\boldsymbol{u}$. As in section II.5.5 we now have from renewal theory for $|r|<1$, Re $\rho_2 \geqq 0$, Re $\rho_3 \geqq 0$,

$$\sum_{n=1}^{\infty} r^n \,\mathrm{E}\{(z_{n+1}=0)\exp(-\rho_2 \boldsymbol{d}_{n+1}-\rho_3 \boldsymbol{k}_{2,n}) \mid z_1=0\}$$

$$=\sum_{n=1}^{\infty} [\mathrm{E}\{r^n \exp(-\rho_2(\boldsymbol{d}_{n+1}-\boldsymbol{d}_1)-\rho_3 \boldsymbol{k}_{2,n})\}]^n = \sum_{n=1}^{\infty} [\mathrm{E}\{r^{\boldsymbol{n}}\,\mathrm{e}^{-\rho_2 \boldsymbol{d}-\rho_3 \boldsymbol{g}}\}]^n$$

$$=\frac{\mathrm{E}\{r^{\boldsymbol{n}} \exp(-\rho_2 \boldsymbol{d}-\rho_3 \boldsymbol{g})\}}{1-\mathrm{E}\{r^{\boldsymbol{n}} \exp(-\rho_2 \boldsymbol{d}-\rho_3 \boldsymbol{g})\}}.$$

Hence by using (2.69) we obtain for $|r| \leqq 1$, Re $\rho_2 \geqq 0$, Re $\rho_3 \geqq 0$,

$$\mathrm{E}\{r^{\boldsymbol{n}} \exp(-\rho_2 \boldsymbol{d}-\rho_3 \boldsymbol{g})\} = 1 - \left[1+\frac{1}{r}\,\zeta_0(r, \infty, -\rho_2, \rho_3)\right]^{-1}, \tag{2.79}$$

with (cf. (2.71), (2.72)),

$$\zeta_0(r, \infty, -\rho_2, \rho_3) = \sum_{n=2}^{\infty} r^n \,\mathrm{E}\{(z_n=0)\exp(-\rho_2 \boldsymbol{d}_n-\rho_3 \boldsymbol{k}_{2,n-1}) \mid z_1=0\}$$

$$= r^2 \beta(1/\alpha)\,\gamma_2(\rho_3+\rho_2) \exp\left\{\sum_{n=1}^{\infty} \frac{r^n}{n}\,\mathrm{E}\{(\boldsymbol{k}_n \leqq 0)\exp(-\rho_3 \boldsymbol{k}_{2,n}+\rho_2 \boldsymbol{k}_n)\}\right\}$$

$$= r^2 \beta(1/\alpha)\,\gamma_2(\rho_3+\rho_2) \exp\left\{\sum_{n=1}^{\infty} \frac{r^n}{n}\,\mathrm{E}\{(\boldsymbol{k}_n \leqq 0)\exp(-\rho_3 \boldsymbol{k}_{3,n}+(\rho_3+\rho_2)\boldsymbol{k}_n)\}\right\}. \tag{2.80}$$

The Laplace-Stieltjes transforms of the distributions of $\boldsymbol{d}$ and $\boldsymbol{g}$ can now be easily obtained; moreover for $|r| \leqq 1$, Re $\rho_2 \geqq 0$, Re $\rho_3 \geqq 0$,

$$\mathrm{E}\{r^{\boldsymbol{n}} \exp(-\rho_2 \boldsymbol{d}-\rho_3 \boldsymbol{g})\} = \mathrm{E}\{r^{\boldsymbol{n}} \exp(\rho_2 \boldsymbol{m}-(\rho_2+\rho_3)\,\boldsymbol{g})\}. \tag{2.81}$$

Since it follows from (2.80) that the latter Laplace-Stieltjes transform exists for Re $\rho_2 \geqq -$Re ρ_3 it follows from the last relation for $|r| \leqq 1$, Re $\rho \geqq 0$ that

$$\mathrm{E}\{r^{\boldsymbol{n}}\,\mathrm{e}^{-\rho \boldsymbol{m}}\} = 1 - \left[1 + r\beta(1/\alpha) \exp\left\{\sum_{n=1}^{\infty} \frac{r^n}{n}\,\mathrm{E}\{\mathrm{e}^{-\rho \boldsymbol{k}_{3,n}}\,(\boldsymbol{k}_n \leqq 0)\}\right\}\right]^{-1},$$

or for $|r_1| \leqq 1$, $|r_2| \leqq 1$,

$$\mathrm{E}\{r_1^{\boldsymbol{n}} r_2^{\boldsymbol{m}}\} = 1 - \left[1 + r_1 \beta(1/\alpha) \exp\left\{\sum_{n=1}^{\infty} \frac{r_1^n}{n}\,\mathrm{E}\{r_2^{\boldsymbol{k}_{3,n}}(\boldsymbol{k}_n \leqq 0)\}\right\}\right]^{-1}. \tag{2.82}$$

The first service batch in the return time $\boldsymbol{u}$ is always empty, and $\boldsymbol{u}$ consists of only one service if no customers arrive during the first service time. It is easily seen that

$$\Pr\{\boldsymbol{n}=1\}=\beta(1/\alpha).$$

The probability that a succession of exactly k empty service batches occurs is

$$\{\beta(1/\alpha)\}^{k-1}\,(1-\beta(1/\alpha)),\qquad k=1,2,\dots .$$

Denote the number of successive empty batches by $\boldsymbol{n}_1$ and the total capacity for service of these empty batches by $\boldsymbol{g}^{(1)}$, then for $|r|<1$, $\mathrm{Re}\,\rho\geqq 0$,

$$\mathrm{E}\{r^{\boldsymbol{n}_1}\,\mathrm{e}^{-\rho \boldsymbol{g}^{(1)}}\} = \frac{r\gamma_2(\rho)\{1-\beta(1/\alpha)\}}{1-r\gamma_2(\rho)\,\beta(1/\alpha)}\,. \tag{2.83}$$

Denote the number of non-empty batches in a return time $\boldsymbol{u}$ by $\boldsymbol{n}_2$, the total unused capacity of these batches by $\boldsymbol{d}^{(2)}$ and the total capacity of these batches by $\boldsymbol{g}^{(2)}$. Since

$$\boldsymbol{n}=\boldsymbol{n}_2+1,\qquad \boldsymbol{d}=\boldsymbol{d}^{(2)}+\boldsymbol{g}_{2,1},\qquad \boldsymbol{g}=\boldsymbol{g}^{(2)}+\boldsymbol{g}_{2,1},$$

we have for $|r|\leqq 1$, $\mathrm{Re}\,\rho_2\geqq 0$, $\mathrm{Re}\,\rho_3\geqq 0$,

$$\mathrm{E}\{r^{\boldsymbol{n}}\exp(-\rho_2\boldsymbol{d}-\rho_3\boldsymbol{g})\}=\mathrm{E}\{r^{\boldsymbol{n}_2+1}\exp(-\rho_2(\boldsymbol{d}^{(2)}+\boldsymbol{g}_{2,1})-\rho_3(\boldsymbol{g}^{(2)}+\boldsymbol{g}_{2,1}))\},$$

so that

$$\mathrm{E}\{r^{\boldsymbol{n}_2}\exp(-\rho_2\boldsymbol{d}^{(2)}-\rho_3\boldsymbol{g}^{(2)})\} = \frac{1}{r\gamma_2(\rho_2+\rho_3)}\,\mathrm{E}\{r^{\boldsymbol{n}}\exp(-\rho_2\boldsymbol{d}-\rho_3\boldsymbol{g})\}. \tag{2.84}$$

The part of the return time $\boldsymbol{u}$ during which non-empty batches are served will be denoted as the busy period $\boldsymbol{p}$ for the transportation problem. Obviously we have

$$\mathrm{E}\{\boldsymbol{p}\}=\beta\,\mathrm{E}\{\boldsymbol{n}_2\}. \tag{2.85}$$

Since

$$\boldsymbol{m}=\boldsymbol{g}^{(2)}-\boldsymbol{d}^{(2)},$$

is the number of customers served during a busy period we obtain from (2.84),

$$\mathrm{E}\{r^{\boldsymbol{n}_2}\exp(-(\rho_2+\rho_3)\,\boldsymbol{d}^{(2)}-\rho_3\boldsymbol{m})\}=\mathrm{E}\{r^{\boldsymbol{n}_2}\exp(-\rho_2\boldsymbol{d}^{(2)}-\rho_3\boldsymbol{g}^{(2)})\}. \tag{2.86}$$

Obviously, $\Pr\{\boldsymbol{d}^{(2)}=0\}$ is the probability that the service capacity of the busy period is fully utilized; from (2.79) and (2.80) it is easily found that

$$\Pr\{\boldsymbol{d}^{(2)}=0\} = \beta(1/\alpha)\exp\left(\sum_{n=1}^{\infty}\frac{1}{n}\Pr\{\boldsymbol{k}_n=0\}\right).$$

Define

$$y_n \overset{\text{def}}{=} [z_n - g_{2,n}]^+, \qquad n=1, 2, \ldots, \tag{2.87}$$

then y_n represents the number of waiting customers immediately after the service of the nth batch has started. Obviously,

$$y_{n+1} = [y_n + g_{3,n} - g_{2,n+1}]^+, \qquad n=1, 2, \ldots, \tag{2.88}$$

and (cf. (2.67)),

$$d_{n+2} - d_{n+1} = -[y_n + g_{3,n} - g_{2,n+1}]^-, \qquad n=1, 2, \ldots . \tag{2.89}$$

The relations (2.88) and (2.89) are essentially the same as the basic relations of section II.5.1. Putting

$$h_{2,n} \overset{\text{def}}{=} \sum_{m=2}^{n+1} g_{2,m}, \qquad h_n \overset{\text{def}}{=} k_{3,n} - h_{2,n}, \qquad n=1, 2, \ldots, \tag{2.90}$$

it immediately follows from section II.5.3 for $|r|<1$, $\operatorname{Re}\rho_1 \geqq 0$, $\operatorname{Re}\rho_2 \leqq 0$, $\operatorname{Re}\rho_3 \geqq 0$, $\operatorname{Re}\rho_4 \geqq 0$ that

$$\begin{aligned}
&\eta(r, \rho_1, \rho_2, \rho_3, \rho_4, 0, 0) \\
&\overset{\text{def}}{=} \sum_{n=1}^{\infty} r^n \operatorname{E}\{\exp(-\rho_1 y_n + \rho_2 d_{n+1} - \rho_3 k_{3,n-1} - \rho_4 h_{2,n-1}) \mid y_1 = 0, d_2 = 0\} \\
&= r \exp\left\{\sum_{n=1}^{\infty} \frac{r^n}{n} \operatorname{E}\{\exp(-\rho_4 k_{3,n} - \rho_3 h_{2,n})(\mathrm{e}^{-\rho_1 h_n}(h_n > 0) + \mathrm{e}^{-\rho_2 h_n}(h_n \leqq 0))\}\right\} \\
&= r \exp\left\{\frac{-1}{2\pi\mathrm{i}} \int_{C_\xi} \left\{\frac{1}{\rho_1 - \xi} + \frac{1}{\xi - \rho_2}\right\} \log\{1 - r\gamma_3(\rho_4 + \xi)\,\gamma_2(\rho_3 - \xi)\}\,\mathrm{d}\xi\right\},
\end{aligned} \tag{2.91}$$

for $\max(-\operatorname{Re}\rho_4, \operatorname{Re}\rho_2) < \operatorname{Re}\xi < \min(\operatorname{Re}\rho_1, \operatorname{Re}\rho_3)$, and

$$\begin{aligned}
&\eta(r, \rho_1, \rho_2, \rho_3, \rho_4, j, d) \\
&\overset{\text{def}}{=} \sum_{n=1}^{\infty} r^n \operatorname{E}\{\exp(-\rho_1 y_n + \rho_2 d_{n+1} - \rho_3 k_{3,n-1} - \rho_4 h_{2,n-1}) \mid y_1 = j, d_2 = d\} \\
&= \frac{1}{2\pi\mathrm{i}} \int_{C_\xi} \mathrm{e}^{-\xi j + \rho_2 d} \left\{\frac{1}{\rho_1 - \xi} + \frac{1}{\xi - \rho_2}\right\} \eta(r, \rho_1, \xi, \rho_3, \rho_4, 0, 0)\,\mathrm{d}\xi.
\end{aligned} \tag{2.92}$$

Obviously, all the results of chapter II.5 may be used immediately. For instance if $a_g<1$ the stationary distribution of the Markov chain $\{\boldsymbol{y}_n,\ n=0, 1, \ldots\}$ is infinitely divisible. It is easily seen that if $a_g<1$ (cf. (2.78) and (2.91)),

$$\sum_{j=0}^{\infty} p^j \lim_{n\to\infty} \Pr\{\boldsymbol{y}_n=j \mid \boldsymbol{y}_n=0\} = \frac{v(p)}{g_3(p)}, \qquad |p|\leqq 1. \tag{2.93}$$

Above the Laplace-Stieltjes transforms and generating functions of the various distributions are given as functions of $g_1(p)$, $g_2(p)$, $g_3(p)$, or of $\gamma_1(\rho)$, $\gamma_2(\rho)$, $\gamma_3(\rho)$. By using the method as described in chapter II.5 we obtain integral expressions which are suitable for evaluating the various transforms (cf. sections II.5.8, II.5.10 and II.5.11).

Next, we shall consider the waiting time distributions. The waiting time of a customer depends on the rule by which the server composes the next batch to be served. The same behaviour of the server as in the preceding section 2.4 is assumed here. We shall only consider the limit for $t\to\infty$ of the waiting time distribution of a customer arriving at time t. It is easily proved that this limit exists by using the theory of regenerative processes (cf. section I.6.4) and by noting that the $\boldsymbol{x}_t$-process ($\boldsymbol{x}_t$ is the number of customers in the system at time t) is regenerative with respect to the renewal process defined by the successive moments at which batch services start.

In the case $a_g<1$ for the stationary situation $Q_j(\zeta)$, denotes the probability that an arriving group of customers meets j waiting customers in the system and that the residual service time at that arrival moment is less than ζ. Supposing that $B(t)$ is not a lattice distribution (cf. (I.6.37)),

$$Q_j(\zeta) = \int_{\sigma=0}^{\infty} \sum_{i=0}^{j} \lim_{n\to\infty} \Pr\{\boldsymbol{y}_n=i\} \sum_{h=0}^{\infty} \mathrm{e}^{-\sigma/\alpha} \frac{(\sigma/\alpha)^h}{h!} g_{1,j-i}^{h*} \frac{1}{\beta} \{B(\zeta+\sigma)-B(\sigma)\}\, \mathrm{d}\sigma.$$

Hence (cf. (2.93)),

$$\sum_{j=0}^{\infty} p^j Q_j(\zeta) = \frac{v(p)}{g_3(p)} \int_{\sigma=0}^{\infty} \exp\{-(1-g_1(p))\sigma/\alpha\}\, \{B(\zeta+\sigma)-B(\sigma)\} \frac{\mathrm{d}\sigma}{\beta},$$
$$|p|\leqq 1. \tag{2.94}$$

It follows that

$$\sum_{j=0}^{\infty} p^j Q_j(\infty) = \frac{v(p)}{g_3(p)} \frac{1-g_3(p)}{\{1-g_1(p)\}\beta/\alpha}, \qquad |p|\leqq 1. \tag{2.95}$$

It is not difficult to show that for $a_g<1$,

$$\lim_{t\to\infty} \Pr\{x_t=j\}=Q_j(\infty), \qquad j=0, 1, \dots . \tag{2.96}$$

As in the preceding section (cf. (2.46)) it now follows for the waiting time distribution $W_m(t)$ of the mth customer of an arrival group, supposing the group contains at least m customers, that

$$W_m(t) = \sum_{j=0}^{\infty} Q_j(t) * \sum_{n=0}^{\infty} B^{n*}(t) \sum_{k=0}^{m+j-1} \{g_{2,k}^{n*}-g_{2,k}^{(n+1)*}\}, \tag{2.97}$$

for $t\geqq 0$, $m=1, 2, \dots$. By the same method as has been used in the derivation of (2.47) we obtain for $\operatorname{Re}\rho\geqq 0$, $|p|<|\xi|=1$,

$$\sum_{m=1}^{\infty} p^m \int_{0-}^{\infty} e^{-\rho t}\, dW_m(t)$$

$$= \frac{1}{2\pi i} \int_{D_\xi} \frac{p\, d\xi}{(\xi-p)(1-\xi)} \frac{1-g_2(\xi)}{1-\beta(\rho)\, g_2(\xi)} \frac{v(1/\xi)}{g_3(1/\xi)} \frac{\beta(\rho)-g_3(1/\xi)}{(1-g_1(1/\xi))\beta/\alpha-\beta\rho}, \tag{2.98}$$

where D_ξ is the unit circle in the complex ξ-plane. For the evaluation of this integral see the remarks on formula (2.47). The above integrand is analytic outside the unit circle except at the zeros of

$$1-\beta(\rho)\, g_2(\xi), \qquad \operatorname{Re}\rho>0,$$

which are poles of the integrand.

III.2.6. The bulk queue $M/G/1$ with accessible batches

In this section we shall discuss briefly the $M/G/1$ bulk queue with accessible batches. Customers arrive in groups and are served in batches. The groups arrive according to a Poisson process. As soon as a batch is completely served it leaves the system and immediately the service time of a new batch starts, even if no customers are present. Hence, once again we have the possibility of empty batches. However, if a batch is being served and does not utilize its full capacity for service, it remains accessible for customers arriving during the service time of this batch until its complete capacity for service is utilized. The service time of the batch is not altered by joining customers, so that these customers only obtain service for the residual service time of the batch, and leave the system together with all other members of the batch.

The queueing model discussed above describes some queueing situations for elevators, but it is particularly useful for queueing at traffic lights (cf. NEWELL [1960, 1965]). In the latter example the cycle of the traffic lights, i.e. the total duration of green and red period (amber is zero), is the service time. The capacity for service of a service period is the number of vehicles that can pass in the green period.

With $\boldsymbol{x}_t$ the number of customers in the system at time t and $\boldsymbol{z}_n$ the number of customers left behind at the moment of the nth batch departure we have

$$\boldsymbol{z}_{n+1}=[\boldsymbol{z}_n-\boldsymbol{g}_{2,n}+\boldsymbol{g}_{3,n}]^+, \qquad n=1, 2, \dots . \tag{2.99}$$

Denoting by $\boldsymbol{d}_{n+1}$ the total unused service capacity of the first n batches, we have

$$\boldsymbol{d}_{n+1}-\boldsymbol{d}_n=-[\boldsymbol{z}_n-\boldsymbol{g}_{2,n}+\boldsymbol{g}_{3,n}]^-, \qquad n=1, 2, \dots; \qquad \boldsymbol{d}_1 \stackrel{\text{def}}{=} 0. \tag{2.100}$$

The set of relations (2.99) and (2.100) has been extensively studied in chapter II.5 (see also the relations (2.89) and (2.99) of the preceding section), and all results found in chapter II.5 may immediately be used here.

III.2.7. The queue $G/M/1$ with batch service

We shall consider the queueing model $G/M1$ with single arrivals and batch service, the behaviour of the server being described by the variant (*ii*)a of section 2.1. The analysis of this queueing problem for the general distribution $\{g_{2,k}, k=1, 2, \dots\}$ of the capacity for service seems rather complicated (cf. LE GALL, [1962] p. 267). Therefore, we shall restrict the analysis here to a special class of distributions $\{g_{2,k}, k=1, 2, \dots\}$ viz. to those distributions represented by

$$g_{2,k}=\frac{\int_0^\infty \frac{(t/\beta)^k}{k!}\,\mathrm{e}^{-t/\beta}\,\mathrm{d}\Psi(t)}{\int_0^\infty \{1-\mathrm{e}^{-t/\beta}\}\,\mathrm{d}\Psi(t)}, \qquad k=1, 2, \dots, \tag{2.101}$$

where β is the average service time. The function $\Psi(t)$ is defined as in chapter I.5. This class of distributions contains quite a number of interesting distributions, e.g.

(*i*) $\Psi(t) = -\int_t^\infty \tau^\nu \,\mathrm{d}\tau, \qquad -2<\nu<1, \quad t\in(0,\infty),$

$$g_{2,k} = \frac{\Gamma(k+\nu+1)}{\Gamma(k+1)\,\Gamma(\nu+1)}, \qquad k=1,2,\dots;$$

(*ii*) $\Psi(t) = -\int_t^\infty \tau^\nu \,\mathrm{e}^{-\mu\tau}\,\mathrm{d}\tau, \qquad \nu>-2, \quad \nu\neq 1, \quad \mu>0, \quad t\in(0,\infty),$

$$g_{2,k} = \begin{cases} \dfrac{\Gamma(k+\nu+1)}{\Gamma(k+1)\,\Gamma(\nu+1)} \, \dfrac{(\beta\mu)^{\nu+1}}{(1+\beta\mu)^{\nu+1}-(\beta\mu)^{\nu+1}} \, \dfrac{1}{(1+\beta\mu)^k}, & k=1,2,\dots, \\[2ex] \dfrac{\beta\mu}{(1+\beta\mu)^k}, & \nu=0, \quad k=1,2,\dots; \end{cases}$$

(*iii*) $\Psi(t) = -\int_t^\infty \tau^{-1} \,\mathrm{e}^{-\mu\tau}\,\mathrm{d}\tau, \qquad \mu>0, \quad t\in(0,\infty),$

$$g_{2,k} = \frac{1}{k}\,\frac{1}{(1+\mu\beta)^k}\left\{\log\frac{1+\mu\beta}{\mu\beta}\right\}^{-1}, \qquad k=1,2,\dots;$$

(*iv*) $\Psi(t) = -\{1-U(t-\delta)\}, \qquad t>0, \quad \delta>0,$

$$g_{2,k} = \frac{(\delta/\beta)^k}{k!}\,\mathrm{e}^{-\delta/\beta}\,(1-\mathrm{e}^{-\delta/\beta})^{-1}, \qquad k=1,2,\dots .$$

Suitable linear combinations of the functions $\Psi(t)$ given above may also be used (cf. COHEN [1963]).

Consider the stochastic process $\boldsymbol{x}_t$ defined as follows. At time t the event "$\boldsymbol{x}_t=-1$" indicates the situation of the system with the server idle; $\boldsymbol{x}_t=i$ represents the event that the server is busy and i waiting customers are present, $i=0,1,\dots$. Define $\boldsymbol{x}_0=k$, and assume that at $t=0$ a customer arrives; $\boldsymbol{t}_n$ denotes the moment of arrival of the nth customer after $t=0$. As in section II.3.1 it is obvious that the process $\{\boldsymbol{x}_t, t\in[0,\infty)\}$ is regenerative with respect to the renewal process with renewal distribution $A(t)$ and with the arrival moments as renewals. Since the service times of the batches are independent, all having the same negative exponential distribution, the $\boldsymbol{x}_t$-process between two successive renewals is a continuous time parameter Markov process with state space the set $(-1, 0, 1, \dots)$. Consequently, the regenerative process $\{\boldsymbol{x}_t, t\in[0,\infty)\}$ is Markovian.

The Q-matrix of this process is determined by the distribution of the capacity for service of the batches. Since we assume this distribution to be of the type given by formula (2.101) it follows from the results of chapter I.5 that the $\boldsymbol{x}_t$-process between two successive renewals can be described by a quasi-derived death process. Denoting the transition probabilities of this death process by ${}_3p_{ij}(t)$, it follows from section I.5.3 that for $t \geqq 0$,

$$
\begin{aligned}
&{}_3p_{-1,-1}(t) = 1, \qquad {}_3p_{ii}(t) = \mathrm{e}^{-t\mu(0)}, \qquad i=0, 1, \ldots, \qquad (2.102)\\
&{}_3p_{0,-1}(t) = 1 - \mathrm{e}^{-t\mu(0)}, \qquad {}_3p_{ij}(t) = 0, \qquad i=-1, 0, 1, \ldots,\\
&\qquad\qquad\qquad\qquad \text{and} \quad j=i+1, i+2, \ldots,
\end{aligned}
$$

$$
{}_3p_{ij}(t) = \int_0^\infty \frac{(\tau/\beta)^{i-j}}{(i-j)!}\, \mathrm{e}^{-\tau/\beta}\, \mathrm{d}_\tau \bar{b}(t, \tau), \qquad i=2, 3, \ldots;\ j=1, \ldots, i-1,
$$

$$
\begin{aligned}
{}_3P_0(t; x) &= \sum_{i=1}^\infty x^i\, {}_3p_{i0}(t)\\
&= \frac{x}{1-x}\, \frac{\mu(x)}{\mu(x)-\mu(0)} \{\mathrm{e}^{-t\mu(0)} - \mathrm{e}^{-t\mu(x)}\}, \qquad |x|<1, \qquad (2.103)
\end{aligned}
$$

$$
\begin{aligned}
{}_3P_{-1}(t; x) &= \sum_{i=1}^\infty x^i\, {}_3p_{i,-1}(t)\\
&= \frac{x}{1-x} \left\{1 - \frac{\mu(x)\, \mathrm{e}^{-t\mu(0)} - \mu(0)\, \mathrm{e}^{-t\mu(x)}}{\mu(x)-\mu(0)}\right\}, \qquad |x|<1,
\end{aligned}
$$

with

$$
\mu(x) = \int_0^\infty \{1 - \exp(-(1-x)t/\beta)\}\, \mathrm{d}\Psi(t), \qquad |x|<1, \qquad (2.104)
$$

$$
\mu(0) = \frac{1}{\beta},
$$

and $b(t, \tau)$ being defined as in section I.5.1 with

$$
\int_{-\infty}^\infty \mathrm{e}^{-\tau\xi}\, \mathrm{d}_\tau b(t, \tau) = \exp\{-t \int_0^\infty \{1-\mathrm{e}^{-\xi\tau}\}\, \mathrm{d}\Psi(\tau)\}, \qquad \mathrm{Re}\, \xi \geqq 0, \quad t \geqq 0. \qquad (2.105)
$$

As the distribution $\{g_{2,k},\ k=1, 2, \ldots\}$ does not change if $\Psi(t)$ is multiplied by a factor, it is always possible to choose this factor so that the second relation of (2.104) holds.

The imbedded process $\{\boldsymbol{y}_n, n=0, 1, \ldots\}$ with

$$\boldsymbol{y}_n \overset{\text{def}}{=} \boldsymbol{x}_{t_n}, \qquad n=1, 2, \ldots, \qquad \boldsymbol{y}_0 \overset{\text{def}}{=} \boldsymbol{x}_0, \tag{2.106}$$

is a discrete time parameter Markov chain with stationary transition probabilities and state space $\{-1, 0, 1, \ldots\}$. Here $\boldsymbol{y}_n$ is the number of waiting customers in the system just before the arrival of the nth arriving customer after $t=0$ if $\boldsymbol{y}_n \geq 0$. For $i, j \in \{-1, 0, 1, \ldots\}$ it follows that

$$p_{ij} \overset{\text{def}}{=} \Pr\{\boldsymbol{y}_{n+1}=j \mid \boldsymbol{y}_n=i\} = \int_0^\infty {}_3p_{i+1,j}(t)\,\mathrm{d}A(t). \tag{2.107}$$

We shall first investigate the necessary and sufficient condition for the Markov chain $\{\boldsymbol{y}_n, n=0, 1, \ldots\}$ to be ergodic. For the present we shall indicate this Markov chain by M. Consider a Markov chain M_2 of which the state space is obtained from that of M by contracting the states E_0 and E_{-1} of M into one single state E_0, so that the state space of M_2 is $\{E_0, E_1, \ldots\}$; its one step transition matrix $({}_2p_{ij})$ is defined by

$$\begin{aligned} {}_2p_{ij} &\overset{\text{def}}{=} p_{ij}, \qquad i=0, 1, \ldots; \quad j=1, 2, \ldots, \\ {}_2p_{i0} &\overset{\text{def}}{=} p_{i0} + p_{i,-1}, \qquad i=0, 1, \ldots . \end{aligned} \tag{2.108}$$

Putting

$$A_2(t) \overset{\text{def}}{=} \int_{-\infty}^{\infty} b(\tau, t)\,\mathrm{d}A(\tau), \qquad -\infty<t<\infty, \tag{2.109}$$

then with some simple algebra we obtain from (2.102), (2.103) and (2.108),

$${}_2p_{ij} = \begin{cases} 0, & i=0,1,\ldots;\ j=i+2, i+3, \ldots, \\ \displaystyle\int_0^\infty \frac{(t/\beta)^{i+1-j}}{(i+1-j)!}\,\mathrm{e}^{-t/\beta}\,\mathrm{d}A_2(t), & i=0,1,\ldots;\ j=1,2,\ldots, i+1, \\ \displaystyle\int_0^\infty \left\{\int_0^t \frac{(\tau/\beta)^i}{i!}\,\mathrm{e}^{-\tau/\beta}\,\frac{\mathrm{d}\tau}{\beta}\right\}\mathrm{d}A_2(t), & i=0,1,\ldots;\ j=0. \end{cases} \tag{2.110}$$

It is easily seen that the transition matrix $({}_2p_{ij})$ above is of exactly the same type as that discussed in chapter II.3 (cf. (II.3.2)) for the $G/M/1$ queue with single arrivals and single service, because it is easily verified from (2.109) and from section I.5.1 that $A_2(t)$ is a distribution function of a non-negative stochastic variable (note that for every fixed $\tau \geqq 0$, $b(\tau, t)$ is a probability distribution function in t).

From the results of section II.3.2 it follows that all states of the chain M_2 are positive recurrent if and only if

$$\beta < \int_0^\infty t \, \mathrm{d}A_2(t).$$

Since

$$\int_0^\infty t \, \mathrm{d}A_2(t) = \alpha \int_0^\infty t \, \mathrm{d}\Psi(t), \qquad \gamma_2 = \sum_{k=1}^\infty k \, g_{2,k} = \int_0^\infty t \, \mathrm{d}\Psi(t),$$

the condition above is equivalent to

$$a_g \stackrel{\text{def}}{=} \frac{\beta}{\alpha\gamma_2} < 1.$$

Further, if $a_g = 1$ then all states of M_2 are null recurrent while they are transient for $a_g > 1$. From the definition of M_2 it is easily verified that the states of M_2 are of the same type as those of M and conversely. Hence $a_g < 1$ is a necessary and sufficient condition for the Markov chain $\{\boldsymbol{y}_n, n=0, 1, \ldots\}$ to be ergodic.

For $a_g < 1$, we define

$$u_i \stackrel{\text{def}}{=} \lim_{n\to\infty} \Pr\{\boldsymbol{y}_n = i \mid \boldsymbol{y}_0 = k\}, \qquad i=0, 1, \ldots,$$

the stationary distribution of the Markov chain $\{\boldsymbol{y}_n, n=0, 1, \ldots\}$; then we have from (2.102), (2.103) and (2.107),

$$u_j = \sum_{i=j-1}^\infty u_i \int_0^\infty \frac{(t/\beta)^{i+1-j}}{(i+1-j)!} \mathrm{e}^{-t/\beta} \, \mathrm{d}A_2(t), \qquad j=1, 2, \ldots, \tag{2.111}$$

$$u_0 = u_{-1} \int_0^\infty {}_3p_{0,0}(t) \, \mathrm{d}A(t) + \sum_{i=0}^\infty u_i \int_0^\infty {}_3p_{i+1,0}(t) \, \mathrm{d}A(t), \tag{2.112}$$

$$u_{-1} = u_{-1} \int_0^\infty {}_3p_{0,-1}(t) \, \mathrm{d}A(t) + \sum_{i=0}^\infty u_i \int_0^\infty {}_3p_{i+1,-1}(t) \, \mathrm{d}A(t). \tag{2.113}$$

The relation (2.111) is satisfied by

$$u_j = C\lambda^j, \qquad j=0, 1, 2, \ldots, \tag{2.114}$$

C being a constant, if λ is a root of

$$z=\alpha(\mu(z)),$$

or equivalently, of (cf. (2.105) and (2.109)),

$$z = \int_0^\infty \exp\{-(1-z)t/\beta\}\, \mathrm{d}A_2(t). \tag{2.115}$$

If $a_g<1$ then by Takács' lemma (cf. app. 6) (2.114) has exactly one root λ_0 inside the unit circle, and in fact $0<\lambda_0<1$. Since

$${}_3p_{00}(t)+{}_3p_{0,-1}(t)=1 \qquad \text{for all } t\geqq 0,$$

it follows by summing (2.112) and (2.113) that

$$u_0 = \sum_{i=0}^\infty u_i \int_0^\infty \{{}_3p_{i+1,0}(t)+{}_3p_{i+1,-1}(t)\}\, \mathrm{d}A(t),$$

and it is easily verified that (2.114) also satisfies the latter relation for λ a root of (2.115). Hence by taking

$$u_j=C\lambda_0^j, \qquad j=0, 1, \ldots, \quad a_g<1, \tag{2.116}$$

and using the normalizing condition

$$\sum_{j=1}^\infty u_j = 1,$$

we can determine C and u_{-1} from (2.113).
The result reads

$$u_{-1} = 1 - \frac{C}{1-\lambda_0}, \tag{2.117}$$

$$C = \frac{(1-\lambda_0)\, \alpha(\mu(0))\{\mu(\lambda_0)-\mu(0)\}}{\mu(\lambda_0)-\mu(0)+\mu(0)\{\lambda_0-\alpha(\mu(0))\}}. \tag{2.118}$$

It is easily verified that $u_{-1}>0$, $C>0$, so that it follows from theorem I.2.5 that the stationary distribution of the Markov chain $\{y_n, n=0, 1, \ldots\}$ is given by (2.116), ..., (2.118).

Next we shall consider the waiting time. Denote the waiting time of the nth arriving customer after $t=0$ by w_n, assuming that customers are served in order of arrival. Since the service process is a quasi-derived death process it follows that

$$_3p_{j+1,j+1}(t)+{}_3p_{j+1,j}(t)+\dots+{}_3p_{j+1,1}(t)$$

is the conditional probability that a customer who meets j waiting customers and the server busy at his arrival, is still waiting at a time t after his arrival. Consequently,

$$\Pr\{\boldsymbol{w}_n>t \mid \boldsymbol{y}_0=k\} = \sum_{j=0}^{\infty} \Pr\{\boldsymbol{y}_n=j \mid \boldsymbol{y}_0=k\} \sum_{i=1}^{j+1} {}_3p_{j+1,i}(t).$$

It follows that if $a_g<1$ then

$$1-W(t) \overset{\text{def}}{=} \lim_{n\to\infty} \Pr\{\boldsymbol{w}_n>t \mid \boldsymbol{y}_0=k\} = \sum_{j=0}^{\infty} u_j \sum_{i=1}^{j+1} {}_3p_{j+1,i}(t)$$

$$= \frac{C}{1-\lambda_0}\,\mathrm{e}^{-t\mu(\lambda_0)}, \qquad t>0.$$

Hence

$$W(t) = \begin{cases} 1-\dfrac{C}{1-\lambda_0}\,\mathrm{e}^{-t\mu(\lambda_0)}, & t>0, \\ 0, & t<0, \end{cases} \tag{2.119}$$

$$\int_0^{\infty} t\,\mathrm{d}W(t) = \frac{C}{(1-\lambda_0)\,\mu(\lambda_0)}.$$

From formula (I.6.38) we have for $a_g<1$,

$$w_j \overset{\text{def}}{=} \lim_{t\to\infty} \Pr\{\boldsymbol{x}_t=j \mid \boldsymbol{y}_0=k\} = \sum_{i=-1}^{\infty} u_i \int_0^{\infty} {}_3p_{i+1,j}(t)\{1-A(t)\}\frac{\mathrm{d}t}{\alpha}. \tag{2.120}$$

It follows that

$$w_j = \frac{1-\lambda_0}{\alpha\mu(\lambda_0)}\,\lambda_0^{j-1}C, \qquad j=1,2,\dots, \tag{2.121}$$

$$w_0 = \frac{1-\alpha(\mu(0))}{\alpha\mu(0)} + \frac{C}{1-\lambda_0}\,\frac{\lambda_0-\alpha(\mu(0))}{\{\mu(\lambda_0)-\mu(0)\}\,\alpha},$$

$$w_{-1} = 1-\sum_{j=0}^{\infty} w_j.$$

The results obtained above are all related to the stationary situation of the Markov chain $\{\boldsymbol{y}_n,\ n=0,1,\dots\}$ and the process $\{\boldsymbol{x}_t,\ t\in[0,\infty)\}$. We shall not analyse these processes here for finite values of n and t. Such an analysis can be given along the same lines as in chapter II.3.

III.2.8. The bulk queue $G/M/1$ in inventory control

Problems in inventory control can often be formulated as queueing problems. In this section we shall discuss an important example. Consider an inventory system, where items are stocked. These items arrive in groups of variable size. Let a group of items arrive at $\boldsymbol{t}_0=0$, and the nth group after $\boldsymbol{t}_0$ at $\boldsymbol{t}_n$. The interarrival times $\boldsymbol{t}_{n+1}-\boldsymbol{t}_n$, $n=0, 1, \ldots$, are supposed to be independent, identically distributed variables with distribution $A(t)$; $\boldsymbol{g}_{1,n+1}$ will represent the number of items in the nth arrival group after $\boldsymbol{t}_0$. Items leave the stock (i.e. are sold) in groups of variable size; the number in the nth departing group after $\boldsymbol{t}_0$ will be $\boldsymbol{g}_{2,n}$. The interdeparture times of successive groups of items sold are assumed to be independent, identically distributed variables all having the negative exponential distribution with mean β. As in section 2.1 it will again be assumed that the families of interarrival times, interdeparture times, arrival group sizes and departure group sizes are independent families of stochastic variables.

Obviously, the inventory model described above is closely related to the $M/G/1$ bulk queue, but in the present case items (customers) do not have a service time but stay until they leave the stock, and at such a departure moment the size of the departing group is determined. The size of the nth departing group is $\boldsymbol{g}_{2,n}$ or the total stock, whichever is smaller. In the following we shall often use the term 'customers' for items.

By $\boldsymbol{x}_t$ we shall again denote the number of customers in the system at time t; $\boldsymbol{x}_0=k$. Define

$$\boldsymbol{y}_n \stackrel{\text{def}}{=} \boldsymbol{x}_{\boldsymbol{t}_n}, \qquad n=0, 1, \ldots,$$

so that $\boldsymbol{y}_n$ represents the number of customers in the system just before the arrival of the nth group after $t=0$. By $\boldsymbol{g}_{4,n+1}$ we shall indicate the total number of customers who may depart during the interarrival time $(\boldsymbol{t}_n, \boldsymbol{t}_{n+1})$, $n=0, 1, \ldots$. Hence, since the interdeparture times are negative exponentially distributed

$$\Pr\{\boldsymbol{g}_{4,n+1}=j\} = \sum_{h=0}^{\infty} \int_0^{\infty} \frac{(t/\beta)^h}{h!}\, \mathrm{e}^{-t/\beta}\, g_{2,j}^{h*}\, \mathrm{d}A(t), \qquad j=0, 1, \ldots, \tag{2.122}$$

with

$$g_{2,0}^{0*} \stackrel{\text{def}}{=} 1, \qquad g_{2,j}^{0*} = 0, \qquad j=1, 2, \ldots,$$

$$g_4(p) \stackrel{\text{def}}{=} \sum_{j=0}^{\infty} p^j \Pr\{\boldsymbol{g}_{4,n+1}=j\} = \alpha\{(1-g_2(p))/\beta\}, \qquad |p| \leqq 1. \tag{2.123}$$

It should be noted that the $\boldsymbol{x}_t$-process is regenerative with respect to the renewal process defined by the successive moments of arrivals; moreover this regenerative process is Markovian since the interdeparture times are negative exponentially distributed and independent.

It is immediately seen that

$$\boldsymbol{y}_{n+1} = [\boldsymbol{y}_n + \boldsymbol{g}_{1,n+1} - \boldsymbol{g}_{4,n+1}]^+, \qquad n=0, 1, \ldots . \tag{2.124}$$

Denote the total demand for items during the time interval $[0, \boldsymbol{t}_n)$, which cannot be satisfied due to shortage of items in the stock, by $\boldsymbol{d}_n$, $n=0, 1, \ldots$, with $\boldsymbol{d}_0 \stackrel{\text{def}}{=} 0$, then

$$\boldsymbol{d}_{n+1} - \boldsymbol{d}_n = -[\boldsymbol{y}_n + \boldsymbol{g}_{1,n+1} - \boldsymbol{g}_{4,n+1}]^-, \qquad n=0, 1, \ldots . \tag{2.125}$$

The set of relations (2.124) and (2.125) have been extensively investigated in chapter II.5, and the results obtained there may be immediately used here.

Define

$$\boldsymbol{k}_{i,n} \stackrel{\text{def}}{=} \begin{cases} \sum_{m=1}^{n} \boldsymbol{g}_{i,m}, & i=1, 4; \quad n=1, 2, \ldots, \\ 0, & n=0, \end{cases} \tag{2.126}$$

$$\boldsymbol{k}_{5,n} \stackrel{\text{def}}{=} \boldsymbol{k}_{1,n} - \boldsymbol{k}_{4,n}, \qquad n=1, 2, \ldots,$$

then we have from section II.5.3,

$$\sum_{n=0}^{\infty} r^n \,\mathrm{E}\{\exp(-\rho_1 \boldsymbol{y}_n + \rho_2 \boldsymbol{d}_n - \rho_3 \boldsymbol{k}_{1,n} - \rho_4 \boldsymbol{k}_{4,n}) \mid \boldsymbol{y}_0 = 0\}$$
$$= \exp\left\{ \sum_{n=1}^{\infty} \frac{r^n}{n} \,\mathrm{E}\{\exp(-\rho_3 \boldsymbol{k}_{1,n} - \rho_4 \boldsymbol{k}_{4,n})((\boldsymbol{k}_{5,n} > 0)\exp(-\rho_1 \boldsymbol{k}_{5,n}) \right.$$
$$\left. + (\boldsymbol{k}_{5,n} \leqq 0)\exp(-\rho_2 \boldsymbol{k}_{5,n}))\} \right\}, \tag{2.127}$$

for $|r|<1$, $\mathrm{Re}\,\rho_1 \geqq 0$, $\mathrm{Re}\,\rho_2 \leqq 0$, $\mathrm{Re}\,\rho_3 \geqq 0$, $\mathrm{Re}\,\rho_4 \geqq 0$.

Define

$$\boldsymbol{n} = \min\{n:\ \boldsymbol{y}_n = 0;\ n=1, 2, \ldots\} \qquad \text{if} \quad \boldsymbol{y}_0 = 0, \tag{2.128}$$

so that $\boldsymbol{n}$ is the number of interarrival times between two successive moments of arrival at which the system is empty; the time $\boldsymbol{c}$ between these moments will again be called the 'busy cycle'. We shall denote the total number of customers arriving during a busy cycle by $\boldsymbol{g}^{(1)}$, i.e. if a busy cycle starts at $\boldsymbol{t}_0=0$ and ends at $\boldsymbol{t}_n$ then

$$\boldsymbol{g}^{(1)} \stackrel{\text{def}}{=} \boldsymbol{g}_{1,1} + \dots + \boldsymbol{g}_{1,\boldsymbol{n}};$$

further

$$\boldsymbol{g}^{(4)} \stackrel{\text{def}}{=} \boldsymbol{g}_{4,1} + \dots + \boldsymbol{g}_{4,\boldsymbol{n}},$$

is the total demand for items during a busy cycle, whereas

$$\boldsymbol{d}^{(1)} \stackrel{\text{def}}{=} \boldsymbol{d}_{\boldsymbol{n}},$$

is the total shortage for items during the busy cycle. As in section II.5.5 we obtain

$$\mathrm{E}\{r^{\boldsymbol{n}} \exp(-\rho_2 \boldsymbol{d}^{(1)} - \rho_3 \boldsymbol{g}^{(1)} - \rho_4 \boldsymbol{g}^{(4)})\}$$

$$= 1 - \exp\left\{ -\sum_{n=1}^{\infty} \frac{r^n}{n} \mathrm{E}\{(\boldsymbol{k}_{5,n} \leqq 0) \exp(-\rho_3 \boldsymbol{k}_{1,n} - \rho_4 \boldsymbol{k}_{4,n} + \rho_2 \boldsymbol{k}_{5,n})\} \right\}, \tag{2.129}$$

for $|r|<1$, $\mathrm{Re}\, \rho_i \geqq 0$, $i=2, 3, 4$.

It should be noted that the formulas given above also apply and are meaningful in the case where $\boldsymbol{g}_{1,n}$ and $\boldsymbol{g}_{2,n}$ are continuous variables.

Next, we determine the time a customer (item) spends in the system (stock). It is assumed that customers belonging to the same arrival group are numbered and that those with the lower numbers are sold first, while items belonging to different arrival groups are sold in the order of arrival.

Denoting the time that the mth customer of the nth arriving group after $\boldsymbol{t}_0=0$ stays in the system by $\boldsymbol{w}_m^{(n)}$ and supposing this arrival group contains at least m customers then (cf. the derivation of (2.46)) for $n=1, 2, \dots$; $t>0$, we have

$$\Pr\{\boldsymbol{w}_m^{(n)} < t \,|\, \boldsymbol{y}_0 = k\}$$

$$= \sum_{j=0}^{\infty} \Pr\{\boldsymbol{y}_n = j \,|\, \boldsymbol{y}_0 = k\} \sum_{r=1}^{\infty} B^{r*}(t) \sum_{i=0}^{m+j-1} \{g_{2,i}^{(r-1)*} - g_{2,i}^{r*}\}, \tag{2.130}$$

with $B(t)$ the negative exponential distribution. It follows for $|p|<1$, $\mathrm{Re}\, \rho>0$ (cf. the derivation of (2.47)) that

$$\sum_{m=1}^{\infty} p^m \, \mathrm{E}\{\exp(-\rho w_m^{(n)}) \mid y_0=k\}$$
$$= \frac{1}{2\pi \mathrm{i}} \int_{D_\xi} \frac{p}{\xi-p} \frac{1}{1-\xi} \mathrm{E}\{\xi^{-y_n} \mid y_0=k\} \frac{1-g_2(\xi)}{1+\beta\rho-g_2(\xi)} \, \mathrm{d}\xi, \qquad (2.131)$$

where D_ξ is the unit circle.

From the results of chapter II.5 it follows that the Markov chain $\{y_n, n=0, 1, \ldots\}$ is ergodic if and only if

$$a_g = \frac{\beta}{\alpha} \frac{\gamma_1}{\gamma_2} < 1.$$

From (2.131) it is then easily shown that the distribution of $w_m^{(n)}$ has for $n \to \infty$ a limit distribution to be denoted by $W_m(t)$. For $a_g<1$, with (cf. (II.5.36)),

$$u(p) \stackrel{\text{def}}{=} \sum_{j=0}^{\infty} p^j \lim_{n\to\infty} \Pr\{y_n=j \mid y_0=k\}$$
$$= \exp\left\{-\sum_{n=1}^{\infty} \frac{1}{n} \sum_{j=1}^{\infty} (1-p^j) \Pr\{k_{5,n}=j\}\right\}, \qquad |p| \leqq 1, \qquad (2.132)$$

it follows from (2.131) for $|p|<1$, $\mathrm{Re}\, \rho>0$ that

$$\sum_{m=1}^{\infty} p^m \int_{0-}^{\infty} \mathrm{e}^{-\rho t} \, \mathrm{d}W_m(t) = \frac{1}{2\pi \mathrm{i}} \int_{D_\xi} \frac{p}{\xi-p} \frac{u(1/\xi)}{1-\xi} \frac{1-g_2(\xi)}{1+\beta\rho-g_2(\xi)} \, \mathrm{d}\xi. \qquad (2.133)$$

Frequently, $u(\xi)$ is analytic for $|\xi|$ slightly larger than 1. In that case the contour D_ξ can be replaced by D'_ξ, i.e. the circle with centre at $\xi=0$ and radius $|\xi|=1-$. Differentiating the integral with respect to ρ and taking $\rho=0$ we obtain (note that the derivative of the integral converges uniformly in ρ for all ρ with $\mathrm{Re}\, \rho \geqq \varepsilon>0$),

$$\sum_{m=1}^{\infty} p^m \int_{0}^{\infty} t \, \mathrm{d}W_m(t) = \frac{1}{2\pi \mathrm{i}} \int_{D'_\xi} \frac{p}{\xi-p} \frac{\beta u(1/\xi)}{(1-\xi)(1-g_2(\xi))} \, \mathrm{d}\xi, \qquad |\xi|=1-, \qquad (2.134)$$

if the integral in the right-hand side is finite.

Finally, a brief comment; if we take for the distribution of $g_{2,n}$ a distribution of the type described by (2.101) we may write (cf. (I.5.2)),

$$\gamma_4(\rho) \stackrel{\text{def}}{=} \mathrm{E}\{\mathrm{e}^{-\rho g_{4,n}}\} = \int_0^{\infty} \exp\{-(1-\mathrm{e}^{-\rho})t/\beta\} \, \mathrm{d}A_2(t), \qquad \mathrm{Re}\, \rho \geqq 0,$$

with $A_2(t)$ as defined in the preceding section (cf. (2.109)). We would also have obtained this expression for $\gamma_4(\rho)$ if we had considered the same inventory model as above but with single departures and with $A_2(t)$ as the distribution function of the interarrival times of the groups. The latter model, however, is the $G/M/1$ queue of section II.3 with group arrivals. To this model we may apply the results of section 2.2.

III.2.9. The $G/G/1$ bulk queue with single arrivals and delayed service

We denote the server's behaviour by *delayed service* if he starts service of the next batch only if the service capacity of this batch can be fully utilized, i.e. if the server becomes idle and is ready for serving the next batch, but the number of available customers in the system is less than the service capacity for this batch, he waits until a sufficient number of new customers has arrived to fill this batch.

It is readily seen that the present model can be immediately reduced to the $G/G/1$ queueing model of chapter II.5 with distribution function of the interarrival times given by

$$\sum_{k=1}^{\infty} g_{2,k} A^{k*}(t). \tag{2.135}$$

Suppose that at $t=0$ the server is idle and no customers are present in the system. The moment of arrival of the nth customer after $t=0$ is as before denoted by $\mathbf{t}_n$, $n=1, 2, \ldots$; $\mathbf{t}_1>0$. Let $\mathbf{g}_{2,n}$, $n=1, 2, \ldots$, be the service capacity of the batch which is the nth one to be served after $t=0$.

Hence the nth service starts at time $\mathbf{t}_{\mathbf{k}_{2,n}}$, where

$$\mathbf{k}_{2,n} = \sum_{m=1}^{n} \mathbf{g}_{2,m}, \qquad n=1, 2, \ldots .$$

Considering the customers who are all served in the same batch as one super customer then our queueing model is the $G/G/1$ queue with distribution function of the interarrival time given by (2.135), because the family of variables $\mathbf{g}_{2,n}$, $n=1, 2, \ldots$, is independent from the interarrival times of the individual customers. For the initial conditions described above all results of chapter II.5 can now be used with $\alpha(\rho)$ replaced by $g_2(\alpha(\rho))$ for the Laplace-Stieltjes transform of the distribution of the interarrival time for the super customers. If the initial conditions are not as described above the distribution function of the interarrival time of the first super customer, and possibly also of a finite number of following super customers, changes. Such modifications can be easily handled.

For the initial conditions 'server idle and no customers present at $t=0$' we shall derive the expression for the waiting time distribution ${}_nW(t)$ of the nth arriving customer. Let $W_m(t)$ denote the distribution function of the waiting time of the mth super customer, so that for the initial conditions above

$$W_1(t) = \begin{cases} 1, & t>0, \\ 0, & t<0, \end{cases}$$

while the generating function of the Laplace-Stieltjes transform $\omega_m(\rho)$ of $W_m(t)$, i.e.

$$\sum_{m=1}^{\infty} r^m \omega_m(\rho), \qquad \mathrm{Re}\,\rho \geqq 0,$$

is given by (II.5.36) with $A(t)$ replaced by the expression (2.135).

It is now easily verified that

$$\sum_{k=0}^{\infty} \sum_{j=0}^{n-1} \Pr\{\boldsymbol{k}_{2,m-1}=j\} \Pr\{\boldsymbol{g}_{2,m}=n+k-j\}\, A^{k*}(t),$$

is the probability that the nth arriving (individual) customer is served in the mth batch and has to wait less than a time t until a sufficient number of customers has arrived to utilize completely the service capacity of the mth batch. Consequently, we have for $t>0$,

$${}_nW(t) = \sum_{m=1}^{\infty} \sum_{k=0}^{\infty} \sum_{j=0}^{n-1} \Pr\{\boldsymbol{k}_{2,m-1}=j\} \Pr\{\boldsymbol{g}_{2,m}=n+k-j\}\, A^{k*}(t) * W_m(t), \tag{2.136}$$

with

$$\Pr\{\boldsymbol{k}_{2,0}=0\} \stackrel{\text{def}}{=} 1, \qquad \Pr\{\boldsymbol{k}_{2,0}=j\} \stackrel{\text{def}}{=} 0, \qquad j=1,2,\dots .$$

From (2.136) we obtain for $|p|<1$, $\mathrm{Re}\,\rho \geqq 0$,

$$\sum_{n=1}^{\infty} p^n \int_{0-}^{\infty} \mathrm{e}^{-\rho t}\, \mathrm{d}\,{}_nW(t) = \frac{p}{p-\alpha(\rho)} \frac{g_2(p)-g_2(\alpha(\rho))}{g_2(p)} \sum_{m=1}^{\infty} g_2^m(p)\, \omega_m(\rho). \tag{2.137}$$

From chapter II.5 it follows that if and only if

$$a_g = \frac{\beta}{\alpha\gamma_2} < 1, \tag{2.138}$$

then $W_m(t)$ has for $m \to \infty$ a proper limit distribution $W(t)$ of which the Laplace-Stieltjes transform is given by (II.5.50) with $A(t)$ replaced by the expression (2.135). It is now easily proved that if (2.138) holds ${}_nW(t)$ has also a proper limit distribution for $n \to \infty$, and it is found by using (II.5.36) that

$$\int_{0-}^{\infty} e^{-\rho t}\, d_t \lim_{n\to\infty} {}_nW(t) = \lim_{p\uparrow 1} (1-p) \sum_{n=1}^{\infty} p^n \int_{0-}^{\infty} e^{-\rho t}\, d\, {}_nW(t)$$

$$= \frac{1-g_2(\alpha(\rho))}{(1-\alpha(\rho))\gamma_2} \int_{0-}^{\infty} e^{-\rho t}\, dW(t), \qquad \text{Re}\, \rho \geqq 0. \qquad (2.139)$$

Other results may be obtained immediately by applying the results of chapter II.5.

A special case of the above model is obtained if the service capacity of a batch is constant, say m (cf. TAKÁCS [1962] p. 81 and also LE GALL [1962] p. 251, BAILEY [1954[b]] and DOWNTON [1955]). If, moreover, $A(t)$ is a negative exponential distribution, the model is equivalent to the queueing system $E_m/G/1$ (cf. also section II.6.5).

II.3. PRIORITY DISCIPLINES FOR THE SINGLE SERVER QUEUE

III.3.1. Introduction

The derivation of the waiting time distributions for the queueing models investigated in part II are based on the assumption that customers are served in order of arrival; the so-called *first come first served* discipline. As already mentioned in section II.1.2 this type of service is only one, though important, variant of a complex of service disciplines. In this chapter we shall discuss a number of queue disciplines of interest for many actual queueing situations. These disciplines concern the choice of the next customer to be served when the server terminates a service.

The *last come first served* discipline occurs when the next customer to be served is the one who arrived last. *Random service* is given when the server chooses the next customer to be served at random out of the waiting customers, i.e. of the waiting customers present at the end of a service time every one has the same probability of being the next one to be served. In the case of *random priority service* the order of service of arriving customers is sampled from a uniform distribution on [0, 1] at the moments of their arrivals (cf. section 3.11).

The last come first served procedure is a special type of a priority discipline, viz. the last arriving customer has priority for service over those waiting customers present at his arrival. However, he looses his priority to those customers arriving during his waiting time. Frequently, a priority discipline does not depend on the state of the queueing system at the arrival of a customer, but is determined by a classification of arriving customers according to some criterion which is independent of the state of the system. Suppose arriving customers are classified into two different types.

If waiting customers of type 1 have always priority for service over those of type 2, while customers of the same type are served in order of arrival, it is said that the single server queue operates according to a *priority discipline with two levels*. Frequently, the term *head of queue* is added here to stress that customers of the same priority level are served in order of arrival.

In a single server queue with two priority levels the server is always available when he becomes idle to start the service of a waiting customer of first priority if any is present. Hence, from the point of view of customers with second priority the availability of the server to them is interrupted by the presence of customers of first priority. Consequently, customers of second priority arrive at a single server queue with *interrupted service*, i.e. for some time the server is not available to provide service and customers have to wait until the interruption has been cleared.

We shall first describe *the general single server queueing process with interrupted service*. The distributions of the interarrival times and the service times of the customers will be indicated by $A^{(2)}(t)$ and $B^{(2)}(t)$, respectively. Interarrival times, and similarly service times, are independent variables. The moments at which interruptions occur will be assumed to be a renewal process with renewal distribution $A^{(1)}(t)$. An *interruption time*, having distribution function $B^{(1)}(t)$, is associated with every interruption. Interruption times are assumed to be independent variables. Moreover, it will be assumed that the four families of stochastic variables, the interarrival times, the service times, the interruption times and the renewal times between interruptions are independent.

Define for $i=1, 2$,

$$\alpha^{(i)} \overset{\text{def}}{=} \int_0^\infty t \, \mathrm{d}A^{(i)}(t), \qquad \alpha_2^{(i)} \overset{\text{def}}{=} \int_0^\infty t^2 \, \mathrm{d}A^{(i)}(t), \quad A^{(i)}(0+) = 0, \tag{3.1}$$

$$\beta^{(i)} \overset{\text{def}}{=} \int_0^\infty t \, \mathrm{d}B^{(i)}(t), \qquad \beta_2^{(i)} \overset{\text{def}}{=} \int_0^\infty t^2 \, \mathrm{d}B^{(i)}(t), \quad B^{(i)}(0+) = 0.$$

An interruption may occur when the server is idle, when he is serving a customer and also during an interruption time. If an interruption occurs when the server serves a customer he may immediately stop the service and pay attention to the interruption or he may first finish the service at hand and then pay attention to the interruption. The first type of interruption will be called a *break-in* or *pre-emptive* interruption, the second type a *non break-in* or a *delayed* or *postponed* interruption. The interruption time starts at the moment when the server pays attention to the interruption. During every interruption time the server is not available to provide service to customers.

An interruption may occur during an interruption time, or before a preceding interruption received attention. If the server disregards such interruptions the interruption discipline will be called *restricted.* The usual discipline will be, however, that the server reacts to every interruption and with every interruption is associated an interruption time, these interruptions take effect one after another. The total time during which the server is not available for service may be the sum of the durations of a number of interruption times.

In the case of delayed interruptions, when the server has dealt with all interruptions, he immediately starts the service of a new customer, if one is present, otherwise he will become idle. In the case of a break-in interruption the customer whose service has been interrupted is again given service. If this service is continued from the point where it was interrupted the interruption discipline is called *pre-emptive resume*; if it is started afresh the discipline is referred to as *pre-emptive repeat.* At an interruption occurring when the server is idle the interruption time starts immediately. The scheme below summarizes the various interruption disciplines discussed above.

$$\begin{array}{l} \text{Interruption} \\ \text{(restricted)} \end{array} \left\{ \begin{array}{l} \text{delayed} \\ \\ \text{break-in (pre-emptive)} \left\{ \begin{array}{l} \text{resume} \\ \\ \text{repeat} \end{array} \right. \end{array} \right.$$

For queueing systems with interrupted service we need the concept of *completion time* of a customer. This is the time between the moment when the service of a customer begins and the first moment thereafter at which the server becomes available to serve another customer. The completion time of a customer will be denoted by $\boldsymbol{c}_{\mathrm{i}}$. The *occupation time* of the server is the time interval during which the server is busy or interrupted and which is preceded and followed by periods during which the server is idle (i.e. not busy and not interrupted).

The priority discipline discussed above is now readily seen to be queueing with delayed interruptions. Customers of first priority are causing interruptions to those of second priority. Priority disciplines with break-in interruptions, so called *pre-emptive priority* disciplines, also arise in actual queueing situations.

So far we have considered only priority queueing with two priority levels.

The generalisation to more priority levels is obvious. Denote the number of priority levels by k. Customers of ith priority level have priority for service over those of jth level if $i<j$. Customers of the same priority level are served in order of arrival. The superscript 'i' refers to customers of ith priority level. $A^{(i)}(t)$ and $B^{(i)}(t)$ are the distributions of interarrival times and service times, respectively; $\boldsymbol{x}_t^{(i)}$ represents the number of customers of ith priority level present in the system at time t. Further

$$\boldsymbol{x}_t = \sum_{i=1}^{k} \boldsymbol{x}_t^{(i)},$$

is the total number of customers present in the system at time t. All service times and all interarrival times are assumed to be independent. Moreover these two families are supposed to be independent.

The generalisation to systems with a non-denumerable class of priority levels is obtained by assigning to every customer a realisation of the non-negative stochastic variable $\boldsymbol{y}$, his *priority level*. Denote the distribution of $\boldsymbol{y}$ by $P(y)$. $P(y)$ is the probability that an arriving customer has priority level $\boldsymbol{y}<y$. For $P(y)$ a pure step-function with k jumps, the priority system with k priority levels is obtained.

An example of a priority system with a non-denumerable class of priorities is obtained when all customers have the same service time distribution $B(t)$ and the customer with the shortest service time of those waiting at the end of a service has priority for service (supposing that $B(t)$ is not a pure step-function). In this case $P(y)=B(y)$, see sections 3.9 and 3.10.

Priority queueing leads to smaller average waiting times for customers with the lower priority levels compared with service in order of arrival. The introduction of priorities in a queueing situation is, therefore, usually motivated by economic arguments; the cost structure in general being related to the costs which result from waiting. Hence, for a given objective function the optimal scheduling of priorities means the determination of an optimal priority distribution $P(y)$. For these aspects of priority queueing see Cox and Smith [1961], Morse [1958], Brosh and Naor [1963].

III.3.2. Waiting time for "last come, first served"

In section II.5.5 we considered the distribution of the residual busy period starting at the arrival of a customer whose waiting time is equal to w (cf. (II.5.82)). It is now easy to see that if with the "last come, first served" discipline an arriving customer finds the server occupied and at his moment of arrival the residual service time of the customer being served equals w, his waiting time distribution is that of a residual busy period with initial waiting time w. Hence, for the determination of the waiting time distribution for the "last come, first served" discipline we have to know the distribution of the residual service time at the moment of arrival of a customer if he finds the server busy.

(*i*) Last come, first served for $M/G/1$.

Denote by ξ_t the residual service time at time t whenever at t a customer arrives and the server is busy. Using the notation of chapter II.4 (cf. (II.4.3)) it follows for $t>0$, $\sigma>0$,

$$\Pr\{\xi_t>\sigma \mid z_0=0\} = \int_{u=0}^{t} \{1-B(t+\sigma-u)\}\, e^{-u/\alpha} \frac{du}{\alpha} + \sum_{n=1}^{\infty} \Pr\{r'_{n+1}>t+\sigma,\ r'_n<t \mid z_0=0\}. \tag{3.2}$$

Since (cf. (II.4.18)) for $n=1, 2, \ldots$,

$$\Pr\{r'_{n+1}>t+\sigma,\ r'_n<t,\ z_n>0 \mid z_0=0\} = \sum_{j=1}^{\infty} \int_{u=0}^{t} \{1-B(t+\sigma-u)\}\, d_u P_{0j}^{(n)}(u), \tag{3.3}$$

$$\Pr\{r'_{n+1}>t+\sigma,\ r'_n<t,\ z_n=0 \mid z_0=0\} = \int_{u=0}^{t} \left\{ \int_{v=0}^{t-u} \frac{1}{\alpha}\, e^{-v/\alpha} \{1-B(t+\sigma-u-v)\}\, dv \right\} d_u P_{00}^{(n)}(u),$$

it follows from (3.2) and (II.4.28) for $t>0$, $\sigma>0$ that

$$\Pr\{\xi_t>\sigma \mid z_0=0\} = \int_{u=0}^{t} \{1-B(t+\sigma-u)\}\, e^{-u/\alpha} \frac{du}{\alpha} + \sum_{j=1}^{\infty} \int_{u=0}^{t} \{1-B(t+\sigma-u)\}\, dm_{0j}(u)$$
$$+ \int_{u=0}^{t} \left\{ \int_{v=0}^{t-u} \frac{1}{\alpha}\, e^{-v/\alpha} \{1-B(t+\sigma-u-v)\}\, dv \right\} dm_{00}(u). \tag{3.4}$$

Using the relations (II.4.23) and (II.4.29) we obtain for Re $\rho>0$,

$$\int_0^\infty e^{-\rho t} \Pr\{\xi_t>\sigma \mid z_0=0\}\, dt$$

$$= \frac{1}{1+\alpha\rho-\mu(\rho, 1)} \frac{1-\mu(\rho, 1)}{1-\beta(\rho)} \int_0^\infty e^{-\rho t} \{1-B(t+\sigma)\}\, dt. \tag{3.5}$$

Since the functions $m_{0j}(t), j=0, 1, \ldots$, are renewal functions it follows easily by applying theorem I.6.2 that $\Pr\{\xi_t>\sigma \mid z_0=0\}$ has a limit for $t\to\infty$. The value of this limit is calculated by applying Abel's theorem for the Laplace transform (cf. app. 2). It follows (cf. (II.4.40)) for $\sigma>0$ that

$$\lim_{t\to\infty} \Pr\{\xi_t>\sigma \mid z_0=0\} = \lim_{\rho\downarrow 0} \rho \int_0^\infty e^{-\rho t} \Pr\{\xi_t>\sigma \mid z_0=1\}\, dt$$

$$= \begin{cases} \dfrac{a}{\beta} \displaystyle\int_\sigma^\infty \{1-B(t)\}\, dt & \text{if } a<1, \\ \dfrac{1}{\beta} \displaystyle\int_\sigma^\infty \{1-B(t)\}\, dt & \text{if } a\geqq 1. \end{cases} \tag{3.6}$$

Compare this result with (II.4.86). It is easily proved that the limit in (3.6) is independent of the initial condition.

From (II.4.94) for the distribution function $G_\sigma(t)$ of the residual busy period with initial waiting time σ, we have

$$\int_0^\infty e^{-st}\, d_t G_\sigma(t) = \exp(-\sigma\{s+(1-\mu(s, 1))/\alpha\}), \qquad \text{Re } s\geqq 0, \quad \sigma>0. \tag{3.7}$$

Hence for the "last come, first served" discipline the distribution of the waiting time $W_{l,t}(\tau)$ for a customer arriving at time t whenever $z_0=0$ is for $t>0$, $-\infty<\tau<\infty$, given by

$$W_{l,t}(\tau) = \Pr\{x_t=0 \mid z_0=0\} - \int_{\sigma=0}^\infty G_\sigma(\tau)\, d_\sigma \Pr\{\xi_t>\sigma \mid z_0=0\}. \tag{3.8}$$

From (II.4.50), (3.6) and (3.8) it follows by applying Helly-Bray's theorem (cf. app. 5), since $G_\sigma(\tau)$ is continuous in σ (cf. (II.4.95)), that $W_{l,t}(\tau)$, for every fixed $\tau \neq 0$, has a limit for $t \to \infty$, and

$$W_l(\tau) \overset{\text{def}}{=} \lim_{t\to\infty} W_{l,t}(\tau)$$

$$= \begin{cases} 1 - a + \dfrac{a}{\beta} \displaystyle\int_{\sigma=0}^{\infty} G_\sigma(\tau)\{1-B(\sigma)\}\,\mathrm{d}\sigma & \text{for } a<1, \\ \dfrac{1}{\beta} \displaystyle\int_{\sigma=0}^{\infty} G_\sigma(\tau)\{1-B(\sigma)\}\,\mathrm{d}\sigma & \text{for } a \geqq 1. \end{cases} \tag{3.9}$$

It is easily shown that $W_l(\tau)$ is independent of the initial state z_0.

From (3.7) and (3.9) and by using (II.4.22), we obtain for $\operatorname{Re} s \geqq 0$,

$$\int_{0-}^{\infty} \mathrm{e}^{-s\tau}\,\mathrm{d}W_l(\tau) = \begin{cases} 1 - a + \dfrac{a}{\beta} \dfrac{1-\mu(s,1)}{s+(1-\mu(s,1))/\alpha} & \text{if } a<1, \\ \dfrac{1}{\beta} \dfrac{1-\mu(s,1)}{s+(1-\mu(s,1))/\alpha} & \text{if } a \geqq 1. \end{cases} \tag{3.10}$$

It follows that

$$W_l(\infty) = \begin{cases} 1 & \text{for } a<1, \\ \dfrac{1}{a} & \text{for } a \geqq 1, \end{cases} \tag{3.11}$$

so that if $a \geqq 1$ the waiting time is finite with probability a^{-1}.

For the average waiting time if $a<1$, we obtain

$$\int_0^{\infty} t\,\mathrm{d}W_l(t) = \frac{1}{2} \frac{a\beta}{1-a} \frac{\beta_2}{\beta^2} \qquad \text{if } \beta_2 = \int_0^{\infty} t^2\,\mathrm{d}B(t) < \infty, \tag{3.12}$$

$$\int_0^{\infty} t^2\,\mathrm{d}W_l(t) = \frac{a}{3(1-a)^2} \frac{\beta_3}{\beta} + \frac{a^2}{2(1-a)^3} \frac{\beta_2^2}{\beta^2} \qquad \text{if } \beta_3 = \int_0^{\infty} t^3\,\mathrm{d}B(t) < \infty,$$

while if $a>1$ the conditional expectation of the waiting time of a customer, whenever his waiting time is finite, is equal to

$$\frac{\alpha}{1-\mu(0,1)}. \tag{3.13}$$

Clearly if $a<1$ the average waiting time is the same as for the "first come, first served" discipline, whereas its variance is larger. Another interesting result is that if $a>1$, the conditional average (3.13) is finite even if $\beta_2=\infty$.

Denote the number of customers to whom a customer arriving at time t must give priority by $\boldsymbol{m}_t$, i.e. $\boldsymbol{m}_t$ is the number of customers arriving later and served before an arriving customer obtains service. Since $\mu(0, r)$ is the generating function of the distribution of the number of customers served during a busy period starting with one customer (cf. (II.4.62)) it follows for $|r|<1$ that

$$\lim_{t\to\infty} \sum_{j=0}^{\infty} r^j \Pr\{\boldsymbol{m}_t=j\}$$

$$= \begin{cases} 1-a+\dfrac{a}{\beta}\displaystyle\int_0^\infty \sum_{j=0}^{\infty} \frac{(t/\alpha)^j}{j!} \mathrm{e}^{-t/\alpha} \mu^j(0, r)\{1-B(t)\}\, \mathrm{d}t, & a<1, \\ \dfrac{1}{\beta}\displaystyle\int_0^\infty \sum_{j=0}^{\infty} \frac{(t/\alpha)^j}{j!} \mathrm{e}^{-t/\alpha} \mu^j(0, r)\{1-B(t)\}\, \mathrm{d}t, & a\geqq 1, \end{cases}$$

$$= \begin{cases} 1-a+\dfrac{1-\mu(0, r)/r}{1-\mu(0, r)}, & a<1, \\ \dfrac{1}{a}\,\dfrac{1-\mu(0, r)/r}{1-\mu(0, r)}, & a\geqq 1. \end{cases} \tag{3.14}$$

It then follows that

$$\lim_{t\to\infty} \mathrm{E}\{\boldsymbol{m}_t\} = \frac{1}{2}\,\frac{a^2}{1-a}\,\frac{\beta_2}{\beta^2} \quad \text{if} \quad a<1, \tag{3.15}$$

$$\lim_{t\to\infty} \mathrm{E}\{\boldsymbol{m}_t \mid \boldsymbol{m}_t<\infty\} = \frac{\mu(0, 1)}{1-\mu(0, 1)} \quad \text{if} \quad a>1.$$

Clearly $a+\lim_{t\to\infty} \mathrm{E}\{\boldsymbol{m}_t\}$ is the average number of customers leaving the system before the arriving customer obtains service. Note that this average is equal to the average number of customers in the system at the moment of arrival of the customer (cf. (II.4.51) and (II.4.53)).

(*ii*) Last come, first served for $G/M/1$.

Since for the present case the distribution of the service time is negative exponential, it follows (cf. lemma II.1.2) that if at some moment the server is busy, the residual service time is also negative exponentially distributed with the same parameter as the service time. Hence, if an arriving customer finds the server busy, the distribution of his waiting time is that of a busy period. Applying the notation of chapter II.3 it is seen from (II.3.89) that

$$\delta_0(\rho, r) = \frac{r-\lambda(\rho, r)}{\beta\rho+1-\lambda(\rho, r)}, \qquad \text{Re } \rho \geqq 0, \quad |r| \leqq 1,$$

is the generating function and Laplace-Stieltjes transform of the joint distribution of the number of customers served in and the duration of a busy period.

Denote the distribution of the waiting time of the nth arriving customer by $W_{l,n}(\tau)$ if $\boldsymbol{y}_0 = k$, and the number of customers to whom the nth arriving customer must give priority by $\boldsymbol{m}_n$. We then have

$$\int_{0-}^{\infty} \mathrm{e}^{-\rho\tau} \, \mathrm{d}_\tau W_{l,n}(\tau) = \Pr\{\boldsymbol{y}_n = 0 \,|\, \boldsymbol{y}_0 = k\} + \Pr\{\boldsymbol{y}_n > 0 \,|\, \boldsymbol{y}_0 = k\} \frac{1-\lambda(\rho, 1)}{\beta\rho+1-\lambda(\rho, 1)},$$

$$\text{Re } \rho \geqq 0,$$

$$\mathrm{E}\{r^{\boldsymbol{m}_n}\} = \Pr\{\boldsymbol{y}_n = 0 \,|\, \boldsymbol{y}_0 = k\} + \Pr\{\boldsymbol{y}_n > 0 \,|\, \boldsymbol{y}_0 = k\} \frac{1-\lambda(0, r)/r}{\beta\rho+1-\lambda(0, r)},$$

$$|r| \leqq 1.$$

Hence, with

$$W_l(\tau) \stackrel{\text{def}}{=} \lim_{n\to\infty} W_{l,n}(\tau),$$

$$\int_{0-}^{\infty} \mathrm{e}^{-\rho\tau} \, \mathrm{d}_\tau W_l(\tau) = \begin{cases} 1 - \lambda(0, 1) + \lambda(0, 1) \dfrac{1-\lambda(\rho, 1)}{\beta\rho+1-\lambda(\rho, 1)}, \\ \qquad\qquad\qquad \text{Re } \rho \geqq 0, \quad a < 1, \\ \dfrac{1-\lambda(\rho, 1)}{\beta\rho+1-\lambda(\rho, 1)}, \qquad \text{Re } \rho \geqq 0, \quad a \geqq 1, \end{cases} \tag{3.16}$$

$$\lim_{n\to\infty} \mathrm{E}\{r^{\boldsymbol{m}_n}\} = \begin{cases} 1 - \lambda(0, 1) + \lambda(0, 1) \dfrac{1-\lambda(0, r)/r}{1-\lambda(0, r)}, & |r| \leqq 1, \quad a < 1, \\ \dfrac{1-\lambda(0, r)/r}{1-\lambda(0, r)}, & |r| \leqq 1, \quad a \geqq 1. \end{cases} \tag{3.17}$$

From (II.3.47) we see that if $a > 1$ the waiting time is finite with probability a^{-1}. It is easy to see that the results (3.16) and (3.17) are independent of the initial condition $\mathbf{y}_0 = k$. Again the average waiting times for "first come, first served" and for "last come, first served" are equal, but the variance for the former is smaller.

(*iii*) Last come, first served for $G/G/1$.

We shall first determine the distribution of the residual service time at the moment of arrival of a customer who finds the server busy. Since the interarrival times, and similarly the service times, are interchangeable stochastic variables, it is obvious that the distribution of the residual service time is independent of the queue discipline. Therefore, in the derivation of the relation for the distribution of the residual service time we consider the $G/G/1$ system with the "first come, first served" discipline. Using the notation of chapter II.5 (cf. section II.5.7) we see that for $j = 1, 2, \ldots; n = 2, 3, \ldots; t > 0$,

$$\{\mathbf{r}_{n-1} < \mathbf{t}_{n+j} \leqq \mathbf{r}_n - t\}$$

represents the event that the $(n+j)$th arriving customer on his arrival finds j customers in the system and that the residual service time of the customer who is served will be at least t.

Since

$$\overline{\{\mathbf{r}_{n-1} < \mathbf{t}_{n+j} \leqq \mathbf{r}_n - t\}} = \{\mathbf{r}_{n-1} \geqq \mathbf{t}_{n+j}\} \cup \{\mathbf{r}_n - t < \mathbf{t}_{n+j}\},$$

it follows for $j = 1, 2, \ldots$, that

$$1 - \Pr\{\mathbf{r}_{n-1} < \mathbf{t}_{n+j} \leqq \mathbf{r}_n - t\}$$

$$= \Pr\{\mathbf{r}_{n-1} \geqq \mathbf{t}_{n+j}\} + \Pr\{\mathbf{r}_n - t < \mathbf{t}_{n+j}\} - \Pr\{\mathbf{r}_{n-1} \geqq \mathbf{t}_{n+j} > \mathbf{r}_n - t\}$$

$$= \int_0^\infty A^{(j+1)*}(u)\,\mathrm{d}\Pr\{\mathbf{w}_{n-1} + \boldsymbol{\tau}_{n-1} < u\} + \int_0^\infty \{1 - A^{j*}(u-t)\}\,\mathrm{d}\Pr\{\mathbf{w}_n + \boldsymbol{\tau}_n < u\}$$

$$- \int_{v=0}^{t} \Big[\int_{u=0}^{\infty} \{A^{(j+1)*}(u) - A^{(j+1)*}(u+v-t)\}\,\mathrm{d}\Pr\{\mathbf{w}_{n-1} + \boldsymbol{\tau}_{n-1} < u\}\Big]\,\mathrm{d}B(v). \tag{3.18}$$

As in section II.5.7 it can be proved that $\Pr\{\mathbf{r}_{n-1} < \mathbf{t}_{n+j} \leqq \mathbf{r}_n - t\}$ has a limit for $n \to \infty$.

Suppose $a < 1$ then from (3.18) and from Hewitt's inversion theorem excluding the case that both $A(t)$ and $B(t)$ have discontinuity points (cf. section II.5.8) we get for $t > 0$,

$$\lim_{n\to\infty} \Pr\{\boldsymbol{r}_{n-1}<\boldsymbol{t}_{n+j}\leqq\boldsymbol{r}_n-t\}$$

$$=\int_0^\infty \{A^{j*}(u)-A^{(j+1)*}(u)\}\,\mathrm{d}(W(u)*B(u))$$

$$-\int_0^\infty \{A^{j*}(u)-A^{j*}(u-t)\}\,\mathrm{d}(W(u)*B(u))$$

$$+\int_{v=0}^{t}\Big[\int_{u=0}^{\infty}\{A^{(j+1)*}(u)-A^{(j+1)*}(u-(t-v))\}\,\mathrm{d}(W(u)*B(u))\Big]\,\mathrm{d}B(v)$$

$$=\frac{1}{2\pi\mathrm{i}}\int_{-\mathrm{i}\infty}^{\mathrm{i}\infty}\mathrm{E}\{\mathrm{e}^{-\xi\boldsymbol{w}}\}\,\beta(\xi)\,\frac{\alpha^j(-\xi)-\alpha^{j+1}(-\xi)}{-\xi}\,\mathrm{d}\xi$$

$$-\frac{1}{2\pi\mathrm{i}}\int_{-\mathrm{i}\infty}^{\mathrm{i}\infty}\mathrm{E}\{\mathrm{e}^{-\xi\boldsymbol{w}}\}\,\beta(\xi)\,\frac{\alpha^j(-\xi)(1-\mathrm{e}^{\xi t})}{-\xi}\,\mathrm{d}\xi$$

$$+\int_{v=0}^{t}\Big[\frac{1}{2\pi\mathrm{i}}\int_{-\mathrm{i}\infty}^{\mathrm{i}\infty}\mathrm{E}\{\mathrm{e}^{-\xi\boldsymbol{w}}\}\,\beta(\xi)\,\frac{\alpha^{j+1}(-\xi)(1-\mathrm{e}^{\xi(t-v)})}{-\xi}\,\mathrm{d}\xi\Big]\,\mathrm{d}B(v). \tag{3.19}$$

For the stationary situation denote the event: "the arriving customer finds the server idle" by $\{\boldsymbol{\zeta}=0\}$, and the event that he finds the server busy and the residual service time is less than t by $\{0<\boldsymbol{\zeta}<t\}$.
Hence

$$1-\Pr\{\boldsymbol{\zeta}<t\}=\sum_{j=1}^{\infty}\lim_{n\to\infty}\Pr\{\boldsymbol{r}_{n-1}<\boldsymbol{t}_{n+j}\leqq\boldsymbol{r}_n-t\},\qquad t>0. \tag{3.20}$$

To calculate the right-hand side of (3.20) from (3.19), we shall suppose that $\beta(\xi)$ exists for Re $\xi>\delta$, with $\delta<0$, so that $\mathrm{E}\{\exp(-\xi\boldsymbol{w})\}$ also exists for Re $\xi=0-$. However, the final result is independent of this assumption. This is easily proved if we truncate $B(t)$ at c ($c>0$) and then take the limit for $c\to\infty$ (cf. for this truncation method section II.5.9). In (3.19) we can replace the path of integration by the path Re $\xi=0-$ and evaluate the sums of the integrands. It follows easily with $a<1$ that (cf. (II.5.139)),

$$\sum_{j=1}^{\infty} \lim_{n\to\infty} \Pr\{\boldsymbol{r}_{n-1}<\boldsymbol{t}_{n+j}\leqq \boldsymbol{r}_n-t\} = 1-u_0$$

$$+\frac{1}{2\pi i}\int_{-i\infty-0}^{i\infty-0} \mathrm{E}\{e^{-\xi\boldsymbol{w}}\} \frac{\beta(\xi)\,\alpha(-\xi)}{1-\alpha(-\xi)}\Big\{1-e^{\xi t}$$

$$-\alpha(-\xi)\int_0^t (1-e^{\xi(t-v)})\,\mathrm{d}B(v)\Big\}\frac{\mathrm{d}\xi}{\xi},\qquad t>0. \tag{3.21}$$

From (3.20) and (3.21) we may determine the Laplace-Stieltjes transform of the distribution of $\boldsymbol{\zeta}$. It is found for Re $\rho\geqq 0$, $a<1$ that

$$\int_{0-}^{\infty} e^{-\rho t}\,\mathrm{d}_t \Pr\{\boldsymbol{\zeta}<t\}$$

$$= u_0 + \frac{1}{2\pi i}\int_{-i\infty-0}^{i\infty-0} \mathrm{E}\{e^{-\xi\boldsymbol{w}}\} \frac{\beta(\xi)\,\alpha(-\xi)}{1-\alpha(-\xi)}\,\frac{1-\beta(\rho)\,\alpha(-\xi)}{\rho-\xi}\,\mathrm{d}\xi, \tag{3.22}$$

where u_0 is the probability that an arriving customer finds the server idle, whereas $\mathrm{E}\{\exp(-\xi\boldsymbol{w})\}$ is the Laplace-Stieltjes transform of the stationary distribution of the waiting time with the "first come, first served" discipline (cf. (II.5.139)). For the second integral in (3.22) we may write

$$\frac{1}{2\pi i}\int_{-i\infty-0}^{i\infty-0} \mathrm{E}\{e^{-\xi\boldsymbol{w}}\} \frac{\beta(\xi)\,\alpha(-\xi)}{\rho-\xi}\,\mathrm{d}\xi$$

$$-\frac{1-\beta(\rho)}{2\pi i}\int_{-i\infty-0}^{i\infty-0} \mathrm{E}\{e^{-\xi\boldsymbol{w}}\}\Big\{\frac{1-\beta(\xi)\,\alpha(-\xi)}{1-\alpha(-\xi)}\alpha(-\xi)+1-\frac{1}{1-\alpha(-\xi)}\Big\}\frac{\mathrm{d}\xi}{\rho-\xi}.$$

From (II.5.8) the first integral in the above expression is equal to

$$\lim_{n\to\infty} \mathrm{E}\{(\boldsymbol{w}_n+\boldsymbol{\tau}_n-\boldsymbol{\sigma}_{n+1}>0)\exp(-\rho(\boldsymbol{w}_n+\boldsymbol{\tau}_n-\boldsymbol{\sigma}_{n+1}))\}$$

$$=\mathrm{E}\{e^{-\rho\boldsymbol{w}}(\boldsymbol{w}>0)\}=\mathrm{E}\{e^{-\rho\boldsymbol{w}}\}-u_0.$$

In the second integral the first term yields zero (cf. the derivation of form. (II.5.180)). Therefore, we obtain from (3.22) for Re $\rho \geqq 0$, $a<1$,

$$\int_{0-}^{\infty} e^{-\rho t}\, d_t \Pr\{\xi<t\} = \beta(\rho)\, E\{e^{-\rho w}\} + \frac{1-\beta(\rho)}{2\pi i} \int_{-i\infty}^{i\infty} \frac{E\{e^{-\xi w}\}}{\rho-\xi} \frac{d\xi}{1-\alpha(-\xi)}. \tag{3.23}$$

Whenever $a \geqq 1$ a similar analysis gives

$$\lim_{n\to\infty} \left\{1 - \sum_{j=1}^{\infty} \Pr\{\boldsymbol{r}_{n-1} < \boldsymbol{t}_{n+j} \leqq \boldsymbol{r}_n - t\}\right\} = \frac{1}{\beta} \int_0^t \{1-B(\tau)\}\, d\tau, \qquad t>0. \tag{3.24}$$

Since the Laplace-Stieltjes transform of the distribution of the residual busy period with an initial waiting time w is given by (II.5.83) it follows that the limit distribution $W_l(t)$ of the waiting time for the "last come, first served" discipline is given by

$$\int_{0-}^{\infty} e^{-\rho t}\, dW_l(t) = u_0 + \int_0^{\infty} E\{e^{-\rho \boldsymbol{p}_1} \mid \boldsymbol{w}_1 = \tau\}\, d\Pr\{0<\xi<\tau\}, \qquad \mathrm{Re}\, \rho \geqq 0,$$

with $\Pr\{\xi<t\}$ given by (3.22) or (3.23) for $a<1$, and by (3.24) for $a \geqq 1$. By using the relation (II.5.83) we obtain an expression for the distribution of the number of customers to whom the arriving customer must give priority.

III.3.3. Random service for $M/G/1$

If at the end of a service time of a customer the server selects the next customer to be served at random from the waiting customers present in the system at that moment, the queueing discipline is called *random service*. For this discipline all waiting customers have the same probability of being the next customer to be served. Suppose at time t_0 a service ends and j customers are waiting. Consider one of these customers and denote the probability that he is not served during the first k service times after t_0, that the total duration of these k service times is less than t and that at the end of these k service times a total of $h+1$ customers are waiting by $r_{jh}^{(k)}(t)$. It follows for $j=1, 2, \ldots$; $h=0, 1, \ldots$; $k=1, 2, \ldots$; $t>0$, that

$$r_{jh}^{(k)}(t) = \int_{u=0-}^{t} \int_{v=0}^{t-u} \sum_{i=0}^{h} \frac{(v/\alpha)^i}{i!} e^{-v/\alpha} dB(v) \frac{h+1-i}{h+2-i} d_u r_{j,h+1-i}^{(k-1)}(u), \tag{3.25}$$

$$r_{jh}^{(0)}(t) \stackrel{\text{def}}{=} U_1(t)\, \delta_{h,j-1},$$

where $\delta_{h,j-1}$ is Kronecker's symbol. Note that

$$r_{jh}^{(1)}(t) = \frac{j-1}{j} \int_0^t e^{-v/\alpha} \frac{(v/\alpha)^{h+2-j}}{(h+2-j)!} dB(v).$$

Obviously, for the customer considered

$$q_{jh}^{(k)}(t) \stackrel{\text{def}}{=} \frac{1}{h+1} r_{jh}^{(k)}(t), \qquad k=1, 2, ..., \tag{3.26}$$

is the probability that he has to wait exactly k service times, that the total duration of these k service times is less than t and that at the start of his service time h other customers are waiting. It follows for $k=1, 2, ...;$ $j=1, 2, ...;$ $h=0, 1, ...;$ $t>0$ that

$$(h+1)\, q_{jh}^{(k)}(t) = \int_{u=0-}^{t} \left\{ \int_{v=0}^{t-u} \sum_{i=0}^{h+1} \frac{(v/\alpha)^i}{i!} e^{-v/\alpha} dB(v) \right\} (h+1-i)\, d_u q_{j,h+1-i}^{(k-1)}(u), \tag{3.27}$$

$$(h+1)\, q_{jh}^{(0)}(t) = U_1(t)\, \delta_{h,j-1}.$$

Defining for $j=1, 2, ...;$ $|p| \leqq 1$, $|r| \leqq 1$,

$$q_j(p, t, r) \stackrel{\text{def}}{=} \sum_{k=0}^{\infty} \sum_{h=0}^{\infty} r^k p^h q_{jh}^{(k)}(t), \tag{3.28}$$

we obtain for $t>0$,

$$\frac{\partial}{\partial p} \{p q_j(p, t, r)\} = r \int_{u=0-}^{t} \left\{ \int_{v=0}^{t-u} e^{-(1-p)v/\alpha} dB(v) \right\} \frac{\partial}{\partial p} d_u q_j(p, u, r) + p^{j-1}.$$

Hence with

$$Q_j(p, \rho, r) \stackrel{\text{def}}{=} \int_{0-}^{\infty} e^{-\rho t} d_t q_j(p, t, r), \qquad \text{Re}\, \rho \geqq 0, \tag{3.29}$$

$$\{p - r\beta\{\rho + (1-p)/\alpha\}\} \frac{\partial}{\partial p} Q_j(p, \rho, r) + Q_j(p, \rho, r) = p^{j-1},$$
$$j=1, 2, ... \tag{3.30}$$

With $\mu(\rho, r)$ as defined in chapter II.4 (cf. (II.4.22)) it follows from (3.30), since $Q_j(p, \rho, r)$ and $(\partial/\partial p)\, Q_j(p, \rho, r)$ are analytic functions of p for $|p|<1$ and fixed ρ and r with $|r|\leqq 1$, Re $\rho\geqq 0$, that

$$Q_j(\mu(\rho, r), \rho, r)=\mu^{j-1}(\rho, r), \qquad j=1, 2, \ldots, \tag{3.31}$$

a result which is obvious since $\mu^k(\rho, r)$ is the Laplace-Stieltjes transform and generating function of the joint distribution of the busy period and the number of customers served in this busy period if it starts with k customers (cf. section II.4.4). The solution of (3.30) satisfying (3.31) reads for $1\geqq |p|>\mu(\rho, r)$, $|r|\leqq 1$, Re $\rho>0$,

$$Q_j(p, \rho, r) = p^{j-1} - \int_{\mu(\rho,r)}^{p} (j-1)\, z^{j-2} \exp\{-\int_z^p \varphi(y, \rho, r)\, \mathrm{d}y\}\, \mathrm{d}z, \tag{3.32}$$

with

$$\varphi(y, \rho, r) \stackrel{\text{def}}{=} [y-r\beta\{\rho+(1-y)/\alpha\}]^{-1}, \qquad \mu(\rho, r)<y<1. \tag{3.33}$$

We determine next the probability $R_j^{(n)}(t)$ that a customer arrives between the nth and $(n+1)$th departure from the queue (cf. section II.4.1), that he finds the server busy, that the residual service time at his moment of arrival is less than t and that immediately before the start of the next service time j customers are waiting, whenever $z_0=i$. Obviously, we have for $|q|\leqq 1$, Re $\rho\geqq 0$, $n=1, 2, \ldots$,

$$\sum_{j=1}^{\infty} q^j \int_0^\infty e^{-\rho t}\, \mathrm{d}_t R_j^{(n)}(t)$$

$$= \sum_{j=1}^{\infty} q^j \int_0^\infty e^{-\rho t} \int_{\sigma=0}^{\infty} \sum_{h=1}^{j} \Bigg[\sum_{k=1}^{h} \Pr\{z_n=k \mid z_0=i\} \frac{(\sigma/\alpha)^{h-k}}{(h-k)!} e^{-\sigma/\alpha} \frac{\mathrm{d}\sigma}{\alpha}$$

$$+ \Pr\{z_n=0 \mid z_0=i\} \frac{(\sigma/\alpha)^{h-1}}{(h-1)!} e^{-\sigma/\alpha} \frac{\mathrm{d}\sigma}{\alpha} \Bigg] \frac{(t/\alpha)^{j-h}}{(j-h)!} e^{-t/\alpha}\, \mathrm{d}_t B(t+\sigma)$$

$$= \frac{1}{\alpha\rho} [\beta\{(1-q)/\alpha\} - \beta\{\rho+(1-q)/\alpha\}]$$

$$\cdot \{ \sum_{k=0}^{\infty} \Pr\{z_n=k \mid z_0=i\}\, q^k - (1-q) \Pr\{z_n=0 \mid z_0=i\}\}.$$

From section II.4.2 it follows that

$$\lim_{n\to\infty}\int_0^\infty e^{-\rho t}\sum_{j=1}^\infty q^j\, d_t R_j^{(n)}(t)$$

$$=\begin{cases} 0 & \text{for } a\geqq 1,\\[2ex] \dfrac{1}{\alpha\rho}\,[\beta\{(1-q)/\alpha\}-\beta\{\rho+(1-q)/\alpha\}]\,\dfrac{(1-a)(1-q)\,q}{\beta\{(1-q)/\alpha\}-q} & \\ & \text{for } a<1.\end{cases}\tag{3.34}$$

Denoting the waiting time distribution of a customer who arrives between the nth and $(n+1)$th departure from the queue by $W_{r,n}(t)$, it is easily seen that $W_{r,n}(t)$ tends to zero for all finite t with $n\to\infty$ if $a\geqq 1$. $W_{r,n}(t)$ has a non-trivial limit for $n\to\infty$ if $a<1$. Denoting this limit distribution by $W_r(t)$, then

$$W_r(t) = 1-a+\sum_{j=1}^\infty q_j(1,t,1) * \lim_{n\to\infty} R_j^{(n)}(t), \qquad a<1. \tag{3.35}$$

From (3.32) and (3.34) (with $r=1$) it follows that for the waiting time distribution,

$$\int_{0-}^\infty e^{-\rho t}\, dW_r(t) = 1-a+a\frac{1-\beta(\rho)}{\beta\rho}$$

$$-\frac{a(1-a)}{\beta\rho}\int_{\mu(\rho,1)}^1 \frac{\partial}{\partial z}\left[(1-z)\left\{1-\frac{z-\beta\{\rho+(1-z)/\alpha\}}{z-\beta\{(1-z)/\alpha\}}\right\}\right]$$

$$\cdot\exp\left\{-\int_z^1 \varphi(y,\rho,1)\,dy\right\}dz, \qquad \text{Re}\,\rho\geqq 0. \tag{3.36}$$

Denote the number of customers who are served before the arriving customer obtains service by $\boldsymbol{m}_r$; it follows from (3.32) and (3.34) for $a<1$ that

$$E\{r^{\boldsymbol{m}_r}\} = 1-(1-a)\int_{\mu(0,r)}^1 \frac{d}{dz}\left\{(1-z)\,\frac{(d/dz)\,\beta\{(1-z)/\alpha\}}{\beta\{(1-z)/\alpha\}-z}\right\}$$

$$\cdot\exp\left\{-\int_z^1 \varphi(y,0,r)\,dy\right\}dz, \qquad |r|\leqq 1.$$

To calculate the average delay note that (cf. (3.35)),

$$\int_0^\infty t\,\mathrm{d}W_r(t) = \sum_{j=1}^\infty \{\int_0^\infty t\,\mathrm{d}q_j(1,t,1) \lim_{n\to\infty} R_j^{(n)}(\infty) + q_j(1,\infty,1)\int_0^\infty t\,\mathrm{d}_t \lim_{n\to\infty} R_j^{(n)}(t)\}.$$

The integrals in these expressions can easily be calculated from the Laplace-Stieltjes transform; for $a<1$ we find that

$$\int_0^\infty t\,\mathrm{d}W_r(t) = \frac{1}{2}\frac{a}{1-a}\frac{\beta_2}{\beta} \qquad \text{if } \beta_2<\infty, \tag{3.37}$$

$$\int_0^\infty t^2\,\mathrm{d}W_r(t) = \frac{2}{3}\frac{1}{1-a}\frac{a}{2-a}\frac{\beta_3}{\beta} + \frac{1}{(1-a)^2}\frac{a^2}{2-a}\left(\frac{\beta_2}{\beta}\right)^2, \qquad \text{if } \beta_3<\infty,$$

hence the average waiting time is equal to that for the "first come, first served" discipline. The higher moments can also be found by starting from (3.35) (note that $q_j(1,t,1)$ and $R_j^{(n)}(t)$ are, for fixed j, probabilities of independent events). The relation (3.36) for $W_r(t)$ is very complicated and difficult to handle even for negative exponentially distributed service times (cf. LE GALL [1962]).

III.3.4. Random service for $G/M/1$

Here we shall denote by $w_i(t)$ the waiting time distribution of a customer who finds at his arrival i customers in the system, $i=0, 1, \dots$. A relation for $w_i(t)$ is found by calculating the probability $w_i^{(1)}(t)$ that an arriving customer, finding i customers present, is served before the next arrival and has a waiting time less than t. If k services end before the next arrival, the arriving customer has a probability

$$\frac{1}{i+1} + \frac{i}{i+1}\frac{1}{i} + \dots + \frac{i}{i+1}\frac{i-1}{i}\cdots\frac{1}{i+2-k} = \frac{k}{i+1},$$

of being served before the next arrival. Since $B(t)$ is a negative exponential distribution the residual service time at any moment is also negative exponentially distributed. Hence for $i=1, 2, \ldots$,

$$w_i^{(1)}(t) = \int_0^t \{1-A(u)\}\left[\sum_{k=0}^{i+1} \frac{k}{i+1} \{\mathrm{d}B^{k*}(u)-\mathrm{d}B^{(k+1)*}(u)\}+\mathrm{d}B^{(i+2)*}(u)\right], \tag{3.38}$$

the last term takes into account the probability that all customers are served before the arrival of the next customer.

Denote by $w_i^{(2)}(t)$ the probability that the arriving customer, finding i customers present, has a waiting time less than t and is not served before the arrival of the next customer. It is easily found that for $i=1, 2, \ldots$,

$$w_i^{(2)}(t) = \int_0^t \mathrm{d}A(u) \sum_{k=0}^{i+1}\left(1 - \frac{k}{i+1}\right)\{B^{k*}(u)-B^{(k+1)*}(u)\}\, w_{i+1-k}(t-u). \tag{3.39}$$

Hence

$$w_i(t)=w_i^{(1)}(t)+w_i^{(2)}(t), \qquad t>0, \quad i=1, 2, \ldots, \tag{3.40}$$

$$w_0(t)=U_1(t).$$

Since

$$B^{k*}(t) = \begin{cases} \displaystyle\int_0^t \frac{(\tau/\beta)^{k-1}}{(k-1)!}\, \mathrm{e}^{-\tau/\beta} \frac{\mathrm{d}\tau}{\beta}, & k=1, 2, \ldots; \quad t>0, \\ U_1(t), & k=0, \end{cases} \tag{3.41}$$

it follows from (3.38), ..., (3.41) with

$$w(p, \rho) \stackrel{\text{def}}{=} \sum_{i=1}^{\infty} \int_0^{\infty} p^i\, \mathrm{e}^{-\rho t}\, \mathrm{d}w_i(t), \qquad |p|<1, \quad \mathrm{Re}\, \rho\geqq 0, \tag{3.42}$$

that

$$[p-\alpha\{\rho+(1-p)/\beta\}] \frac{\partial}{\partial p} w(p, \rho) + w(p, \rho) + \alpha\{\rho+1/\beta\}\, w_1(\rho)$$

$$= \frac{1-\alpha\{\rho+(1-p)/\beta\}}{(\beta\rho+1-p)(1-p)} - \frac{1-\alpha\{\rho+1/\beta\}}{\beta\rho+1}. \tag{3.43}$$

With $\lambda(\rho, 1)$ as defined in chapter II.3 (cf. (II.3.43)) it follows from (3.43), since $(\partial/\partial p)\, w(p, \rho)$ and $w(p, \rho)$ are analytic functions of p for $|p|<1$ and $\text{Re}\, \rho>0$, that for $\text{Re}\, \rho>0$,

$$w(\lambda(\rho, 1), \rho) + \alpha\{\rho+1/\beta\}\, w_1(\rho) = \frac{1}{\beta\rho+1-\lambda(\rho, 1)} - \frac{1-\alpha\{\rho+1/\beta\}}{1+\beta\rho}. \tag{3.44}$$

Putting

$$\begin{aligned} \Psi(p, \rho) &\overset{\text{def}}{=} [p-\alpha\{\rho+(1-p)/\beta\}]^{-1}, && p\neq\lambda(\rho, 1), \quad \text{Re}\, \rho>0, \\ \zeta(p, \rho) &\overset{\text{def}}{=} \frac{1-\alpha\{\rho+(1-p)/\beta\}}{(\beta\rho+1-p)(1-p)} - \frac{1-\alpha\{\rho+1/\beta\}}{1+\beta\rho}, && |p|<1, \quad \text{Re}\, \rho>0, \end{aligned} \tag{3.45}$$

it follows from (3.43) for $\text{Re}\, \rho>0$, $|p|<1$ that

$$\begin{aligned} w(p, \rho) &= \exp\left\{-\int_{G_p} \Psi(y, \rho)\, \mathrm{d}y\right\} \\ &\cdot\left\{C + \int_{G_p} \Psi(z, \rho)\, \{\zeta(z, \rho) - \alpha\{\rho + 1/\beta\}\, w_1(\rho)\} \exp\left[\int_{G_z} \Psi(y, \rho)\, \mathrm{d}y\right] \mathrm{d}z\right\}, \end{aligned} \tag{3.46}$$

where C is the integration constant and G_p is a simple curve in the p-plane terminating at point p and starting at $\lambda(\rho, 1)+\varepsilon$ with $|\varepsilon|>0$ but sufficiently small, and leading to the point p, so that this path lies outside the circle with radius $|\varepsilon|$ and centre at $\lambda(\rho, 1)$.

Taking $p=\lambda(\rho, 1)+\varepsilon$ and letting $\varepsilon\to0$ it follows from (3.46) and (3.44) that

$$C = \zeta(\lambda(\rho, 1), \rho) - \alpha\{\rho + 1/\beta\}\, w_1(\rho). \tag{3.47}$$

Elimination of C from (3.46) and (3.47) gives after partial integration

$$\begin{aligned} w(p, \rho) &= \zeta(p, \rho) - \alpha\{\rho + 1/\beta\}\, w_1(\rho) \\ &\quad - \lim_{|\varepsilon|\to0} \int_{G_p} \left\{\frac{\partial}{\partial z}\zeta(z, \rho)\right\} \exp\left[-\int_z^p \Psi(y, \rho)\, \mathrm{d}y\right] \mathrm{d}z, \end{aligned} \tag{3.48}$$

for $|p|<1$, $\text{Re}\, \rho>0$. Taking in (3.48) $p=0$ it follows from (3.42) for $|\varepsilon|\to0$,

$$\alpha\{\rho+1/\beta\}\, w_1(\rho)$$

$$= - \lim_{|\varepsilon|\to 0} \int_{G_0} \left\{\frac{\partial}{\partial z}\zeta(z,\rho)\right\} \exp\left[\int_0^z \Psi(y,\rho)\,\mathrm{d}y\right] \mathrm{d}z, \qquad \mathrm{Re}\,\rho>0. \tag{3.49}$$

It is easily proved from the behaviour of the function $\Psi(p, \rho)$ in the neighbourhood of $p=\lambda(\rho, 1)$ that the limits in (3.48) and (3.49) exist. The function $w(p, \rho)$ is now completely known for $|p|<1$, $\mathrm{Re}\,\rho>0$.

With $\boldsymbol{y}_n$ denoting the number of customers in the system just before the arrival of the nth customer (cf. section II.3.2) it follows that

$$U_1(t)\,\Pr\{\boldsymbol{y}_n=0 \mid \boldsymbol{y}_0=j\} + \sum_{i=1}^{\infty} \Pr\{\boldsymbol{y}_n=i \mid \boldsymbol{y}_0=j\}\, w_i(t)$$

is the distribution function of the waiting time for the nth customer if $\boldsymbol{y}_0=j$ in the case of random service. If $a<1$ it follows easily from (II.3.15) that this waiting time distribution converges in distribution to a limit distribution $W_r(t)$ of which the Laplace-Stieltjes transform is given by

$$\int_{0-}^{\infty} \mathrm{e}^{-\rho t}\,\mathrm{d}W_r(t) = 1-\lambda_0+(1-\lambda_0)\,w(\lambda_0,\rho), \qquad \mathrm{Re}\,\rho\geqq 0, \quad \lambda_0=\lambda(0,1). \tag{3.50}$$

III.3.5. Completion times

It will be assumed that interruptions occur according to a Poisson process, hence (cf. (3.1)),

$$A^{(1)}(t)=1-\mathrm{e}^{-t/\alpha^{(1)}}, \qquad t>0.$$

Since the negative exponential distribution has no "memory" it follows that completion times of customers are independent, identically distributed variables.

During every interruption time new interruptions may occur. Hence, the time between an interruption and the first moment at which the server becomes available again is a busy period for a $M/G/1$ queueing system with average interarrival time $\alpha^{(1)}$ and service time distribution $B^{(1)}(t)$. For such a queueing system the Laplace-Stieltjes transform of the busy period is given by $\mu^{(1)}(\rho, 1)$ (cf. section II.4.4), where $\mu^{(1)}(\rho, 1)$ is the zero with smallest absolute value of

$$p-\beta^{(1)}\{\rho+(1-p)/\alpha^{(1)}\}, \qquad \operatorname{Re}\rho\geqq 0,$$

with

$$\beta^{(1)}(\rho)=\int_0^\infty e^{-\rho t}\,dB^{(1)}(t).$$

(*i*) Delayed interruptions.

The probability of k interruptions during a service time τ of a customer is given by

$$\frac{\{\tau/\alpha^{(1)}\}^k}{k!}\,e^{-\tau/\alpha^{(1)}}.$$

Since interruption times and service times are independent it follows for the completion time $\boldsymbol{c}_i$ that

$$\begin{aligned}
E\{e^{-\rho \boldsymbol{c}_i}\} &= \sum_{k=0}^{\infty}\int_0^\infty e^{-\rho t}\,\frac{\{t/\alpha^{(1)}\}^k}{k!}\,e^{-t/\alpha^{(1)}}\,dB^{(2)}(t)\{\mu^{(1)}(\rho,1)\}^k\\
&=\beta^{(2)}\{\rho+(1-\mu^{(1)}(\rho,1))/\alpha^{(1)}\}, \qquad \operatorname{Re}\rho\geqq 0, \qquad (3.51)
\end{aligned}$$

with

$$\beta^{(2)}(\rho)=\int_0^\infty e^{-\rho t}\,dB^{(2)}(t), \qquad \operatorname{Re}\rho\geqq 0.$$

It follows (cf. (II.4.66)) that

$$E\{\boldsymbol{c}_i\}=\frac{\beta^{(2)}}{1-a^{(1)}} \qquad \text{if } a^{(1)}<1, \tag{3.52}$$

where

$$a^{(1)}\overset{\text{def}}{=}\frac{\beta^{(1)}}{\alpha^{(1)}}, \qquad a^{(2)}\overset{\text{def}}{=}\frac{\beta^{(2)}}{\alpha^{(2)}}, \tag{3.53}$$

and

$$E\{\boldsymbol{c}_i^2\}=\frac{\beta_2^{(2)}}{(1-a^{(1)})^2}+\beta_2^{(1)}\frac{\beta^{(2)}}{\beta^{(1)}}\frac{a^{(1)}}{\{1-a^{(1)}\}^3}, \qquad a^{(1)}<1. \tag{3.54}$$

(*ii*) Pre-emptive resume interruptions.

Since time intervals between successive interruptions are negative exponentially distributed and independent, the time interval between resumption of service and the next interruption is also negative exponentially distributed with the same parameter. Hence the probability that a service time of duration τ has k interruptions is also given by

$$\frac{(\tau/\alpha^{(1)})^k}{k!}\,\mathrm{e}^{-\tau/\alpha^{(1)}}.$$

As above we again obtain relation (3.51) for $\mathrm{E}\{\exp(-\rho \boldsymbol{c}_{\mathrm{i}})\}$.

(*iii*) Pre-emptive repeat interruptions.

Obviously, the probability of exactly k interruptions in a completion time $\boldsymbol{c}_{\mathrm{i}}$ of which the service time $\boldsymbol{\tau}$ has duration τ is given by

$$\{1-\mathrm{e}^{-\tau/\alpha^{(1)}}\}^k\,\mathrm{e}^{-\tau/\alpha^{(1)}}.$$

Denote the past service time between a restart and the next interruption by $\boldsymbol{t}$ then

$$\Pr\{\boldsymbol{t}<t \mid \boldsymbol{t}<\tau\} = \frac{1-\mathrm{e}^{-t/\alpha^{(1)}}}{1-\mathrm{e}^{-\tau/\alpha^{(1)}}}, \qquad 0<t<\tau,$$

so that

$$\mathrm{E}\{\mathrm{e}^{-\rho \boldsymbol{t}} \mid \boldsymbol{\tau}=\tau\} = \frac{1}{1+\rho\alpha^{(1)}}\,\frac{1-\exp(-(1+\rho\alpha^{(1)})\,\tau/\alpha^{(1)})}{1-\exp(-\tau/\alpha^{(1)})}.$$

Hence

$$\mathrm{E}\{\mathrm{e}^{-\rho \boldsymbol{c}_{\mathrm{i}}} \mid \boldsymbol{\tau}=\tau\}$$

$$= \sum_{k=0}^{\infty} [\mathrm{E}\{\mathrm{e}^{-\rho \boldsymbol{t}} \mid \boldsymbol{\tau}=\tau\}\mu^{(1)}(\rho,1)\{1-\mathrm{e}^{-\tau/\alpha^{(1)}}\}]^k \exp(-(1+\rho\alpha^{(1)})\,\tau/\alpha^{(1)}).$$

It follows for $\mathrm{Re}\,\rho \geqq 0$ that

$$\mathrm{E}\{\mathrm{e}^{-\rho \boldsymbol{c}_{\mathrm{i}}}\} = \int_0^\infty \exp\{-(\rho+1/\alpha^{(1)})\tau\} \cdot\left[1-\frac{\mu^{(1)}(\rho,1)}{1+\rho\alpha^{(1)}}\{1-\exp\{-(\rho+1/\alpha^{(1)})\tau\}\}\right]^{-1} \mathrm{d}B^{(2)}(\tau), \tag{3.55}$$

and if $\beta^{(2)}(-1/\alpha^{(1)})$ exists

$$\mathrm{E}\{\boldsymbol{c}_{\mathrm{i}}\} = \frac{\alpha^{(1)}}{1-a^{(1)}}\{\beta^{(2)}(-1/\alpha^{(1)})-1\} \qquad \text{if} \quad a^{(1)}<1. \tag{3.56}$$

(*iv*) Restricted interruptions.

For these interruption processes again pre-emptive and delayed disciplines can be distinguished. It is easy to see that for the various possibilities the Laplace-Stieltjes transform of the distribution of the completion time follows from the above expressions with $\mu^{(1)}(\rho, 1)$ replaced by $\beta^{(1)}(\rho)$.

III.3.6. Pre-emptive resume priority for $M/G/1$ with two priority levels

Customers of first priority level cause break-in interruptions to customers of second priority. The interruption discipline is pre-emptive resume. The arrival processes for both types of customers are Poisson processes, i.e.

$$A^{(i)}(t)=1-e^{-t/\alpha^{(i)}}, \qquad i=1,2; \quad t>0.$$

Since first priority customers have break-in priority the process $\{x_t^{(1)}, t\in[0, \infty)\}$, $x_t^{(1)}$ being the number of first priority customers present in the system at time t, is the queueing process $M/G/1$ of chapter II.4 with service time distribution $B^{(1)}(t)$. Hence all probabilities of events relating to $x_t^{(1)}$ can be immediately obtained from chapter II.4.

The stochastic process $\{x_t, t\in[0, \infty)\}$ with x_t the total number of customers in the system at time t is also a queueing process $M/G/1$ because the superposition of two Poisson processes is again a Poisson process. For this process x_t, the service time distribution $B(t)$ is given by

$$B(t) = \frac{\alpha^{(2)}}{\alpha^{(1)}+\alpha^{(2)}} B^{(1)}(t) + \frac{\alpha^{(1)}}{\alpha^{(1)}+\alpha^{(2)}} B^{(2)}(t), \tag{3.57}$$

and the average interarrival time α by

$$\alpha = \frac{\alpha^{(1)}\alpha^{(2)}}{\alpha^{(1)}+\alpha^{(2)}}. \tag{3.58}$$

Obviously, the "occupation time" of the server is now the busy period of the x_t-process. It follows that the occupation time is finite with probability one, if (cf. section II.4.4),

$$a = \frac{1}{\alpha}\int_0^\infty t\, dB(t) = \frac{\beta^{(1)}}{\alpha^{(1)}} + \frac{\beta^{(2)}}{\alpha^{(2)}} = a^{(1)} + a^{(2)} < 1. \tag{3.59}$$

From

$$\Pr\{\boldsymbol{x}_t=0\}=\Pr\{\boldsymbol{x}_t^{(2)}=0 \mid \boldsymbol{x}_t^{(1)}=0\}\,\Pr\{\boldsymbol{x}_t^{(1)}=0\},$$

it follows that (cf. (II.4.50)),

$$\lim_{t\to\infty}\Pr\{\boldsymbol{x}_t^{(2)}=0 \mid \boldsymbol{x}_t^{(1)}=0\} = \frac{1-a}{1-a^{(1)}} \quad \text{if} \quad a<1, \tag{3.60}$$

$$\lim_{t\to\infty}\Pr\{\boldsymbol{x}_t^{(2)}>0 \mid \boldsymbol{x}_t^{(1)}=0\} = \frac{a^{(2)}}{1-a^{(1)}} \quad \text{if} \quad a<1,$$

since $a<1$ implies $a^{(1)}<1$.

To determine the distribution of the waiting time $\boldsymbol{w}_t^{(2)}$ of a second priority customer arriving at time t, note that if $\boldsymbol{v}_t$ is the virtual waiting time for the queueing process $\boldsymbol{x}_t$, $\boldsymbol{w}_t^{(2)}$ equals the length of the initial busy period $\boldsymbol{\theta}_v$ of the $\boldsymbol{x}_t^{(1)}$-process with initial waiting time v equal to $\boldsymbol{v}_t$. Hence

$$\Pr\{\boldsymbol{w}_t^{(2)}<\tau\} = \int_{0-}^{\infty} \Pr\{\boldsymbol{\theta}_v<\tau\}\,\mathrm{d}_v \Pr\{\boldsymbol{v}_t<v\}. \tag{3.61}$$

For $a<1$ the distribution of $\boldsymbol{v}_t$ has a limit for $t\to\infty$ (cf. II.4.90), hence from Helly-Bray's theorem (cf. app. 5) it is seen that if $a<1$, $\Pr\{\boldsymbol{w}_t^{(2)}<\tau\}$ also has a limit distribution $W^{(2)}(\tau)$ for $t\to\infty$. From (II.4.94) we now obtain for $\operatorname{Re}\rho\geqq 0$,

$$\int_0^\infty \mathrm{e}^{-\rho t}\,\mathrm{d}W^{(2)}(t) = \int_0^\infty \exp[-\{\rho+(1-\mu^{(1)}(\rho,1))/\alpha^{(1)}\}\,\tau]\,\mathrm{d}_\tau \lim_{t\to\infty}\Pr\{\boldsymbol{v}_t<\tau\}, \tag{3.62}$$

with $\mu^{(1)}(\rho, 1)$ the Laplace-Stieltjes transform of the busy period for the $\boldsymbol{x}_t^{(1)}$-process. From (3.62) and (II.4.90) it follows for $\operatorname{Re}\rho\geqq 0$ if $a<1$ that

$$\int_{0-}^{\infty} \mathrm{e}^{-\rho t}\,\mathrm{d}W^{(2)}(t) = (1-a)\,\alpha^{(2)}\,\frac{\rho+\{1-\mu^{(1)}(\rho,1)\}/\alpha^{(1)}}{-1+\rho\alpha^{(2)}+\beta^{(2)}\{\rho+(1-\mu^{(1)}(\rho,1))/\alpha^{(1)}\}}, \tag{3.63}$$

$W(0+)=1-a.$

For the first and second moments we obtain from (3.62),

$$\int_0^\infty t\,\mathrm{d}W^{(2)}(t) = \frac{1}{1-a^{(1)}}\lim_{t\to\infty}\mathrm{E}\{\boldsymbol{v}_t\} = \frac{1}{2(1-a^{(1)})(1-a)}\left\{\frac{\beta_2^{(1)}}{\alpha^{(1)}}+\frac{\beta_2^{(2)}}{\alpha^{(2)}}\right\}, \tag{3.64}$$

$$\int_0^\infty t^2\,\mathrm{d}W^{(2)}(t) = \frac{1}{(1-a^{(1)})^2}\lim_{t\to\infty}\mathrm{E}\{\boldsymbol{v}_t^2\} + \frac{1}{\alpha^{(1)}}\,\frac{\beta_2^{(1)}}{(1-a^{(1)})^3}\lim_{t\to\infty}\mathrm{E}\{\boldsymbol{v}_t\}$$

$$= \frac{1}{3(1-a^{(1)})^2\,(1-a)}\left\{\frac{\beta_3^{(1)}}{\alpha^{(1)}} + \frac{\beta_3^{(2)}}{\alpha^{(2)}}\right\}$$

$$+ \frac{1}{2(1-a^{(1)})^2\,(1-a)}\left\{\frac{\beta_2^{(1)}}{\alpha^{(1)}} + \frac{\beta_2^{(2)}}{\alpha^{(2)}}\right\}$$

$$\cdot\left\{\frac{1}{1-a}\left(\frac{\beta_2^{(1)}}{\alpha^{(1)}} + \frac{\beta_2^{(2)}}{\alpha^{(2)}}\right) + \frac{1}{1-a^{(1)}}\,\frac{\beta_2^{(1)}}{\alpha^{(1)}}\right\}.$$

For the stationary situation the distribution function of the time spent in the system by a second priority customer is given by the convolution of $W^{(2)}(t)$ and the completion time distribution (cf. (3.51)).

Next, we shall determine for the stationary situation the distribution $\{\boldsymbol{v}_j^{(2)},\ j=0, 1, \dots\}$ of the number of second priority customers present in the system immediately after the departure of a second priority customer. Since the number of second priority customers present in the system at the departure of a second priority customer is equal to the number of customers arriving during the waiting time and the completion time of the departing customer it follows for $|p|\leqq 1$, $a<1$ that

$$\sum_{j=0}^{\infty} v_j^{(2)}p^j = \sum_{j=0}^{\infty} p^j \int_{0-}^{\infty} \frac{\{t/\alpha^{(2)}\}^j}{j!}\,\mathrm{e}^{-t/\alpha^{(2)}}\,\mathrm{d}_t\{W^{(2)}(t) * \Pr\{\boldsymbol{c}_{\mathrm{i}}<t\}\}$$

$$= \int_{0-}^{\infty} \mathrm{e}^{-(1-p)t/\alpha^{(2)}}\mathrm{E}\{\mathrm{e}^{-(1-p)\boldsymbol{c}_{\mathrm{i}}/\alpha^{(2)}}\}\,\mathrm{d}_t W^{(2)}(t), \tag{3.65}$$

where $\boldsymbol{c}_{\mathrm{i}}$ is the completion time of a second priority customer, the transform of its distribution is given by (3.51), that of $W^{(2)}(t)$ by (3.62). It follows that

$$v_0^{(2)} = (1-a)\left[1 + \frac{\alpha^{(2)}}{\alpha^{(1)}}\{1 - \mu^{(1)}(1/\alpha^{(2)}, 1)\}\right], \tag{3.66}$$

$$\sum_{j=0}^{\infty} j v_j^{(2)} = \frac{1}{1-a^{(1)}}\left\{a^{(2)} + \frac{1}{2(1-a)\,\alpha^{(2)}}\left(\frac{\beta_2^{(1)}}{\alpha^{(1)}} + \frac{\beta_2^{(2)}}{\alpha^{(2)}}\right)\right\}.$$

Note that

$$\mathrm{E}\{\boldsymbol{x}_t^{(2)}\} = \mathrm{E}\{\boldsymbol{x}_t\} - \mathrm{E}\{\boldsymbol{x}_t^{(1)}\},$$

the right-hand side can be found by using the results of chapter II.4 with α and $B(t)$ given by (3.57) and (3.58).

It should be noted that the moments at which a second priority customer leaves the system are regeneration points of the $\boldsymbol{x}_t^{(2)}$-process, because at such moments no first priority customers are present in the system. We shall determine for the imbedded Markov chain of the $\boldsymbol{x}_t^{(2)}$-process at these regeneration points the one step transition matrix (p_{ij}).

It is easy to see that

$$p_{ij} = \begin{cases} 0, & j=0, 1, \dots, i-2; \quad i=2, 3, \dots, \\ \displaystyle\int_0^\infty \frac{\{t/\alpha^{(2)}\}^{j+1-i}}{(j+1-i)!} \, \mathrm{e}^{-t/\alpha^{(2)}} \, \mathrm{d}_t \Pr\{\boldsymbol{c}_\mathrm{i} < t\}, & \\ & j=i-1, i, \dots; \quad i=1, 2, \dots . \end{cases} \tag{3.67}$$

If a second priority customer leaves the system empty, the time which the next arriving second priority customer will spend in the system is the sum of his completion time and the duration of a residual busy period of the $\boldsymbol{x}_t^{(1)}$-system. This busy period $\boldsymbol{\theta}$ starts with an inital waiting time $\boldsymbol{v}_\tau^{(1)}$ which is equal to the virtual waiting time of the $\boldsymbol{x}_t^{(1)}$-system at the moment τ when the second priority customer arrives. From (II.4.94) and (II.4.95) it follows for $\mathrm{Re}\, \rho \geqq 0$ that

$$\begin{aligned} \mathrm{E}\{\mathrm{e}^{-\rho\boldsymbol{\theta}}\} &= \int_{t=0}^\infty \int_{\sigma=0}^\infty \exp[-\{\rho + (1-\mu^{(1)}(\rho, 1))/\alpha^{(1)}\}\,\sigma] \\ &\qquad \cdot \mathrm{d}_\sigma \Pr\{\boldsymbol{v}_t^{(1)} < \sigma \mid \boldsymbol{v}_0^{(1)} = 0\} \, \mathrm{e}^{-t/\alpha^{(2)}} \, \mathrm{d}t/\alpha^{(2)} \\ &= \frac{\alpha^{(1)} - \alpha^{(1)}\alpha^{(2)}\rho + (\mu^{(1)}(\rho, 1) - \mu^{(1)}(1/\alpha^{(2)}, 1))\alpha^{(2)}}{[\alpha^{(1)} + \alpha^{(2)} - \alpha^{(2)}\mu^{(1)}(1/\alpha^{(2)}, 1)](1-\alpha^{(2)}\rho)}, \end{aligned} \tag{3.68}$$

where we used (II.4.91). It follows that

$$p_{0j} = \int_{t=0-}^\infty \mathrm{e}^{-t/\alpha^{(2)}} \frac{(t/\alpha^{(2)})^j}{j!} \, \mathrm{d}_t[\Pr\{\boldsymbol{\theta} < t\} * \Pr\{\boldsymbol{c}_\mathrm{i} < t\}], \qquad j=0, 1, 2, \dots, \tag{3.69}$$

with the distribution of $\boldsymbol{c}_\mathrm{i}$ given by (3.51).

For $a < 1$ it is easily verified that $\{v_j^{(2)}, j=0, 1, \dots\}$ as given by (3.65) satisfies

$$v_j = \sum_{i=0}^\infty v_i p_{ij}.$$

III.3.7. Pre-emptive repeat priority for $M/G/1$ with two priority levels

Customers of first priority cause break-in interruptions to customers of second priority. The interruption discipline is pre-emptive repeat. Arrival processes of both types of customers are Poisson processes,

$$A^{(i)}(t)=1-\mathrm{e}^{-t/\alpha^{(i)}}, \qquad t>0, \quad i=1,2.$$

The $\boldsymbol{x}_t^{(1)}$-process is the $M/G/1$ queueing process with service time distribution $B^{(1)}(t)$ and average interarrival time $\alpha^{(1)}$. It is easily seen that here the $\boldsymbol{x}_t^{(2)}$-process is of essentially the same structure as the $\boldsymbol{x}_t^{(2)}$-process for pre-emptive resume of the preceding section, only the distribution of the completion time is different. For the imbedded Markov chain of the $\boldsymbol{x}_t^{(2)}$-process the one step transition matrix (p_{ij}) is given by (3.67) and (3.69) with the distribution function of the completion time $\boldsymbol{c}_\mathrm{i}$ given by (3.55). Using Foster's criteria (cf. section I.2.4) it is easily proved that the imbedded Markov chain is ergodic if and only if

$$a^{(1)}<1, \qquad \mathrm{E}\{\boldsymbol{c}_\mathrm{i}\}<\alpha^{(2)}, \tag{3.70}$$

with $\mathrm{E}\{\boldsymbol{c}_\mathrm{i}\}$ given by (3.56). Assuming that (3.70) is satisfied, it follows for the stationary situation that the distribution $\{v_j^{(2)},\ j=0,1,2,\dots\}$ of the number of second priority customers left behind in the system by a departing second priority customer satisfies

$$\sum_{j=0}^{\infty} v_j = 1, \qquad v_j = \sum_{i=0}^{\infty} v_i p_{ij}. \tag{3.71}$$

From (3.71) it is found for $|p|\leqq 1$,

$$\sum_{j=0}^{\infty} v_j^{(2)} p^j = v_0^{(2)} \frac{\mathrm{E}\{\mathrm{e}^{-q_2\boldsymbol{c}_\mathrm{i}}\}}{\mathrm{E}\{\mathrm{e}^{-q_2\boldsymbol{c}_\mathrm{i}}\}-p}\; \frac{1+\alpha^{(1)}q_2-\mu^{(1)}(q_2,1)}{1+\alpha^{(1)}/\alpha^{(2)}-\mu^{(1)}(1/\alpha^{(2)},1)}$$

$$= \mathrm{E}\{\mathrm{e}^{-q_2\boldsymbol{c}_\mathrm{i}}\} \int_{0-}^{\infty} \mathrm{e}^{-q_2 t}\, \mathrm{d}W^{(2)}(t), \tag{3.72}$$

$$v_0^{(2)} = \frac{\alpha^{(2)}-\mathrm{E}\{\boldsymbol{c}_\mathrm{i}\}}{\alpha^{(1)}}(1-a^{(1)})\left\{1+\frac{\alpha^{(1)}}{\alpha^{(2)}}-\mu^{(1)}(1/\alpha^{(2)},1)\right\}$$

$$=\left\{1-a^{(1)}+\frac{\alpha^{(1)}}{\alpha^{(2)}}\{1-\beta^{(2)}(-1/\alpha^{(2)})\}\right\}\left\{1+\frac{\alpha^{(2)}}{\alpha^{(1)}}\{1-\mu^{(1)}(1/\alpha^{(2)},1)\}\right\}, \tag{3.73}$$

$$W^{(2)}(0+) = 1 - a^{(1)} + \frac{\alpha^{(1)}}{\alpha^{(2)}}\{1 - \beta^{(2)}(-1/\alpha^{(2)})\}, \tag{3.74}$$

where

$$q_2 \overset{\text{def}}{=} \frac{1}{\alpha^{(2)}}(1-p),$$

and $W^{(2)}(t)$ is the waiting time distribution of a second priority customer, while $\mathrm{E}\{\boldsymbol{c}_1\}$ and $\mathrm{E}\{\mathrm{e}^{-\rho \boldsymbol{c}_1}\}$ are given by (3.56) and (3.55). The relation between $W^{(2)}(t)$ and $v_j^{(2)}$ is based on the same argument as in the preceding section.

III.3.8. Non break-in priority for $M/G/1$ with two and three priority levels

(*i*) Two priority levels.

The interruption discipline is that of delayed interruptions. Arrival processes of both types of customers are Poisson processes,

$$A^{(i)}(t) = 1 - \mathrm{e}^{-t/\alpha^{(i)}}, \qquad t > 0, \quad i = 1, 2, \dots .$$

The $\boldsymbol{x}_t$-process with $\boldsymbol{x}_t$ the total number of customers in the system at time t, is again the $M/G/1$ queueing process with service time distribution $B(t)$ given by (3.57) and with average interarrival time α given by (3.58). Denote the stationary virtual waiting time distribution of the $\boldsymbol{x}_t$-process by $V(t)$ if

$$a^{(1)} + a^{(2)} = \frac{\beta^{(1)}}{\alpha^{(1)}} + \frac{\beta^{(2)}}{\alpha^{(2)}} < 1.$$

From (II.4.81),

$$\int_{0-}^{\infty} \mathrm{e}^{-\rho t}\,\mathrm{d}V(t) = (1 - a^{(1)} - a^{(2)})\frac{\alpha\rho}{\beta(\rho) + \alpha\rho - 1}, \qquad \mathrm{Re}\,\rho \geqq 0, \tag{3.75}$$

$$\beta(\rho) = \int_0^{\infty} \mathrm{e}^{-\rho t}\,\mathrm{d}B(t).$$

For an arriving customer of second priority the waiting time is the sum of the time $\boldsymbol{v}$ needed to serve all the customers present in the system at the moment of his arrival and the time needed to clear all first priority customers arriving during $\boldsymbol{v}$. The latter time is a residual busy period, starting with the number of first priority customers arriving during $\boldsymbol{v}$ of a $M/G/1$ queueing system with service time distribution $B^{(1)}(t)$ and average interarrival time $\alpha^{(1)}$. From the results of chapter II.4 it is easily proved that the distribution of the waiting time of a second priority customer arriving at time t has a limit for $t\to\infty$, which is a proper probability distribution if $a^{(1)}+a^{(2)}<1$. Denote this limit distribution by $W^{(2)}(t)$ if $a=a^{(1)}+a^{(2)}<1$; then (cf. section II.4.4),

$$\int_{0-}^{\infty} e^{-\rho t}\,dW^{(2)}(t) = \int_{0-}^{\infty} \sum_{j=0}^{\infty} \frac{(t/\alpha^{(1)})^j}{j!} e^{-t/\alpha^{(1)}} e^{-\rho t}\,dV(t)\{\mu^{(1)}(\rho,1)\}^j$$

$$= \int_{0}^{\infty} \exp\left[-t\{\rho+(1-\mu^{(1)}(\rho,1))/\alpha^{(1)}\}\right] dV(t)$$

$$= (1-a)\frac{\{\rho+(1-\mu^{(1)}(\rho,1))/\alpha^{(1)}\}\,\alpha^{(2)}}{-1+\rho\alpha^{(2)}+\beta^{(2)}\{\rho+(1-\mu^{(1)}(\rho,1))/\alpha^{(1)}\}},$$

$$\operatorname{Re}\rho\geqq 0, \qquad (3.76)$$

where $\mu^{(1)}(\rho,1)$ is the zero with smallest absolute value of

$$z-\beta^{(1)}\{\rho+(1-z)/\alpha^{(1)}\}, \qquad \operatorname{Re}\rho\geqq 0.$$

Note that in the present case the distribution of the waiting time for second priority customers is identical with that of second priority customers for pre-emptive resume priority (cf. (3.63)). The expressions for the first and second moment of $W^{(2)}(t)$ are given by (3.64).

Next we determine the distribution of the waiting time of first priority customers. Denote the number of first and second priority customers left behind in the system by the nth departing customer after $t=0$ by $z_n^{(1)}$ and $z_n^{(2)}$. Further $\boldsymbol{v}_j^{(i)}$ will denote the number of ith priority customers arriving during a service time of a jth priority customer, $i=1,2$; $j=1,2$. We note that, if at some moment t_0 the system is empty,

$$\frac{\alpha^{(i)}}{\alpha^{(1)}+\alpha^{(2)}}, \qquad (3.77)$$

is the probability that the first customer arriving after t_0 is not an ith priority customer. We have for $|p|\leqq 1$,

$$\begin{aligned}
\mathrm{E}\{p^{z_{n+1}^{(1)}} \mid \boldsymbol{x}_0=0\} = {} & \mathrm{E}\{p^{z_n^{(1)}-1+\nu_1^{(1)}}(z_n^{(1)}>0) \mid \boldsymbol{x}_0=0\} \\
& + \mathrm{E}\{(z_n^{(1)}=0)(z_n^{(2)}>0)\, p^{\nu_2^{(1)}} \mid \boldsymbol{x}_0=0\} \\
& + \mathrm{E}\{(z_n^{(1)}=0)(z_n^{(2)}=0)\, p^{\nu_1} \mid \boldsymbol{x}_0=0\} \frac{\alpha^{(2)}}{\alpha^{(1)}+\alpha^{(2)}} \\
& + \mathrm{E}\{(z_n^{(1)}=0)(z_n^{(2)}=0)\, p^{\nu_2^{(1)}} \mid \boldsymbol{x}_0=0\} \frac{\alpha^{(1)}}{\alpha^{(1)}+\alpha^{(2)}}. \qquad (3.78)
\end{aligned}$$

For the vector Markov chain $\{(z_n^{(1)},\ z_n^{(2)}),\ n=1, 2, \ldots\}$ the empty state is positive recurrent if and only if $a^{(1)}+a^{(2)}<1$, because the empty state of the $\boldsymbol{x}_t$-process is positive recurrent if and only if $a^{(1)}+a^{(2)}<1$. Since the vector Markov chain is irreducible all its states are positive recurrent if and only if $a^{(1)}+a^{(2)}<1$. Moreover, its states are aperiodic so that the joint distribution of $z_n^{(1)}$ and $z_n^{(2)}$ has a limit distribution for $n\to\infty$, which is a proper probability distribution. Denote the stationary distribution of the number of first priority customers left behind in the system by a departing customer by $\{v_j^{(1)}, j=0, 1, \ldots\}$. It follows for $a^{(1)}+a^{(2)}<1$ that

$$\begin{aligned}
\lim_{n\to\infty} \Pr\{z_n^{(1)}=0, z_n^{(2)}=0 \mid \boldsymbol{x}_0=0\} &= \lim_{t\to\infty} \Pr\{\boldsymbol{x}_t=0 \mid \boldsymbol{x}_0=0\} = 1 - a^{(1)} - a^{(2)}, \\
\lim_{n\to\infty} \Pr\{z_n^{(1)}=0, z_n^{(2)}>0 \mid \boldsymbol{x}_0=0\} &= v_0^{(1)} - (1-a^{(1)}-a^{(2)}). \qquad (3.79)
\end{aligned}$$

From (3.78) and (3.79) it now follows by letting $n\to\infty$ and defining

$$v^{(1)}(p) \overset{\text{def}}{=} \sum_{j=0}^{\infty} p^j v_j^{(1)} = \lim_{n\to\infty} \mathrm{E}\{p^{z_n^{(1)}} \mid \boldsymbol{x}_0=0\}, \qquad |p|\leqq 1,$$

that for $a^{(1)}+a^{(2)}<1$,

$$\begin{aligned}
v^{(1)}(p) = {} & v_0^{(1)} \frac{\beta^{(1)}(q)-p\beta^{(2)}(q)}{\beta^{(1)}(q)-p} \\
& + (1-a^{(1)}-a^{(2)}) \frac{\alpha^{(2)}}{\alpha^{(1)}+\alpha^{(2)}} \frac{\beta^{(2)}(q)-\beta^{(1)}(q)}{\beta^{(1)}(q)-p}\, p, \qquad (3.80)
\end{aligned}$$

where

$$q \overset{\text{def}}{=} \frac{1}{\alpha^{(1)}} (1-p).$$

The norming condition determines $v_0^{(1)}$,

$$v_0^{(1)} = 1 - \frac{\alpha^{(2)}}{\alpha^{(1)}+\alpha^{(2)}} (a^{(1)}+a^{(2)}), \qquad a^{(1)}+a^{(2)}<1. \qquad (3.81)$$

Denoting the indicator function of the event that the nth departing customer is an ith priority customer by $U_n^{(i)}$, $i=1, 2$, we have

$$\mathrm{E}\{p^{z_{n+1}^{(1)}}\, U_{n+1}^{(1)} \mid x_0=0\} = \mathrm{E}\{p^{z_n^{(1)}-1+\nu_1^{(1)}}\, (z_n^{(1)}>0) \mid x_0=0\}$$

$$+\, \mathrm{E}\{(z_n^{(1)}=0)(z_n^{(2)}=0)\, p^{\nu_1^{(1)}} \mid x_0=0\} \frac{\alpha^{(2)}}{\alpha^{(1)}+\alpha^{(2)}}. \tag{3.82}$$

Hence for $|p|\leqq 1$, $a^{(1)}+a^{(2)}<1$,

$$v_1^{(1)}(p) \stackrel{\text{def}}{=} \lim_{n\to\infty} \mathrm{E}\{p^{z_n^{(1)}}\, U_n^{(1)} \mid x_0=0\}$$

$$= \left[\frac{1}{p}(v^{(1)}(p)-v^{(1)}(0)) + \frac{\alpha^{(2)}}{\alpha^{(1)}+\alpha^{(2)}}(1-a^{(1)}-a^{(2)})\right]\beta^{(1)}(q)$$

$$= \left[v_0^{(1)}(1-\beta^{(2)}(q)) + (1-a^{(1)}-a^{(2)})\frac{\alpha^{(2)}}{\alpha^{(1)}+\alpha^{(2)}}(\beta^{(2)}(q)-p)\right]\frac{\beta^{(1)}(q)}{\beta^{(1)}(q)-p}. \tag{3.83}$$

It follows for $a^{(1)}+a^{(2)}<1$,

$$\lim_{n\to\infty} \mathrm{E}\{U_n^{(1)} \mid x_0=0\} = \frac{\alpha^{(2)}}{\alpha^{(1)}+\alpha^{(2)}}. \tag{3.84}$$

From (3.83) and (3.84) it now follows that

$$\frac{\alpha^{(1)}+\alpha^{(2)}}{\alpha^{(2)}}\, v_1^{(1)}(p),$$

is the generating function of the distribution of the number of first priority customers left behind in the system if the departing customer is a first priority customer. Hence for the stationary situation the waiting time distribution $W^{(1)}(t)$ of first priority customers is given by

$$\int_{0-}^{\infty} \mathrm{e}^{-\rho t}\, \mathrm{d}W^{(1)}(t) = \frac{\alpha^{(1)}+\alpha^{(2)}}{\alpha^{(2)}}\, \frac{v_1^{(1)}(1-\alpha^{(1)}\rho)}{\beta^{(1)}(\rho)}$$

$$= \frac{1-a^{(1)}-a^{(2)}+a^{(2)}(1-\beta^{(2)}(\rho))/\rho\beta^{(2)}}{1-a^{(1)}(1-\beta^{(1)}(\rho))/\rho\beta^{(1)}},$$

$$\operatorname{Re}\rho\geqq 0;\quad a^{(1)}+a^{(2)}<1. \tag{3.85}$$

With

$$\beta_3^{(i)} \stackrel{\text{def}}{=} \int_0^\infty t^3\, \mathrm{d}B^{(i)}(t),$$

we obtain for the moments of $W^{(1)}(t)$ if $a^{(1)}+a^{(2)}<1$,

$$\int_0^\infty t\,\mathrm{d}W^{(1)}(t) = \frac{1}{2(1-a^{(1)})}\left\{\frac{\beta_2^{(1)}}{\alpha^{(1)}}+\frac{\beta_2^{(2)}}{\alpha^{(2)}}\right\}, \tag{3.86}$$

$$\int_0^\infty t^2\,\mathrm{d}W^{(1)}(t) = \frac{1}{3(1-a^{(1)})}\left\{\frac{\beta_3^{(1)}}{\alpha^{(1)}}+\frac{\beta_3^{(2)}}{\alpha^{(2)}}\right\}$$

$$+\frac{1}{2(1-a^{(1)})^2}\frac{\beta_2^{(1)}}{\alpha^{(1)}}\left\{\frac{\beta_2^{(1)}}{\alpha^{(1)}}+\frac{\beta_2^{(2)}}{\alpha^{(2)}}\right\}.$$

Above we obtained the stationary distribution of the waiting time of first priority customers if $a^{(1)}+a^{(2)}<1$. Next, we shall discuss the waiting time of first priority customers if $a^{(1)}<1$, $a^{(1)}+a^{(2)}\geqq 1$. In this case the $\boldsymbol{x}_t$-process is not ergodic. Obviously $\boldsymbol{x}_t^{(2)}\to\infty$ for $t\to\infty$ with probability one. The queueing process for the first priority customers is now a variant of the $M/G/1$ queue (cf. FINCH [1959a]). An arriving first priority customer, who finds no first priority customers present in the system on his arrival has to wait a random time, which is the residual service time of a second priority service at the moment of his arrival. This queueing process is ergodic if the random time above has a finite first moment.

To obtain the distribution of the waiting time for first priority customers in the stationary situation we may start again from (3.78) but now with

$$\lim_{n\to\infty}\Pr\{\boldsymbol{z}_n^{(1)}=0,\ \boldsymbol{z}_n^{(2)}=0\}=0.$$

As above we obtain for $|p|\leqq 1$, $a^{(1)}<1$, $a^{(1)}+a^{(2)}\geqq 1$,

$$v^{(1)}(p) = v_0^{(1)}\frac{\beta^{(1)}(q)-\beta^{(2)}(q)\,p}{\beta^{(1)}(q)-p},\qquad q=(1-p)/\alpha^{(1)}, \tag{3.87}$$

$$v_0^{(1)} = \frac{1-a^{(1)}}{1-a^{(1)}+\alpha^{(2)}a^{(2)}/\alpha^{(1)}}.$$

Further

$$v_1^{(1)}(p) = v_0^{(1)}\frac{1-\beta^{(2)}(q)}{\beta^{(1)}(q)-p}\,\beta^{(1)}(q), \tag{3.88}$$

so that

$$\lim_{n\to\infty}\mathrm{E}\{\boldsymbol{U}_n^{(1)}\mid \boldsymbol{x}_0=0\} = \frac{\alpha^{(2)}a^{(2)}/\alpha^{(1)}}{1-a^{(1)}+\alpha^{(2)}a^{(2)}/\alpha^{(1)}}.$$

Hence

$$\lim_{n\to\infty} \mathrm{E}\{p^{z_n(1)} \mid U_n^{(1)}=1, x_0=0\} = \frac{1-a^{(1)}}{\alpha^{(2)}a^{(2)}/\alpha^{(1)}} \frac{1-\beta^{(2)}(q)}{\beta^{(1)}(q)-p}\, \beta^{(1)}(q), \qquad (3.89)$$

from which we obtain for the waiting time distribution of first priority customers

$$\int_0^\infty \mathrm{e}^{-\rho t}\, \mathrm{d}W^{(1)}(t) = \frac{\{1-a^{(1)}\}(1-\beta^{(2)}(\rho))/(\rho\beta^{(2)})}{1-a^{(1)}(1-\beta^{(1)}(\rho))/(\rho\beta^{(1)})},$$

$$\operatorname{Re}\rho \geqq 0, \quad a^{(1)}<1, \quad a^{(1)}+a^{(2)} \geqq 1, \qquad (3.90)$$

$$\int_0^\infty t\, \mathrm{d}W^{(1)}(t) = \frac{1}{2}\left\{\frac{1}{1-a^{(1)}}\frac{\beta_2^{(1)}}{\alpha^{(1)}} + \frac{1}{a^{(2)}}\frac{\beta_2^{(2)}}{\alpha^{(2)}}\right\},$$

$$\int_0^\infty t^2\, \mathrm{d}W^{(1)}(t) = \frac{1}{3}\left\{\frac{1}{1-a^{(1)}}\frac{\beta_3^{(1)}}{\alpha^{(1)}} + \frac{1}{a^{(2)}}\frac{\beta_3^{(2)}}{\alpha^{(2)}}\right\} + \frac{1}{2(1-a^{(1)})}\frac{\beta_2^{(1)}}{\alpha^{(1)}}\left\{\frac{1}{1-a^{(1)}}\frac{\beta_2^{(1)}}{\alpha^{(1)}} + \frac{1}{a^{(2)}}\frac{\beta_2^{(2)}}{\alpha^{(2)}}\right\}.$$

If $a^{(1)}+a^{(2)}<1$ and customers are served in order of arrival the average waiting time of a customer is given by

$$\frac{1}{2(1-a)}\left\{\frac{\beta_2^{(1)}}{\alpha^{(1)}} + \frac{\beta_2^{(2)}}{\alpha^{(2)}}\right\}.$$

With priority serving, however, the average waiting time of a customer is given by

$$\frac{\alpha^{(2)}}{\alpha^{(1)}+\alpha^{(2)}}\int_0^\infty t\, \mathrm{d}W^{(1)}(t) + \frac{\alpha^{(1)}}{\alpha^{(1)}+\alpha^{(2)}}\int_0^\infty t\, \mathrm{d}W^{(2)}(t)$$

$$= \frac{1}{2(1-a^{(1)}-a^{(2)})}\left\{\frac{\beta_2^{(1)}}{\alpha^{(1)}} + \frac{\beta_2^{(2)}}{\alpha^{(2)}}\right\} \cdot \frac{1}{1-a^{(1)}}\left\{1 - (a^{(1)}+a^{(2)})\frac{\alpha^{(2)}}{\alpha^{(1)}+\alpha^{(2)}}\right\}. \qquad (3.91)$$

Since

$$\frac{1}{1-a^{(1)}}\left\{1 - (a^{(1)}+a^{(2)})\frac{\alpha^{(2)}}{\alpha^{(1)}+\alpha^{(2)}}\right\} < 1 \qquad \text{if} \quad \beta^{(1)}<\beta^{(2)},$$

it follows that if first priority is given to customers with the smaller average service time the overall average delay decreases.

(*ii*) Three priority levels.

We shall discuss here only the situation with

$$a^{(1)}+a^{(2)}<1, \qquad a^{(1)}+a^{(2)}+a^{(3)}\geqq 1, \tag{3.92}$$

as the general case will be considered in section 3.10. All definitions given above will also hold here, but now for $i=1, 2, 3$. As before it is seen that the $\boldsymbol{x}_t$-process is not ergodic; the $\boldsymbol{x}_t^{(1)}+\boldsymbol{x}_t^{(2)}$-process, however, is ergodic. Hence, $\boldsymbol{z}_n^{(3)}\to\infty$ for $n\to\infty$. It follows (cf. (3.78)) that for $|p|\leqq 1$,

$$\begin{aligned}\mathrm{E}\{p^{\boldsymbol{z}_{n+1}{}^{(1)}} \mid \boldsymbol{x}_0=0\} &= \mathrm{E}\{p^{\boldsymbol{z}_n{}^{(1)}-1+\boldsymbol{\nu}_1{}^{(1)}}\,(\boldsymbol{z}_n^{(1)}>0) \mid \boldsymbol{x}_0=0\}\\ &\quad+\mathrm{E}\{(\boldsymbol{z}_n^{(1)}=0)(\boldsymbol{z}_n^{(2)}>0)\,p^{\boldsymbol{\nu}_2{}^{(1)}} \mid \boldsymbol{x}_0=0\}\\ &\quad+\mathrm{E}\{(\boldsymbol{z}_n^{(1)}=0)(\boldsymbol{z}_n^{(2)}=0)(\boldsymbol{z}_n^{(3)}>0)\,p^{\boldsymbol{\nu}_3{}^{(1)}} \mid \boldsymbol{x}_0=0\}\\ &\quad+\Pr\{\boldsymbol{z}_n^{(1)}+\boldsymbol{z}_n^{(2)}+\boldsymbol{z}_n^{(3)}=0\}\,D(p),\end{aligned}$$

where $D(p)$ is some function of p. As above it can be shown that all terms in the above relation have limits for $n\to\infty$, the limit of the last term being zero (cf. (3.92)). Letting $n\to\infty$ we obtain for $|p|\leqq 1$,

$$v^{(1)}(p)=\frac{1}{\beta^{(1)}(q)-p}\,[v^{(1)}(0)\{\beta^{(1)}(q)-p\,\beta^{(2)}(q)\}+v^{(2)}(0)\,p\{\beta^{(2)}(q)-\beta^{(3)}(q)\}],$$

with

$$q=\frac{1}{\alpha^{(1)}}(1-p), \qquad v^{(2)}(0)\overset{\text{def}}{=}\lim_{n\to\infty}\Pr\{\boldsymbol{z}_n^{(1)}=0,\ \boldsymbol{z}_n^{(2)}=0 \mid \boldsymbol{x}_0=0\}.$$

The norming condition $v^{(1)}(1)=1$ yields

$$\alpha^{(1)}-\beta^{(1)}=v^{(1)}(0)\{\alpha^{(1)}-\beta^{(1)}+\beta^{(2)}\}+v^{(2)}(0)\{\beta^{(3)}-\beta^{(2)}\}. \tag{3.93}$$

The constant $v^{(2)}(0)$ can be obtained from (3.87) if we note that a two level priority system is obtained when pooling the customers of priority level 1 and 2. Hence from (3.87),

$$v^{(2)}(0)=\{1-a^{(1)}-a^{(2)}\}\left\{1-a^{(1)}-a^{(2)}+\left(\frac{1}{\alpha^{(1)}}+\frac{1}{\alpha^{(2)}}\right)\beta^{(3)}\right\}^{-1},$$

so that from (3.93),

$$v^{(1)}(0)=\left\{1-a^{(1)}-a^{(2)}+\frac{\beta^{(3)}}{\alpha^{(2)}}\right\}\left\{1-a^{(1)}-a^{(2)}+\left(\frac{1}{\alpha^{(1)}}+\frac{1}{\alpha^{(2)}}\right)\beta^{(3)}\right\}^{-1}.$$

Further for $|p|\leqq 1$,

$$\lim_{n\to\infty} \mathrm{E}\{p^{z_n^{(1)}}\, U_n^{(1)} \mid x_0=0\} = \beta^{(1)}(q)\{v^{(1)}(p)-v^{(1)}(0)\}\, p^{-1}$$

$$= \frac{\beta^{(1)}(q)}{\beta^{(1)}(q)-p}\, [v^{(1)}(0)\{1-\beta^{(2)}(q)\}+v^{(2)}(0)\{\beta^{(2)}(q)-\beta^{(3)}(q)\}].$$

Hence

$$\lim_{n\to\infty} \mathrm{E}\{U_n^{(1)} \mid x_0=0\} = \frac{\beta^{(3)}}{\alpha^{(1)}} \Big/ \left\{1 - a^{(1)} - a^{(2)} + \left(\frac{1}{\alpha^{(1)}} + \frac{1}{\alpha^{(2)}}\right)\beta^{(3)}\right\}.$$

It now follows using the argument which led to (3.90) that the stationary distribution of the waiting time $W^{(1)}(t)$ for first priority customers is given by

$$\int_{0-}^{\infty} \mathrm{e}^{-\rho t}\, \mathrm{d}W^{(1)}(t)$$

$$= \frac{1}{1-a^{(1)}(1-\beta^{(1)}(\rho))/(\rho\beta^{(1)})}\left\{a^{(2)}\,\frac{1-\beta^{(2)}(\rho)}{\rho\beta^{(2)}} + \{1-a^{(1)}-a^{(2)}\}\,\frac{1-\beta^{(3)}(\rho)}{\rho\beta^{(3)}}\right\}, \tag{3.94}$$

for $a^{(1)}+a^{(2)}<1$, $a^{(1)}+a^{(2)}+a^{(3)}\geqq 1$, $\mathrm{Re}\,\rho\geqq 0$,

$$\int_0^{\infty} t\, \mathrm{d}W^{(1)}(t) = \frac{a^{(1)}}{2(1-a^{(1)})}\,\frac{\beta_2^{(1)}}{\beta^{(1)}} + \frac{a^{(2)}}{2(1-a^{(1)})}\,\frac{\beta_2^{(2)}}{\beta^{(2)}} + \frac{1-a^{(1)}-a^{(2)}}{2(1-a^{(1)})}\,\frac{\beta_2^{(3)}}{\beta^{(3)}}, \tag{3.95}$$

$$\int_0^{\infty} t^2\, \mathrm{d}W^{(1)}(t) = \frac{a^{(1)}}{3(1-a^{(1)})}\,\frac{\beta_3^{(1)}}{\beta^{(1)}} + \frac{a^{(2)}}{3(1-a^{(1)})}\,\frac{\beta_3^{(2)}}{\beta^{(2)}} + \frac{1-a^{(1)}-a^{(2)}}{3(1-a^{(1)})}\,\frac{\beta_3^{(3)}}{\beta^{(3)}} + \frac{a^{(1)}}{2(1-a^{(1)})^2}\,\frac{\beta_2^{(1)}}{\beta^{(1)}} \cdot \left\{a^{(1)}\,\frac{\beta_2^{(1)}}{\beta^{(1)}} + a^{(2)}\,\frac{\beta_2^{(2)}}{\beta^{(2)}} + (1-a^{(1)}-a^{(2)})\,\frac{\beta_2^{(3)}}{\beta^{(3)}}\right\}.$$

III.3.9. General pre-emptive resume priority for $M/G/1$

Interarrival times of successive arriving customers are independent, identically distributed variables with distribution

$$A(t) = 1 - e^{-t/\alpha}, \qquad t>0, \qquad \Lambda \overset{\text{def}}{=} \alpha^{-1}. \tag{3.96}$$

The priority level $\boldsymbol{y}$ of an arriving customer is a non-negative stochastic variable. It will be assumed that the priority levels of the various customers are independent, identically distributed variables with distribution

$$P(y) \overset{\text{def}}{=} \begin{cases} \Pr\{\boldsymbol{y}<y\}, & y>0, \\ 0, & y<0, \end{cases} \tag{3.97}$$
$$P(\infty) = 1.$$

$B_y(t)$ will denote the service time distribution of a customer of priority level $\boldsymbol{y}=y$. The service times are supposed to be independent variables;

$$\beta_y(\rho) = \int_0^\infty e^{-\rho t}\, dB_y(t), \qquad \operatorname{Re}\rho \geqq 0; \qquad \beta_{y,i} \overset{\text{def}}{=} \int_0^\infty t^i\, dB_y(t),$$
$$i=1, 2, \ldots, \tag{3.98}$$

for every i for which the integral is finite. The families of interarrival times and of service times are independent, and the families of interarrival times and of priority levels are assumed to be independent.

Customers of priority level x have priority for obtaining service over those of priority level y if $y>x$; customers of the same priority level are served in order of arrival.

The interruption discipline is pre-emptive resume.

Define

$$\Lambda(y) \overset{\text{def}}{=} \Lambda P(y), \qquad \lambda(y) \overset{\text{def}}{=} \Lambda(y+) - \Lambda(y), \tag{3.99}$$
$$a(y) \overset{\text{def}}{=} \Lambda \int_{0-}^{y} \beta_{x,1}\, dP(x), \qquad a \overset{\text{def}}{=} \lim_{y\to\infty} a(y),$$

and denote the stationary distribution of the waiting time of a customer with priority level $\boldsymbol{y}=y$ by $W_y(t)$, whenever this stationary waiting time distribution exists.

Customers with priority level $\boldsymbol{y}>y$ do not influence the queueing of customers with priority level $\boldsymbol{y}\leqq y$. Hence the queueing process of customers with priority level $\mathbf{y}=y$ is the pre-emptive resume priority $M/G/1$ system with two priority levels. We may therefore use the results of section 3.6 with

$$\Lambda P(y) = \{\alpha^{(1)}\}^{-1}, \qquad \lambda_y = \{\alpha^{(2)}\}^{-1}, \tag{3.100}$$

$$B_y(t) = B^{(2)}(t), \qquad \frac{1}{P(y)} \int_{0-}^{y} B_x(t)\, \mathrm{d}P(x) = B^{(1)}(t).$$

Hence, if $a(y+)<1$ then customers with priority level $\mathbf{y}=y$ have a stationary waiting time distribution. It follows from (3.63) for $\operatorname{Re} \rho \geqq 0$, $a(y+)<1$, $y>0$ that

$$\int_{0-}^{\infty} \mathrm{e}^{-\rho t}\, \mathrm{d}W_y(t) = \{1-a(y+)\} \frac{\rho+\Lambda(y)\{1-\mu_y(\rho, 1)\}}{-\lambda_y+\rho+\lambda_y \beta_y\{\rho+\Lambda(y)(1-\mu_y(\rho, 1))\}}, \tag{3.101}$$

if y is a discontinuity point of $P(y)$, and for $\operatorname{Re} \rho \geqq 0$ that

$$\int_{0-}^{\infty} \mathrm{e}^{-\rho t}\, \mathrm{d}W_y(t) = \{1-a(y)\} \left\{1 + \frac{\Lambda(y)}{\rho} \{1-\mu_y(\rho, 1)\}\right\}$$

$$= 1 - a(y) + a(y) \frac{1-\mu_y(\rho, 1)}{\rho c(y)}, \tag{3.102}$$

if y is a continuity point of $P(y)$; here $\mu_y(\rho, 1)$ is the zero with smallest absolute value of

$$z - \frac{1}{P(y)} \int_{0-}^{y} \beta_x\{\rho+\Lambda(y)(1-z)\}\, \mathrm{d}P(x), \qquad \operatorname{Re} \rho \geqq 0,$$

and

$$c(y) \stackrel{\text{def}}{=} \frac{1}{\{1-a(y)\}\, P(y)} \int_{0-}^{y} \beta_{x,1}\, \mathrm{d}P(x), \qquad y \geqq 0. \tag{3.103}$$

For y a continuity point of $P(y)$ the transform $\mu_y(\rho, 1)$ is the Laplace-Stieltjes transform of the busy period of a $M/G/1$ queueing system without priority; $c(y)$ is the average duration of this busy period. The relation (3.102) is easily interpreted since $a(y)$ is the probability that the server is busy and $\{1-\mu_y(\rho, 1)\}/\{\rho c(y)\}$ is the Laplace-Stieltjes transform of the residual busy period.

The expressions for the moments of $W_y(t)$ for y a discontinuity point follow from (3.64). For y a continuity point of $P(y)$ we obtain from (3.102) and (II.4.66) if $a(y)<1$,

$$\int_0^\infty t\,\mathrm{d}W_y(t) = \frac{\Lambda}{2(1-a(y))^2}\int_{0-}^{y}\beta_{x,2}\,\mathrm{d}P(x), \tag{3.104}$$

$$\int_0^\infty t^2\,\mathrm{d}W_y(t) = \frac{\Lambda}{3(1-a(y))^3}\int_{0-}^{y}\beta_{x,3}\,\mathrm{d}P(x) + \frac{\Lambda^2}{(1-a(y))^4}\left\{\int_{0-}^{y}\beta_{x,2}\,\mathrm{d}P(x)\right\}^2.$$

An interesting application of the relations (3.104) is obtained for the $M/G/1$ queueing system with pre-emptive resume priority and all customers having the same service time distribution $B(t)$; the priority discipline being: *customers with shorter service time have priority*. In this case

$$B_y(t) = \begin{cases} 1 & \text{for} \quad t>y, \\ 0 & \text{for} \quad t<y, \end{cases} \tag{3.105}$$

and

$$P(y)=B(y). \tag{3.106}$$

It follows that

$$a(y) = \Lambda\int_0^y x\,\mathrm{d}B(x),$$

so that

$$\int_0^\infty t\,\mathrm{d}W_y(t) = \frac{\Lambda}{2\{1-\Lambda\int_0^y x\,\mathrm{d}B(x)\}^2}\int_0^y x^2\,\mathrm{d}B(x).$$

III.3.10. General non break-in priority for $M/G/1$

In this section we shall consider the same model as in the preceding section, the priority discipline, however, being that of delayed interruptions. The notations and definitions of that section are also used here. Denote the number of customers of priority level $\boldsymbol{y}\leq y$ and of priority level $\boldsymbol{y}>y$, present in the system at time t by $\boldsymbol{x}_t^{(1)}(y)$ and $\boldsymbol{x}_t^{(2)}(y)$, respectively. Suppose for the moment that customers of priority level $\boldsymbol{y}\leq y$ are served in order of arrival, similarly for those of level $\boldsymbol{y}>y$, but for two customers one of level

$\boldsymbol{y} \leqq y$ and one of level $\boldsymbol{y} > y$ the priority discipline is maintained. Clearly the two processes $\boldsymbol{x}_t^{(1)}(y)$ and $\boldsymbol{x}_t^{(2)}(y)$ constitute a non break-in priority system with two levels. Consequently, we may apply the results of section 3.8. In case $a<1$ (cf. (3.99)) a customer of priority level $\boldsymbol{y} \leqq y$ has a stationary waiting time distribution $W_y^{(1)}(t)$ for the modified system, which is given by (3.85) with

$$a = \lim_{y\to\infty} a(y), \qquad a^{(2)} = a - a(y+), \qquad a^{(1)} = a(y+), \tag{3.107}$$

$$\beta^{(1)}(\rho) = \frac{1}{P(y+)} \int_{0-}^{y+} \beta_x(\rho)\, \mathrm{d}P(x), \quad \beta^{(2)}(\rho) = \frac{1}{1-P(y+)} \int_{y+}^{\infty} \beta_x(\rho)\, \mathrm{d}P(x),$$

$$\beta^{(1)} = \frac{1}{P(y+)} \int_{0-}^{y+} \beta_{x,1}\, \mathrm{d}P(x), \qquad \beta^{(2)} = \frac{1}{1-P(y+)} \int_{y+}^{\infty} \beta_{x,1}\, \mathrm{d}P(x).$$

Restoring only the priority discipline between customers of priority level $\boldsymbol{y}<y$ and customers of level $\boldsymbol{y}=y$ for the $\boldsymbol{x}_t^{(1)}(y)$-process, we again obtain a two level priority system and if $a<1$ we may use the second part of formula (3.76) for the stationary waiting time distribution $W_y(t)$ of a customer of level $\boldsymbol{y}=y$. Hence

$$\int_{0-}^{\infty} \mathrm{e}^{-\rho t}\, \mathrm{d}W_y(t) = \frac{1-a+a^{(2)}(1-\beta^{(2)}(q_y))/(q_y\beta^{(2)})}{1-a^{(1)}(1-\beta^{(1)}(q_y))/(q_y\beta^{(1)})}, \qquad \operatorname{Re}\rho \geqq 0, \quad a<1, \tag{3.108}$$

with (cf. (3.102) for $\mu_y(\rho, 1)$),

$$q_y \stackrel{\text{def}}{=} \rho + \Lambda(y)\{1-\mu_y(\rho, 1)\}, \tag{3.109}$$

and a, $\beta^{(1)}(\rho)$, ..., given by (3.107). From (3.107), (3.108) and by using the definition of $\mu_y(\rho, 1)$ it follows for $a<1$,

$$\int_{0-}^{\infty} \mathrm{e}^{-\rho t}\, \mathrm{d}W_y(t) = \frac{(1-a)\,q_y + (1-\beta^{(2)}(q_y))(1-P(y+))\,\Lambda}{\rho-\lambda(y)\{1-\beta_y(q_y)\}}, \quad \operatorname{Re}\rho \geqq 0, \tag{3.110}$$

if y is a discontinuity point of $P(y)$, and

$$\int_{0-}^{\infty} \mathrm{e}^{-\rho t}\, \mathrm{d}W_y(t) = \frac{1}{\rho}\,[(1-a)\,q_y + (1-\beta^{(2)}(q_y))(1-P(y))\,\Lambda], \quad \operatorname{Re}\rho \geqq 0, \tag{3.111}$$

if y is a continuity point of $P(y)$.

It follows for y a discontinuity point of $P(y)$ that

$$\int_0^\infty t\,\mathrm{d}W_y(t) = \frac{\Lambda\beta_2}{2(1-a(y))(1-a(y+))}, \tag{3.112}$$

$$\int_0^\infty t^2\,\mathrm{d}W_y(t) = \frac{\Lambda\beta_3}{3(1-a(y))^2\,(1-a(y+))}$$

$$+\frac{\Lambda^2\beta_2}{2(1-a(y))^2\,(1-a(y+))}\left\{\frac{\int_{0-}^{y+}\beta_{x,2}\,\mathrm{d}P(x)}{1-a(y+)}+\frac{\int_{0-}^{y}\beta_{x,2}\,\mathrm{d}P(x)}{1-a(y)}\right\},$$

and for a continuity point of $P(y)$ that

$$\int_0^\infty t\,\mathrm{d}W_y(t) = \frac{\Lambda\beta_2}{2(1-a(y))^2}, \tag{3.113}$$

$$\int_0^\infty t^2\,\mathrm{d}W_y(t) = \frac{\Lambda\beta_3}{3(1-a(y))^3}+\frac{\Lambda^2\beta_2}{(1-a(y))^4}\int_{0-}^{y}\beta_{x,2}\,\mathrm{d}P(x),$$

with

$$\beta_2 \stackrel{\text{def}}{=} \int_{0-}^{\infty}\beta_{x,2}\,\mathrm{d}P(x), \qquad \beta_3 \stackrel{\text{def}}{=} \int_{0-}^{\infty}\beta_{x,3}\,\mathrm{d}P(x). \tag{3.114}$$

The above results hold only if $a<1$. We shall now treat the situation for $a>1$. Let r be defined by

$$r \stackrel{\text{def}}{=} \sup\{y\colon\ a(y)<1\},$$

and suppose r is finite. Denote the number of customers of priority level $\boldsymbol{y}\leqq y$, of priority level $\boldsymbol{y}$ with $y<\boldsymbol{y}<r$, and of priority level $\boldsymbol{y}\geqq r$, who are in the system at time t by $\boldsymbol{x}_t^{(1)}$, $\boldsymbol{x}_t^{(2)}$ and $\boldsymbol{x}_t^{(3)}$ respectively. Supposing for the moment that customers of the $\boldsymbol{x}_t^{(i)}$-process are served in order of arrival, $i=1, 2, 3$, but with the priority discipline between the three processes maintained, then we encounter the three priority queueing system of section 3.8. By the same argument as that used in the beginning of this section we see that the stationary waiting time distribution $W_y(t)$ of a customer of priority

level $y=y<r$ follows from (3.94). For Re $\rho \geqq 0$, $y<r$,

$$\int_{0-}^{\infty} e^{-\rho t}\, dW_y(t) = \left\{1 - d^{(1)} \frac{1-\gamma^{(1)}(q_y)}{q_y \gamma^{(1)}}\right\}^{-1}$$

$$\cdot \left\{d^{(2)} \frac{1-\gamma^{(2)}(q_y)}{q_y \gamma^{(2)}} + \{1-d^{(1)}-d^{(2)}\} \frac{1-\gamma^{(3)}(q_y)}{q_y \gamma^{(3)}}\right\}, \qquad (3.115)$$

with q_y defined by (3.109), and

$$d^{(1)} \stackrel{\text{def}}{=} \Lambda \int_{0-}^{y+} \beta_{x,1}\, dP(x), \qquad d^{(1)} + d^{(2)} \stackrel{\text{def}}{=} \Lambda \int_{0-}^{r} \beta_{x,1}\, dP(x), \qquad (3.116)$$

$$\gamma^{(1)}(\rho) \stackrel{\text{def}}{=} \frac{1}{P(y+)} \int_{0-}^{y+} \beta_x(\rho)\, dP(x),$$

$$\gamma^{(2)}(\rho) \stackrel{\text{def}}{=} \frac{1}{P(r)-P(y+)} \int_{y+}^{r} \beta_x(\rho)\, dP(x),$$

$$\gamma^{(3)}(\rho) \stackrel{\text{def}}{=} \frac{1}{1-P(r)} \int_{r}^{\infty} \beta_x(\rho)\, dP(x),$$

$$\gamma^{(i)} \stackrel{\text{def}}{=} - \frac{d}{d\rho} \gamma^{(i)}(\rho) \bigg|_{\rho=0}, \qquad i=1, 2, 3.$$

It should be noted that if r is a continuity point of $a(y)$ then

$$1-d^{(1)}-d^{(2)}=0. \qquad (3.117)$$

For y a continuity point of $P(y)$ we have for Re $\rho \geqq 0$,

$$\int_{0-}^{\infty} e^{-\rho t}\, dW_y(t)$$

$$= \frac{1}{\rho} [\{1-\gamma^{(2)}(q_y)\}\, d^{(2)}/\gamma^{(2)} + \{1-\gamma^{(3)}(q_y)\}(1-d^{(1)}-d^{(2)})/\gamma^{(3)}], \qquad (3.118)$$

and for y a discontinuity point of $P(y)$,

$$\int_{0-}^{\infty} e^{-\rho t}\,dW_y(t)$$

$$= \frac{\{1-\gamma^{(2)}(q_y)\}\,d^{(2)}/\gamma^{(2)}+\{1-\gamma^{(3)}(q_y)\}(1-d^{(1)}-d^{(2)})/\gamma^{(3)}}{\rho-\lambda(y)\{1-\beta_y(q_y)\}}. \tag{3.119}$$

For the moments we have (cf. (3.114)),

$$\int_0^{\infty} t\,dW_y(t) = \frac{\Lambda}{2(1-a(y))(1-a(y+))}\left\{\beta_2 - \frac{a-1}{a-a(r)}\int_r^{\infty}\beta_{x,2}\,dP(x)\right\},$$

(3.120)

$$\int_0^{\infty} t^2\,dW_y(t) = \frac{\Lambda}{3(1-a(y))^2\,(1-a(y+))}\left\{\beta_3 - \frac{a-1}{a-a(r)}\int_r^{\infty}\beta_{x,3}\,dP(x)\right\}$$

$$+\frac{\Lambda^2}{2(1-a(y))^2\,(1-a(y+))}\left\{\beta_2 - \frac{a-1}{a-a(r)}\int_r^{\infty}\beta_{x,2}\,dP(x)\right\}$$

$$\cdot\left\{\frac{\int_{0-}^{y+}\beta_{x,2}\,dP(x)}{1-a(y+)} + \frac{\int_{0-}^{y}\beta_{x,2}\,dP(x)}{1-a(y)}\right\}.$$

Applying these results to the queueing system $M/G/1$ with all customers having the same service time distribution $B(t)$ and with priority discipline: "customers with shorter service time have priority" (cf. end of section 3.9) then if it is assumed that $B(t)$ is continuous we obtain

$$\int_0^{\infty} t\,dW_y(t) = \frac{\Lambda\beta_2}{2(1-a(y))^2}, \qquad \beta_i = \int_0^{\infty} x^i\,dB(x), \qquad i=1,2,3, \tag{3.121}$$

$$\int_0^{\infty} t^2\,dW_y(t) = \frac{\Lambda\beta_3}{3(1-a(y))^3} + \frac{\Lambda^2\beta_2}{(1-a(y))^4}\int_0^{y} x^2\,dB(x),$$

$$a(y) = \Lambda\int_0^{y} x\,dB(x),$$

for $y>0$, $a<1$, and

$$\int_0^\infty t\,\mathrm{d}W_y(t) = \frac{\Lambda \int_0^r x^2\,\mathrm{d}B(x)}{2(1-a(y))^2},$$

$$\int_0^\infty t^2\,\mathrm{d}W_y(t) = \frac{\Lambda \int_0^r x^3\,\mathrm{d}B(x)}{3(1-a(y))^3} + \frac{\Lambda^2 \int_0^r x^2\,\mathrm{d}B(x)}{(1-a(y))^4} \int_0^y x^2\,\mathrm{d}B(x),$$

for $0<y<r$, $a>1$; Λ being the arrival rate of the customers ($\Lambda=\alpha^{-1}$).

From the relations (3.121) it follows easily that for the total average waiting time for all customers we have

$$\int_0^\infty \left\{\int_0^\infty t\,\mathrm{d}W_y(t)\right\}\mathrm{d}P(y) = \tfrac{1}{2}\Lambda\beta_2 \int_0^\infty \frac{\mathrm{d}B(y)}{(1-a(y))^2} < \tfrac{1}{2}\Lambda\beta_2 \frac{1}{1-a}, \qquad a<1,$$

where the right-hand side is the average waiting time for the $M/G/1$ system with all customers having the same service time distribution and with service in order of arrival (cf. (II.4.82)). To prove the inequality note that $B(t)$ is continuous and that

$$a(y) = aB(y) \int_0^y \frac{t}{\beta}\,\frac{\mathrm{d}B(t)}{B(y)} < aB(y) \qquad \text{if} \quad B(y)>0.$$

Hence

$$\int_0^\infty \frac{\mathrm{d}B(y)}{\{1-a(y)\}^2} < \int_0^\infty \frac{\mathrm{d}B(y)}{\{1-aB(y)\}^2} = \frac{1}{1-a}.$$

Introduction of two priority levels depending on the service time of the arriving customers is a less refined but also very effective method of reducing the average waiting time for all customers in a $M/G/1$ system with all customers having the same service time distribution. Customers have priority level 1 if their service time is less than $w\beta$ and priority level 2 if it is at least $w\beta$, $0<w<\infty$. If $a<1$, the total average waiting time for all customers is now given by (cf. (3.112)),

$$\int_0^\infty \left\{ \int_0^\infty t \, \mathrm{d}W_y(t) \right\} \mathrm{d}P(y) = \tfrac{1}{2}\Lambda\beta_2 \frac{B(w\beta) + (1 - B(w\beta))/(1-a)}{1 - \Lambda \int_0^{w\beta} t \, \mathrm{d}B(t)}$$

$$= \frac{\Lambda\beta_2}{2(1-a)} \frac{1 - aB(w\beta)}{1 - a \int_0^{w\beta} (t/\beta) \, \mathrm{d}B(t)} \leqq \frac{\Lambda\beta_2}{2(1-a)}.$$

For $w=0$ and $w \to \infty$ we of course again obtain the average waiting time for service in order of arrival. Hence if $B(t)$ is not degenerate at β (constant service time) a value of w exists for which the total average waiting time has a minimum (cf. Cox and Smith [1961]).

III.3.11. Non break-in random priorities for $M/G/1$

A queueing system with random priorities is a system for which the order of service of the customers is sampled from a uniform distribution on [0, 1], i.e. the priority variable $\boldsymbol{y}$ has a uniform distribution on [0, 1]. The sampling is effected at the moment the customer arrives. The priority variables and the service times are independent families of variables.

Before discussing the $M/G/1$ queueing system with non break-in random priorities we shall first consider the $M/G/1$ queueing system with general non break-in priorities and with all customers having the same service time distribution $B(t)$.

For the present it will be assumed that $a<1$. For the stationary situation denote the distribution of the waiting time of a customer by $W(t)$ then

$$W(t) = \int_{0-}^{\infty} W_y(t) \, \mathrm{d}P(y), \tag{3.122}$$

with $W_y(t)$ being the stationary waiting time distribution of a customer of priority level $\boldsymbol{y}=y$. $W_y(t)$ is given by (3.108). From (3.112) we have

$$\int_0^\infty t \, \mathrm{d}W_y(t) = \begin{cases} \dfrac{\beta_2}{2\alpha} \dfrac{1}{(1-aP(y))^2} & \text{if } P(y+)=P(y), \\[2ex] \dfrac{\beta_2}{2\alpha} \dfrac{1}{a\{P(y+)-P(y)\}} \left\{ \dfrac{1}{1-aP(y+)} - \dfrac{1}{1-aP(y)} \right\} & \\ & \text{if } P(y+)>P(y). \end{cases}$$

It follows easily for the average waiting time

$$\int_0^\infty t\,\mathrm{d}W(t) = \int_0^\infty \int_{0-}^\infty t\,\mathrm{d}W_y(t)\,\mathrm{d}P(y) = \frac{a}{2(1-a)}\,\frac{\beta_2}{\beta}, \qquad a<1, \qquad (3.123)$$

which is equal to that for "service in order of arrival", and hence also equal to that for "last come, first served" and to that for "random service".

From (3.112) or from the results of section 3.7 it is easily derived that $\int_0^\infty t^2\,\mathrm{d}W(t)$, for a two priority level system, is larger than the second moment of the waiting time for "service in order of arrival" if all customers have the same service time distribution and if the priority level variable y is independent of the service time distribution. For these conditions it follows that if, in a k-priority level system $M/G/1$, one of the priority groups is split up into two new priority groups so that a $k+1$ priority level system is obtained then the second moment of $W(t)$ as defined by (3.122) will not decrease. Hence this second moment is a non-decreasing function of $k=2, 3, \ldots$.

Let $P(y)$ (cf. (3.97)) be a continuous distribution and let $P_n(y)$ be the distribution of the priority levels for a $M/G/1$ system with n priority levels, so that $P_n(y)$ has exactly n points of increase. It is supposed that all customers have the same service time distribution. Moreover it is assumed that the sequence $P_n(y)$, $n=2, 3, \ldots$, converges completely for $n\to\infty$ to $P(y)$. It is well-known that for a given $P(y)$ such a sequence always exists. From (3.112) we easily derive that for every y and every $n=2, 3, \ldots$,

$$\frac{\Lambda\beta_3}{3(1-aP_n(y+))^3} + \frac{\Lambda^2\beta_2^2 P_n(y+)}{(1-aP_n(y+))^4} \geq \int_0^\infty t^2\,\mathrm{d}W_y^{(n)}(t) \geq \frac{\Lambda\beta_3}{3(1-aP_n(y))^3} + \frac{\Lambda^2\beta_2^2 P_n(y)}{(1-aP_n(y))^4}.$$

Integrate all terms of this inequality with respect to $P_n(y)$. The integrals of the first and last terms have limits for $n\to\infty$, these limits being equal. Hence,

$$\lim_{n\to\infty} \int_0^\infty \int_0^\infty t^2\,\mathrm{d}W_y^{(n)}(t)\,\mathrm{d}P_n(y) = \tfrac{1}{3}\Lambda\beta_3 \int_0^\infty \frac{\mathrm{d}P(y)}{(1-aP(y))^3} + \Lambda^2\beta_2^2 \int_0^\infty \frac{P(y)\,\mathrm{d}P(y)}{(1-aP(y))^4}$$

$$= \frac{2a-a^2}{6(1-a)^2}\,\frac{\beta_3}{\beta} + \frac{3a^2-a^3}{6(1-a)^3}\left(\frac{\beta_2}{\beta}\right)^2, \qquad a<1. \qquad (3.124)$$

By a similar argument it can be shown that for every t,

$$\lim_{n\to\infty} \int_0^\infty W_y^{(n)}(t)\,\mathrm{d}P_n(y) = \int_0^\infty W_y(t)\,\mathrm{d}P(y),$$

with $W_y(t)$ the waiting time distribution for a customer with priority level $\mathbf{y}=y$ and $P(y)$ the distribution of $\mathbf{y}$. Since all customers have the same service time distribution it follows from (3.111) that

$$\int_{0-}^{\infty} \mathrm{e}^{-\rho t}\,\mathrm{d}W_y(t) = 1-a+a\frac{1-\mu_y(\rho,1)}{\rho\beta/(1-aP(y))}, \qquad \operatorname{Re}\rho\geqq 0. \qquad (3.125)$$

From section II.4.4 it is known that $\mu_y(\rho, 1)$ is the Laplace-Stieltjes transform of the busy period distribution $D_y(t)$ of a $M/G/1$ queueing system with $\alpha^{-1}P(y)$ as the average arrival rate of the customers. Hence (cf. (II.4.63)),

$$D_y(t) = \sum_{j=1}^{\infty} \int_0^t \mathrm{e}^{-\tau P(y)/\alpha} \frac{\{P(y)\,\tau/\alpha\}^{j-1}}{j!}\,\mathrm{d}B^{j*}(\tau), \qquad t>0. \qquad (3.126)$$

From (3.125) and (3.126) it is now easily found that

$$W(t) = \int_0^1 W_y(t)\,\mathrm{d}P(y) = 1-a+a\int_0^t \Bigg\{\int_0^1 (1-ax)\,\mathrm{d}x$$

$$-\int_{\tau=0}^{\sigma}\int_{x=0}^{1} \sum_{j=1}^{\infty}(1-ax)\frac{(ax\tau/\beta)^{j-1}}{j!}\,\mathrm{e}^{-ax\tau/\beta}\,\mathrm{d}x\,\mathrm{d}B^{j*}(\tau)\Bigg\}\frac{\mathrm{d}\sigma}{\beta}, \qquad (3.127)$$

for $t>0$ and $a<1$, and with first and second moments given by (3.123) and (3.124).

From (3.127) (cf. also (3.124)) it is seen that the distribution function $W(t)$ is independent of the form of $P(y)$ (for $P(y)$ a continuous function). This fact is easily explained. Let $f(x)$ be an increasing and continuous function of x with $f(0)=0$, $f(\infty)=1$. If the priority level $\mathbf{z}=f(\mathbf{y})$ is assigned to an arriving customer instead of the priority level $\mathbf{y}$ then the relative ordering of the priority levels $\mathbf{z}$ of arriving customers is not disturbed, and consequently the order in which they are served is not altered.

Clearly

$$\Pr\{z<x\}=\Pr\{f(y)<x\}=\Pr\{y<f_*(x)\}=P(f_*(x)),$$

where $f_*(x)$ is the inverse function of $f(x)$, it is single-valued. By taking $f(y)=P(y)$, if $P(y)$ is everywhere increasing, it follows that

$$\Pr\{z<x\} = \begin{cases} x, & 0\leqq x\leqq 1, \\ 1, & x\geqq 1, \end{cases}$$

so that the priority level z has a uniform distribution on [0, 1]. This explains the independence of $W(t)$ of the form of $P(y)$ if $P(y)$ is increasing everywhere. If $P(y)$ is non-decreasing and constant on an interval $[y_0, y_1]$ then priority levels y with $y_0\leqq y\leqq y_1$ have probability zero; we may define $P(P_*(y)) \overset{\text{def}}{=} y$ for $y_0\leqq y<y_1$ and obtain the same result.

Next we consider the case $a>1$ for general non break-in priorities with all customers having the same service time distribution and with $P(y)$ a continuous function. It follows from (3.118) that

$$\int_0^\infty e^{-\rho t}\,dW_y(t) = \frac{1-\mu_y(\rho, 1)}{\rho\beta/(1-aP(y))} \qquad \text{for } 0<y<r, \quad a>1,$$

with r defined by

$$aP(r)=1,$$

and for $y<r$,

$$\int_0^\infty t\,dW_y(t) = \frac{1}{2(1-aP(y))^2}\,\frac{\beta_2}{\beta},$$

$$\int_0^\infty t^2\,dW_y(t) = \frac{1}{3(1-aP(y))^3}\,\frac{\beta_3}{\beta} + \frac{aP(y)}{(1-aP(y))^4}\left(\frac{\beta_2}{\beta}\right)^2.$$

Note that a^{-1} is the probability that an arriving customer has a finite average waiting time.

Denote the probability that an arriving customer will have a priority level $y<y$ and a waiting time less than t by $W(t, y)$. It follows (cf. (3.126)) that for $a>1$, $y<r$, $t>0$,

$$W(t, y) = \int_0^y W_x(t)\,\mathrm{d}P(x)$$

$$= \int_0^t \Big\{ \int_0^y (1-aP(x))\,\mathrm{d}P(x)$$

$$- \int_{\tau=0}^{\sigma} \int_{x=0}^{y} \sum_{j=1}^{\infty} (1-aP(x)) \frac{(a\tau P(x)/\beta)^{j-1}}{j!} \mathrm{e}^{-a\tau P(x)/\beta}\,\mathrm{d}P(x)\mathrm{d}B^{j*}(\tau) \Big\} \frac{\mathrm{d}\sigma}{\beta},$$

and

$$\int_0^\infty t\,\mathrm{d}_t W(t, y) = \frac{P(y)}{2(1-aP(y))} \frac{\beta_2}{\beta},$$

$$\int_0^\infty t^2\,\mathrm{d}_t W(t, y) = \frac{2-aP(y)}{6(1-aP(y))^2} P(y) \frac{\beta_3}{\beta} + \frac{a(3-aP(y))}{6(1-aP(y))^3} \left\{ \frac{\beta_2}{\beta} P(y) \right\}^2.$$

Since a^{-1} is the probability that an arriving customer will obtain a priority level $y<r$ it follows that

$$a \int_0^\infty t\,\mathrm{d}_t W(t, y) = \frac{aP(y)}{2(1-aP(y))} \frac{\beta_2}{\beta}, \qquad y<r,$$

$$a \int_0^\infty t^2\,\mathrm{d}_t W(t, y) = \frac{aP(y)(2-aP(y))}{6(1-aP(y))^2} \frac{\beta_3}{\beta}$$

$$+ \frac{(3-aP(y))(aP(y))^2}{6(1-aP(y))^3} \left(\frac{\beta_2}{\beta} \right)^2, \qquad y<r,$$

are the first and second moments of the waiting time of a customer and the probability that he has a priority level $\boldsymbol{y}<y<r$ given that his priority level is less than r. It is interesting to compare these moments with those given by (3.123) and (3.124) for the case $a<1$.

For the system with random priority discipline and all customers having the same service time distribution the waiting time distribution $W(t)$ of a customer can now be immediately obtained from the results above by substituting

$$P(y) = \begin{cases} y & \text{for } y<1, \\ 1 & \text{for } y \geqq 1. \end{cases}$$

In particular it should be mentioned that for $a<1$ the waiting time $W(t)$ is independent of $P(y)$ if $P(y)$ is a continuous function.

Random priority discipline and random service have in common that for both disciplines the order of service of customers is determined by a purely random scheme.

III.3.12. Comparison of variances for various service disciplines for $M/G/1$

In the preceding sections a number of service disciplines have been investigated. To have at least some evaluation of these disciplines available the variances of the stationary waiting time distributions for these disciplines will be compared for the $M/G/1$ queueing system with all customers having the same service time distribution.

It is easily shown that for the stationary waiting time distributions the variance for service in order of arrival (cf. (II.4.82)) is smaller than that for random service (cf. (3.37)), the latter being smaller than that for random priorities (cf. (3.124)). Whereas for the "last come, first served" discipline the variance is largest (cf. (3.12)). These disciplines as well as the "non break-in" priority discipline with a finite number of priority levels have the same average waiting time (cf. (3.123)). For these disciplines the variance may be found from

$$\int_0^\infty t^2 \, \mathrm{d}W(t) = \int_{0-}^\infty \{\int_0^\infty t^2 \, \mathrm{d}W_y(t)\} \, \mathrm{d}P(y),$$

with $\int_0^\infty t^2 \, \mathrm{d}W_y(t)$ given by (3.112). This variance is always larger than that for service in the order of arrival, but smaller than that for random priorities. Whether this variance is larger or smaller than that for random service depends on $a(=\alpha^{-1}\beta)$ and on $P(y)$.

In the table below the general expressions for the variances for the various disciplines are listed for some values of a. For the two-priority discipline $P(y)$ has been chosen with two jumps each of magnitude $\frac{1}{2}$; for the three-priority discipline $P(y)$ has three jumps each equal to $\frac{1}{3}$. For negative exponentially distributed service times and for constant service times the values of the variances are specified in the table. It is seen that the variances for random service and for the two-priority discipline differ only by a few percents. Possibly the two-priority discipline may be a good approximation for the random service discipline. A good approximation method for the latter case is much needed, since the results obtained for random service are difficult to handle in applications.

Table of variances for various service disciplines for $M/G/1$.

β, β_2, β_3 respectively the first, second and third moments of service time distribution, $a=\Lambda\beta$, Λ the average arrival rate, $t=\beta_2/\beta$, $d=\beta_3/\beta$

	a	order of arr.	2 prior. ($\frac{1}{2}$, $\frac{1}{2}$)	3 prior. ($\frac{1}{3}$, $\frac{1}{3}$, $\frac{1}{3}$)	rand. serv.	rand. prior.	l. come, f. serv.
first moment	0.50	$0.50t$ ←	······	······	······	······	→ $0.50t$
	0.75	$1.50t$ ←	······	······	······	······	→ $1.50t$
	0.90	$4.50t$ ←	······	······	······	······	→ $4.50t$
variance 'general'	0.50	$0.33d+\ 0.25t^2$	$0.41d+\ 0.40t^2$	$0.44d+\ 0.46t^2$	$0.44d+\ 0.42t^2$	$0.50d+\ 0.58t^2$	$0.67d+\ 0.75t^2$
	0.75	1.00 2.25	1.48 4.84	1.74 6.40	1.60 4.95	2.50 11.25	4.00 15.75
	0.90	3.00 20.25	5.23 53.11	6.90 79.76	5.45 53.39	16.50 263.25	30.00 384.75
variance 'neg. exp.'	0.50	$3.00\beta^2$	$4.04\beta^2$	$4.45\beta^2$	$4.33\beta^2$	$5.33\beta^2$	$7.00\beta^2$
	0.75	15.00	28.25	36.04	29.40	60.00	87.00
	0.90	99.00	243.84	360.44	246.24	1152.00	1719.00
variance 'constant'	0.50	$0.58\beta^2$	$0.81\beta^2$	$0.90\beta^2$	$0.86\beta^2$	$1.08\beta^2$	$1.42\beta^2$
	0.75	3.25	6.32	8.14	6.55	13.75	19.75
	0.90	23.25	58.34	86.66	58.84	279.75	414.75

III.4. UNIFORMLY BOUNDED ACTUAL WAITING TIME

III.4.1. Introduction

In this chapter we shall study the stochastic process $\{\boldsymbol{v}_n, n=1, 2, ...\}$ which is recursively defined by the relations

$$\begin{aligned} \boldsymbol{v}_{n+1} &= \min(K, [\boldsymbol{v}_n+\boldsymbol{\tau}_n-\boldsymbol{\sigma}_{n+1}]^+), \qquad n=1, 2, ..., \\ \boldsymbol{v}_1 &= v, \qquad 0 \leqq v \leqq K, \end{aligned} \tag{4.1}$$

with K a positive constant.

The variable $\boldsymbol{v}_n$ may be interpreted as the actual waiting time of the nth arriving customer if the server acts in such a way that the actual waiting time never exceeds K. A queueing system of this type will be called a single server queue with uniformly bounded actual waiting time. It should be noted that for this system $\boldsymbol{v}_n+\boldsymbol{\tau}_n$ is in general not a bounded variable. In fact the queueing model, determined by the relations (4.1) and with customers served in order of arrival, may be described as follows. A customer obtains full service if the next customer arrives at a moment at which the work still to be handled by the server is less than K; if it exceeds K his service time is so much shortened as to make the actual waiting time of the next arriving customer equal to K. The decision to shorten the service time of a customer is taken by the server at the moment of arrival of the next customer. Once the mathematical description of the process $\{\boldsymbol{v}_n, n=1, 2, ...\}$ is known it is possible to obtain from this solution new results for the $G/G/1$ queueing model. In fact it is possible to obtain for the $G/G/1$ model the joint distribution of the busy cycle and of the supremum of the actual waiting time of the customers arriving during this busy cycle. The distribution of the number of the customer who is the first one with an actual waiting time exceeding K may also be obtained. Moreover, it should be noted that the description of the process $\{\boldsymbol{v}_n, n=1, 2, \dots\}$ leads to interesting results in fluctuation theory (cf. I.6).

To simplify the analysis of the process $\{\boldsymbol{v}_n, n=1, 2, \dots\}$ it will be assumed that the Laplace-Stieltjes transforms $\alpha(\rho)$ and $\beta(\rho)$ of $A(.)$ and $B(.)$, respectively, are rational functions of ρ, i.e.

$$\int_0^\infty \mathrm{e}^{-\rho t}\,\mathrm{d}A(t) = \alpha(\rho) = \frac{\alpha_1(\rho)}{\alpha_2(\rho)}, \qquad \int_0^\infty \mathrm{e}^{-\rho t}\,\mathrm{d}B(t) = \beta(\rho) = \frac{\beta_1(\rho)}{\beta_2(\rho)}, \tag{4.2}$$

with $\alpha_1(\rho)$, $\alpha_2(\rho)$, $\beta_1(\rho)$ and $\beta_2(\rho)$ polynomials in ρ; the degree of $\alpha_2(\rho)$ will be m, that of $\beta_2(\rho)$ will be n. Since it is assumed that $A(0+)=B(0+)=0$ (cf. II.1.2) it follows that the degree of $\alpha_1(\rho)$ is at most $m-1$, and the degree of $\beta_1(\rho)$ is at most $n-1$. In sections 4.5 and 4.6 where the results of the general theory will be applied to the queueing models $M/G/1$ and $G/M/1$ it will be seen that for these the assumption of rationality is superfluous.

III.4.2. The time dependent solution

From section II.5.1 we have the following relations:
for real values of x:

$$\exp(-\rho[x]^-) = 1 - \frac{1}{2\pi\mathrm{i}} \int_{C_\xi} \left(\frac{1}{\rho-\xi} + \frac{1}{\xi}\right) \mathrm{e}^{-\xi x}\,\mathrm{d}\xi,$$

$$\operatorname{Re}\rho < \operatorname{Re}\xi, \quad \operatorname{Re}\xi > 0; \tag{4.3}$$

$$\exp(-\rho[x]^+) = \frac{1}{2\pi\mathrm{i}} \int_{C_\eta} \left(\frac{1}{\rho-\eta} + \frac{1}{\eta}\right) \mathrm{e}^{-\eta x}\,\mathrm{d}\eta,$$

$$\operatorname{Re}\rho > \operatorname{Re}\eta > 0.$$

From (4.1) and (4.3) we obtain for $\operatorname{Re}\xi > \operatorname{Re}\rho$, $\operatorname{Re}\xi > \operatorname{Re}\eta > 0$, and $n=1, 2, \dots$,

$$\exp(-\rho\boldsymbol{v}_{n+1}) = \exp(-\rho K - \rho[[\boldsymbol{v}_n + \boldsymbol{\tau}_n - \boldsymbol{\sigma}_{n+1}]^+ - K]^-)$$

$$= \mathrm{e}^{-\rho K} - \frac{\mathrm{e}^{-\rho K}}{2\pi\mathrm{i}} \int_{C_\xi} \left(\frac{1}{\rho-\xi} + \frac{1}{\xi}\right) \exp(\xi K - \xi[\boldsymbol{v}_n + \boldsymbol{\tau}_n - \boldsymbol{\sigma}_{n+1}]^+)\,\mathrm{d}\xi =$$

$$= \mathrm{e}^{-\rho K} - \frac{\mathrm{e}^{-\rho K}}{2\pi \mathrm{i}} \int_{C_\xi} \left(\frac{1}{\rho - \xi} + \frac{1}{\xi} \right) \frac{\mathrm{e}^{\xi K}}{2\pi \mathrm{i}}$$

$$\cdot \left\{ \int_{C_\eta} \left(\frac{1}{\xi - \eta} + \frac{1}{\eta} \right) \exp(-\eta(\boldsymbol{v}_n + \boldsymbol{\tau}_n - \boldsymbol{\sigma}_{n+1}))\, \mathrm{d}\eta \right\} \mathrm{d}\xi.$$

From (4.1) it follows that $\boldsymbol{v}_n$ is independent of $\boldsymbol{\tau}_n$ and of $\boldsymbol{\sigma}_{n+1}$, hence we obtain from the above relation

$$\mathrm{E}\{\exp(-\rho \boldsymbol{v}_{n+1} - \rho_3 \boldsymbol{a}_n - \rho_4 \boldsymbol{b}_n) \mid \boldsymbol{v}_1 = v\} = \mathrm{e}^{-\rho K}\, \alpha^n(\rho_3)\, \beta^n(\rho_4)$$

$$- \frac{\mathrm{e}^{-\rho K}}{2\pi \mathrm{i}} \int_{C_\xi} \left(\frac{1}{\rho - \xi} + \frac{1}{\xi} \right) \frac{\mathrm{e}^{\xi K}}{2\pi \mathrm{i}} \left\{ \int_{C_\eta} \left(\frac{1}{\xi - \eta} + \frac{1}{\eta} \right) \right.$$

$$\left. \cdot\, \alpha(\rho_3 - \eta)\, \beta(\rho_4 + \eta)\, \mathrm{E}\{\exp(-\eta \boldsymbol{v}_n - \rho_3 \boldsymbol{a}_{n-1} - \rho_4 \boldsymbol{b}_{n-1}) \mid \boldsymbol{v}_1 = v\}\, \mathrm{d}\eta \right\} \mathrm{d}\xi, \tag{4.4}$$

for $\mathrm{Re}\,\xi > \mathrm{Re}\,\rho$, $\mathrm{Re}\,\xi > \mathrm{Re}\,\eta > 0$, $\mathrm{Re}(\rho_3 - \eta) > 0$, $\mathrm{Re}\,\rho_4 \geqq 0$, $n = 1, 2, \ldots,$ with

$$\boldsymbol{a}_n \stackrel{\text{def}}{=} \begin{cases} \sum_{i=1}^{n} \boldsymbol{\sigma}_{i+1}, \\ 0, \end{cases} \qquad \boldsymbol{b}_n \stackrel{\text{def}}{=} \begin{cases} \sum_{i=1}^{n} \boldsymbol{\tau}_i, & n = 1, 2, \ldots, \\ 0, & n = 0. \end{cases}$$

The permutations of the operators E and $\int_{C_\xi} \ldots$, and $\int_{C_\eta} \ldots$ in the derivation of (4.4) are easily justified since in the domain of integration $\beta(\rho_4 + \eta)$ and $\alpha(\rho_3 - \eta)$ and the conditional expectation $\mathrm{E}\{\ldots \mid \boldsymbol{v}_1 = v\}$ are bounded, note $0 \leqq \boldsymbol{v}_n \leqq K$, and since

$$\int_{C_\xi} \left| \frac{1}{\rho - \xi} + \frac{1}{\xi} \right| |\mathrm{d}\xi| < \infty.$$

We define for $|r| < 1$, $|\rho| < \infty$, $\mathrm{Re}\,\rho_3 \geqq 0$, $\mathrm{Re}\,\rho_4 \geqq 0$, $0 \leqq v \leqq K$,

$$\varphi(r, \rho, \rho_3, \rho_4, v) \stackrel{\text{def}}{=} \sum_{n=1}^{\infty} r^n\, \mathrm{E}\{\exp(-\rho \boldsymbol{v}_n - \rho_3 \boldsymbol{a}_{n-1} - \rho_4 \boldsymbol{b}_{n-1}) \mid \boldsymbol{v}_1 = v\}, \tag{4.5}$$

since

$$0 \leqq \boldsymbol{v}_n \leqq K \quad \text{for all } n = 1, 2, \ldots, \quad \text{with probability one,} \tag{4.6}$$

it follows that for $|\rho| < \infty$, and the other conditions just mentioned, the right-hand side of (4.5) exists. From (4.4) and (4.5) it follows easily that

$$\varphi(r,\rho,\rho_3,\rho_4,v)-r\,\mathrm{e}^{-\rho v}-\frac{r^2\alpha(\rho_3)\,\beta(\rho_4)}{1-r\alpha(\rho_3)\,\beta(\rho_4)}\,\mathrm{e}^{-\rho K}$$

$$=-\frac{r\,\mathrm{e}^{-\rho K}}{2\pi\mathrm{i}}\int_{C_\xi}\left(\frac{1}{\rho-\xi}+\frac{1}{\xi}\right)\frac{\mathrm{e}^{\xi K}}{2\pi\mathrm{i}}$$

$$\cdot\left\{\int_{C_\eta}\left(\frac{1}{\xi-\eta}+\frac{1}{\eta}\right)\varphi(r,\eta,\rho_3,\rho_4,v)\,\alpha(\rho_3-\eta)\,\beta(\rho_4+\eta)\,\mathrm{d}\eta\right\}\mathrm{d}\xi, \tag{4.7}$$

for $|r|<1$, $\mathrm{Re}\,\xi>\mathrm{Re}\,\rho$, $\mathrm{Re}\,\xi>\mathrm{Re}\,\eta>0$, $\mathrm{Re}(\rho_3-\eta)>0$, $\mathrm{Re}\,\rho_4\geq 0$, $0\leq v\leq K$.

The latter relation is the integral equation for the process $\{\boldsymbol{v}_n,\ n=1,\ldots\}$ with $\boldsymbol{v}_1=v$ defined by (4.1) (cf. COHEN [1967b]). Before studying the solution of (4.7) we shall first discuss some properties of the function $\varphi(r,\rho,\rho_3,\rho_4,v)$.

From (4.6) it follows that $\mathrm{E}\{\exp(-\rho\boldsymbol{v}_n)\mid\boldsymbol{v}_1=v\}$ is an entire function of exponential type and order one, for all $n=2,3,\ldots$; in fact

$$|\mathrm{E}\{\mathrm{e}^{-\rho\boldsymbol{v}_n}\mid\boldsymbol{v}_1=v\}|\leq\mathrm{e}^{-K\mathrm{Re}\,\rho}\qquad\text{for}\quad 1\tfrac{1}{2}\pi>\arg\rho>\tfrac{1}{2}\pi. \tag{4.8}$$

Further it is well known that for $n=2,3,\ldots,$

$$\lim_{|\rho|\to\infty}\mathrm{E}\{\mathrm{e}^{-\rho\boldsymbol{v}_n}\mid\boldsymbol{v}_1=v\}=\Pr\{\boldsymbol{v}_n=0\mid\boldsymbol{v}_1=v\},\qquad -\tfrac{1}{2}\pi<\arg\rho<\tfrac{1}{2}\pi. \tag{4.9}$$

Hence it follows from (4.5) that for $|r|<1$, $\mathrm{Re}\,\rho_3\geq 0$, $\mathrm{Re}\,\rho_4\geq 0$, $0\leq v\leq K$, two bounded functions $C_1(r,\rho_3,\rho_4,v)$ and $C_2(r,\rho_3,\rho_4,v)$ exist such that

$$\lim_{|\rho|\to\infty}|\mathrm{e}^{\rho K}\,\varphi(r,\rho,\rho_3,\rho_4,v)|=C_1(r,\rho_3,\rho_4,v),\qquad 1\tfrac{1}{2}\pi>\arg\rho>\tfrac{1}{2}\pi, \tag{4.10}$$

$$\lim_{|\rho|\to\infty}|\varphi(r,\rho,\rho_3,\rho_4,v)|=C_2(r,\rho_3,\rho_4,v),\qquad \tfrac{1}{2}\pi>\arg\rho>-\tfrac{1}{2}\pi.$$

As already stated above (cf. (4.5)) we have

$$|\varphi(r,\rho,\rho_3,\rho_4,v)|<\infty,\qquad\text{for}\quad |r|<1,\ \ \mathrm{Re}\,\rho_3\geq 0,\ \ \mathrm{Re}\,\rho_4\geq 0,\ \ |\rho|<\infty. \tag{4.11}$$

We now transform the integral equation (4.7) by contour integration into an equivalent integral equation. For the right-hand side of (4.7) we may write by using contour integration

$$-\frac{r\,\mathrm{e}^{-\rho K}}{2\pi\mathrm{i}}\int_{C_\xi}\left(\frac{1}{\rho-\xi}+\frac{1}{\xi}\right)\frac{\mathrm{e}^{\xi K}}{2\pi\mathrm{i}}$$

$$\cdot\left\{\int_{C_\eta}\left(\frac{1}{\xi-\eta}+\frac{1}{\eta}\right)\varphi(r,\eta,\rho_3,\rho_4,v)\,\alpha(\rho_3-\eta)\,\beta(\rho_4+\eta)\,\mathrm{d}\eta\right\}\mathrm{d}\xi=$$

$$
\begin{aligned}
&= -\frac{r\,\mathrm{e}^{-\rho K}}{2\pi\mathrm{i}} \int_{C_\xi} \left(\frac{1}{\rho-\xi}+\frac{1}{\xi}\right) \mathrm{e}^{\xi K}\, \varphi(r,\xi,\rho_3,\rho_4,v)\, \alpha(\rho_3-\xi)\, \beta(\rho_4+\xi)\, \mathrm{d}\xi \\
&\quad - \frac{r\,\mathrm{e}^{-\rho K}}{2\pi\mathrm{i}} \int_{C_\xi} \left(\frac{1}{\rho-\xi}+\frac{1}{\xi}\right) \frac{\mathrm{e}^{\xi K}}{2\pi\mathrm{i}} \\
&\quad \cdot \left\{ \int_{C_\omega} \left(\frac{1}{\xi-\omega}+\frac{1}{\omega}\right) \varphi(r,\omega,\rho_3,\rho_4,v)\, \alpha(\rho_3-\omega)\, \beta(\rho_4+\omega)\, \mathrm{d}\omega \right\} \mathrm{d}\xi \\
&= r\varphi(r,\rho,\rho_3,\rho_4,v)\, \alpha(\rho_3-\rho)\, \beta(\rho_4+\rho) - \frac{r^2\alpha(\rho_3)\, \beta(\rho_4)}{1-r\alpha(\rho_3)\, \beta(\rho_4)}\, \mathrm{e}^{-\rho K} \\
&\quad - \frac{r\,\mathrm{e}^{-\rho K}}{2\pi\mathrm{i}} \int_{C_\zeta} \left(\frac{1}{\rho-\zeta}+\frac{1}{\zeta}\right) \mathrm{e}^{\zeta K}\, \varphi(r,\zeta,\rho_3,\rho_4,v)\, \alpha(\rho_3-\zeta)\, \beta(\rho_4+\zeta)\, \mathrm{d}\zeta \\
&\quad + r \int_{C_\omega} \left(\frac{1}{\rho-\omega}+\frac{1}{\omega}\right) \varphi(r,\omega,\rho_3,\rho_4,v)\, \alpha(\rho_3-\omega)\, \beta(\rho_4+\omega)\, \mathrm{d}\omega \\
&\quad - \frac{r\,\mathrm{e}^{-\rho K}}{2\pi\mathrm{i}} \int_{C_\zeta} \left(\frac{1}{\rho-\zeta}+\frac{1}{\zeta}\right) \frac{\mathrm{e}^{\zeta K}}{2\pi\mathrm{i}} \\
&\quad \cdot \left\{ \int_{C_\omega} \left(\frac{1}{\zeta-\omega}+\frac{1}{\omega}\right) \varphi(r,\omega,\rho_3,\rho_4,v)\, \alpha(\rho_3-\omega)\, \beta(\rho_4+\omega)\, \mathrm{d}\omega \right\} \mathrm{d}\zeta,
\end{aligned}
\tag{4.12}
$$

for $\mathrm{Re}\,\xi > \mathrm{Re}\,\rho > 0$, $\mathrm{Re}\,\omega > \mathrm{Re}\,\xi$, $\mathrm{Re}(\rho_3-\omega) \geqq 0$, $\mathrm{Re}\,\rho_4 \geqq 0$,
$\mathrm{Re}\,\zeta < 0 < \mathrm{Re}\,\rho < \mathrm{Re}\,\omega$, $|r|<1$, $\mathrm{Re}\,\xi > \mathrm{Re}\,\eta > 0$.
From (4.7), (4.10), (4.11) and (4.12) it is easily seen that we have for $|r|<1$, $\mathrm{Re}\,\zeta < 0 < \mathrm{Re}\,\rho < \mathrm{Re}\,\omega$, $\mathrm{Re}(\rho_3-\omega) \geqq 0$, $\mathrm{Re}(\rho_4+\zeta) \geqq 0$,

$$
\begin{aligned}
&\{1-r\alpha(\rho_3-\rho)\, \beta(\rho_4+\rho)\}\, \varphi(r,\rho,\rho_3,\rho_4,v) - r\,\mathrm{e}^{-\rho v} \\
&= -\frac{1}{2\pi\mathrm{i}} \int_{C_\omega} \left(\frac{1}{\rho-\omega}+\frac{1}{\omega}\right) \{1-r\alpha(\rho_3-\omega)\, \beta(\rho_4+\omega)\}\, \varphi(r,\omega,\rho_3,\rho_4,v)\, \mathrm{d}\omega \\
&\quad + \frac{\mathrm{e}^{-\rho K}}{2\pi\mathrm{i}} \int_{C_\zeta} \left(\frac{1}{\rho-\zeta}+\frac{1}{\zeta}\right) \mathrm{e}^{\zeta K} \{1-r\alpha(\rho_3-\zeta)\, \beta(\rho_4+\zeta)\}\, \varphi(r,\zeta,\rho_3,\rho_4,v)\, \mathrm{d}\zeta
\end{aligned}
$$

$$+\frac{\mathrm{e}^{-\rho K}}{2\pi\mathrm{i}}\int\limits_{C_\zeta}\left(\frac{1}{\rho-\zeta}+\frac{1}{\zeta}\right)\frac{\mathrm{e}^{\zeta K}}{2\pi\mathrm{i}}$$

$$\cdot\left\{\int\limits_{C_\omega}\left(\frac{1}{\zeta-\omega}+\frac{1}{\omega}\right)\{1-r\alpha(\rho_3-\omega)\,\beta(\rho_4+\omega)\}\,\varphi(r,\omega,\rho_3,\rho_4,v)\,\mathrm{d}\omega\right\}\mathrm{d}\zeta. \tag{4.13}$$

It should be noted that for the derivation of the integral equation (4.13) no use has been made of the assumption that $\alpha(\rho)$ and $\beta(\rho)$ are rational functions. However, for the construction of the solution of (4.13) this assumption will be used.

Application of Rouché's theorem to the function of ρ,

$$1-r\,\alpha(\rho_3-\rho)\,\beta(\rho_4+\rho),\qquad\text{with}\quad |r|<1,\quad \mathrm{Re}\,\rho_3\geqq 0,\quad \mathrm{Re}\,\rho_4\geqq 0, \tag{4.14}$$

leads readily to the conclusion that this function has exactly $m+n$ zeros (cf. (4.2)) $\delta_j(r,\rho_3,\rho_4)$, $j=1,\dots,m$, and $\varepsilon_i(r,\rho_3,\rho_4)$, $i=1,\dots,n$. It is easily seen that these zeros are continuous functions of r, ρ_3 and ρ_4, moreover of the $m+n$ zeros n are lying in the left halfplane and m in the right halfplane, i.e. for $|r|<1$, $\mathrm{Re}\,\rho_3\geqq 0$, $\mathrm{Re}\,\rho_4\geqq 0$, or $|r|\leqq 1$, $\mathrm{Re}\,\rho_3>0$, $\mathrm{Re}\,\rho_4\geqq 0$, or $|r|\leqq 1$, $\mathrm{Re}\,\rho_3\geqq 0$, $\mathrm{Re}\,\rho_4>0$,

$$\mathrm{Re}\,\varepsilon_i(r,\rho_3,\rho_4)<0,\quad i=1,\dots,n;\qquad \mathrm{Re}\,\delta_j(r,\rho_3,\rho_4)>0,\quad j=1,\dots,m. \tag{4.15}$$

We define

$$\nu_1\equiv\nu_1(r,\rho_3,\rho_4)\overset{\text{def}}{=}\min_{1\leqq j\leqq m}\mathrm{Re}\,\delta_j(r,\rho_3,\rho_4), \tag{4.16}$$

$$\nu_2\equiv\nu_2(r,\rho_3,\rho_4)\overset{\text{def}}{=}\max_{1\leqq i\leqq n}\mathrm{Re}\,\varepsilon_i(r,\rho_3,\rho_4),$$

and

$$H_-(r,\rho,\rho_3,\rho_4)\overset{\text{def}}{=}\frac{1-r\alpha(\rho_3-\rho)\,\beta(\rho_4+\rho)}{\prod\limits_{i=1}^{n}(\rho-\varepsilon_i(r,\rho_3,\rho_4))}\,\beta_2(\rho_4+\rho), \tag{4.17}$$

$$H_+(r,\rho,\rho_3,\rho_4)\overset{\text{def}}{=}\frac{1-r\alpha(\rho_3-\rho)\,\beta(\rho_4+\rho)}{\prod\limits_{j=1}^{m}(\rho-\delta_j(r,\rho_3,\rho_4))}\,\alpha_2(\rho_3-\rho),$$

for all those values of r, ρ, ρ_3, ρ_4, for which the right-hand sides exist.

Note that $H_+(\dots)$ and $H_-(\dots)$ are rational functions of ρ. Clearly for $|r|<1$, $\operatorname{Re}\rho_3 \geqq 0$, $\operatorname{Re}\rho_4 \geqq 0$, $H_-(\dots)$ is analytic for $\operatorname{Re}\rho < \operatorname{Re}\rho_3$ and has no zeros for $\operatorname{Re}\rho < \nu_1$, while $H_+(\dots)$ is analytic for $\operatorname{Re}\rho > -\operatorname{Re}\rho_4$ and has no zeros for $\operatorname{Re}\rho > \nu_2$.

Next, we define two polynomials $f(r, \rho, \rho_3, \rho_4, v)$ and $g(r, \rho, \rho_3, \rho_4, v)$ in ρ of which the coefficients are functions of r, ρ_3, ρ_4 and v, and such that
(*i*) $f(r, \rho, \rho_3, \rho_4, v)$ is a polynomial of degree $m-1$ in ρ and $g(r, \rho, \rho_3, \rho_4, v)$ a polynomial of degree $n-1$ in ρ;
(*ii*) for $|r|<1$, $\operatorname{Re}\rho_3 \geqq 0$, $\operatorname{Re}\rho_4 \geqq 0$, $0 \leqq v \leqq K$,

$$\frac{\rho f(r, \rho, \rho_3, \rho_4, v)}{\alpha_2(\rho_3-\rho)} + \frac{\rho g(r, \rho, \rho_3, \rho_4, v)}{\beta_2(\rho_4+\rho)} \mathrm{e}^{-\rho K} + r\,\mathrm{e}^{-\rho v} = 0, \tag{4.18}$$

for $\rho = \delta_j(r, \rho_3, \rho_4)$, $j=1, \dots, m$, and for $\rho = \varepsilon_i(r, \rho_3, \rho_4)$, $i=1, \dots, n$.

It is easily verified that for the conditions mentioned the functions $f(r, \rho, \rho_3, \rho_4, v)$ and $g(r, \rho, \rho_3, \rho_4, v)$ are uniquely determined.

We prove that the following pair of integral equations determine the polynomials $f(\dots)$ and $g(\dots)$ equally well:

$$f(r, \rho, \rho_3, \rho_4, v) = \frac{-\alpha_2(\rho_3-\rho)\, H_-(r, \rho, \rho_3, \rho_4)}{2\pi \mathrm{i}}$$

$$\cdot \int_{C_\xi} \left\{ \frac{g(r, \xi, \rho_3, \rho_4, v)}{\beta_2(\rho_4+\xi)} \mathrm{e}^{-\xi K} + \frac{r\,\mathrm{e}^{-\xi v}}{\xi} \right\} \frac{1}{H_-(r, \xi, \rho_3, \rho_4)} \frac{\mathrm{d}\xi}{\xi-\rho}, \tag{4.19}$$

$$g(r, \rho, \rho_3, \rho_4, v) = \frac{\beta_2(\rho_4+\rho)\, H_+(r, \rho, \rho_3, \rho_4)}{2\pi \mathrm{i}}$$

$$\cdot \int_{C_\eta} \left\{ \frac{f(r, \eta, \rho_3, \rho_4, v)}{\alpha_2(\rho_3-\eta)} \mathrm{e}^{\eta K} + \frac{r \mathrm{e}^{(K-v)\eta}}{\eta} \right\} \frac{1}{H_+(r, \eta, \rho_3, \rho_4)} \frac{\mathrm{d}\eta}{\eta-\rho},$$

for $|r|<1$, $\nu_1 > \operatorname{Re}\xi > 0$, $\operatorname{Re}\xi > \operatorname{Re}\rho$, $\nu_2 < \operatorname{Re}\eta < 0$, $\operatorname{Re}\eta < \operatorname{Re}\rho$, $\operatorname{Re}\rho_3 \geqq 0$, $\operatorname{Re}\rho_4 \geqq 0$, $0 \leqq v \leqq K$.

For $g(r, \rho, \rho_3, \rho_4, v)$ a polynomial in ρ it follows from (4.19) and (4.17) that $f(r, \rho, \rho_3, \rho_4, v)$ is a polynomial of degree $m-1$, similarly it is seen from (4.19) that $g(r, \rho, \rho_3, \rho_4, v)$ is a polynomial of degree $n-1$. It is easily verified that the integral in the first relation of (4.19) is equal to the sum of the residues of the integrand at its poles $\xi = \delta_j(r, \rho_3, \rho_4)$, $j=1, \dots, m$, apart from a factor -1. Evaluation of this integral and then taking $\rho=\delta_j(r, \rho_3, \rho_4)$, $j=1, \ldots, m$, yields the condition (4.18) for $\rho=\delta_j(r, \rho_3, \rho_4)$, $j=1, \ldots, m$. Application of a similar procedure to the second integral relation of (4.19) yields the condition (4.18) for $\rho=\varepsilon_i(r, \rho_3, \rho_4)$, $i=1, \ldots, n$. The proof is complete.

THEOREM 4.1. *The function* $\varphi(r, \rho, \rho_3, \rho_4, v)$ *defined by* (4.5) *for* $|r|<1$, $|\rho|<\infty$, $\mathrm{Re}\,\rho_3\geqq 0$, $\mathrm{Re}\,\rho_4\geqq 0$, $0\leqq v\leqq K$ *is given by*

$$\varphi(r, \rho, \rho_3, \rho_4, v) = \frac{1}{1-r\alpha(\rho_3-\rho)\,\beta(\rho_4+\rho)}$$

$$\cdot\left\{\frac{\rho f(r, \rho, \rho_3, \rho_4, v)}{\alpha_2(\rho_3-\rho)} + \frac{\rho g(r, \rho, \rho_3, \rho_4, v)}{\beta_2(\rho_4+\rho)}\,\mathrm{e}^{-\rho K} + r\,\mathrm{e}^{-\rho v}\right\}, \tag{4.20}$$

with the polynomials $f(r, \rho, \rho_3, \rho_4, v)$ *and* $g(r, \rho, \rho_3, \rho_4, v)$ *in* ρ *given by* (4.19) *or equivalently by* (4.18).

Proof. By writing for the present

$$F(\rho) = \frac{f(r, \rho, \rho_3, \rho_4, v)}{\alpha_2(\rho_3-\rho)}, \qquad G(\rho) = \frac{g(r, \rho, \rho_3, \rho_4, v)}{\beta_2(\rho_4+\rho)}, \tag{4.21}$$

it follows by substituting the expression (4.20) into (4.13) that

$$\rho F(\rho) + \rho G(\rho)\,\mathrm{e}^{-\rho K}$$

$$= -\frac{1}{2\pi\mathrm{i}}\int_{C_\omega}\left(\frac{1}{\rho-\omega}+\frac{1}{\omega}\right)\{\omega F(\omega)+\omega G(\omega)\,\mathrm{e}^{-\omega K}+r\,\mathrm{e}^{-\omega v}\}\,\mathrm{d}\omega$$

$$+\frac{\mathrm{e}^{-\rho K}}{2\pi\mathrm{i}}\int_{C_\zeta}\left(\frac{1}{\rho-\zeta}+\frac{1}{\zeta}\right)\mathrm{e}^{\zeta K}\{\zeta F(\zeta)+\zeta G(\zeta)\,\mathrm{e}^{-\zeta K}+r\,\mathrm{e}^{-\zeta v}\}\,\mathrm{d}\zeta$$

$$+\frac{\mathrm{e}^{-\rho K}}{2\pi\mathrm{i}}\int_{C_\zeta}\left(\frac{1}{\rho-\zeta}+\frac{1}{\zeta}\right)\frac{\mathrm{e}^{\zeta K}}{2\pi\mathrm{i}}$$

$$\cdot\left\{\int_{C_\omega}\left(\frac{1}{\zeta-\omega}+\frac{1}{\omega}\right)\{\omega F(\omega)+\omega G(\omega)\,\mathrm{e}^{-\omega K}+r\,\mathrm{e}^{-\omega v}\}\,\mathrm{d}\omega\right\}\mathrm{d}\zeta, \tag{4.22}$$

for $\mathrm{Re}\,\zeta<0<\mathrm{Re}\,\rho<\mathrm{Re}\,\omega$, $\mathrm{Re}(\rho_3-\omega)>0$, $\mathrm{Re}(\rho_4+\zeta)>0$.

Since $F(\rho)$ and $G(\rho)$ are of $O(|\rho|^{-1})$ for $|\rho|\to\infty$ it follows from (4.21) that

$$\frac{1}{2\pi i}\int_{C_\omega} F(\omega)\frac{d\omega}{\omega-\rho}=F(\rho),\qquad \frac{1}{2\pi i}\int_{C_\omega} G(\omega)\frac{e^{-\omega K}}{\omega-\rho}\,d\omega=0,$$
$$\frac{1}{2\pi i}\int_{C_\zeta} F(\zeta)\frac{e^{\zeta K}}{\rho-\zeta}\,d\zeta=0,\qquad \frac{1}{2\pi i}\int_{C_\zeta} G(\zeta)\frac{d\zeta}{\rho-\zeta}=G(\rho), \tag{4.23}$$

since $\alpha(\rho)$ and $\beta(\rho)$ are analytic for $\operatorname{Re}\rho>0$. Using the relations (4.23) it is easily verified that the expression (4.22) is an identity.

It follows from (4.18) that the function expressed by (4.20) satisfies the conditions (4.10) and (4.11).

From (4.14) it is seen that for $\operatorname{Re}\rho_3\geqq 0$, $\operatorname{Re}\rho_4\geqq 0$ the zeros $\varepsilon_i(r,\rho_3,\rho_4)$, $i=1,\dots,n$, and $\delta_j(r,\rho_3,\rho_4)$, $j=1,\dots,m$, all have series expansions in r, which are all absolutely convergent for $|r|$ sufficiently small, say $|r|<r_0$, $1>r_0>0$. From the system of linear equations (4.18) for the coefficients of the powers of ρ in the polynomials $f(\dots)$ and $g(\dots)$ it is readily seen that these polynomials have absolutely convergent series expansions with respect to r for $|r|<r_0$. Consequently, the right-hand side of (4.20) possesses a power series expansion with respect to r, which is absolutely convergent for $|r|<r_0$. Hence, we may write for $|\rho|<\infty$, $\operatorname{Re}\rho_3\geqq 0$, $\operatorname{Re}\rho_4\geqq 0$, $0\leqq v\leqq K$,

$$\varphi(r,\rho,\rho_3,\rho_4,v)=\sum_{n=1}^{\infty} r^n a_n(\rho,\rho_3,\rho_4,v),\qquad |r|<r_0, \tag{4.24}$$

with $\varphi(\dots)$ given by (4.20). It follows from

$$a_n(\rho,\rho_3,\rho_4,v)=\frac{1}{2\pi i}\int_{D_{r_1}}\frac{\varphi(r,\rho,\rho_3,\rho_4,v)}{r^{n+1}}\,dr, \tag{4.25}$$

with D_{r_1} a circle in the r-plane with radius $r_1<r_0$ and centre at $r=0$. Moreover, since $\varphi(r,\rho,\rho_3,\rho_4,v)$ as given by (4.20) is uniformly bounded in r for $|r|<r_0$, say by $c(\rho,\rho_3,\rho_4,v)$ with $|\rho|<\infty$, $\operatorname{Re}\rho_3\geqq 0$, $\operatorname{Re}\rho_4\geqq 0$, $0\leqq v\leqq K$, we have

$$|a_n(\rho,\rho_3,\rho_4,v)|<r_0^{-n}c(\rho,\rho_3,\rho_4,v). \tag{4.26}$$

Since $\varphi(r,\rho,\rho_3,\rho_4,v)$ as given by (4.20) satisfies (4.13) or equivalently (4.7), it follows by substituting (4.24) into (4.7), by multiplying the resulting relation by $r^{-(n+1)}$ and then integrating term by term along the contour D_{r_1}

that

$$a_{n+1}(\rho, \rho_3, \rho_4, v) = \mathrm{e}^{-\rho K} \alpha^n(\rho_3)\, \beta^n(\rho_4) - \frac{\mathrm{e}^{-\rho K}}{2\pi\mathrm{i}} \int\limits_{C_\xi} \left(\frac{1}{\rho-\xi} + \frac{1}{\xi}\right) \frac{\mathrm{e}^{\xi K}}{2\pi\mathrm{i}}$$

$$\cdot \left\{ \int\limits_{C_\eta} \left(\frac{1}{\xi-\eta} + \frac{1}{\eta}\right) \alpha(\rho_3-\eta)\, \beta(\rho_4+\eta)\, a_n(\eta, \rho_3, \rho_4, v)\, \mathrm{d}\eta \right\} \mathrm{d}\xi, \tag{4.27}$$

for $n=1, 2, \ldots$; $\mathrm{Re}\,\xi > \mathrm{Re}\,\rho$, $\mathrm{Re}\,\xi > \mathrm{Re}\,\eta > 0$, $\mathrm{Re}(\rho_3-\eta)>0$, $\mathrm{Re}(\rho_4+\eta)>0$, and

$$a_1(\rho, \rho_3, \rho_4, v) = \mathrm{e}^{-\rho v}.$$

This set of recurrence relations determines $a_{n+1}(\ldots)$ uniquely; it is the same set of relations as (4.4). Hence, the expression (4.20) represents the function defined in (4.5). The proof is complete.

III.4.3. The stationary distribution

In this section we shall prove the existence of a unique stationary distribution for the process $\{\boldsymbol{v}_n, n=1, 2, \ldots\}$ defined by (4.1). For this proof the assumption of rationality of the transforms $\alpha(\rho)$ and $\beta(\rho)$ will not be used.

We shall exclude the trivial case that $\boldsymbol{\sigma}_{n+1} = \boldsymbol{\tau}_n = \text{constant}$ for all $n=1, 2, \ldots$; hence it is assumed that

$$\Pr\{\boldsymbol{\sigma}_{n+1} \neq \boldsymbol{\tau}_n\} > 0, \qquad n=1, 2, \ldots . \tag{4.28}$$

It is not difficult to verify by an enumeration of the possible cases that the variables $\boldsymbol{v}_n$, $n=1, 2, \ldots$, defined by (4.1) also satisfy the set of recurrence relations

$$\begin{aligned} K - \boldsymbol{v}_{n+1} &= \min(K, [K - \boldsymbol{v}_n + \boldsymbol{\sigma}_{n+1} - \boldsymbol{\tau}_n]^+), \qquad n=1, 2, \ldots, \\ K - \boldsymbol{v}_1 &= K - v. \end{aligned} \tag{4.29}$$

Consequently, $\boldsymbol{v}_n$ and $K - \boldsymbol{v}_n$ are dual variables. Define the sequence $\boldsymbol{w}_n$, $n=1, 2, \ldots$, by

$$\begin{aligned} \boldsymbol{w}_{n+1} &= [\boldsymbol{w}_n + \boldsymbol{\tau}_n - \boldsymbol{\sigma}_{n+1}]^+, \qquad n=1, 2, \ldots, \\ \boldsymbol{w}_1 &= w. \end{aligned} \tag{4.30}$$

From the relations (4.1) and (4.30) it is now easily proved that if $w=v$ then for every $n=1, 2, \ldots$

$$\boldsymbol{w}_n \geqq \boldsymbol{v}_n \qquad \text{with probability 1.} \tag{4.31}$$

By the same argument as that used in 5.3 it is now proved that

$$\begin{aligned} &\Pr\{\boldsymbol{v}_n=0, \text{ i.o.}\}=1 \qquad \text{if} \quad a \leqq 1, \\ &\Pr\{\boldsymbol{v}_n=K, \text{ i.o.}\}=1 \qquad \text{if} \quad a \leqq 1, \end{aligned} \tag{4.32}$$

with

$$a = \beta/\alpha, \qquad \beta \stackrel{\text{def}}{=} \mathrm{E}\{\boldsymbol{\tau}_n\}, \qquad \alpha \stackrel{\text{def}}{=} \mathrm{E}\{\boldsymbol{\sigma}_n\}, \tag{4.33}$$

if it is assumed that α and β are finite. The same applies for the existence of a unique stationary distribution $V(x)$ for the process $\{\boldsymbol{v}_n, n=1, 2, \ldots\}$, as well as for the convergence of the distribution of $\boldsymbol{v}_n$ for $n \to \infty$ to this stationary distribution, independently of the initial conditions of the process.

From Helly-Bray's theorem and a well-known Abelian theorem we have for $|\rho| < \infty$,

$$\varphi(\rho) \stackrel{\text{def}}{=} \int_{0-}^{\infty} \mathrm{e}^{-\rho x} \, \mathrm{d}V(x) = \lim_{r \to 1} (1-r)\, \varphi(r, \rho, 0, 0, v). \tag{4.34}$$

Clearly if $\boldsymbol{v}_1$ has the distribution function $V(x)$, then all $\boldsymbol{v}_n$, $n=1, 2, \ldots$, have this distribution and satisfy

$$\boldsymbol{v}_{n+1} = \min(K, [\boldsymbol{v}_n + \boldsymbol{\tau}_n - \boldsymbol{\sigma}_{n+1}]^+), \qquad n=1, 2, \ldots . \tag{4.35}$$

Since $0 \leqq \boldsymbol{v}_n \leqq K$ with probability one, it follows that $\varphi(\rho)$ is an entire function of exponential type and order one, hence

$$\begin{aligned} &|\varphi(\rho)| < \infty \qquad \text{for} \quad |\rho| < \infty, \\ &\lim_{|\rho| \to \infty} |\varphi(\rho)| < \infty, \qquad \tfrac{1}{2}\pi > \arg \rho > -\tfrac{1}{2}\pi, \\ &\lim_{|\rho| \to \infty} |\mathrm{e}^{\rho K} \varphi(\rho)| < \infty, \qquad 1\tfrac{1}{2}\pi > \arg \rho > \tfrac{1}{2}\pi. \end{aligned} \tag{4.36}$$

Define μ_1 as the real part of the singularity of $\alpha(-\rho)$ nearest to the imaginary axis in the ρ-plane, and μ_2 that of $\beta(\rho)$; hence $\mu_1 > 0$, $\mu_2 < 0$.
As in section 4.2 it follows from (4.35) that for

$$\mu_2 < \operatorname{Re} \zeta < 0 < \operatorname{Re} \rho < \operatorname{Re} \omega < \mu_1,$$

$$\{1-\alpha(-\rho)\,\beta(\rho)\}\,\varphi(\rho) = -\frac{1}{2\pi i}\int_{C_\omega}\left(\frac{1}{\rho-\omega}+\frac{1}{\omega}\right)\{1-\alpha(-\omega)\,\beta(\omega)\}\,\varphi(\omega)\,d\omega$$

$$+\frac{e^{-\rho K}}{2\pi i}\int_{C_\zeta}\left(\frac{1}{\rho-\zeta}+\frac{1}{\zeta}\right)e^{\zeta K}\,\{1-\alpha(-\zeta)\,\beta(\zeta)\}\,\varphi(\zeta)\,d\zeta$$

$$+\frac{e^{-\rho K}}{2\pi i}\int_{C_\zeta}\left(\frac{1}{\rho-\zeta}+\frac{1}{\zeta}\right)\frac{e^{\zeta K}}{2\pi i}$$

$$\cdot\left\{\int_{C_\omega}\left(\frac{1}{\zeta-\omega}+\frac{1}{\omega}\right)\{1-\alpha(-\omega)\,\beta(\omega)\}\,\varphi(\omega)\,d\omega\right\}d\zeta. \tag{4.37}$$

Applying Rouché's theorem to the function

$$1-\alpha(-\rho)\,\beta(\rho), \tag{4.38}$$

it follows that this function has $m+n$ zeros of which $\rho=0$ is always a simple zero if $a\neq 1$, a double zero if $a=1$. Denote by $\varepsilon_i(1,0,0)$, $i=1,\dots,n_0$, the zeros in the left halfplane, and by $\delta_j(1,0,0)$, $j=1,\dots,m_0$, those in the right halfplane, then

$$n_0=\begin{cases} n, \\ n-1, \\ n-1, \end{cases}\qquad m_0=\begin{cases} m-1 & \text{if } a<1, \\ m-1 & \text{if } a=1, \\ m & \text{if } a>1. \end{cases} \tag{4.39}$$

Define

$$\nu_{01}\overset{\text{def}}{=}\min(\mu_1,\ \operatorname{Re}\delta_j(1,0,0),\ j=1,\dots,m_0),$$

$$\nu_{02}\overset{\text{def}}{=}\max(\mu_2,\ \operatorname{Re}\varepsilon_i(1,0,0),\ i=1,\dots,n_0),$$

so that

$$\nu_{01}>0,\qquad \nu_{02}<0.$$

Further

$$H_-(\rho)\overset{\text{def}}{=}\frac{\{1-\alpha(-\rho)\,\beta(\rho)\}\,\beta_2(\rho)}{\prod_{i=1}^{n_0}(\rho-\varepsilon_i(1,0,0))}\,\frac{1}{\rho^{n-n_0}}, \tag{4.40}$$

$$H_+(\rho)\overset{\text{def}}{=}\frac{\{1-\alpha(-\rho)\,\beta(\rho)\}\,\alpha_2(-\rho)}{\prod_{j=1}^{m_0}(\rho-\delta_j(1,0,0))}\,\frac{1}{\rho^{m-m_0}},$$

for all those ρ for which the right-hand sides exist; by definition an empty product is equal to 1.

Next, we define two polynomials $f(\rho)$ and $g(\rho)$ such that

(*i*) $f(\rho)$ is of degree $m-1$, and $g(\rho)$ is of degree $n-1$;

(*ii*)

$$\frac{\rho}{1-\alpha(-\rho)\,\beta(\rho)}\left\{\frac{f(\rho)}{\alpha_2(-\rho)}+\frac{g(\rho)}{\beta_2(\rho)}\right\}=1 \quad \text{for} \quad \rho\to 0, \tag{4.41}$$

$$\frac{f(\rho)}{\alpha_2(-\rho)}+\frac{g(\rho)}{\beta_2(\rho)}\,\mathrm{e}^{-\rho K}=0, \tag{4.42}$$

for $\rho=\varepsilon_i(1,0,0)$, $i=1,\dots,n_0$; and $\rho=\delta_j(1,0,0)$, $j=1,\dots,m_0$, if $a\neq 1$, while if $a=1$ the condition (4.42) should also hold for $\rho=0$.

The conditions above yield $m+n$ linear equations for the $m+n$ coefficients in the polynomials $f(\rho)$ and $g(\rho)$, and determine these polynomials uniquely. As in section 4.2 it can be proved that if $a\neq 1$ then the conditions (4.42) may be replaced by

$$f(\rho)=-\frac{\alpha_2(-\rho)\,H_-(\rho)}{2\pi\mathrm{i}}\int_{C_\xi}\frac{g(\xi)}{\beta_2(\xi)}\,\frac{\mathrm{e}^{-\xi K}}{H_-(\xi)}\,\frac{\mathrm{d}\xi}{\xi-\rho}, \tag{4.43}$$

$$g(\rho)=\frac{\beta_2(\rho)\,H_+(\rho)}{2\pi\mathrm{i}}\int_{C_\eta}\frac{f(\eta)}{\alpha_2(-\eta)}\,\frac{\mathrm{e}^{\eta K}}{H_+(\eta)}\,\frac{\mathrm{d}\eta}{\eta-\rho},$$

with $\nu_{01}>\operatorname{Re}\xi>0$, $\operatorname{Re}\xi>\operatorname{Re}\rho$, $\nu_{02}<\operatorname{Re}\eta<0$, $\operatorname{Re}\eta<\operatorname{Re}\rho$.

As in section 4.2 it is now shown that for $|\rho|<\infty$,

$$\varphi(\rho)=\frac{\rho}{1-\alpha(-\rho)\,\beta(\rho)}\left\{\frac{f(\rho)}{\alpha_2(-\rho)}+\frac{g(\rho)}{\beta_2(\rho)}\mathrm{e}^{-\rho K}\right\}, \tag{4.44}$$

with the polynomials $f(\rho)$ and $g(\rho)$ determined by the above conditions, is a solution of (4.37) and has the properties (4.36). Moreover, it is easily verified from (4.18) and (4.20) that for the expression for $\varphi(\rho)$ given by (4.44) the following holds:

$$\varphi(\rho)=\lim_{r\to 1}(1-r)\,\varphi(r,\rho,0,0,v), \qquad |\rho|<\infty;$$

note that for $a\neq 1$, $\rho_3=\rho_4=0$, one of the zeros of (4.14) behaves as $O(|1-r|)$ for $r\to 1$ (cf. (II.5.203)). Consequently, $\varphi(\rho)$ as given by (4.44) represents the Laplace-Stieltjes transform of the stationary distribution of the process $\{\boldsymbol{v}_n,\ n=1,2,\dots\}$.

Applying the inversion formula for the Laplace-Stieltjes transform we have for $\mathrm{Re}\,\rho>0$,

$$V(z)=\begin{cases}0, & z<0,\\ \dfrac{1}{2\pi i}\displaystyle\int_{C_\rho}\frac{e^{\rho z}}{\rho}\,\frac{\rho}{1-\alpha(-\rho)\,\beta(\rho)}\left\{\frac{f(\rho)}{\alpha_2(-\rho)}+\frac{g(\rho)}{\beta_2(\rho)}\,e^{-\rho K}\right\}d\rho, & \\ & z>0,\end{cases} \quad (4.45)$$

if z is a continuity point of $V(z)$, otherwise the left-hand side should be replaced by $\frac{1}{2}\{V(z+)+V(z-)\}$. The integrand in (4.45) has only one singularity at $\rho=0$, since $\varphi(\rho)$ is an entire function. It follows that for

$$\mathrm{Re}\,\eta>\max_{1\leqq j\leqq m_0}\mathrm{Re}\,\delta_j(1,0,0),\qquad \mathrm{Re}\,\zeta<\min_{1\leqq i\leqq n_0}\mathrm{Re}\,\varepsilon_i(1,0,0),$$

$$V(z)=\begin{cases}\dfrac{1}{2\pi i}\displaystyle\int_{C_\eta}\frac{e^{\eta z}}{1-\alpha(-\eta)\,\beta(\eta)}\,\frac{f(\eta)}{\alpha_2(-\eta)}\,d\eta, & 0<z<K,\\ 1+\dfrac{1}{2\pi i}\displaystyle\int_{C_\zeta}\frac{e^{-\zeta(K-z)}}{1-\alpha(-\zeta)\,\beta(\zeta)}\,\frac{g(\zeta)}{\beta_2(\zeta)}\,d\zeta, & 0<z<K.\end{cases} \quad (4.46)$$

It is not difficult to show that $V(z)$ has discontinuity points at $z=0$ and $z=K$. Clearly

$$V(0+)=\lim_{\rho\to\infty}\varphi(\rho)=\lim_{\rho\to\infty}\frac{\rho f(\rho)}{\alpha_2(-\rho)},\qquad \rho\text{ real}, \quad (4.47)$$

$$1-V(K-)=\lim_{\rho\to-\infty}e^{\rho K}\varphi(\rho)=\lim_{\rho\to-\infty}\frac{\rho g(\rho)}{\beta_2(\rho)},\qquad \rho\text{ real}.$$

III.4.4. Entrance times and return times

We shall consider entrance and return time distributions for the process $\{\boldsymbol{v}_n,\ n=1,2,\ldots\}$. Obviously the process $\{\boldsymbol{v}_n,\ n=1,2,\ldots\}$ is a discrete time parameter Markov process with state space the closed interval $[0,K]$. We shall say that at time n the process is in state E_0 if $\boldsymbol{v}_n=0$ and in E_K if $\boldsymbol{v}_n=K$.

Clearly the successive occurrences of the state E_0, and similarly of state E_K, are imbedded renewal processes of the process $\{\boldsymbol{v}_n,\ n=1,2,\ldots\}$.

Define for $0 \leqq v \leqq K$,

$$\boldsymbol{\alpha}_{v0} \stackrel{\text{def}}{=} \begin{cases} 1 & \text{if} \quad \boldsymbol{v}_2 = 0, \quad \boldsymbol{v}_1 = v, \\ \min\limits_{n=2,3,\dots} \{n: \boldsymbol{v}_{n+1} = 0, \boldsymbol{v}_i > 0, i = 2, \dots, n\} & \text{if} \quad \boldsymbol{v}_1 = v; \end{cases} \tag{4.48}$$

$$\boldsymbol{\alpha}_{vK} \stackrel{\text{def}}{=} \begin{cases} 1 & \text{if} \quad \boldsymbol{v}_2 = K, \quad \boldsymbol{v}_1 = v, \\ \min\limits_{n=2,3,\dots} \{n: \boldsymbol{v}_{n+1} = K, \boldsymbol{v}_i < K, i = 2, \dots, n\} & \text{if} \quad \boldsymbol{v}_1 = v; \end{cases}$$

$$\begin{aligned} \boldsymbol{\gamma}_{v0} &\stackrel{\text{def}}{=} \boldsymbol{a}_{\boldsymbol{\alpha}_{v0}}, \qquad \boldsymbol{\gamma}_{vK} \stackrel{\text{def}}{=} \boldsymbol{a}_{\boldsymbol{\alpha}_{vK}}, \\ \boldsymbol{p}_{v0} &\stackrel{\text{def}}{=} \boldsymbol{b}_{\boldsymbol{\alpha}_{v0}}, \qquad \boldsymbol{p}_{vK} \stackrel{\text{def}}{=} \boldsymbol{b}_{\boldsymbol{\alpha}_{vK}}. \end{aligned} \tag{4.49}$$

Clearly $\boldsymbol{\alpha}_{00}$ is the return time in discrete time of E_0, and $\boldsymbol{\gamma}_{00}$ the return time in continuous time of E_0; $\boldsymbol{p}_{00}$ is the total amount of work for the server brought into the system by the customers arriving during the return time $\boldsymbol{\gamma}_{00}$. For the state E_K the variables $\boldsymbol{\alpha}_{KK}$, $\boldsymbol{\gamma}_{KK}$ and $\boldsymbol{p}_{KK}$ have a similar interpretation.

Further we define for $z_1 \geqq 0$, $z_2 \geqq 0$, $0 \leqq v \leqq K$, $n = 1, 2, \dots,$

$$\begin{aligned} p_{v0}^{(n)}(z_1, z_2) &\stackrel{\text{def}}{=} \Pr\{\boldsymbol{v}_{n+1} = 0, \boldsymbol{a}_n < z_1, \boldsymbol{b}_n < z_2 \mid \boldsymbol{v}_1 = v\}, \\ p_{vK}^{(n)}(z_1, z_2) &\stackrel{\text{def}}{=} \Pr\{\boldsymbol{v}_{n+1} = K, \boldsymbol{a}_n < z_1, \boldsymbol{b}_n < z_2 \mid \boldsymbol{v}_1 = v\}, \end{aligned} \tag{4.50}$$

and

$$\begin{aligned} rc_f(r, \rho_3, \rho_4, v) &\stackrel{\text{def}}{=} \lim_{\rho \to \infty} \frac{\rho f(r, \rho, \rho_3, \rho_4, v)}{\alpha_2(\rho_3 - \rho)}, \qquad \rho \text{ real}, \\ rc_g(r, \rho_3, \rho_4, v) &\stackrel{\text{def}}{=} \lim_{\rho \to -\infty} \frac{\rho g(r, \rho, \rho_3, \rho_4, v)}{\beta_2(\rho_4 + \rho)}, \qquad \rho \text{ real}, \end{aligned} \tag{4.51}$$

for $|r| < 1$, $\operatorname{Re} \rho_3 \geqq 0$, $\operatorname{Re} \rho_4 \geqq 0$, $0 \leqq v \leqq K$.

It follows from (4.20) that for $|r| < 1$, $\operatorname{Re} \rho_3 \geqq 0$, $\operatorname{Re} \rho_4 \geqq 0$,

$$\begin{aligned} &\sum_{n=1}^{\infty} r^n \operatorname{E}\{(\boldsymbol{v}_n = 0)\exp(-\rho_3 \boldsymbol{a}_{n-1} - \rho_4 \boldsymbol{b}_{n-1}) \mid \boldsymbol{v}_1 = v\} \\ &\qquad = \begin{cases} rc_f(r, \rho_3, \rho_4, 0) + r & \text{if} \quad v = 0, \\ rc_f(r, \rho_3, \rho_4, v), & 0 < v \leqq K, \end{cases} \\ &\sum_{n=1}^{\infty} r^n \operatorname{E}\{(\boldsymbol{v}_n = K)\exp(-\rho_3 \boldsymbol{a}_{n-1} - \rho_4 \boldsymbol{b}_{n-1}) \mid \boldsymbol{v}_1 = v\} \\ &\qquad = \begin{cases} rc_g(r, \rho_3, \rho_4, v), & 0 \leqq v < K, \\ rc_g(r, \rho_3, \rho_4, K) + r, & v = K; \end{cases} \end{aligned} \tag{4.52}$$

here $(\boldsymbol{v}_n=0)$ and $(\boldsymbol{v}_n=K)$ are the indicator functions of the events $\{\boldsymbol{v}_n=0\}$ and $\{\boldsymbol{v}_n=K\}$, respectively.

Introduce the Laplace-Stieltjes transforms of the functions defined in (4.50). These transforms are indicated by the sign "$\sim$", so that for instance

$$\tilde{p}_{v0}^{(n)}(\rho_3, \rho_4) \overset{\text{def}}{=} \int_{0-}^{\infty} \int_{0-}^{\infty} \exp(-\rho_3 z_1 - \rho_4 z_2)\, \mathrm{d}_{z_1}\, \mathrm{d}_{z_2}\, p_{v0}^{(n)}(z_1, z_2), \tag{4.53}$$
$$\operatorname{Re} \rho_3 \geqq 0, \quad \operatorname{Re} \rho_4 \geqq 0.$$

It follows from (4.50), (4.52) and (4.53) for $|r|<1$, $\operatorname{Re} \rho_3 \geqq 0$, $\operatorname{Re} \rho_4 \geqq 0$, $0 \leqq v \leqq K$ that

$$\sum_{n=1}^{\infty} r^n \tilde{p}_{v0}^{(n)}(\rho_3, \rho_4) = c_f(r, \rho_3, \rho_4, v), \tag{4.54}$$
$$\sum_{n=1}^{\infty} r^n \tilde{p}_{vK}^{(n)}(\rho_3, \rho_4) = c_g(r, \rho_3, \rho_4, v).$$

Applying the fundamental relations from renewal theory to the two imbedded renewal processes of the process $\{\boldsymbol{v}_n, n=1, 2, \ldots\}$ we immediately obtain the relations

$$\mathrm{E}\{r^{\boldsymbol{\alpha}_{v0}} \exp(-\rho_3 \boldsymbol{\gamma}_{v0} - \rho_4 \boldsymbol{p}_{v0})\} = \frac{c_f(r, \rho_3, \rho_4, v)}{1 + c_f(r, \rho_3, \rho_4, 0)}, \qquad 0 \leqq v \leqq K, \tag{4.55}$$
$$\mathrm{E}\{r^{\boldsymbol{\alpha}_{vK}} \exp(-\rho_3 \boldsymbol{\gamma}_{vK} - \rho_4 \boldsymbol{p}_{vK})\} = \frac{c_g(r, \rho_3, \rho_4, v)}{1 + c_g(r, \rho_3, \rho_4, K)}, \qquad 0 \leqq v \leqq K,$$

for $|r|<1$, $\operatorname{Re} \rho_3 \geqq 0$, $\operatorname{Re} \rho_4 \geqq 0$.

The relations (4.55) contain all the information about entrance times and return times.

Define for $z_1 \geqq 0$, $z_2 \geqq 0$, $0 \leqq v \leqq K$, $n=1, 2, \ldots,$

$$f_{v0}^{(n)}(z_1, z_2) \overset{\text{def}}{=} \Pr\{\boldsymbol{\alpha}_{v0}=n, \boldsymbol{\gamma}_{v0}<z_1, \boldsymbol{p}_{v0}<z_2\}, \tag{4.56}$$
$$f_{vK}^{(n)}(z_1, z_2) \overset{\text{def}}{=} \Pr\{\boldsymbol{\alpha}_{vK}=n, \boldsymbol{\gamma}_{vK}<z_1, \boldsymbol{p}_{vK}<z_2\},$$

$$p_{K;v0}^{(n)}(z_1, z_2) \overset{\text{def}}{=} \begin{cases} p_{v0}^{(1)}(z_1, z_2), & n=1, \\ \Pr\{\boldsymbol{v}_{n+1}=0, \boldsymbol{v}_j<K, j=2, \ldots, n;\ \boldsymbol{a}_n<z_1, \boldsymbol{b}_n<z_2 \mid \boldsymbol{v}_1=v\}, & \\ & n=2, 3, \ldots; \end{cases}$$

$$p_{0;vK}^{(n)}(z_1, z_2) \overset{\text{def}}{=} \begin{cases} p_{vK}^{(1)}(z_1, z_2), & n=1, \\ \Pr\{\boldsymbol{v}_{n+1}=K, \boldsymbol{v}_j>0, j=2, \ldots, n;\ \boldsymbol{a}_n<z_1, \boldsymbol{b}_n<z_2 \mid \boldsymbol{v}_1=v\}, & \\ & n=2, 3, \ldots; \end{cases}$$

$$f_{K;v0}^{(n)}(z_1, z_2) \overset{\text{def}}{=} \begin{cases} f_{v0}^{(1)}(z_1, z_2), & n=1, \\ \Pr\{\boldsymbol{\alpha}_{v0}=n, \boldsymbol{\gamma}_{v0}<z_1, \boldsymbol{p}_{v0}<z_2, \boldsymbol{v}_j<K, j=2, \dots, n\}, & \\ & n=2, 3 \dots; \end{cases}$$

$$f_{0;vK}^{(n)}(z_1, z_2) \overset{\text{def}}{=} \begin{cases} f_{vK}^{(1)}(z_1, z_2), & n=1, \\ \Pr\{\boldsymbol{\alpha}_{vK}=n, \boldsymbol{\gamma}_{vK}<z_1, \boldsymbol{p}_{vK}<z_2, \boldsymbol{v}_j>0, j=2, \dots, n\}, & \\ & n=2, 3, \dots . \end{cases}$$

Since E_0 and E_K are regenerative events of the process $\{\boldsymbol{v}_n, n=1, 2, \dots\}$ it follows for $\operatorname{Re} \rho_3 \geqq 0$, $\operatorname{Re} \rho_4 \geqq 0$, $n=2, 3, \dots$, and using the notation introduced in (4.53) that

$$\tilde{p}_{v0}^{(n)}(\rho_3, \rho_4) = \tilde{p}_{K;v0}^{(n)}(\rho_3, \rho_4) + \sum_{m=1}^{n-1} \tilde{f}_{vK}^{(m)}(\rho_3, \rho_4)\, \tilde{p}_{K0}^{(n-m)}(\rho_3, \rho_4), \qquad n=2, 3, \dots, \tag{4.57}$$

$$\tilde{p}_{vK}^{(n)}(\rho_3, \rho_4) = \tilde{p}_{0;vK}^{(n)}(\rho_3, \rho_4) + \sum_{m=1}^{n-1} \tilde{f}_{v0}^{(m)}(\rho_3, \rho_4)\, \tilde{p}_{0K}^{(n-m)}(\rho_3, \rho_4), \qquad n=2, 3, \dots,$$

$$\tilde{p}_{K;v0}^{(n)}(\rho_3, \rho_4) = \tilde{f}_{K;v0}^{(n)}(\rho_3, \rho_4) + \sum_{m=1}^{n-1} \tilde{f}_{K;v0}^{(m)}(\rho_3, \rho_4)\, \tilde{p}_{K;00}^{(n-m)}(\rho_3, \rho_4), \qquad n=2, 3, \dots,$$

$$\tilde{p}_{0;vK}^{(n)}(\rho_3, \rho_4) = \tilde{f}_{0;vK}^{(n)}(\rho_3, \rho_4) + \sum_{m=1}^{n-1} \tilde{f}_{0;vK}^{(m)}(\rho_3, \rho_4)\, \tilde{p}_{0;KK}^{(n-m)}(\rho_3, \rho_4), \qquad n=2, 3, \dots .$$

All the taboo probabilities defined above may be expressed in the functions $c_f(\dots)$ and $c_g(\dots)$ by using generating functions. These taboo probabilities are not only of interest for the process $\{\boldsymbol{v}_n, n=1, 2, \dots\}$, but also for the general queueing system $G/G/1$.

For instance $\sum_{n=1}^{\infty} f_{K;00}^{(n)}(z_1, z_2)$ is the probability that for the general single server queue $G/G/1$ the busy cycle is less than z_1, the busy period is less than z_2, and the supremum of the actual waiting time during this busy cycle is less than K; $f_{0K}^{(n)}(\infty, \infty)$ is the probability that the nth arriving customer is the first one whose actual waiting time exceeds K if the waiting time of the first arriving customer is zero.

III.4.5. Application to the system $M/G/1$

In this section we shall apply the results obtained in the previous sections to the queueing model $M/K_n/1$, i.e. we have

$$\alpha(\rho) = \frac{1}{1+\alpha\rho}, \qquad \beta(\rho) = \frac{\beta_1(\rho)}{\beta_2(\rho)}, \tag{4.58}$$

with $\beta_1(\rho)$ and $\beta_2(\rho)$ polynomials in ρ; $\beta_2(\rho)$ is of degree n, $\beta_1(\rho)$ of degree $n-1$ at most. At the end of this section it will be shown that the assumption of rationality of $\beta(\rho)$ is superfluous, so that all results hold for general $B(.)$.

Since for the present case $m=1$ it follows from the definition of $f(r, \rho, \rho_3, \rho_4, v)$ that this function is independent of ρ. Define for $|r|>1$, $\text{Re}\, \rho_3 \geqq 0$, $\text{Re}\, \rho_4 \geqq 0$, $0 \leqq v \leqq K$,

$$f_0(r, \rho_3, \rho_4, v) \stackrel{\text{def}}{=} f(r, \rho, \rho_3, \rho_4, v). \tag{4.59}$$

From (4.18) we obtain for $|r|<1$, $\text{Re}\, \rho_3 \geqq 0$, $\text{Re}\, \rho_4 \geqq 0$, $0 \leqq v \leqq K$,

$$f_0(r, \rho_3, \rho_4, v) = -\frac{g(r, \delta_1, \rho_3, \rho_4, v)}{\beta_2(\rho_4+\delta_1)} \cdot \{1-(\delta_1-\rho_3)\,\alpha\}\, \mathrm{e}^{-\delta_1 K} - r\,\frac{1-(\delta_1-\rho_3)\,\alpha}{\delta_1}\,\mathrm{e}^{-\delta_1 v}, \tag{4.60}$$

with

$$\delta_1 = \delta_1(r, \rho_3, \rho_4).$$

Further from (4.17) and (4.19),

$$\frac{g(r, \rho, \rho_3, \rho_4, v)}{\beta_2(\rho+\rho_4)} = \frac{1-(\rho-\rho_3)\,\alpha - r\beta(\rho+\rho_4)}{2\pi\mathrm{i}(\rho-\delta_1)} \cdot \int_{C_\eta} \left\{ f_0(r, \rho_3, \rho_4, v)\,\frac{\mathrm{e}^{\eta K}}{1-(\eta-\rho_3)\,\alpha} + r\,\frac{\mathrm{e}^{\eta(K-v)}}{\eta} \right\} \cdot \frac{\eta-\delta_1}{1-(\eta-\rho_3)\,\alpha - r\beta(\eta+\rho_4)}\,\frac{\mathrm{d}\eta}{\eta-\rho}, \tag{4.61}$$

for $|r|<1$, $\text{Re}\, \rho_3 \geqq 0$, $\text{Re}\, \rho_4 \geqq 0$, $\nu_2 < \text{Re}\, \eta < 0$, $\text{Re}\, \eta < \text{Re}\, \rho$.
In (4.61) we take $\rho=\delta_1$ and eliminate $g(r, \delta_1, \rho_3, \rho_4, v)$ from the resulting relation and from (4.60). It follows that

$$
-\lim_{\rho\to\delta_1} \frac{\rho-\delta_1}{1-(\rho-\rho_3)\,\alpha-r\beta(\rho_4+\rho)}
$$

$$
\cdot\left\{\frac{f_0(r,\rho_3,\rho_4,v)}{1-(\delta_1-\rho_3)\,\alpha}\,\mathrm{e}^{\delta_1 K}+\frac{r}{\delta_1}\,\mathrm{e}^{\delta_1(K-v)}\right\}
$$

$$
=\frac{1}{2\pi\mathrm{i}}\int_{C_\eta}\left\{f_0(r,\rho_3,\rho_4,v)\,\frac{\mathrm{e}^{\eta K}}{1-(\eta-\rho_3)\,\alpha}+r\,\frac{\mathrm{e}^{\eta(K-v)}}{\eta}\right\}
$$

$$
\cdot\,\frac{\mathrm{d}\eta}{1-(\eta-\rho_3)\,\alpha-r\beta(\eta+\rho_4)}, \tag{4.62}
$$

for $\nu_2<\operatorname{Re}\eta<0$, $|r|<1$, $\operatorname{Re}\rho_3\geqq 0$, $\operatorname{Re}\rho_4\geqq 0$.
The relation (4.62) determines $f_0(r,\rho_3,\rho_4,v)$. Obviously the left-hand side of (4.62) is equal to minus the residue at $\eta=\delta_1$ of the integrand. Hence by contour integration (4.62) can be simplified. Since $\alpha^{-1}+\operatorname{Re}\rho_3$ does not always exceed $\operatorname{Re}\delta_1$ we shall use for the contour integration of (4.62) the contour A_η defined as follows.

If $\operatorname{Re}(\delta_1-\rho_3)\,\alpha<1$ the contour A_η is the line C_η with $\operatorname{Re}\delta_1<\operatorname{Re}\eta<1/\alpha+\operatorname{Re}\rho_3$; note that $(\delta_1-\rho_3)\,\alpha<1$ if $0<r\leqq 1$, $\rho_3\geqq 0$, $\rho_4\geqq 0$.

If $\operatorname{Re}(\delta_1-\rho_3)\,\alpha>1$ denote by A and B the points $\gamma_0+\mathrm{i}\gamma_1$, and $\gamma_0+\mathrm{i}\gamma_2$ with $0<\gamma_0<1/\alpha+\operatorname{Re}\rho_3$, $\gamma_1<\gamma_2$. The contour A_η consists of the half-lines $\operatorname{Re}\eta=\gamma_0$, $\operatorname{Im}\eta\leqq\gamma_1$, and $\operatorname{Re}\eta=\gamma_0$, $\operatorname{Im}\eta\geqq\gamma_2$, and of a loop starting at A, ending at B and which encloses the point δ_1 but not the point $1/\alpha+\rho_3$ (see fig. 2).

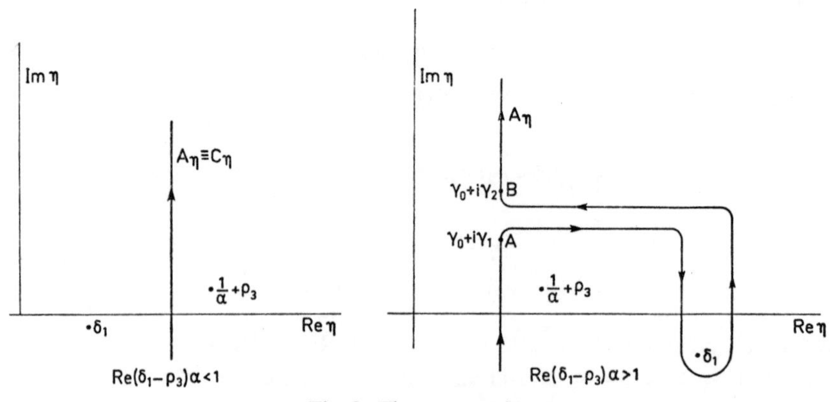

Fig. 2. The contour A_η.

Hence, when moving along A_η from $\gamma_0 - i\infty$ to $\gamma_0 + i\infty$ the point $1/\alpha + \rho_3$ lies to the right of A_η, while the point δ_1 lies to the left.

Replacing the contour C_η with $\text{Re}\,\eta < 0$ by A_η in (4.62) we have to account for the residues at $\eta = 0$ and at $\eta = \delta_1$. By contour integration we obtain

$$\frac{f_0(r, \rho_3, \rho_4, v)}{2\pi i} \int\limits_{A_\eta} \frac{e^{\eta K}}{1-(\eta-\rho_3)\,\alpha} \frac{d\eta}{1-(\eta-\rho_3)\,\alpha - r\beta(\eta+\rho_4)}$$
$$= \frac{r}{1+\alpha\rho_3 - r\beta(\rho_4)} - \frac{r}{2\pi i} \int\limits_{C_\zeta} \frac{e^{\zeta(K-v)}}{\zeta} \frac{d\zeta}{1-(\zeta-\rho_3)\,\alpha - r\beta(\rho_4+\zeta)}, \tag{4.63}$$

for $|r| < 1$, $\text{Re}\,\rho_3 \geqq 0$, $\text{Re}\,\rho_4 \geqq 0$, $\text{Re}\,\delta_1 < \text{Re}\,\zeta$, $0 \leqq v \leqq K$.

From (4.61) and (4.51) we determine the function $c_g(r, \rho_3, \rho_4, v)$. The integrand in (4.61) has a pole at $\eta = 0$ and at $\eta = \rho$. We obtain from (4.61) for ρ real

$$rc_g(r, \rho_3, \rho_4, v) = \lim_{\rho \to -\infty} \rho \frac{1-(\rho-\rho_3)\,\alpha - r\beta(\rho+\rho_4)}{\rho - \delta_1}$$
$$\cdot \Bigg[\Bigg\{ -f_0(r, \rho_3, \rho_4, v) \frac{e^{\rho K}}{1-(\rho-\rho_3)\,\alpha} - \frac{r}{\rho} e^{(K-v)\rho} \Bigg\}$$
$$\cdot \frac{\rho - \delta_1}{1-(\rho-\rho_3)\,\alpha - r\beta(\rho_4+\rho)} - \frac{r\delta_1}{\rho} \frac{1}{1+\alpha\rho_3 - r\beta(\rho_4)}$$
$$+ \frac{1}{2\pi i} \int\limits_{A_\eta} \Bigg\{ \frac{f_0(r, \rho_3, \rho_4, v)}{1-(\eta-\rho_3)\,\alpha} e^{\eta K} + \frac{r}{\eta} e^{(K-v)\eta} \Bigg\} \frac{\eta - \delta_1}{1-(\eta-\rho_3)\,\alpha - r\beta(\rho_4+\eta)} \frac{d\eta}{\eta - \rho} \Bigg].$$

From this relation and from (4.63) it follows easily that for $|r| < 1$, $\text{Re}\,\rho_3 \geqq 0$, $\text{Re}\,\rho_4 \geqq 0$, $\text{Re}\,\delta_1 < \text{Re}\,\zeta$,

$$rc_g(r, \rho_3, \rho_4, v)$$
$$\begin{cases} = \dfrac{1}{2\pi i} \displaystyle\int\limits_{A_\eta} f_0(r, \rho_3, \rho_4, v) \frac{e^{\eta K}}{1-(\eta-\rho_3)\,\alpha} \frac{\alpha\eta\, d\eta}{1-(\eta-\rho_3)\,\alpha - r\beta(\eta+\rho_4)} \\ \quad + \dfrac{1}{2\pi i} \displaystyle\int\limits_{C_\zeta} e^{\zeta(K-v)} \frac{\alpha r\, d\zeta}{1-(\zeta-\rho_3)\,\alpha - r\beta(\zeta+\rho_4)}, \qquad 0 \leqq v < K, \\ = -r + \dfrac{1}{2\pi i} \displaystyle\int\limits_{A_\eta} f_0(r, \rho_3, \rho_4, K) \frac{e^{\eta K}}{1-(\eta-\rho_3)\,\alpha} \frac{\alpha\eta\, d\eta}{1-(\eta-\rho_3)\,\alpha - r\beta(\eta+\rho_4)}, \\ \qquad \text{for} \quad v = K. \end{cases} \tag{4.64}$$

Further

$$rc_f(r, \rho_3, \rho_4, v) = -\frac{1}{\alpha} f_0(r, \rho_3, \rho_4, v). \tag{4.65}$$

By using the relations (4.61), ..., (4.65) we may now calculate all quantities of interest for the present queueing problem. We start with the stationary distribution of the waiting time $V(x)$.
From (4.63) it follows (cf. (4.34) and (4.44)) that

$$\begin{aligned}\frac{f(\rho)}{\alpha_2(-\rho)} &= \frac{1}{1-\alpha\rho} \lim_{r\uparrow 1} (1-r) f_0(r, 0, 0, v)\\ &= \frac{1}{1-\alpha\rho}\left[\frac{1}{2\pi i}\int_{C_\zeta} \frac{e^{\zeta K}}{1-\zeta\alpha}\,\frac{d\zeta}{1-\alpha\zeta-\beta(\zeta)}\right]^{-1},\end{aligned} \tag{4.66}$$

for $\rho \neq 1/\alpha$, $\delta_1(1, 0, 0) < \operatorname{Re}\zeta < 1/\alpha$. Note that $(\delta_1 - \rho_3)\,\alpha < 1$ if $0 < r \leqq 1$, $\rho_3 \geqq 0$, $\rho_4 \geqq 0$. Consequently from (4.46) and (4.47),

$$V(z) = \frac{\dfrac{1}{2\pi i}\displaystyle\int_{C_\eta} \frac{e^{\eta z}}{\beta(\eta)+\alpha\eta-1}\,d\eta}{\dfrac{1}{2\pi i}\displaystyle\int_{C_\zeta} \frac{e^{\zeta K}}{(1-\alpha\zeta)(\beta(\zeta)+\alpha\zeta-1)}\,d\zeta}, \qquad 0<z<K, \tag{4.67}$$

$$V(0+) = \left\{\frac{1}{2\pi i}\int_{C_\zeta} \frac{\alpha\, e^{\zeta K}}{(1-\alpha\zeta)(\beta(\zeta)+\alpha\zeta-1)}\,d\zeta\right\}^{-1},$$

$$1 - V(K-) = \frac{\dfrac{1}{2\pi i}\displaystyle\int_{C_\zeta} \frac{\alpha\zeta\, e^{\zeta K}}{(1-\alpha\zeta)(\beta(\zeta)+\alpha\zeta-1)}\,d\zeta}{\dfrac{1}{2\pi i}\displaystyle\int_{C_\zeta} \frac{e^{\zeta K}}{(1-\alpha\zeta)(\beta(\zeta)+\alpha\zeta-1)}\,d\zeta},$$

for $\operatorname{Re}\eta > \delta_1(1, 0, 0)$, $\delta_1(1, 0, 0) < \operatorname{Re}\zeta < 1/\alpha$.
From (4.55), (4.63), (4.64) and (4.65) we obtain the expressions for return times and entrance times distributions.

For $|r|\leqq 1$, $\operatorname{Re}\rho_3\geqq 0$, $\operatorname{Re}\rho_4\geqq 0$, $\operatorname{Re}\delta_1<\operatorname{Re}\zeta$, $0\leqq v\leqq K$,

$$\mathrm{E}\{r^{\boldsymbol{\alpha}_{v0}}\,\mathrm{e}^{-\rho_3\boldsymbol{\gamma}_{v0}-\rho_4\boldsymbol{p}_{v0}}\} = \frac{1+\dfrac{1+\alpha\rho_3-r\beta(\rho_4)}{2\pi\mathrm{i}}\displaystyle\int_{C_\zeta}\frac{\mathrm{e}^{\zeta(K-v)}}{\zeta}\,\frac{\mathrm{d}\zeta}{r\beta(\rho_4+\zeta)+(\zeta-\rho_3)\,\alpha-1}}{1+(1+\alpha\rho_3)\dfrac{1+\alpha\rho_3-r\beta(\rho_4)}{2\pi\mathrm{i}}\displaystyle\int_{A_\eta}\frac{\mathrm{e}^{\eta K}}{1-(\eta-\rho_3)\,\alpha}\,\frac{1}{r\beta(\rho_4+\eta)+(\eta-\rho_3)\,\alpha-1}\,\frac{\mathrm{d}\eta}{\eta}}, \tag{4.68}$$

$$\mathrm{E}\{r^{\boldsymbol{\alpha}_{KK}}\,\mathrm{e}^{-\rho_3\boldsymbol{\gamma}_{KK}-\rho_4\boldsymbol{p}_{KK}}\} = 1-(1+\alpha\rho_3-r\beta(\rho_4))\frac{\dfrac{1}{2\pi\mathrm{i}}\displaystyle\int_{A_\eta}\frac{\mathrm{e}^{\eta K}}{1-(\eta-\rho_3)\,\alpha}\,\frac{\mathrm{d}\eta}{r\beta(\rho_4+\eta)+(\eta-\rho_3)\,\alpha-1}}{\dfrac{1}{2\pi\mathrm{i}}\displaystyle\int_{A_\eta}\frac{\mathrm{e}^{\eta K}}{1-(\eta-\rho_3)\,\alpha}\,\frac{\alpha\eta\,\mathrm{d}\eta}{r\beta(\rho_4+\eta)+(\eta-\rho_3)\,\alpha-1}}, \tag{4.69}$$

the expression for $\mathrm{E}\{r^{\boldsymbol{\alpha}_{vK}}\,\mathrm{e}^{-\rho_3\boldsymbol{\gamma}_{vK}-\rho_4\boldsymbol{p}_{vK}}\}$ can also be obtained without difficulty but this is rather lengthy and it is therefore omitted.

Next we shall derive expressions for the taboo probabilities. From (4.54), ..., (4.57) we obtain for $\operatorname{Re}\rho_3\geqq 0$, $\operatorname{Re}\rho_4\geqq 0$, $0\leqq v\leqq K$,

$$\sum_{n=1}^{\infty} r^n\tilde{p}^{(n)}_{K;v0}(\rho_3,\rho_4) = c_f(r,\rho_3,\rho_4,v) - \frac{c_g(r,\rho_3,\rho_4,v)}{1+c_g(r,\rho_3,\rho_4,K)}\,c_f(r,\rho_3,\rho_4,K), \qquad |r|<1; \tag{4.70}$$

$$\sum_{n=1}^{\infty} r^n\tilde{f}^{(n)}_{K;v0}(\rho_3,\rho_4) = \frac{\displaystyle\sum_{n=1}^{\infty} r^n\tilde{p}^{(n)}_{K;v0}(\rho_3,\rho_4)}{1+\displaystyle\sum_{n=1}^{\infty} r^n\tilde{p}^{(n)}_{K;00}(\rho_3,\rho_4)}, \qquad |r|\leqq 1.$$

From these relations we obtain

$$\sum_{n=1}^{\infty} r^n\tilde{f}^{(n)}_{K;v0}(\rho_3,\rho_4) = \frac{\dfrac{1}{2\pi\mathrm{i}}\displaystyle\int_{C_\zeta}\frac{\mathrm{e}^{\zeta(K-v)}}{1-(\zeta-\rho_3)\,\alpha-r\beta(\rho_4+\zeta)}\,\mathrm{d}\zeta}{\dfrac{1+\alpha\rho_3}{2\pi\mathrm{i}}\displaystyle\int_{A_\eta}\frac{\mathrm{e}^{\eta K}}{1-(\eta-\rho_3)\,\alpha}\,\frac{\mathrm{d}\eta}{1-(\eta-\rho_3)\,\alpha-r\beta(\eta+\rho_4)}}, \qquad 0\leqq v<K, \tag{4.71}$$

$$\sum_{n=1}^{\infty} r^n \tilde{f}^{(n)}_{K;KO}(\rho_3, \rho_4)$$

$$= \left\{ \frac{1+\alpha\rho_3}{2\pi i} \int_{A_\eta} \frac{e^{\eta K}}{1-(\eta-\rho_3)\,\alpha} \frac{\alpha \,d\eta}{r\beta(\rho_4+\eta)+(\eta-\rho_3)\,\alpha-1} \right\}^{-1}, \tag{4.72}$$

for $|r| \leqq 1$, $\operatorname{Re} \rho_3 \geqq 0$, $\operatorname{Re} \rho_4 \geqq 0$, $\operatorname{Re} \delta_1 < \operatorname{Re} \zeta$.

In particular we have for $\delta_1(1, 0, 0) < \operatorname{Re} \zeta < 1/\alpha$,

$$\sum_{n=1}^{\infty} f^{(n)}_{K;00} = \frac{\dfrac{1}{2\pi i} \displaystyle\int_{C_\zeta} \frac{e^{\zeta K}}{\beta(\zeta)+\alpha\zeta-1}\, d\zeta}{\dfrac{1}{2\pi i} \displaystyle\int_{C_\zeta} \frac{e^{\zeta K}}{1-\zeta\alpha} \frac{d\zeta}{\beta(\zeta)+\alpha\zeta-1}}, \tag{4.73}$$

$$\sum_{n=1}^{\infty} f^{(n)}_{K;KO} = \left\{ \frac{1}{2\pi i} \int_{C_\zeta} \frac{e^{\zeta K}}{1-\alpha\zeta} \frac{\alpha\, d\zeta}{\beta(\zeta)+\alpha\zeta-1} \right\}^{-1}. \tag{4.74}$$

Note that $\sum_{n=1}^{\infty} f^{(n)}_{K;00}$ represents for the $M/G/1$ queueing system the probability that the supremum of the actual waiting time during a busy period is less than K. In chapter III.7 the distribution (4.73) is investigated.

From (4.54), ..., (4.57) we obtain for $\operatorname{Re} \rho_3 \geqq 0$, $\operatorname{Re} \rho_4 \geqq 0$,

$$\sum_{n=1}^{\infty} r^n \tilde{p}^{(n)}_{0;KK}(\rho_3, \rho_4) = c_g(r, \rho_3, \rho_4, K) - \frac{c_f(r, \rho_3, \rho_4, K)}{1+c_f(r, \rho_3, \rho_4, 0)}\, c_g(r, \rho_3, \rho_4, 0), \qquad |r|<1, \tag{4.75}$$

$$\sum_{n=1}^{\infty} r^n \tilde{f}^{(n)}_{0;KK}(\rho_3, \rho_4) = \frac{\sum_{n=1}^{\infty} r^n p^{(n)}_{0;KK}(\rho_3, \rho_4)}{1+\sum_{n=1}^{\infty} r^n p^{(n)}_{0;KK}(\rho_3, \rho_4)}, \qquad |r| \leqq 1.$$

As above it follows for $|r| \leqq 1$, $\operatorname{Re} \rho_3 \geqq 0$, $\operatorname{Re} \rho_4 \geqq 0$, $\operatorname{Re} \delta_1 < \operatorname{Re} \zeta$ that

$$\sum_{n=1}^{\infty} r^n \tilde{f}^{(n)}_{0;KK}(\rho_3, \rho_4) = 1$$

$$- \frac{\dfrac{1}{1+\alpha\rho_3} - \dfrac{1+\alpha\rho_3 - r\beta(\rho_4)}{2\pi i} \displaystyle\int_{A_\eta} \frac{e^{\eta K}}{1-(\eta-\rho_3)\,\alpha} \frac{1}{r\beta(\rho_4+\eta)+(\eta-\rho_3)\,\alpha-1} \frac{d\eta}{\eta}}{\dfrac{1}{2\pi i} \displaystyle\int_{A_\eta} \frac{e^{\eta K}}{1-(\eta-\rho_3)\,\alpha} \frac{\alpha\, d\eta}{r\beta(\rho_4+\eta)+(\eta-\rho_3)\,\alpha-1}}. \tag{4.76}$$

Hence for $\delta_1(1, 0, 0) < \operatorname{Re} \zeta < 1/\alpha$,

$$\sum_{n=1}^{\infty} f_{0;KK}^{(n)} = 1 - \left\{ \frac{1}{2\pi i} \int_{C_\zeta} \frac{e^{\zeta K}}{1-\zeta\alpha} \frac{\alpha \, d\zeta}{\beta(\zeta)+\zeta\alpha-1} \right\}^{-1}. \tag{4.77}$$

Finally, we shall prove that the results obtained above are independent of the assumption of rationality of $\beta(\rho)$ (cf. (4.58)), so that they hold for general $B(.)$.

From the expressions obtained above for

$$f(r, \rho, \rho_3, \rho_4, v) \quad \text{and} \quad g(r, \rho, \rho_3, \rho_4, v)/\beta_2(\rho_4+\rho),$$

(cf. (4.63) and (4.61)) it is seen that these expressions involve only $\beta(\rho)$ and not $\beta_1(\rho)$ nor $\beta_2(\rho)$. Consequently, the expression $\varphi(r, \rho, \rho_3, \rho_4, v)$ (cf. (4.20)) contains for the present case $M/G/1$ only $\beta(\rho)$ if the terms mentioned above are eliminated. The resulting expression for $\varphi(r, \rho, \rho_3, \rho_4, v)$ can easily be shown to satisfy the integral equation (4.7) and the other conditions (cf. (4.9), ..., (4.11)). The integral equation (4.7) has been derived without the assumption of rationality. Hence the expression for $\varphi(r, \rho, \rho_3, \rho_4, v)$ obtained in this section holds for general $\beta(\rho)$. The assumption of rationality of $\beta(\rho)$ has been used in (4.51), (4.52) and in the derivation of (4.64) to obtain from $\varphi(r, \rho, \rho_3, \rho_4, v)$ the expressions (4.69), ..., (4.77) for the return and entrance times and the taboo probabilities. It is not difficult to see, however, that only minor changes are needed in (4.51), (4.52) and in the derivation of (4.64) to justify the validity of the relations (4.69), ..., (4.77) for general $\beta(\rho)$.

III.4.6. Application to the system $G/M/1$

In this section we shall apply the results obtained in the previous sections to the queueing model $K_m/M/1$, i.e. we have

$$\alpha(\rho) = \frac{\alpha_1(\rho)}{\alpha_2(\rho)}, \qquad \beta(\rho) = \frac{1}{1+\beta\rho}, \tag{4.78}$$

with $\alpha_1(\rho)$ and $\alpha_2(\rho)$ polynomials in ρ, $\alpha_2(\rho)$ is of degree m, $\alpha_1(\rho)$ of degree $m-1$ at most. The assumption of rationality of $\alpha(\rho)$ is superfluous, so that all results hold for general $A(.)$.

Since for the present case $n=1$ it follows from the definition of $g(r, \rho, \rho_3, \rho_4, v)$ that this function is independent of ρ. Define for $|r|<1$,

$\operatorname{Re} \rho_3 \geqq 0$, $\operatorname{Re} \rho_4 \geqq 0$, $0 \leqq v \leqq K$,

$$g_0(r, \rho_3, \rho_4, v) \stackrel{\text{def}}{=} g(r, \rho, \rho_3, \rho_4, v). \tag{4.79}$$

In exactly the same way as in the previous section we obtain with

$$\varepsilon_1 = \varepsilon_1(r, \rho_3, \rho_4), \qquad |r|<1, \qquad \operatorname{Re} \rho_3 \geqq 0, \quad \operatorname{Re} \rho_4 \geqq 0,$$

for $|r|<1$, $\operatorname{Re} \rho_3 \geqq 0$, $\operatorname{Re} \rho_4 \geqq 0$, $\nu_1 > \operatorname{Re} \xi > 0$, $\operatorname{Re} \xi > \operatorname{Re} \rho$, $0 \leqq v \leqq K$,

$$\begin{aligned}\frac{f(r, \rho, \rho_3, \rho_4, v)}{\alpha_2(\rho_3-\rho)} = &-\frac{1+(\rho+\rho_4)\,\beta - r\alpha(\rho_3-\rho)}{2\pi \mathrm{i}(\rho-\varepsilon_1)} \\ &\cdot \int_{C_\xi} \left\{ g_0(r, \rho_3, \rho_4, v) \frac{\mathrm{e}^{-\xi K}}{1+(\rho_4+\xi)\,\beta} + \frac{r\,\mathrm{e}^{-\xi v}}{\xi} \right\} \\ &\cdot \frac{\xi-\varepsilon_1}{1+(\rho_4+\xi)\,\beta - r\alpha(\rho_3-\xi)} \frac{\mathrm{d}\xi}{\xi-\rho},\end{aligned} \tag{4.80}$$

and for $\operatorname{Re} \varepsilon_1 > \operatorname{Re} \zeta$,

$$\begin{aligned}&\frac{g_0(r, \rho_3, \rho_4, v)}{2\pi \mathrm{i}} \int_{B_\xi} \frac{\mathrm{e}^{-\xi K}}{1+(\rho_4+\xi)\,\beta} \frac{\mathrm{d}\xi}{1+(\rho_4+\xi)\,\beta - r\alpha(\rho_3-\xi)} \\ &= -\frac{r}{1+\beta\rho_4 - r\alpha(\rho_3)} - \frac{r}{2\pi \mathrm{i}} \int_{C_\zeta} \frac{\mathrm{e}^{-\zeta v}}{\zeta} \frac{\mathrm{d}\zeta}{1+(\rho_4+\zeta)\,\beta - r\alpha(\rho_3-\zeta)},\end{aligned} \tag{4.81}$$

$$rc_g(r, \rho_3, \rho_4, v) = \frac{1}{\beta} g_0(r, \rho_3, \rho_4, v), \tag{4.82}$$

$$rc_f(r, \rho_3, \rho_4, v)$$

$$\begin{cases} = \dfrac{1}{2\pi \mathrm{i}} \displaystyle\int_{B_\xi} g_0(r, \rho_3, \rho_4, v) \frac{\mathrm{e}^{-\xi K}}{1+(\xi+\rho_4)\,\beta} \frac{\beta\xi\,\mathrm{d}\xi}{1+(\xi+\rho_4)\,\beta - r\alpha(\rho_3-\xi)} \\ \quad + \dfrac{1}{2\pi \mathrm{i}} \displaystyle\int_{C_\zeta} \mathrm{e}^{-\zeta v} \frac{\beta r\,\mathrm{d}\zeta}{1+(\zeta+\rho_4)\,\beta - r\alpha(\rho_3-\zeta)}, \qquad \text{for} \quad 0<v \leqq K, \\ = -r + \dfrac{1}{2\pi \mathrm{i}} \displaystyle\int_{B_\xi} g_0(r, \rho_3, \rho_4, 0) \frac{\mathrm{e}^{-\xi K}}{1+(\xi+\rho_4)\,\beta} \frac{\beta\xi\,\mathrm{d}\xi}{1+(\xi+\rho_4)\,\beta - r\alpha(\rho_3-\xi)}, \\ \qquad\qquad \text{for} \quad v=0, \end{cases}$$

with the contour B_ξ (see fig. 3) defined analogously to the contour A_η of the preceding section (see fig. 2). In fig. 3 we have taken $-1/\beta<\gamma_3<0$, $\gamma_1<\gamma_2$.

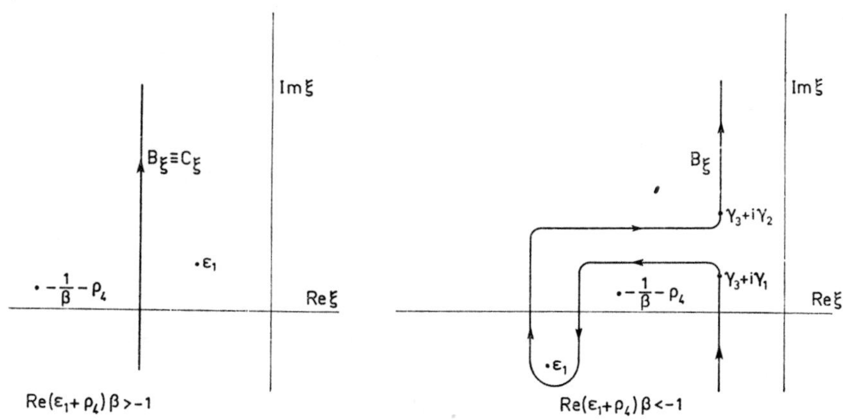

Fig. 3. The contour B_ξ.

From the above relations and using the same type of derivations as in the preceding section we obtain the expressions:

$$\frac{g(\rho)}{\beta_2(\rho)}=\frac{-1}{1+\beta\rho}\left[\frac{1}{2\pi i}\int_{C_\zeta}\frac{e^{-\zeta K}}{1+\zeta\beta}\,\frac{d\zeta}{1+\zeta\beta-\alpha(-\zeta)}\right]^{-1},$$

for $\rho\neq-1/\beta$, $\varepsilon_1(1,0,0)>\operatorname{Re}\zeta>-1/\beta$. Hence from (4.46),

$$V(z)=1-\frac{\dfrac{1}{2\pi i}\displaystyle\int_{C_\zeta}\frac{e^{-\zeta(K-z)}}{1+\beta\zeta-\alpha(-\zeta)}\,d\zeta}{\dfrac{1}{2\pi i}\displaystyle\int_{C_\zeta}\frac{e^{-\zeta K}}{1+\zeta\beta}\,\frac{d\zeta}{1+\beta\zeta-\alpha(-\zeta)}},\qquad 0<z<K, \quad (4.83)$$

$$V(0+)=1-\frac{\dfrac{1}{2\pi i}\displaystyle\int_{C_\zeta}\frac{e^{-\zeta K}\,d\zeta}{1+\beta\zeta-\alpha(-\zeta)}}{\dfrac{1}{2\pi i}\displaystyle\int_{C_\zeta}\frac{e^{-\zeta K}}{1+\zeta\beta}\,\frac{d\zeta}{1+\beta\zeta-\alpha(-\zeta)}},$$

$$1-V(K-)=-\left\{\frac{1}{2\pi i}\int_{C_\zeta}\frac{e^{-\zeta K}}{1+\zeta\beta}\,\frac{\beta\,d\zeta}{1+\beta\zeta-\alpha(-\zeta)}\right\}^{-1},$$

for $\varepsilon_1(1,0,0)>\operatorname{Re}\zeta>-1/\beta$.

For the transforms of the distributions of return times and entrance times we obtain for $|r|\leqq 1$, $\operatorname{Re}\rho_3\geqq 0$, $\operatorname{Re}\rho_4\geqq 0$, $0\leqq v\leqq K$ (cf. (4.55)), $\operatorname{Re}\varepsilon_1>\operatorname{Re}\zeta$,

$$\mathrm{E}\{r^{\boldsymbol{\alpha}_{00}}\,\mathrm{e}^{-\rho_3\boldsymbol{\gamma}_{00}-\rho_4\boldsymbol{p}_{00}}\} = 1+(1+\beta\rho_4-r\alpha(\rho_3))\frac{\dfrac{1}{2\pi\mathrm{i}}\displaystyle\int_{B_\xi}\frac{\mathrm{e}^{-\xi K}}{1+(\xi+\rho_4)\beta}\,\frac{\mathrm{d}\xi}{1+(\xi+\rho_4)\beta-r\alpha(\rho_3-\xi)}}{\dfrac{1}{2\pi\mathrm{i}}\displaystyle\int_{B_\xi}\frac{\mathrm{e}^{-\xi K}}{1+(\xi+\rho_4)\beta}\,\frac{\beta\xi\,\mathrm{d}\xi}{1+(\xi+\rho_4)\beta-r\alpha(\rho_3-\xi)}},$$

$$\mathrm{E}\{r^{\boldsymbol{\alpha}_{vK}}\,\mathrm{e}^{-\rho_3\boldsymbol{\gamma}_{vK}-\rho_4\boldsymbol{p}_{vK}}\} = \frac{1+\dfrac{1+\beta\rho_4-r\alpha(\rho_3)}{2\pi\mathrm{i}}\displaystyle\int_{C_\zeta}\frac{\mathrm{e}^{-\zeta v}}{\zeta}\,\frac{\mathrm{d}\zeta}{1+(\zeta+\rho_4)\beta-r\alpha(\rho_3-\zeta)}}{1+(1+\beta\rho_4)\dfrac{1+\beta\rho_4-r\alpha(\rho_3)}{2\pi\mathrm{i}}\displaystyle\int_{B_\xi}\frac{\mathrm{e}^{-\xi K}}{1+(\xi+\rho_4)\beta}\,\frac{1}{1+(\xi+\rho_4)\beta-r\alpha(\rho_3-\xi)}\,\frac{\mathrm{d}\xi}{\xi}}. \tag{4.84}$$

As in the previous section we obtain for $|r|\leqq 1$, $\operatorname{Re}\rho_3\geqq 0$, $\operatorname{Re}\rho_4\geqq 0$, $\operatorname{Re}\varepsilon_1>\operatorname{Re}\zeta$,

$$\sum_{n=1}^{\infty} r^n\tilde{f}^{(n)}_{K;00}(\rho_3,\rho_4) = 1+\frac{\dfrac{1}{1+\beta\rho_4}+\dfrac{1+\beta\rho_4-r\alpha(\rho_3)}{2\pi\mathrm{i}}\displaystyle\int_{B_\xi}\frac{\mathrm{e}^{-\xi K}}{1+(\xi+\rho_4)\beta}\,\frac{1}{1+(\xi+\rho_4)\beta-r\alpha(\rho_3-\xi)}\,\frac{\mathrm{d}\xi}{\xi}}{\dfrac{1}{2\pi\mathrm{i}}\displaystyle\int_{B_\xi}\frac{\mathrm{e}^{-\xi K}}{1+(\xi+\rho_4)\beta}\,\frac{\beta\,\mathrm{d}\xi}{1+(\xi+\rho_4)\beta-r\alpha(\rho_3-\xi)}},$$

$$\sum_{n=1}^{\infty} r^n\tilde{f}^{(n)}_{0;vK}(\rho_3,\rho_4) = \frac{\dfrac{1}{2\pi\mathrm{i}}\displaystyle\int_{C_\zeta}\frac{\mathrm{e}^{-\zeta v}\,\mathrm{d}\zeta}{1+(\zeta+\rho_4)\beta-r\alpha(\rho_3-\zeta)}}{\dfrac{1+\rho_4\beta}{2\pi\mathrm{i}}\displaystyle\int_{B_\xi}\frac{\mathrm{e}^{-\xi K}}{1+(\xi+\rho_4)\beta}\,\frac{\mathrm{d}\xi}{1+(\xi+\rho_4)\beta-r\alpha(\rho_3-\xi)}}, \qquad 0<v\leqq K, \tag{4.85}$$

$$\sum_{n=1}^{\infty} r^n\tilde{f}^{(n)}_{0;0K}(\rho_3,\rho_4) = -\left\{\frac{1+\rho_4\beta}{2\pi\mathrm{i}}\int_{B_\xi}\frac{\mathrm{e}^{-\xi K}}{1+(\xi+\rho_4)\beta}\,\frac{\beta\,\mathrm{d}\xi}{1+(\xi+\rho_4)\beta-r\alpha(\rho_3-\xi)}\right\}^{-1}.$$

It should be noted that all the relations of the present section can be obtained immediately from those of the preceding section by using the duality relation between $\boldsymbol{v}_n$ and $K-\boldsymbol{v}_n$, cf. (4.29). From this, or by the same argument as used in the previous section, it is easily seen that the results obtained above hold for general distributions $A(.)$.

III.4.7. Application to the system $M/M/1$

For the $M/M/1$ system, i.e. for

$$\alpha(\rho) = \frac{1}{1+\alpha\rho}, \qquad \beta(\rho) = \frac{1}{1+\beta\rho},$$

we shall specify here some of the results obtained in the previous sections. We shall only give the simpler expressions. All results obtained in the previous section can easily be evaluated by contour integration, but since the more general relations are rather lengthy they are omitted here. It is found that

$$V(0+) = \begin{cases} \dfrac{1-a}{1-a^2\exp\{-(1-a)\,K/\beta\}}, & a\neq 1, \\[2ex] \dfrac{1}{2+K/\beta}, & a=1, \end{cases}$$

$$1 - V(K-) = \begin{cases} \dfrac{a(1-a)\exp\{-(1-a)\,K/\beta\}}{1-a^2\exp\{-(1-a)\,K/\beta\}}, & a\neq 1, \\[2ex] \dfrac{1}{2+K/\beta}, & a=1, \end{cases}$$

$$V(x) = \begin{cases} 0, & & x<0, \\[1ex] \dfrac{1-a\exp\{-(1-a)\,x/\beta\}}{1-a^2\exp\{-(1-a)\,K/\beta\}}, & a\neq 1, & 0<x<K, \\[2ex] \dfrac{1+x/\beta}{2+K/\beta}, & a=1, & 0<x<K, \\[2ex] 1, & & x>K. \end{cases}$$

$$\sum_{n=1}^{\infty} f_{K;00}^{(n)} = \begin{cases} \dfrac{1-a\exp\{-(1-a)\,K/\beta\}}{1-a^2\exp\{-(1-a)\,K/\beta\}}, & a \neq 1, \\[2ex] \dfrac{1+K/\beta}{2+K/\beta}, & a=1, \end{cases}$$

$$\sum_{n=1}^{\infty} f_{0;KK}^{(n)} = \begin{cases} a\,\dfrac{1-a\exp\{-(1-a)\,K/\beta\}}{1-a^2\exp\{-(1-a)\,K/\beta\}}, & a \neq 1, \\[2ex] \dfrac{1+K/\beta}{2+K/\beta}, & a=1. \end{cases}$$

III.5. THE FINITE DAM; UNIFORMLY BOUNDED VIRTUAL WAITING TIME

III.5.1. Introduction

The $G/G/1$ queueing model described in chapter II.5 is often used as a storage model for a dam or water reservoir with infinite capacity and with constant release of water per unit time if it is not dry. The arrival moments $\boldsymbol{t}_n$, $n=1, 2, \ldots$, for this model, are the successive moments at which the dam receives inputs, while $\boldsymbol{\tau}_n$, $n=1, 2, \ldots$, is the quantity of water brought into the dam at time $\boldsymbol{t}_n$. It is assumed that the water supply is instantaneous. Since between two successive inputs the total quantity of water decreases linearly if the dam is not dry, the equivalence of the $G/G1$ queueing model and that of the dam with infinite capacity, constant release and instantaneous water supply is apparent. In the terminology of dam theory the busy period is called the wet period of the dam.

Clearly many of the results obtained in previous chapters have an interpretation in dam theory. The dam model with infinite capacity has been the subject of many studies. (For an extensive list of references see PRABHU [1965[a]].)

An important model in dam theory is the model with finite capacity K. This model is the same as that above, except that the capacity is finite and equal to K, so that if at some input moment $\boldsymbol{t}_n$ the supply is so large that the content would exceed K, the excess water overflows. To be more precise, suppose $\boldsymbol{w}_n$ is the content of the dam just before $\boldsymbol{t}_n$ then if $\boldsymbol{w}_n+\boldsymbol{\tau}_n<K$ no overflow occurs, while if $\boldsymbol{w}_n+\boldsymbol{\tau}_n>K$ a quantity of water equal to $\boldsymbol{w}_n+\boldsymbol{\tau}_n-K$ overflows.

In this chapter we shall mainly investigate the finite dam model as described above. Since the dam content at some time t is equivalent to the virtual waiting time $\boldsymbol{v}_t$ of the $G/G/1$ model, we may also consider the finite dam as a queueing system with uniformly bounded virtual waiting time.

It is now easily seen that for the stochastic process $\{\boldsymbol{w}_n,\ n=1, 2, ...\}$ we have

$$\boldsymbol{w}_{n+1} = \begin{cases} [\boldsymbol{w}_n+\boldsymbol{\tau}_n-\boldsymbol{\sigma}_{n+1}]^+ & \text{if } \boldsymbol{w}_n+\boldsymbol{\tau}_n<K, \\ [K-\boldsymbol{\sigma}_{n+1}]^+ & \text{if } \boldsymbol{w}_n+\boldsymbol{\tau}_n\geqq K, \end{cases}$$

for $n=1, 2, \dots$, and $\boldsymbol{w}_1=w$, or equivalently,

$$\begin{aligned} \boldsymbol{w}_{n+1} &= [K-\boldsymbol{\sigma}_{n+1}+[\boldsymbol{w}_n+\boldsymbol{\tau}_n-K]^-]^+, \qquad && n=1, 2, \dots, \\ \boldsymbol{w}_1 &= w, && 0\leqq w\leqq K. \end{aligned} \tag{5.1}$$

Obviously the process $\{\boldsymbol{w}_n,\ n=1, 2, \dots\}$ is a discrete time parameter Markov process with state space the closed interval $[0, K]$.

Once the mathematical description of the process $\{\boldsymbol{w}_n,\ n=1, 2, \dots\}$ has been obtained we may derive new results for the infinite model $G/G/1$. For instance if we know for the finite model the probability that during a busy period or wet period no overflow occurs this probability for the infinite model is the probability that during a busy period the supremum of the virtual waiting time is less than K. Moreover, it will be shown in section 5.8 that from the solution for the finite model we may obtain for the infinite model the distribution of the number of crossings from above of a level K during a busy period and also the distribution of time between two such successive crossings.

To simplify the analysis, it will be assumed here, as in section 4.1 that $\alpha(\rho)$ and $\beta(\rho)$, the Laplace-Stieltjes transforms of $A(.)$ and $B(.)$, respectively, are rational functions of ρ, i.e.

$$\alpha(\rho) = \frac{\alpha_1(\rho)}{\alpha_2(\rho)}, \qquad \beta(\rho) = \frac{\beta_1(\rho)}{\beta_2(\rho)},$$

with $\alpha_1(\rho)$, $\beta_1(\rho)$, $\alpha_2(\rho)$, and $\beta_2(\rho)$ polynomials in ρ; the degree of $\alpha_2(\rho)$ being m and that of $\beta_2(\rho)$ being n. The degree of $\alpha_1(\rho)$ is at most $m-1$ and that of $\beta_1(\rho)$ at most $n-1$. In sections 5.5 and 5.6 where the results of the general theory will be applied to the models $M/G/1$ and $G/M/1$ it will be seen that the assumption of rationality is superfluous.

It should be noted that (5.1) implies that all probability distributions concerning the process $\{\boldsymbol{w}_n,\ n=1, 2, \dots\}$ are independent of the form of the tails of the distributions $A(t)$ and $B(t)$ for $t>K$.

III.5.2. The time dependent solution

In this section we shall discuss the integral equation for the process $\{\boldsymbol{w}_n, n=1, 2, ...\}$.

Define for $|r|<1$, $|\rho|<\infty$, $\mathrm{Re}\,\rho_3 \geqq 0$, $\mathrm{Re}\,\rho_4 \geqq 0$, $0 \leqq w \leqq K$,

$$\omega(r, \rho, \rho_3, \rho_4, w) \overset{\mathrm{def}}{=} \sum_{n=1}^{\infty} r^n \, \mathrm{E}\{\exp(-\rho \boldsymbol{w}_n - \rho_3 \boldsymbol{a}_{n-1} - \rho_4 \boldsymbol{b}_{n-1}) \mid \boldsymbol{w}_1 = w\}. \tag{5.2}$$

Since

$$0 \leqq \boldsymbol{w}_n \leqq K \qquad \text{with probability one for all } n=1, 2, ..., \tag{5.3}$$

it follows that

$$\mathrm{E}\{\mathrm{e}^{-\rho \boldsymbol{w}_n} \mid \boldsymbol{w}_1 = w\},$$

is an entire function of ρ of exponential type and order 1; in fact

$$|\mathrm{E}\{\mathrm{e}^{-\rho \boldsymbol{w}_n} \mid \boldsymbol{w}_1 = w\}| \leqq \mathrm{e}^{-K \mathrm{Re}\,\rho} \qquad \text{for} \quad 1\tfrac{1}{2}\pi > \arg \rho > \tfrac{1}{2}\pi. \tag{5.4}$$

As in section 4.2 it follows that

$$|\omega(r, \rho, \rho_3, \rho_4, w)| < \infty \qquad \text{for } |r|<1, \ \mathrm{Re}\,\rho_3 \geqq 0, \ \mathrm{Re}\,\rho_4 \geqq 0, \ |\rho|<\infty, \tag{5.5}$$

and that bounded functions $D_1(r, \rho_3, \rho_4, w)$ and $D_2(r, \rho_3, \rho_4, w)$ exist such that

$$\lim_{|\rho| \to \infty} |\mathrm{e}^{\rho K} \omega(r, \rho, \rho_3, \rho_4, w)| = D_1(r, \rho_3, \rho_4, w), \qquad 1\tfrac{1}{2}\pi > \arg \rho > \tfrac{1}{2}\pi, \tag{5.6}$$

$$\lim_{|\rho| \to \infty} |\omega(r, \rho, \rho_3, \rho_4, w)| = D_2(r, \rho_3, \rho_4, w), \qquad \tfrac{1}{2}\pi > \arg \rho > -\tfrac{1}{2}\pi,$$

for $|r|<1$, $\mathrm{Re}\,\rho_3 \geqq 0$, $\mathrm{Re}\,\rho_4 \geqq 0$, $0 \leqq w \leqq K$.

Using the relations (4.3) it follows from (5.1) that for $\mathrm{Re}\,\rho > \mathrm{Re}\,\xi > 0$, $\mathrm{Re}\,\eta > \mathrm{Re}\,\xi > 0$, and $n=1, 2, ...,$

$$\exp\{-\rho \boldsymbol{w}_{n+1} - \rho_3 \boldsymbol{a}_n - \rho_4 \boldsymbol{b}_n\}$$

$$= \frac{1}{2\pi \mathrm{i}} \int_{C_\xi} \left(\frac{1}{\rho - \xi} + \frac{1}{\xi} \right) \cdot \exp\{-\xi K + \xi \boldsymbol{\sigma}_{n+1} - \rho_3 \boldsymbol{a}_n - \rho_4 \boldsymbol{b}_n - \xi[\boldsymbol{w}_n + \boldsymbol{\tau}_n - K]^-\} \, \mathrm{d}\xi =$$

$$= \frac{1}{2\pi i} \int_{C_\xi} \left(\frac{1}{\rho - \xi} + \frac{1}{\xi} \right) \exp\{-\xi K + \xi \boldsymbol{\sigma}_{n+1} - \rho_3 \boldsymbol{a}_n - \rho_4 \boldsymbol{b}_n\} \, d\xi$$

$$- \frac{1}{2\pi i} \int_{C_\xi} \left(\frac{1}{\rho - \xi} + \frac{1}{\xi} \right) \exp\{-\xi K + \xi \boldsymbol{\sigma}_{n+1}\} \frac{d\xi}{2\pi i}$$

$$\cdot \left\{ \int_{C_\eta} \left(\frac{1}{\xi - \eta} + \frac{1}{\eta} \right) \exp\{\eta K - \eta \boldsymbol{w}_n - \eta \boldsymbol{\tau}_n - \rho_3 \boldsymbol{a}_n - \rho_4 \boldsymbol{b}_n\} \, d\eta \right\}. \qquad (5.7)$$

Taking expectations in (5.7), then multiplying by r^n and summing over $n=1, 2, \ldots$, we find for $|r|<1$, $0 \leqq w < K$, $\operatorname{Re} \rho > \operatorname{Re} \xi > 0$, $\operatorname{Re} \eta > \operatorname{Re} \xi > 0$, $\operatorname{Re}(\rho_3 - \xi) > 0$, $\operatorname{Re} \rho_4 \geqq 0$,

$$\omega(r, \rho, \rho_3, \rho_4, w) - r\, e^{-\rho w}$$

$$= \frac{1}{2\pi i} \int_{C_\xi} \left(\frac{1}{\rho - \xi} + \frac{1}{\xi} \right) e^{-\xi K} \frac{r^2 \beta(\rho_4)\, \alpha(\rho_3 - \xi)}{1 - r\alpha(\rho_3)\, \beta(\rho_4)} \, d\xi$$

$$- \frac{r}{2\pi i} \int_{C_\xi} \left(\frac{1}{\rho - \xi} + \frac{1}{\xi} \right) e^{-\xi K} \alpha(\rho_3 - \xi) \frac{d\xi}{2\pi i}$$

$$\cdot \left\{ \int_{C_\eta} \left(\frac{1}{\xi - \eta} + \frac{1}{\eta} \right) e^{\eta K} \omega(r, \eta, \rho_3, \rho_4, w)\, \beta(\rho_4 + \eta) \, d\eta \right\}, \qquad (5.8)$$

since it is easily verified that summation, expectation and integrations may be interchanged. The last relation is the integral equation for the finite dam. We shall replace the relation (5.8) by an equivalent one.

In (5.8) we move the path of integration with respect to η to a path just left of the imaginary axis in the η-plane, so that we obtain for $\operatorname{Re} \eta = 0-$,

$$\omega(r, \rho, \rho_3, \rho_4, w) - r\, e^{-\rho w}$$

$$= \frac{r}{2\pi i} \int_{C_\xi} \left(\frac{1}{\rho - \xi} + \frac{1}{\xi} \right) \alpha(\rho_3 - \xi)\, \beta(\rho_4 + \xi)\, \omega(r, \xi, \rho_3, \rho_4, w) \, d\xi$$

$$- \frac{r}{2\pi i} \int_{C_\xi} \left(\frac{1}{\rho - \xi} + \frac{1}{\xi} \right) e^{-\xi K} \alpha(\rho_3 - \xi) \frac{d\xi}{2\pi i}$$

$$\cdot \left\{ \int_{C_\eta} \left(\frac{1}{\xi - \eta} + \frac{1}{\eta} \right) e^{\eta K} \omega(r, \eta, \rho_3, \rho_4, w)\, \beta(\rho_4 + \eta) \, d\eta \right\};$$

the latter relation is equivalent to (cf. (5.5), (5.6)),

$$r\,\mathrm{e}^{-\rho w} = \frac{1}{2\pi\mathrm{i}} \int\limits_{C_\xi} \left(\frac{1}{\rho-\xi} + \frac{1}{\xi}\right) \cdot \{1 - r\alpha(\rho_3-\xi)\,\beta(\rho_4+\xi)\}\,\omega(r, \xi, \rho_3, \rho_4, w)\,\mathrm{d}\xi + \frac{r}{2\pi\mathrm{i}} \int\limits_{C_\xi} \left(\frac{1}{\rho-\xi} + \frac{1}{\xi}\right) \mathrm{e}^{-\xi K}\,\alpha(\rho_3-\xi)\,\frac{\mathrm{d}\xi}{2\pi\mathrm{i}} \cdot \left\{ \int\limits_{C_\eta} \left(\frac{1}{\xi-\eta} + \frac{1}{\eta}\right) \mathrm{e}^{\eta K}\,\omega(r, \eta, \rho_3, \rho_4, w)\,\beta(\rho_4+\eta)\,\mathrm{d}\eta \right\}, \tag{5.9}$$

for $|r|<1$, $\mathrm{Re}\,\rho > \mathrm{Re}\,\xi > 0$, $\mathrm{Re}\,\eta = 0-$, $\mathrm{Re}(\rho_3-\xi) \geqq 0$, $\mathrm{Re}(\rho_4+\eta) \geqq 0$, $0 \leqq w \leqq K$. It should be noted that in the derivation of the integral equation (5.9) the assumption of rationality of the transforms $\alpha(\rho)$ and $\beta(\rho)$ has not been used.

We define two polynomials $x(r, \rho, \rho_3, \rho_4, w)$ and $y(r, \rho, \rho_3, \rho_4, w)$ in ρ of which the coefficients are functions of r, ρ_3, ρ_4 and w, and such that

(*i*) $x(r, \rho, \rho_3, \rho_4, w)$ is a polynomial of degree $m-1$ in ρ and $y(r, \rho, \rho_3, \rho_4, w)$ is a polynomial of degree $n-1$ in ρ;

(*ii*) for $|r|<1$, $\mathrm{Re}\,\rho_3 \geqq 0$, $\mathrm{Re}\,\rho_4 \geqq 0$, $0 \leqq w \leqq K$,

$$\frac{\rho x(r, \rho, \rho_3, \rho_4, w)}{\alpha_2(\rho_3-\rho)} + \alpha(\rho_3-\rho)\,\frac{\rho y(r, \rho, \rho_3, \rho_4, w)}{\beta_2(\rho_4+\rho)}\,\mathrm{e}^{-\rho K} + r\,\mathrm{e}^{-\rho w} = 0, \tag{5.10}$$

for $\rho = \varepsilon_i(r, \rho_3, \rho_4)$, $i = 1, \ldots, n$, and for $\rho = \delta_j(r, \rho_3, \rho_4)$, $j = 1, \ldots, m$ (cf. (4.14) and (4.15)).

Obviously these conditions determine the polynomials $x(\ldots)$ and $y(\ldots)$ uniquely.

The following pair of integral equations determine the polynomials $x(\ldots)$ and $y(\ldots)$ just introduced equally well. For $|r|<1$, $\mathrm{Re}\,\rho_3 \geqq 0$, $\mathrm{Re}\,\rho_4 \geqq 0$, $0 \leqq w \leqq K$ (cf. (4.16) and (4.17)),

$$x(r, \rho, \rho_3, \rho_4, w) = -\frac{\alpha_2(\rho_3-\rho)\,H_-(r, \rho, \rho_3, \rho_4)}{2\pi\mathrm{i}} \cdot \int\limits_{C_\xi} \left\{ \alpha(\rho_3-\xi)\,\frac{y(r, \xi, \rho_3, \rho_4, w)}{\beta_2(\rho_4+\xi)}\,\mathrm{e}^{-\xi K} + \frac{r}{\xi}\,\mathrm{e}^{-\xi w} \right\} \cdot \frac{1}{H_-(r, \xi, \rho_3, \rho_4)}\,\frac{\mathrm{d}\xi}{\xi-\rho}, \tag{5.11}$$

for $\nu_1 > \text{Re}\,\xi > 0$, $\text{Re}\,\xi > \text{Re}\,\rho$; and for $\nu_2 < \text{Re}\,\eta < 0$, $\text{Re}\,\eta < \text{Re}\,\rho$,

$$y(r, \rho, \rho_3, \rho_4, w) = r\,\frac{\beta_2(\rho_4+\rho)\,H_+(r, \rho, \rho_3, \rho_4)}{2\pi\mathrm{i}}$$

$$\cdot \int_{C_\eta} \left\{ \frac{x(r, \eta, \rho_3, \rho_4, w)}{\alpha_2(\rho_3-\eta)}\,\mathrm{e}^{\eta K} + \frac{r}{\eta}\,\mathrm{e}^{(K-w)\eta} \right\}$$

$$\cdot \frac{\beta(\rho_4+\eta)}{H_+(r, \eta, \rho_3, \rho_4)}\,\frac{\mathrm{d}\eta}{\eta-\rho}.$$

The proof that the integral relations (5.11) determine the same polynomials as the conditions (5.10) is completely analogous to that of section 4.2 (cf. (4.19)), and is, therefore, omitted.

THEOREM 5.1. *The function* $\omega(r, \rho, \rho_3, \rho_4, w)$ *defined by* (5.8) *for* $|r|<1$, $|\rho|<\infty$, $\text{Re}\,\rho_3 \geqq 0$, $\text{Re}\,\rho_4 \geqq 0$, $0 \leqq w \leqq K$, *is given by*

$$\omega(r, \rho, \rho_3, \rho_4, w) = \frac{1}{1-r\alpha(\rho_3-\rho)\,\beta(\rho_4+\rho)}$$

$$\cdot \left\{ \frac{\rho x(r, \rho, \rho_3, \rho_4, w)}{\alpha_2(\rho_3-\rho)} + \alpha(\rho_3-\rho)\,\frac{\rho y(r, \rho, \rho_3, \rho_4, w)}{\beta_2(\rho_4+\rho)}\,\mathrm{e}^{-\rho K} + r\,\mathrm{e}^{-\rho w} \right\}, \qquad (5.12)$$

with the polynomials $x(r, \rho, \rho_3, \rho_4, w)$ *and* $y(r, \rho, \rho_3, \rho_4, w)$ *in* ρ *given by* (5.10) *or equivalently, by* (5.11).

Proof. It is easily verified that the right-hand side of (5.12) is an entire function of ρ, and that it satisfies the conditions (5.5) and (5.6). Substitution of the expression (5.12) into (5.9) leads to

$$0 = \frac{1}{2\pi\mathrm{i}} \int_{C_\xi} \left(\frac{1}{\rho-\xi} + \frac{1}{\xi}\right) \left\{ \frac{\xi x(r, \xi)}{\alpha_2(\rho_3-\xi)} + \alpha(\rho_3-\xi)\,\frac{\xi y(r, \xi)}{\beta_2(\rho_4+\xi)}\,\mathrm{e}^{-\xi K} \right\} \mathrm{d}\xi$$

$$+ \frac{r}{2\pi\mathrm{i}} \int_{C_\xi} \left(\frac{1}{\rho-\xi} + \frac{1}{\xi}\right) \mathrm{e}^{-\xi K}\,\alpha(\rho_3-\xi)\,\frac{\mathrm{d}\xi}{2\pi\mathrm{i}}$$

$$\cdot \left\{ \int_{C_\eta} \left(\frac{1}{\xi-\eta} + \frac{1}{\eta}\right) \mathrm{d}\eta \left[-\frac{\eta y(r, \eta)}{r\beta_2(\rho_4+\eta)} + \right.\right.$$

$$+\left\{\beta(\rho_4+\eta)\frac{\eta x(r,\eta)}{\alpha_2(\rho_3-\eta)}+\frac{\eta y(r,\eta)}{r\beta_2(\rho_4+\eta)}\,e^{-\eta K}+r\beta(\rho_4+\eta)\,e^{-\eta w}\right\}$$

$$\left.\left.\cdot\frac{e^{\eta K}}{1-r\alpha(\rho_3-\eta)\,\beta(\rho_4+\eta)}\right]\right\}, \tag{5.13}$$

for $\mathrm{Re}\,\rho>\mathrm{Re}\,\xi>0$, $\mathrm{Re}\,\eta=0-$, $\mathrm{Re}(\rho_3-\xi)\geqq 0$, $\mathrm{Re}(\rho_4+\eta)\geqq 0$, $0\leqq w\leqq K$, and where for the present $x(r, \rho)$ and $y(r, \rho)$ stand for $x(r, \rho, \rho_3, \rho_4, w)$ and $y(r, \rho, \rho_3, \rho_4, w)$, respectively. From (5.10) and (4.14) it is readily verified that the second term in the last integral of (5.13) is an entire function. It is therefore easily seen that this term does not contribute to the last integral. Consequently, the last integral may be written as

$$\frac{-1}{2\pi i}\int_{C_\eta}\frac{\xi}{\xi-\eta}\,\frac{y(r,\eta)}{r\beta_2(\rho_4+\eta)}\,d\eta=-\frac{\xi y(r,\xi)}{r\beta_2(\rho_4+\xi)}$$

$$\text{for}\quad \mathrm{Re}\,\xi>\mathrm{Re}\,\eta=0-. \tag{5.14}$$

From (5.14) it is seen that (5.13) reduces to

$$0=\frac{1}{2\pi i}\int_{C_\xi}\left(\frac{1}{\rho-\xi}+\frac{1}{\xi}\right)\frac{\xi x(r,\xi)}{\alpha_2(\rho_3-\xi)}\,d\xi,\qquad \mathrm{Re}\,\rho>\mathrm{Re}\,\xi>0;$$

but this relation is an identity since $\alpha_2(\rho_3-\xi)$ has no zeros for $\mathrm{Re}\,\xi\leqq 0$. Consequently, $\omega(r, \rho, \rho_3, \rho_4, w)$ as given by (5.12) is a solution of (5.9) and hence of (5.8). To show that the expression (5.12) is indeed the function defined in (5.2) it remains to be shown that the coefficients of the powers of r in the series expansion of the right-hand side of (5.12) with respect to r satisfy a system of recurrence relations with a unique solution and which is identical with such a system for

$$\mathrm{E}\{\exp(-\rho \boldsymbol{w}_n-\rho_3\boldsymbol{a}_{n-1}-\rho_4\boldsymbol{b}_{n-1})\mid \boldsymbol{w}_1=w\},\qquad n=1,2,\dots,$$

and having the same initial condition. This proof is completely analogous to that of theorem 4.1 and is, therefore, omitted. The proof is complete.

Define the content of the dam immediately after the nth input by $\boldsymbol{u}_n$, i.e.

$$\boldsymbol{u}_n\overset{\text{def}}{=}\min(K,\boldsymbol{w}_n+\boldsymbol{\tau}_n),\qquad n=1,2,\dots, \tag{5.15}$$

or

$$\boldsymbol{u}_n=K+[\boldsymbol{w}_n+\boldsymbol{\tau}_n-K]^-,\qquad n=1,2,\dots .$$

Since

$$\boldsymbol{w}_{n+1} = [\boldsymbol{u}_n - \boldsymbol{\sigma}_{n+1}]^+, \qquad n=1, 2, \ldots,$$

it follows that

$$\boldsymbol{u}_{n+1} = K + [[\boldsymbol{u}_n - \boldsymbol{\sigma}_{n+1}]^+ + \boldsymbol{\tau}_{n+1} - K]^-, \qquad n=1, 2, \ldots,$$

or

$$-\boldsymbol{u}_{n+1} = -K + [-[\boldsymbol{u}_n - \boldsymbol{\sigma}_{n+1}]^+ - \boldsymbol{\tau}_{n+1} + K]^+,$$

so that

$$\begin{aligned} K - \boldsymbol{u}_{n+1} &= [K - \boldsymbol{\tau}_{n+1} + [\boldsymbol{\sigma}_{n+1} + (K - \boldsymbol{u}_n) - K]^-]^+, \qquad n=1, 2, \ldots, \\ K - \boldsymbol{u}_1 &= K - u, \qquad 0 \leqq u \leqq K, \end{aligned} \tag{5.16}$$

if u is the quantity of water in the system just after the first arrival.

It should be noted that the sequences $\boldsymbol{w}_n$, $n=1, 2, \ldots$, and $K-\boldsymbol{u}_n$, $n=1, 2, \ldots$, satisfy the same type of recurrence relations. The duality of $\boldsymbol{w}_n$ and $K-\boldsymbol{u}_n$ has been pointed out by DALEY [1964].

Obviously, the process $\{\boldsymbol{u}_n, n=1, 2, \ldots\}$ is a discrete time parameter Markov process with state space the closed interval $[0, K]$.

Since the set of recurrence relations for $K-\boldsymbol{u}_n$, $n=1, 2, \ldots$, is the same as that for $\boldsymbol{w}_n$, $n=1, 2, \ldots$, except for the replacement of $\boldsymbol{\tau}_{n+1}$ by $\boldsymbol{\sigma}_{n+1}$ and $\boldsymbol{\sigma}_{n+1}$ by $\boldsymbol{\tau}_n$ we may immediately write down the time dependent solution for the process $\{K-\boldsymbol{u}_n, n=1, 2, \ldots\}$. From that result it is not difficult to find the time dependent description of the process $\{\boldsymbol{u}_n, n=1, 2, \ldots\}$.

Defining for $|r|<1$, $|\rho|<\infty$, $\operatorname{Re}\rho_3 \geqq 0$, $\operatorname{Re}\rho_4 \geqq 0$, $0 \leqq u \leqq K$,

$$\Pi(r, \rho, \rho_3, \rho_4, u) \stackrel{\text{def}}{=} \sum_{n=1}^{\infty} r^n \operatorname{E}\{\exp\{-\rho \boldsymbol{u}_n - \rho_3 \boldsymbol{a}_{n-1} - \rho_4(\boldsymbol{b}_n - \boldsymbol{b}_1)\} \mid \boldsymbol{u}_1 = u\}, \tag{5.17}$$

it is found that

$$\Pi(r, \rho, \rho_3, \rho_4, u) = \frac{1}{1 - r\alpha(\rho_3 - \rho)\,\beta(\rho_4 + \rho)}$$

$$\cdot \left\{ \beta(\rho_4 + \rho) \frac{\rho k(r, \rho, \rho_3, \rho_4, u)}{\alpha_2(\rho_3 - \rho)} + \frac{\rho h(r, \rho, \rho_3, \rho_4, u)}{\beta_2(\rho_4 + \rho)} \mathrm{e}^{-\rho K} + r\,\mathrm{e}^{-\rho u} \right\}, \tag{5.18}$$

with $k(r, \rho, \rho_3, \rho_4, u)$ and $h(r, \rho, \rho_3, \rho_4, u)$ polynomials in ρ of degree $m-1$ and $n-1$, respectively, and which are determined by

$$\beta(\rho_4 + \rho) \frac{k(r, \rho, \rho_3, \rho_4, u)}{\alpha_2(\rho_3 - \rho)} + \frac{h(r, \rho, \rho_3, \rho_4, u)}{\beta_2(\rho_4 + \rho)} \mathrm{e}^{-\rho K} + \frac{r}{\rho} \mathrm{e}^{-\rho u} = 0, \tag{5.19}$$

for $\rho=\varepsilon_i(r, \rho_3, \rho_4)$, $i=1, \ldots, n$, and $\rho=\delta_j(r, \rho_3, \rho_4)$, $j=1, \ldots, m$, with $|r|<1$, $\operatorname{Re}\rho_3 \geqq 0$, $\operatorname{Re}\rho_4 \geqq 0$.

As before we have a pair of integral relations which also determine these polynomials. These relations read:
for $|r|<1$, $\operatorname{Re}\rho_3 \geqq 0$, $\operatorname{Re}\rho_4 \geqq 0$, $|\rho|<\infty$, $0 \leqq u \leqq K$,

$$k(r, \rho, \rho_3, \rho_4, u) = -r\,\frac{\alpha_2(\rho_3-\rho)\,H_-(r, \rho, \rho_3, \rho_4)}{2\pi\mathrm{i}} \cdot \int_{C_\xi} \left\{ \frac{h(r, \xi, \rho_3, \rho_4, u)}{\beta_2(\rho_4+\xi)}\,\mathrm{e}^{-\xi K} + \frac{r}{\xi}\,\mathrm{e}^{-\xi u} \right\} \cdot \frac{\alpha(\rho_3-\xi)}{H_-(r, \xi, \rho_3, \rho_4)}\,\frac{\mathrm{d}\xi}{\xi-\rho}, \tag{5.20}$$

for $\nu_1 > \operatorname{Re}\xi > 0$, $\operatorname{Re}\xi > \operatorname{Re}\rho$, and

$$h(r, \rho, \rho_3, \rho_4, u) = \frac{\beta_2(\rho_4+\rho)\,H_+(r, \rho, \rho_3, \rho_4)}{2\pi\mathrm{i}} \cdot \int_{C_\eta} \left\{ \beta(\rho_4+\eta)\,\frac{k(r, \eta, \rho_3, \rho_4, u)}{\alpha_2(\rho_3-\eta)}\,\mathrm{e}^{\eta K} + \frac{r}{\eta}\,\mathrm{e}^{(K-u)\eta} \right\} \cdot \frac{1}{H_+(r, \eta, \rho_3, \rho_4)}\,\frac{\mathrm{d}\eta}{\eta-\rho}, \tag{5.21}$$

for $\nu_2 < \operatorname{Re}\eta < 0$, $\operatorname{Re}\eta < \operatorname{Re}\rho$.

To complete the time dependent description of the processes

$$\{\boldsymbol{w}_n,\ n=1, 2, \ldots\} \quad \text{and} \quad \{\boldsymbol{u}_n,\ n=1, 2, \ldots\}$$

we define for $|r|<1$, $\operatorname{Re}\rho_3 \geqq 0$, $\operatorname{Re}\rho_4 \geqq 0$, $0 \leqq w \leqq K$, $0 \leqq u \leqq K$, $|\rho|<\infty$,

$$\omega_1(r, \rho, \rho_3, \rho_4, w) \stackrel{\text{def}}{=} \sum_{n=1}^{\infty} r^n\,\mathrm{E}\{\exp(-\rho\boldsymbol{u}_n - \rho_3\boldsymbol{a}_{n-1} - \rho_4\boldsymbol{b}_n) \mid \boldsymbol{w}_1 = w\}, \tag{5.22}$$

$$\Pi_1(r, \rho, \rho_3, \rho_4, u) \stackrel{\text{def}}{=} \sum_{n=2}^{\infty} r^{n-1}\,\mathrm{E}\{\exp\{-\rho\boldsymbol{w}_n - \rho_3\boldsymbol{a}_{n-1} - \rho_4(\boldsymbol{b}_{n-1} - \boldsymbol{b}_1)\} \mid \boldsymbol{u}_1 = u\}. \tag{5.23}$$

We shall show that for $|r|<1$, $|\rho|<\infty$, $\operatorname{Re}\rho_3 \geqq 0$, $\operatorname{Re}\rho_4 \geqq 0$, $0 \leqq w \leqq K$, $0 \leqq u \leqq K$,

$$\omega_1(r,\rho,\rho_3,\rho_4,w) = \frac{1}{1-r\alpha(\rho_3-\rho)\,\beta(\rho_4+\rho)} \cdot \left\{\beta(\rho_4+\rho)\,\frac{\rho x(r,\rho,\rho_3,\rho_4,w)}{\alpha_2(\rho_3-\rho)} + \frac{\rho y(r,\rho,\rho_3,\rho_4,w)}{r\beta_2(\rho_4+\rho)}\,\mathrm{e}^{-\rho K} + r\beta(\rho_4+\rho)\,\mathrm{e}^{-\rho w}\right\}, \tag{5.24}$$

$$\Pi_1(r,\rho,\rho_3,\rho_4,u) = \frac{1}{1-r\alpha(\rho_3-\rho)\,\beta(\rho_4+\rho)} \cdot \left\{\frac{\rho k(r,\rho,\rho_3,\rho_4,u)}{r\alpha_2(\rho_3-\rho)} + \alpha(\rho_3-\rho)\,\frac{\rho h(r,\rho,\rho_3,\rho_4,u)}{\beta_2(\rho_4+\rho)}\,\mathrm{e}^{-\rho K} + r\alpha(\rho_3-\rho)\,\mathrm{e}^{-\rho u}\right\}. \tag{5.25}$$

To prove (5.24) note that it follows from (5.22) and (5.15) that

$$\omega_1(r,\rho,\rho_3,\rho_4,w) = \sum_{n=1}^{\infty} r^n\,\mathrm{E}\{\exp\{-\rho\min(K,\boldsymbol{w}_n+\boldsymbol{\tau}_n)-\rho_3\boldsymbol{a}_{n-1}-\rho_4\boldsymbol{b}_n\}\mid \boldsymbol{w}_1=w\}.$$

Since for real values of x,

$$\exp\{-\rho[x]^-\} = \mathrm{e}^{-\rho x} - \frac{1}{2\pi\mathrm{i}}\int_{C_\eta}\left(\frac{1}{\rho-\eta}+\frac{1}{\eta}\right)\mathrm{e}^{-\eta x}\,\mathrm{d}\eta, \qquad \operatorname{Re}\eta<0<\operatorname{Re}\rho,$$

it follows from the above relation that for $\operatorname{Re}\eta>\operatorname{Re}\rho>0$, $\operatorname{Re}\zeta=0-$, $\operatorname{Re}\rho_3\geqq 0$, $\operatorname{Re}\rho_4\geqq 0$, $|r|<1$, $0\leqq w\leqq K$,

$$\begin{aligned}\omega_1(r,\rho,\rho_3,\rho_4,w) &= \sum_{n=1}^{\infty} r^n\,\mathrm{E}\left\{\left(\mathrm{e}^{-\rho K} - \frac{\mathrm{e}^{-\rho K}}{2\pi\mathrm{i}}\int_{C_\eta}\left(\frac{1}{\rho-\eta}+\frac{1}{\eta}\right)\right.\right.\\ &\qquad \left.\left.\cdot\exp\{-\eta(\boldsymbol{w}_n+\boldsymbol{\tau}_n-K)\}\,\mathrm{d}\eta\right)\exp\{-\rho_3\boldsymbol{a}_{n-1}-\rho_4\boldsymbol{b}_n\}\mid \boldsymbol{w}_1=w\right\}\\ &= \beta(\rho+\rho_4)\,\omega(r,\rho,\rho_3,\rho_4,w) - \frac{\mathrm{e}^{-\rho K}}{2\pi\mathrm{i}}\int_{C_\zeta}\left(\frac{1}{\rho-\zeta}+\frac{1}{\zeta}\right)\\ &\qquad \cdot\mathrm{e}^{\zeta K}\,\beta(\zeta+\rho_4)\,\omega(r,\zeta,\rho_3,\rho_4,w)\,\mathrm{d}\zeta. \end{aligned} \tag{5.26}$$

Since

$$\beta(\rho_4+\rho)\,\omega(r,\rho,\rho_3,\rho_4,w) = -\frac{\rho y(r,\rho,\rho_3,\rho_4,w)}{r\beta_2(\rho_4+\rho)}\,\mathrm{e}^{-\rho K}$$

$$+\frac{1}{1-r\alpha(\rho_3-\rho)\,\beta(\rho_4+\rho)}\left\{\beta(\rho_4+\rho)\,\frac{\rho x(r,\rho,\rho_3,\rho_4,w)}{\alpha_2(\rho_3-\rho)}\right.$$

$$\left.+\frac{\rho y(r,\rho,\rho_3,\rho_4,w)}{r\beta_2(\rho_4+\rho)}\,\mathrm{e}^{-\rho K}+r\beta(\rho_4+\rho)\,\mathrm{e}^{-\rho w}\right\},$$

and since the second term in the last relation is an entire function of ρ (cf. (5.10)), it follows immediately from (5.26) that

$$\omega_1(r,\rho,\rho_3,\rho_4,w)$$

$$=\beta(\rho_4+\rho)\,\omega(r,\rho,\rho_3,\rho_4,w)+\frac{\mathrm{e}^{-\rho K}}{2\pi\mathrm{i}}\int_{C_\zeta}\frac{\rho}{\rho-\zeta}\,\frac{y(r,\zeta,\rho_3,\rho_4,w)}{r\beta_2(\zeta+\rho_4)}\,\mathrm{d}\zeta$$

$$=\beta(\rho_4+\rho)\,\omega(r,\rho,\rho_3,\rho_4,w)+\frac{\rho y(r,\rho,\rho_3,\rho_4,w)}{r\beta_2(\rho_4+\rho)}\,\mathrm{e}^{-\rho K}.$$

From the latter relation and (5.12) the relation (5.24) follows. Since

$$\boldsymbol{w}_{n+1}=[\boldsymbol{u}_n-\boldsymbol{\sigma}_{n+1}]^+,\qquad n=1,2,\ldots,$$

and for real x,

$$\exp\{-\rho[x]^+\}=\frac{1}{2\pi\mathrm{i}}\int_{C_\xi}\left(\frac{1}{\rho-\xi}+\frac{1}{\xi}\right)\mathrm{e}^{-\xi x}\,\mathrm{d}\xi,\qquad \operatorname{Re}\rho>\operatorname{Re}\xi>0,$$

it follows from (5.23) for $\operatorname{Re}\rho>\operatorname{Re}\xi>0$, $\operatorname{Re}\eta>\operatorname{Re}\rho$ that

$$\Pi_1(r,\rho,\rho_3,\rho_4,u)$$

$$=\frac{1}{r}\sum_{n=1}^{\infty}r^{n+1}\,\mathrm{E}\{\exp\{-\rho[\boldsymbol{u}_n-\boldsymbol{\sigma}_{n+1}]^+-\rho_3\boldsymbol{a}_n-\rho_4(\boldsymbol{b}_n-\boldsymbol{b}_1)\}\mid\boldsymbol{u}_1=u\}$$

$$=\frac{1}{r}\sum_{n=1}^{\infty}r^{n+1}\,\mathrm{E}\left\{\frac{1}{2\pi\mathrm{i}}\int_{C_\xi}\left(\frac{1}{\rho-\xi}+\frac{1}{\xi}\right)\right.$$

$$\left.\cdot\exp\{-\xi(\boldsymbol{u}_n-\boldsymbol{\sigma}_{n+1})\}\,\mathrm{d}\xi\,\exp\{-\rho_3\boldsymbol{a}_n-\rho_4(\boldsymbol{b}_n-\boldsymbol{b}_1)\}\mid\boldsymbol{u}_1=u\right\}=$$

$$= \frac{1}{2\pi i} \int_{C_\xi} \left(\frac{1}{\rho - \xi} + \frac{1}{\xi} \right) \alpha(\rho_3 - \xi)\, \Pi(r, \xi, \rho_3, \rho_4, u)\, d\xi$$

$$= \alpha(\rho_3 - \rho)\, \Pi(r, \rho, \rho_3, \rho_4, u)$$

$$+ \frac{1}{2\pi i} \int_{C_\eta} \left(\frac{1}{\rho - \eta} + \frac{1}{\eta} \right) \alpha(\rho_3 - \eta)\, \Pi(r, \eta, \rho_3, \rho_4, u)\, d\eta,$$

if ρ and η are chosen in such a way that all poles of $\alpha(\rho_3 - \eta)$ lie to the right of the path of integration C_η, this is always possible. Noting that $\Pi(r, \rho, \rho_3, \rho_4, u)$ is an entire function of ρ, and that the only poles of the last integrand are the poles of $\alpha(\rho_3 - \eta)$, it follows easily that with $0 < \text{Re}\, \rho < \text{Re}\, \eta$, since $\alpha(\rho_3 - \zeta)$ is analytic for $\text{Re}\, \zeta \leqq \text{Re}\, \eta$,

$$\Pi_1(r, \rho, \rho_3, \rho_4, u) = \alpha(\rho_3 - \rho)\, \Pi(r, \rho, \rho_3, \rho_4, u)$$

$$- \frac{1}{2\pi i} \int_{C_\eta} \frac{\rho}{\rho - \eta} \frac{k(r, \eta, \rho_3, \rho_4, u)}{r \alpha_2(\rho_3 - \eta)}\, d\eta$$

$$= \alpha(\rho_3 - \rho)\, \Pi(r, \rho, \rho_3, \rho_4, u) + \frac{\rho k(r, \rho, \rho_3, \rho_4, u)}{r \alpha_2(\rho_3 - \rho)}.$$

From the last relation and from (5.18) we obtain (5.25). In the derivation of the relations (5.24) and (5.25) expectations and integrations have been interchanged. That this is permitted is easily verified since the integrals converge absolutely.

III.5.3. The stationary distributions

In this section we shall study the stationary distributions of the processes $\{\boldsymbol{w}_n, n=1, 2, \ldots\}$ and $\{\boldsymbol{u}_n, n=1, 2, \ldots\}$. For the proof of the existence of these stationary distributions our assumption of rationality of the transforms $\alpha(\rho)$ and $\beta(\rho)$ (cf. section 5.1) will not be used.

First we see that if the process $\{\boldsymbol{w}_n, n=1, 2, \ldots\}$ has a stationary distribution then the process $\{\boldsymbol{u}_n, n=1, 2, \ldots\}$ also has a stationary distribution and conversely. This follows from the relation (5.15).

We shall exclude the trivial case that $\boldsymbol{\sigma}_{n+1} = \boldsymbol{\tau}_n = \text{constant}$, i.e. we assume that

$$\Pr\{\boldsymbol{\sigma}_{n+1} \neq \boldsymbol{\tau}_n\} > 0, \qquad n = 1, 2, \ldots . \tag{5.27}$$

Consider the stochastic process $\{\boldsymbol{w}_n^{(1)},\ n=1, 2, \ldots\}$ defined by

$$\boldsymbol{w}_{n+1}^{(1)}=[\boldsymbol{w}_n^{(1)}+\boldsymbol{\tau}_n-\boldsymbol{\sigma}_{n+1}]^+, \qquad n=1, 2, \ldots, \tag{5.28}$$
$$\boldsymbol{w}_1^{(1)}=w.$$

We first show that

$$\boldsymbol{w}_n^{(1)}\geqq \boldsymbol{w}_n \qquad \text{with probability 1 for all } n=1, 2, \ldots . \tag{5.29}$$

The proof is given by induction on n. Suppose $\boldsymbol{w}_n^{(1)}\geqq \boldsymbol{w}_n$. If $\boldsymbol{w}_n+\boldsymbol{\tau}_n\geqq K$ then $\boldsymbol{w}_n^{(1)}+\boldsymbol{\tau}_n\geqq K$ and hence from (5.28) and (5.1)

$$\boldsymbol{w}_{n+1}^{(1)}\geqq[K-\boldsymbol{\sigma}_{n+1}]^+=\boldsymbol{w}_{n+1}.$$

If $\boldsymbol{w}_n+\boldsymbol{\tau}_n<K$ then from (5.1) and (5.28),

$$\boldsymbol{w}_{n+1}=[\boldsymbol{w}_n+\boldsymbol{\tau}_n-\boldsymbol{\sigma}_{n+1}]^+\leqq[\boldsymbol{w}_n^{(1)}+\boldsymbol{\tau}_n-\boldsymbol{\sigma}_{n+1}]^+=\boldsymbol{w}_{n+1}^{(1)}.$$

Hence, since $\boldsymbol{w}_1^{(1)}=\boldsymbol{w}_1$, the relation (5.29) is proved.

The process $\{\boldsymbol{w}_n^{(1)},\ n=1, 2, \ldots\}$ defined by (5.28) is the waiting time process for the general single server queueing system $G/G/1$. For this process it is known (cf. section II.5.4) that if

$$\sum_{n=1}^{\infty}\frac{1}{n}\Pr\{\boldsymbol{b}_n-\boldsymbol{a}_n<0\} = \infty, \qquad \sum_{n=1}^{\infty}\frac{1}{n}\Pr\{\boldsymbol{b}_n-\boldsymbol{a}_n>0\} < \infty, \tag{5.30}$$

or

$$\sum_{n=1}^{\infty}\frac{1}{n}\Pr\{\boldsymbol{b}_n-\boldsymbol{a}_n<0\} = \infty, \qquad \sum_{n=1}^{\infty}\frac{1}{n}\Pr\{\boldsymbol{b}_n-\boldsymbol{a}_n>0\} = \infty, \tag{5.31}$$

then

$$\Pr\{\boldsymbol{w}_2^{(1)}=0 \mid \boldsymbol{w}_1^{(1)}=0\}>0, \tag{5.32}$$
$$\Pr\{\boldsymbol{w}_n^{(1)}=0 \mid \boldsymbol{w}_1^{(1)}=w\}>0, \qquad \text{for } n \text{ sufficiently large.}$$

Note that if

$$\alpha \stackrel{\text{def}}{=} \mathrm{E}\{\boldsymbol{\sigma}_n\} < \infty, \qquad \beta \stackrel{\text{def}}{=} \mathrm{E}\{\boldsymbol{\tau}_n\} < \infty, \qquad a \stackrel{\text{def}}{=} \beta/\alpha, \tag{5.33}$$

then (5.30) is equivalent with $a<1$ and (5.31) with $a=1$.

Define

$$n_1 \stackrel{\text{def}}{=} \min_{n=2,3,\ldots} \{n\colon \Pr\{\boldsymbol{w}_n^{(1)}=0 \mid \boldsymbol{w}_1^{(1)}=K\}>0\},$$

and

$$b_1 \stackrel{\text{def}}{=} \Pr\{\boldsymbol{w}_{n_1}^{(1)}=0 \mid \boldsymbol{w}_1^{(1)}=K\}, \qquad b_2 \stackrel{\text{def}}{=} \Pr\{\boldsymbol{w}_2^{(1)}=0 \mid \boldsymbol{w}_1^{(1)}=0\}. \tag{5.34}$$

Since the event $\{\boldsymbol{w}_n^{(1)}=0\}$ is a regenerative event for the process $\{\boldsymbol{w}_n^{(1)}, n=1, 2, ...\}$ it follows that

$$\Pr\{\boldsymbol{w}_N^{(1)}=0 \mid \boldsymbol{w}_1^{(1)}=K\} \geqq b_1 b_2^{N-n_1}, \qquad \text{for } N \text{ sufficiently large.} \tag{5.35}$$

From (5.29) and (5.35) it now follows for N sufficiently large and $0 \leqq w \leqq K$ that

$$\begin{aligned}\Pr\{\boldsymbol{w}_{n+N}=0 \mid \boldsymbol{w}_n=w\} &\geqq \Pr\{\boldsymbol{w}_{n+N}=0 \mid \boldsymbol{w}_n=K\} \\ &\geqq \Pr\{\boldsymbol{w}_{n+N}^{(1)}=0 \mid \boldsymbol{w}_n^{(1)}=K\} \geqq b_1 b_2^{N-n_1+1}.\end{aligned} \tag{5.36}$$

If the conditions (5.30) or (5.31) hold it is well-known that

$$\sum_{n=1}^{\infty} \Pr\{\boldsymbol{w}_n^{(1)}=0 \mid \boldsymbol{w}_1^{(1)}=w\} = \infty,$$

so that

$$\sum_{n=1}^{\infty} \Pr\{\boldsymbol{w}_n=0 \mid \boldsymbol{w}_1=w\} = \infty.$$

Since the event $\{\boldsymbol{w}_n=0\}$ is a regenerative event for the process $\{\boldsymbol{w}_n, n=1,2, ...\}$ it follows that if the conditions (5.30) or (5.31) apply, i.e. if $a \leqq 1$ in case α and β are finite then

$$\Pr\{\boldsymbol{w}_n=0, \text{ i.o.}\}=1. \tag{5.37}$$

From the duality between the processes $\{\boldsymbol{w}_n, n=1,2, ...\}$ and $\{\boldsymbol{u}_n, n=1,2, ...\}$ it follows that if $a \geqq 1$ then

$$\Pr\{\boldsymbol{u}_n=K, \text{ i.o.}\}=1. \tag{5.38}$$

To prove the existence of a stationary distribution for the process $\{\boldsymbol{w}_n, n=1, 2, ...\}$ we now apply an idea of DALEY [1964]. Let μ be a measure on the Borelfield $\mathscr{B}$ generated by all open intervals with rational end-points of the point set $[0, K]$, and such that $\mu(A)=1$ if $0 \in A$, $\mu(A)=0$ if $0 \notin A$, with $A \in \mathscr{B}$; hence $\mu([0, K])=1$. Suppose (5.30) or (5.31) holds or, equivalently $a \leqq 1$, if α and β are finite. For $0<\varepsilon<b_1 b_2^{N-n_1+1}$ and any set $B \in \mathscr{B}$ with $\mu(B) \leqq \varepsilon$ we have for every $w \in [0, K]$ from (5.36),

$$\Pr\{\boldsymbol{w}_{n+N} \in B \mid \boldsymbol{w}_n=w\}=1-\Pr\{\boldsymbol{w}_{n+N} \notin B \mid \boldsymbol{w}_n=w\},$$

$$\Pr\{\boldsymbol{w}_{n+N} \notin B \mid \boldsymbol{w}_n=w\} \geqq \Pr\{\boldsymbol{w}_{n+N}=0 \mid \boldsymbol{w}_n=w\} \geqq b_1 b_2^{N-n_1+1},$$

so that

$$\Pr\{\boldsymbol{w}_{n+N} \in B \mid \boldsymbol{w}_n=w\} \leqq 1-b_1 b_2^{N-n_1+1}<1-\varepsilon,$$

for N sufficiently large. The latter inequality implies that Doeblin's condition for the Markov process $\{\boldsymbol{w}_n,\ n=1, 2, \ldots\}$ is satisfied (cf. DOOB [1953]). Hence, if the conditions (5.30) or (5.31) apply, the process $\{\boldsymbol{w}_n,\ n=1, 2, \ldots\}$ has a unique stationary distribution and as $n\to\infty$ the distribution $\Pr\{\boldsymbol{w}_n<x \mid \boldsymbol{w}_1=w\}$ converges to it for every w, with $0\leqq w\leqq K$. The same holds for the process $\{\boldsymbol{u}_n,\ n=1, 2, \ldots\}$. In the case in which

$$\sum_{n=1}^{\infty} \frac{1}{n} \Pr\{\boldsymbol{b}_n-\boldsymbol{a}_n<0\} < \infty, \tag{5.39}$$

or equivalently, $a>1$, if α and β are finite, we obtain by the same argument as above that the process $\{K-\boldsymbol{u}_n,\ n=1, 2, \ldots\}$ has a stationary distribution. Hence we reach the same results as before, but now for $a>1$. The following statement, first formulated and proved by DALEY [1964], has been established. If $\Pr\{\boldsymbol{\sigma}_{n+1}\neq\boldsymbol{\tau}_n\}>0$, the processes $\{\boldsymbol{w}_n,\ n=1, 2, \ldots\}$ and $\{\boldsymbol{u}_n,\ n=1, 2, \ldots\}$ each have a unique stationary distribution, the distribution of $\boldsymbol{w}_n$ converges for $n\to\infty$ to the stationary distribution of the process $\{\boldsymbol{w}_n,\ n=1, 2, \ldots\}$, independently of the initial condition; similarly for the process $\{\boldsymbol{u}_n,\ n=1, 2, \ldots\}$.

Next, we shall investigate these stationary distributions in the case where $\alpha(\rho)$ and $\beta(\rho)$ are rational transforms. Note that under this assumption α and β are always finite.

The stationary distributions of the processes $\{\boldsymbol{w}_n,\ n=1, 2, \ldots\}$ and $\{\boldsymbol{u}_n,\ n=1, 2, \ldots\}$ will be indicated by $W(x)$ and $U(x)$, respectively. From Helly-Bray's theorem and a well-known Abelian theorem (cf. app. 1) it follows for $|\rho|<\infty$ that

$$\omega(\rho) \stackrel{\text{def}}{=} \int_{0-}^{\infty} e^{-\rho x}\, dW(x) = \lim_{r\uparrow 1} (1-r)\, \omega(r, \rho, 0, 0, w), \tag{5.40}$$

$$\Pi(\rho) \stackrel{\text{def}}{=} \int_{0-}^{\infty} e^{-\rho x}\, dU(x) = \lim_{r\uparrow 1} (1-r)\, \Pi(r, \rho, 0, 0, w).$$

Since $W(x)$ is the stationary distribution of the process $\{\boldsymbol{w}_n,\ n=1, 2, \ldots\}$ it follows that if $\boldsymbol{w}_1$ has the distribution $W(x)$, then all $\{\boldsymbol{w}_n,\ n=1, 2, \ldots\}$ have $W(x)$ as distribution function and satisfy

$$\boldsymbol{w}_{n+1}=[K-\boldsymbol{\sigma}_{n+1}+[\boldsymbol{w}_n+\boldsymbol{\tau}_n-K]^-]^+, \qquad n=1, 2, \ldots . \tag{5.41}$$

Obviously we now have, since $0\leqq\boldsymbol{w}_n\leqq K$ with probability one, that $\omega(\rho)$ is an entire function of exponential type and order 1, hence

$$|\omega(\rho)| < \infty \qquad \text{for} \quad \rho| < \infty, \tag{5.42}$$

$$\lim_{|\rho|\to\infty} |\omega(\rho)| < \infty, \qquad \tfrac{1}{2}\pi > \arg\rho > -\tfrac{1}{2}\pi,$$

$$\lim_{|\rho|\to\infty} |\mathrm{e}^{\rho K}\,\omega(\rho)| < \infty, \qquad 1\tfrac{1}{2}\pi > \arg\rho > \tfrac{1}{2}\pi.$$

We define the numbers μ_1 and μ_2 by the condition that μ_1 is the real part of the singularity of $\alpha(-\rho)$ nearest to the imaginary axis in the ρ plane, and μ_2 that of $\beta(\rho)$; hence $\mu_1 > 0$, $\mu_2 < 0$.

As in section 4.3, it follows from (5.40) and (5.41) that for $\operatorname{Re}\rho > \operatorname{Re}\xi$, $\mu_1 > \operatorname{Re}\xi > 0$, $\operatorname{Re}\eta = 0-$,

$$0 = \frac{1}{2\pi\mathrm{i}} \int_{C_\xi} \left(\frac{1}{\rho-\xi} + \frac{1}{\xi}\right)\{1-\alpha(-\xi)\,\beta(\xi)\}\,\omega(\xi)\,\mathrm{d}\xi$$
$$+ \frac{1}{2\pi\mathrm{i}} \int_{C_\xi} \left(\frac{1}{\rho-\xi} + \frac{1}{\xi}\right)\mathrm{e}^{-\xi K}\,\alpha(-\xi)\,\frac{\mathrm{d}\xi}{2\pi\mathrm{i}}$$
$$\cdot \left\{ \int_{C_\eta} \left(\frac{1}{\xi-\eta} + \frac{1}{\eta}\right)\mathrm{e}^{\eta K}\,\omega(\eta)\,\beta(\eta)\,\mathrm{d}\eta \right\}. \tag{5.43}$$

Next, we define two polynomials $x(\rho)$ and $y(\rho)$ in ρ such that
(*i*) $x(\rho)$ is of degree $m-1$, and $y(\rho)$ of degree $n-1$;
(*ii*)

$$\frac{\rho}{1-\alpha(-\rho)\,\beta(\rho)} \left\{ \frac{x(\rho)}{\alpha_2(-\rho)} + \alpha(-\rho)\,\frac{y(\rho)}{\beta_2(\rho)} \right\} = 1 \qquad \text{for} \quad \rho \to 0, \tag{5.44}$$

$$\frac{x(\rho)}{\alpha_2(-\rho)} + \alpha(-\rho)\,\frac{y(\rho)}{\beta_2(\rho)}\,\mathrm{e}^{-\rho K} = 0, \tag{5.45}$$

for $\rho = \varepsilon_i(1, 0, 0)$, $i = 1, \ldots, n_0$, and $\rho = \delta_j(1, 0, 0)$, $j = 1, \ldots, m_0$, if $a \neq 1$ (cf. (4.39)), while for $a = 1$ the condition (5.45) should also hold for $\rho = 0$.

The conditions above yield $m+n$ linear equations for the $m+n$ coefficients in the polynomials $x(\rho)$ and $y(\rho)$, and determine these polynomials uniquely. As in sections 5.2 and 4.3 it is proved that if $a \neq 1$ the conditions (5.44) and (5.45) may be replaced by

$$x(\rho) = -\frac{\alpha_2(-\rho)\,H_-(\rho)}{2\pi i}\int_{C_\xi} \alpha(-\xi)\,\frac{y(\xi)}{\beta_2(\xi)}\,\frac{e^{-\xi K}}{H_-(\xi)}\,\frac{d\xi}{\xi-\rho}, \tag{5.46}$$

$$y(\rho) = r\,\frac{\beta_2(\rho)\,H_+(\rho)}{2\pi i}\int_{C_\eta} \beta(\eta)\,\frac{x(\eta)}{\alpha_2(-\eta)}\,\frac{e^{\eta K}}{H_+(\eta)}\,\frac{d\eta}{\eta-\rho},$$

with $\nu_{01} > \text{Re}\,\xi > 0$, $\text{Re}\,\xi > \text{Re}\,\rho$, $\nu_{02} < \text{Re}\,\eta < 0$, $\text{Re}\,\eta < \text{Re}\,\rho$.

In the same way as in section 5.2 it can be proved that for $|\rho| < \infty$,

$$\omega(\rho) = \frac{\rho}{1-\alpha(-\rho)\,\beta(\rho)}\left\{\frac{x(\rho)}{\alpha_2(-\rho)} + \alpha(-\rho)\,\frac{y(\rho)}{\beta_2(\rho)}\,e^{-\rho K}\right\}, \tag{5.47}$$

with the polynomials $x(\rho)$ and $y(\rho)$ determined by the above conditions, is a solution of (5.43), and has the properties (5.42). Moreover, it is easily verified from (5.12) that for $\omega(\rho)$ given by (5.47) we have

$$\omega(\rho) = \lim_{r\uparrow 1}\,(1-r)\,\omega(r, \rho, 0, 0, w) \qquad \text{for all } 0 \leq w \leq K.$$

Consequently, $\omega(\rho)$ as given by (5.47) represents the Laplace-Stieltjes transform of the stationary distribution of the process $\{\boldsymbol{w}_n, n=1, 2, \ldots\}$.

In exactly the same way it is shown that the Laplace-Stieltjes transform of the stationary distribution of the process $\{\boldsymbol{u}_n, n=1, 2, \ldots\}$ is given by

$$\Pi(\rho) = \frac{\rho}{1-\alpha(-\rho)\,\beta(\rho)}\left\{\beta(\rho)\,\frac{x(\rho)}{\alpha_2(-\rho)} + \frac{y(\rho)}{\beta_2(\rho)}\,e^{-\rho K}\right\}. \tag{5.48}$$

It should be noted that (cf. (5.10) and (5.19)),

$$x(\rho) = \lim_{r\uparrow 1}\,(1-r)\,x(r, \rho, 0, 0, w) = \lim_{r\uparrow 1}\,(1-r)\,k(r, \rho, 0, 0, u), \tag{5.49}$$

$$y(\rho) = \lim_{r\uparrow 1}\,(1-r)\,y(r, \rho, 0, 0, w) = \lim_{r\uparrow 1}\,(1-r)\,h(r, \rho, 0, 0, u).$$

Applying the inversion formula for the Laplace-Stieltjes transform we have for $\text{Re}\,\rho > 0$,

$$W(z) = \begin{cases} \dfrac{1}{2\pi i}\displaystyle\int_{C_\rho} \frac{e^{\rho z}}{\rho}\,\frac{\rho}{1-\alpha(-\rho)\,\beta(\rho)}\left\{\frac{x(\rho)}{\alpha_2(-\rho)} + \alpha(-\rho)\,\frac{y(\rho)}{\beta_2(\rho)}\,e^{-\rho K}\right\} d\rho, & z \geq 0, \\ 0, & z < 0, \end{cases} \tag{5.50}$$

$$U(z) = \begin{cases} \dfrac{1}{2\pi i} \displaystyle\int_{C_\rho} \dfrac{e^{\rho z}}{\rho} \dfrac{\rho}{1-\alpha(-\rho)\,\beta(\rho)} \left\{\beta(\rho)\dfrac{x(\rho)}{\alpha_2(-\rho)} + \dfrac{y(\rho)}{\beta_2(\rho)} e^{-\rho K}\right\} d\rho, & z \geqq 0, \\ 0, & z<0, \end{cases}$$

if z is a continuity point of $W(z)$ and $U(z)$, respectively. If z is not a continuity point then the left-hand sides should be replaced by $\frac{1}{2}\{W(z+)+W(z-)\}$ and $\frac{1}{2}\{U(z+)+U(z-)\}$, respectively.

The integrands in (5.50) have only one singularity at $\rho=0$ since $\omega(\rho)$ and $\Pi(\rho)$ are entire functions. It follows easily from (5.50) that for

$$\operatorname{Re}\eta > \max_{1\leqq j\leqq m_0} \operatorname{Re}\delta_j(1,0,0), \qquad \operatorname{Re}\zeta < \min_{1\leqq i\leqq n_0} \operatorname{Re}\varepsilon_i(1,0,0),$$

$$W(z) = \begin{cases} \dfrac{1}{2\pi i} \displaystyle\int_{C_\eta} \dfrac{e^{\eta z}}{1-\alpha(-\eta)\,\beta(\eta)} \dfrac{x(\eta)}{\alpha_2(-\eta)} d\eta, & 0\leqq z<K, \\ 1 + \dfrac{1}{2\pi i} \displaystyle\int_{C_\zeta} \dfrac{\alpha(-\zeta)\, e^{-\zeta(K-z)}}{1-\alpha(-\zeta)\,\beta(\zeta)} \dfrac{y(\zeta)}{\beta_2(\zeta)} d\zeta, & 0<z\leqq K, \end{cases} \tag{5.51}$$

$$U(z) = \begin{cases} \dfrac{1}{2\pi i} \displaystyle\int_{C_\eta} \dfrac{\beta(\eta)\, e^{\eta z}}{1-\alpha(-\eta)\,\beta(\eta)} \dfrac{x(\eta)}{\alpha_2(-\eta)} d\eta, & 0\leqq z<K, \\ 1 + \dfrac{1}{2\pi i} \displaystyle\int_{C_\zeta} \dfrac{e^{-\zeta(K-z)}}{1-\alpha(-\zeta)\,\beta(\zeta)} \dfrac{y(\zeta)}{\beta_2(\zeta)} d\zeta, & 0<z\leqq K. \end{cases}$$

It is easily verified that $W(z)$ has only one discontinuity point viz. $z=0$, and that $U(z)$ has also one discontinuity point, viz. $z=K$ (cf. also (5.37) and (5.38)). Clearly

$$W(0+) = \lim_{\rho\to\infty} \omega(\rho) = \lim_{\rho\to\infty} \frac{\rho x(\rho)}{\alpha_2(-\rho)}, \qquad \rho \text{ real}, \tag{5.52}$$

$$1 - U(K-) = \lim_{\rho\to-\infty} e^{\rho K}\,\Pi(\rho) = \lim_{\rho\to-\infty} \frac{\rho y(\rho)}{\beta_2(\rho)}, \qquad \rho \text{ real}.$$

III.5.4. Entrance times and return times

In this section we shall derive expressions for the distributions of entrance times and return times for the processes $\{\boldsymbol{w}_n, n=1,2,\dots\}$ and $\{\boldsymbol{u}_n, n=1,2,\dots\}$.

We shall say that at time n the process $\{(\boldsymbol{w}_n, \boldsymbol{u}_n), n=1, 2, \ldots\}$ is in state E_0 if $\boldsymbol{w}_n=0$, and in state E_K if $\boldsymbol{u}_n=K$. Obviously the successive occurrences of states E_0 constitute an imbedded renewal process for $\{\boldsymbol{w}_n, n=1, 2, \ldots\}$ and similarly, the successive occurrences of E_K are an imbedded renewal process for $\{\boldsymbol{u}_n, n=1, 2, \ldots\}$.

Define for $0 \leqq w < K$, $0 < u \leqq K$,

$$\boldsymbol{\alpha}_{w0} \stackrel{\text{def}}{=} \begin{cases} 1 & \text{if } \boldsymbol{w}_2=0, \ \boldsymbol{w}_1=w, \\ \min\limits_{n=2,3,\ldots} \{n: \boldsymbol{w}_{n+1}=0, \boldsymbol{w}_i>0, \ i=2, \ldots, n\} & \text{if } \boldsymbol{w}_1=w, \end{cases} \tag{5.53}$$

$$\boldsymbol{\alpha}_{wK} \stackrel{\text{def}}{=} \begin{cases} 1 & \text{if } \boldsymbol{u}_1=K, \ \boldsymbol{w}_1=w, \\ \min\limits_{n=2,3,\ldots} \{n: \boldsymbol{u}_n=K, \boldsymbol{u}_i<K, \ i=1, \ldots, n-1\} & \text{if } \boldsymbol{w}_1=w, \end{cases}$$

$$\boldsymbol{\beta}_{uK} \stackrel{\text{def}}{=} \begin{cases} 1 & \text{if } \boldsymbol{u}_2=K, \ \boldsymbol{u}_1=u, \\ \min\limits_{n=2,3,\ldots} \{n: \boldsymbol{u}_{n+1}=K, \boldsymbol{u}_i<K, \ i=2, \ldots, n\} & \text{if } \boldsymbol{u}_1=u, \end{cases}$$

$$\boldsymbol{\beta}_{u0} \stackrel{\text{def}}{=} \begin{cases} 0 & \text{if } \boldsymbol{w}_2=0, \ \boldsymbol{u}_1=u, \\ -1+\min\limits_{n=2,3,\ldots} \{n: \boldsymbol{w}_{n+1}=0, \boldsymbol{w}_i>0, \ i=2, \ldots, n\} & \text{if } \boldsymbol{u}_1=u, \end{cases}$$

$$\begin{aligned} \boldsymbol{\gamma}_{w0} &\stackrel{\text{def}}{=} \boldsymbol{a}_{\boldsymbol{\alpha}_{w0}}, & \boldsymbol{\gamma}_{wK} &\stackrel{\text{def}}{=} \boldsymbol{a}_{\boldsymbol{\alpha}_{wK}-1}, \\ \boldsymbol{\delta}_{uK} &\stackrel{\text{def}}{=} \boldsymbol{a}_{\boldsymbol{\beta}_{uK}}, & \boldsymbol{\delta}_{u0} &\stackrel{\text{def}}{=} \boldsymbol{a}_{\boldsymbol{\beta}_{u0}+1}, \\ \boldsymbol{p}_{w0} &\stackrel{\text{def}}{=} \boldsymbol{b}_{\boldsymbol{\alpha}_{w0}}, & \boldsymbol{p}_{wK} &\stackrel{\text{def}}{=} \boldsymbol{b}_{\boldsymbol{\alpha}_{wK}}, \\ \boldsymbol{q}_{uK} &\stackrel{\text{def}}{=} \boldsymbol{b}_{\boldsymbol{\beta}_{uK}+1} - \boldsymbol{b}_1, & \boldsymbol{q}_{u0} &\stackrel{\text{def}}{=} \boldsymbol{b}_{\boldsymbol{\beta}_{u0}+1} - \boldsymbol{b}_1. \end{aligned} \tag{5.54}$$

It is easily seen that $\boldsymbol{\alpha}_{00}$ is the return time in discrete time of E_0, while $\boldsymbol{\gamma}_{00}$ is the return time of E_0 in continuous time; $\boldsymbol{\alpha}_{w0}$ and $\boldsymbol{\gamma}_{w0}$ are the entrance times from $\boldsymbol{w}_1=w$ into E_0, in discrete and continuous time, respectively. Further $\boldsymbol{\alpha}_{wK}$ and $\boldsymbol{\gamma}_{wK}$ are entrance times from $\boldsymbol{w}_1=w$ into E_K, discrete and continuous, respectively. The variables $\boldsymbol{p}_{w0}$ and $\boldsymbol{p}_{wK}$ represent the amount of work brought into the system by the arriving customers during the entrance time from $\boldsymbol{w}_1=w$ into E_0 and into E_K, respectively.

The variables $\boldsymbol{\beta}_{KK}$ and $\boldsymbol{\delta}_{KK}$ are the discrete and continuous return times between successive overflows; $\boldsymbol{q}_{KK}$ is the amount of water which flowed into the system between two successive overflows of the system. Note that by definition $\boldsymbol{a}_0=0$, $\boldsymbol{b}_0=0$.

Further, we define for $z_1 \geqq 0$, $z_2 \geqq 0$, $0 \leqq w < K$, $0 < u \leqq K$,

$$p_{w0}^{(n)}(z_1, z_2) \stackrel{\text{def}}{=} \Pr\{\boldsymbol{w}_{n+1}=0, \boldsymbol{a}_n < z_1, \boldsymbol{b}_n < z_2 \mid \boldsymbol{w}_1 = w\}, \quad n=1, 2, \ldots, \tag{5.55}$$

$$p_{wK}^{(n)}(z_1, z_2) \stackrel{\text{def}}{=} \Pr\{\boldsymbol{u}_n = K, \boldsymbol{a}_{n-1} < z_1, \boldsymbol{b}_n < z_2 \mid \boldsymbol{w}_1 = w\}, \quad n=1, 2, \ldots,$$

$$q_{uK}^{(n)}(z_1, z_2) \stackrel{\text{def}}{=} \Pr\{\boldsymbol{u}_{n+1} = K, \boldsymbol{a}_n < z_1, \boldsymbol{b}_{n+1} - \boldsymbol{b}_1 < z_2 \mid \boldsymbol{u}_1 = u\}, \quad n=1, 2, \ldots,$$

$$q_{u0}^{(n)}(z_1, z_2) \stackrel{\text{def}}{=} \Pr\{\boldsymbol{w}_{n+2} = 0, \boldsymbol{a}_{n+1} < z_1, \boldsymbol{b}_{n+1} - \boldsymbol{b}_1 < z_2 \mid \boldsymbol{u}_1 = u\}, \quad n=0, 1, \ldots .$$

To obtain expressions for the probabilities defined in (5.55) we introduce the functions

$$rc_x(r, \rho_3, \rho_4, w) \stackrel{\text{def}}{=} \lim_{\rho \to \infty} \frac{\rho x(r, \rho, \rho_3, \rho_4, w)}{\alpha_2(\rho_3 - \rho)}, \quad \rho \text{ real}, \tag{5.56}$$

$$r^2 c_y(r, \rho_3, \rho_4, w) \stackrel{\text{def}}{=} \lim_{\rho \to -\infty} \frac{\rho y(r, \rho, \rho_3, \rho_4, w)}{\beta_2(\rho_4 + \rho)}, \quad \rho \text{ real},$$

$$r^2 c_k(r, \rho_3, \rho_4, u) \stackrel{\text{def}}{=} \lim_{\rho \to \infty} \frac{\rho k(r, \rho, \rho_3, \rho_4, u)}{\alpha_2(\rho_3 - \rho)}, \quad \rho \text{ real},$$

$$rc_h(r, \rho_3, \rho_4, u) \stackrel{\text{def}}{=} \lim_{\rho \to -\infty} \frac{\rho h(r, \rho, \rho_3, \rho_4, u)}{\beta_2(\rho_4 + \rho)}, \quad \rho \text{ real},$$

for $|r| < 1$, $\operatorname{Re} \rho_3 \geqq 0$, $\operatorname{Re} \rho_4 \geqq 0$, $0 \leqq w < K$, $0 < u \leqq K$.

It follows from (5.12), (5.18), (5.24) and (5.25) that for $|r| < 1$, $\operatorname{Re} \rho_3 \geqq 0$, $\operatorname{Re} \rho_4 \geqq 0$, $w < K$, $0 < u$,

$$\sum_{n=1}^{\infty} r^n \operatorname{E}\{(\boldsymbol{w}_n = 0) \exp\{-\rho_3 \boldsymbol{a}_{n-1} - \rho_4 \boldsymbol{b}_{n-1}\} \mid \boldsymbol{w}_1 = w\} = \begin{cases} rc_x(r, \rho_3, \rho_4, 0) + r & \text{if } w = 0, \\ rc_x(r, \rho_3, \rho_4, w) & \text{if } w > 0, \end{cases} \tag{5.57}$$

$$\sum_{n=1}^{\infty} r^n \operatorname{E}\{(\boldsymbol{u}_n = K) \exp\{-\rho_3 \boldsymbol{a}_{n-1} - \rho_4 \boldsymbol{b}_n\} \mid \boldsymbol{w}_1 = w\} = rc_y(r, \rho_3, \rho_4, w), \quad w \geqq 0,$$

$$\sum_{n=1}^{\infty} r^n \operatorname{E}\{(\boldsymbol{u}_n = K) \exp\{-\rho_3 \boldsymbol{a}_{n-1} - \rho_4 (\boldsymbol{b}_n - \boldsymbol{b}_1)\} \mid \boldsymbol{u}_1 = u\} = \begin{cases} rc_h(r, \rho_3, \rho_4, K) + r, & u = K, \\ rc_h(r, \rho_3, \rho_4, u), & u < K, \end{cases}$$

$$\sum_{n=2}^{\infty} r^{n-1} \mathrm{E}\{(\boldsymbol{w}_n=0)\exp\{-\rho_3\boldsymbol{a}_{n-1}-\rho_4(\boldsymbol{b}_{n-1}-\boldsymbol{b}_1)\} \mid \boldsymbol{u}_1=u\} = rc_k(r, \rho_3, \rho_4, u), \quad u \leqq K,$$

here $(\boldsymbol{w}_n=0)$ and $(\boldsymbol{u}_n=K)$ are the indicator functions of the events $\{\boldsymbol{w}_n=0\}$ and $\{\boldsymbol{u}_n=K\}$, respectively.

Introduce the Laplace-Stieltjes transforms of the functions defined in (5.55) with respect to z_1 and z_2. These transforms are indicated by the sign "$\sim$", so that for instance for $\mathrm{Re}\,\rho_3 \geqq 0$, $\mathrm{Re}\,\rho_4 \geqq 0$,

$$\tilde{p}_{w0}^{(n)}(\rho_3, \rho_4) \stackrel{\text{def}}{=} \int_{0-}^{\infty}\int_{0-}^{\infty} \exp\{-\rho_3 z_1-\rho_4 z_2\}\,\mathrm{d}_{z_1}\,\mathrm{d}_{z_2}\,p_{w0}^{(n)}(z_1, z_2).$$

It follows from (5.55) and (5.57) for $|r|<1$, $\mathrm{Re}\,\rho_3 \geqq 0$, $\mathrm{Re}\,\rho_4 \geqq 0$, $0 \leqq w < K$, $0 < u \leqq K$ that

$$\sum_{n=1}^{\infty} r^n \tilde{p}_{w0}^{(n)}(\rho_3, \rho_4) = c_x(r, \rho_3, \rho_4, w), \tag{5.58}$$

$$\sum_{n=1}^{\infty} r^n \tilde{p}_{wK}^{(n)}(\rho_3, \rho_4) = rc_y(r, \rho_3, \rho_4, w),$$

$$\sum_{n=1}^{\infty} r^n \tilde{q}_{uK}^{(n)}(\rho_3, \rho_4) = c_h(r, \rho_3, \rho_4, u),$$

$$\sum_{n=0}^{\infty} r^{n+1} \tilde{q}_{u0}^{(n)}(\rho_3, \rho_4) = rc_k(r, \rho_3, \rho_4, u).$$

From (5.54) and (5.58) we can immediately write down the expressions for the generating functions and Laplace-Stieltjes transforms of the distributions of entrance times and return times by applying well-known relations of renewal theory. It follows for $|r| \leqq 1$, $\mathrm{Re}\,\rho_3 \geqq 0$, $\mathrm{Re}\,\rho_4 \geqq 0$, $0 \leqq w < K$, $0 < u \leqq K$ that

$$\mathrm{E}\{r^{\boldsymbol{\alpha}_{w0}} \exp(-\rho_3 \boldsymbol{\gamma}_{w0}-\rho_4 \boldsymbol{p}_{w0})\} = \frac{c_x(r, \rho_3, \rho_4, w)}{1+c_x(r, \rho_3, \rho_4, 0)}, \tag{5.59}$$

$$\mathrm{E}\{r^{\boldsymbol{\alpha}_{wK}} \exp(-\rho_3 \boldsymbol{\gamma}_{wK}-\rho_4 \boldsymbol{p}_{wK})\} = \frac{rc_y(r, \rho_3, \rho_4, w)}{1+c_h(r, \rho_3, \rho_4, K)},$$

$$\mathrm{E}\{r^{\boldsymbol{\beta}_{uK}} \exp(-\rho_3 \boldsymbol{\delta}_{uK}-\rho_4 \boldsymbol{q}_{uK})\} = \frac{c_h(r, \rho_3, \rho_4, u)}{1+c_h(r, \rho_3, \rho_4, K)},$$

$$\mathrm{E}\{r^{\boldsymbol{\beta}_{u0}} \exp(-\rho_3 \boldsymbol{\delta}_{u0}-\rho_4 \boldsymbol{q}_{u0})\} = \frac{c_k(r, \rho_3, \rho_4, u)}{1+c_x(r, \rho_3, \rho_4, 0)}.$$

Define for $z_1 \geqq 0$, $z_2 \geqq 0$, $0 \leqq w < K$, $0 < u \leqq K$,

$$f_{w0}^{(n)}(z_1, z_2) \stackrel{\text{def}}{=} \Pr\{\boldsymbol{\alpha}_{w0} = n,\ \boldsymbol{\gamma}_{w0} < z_1,\ \boldsymbol{p}_{w0} < z_2\}, \qquad n = 1, 2, \ldots, \tag{5.60}$$

$$f_{wK}^{(n)}(z_1, z_2) \stackrel{\text{def}}{=} \Pr\{\boldsymbol{\alpha}_{wK} = n,\ \boldsymbol{\gamma}_{wK} < z_1,\ \boldsymbol{p}_{wK} < z_2\}, \qquad n = 1, 2, \ldots,$$

$$h_{uK}^{(n)}(z_1, z_2) \stackrel{\text{def}}{=} \Pr\{\boldsymbol{\beta}_{uK} = n,\ \boldsymbol{\delta}_{uK} < z_1,\ \boldsymbol{q}_{uK} < z_2\}, \qquad n = 1, 2, \ldots,$$

$$h_{u0}^{(n)}(z_1, z_2) \stackrel{\text{def}}{=} \Pr\{\boldsymbol{\beta}_{u0} = n,\ \boldsymbol{\delta}_{u0} < z_1,\ \boldsymbol{q}_{u0} < z_2\}, \qquad n = 0, 1, \ldots .$$

$$p_{K;w0}^{(n)}(z_1, z_2) \stackrel{\text{def}}{=} \Pr\{\boldsymbol{w}_{n+1} = 0,\ \boldsymbol{u}_j < K,\ j = 1, \ldots, n;\ \boldsymbol{a}_n < z_1, \boldsymbol{b}_n < z_2 \mid \boldsymbol{w}_1 = w\},$$
$$n = 1, 2, \ldots; \tag{5.61}$$

$$p_{0;wK}^{(n)}(z_1, z_2) \stackrel{\text{def}}{=} \begin{cases} p_{wK}^{(1)}(z_1, z_2), & n = 1, \\ \Pr\{\boldsymbol{u}_n = K,\ \boldsymbol{w}_j > 0,\ j = 2, \ldots, n;\ \boldsymbol{a}_{n-1} < z_1, \boldsymbol{b}_n < z_2 \mid \boldsymbol{w}_1 = w\}, & \\ & n = 2, 3, \ldots; \end{cases}$$

$$q_{0;uK}^{(n)}(z_1, z_2) \stackrel{\text{def}}{=} \Pr\{\boldsymbol{u}_{n+1} = K,\ \boldsymbol{w}_j > 0,\ j = 2, \ldots, n+1;$$
$$\boldsymbol{a}_n < z_1, \boldsymbol{b}_{n+1} - \boldsymbol{b}_1 < z_2 \mid \boldsymbol{u}_1 = u\}, \qquad n = 1, 2, \ldots;$$

$$q_{K;u0}^{(n)}(z_1, z_2) \stackrel{\text{def}}{=} \begin{cases} q_{u0}^{(0)}(z_1, z_2), & n = 0, \\ \Pr\{\boldsymbol{w}_{n+2} = 0,\ \boldsymbol{u}_j < K,\ j = 2, \ldots, n+2; & \\ \boldsymbol{a}_{n+1} < z_1, \boldsymbol{b}_{n+1} - \boldsymbol{b}_1 < z_2 \mid \boldsymbol{u}_1 = u\}, & n = 1, 2, \ldots; \end{cases}$$

$$f_{K;w0}^{(n)}(z_1, z_2) \stackrel{\text{def}}{=} \Pr\{\boldsymbol{\alpha}_{w0} = n,\ \boldsymbol{\gamma}_{w0} < z_1, \boldsymbol{p}_{w0} < z_2,\ \boldsymbol{u}_j < K,\ j = 1, \ldots, n\},$$
$$n = 1, 2, \ldots; \tag{5.62}$$

$$f_{0;wK}^{(n)}(z_1, z_2) \stackrel{\text{def}}{=} \begin{cases} f_{wK}^{(1)}(z_1, z_2), & n = 1, \\ \Pr\{\boldsymbol{\alpha}_{wK} = n,\ \boldsymbol{\gamma}_{wK} < z_1, \boldsymbol{p}_{wK} < z_2,\ \boldsymbol{w}_j > 0,\ j = 2, \ldots, n\}, & \\ & n = 2, 3, \ldots; \end{cases}$$

$$h_{0;uK}^{(n)}(z_1, z_2) \stackrel{\text{def}}{=} \Pr\{\boldsymbol{\beta}_{uK} = n,\ \boldsymbol{\delta}_{uK} < z_1, \boldsymbol{q}_{uK} < z_2,\ \boldsymbol{w}_j > 0,\ j = 2, \ldots, n+1\},$$
$$n = 1, 2, \ldots;$$

$$h_{K;u0}^{(n)}(z_1, z_2) \stackrel{\text{def}}{=} \begin{cases} h_{u0}^{(0)}(z_1, z_2), & n = 0, \\ \Pr\{\boldsymbol{\beta}_{u0} = n,\ \boldsymbol{\delta}_{u0} < z_1, \boldsymbol{q}_{u0} < z_2,\ \boldsymbol{u}_j < K,\ j = 2, \ldots, n+1\}, & \\ & n = 1, 2, \ldots . \end{cases}$$

Since E_0 and E_K are regenerative events it is easily seen that the definitions given above imply that for Re $\rho_3 \geqq 0$, Re $\rho_4 \geqq 0$, $0 \leqq w < K$, $0 < u \leqq K$,

$$\tilde{p}_{w0}^{(n)}(\rho_3, \rho_4) = \tilde{p}_{K;w0}^{(n)}(\rho_3, \rho_4) + \sum_{m=1}^{n} \tilde{f}_{wK}^{(m)}(\rho_3, \rho_4)\, \tilde{q}_{K0}^{(n-m)}(\rho_3, \rho_4), \quad n=1, 2, \ldots; \tag{5.63}$$

$$\tilde{p}_{wK}^{(n)}(\rho_3, \rho_4) = \tilde{p}_{0;wK}^{(n)}(\rho_3, \rho_4) + \sum_{m=1}^{n-1} \tilde{f}_{w0}^{(m)}(\rho_3, \rho_4)\, \tilde{p}_{0K}^{(n-m)}(\rho_3, \rho_4), \quad n=2, 3, \ldots;$$

$$\tilde{q}_{uK}^{(n)}(\rho_3, \rho_4) = \tilde{q}_{0;uK}^{(n)}(\rho_3, \rho_4) + \sum_{m=0}^{n-1} \tilde{h}_{u0}^{(m)}(\rho_3, \rho_4)\, \tilde{p}_{0K}^{(n-m)}(\rho_3, \rho_4), \quad n=1, 2, \ldots;$$

$$\tilde{q}_{u0}^{(n)}(\rho_3, \rho_4) = \tilde{q}_{K;u0}^{(n)}(\rho_3, \rho_4) + \sum_{m=1}^{n} \tilde{h}_{uK}^{(m)}(\rho_3, \rho_4)\, \tilde{q}_{K0}^{(n-m)}(\rho_3, \rho_4), \quad n=1, 2, \ldots;$$

$$\tilde{p}_{K;w0}^{(n)}(\rho_3, \rho_4) = \tilde{f}_{K;w0}^{(n)}(\rho_3, \rho_4) + \sum_{m=1}^{n-1} \tilde{f}_{K;w0}^{(m)}(\rho_3, \rho_4)\, \tilde{p}_{K;00}^{(n-m)}(\rho_3, \rho_4), \quad n=2, 3, \ldots; \tag{5.64}$$

$$\tilde{p}_{0;wK}^{(n)}(\rho_3, \rho_4) = \tilde{f}_{0;wK}^{(n)}(\rho_3, \rho_4) + \sum_{m=1}^{n-1} \tilde{f}_{0;wK}^{(m)}(\rho_3, \rho_4)\, \tilde{q}_{0;KK}^{(n-m)}(\rho_3, \rho_4), \quad n=2, 3, \ldots;$$

$$\tilde{q}_{0;uK}^{(n)}(\rho_3, \rho_4) = \tilde{h}_{0;uK}^{(n)}(\rho_3, \rho_4) + \sum_{m=1}^{n-1} \tilde{h}_{0;uK}^{(m)}(\rho_3, \rho_4)\, \tilde{q}_{0;KK}^{(n-m)}(\rho_3, \rho_4), \quad n=2, 3, \ldots;$$

$$\tilde{q}_{K;u0}^{(n)}(\rho_3, \rho_4) = \tilde{h}_{K;u0}^{(n)}(\rho_3, \rho_4) + \sum_{m=0}^{n-1} \tilde{h}_{K;u0}^{(m)}(\rho_3, \rho_4)\, \tilde{p}_{K;00}^{(n-m)}(\rho_3, \rho_4), \quad n=1, 2, \ldots .$$

The quantities defined above can easily be expressed in the functions $c_x(\ldots)$, $c_y(\ldots)$, $c_k(\ldots)$ and $c_h(\ldots)$ of (5.57). Several are not only of interest for the finite dam or queueing system with uniformly bounded virtual waiting time, but also for the infinite dam and the general queueing system. We shall give some examples. For the $G/G/1$ queue: $\sum_{n=1}^{\infty} f_{K;00}^{(n)}(z_1, z_2)$ is the probability that a busy cycle has a duration less than z_1, that the amount of work brought into the system during the busy cycle is less than z_2 and that the supremum of the virtual waiting time during this busy cycle is less than K. For the infinite dam: $\sum_{n=1}^{\infty} f_{K;00}^{(n)}(\infty, z_2)$ is the probability that the duration of the wet period is less than z_2 and that during this wet period the content never exceeds K. For the $G/G/1$ queue: $\sum_{n=1}^{\infty} h^{(n)}_{0;KK}(z_1, z_2)$ is the probability that if at t_1 the virtual waiting time is equal to K, the time at which for the first time thereafter the virtual waiting time is not less than K is less than z_1, that the total amount of work brought into the system during this time is less than z_2 and that during this time the virtual waiting time has never been zero.

Finally, we derive a relation for the distribution of the number of times that an overflow occurs during a busy period or wet period of the dam. Denote this number by $\boldsymbol{\Phi}_K$, it then follows with

$$f^{(n)}_{K;00} \stackrel{\text{def}}{=} f^{(n)}_{K;00}(\infty, \infty), \qquad f^{(n)}_{0;0K} \stackrel{\text{def}}{=} f^{(n)}_{0;0K}(\infty, \infty),$$

$$h^{(n)}_{0;KK} \stackrel{\text{def}}{=} h^{(n)}_{0;KK}(\infty, \infty), \qquad h^{(n)}_{K;K0} \stackrel{\text{def}}{=} h^{(n)}_{K;K0}(\infty, \infty),$$

that

$$\Pr\{\boldsymbol{\Phi}_K=0\} = \sum_{n=1}^{\infty} f^{(n)}_{K;00}, \tag{5.65}$$

$$\Pr\{\boldsymbol{\Phi}_K=m\} = \{1 - \sum_{n=1}^{\infty} f^{(n)}_{K;00}\}\{\sum_{n=1}^{\infty} h^{(n)}_{0;KK}\}^{m-1} \sum_{n=0}^{\infty} h^{(n)}_{K;K0}, \qquad m=1, 2, \ldots .$$

Note that

$$1 - \sum_{n=1}^{\infty} h^{(n)}_{0;KK} = \sum_{n=0}^{\infty} h^{(n)}_{K;K0}.$$

It follows that

$$\mathrm{E}\{\boldsymbol{\Phi}_K\} = \frac{1 - \sum_{n=1}^{\infty} f^{(n)}_{K;00}}{1 - \sum_{n=1}^{\infty} h^{(n)}_{0;KK}}. \tag{5.66}$$

III.5.5. The finite dam model $M/G/1$

In this section we shall apply the results obtained in the previous sections to the queueing model $M/K_n/1$, i.e. we have

$$\alpha(\rho) = \frac{1}{1+\alpha\rho}, \qquad \beta(\rho) = \frac{\beta_1(\rho)}{\beta_2(\rho)}, \tag{5.67}$$

with $\beta_1(\rho)$ and $\beta_2(\rho)$ polynomials in ρ, $\beta_2(\rho)$ is of degree n and $\beta_1(\rho)$ of degree $n-1$ at most. At the end of this section it will be shown that the assumption of rationality of $\beta(\rho)$ is superfluous, so that all results hold for general $B(.)$. Since in the present case $m=1$ it follows from the definitions of $x(r, \rho, \rho_3, \rho_4, w)$ and $k(r, \rho, \rho_3, \rho_4, u)$ (cf. (5.12) and (5.18)) that these functions are independent of ρ.

Define for $|r|<1$, Re $\rho_3 \geqq 0$, Re $\rho_4 \geqq 0$, $0 \leqq w < K$, $0 < u \leqq K$,

$$x_0(r, \rho_3, \rho_4, w) \overset{\text{def}}{=} x(r, \rho, \rho_3, \rho_4, w), \qquad (5.68)$$

$$k_0(r, \rho_3, \rho_4, u) \overset{\text{def}}{=} k(r, \rho, \rho_3, \rho_4, u).$$

From (5.10) it follows for $|r|<1$, Re $\rho_3 \geqq 0$, Re $\rho_4 \geqq 0$, $0 \leqq w < K$ with

$$\delta_1 \equiv \delta_1(r, \rho_3, \rho_4), \qquad (5.69)$$

that

$$x_0(r, \rho_3, \rho_4, w) = -\frac{y(r, \delta_1, \rho_3, \rho_4, w)}{\beta_2(\rho_4+\delta_1)} \mathrm{e}^{-\delta_1 K} - \frac{r}{\delta_1}\{1+(\rho_3-\delta_1)\,\alpha\}\,\mathrm{e}^{-\delta_1 w}. \qquad (5.70)$$

From (5.11) we have for $|r|<1$, Re $\rho_3 \geqq 0$, Re $\rho_4 \geqq 0$, $\nu_2 < \operatorname{Re} \eta < 0$, $\operatorname{Re} \eta < \operatorname{Re} \rho$, $0 \leqq w < K$, that

$$\frac{y(r, \rho, \rho_3, \rho_4, w)}{r\beta_2(\rho_4+\rho)} = \frac{1+(\rho_3-\rho)\,\alpha - r\beta(\rho_4+\rho)}{2\pi\mathrm{i}(\rho-\delta_1)} \int_{C_\eta} \left\{ \frac{x_0(r, \rho_3, \rho_4, w)}{1+(\rho_3-\eta)\,\alpha}\,\mathrm{e}^{\eta K} + \frac{r}{\eta}\mathrm{e}^{(K-w)\eta} \right\} \frac{(\eta-\delta_1)\,\beta(\rho_4+\eta)}{1+(\rho_3-\eta)\,\alpha - r\beta(\rho_4+\eta)} \frac{\mathrm{d}\eta}{\eta-\rho}$$

$$= \frac{1+(\rho_3-\rho)\,\alpha - r\beta(\rho_4+\rho)}{2\pi\mathrm{i}(\rho-\delta_1)} \int_{C_\eta} \left\{ x_0(r, \rho_3, \rho_4, w)\,\frac{\mathrm{e}^{\eta K}}{r} + \frac{r}{\eta}\beta(\rho_4+\eta)\,\mathrm{e}^{(K-w)\eta} \right\} \frac{\eta-\delta_1}{1+(\rho_3-\eta)\,\alpha - r\beta(\rho_4+\eta)} \frac{\mathrm{d}\eta}{\eta-\rho}, \qquad (5.71)$$

since

$$\frac{r\beta(\rho_4+\eta)}{\{1+(\rho_3-\eta)\,\alpha\}\{1+(\rho_3-\eta)\,\alpha - r\beta(\rho_4+\eta)\}} = \frac{-1}{1+(\rho_3-\eta)\,\alpha} + \frac{1}{1+(\rho_3-\eta)\,\alpha - r\beta(\rho_4+\eta)}.$$

Inserting $\rho=\delta_1$ in (5.71) and eliminating $y(r, \delta_1, \rho_3, \rho_4, w)$ from (5.70) and (5.71) we obtain for $|r|<1$, Re $\rho_3 \geqq 0$, Re $\rho_4 \geqq 0$, $\nu_2 < \operatorname{Re} \eta < 0$, $0 \leqq w < K$,

$$\left[-\frac{1}{r}\,x_0(r,\rho_3,\rho_4,w)\,\mathrm{e}^{\delta_1 K}-\{1+(\rho_3-\delta_1)\,\alpha\}\,\frac{\mathrm{e}^{(K-w)\delta_1}}{\delta_1}\right]$$

$$\cdot\lim_{\rho\to\delta_1}\frac{\rho-\delta_1}{1+(\rho_3-\rho)\,\alpha-r\beta(\rho_4+\rho)}$$

$$=\frac{1}{2\pi\mathrm{i}}\int_{\mathrm{C}_\eta}\left\{\frac{\mathrm{e}^{\eta K}}{r}\,x_0(r,\rho_3,\rho_4,w)+\frac{r}{\eta}\,\beta(\rho_4+\eta)\,\mathrm{e}^{(K-w)\eta}\right\}$$

$$\cdot\frac{\mathrm{d}\eta}{1+(\rho_3-\eta)\,\alpha-r\beta(\rho_4+\eta)}. \tag{5.72}$$

In the last integral we replace by contour integration the path C_η with $\nu_2<\mathrm{Re}\,\eta<0$ by a path just right of $\eta=\delta_1$. Hence we have to account for the residues at $\eta=0$ and at $\eta=\delta_1$.
It follows easily that

$$\frac{x_0(r,\rho_3,\rho_4,w)}{2\pi\mathrm{i}}\int_{\mathrm{C}_\eta}\frac{\mathrm{e}^{\eta K}}{1+(\rho_3-\eta)\,\alpha-r\beta(\rho_4+\eta)}\,\mathrm{d}\eta$$

$$=\frac{r^2\beta(\rho_4)}{1+\alpha\rho_3-r\beta(\rho_4)}-\frac{r^2}{2\pi\mathrm{i}}\int_{\mathrm{C}_\eta}\frac{\mathrm{e}^{\eta(K-w)}}{\eta}\,\frac{\beta(\rho_4+\eta)}{1+(\rho_3-\eta)\,\alpha-r\beta(\rho_4+\eta)}\,\mathrm{d}\eta, \tag{5.73}$$

for $|r|<1$, $\mathrm{Re}\,\rho_3\geqq 0$, $\mathrm{Re}\,\rho_4\geqq 0$, $\mathrm{Re}\,\eta>\mathrm{Re}\,\delta_1$, $0\leqq w<K$.
From (5.19) and (5.21) we obtain

$$k_0(r,\rho_3,\rho_4,u)\,\frac{\beta(\rho_4+\delta_1)}{1+(\rho_3-\delta_1)\,\alpha}+\frac{h(r,\delta_1,\rho_3,\rho_4,u)}{\beta_2(\rho_4+\delta_1)}\,\mathrm{e}^{-\delta_1 K}+\frac{r}{\delta_1}\,\mathrm{e}^{-\delta_1 u}=0,$$

$$\frac{h(r,\rho,\rho_3,\rho_4,u)}{\beta_2(\rho_4+\rho)}=\frac{1+(\rho_3-\rho)\,\alpha-r\beta(\rho_4+\rho)}{2\pi\mathrm{i}(\rho-\delta_1)} \tag{5.74}$$

$$\cdot\int_{\mathrm{C}_\eta}\left\{\beta(\rho_4+\eta)\,\frac{k_0(r,\rho_3,\rho_4,u)}{1+(\rho_3-\eta)\,\alpha}\,\mathrm{e}^{\eta K}+\frac{r}{\eta}\,\mathrm{e}^{(K-u)\eta}\right\}$$

$$\cdot\frac{\eta-\delta_1}{1+(\rho_3-\eta)\,\alpha-r\beta(\rho_4+\eta)}\,\frac{\mathrm{d}\eta}{\eta-\rho},$$

for $|r|<1$, $\mathrm{Re}\,\rho_3\geqq 0$, $\mathrm{Re}\,\rho_4\geqq 0$, $\nu_2<\mathrm{Re}\,\eta<0$, $\mathrm{Re}\,\eta<\mathrm{Re}\,\rho$, $0<u\leqq K$.

An analysis completely similar to that above yields

$$\frac{k_0(r, \rho_3, \rho_4, u)}{2\pi i} \int_{C_\eta} \frac{e^{\eta K}}{1+(\rho_3-\eta)\alpha - r\beta(\rho_4+\eta)} d\eta$$

$$= \frac{r^2}{1+\alpha\rho_3 - r\beta(\rho_4)} - \frac{r^2}{2\pi i} \int_{C_\eta} \frac{e^{\eta(K-u)}}{\eta} \cdot \frac{d\eta}{1+(\rho_3-\eta)\alpha - r\beta(\rho_4+\eta)}, \tag{5.75}$$

for $|r|<1$, Re $\rho_3 \geqq 0$, Re $\rho_4 \geqq 0$, Re $\delta_1 <$ Re η, $0<u\leqq K$.

The relations (5.71), (5.73), (5.74) and (5.75) determine all the functions $x(\ldots)$, $y(\ldots)$, $k(\ldots)$ and $h(\ldots)$ which are needed for the calculation of the functions $\omega(\ldots)$, $\omega_1(\ldots)$, $\Pi(\ldots)$ and $\Pi_1(\ldots)$.

To determine the function $c_y(r, \rho_3, \rho_4, w)$ (cf. (5.56)) we multiply the expression (5.71) by $r\rho$ and move the path of integration C_η to the path A_ζ as defined in section 4.5. We then obtain

$$r^2 c_y(r, \rho_3, \rho_4, w) = \lim_{\rho\to-\infty} r\rho \frac{1+(\rho_3-\rho)\alpha - r\beta(\rho_4+\rho)}{\rho-\delta_1}$$

$$\cdot \Bigg[\left\{ -\frac{x_0(r, \rho_3, \rho_4, w)}{1+(\rho_3-\rho)\alpha} e^{\rho K} - \frac{r}{\rho} e^{(K-w)\rho} \right\}$$

$$\cdot \frac{(\rho-\delta_1)\beta(\rho_4+\rho)}{1+(\rho_3-\rho)\alpha - r\beta(\rho_4+\rho)} - \frac{r\delta_1}{\rho} \frac{\beta(\rho_4)}{1+\alpha\rho_3 - r\beta(\rho_4)}$$

$$+ \frac{1}{2\pi i} \int_{A_\zeta} \left\{ \frac{x_0(r, \rho_3, \rho_4, w)}{1+(\rho_3-\zeta)\alpha} e^{\zeta K} + \frac{r}{\zeta} e^{(K-w)\zeta} \right\}$$

$$\cdot \frac{(\zeta-\delta_1)\beta(\rho_4+\zeta)}{1+(\rho_3-\zeta)\alpha - r\beta(\rho_4+\zeta)} \frac{d\zeta}{\zeta-\rho} \Bigg].$$

Using (5.73) for $w<K$ this relation reduces to

$$r^2 c_y(r, \rho_3, \rho_4, w) = \lim_{\rho\to-\infty} \frac{1}{2\pi i} \int_{A_\zeta} \left\{ \frac{x_0(r, \rho_3, \rho_4, w)}{1+(\rho_3-\zeta)\alpha} e^{\zeta K} + \frac{r}{\zeta} e^{(K-w)\zeta} \right\}$$

$$\cdot \frac{\alpha\zeta r\beta(\rho_4+\zeta)}{1+(\rho_3-\zeta)\alpha - r\beta(\rho_4+\zeta)} \frac{\rho}{\rho-\zeta} d\zeta.$$

The limit in the expression above can easily be calculated since it is readily verified that limit operation and integration may be interchanged if $w<K$.

Calculating the limit, using (5.73) and from

$$\frac{\alpha\zeta r\beta(\rho_4+\zeta)}{\{1+(\rho_3-\zeta)\,\alpha\}\{1+(\rho_3-\zeta)\,\alpha-r\beta(\rho_4+\zeta)\}}$$

$$= -\frac{1+\alpha\rho_3}{1+(\rho_3-\zeta)\,\alpha}+\frac{1+\alpha\rho_3-r\beta(\rho_4+\zeta)}{1+(\rho_3-\zeta)\,\alpha-r\beta(\rho_4+\zeta)},$$

it is found that for $|r|<1$, $\operatorname{Re}\rho_3\geqq 0$, $\operatorname{Re}\rho_4\geqq 0$, $\operatorname{Re}\delta_1<\operatorname{Re}\eta$, $0\leqq w<K$,

$$r^2 c_y(r,\rho_3,\rho_4,w) = \frac{x_0(r,\rho_3,\rho_4,w)}{2\pi\mathrm{i}}\int_{C_\eta}\left\{\frac{\alpha\eta}{1+(\rho_3-\eta)\,\alpha-r\beta(\rho_4+\eta)}+1\right\}\mathrm{e}^{\eta K}\,\mathrm{d}\eta$$
$$+\frac{r^2}{2\pi\mathrm{i}}\int_{C_\eta}\mathrm{e}^{(K-w)\eta}\,\frac{\alpha\beta(\rho_4+\eta)}{1+(\rho_3-\eta)\,\alpha-r\beta(\rho_4+\eta)}\,\mathrm{d}\eta. \qquad (5.76)$$

From (5.56) we obtain for $|r|<1$, $\operatorname{Re}\rho_3\geqq 0$, $\operatorname{Re}\rho_4\geqq 0$,

$$r c_x(r,\rho_3,\rho_4,w) = -\frac{1}{\alpha}\,x_0(r,\rho_3,\rho_4,w), \qquad 0\leqq w<K, \qquad (5.77)$$

$$r^2 c_k(r,\rho_3,\rho_4,u) = -\frac{1}{\alpha}\,k_0(r,\rho_3,\rho_4,u), \qquad 0<u\leqq K,$$

and from (5.56) and (5.74) in the same way that (5.76) has been derived we find that

$$r^2 c_h(r,\rho_3,\rho_4,u)$$
$$=\begin{cases} \dfrac{k_0(r,\rho_3,\rho_4,u)}{2\pi\mathrm{i}}\displaystyle\int_{C_\eta}\left\{\frac{\alpha\eta}{1+(\rho_3-\eta)\,\alpha-r\beta(\rho_4+\eta)}+1\right\}\mathrm{e}^{\eta K}\,\mathrm{d}\eta \\ \quad +\dfrac{r^2}{2\pi\mathrm{i}}\displaystyle\int_{C_\eta}\mathrm{e}^{(K-u)\eta}\,\frac{\alpha\,\mathrm{d}\eta}{1+(\rho_3-\eta)\,\alpha-r\beta(\rho_4+\eta)}, \qquad 0<u<K, \\ -r^2+\dfrac{k_0(r,\rho_3,\rho_4,K)}{2\pi\mathrm{i}}\displaystyle\int_{C_\eta}\left\{\frac{\alpha\eta}{1+(\rho_3-\eta)\,\alpha-r\beta(\rho_4+\eta)}+1\right\}\mathrm{e}^{\eta K}\,\mathrm{d}\eta, \\ \qquad\qquad u=K, \end{cases} \qquad (5.78)$$

for $|r|<1$, $\operatorname{Re}\rho_3\geqq 0$, $\operatorname{Re}\rho_4\geqq 0$, $\operatorname{Re}\eta>\operatorname{Re}\delta_1$.

We now derive expressions for the limit distributions $W(x)$ and $U(x)$, cf. (5.40) and (5.51). From (5.49) and (5.73) we obtain

$$\frac{x(\rho)}{\alpha_2(\rho)}=\frac{1}{1-\alpha\rho}\left[\frac{1}{2\pi i}\int_{C_\eta}\frac{e^{\eta K}}{1-\alpha\eta-\beta(\eta)}\,d\eta\right]^{-1},\qquad \rho\neq\frac{1}{\alpha},$$
$$\operatorname{Re}\eta>\delta_1(1,0,0). \tag{5.79}$$

Hence from (5.51) and (5.52) for $\operatorname{Re}\eta>\delta_1(1,0,0)$,

$$W(z)=\begin{cases}0, & z\leqq 0,\\[2mm] \dfrac{\dfrac{1}{2\pi i}\displaystyle\int_{C_\eta}\frac{e^{\eta z}}{\beta(\eta)+\alpha\eta-1}\,d\eta}{\dfrac{1}{2\pi i}\displaystyle\int_{C_\eta}\frac{e^{\eta K}}{\beta(\eta)+\alpha\eta-1}\,d\eta}, & 0<z\leqq K,\\[2mm] 1, & z\geqq K,\end{cases} \tag{5.80}$$

$$U(z)=\begin{cases}0, & z<0,\\[2mm] \dfrac{\dfrac{1}{2\pi i}\displaystyle\int_{C_\eta}\frac{\beta(\eta)\,e^{\eta z}}{\beta(\eta)+\alpha\eta-1}\,d\eta}{\dfrac{1}{2\pi i}\displaystyle\int_{C_\eta}\frac{e^{\eta K}}{\beta(\eta)+\alpha\eta-1}\,d\eta}, & 0\leqq z<K,\\[2mm] 1, & z\geqq K,\end{cases}$$

$$W(0+)=\left[\frac{1}{2\pi i}\int_{C_\eta}\frac{e^{\eta K}\,\alpha\,d\eta}{\beta(\eta)+\alpha\eta-1}\right]^{-1},$$

$$1-U(K-)=1-\frac{\dfrac{1}{2\pi i}\displaystyle\int_{C_\eta}\frac{\beta(\eta)\,e^{\eta K}}{\beta(\eta)+\alpha\eta-1}\,d\eta}{\dfrac{1}{2\pi i}\displaystyle\int_{C_\eta}\frac{e^{\eta K}}{\beta(\eta)+\alpha\eta-1}\,d\eta}.$$

If $a<1$ then

$$(1-a)\frac{\alpha\rho}{\beta(\rho)+\alpha\rho-1},$$

is the Laplace-Stieltjes transform of the stationary waiting time distribution of the $M/G/1$ queueing system. Denote by $\boldsymbol{w}$ a stochastic variable with this stationary distribution as distribution function, and let $\boldsymbol{\tau}$ be a variable with distribution function $B(.)$. Suppose that $\boldsymbol{w}$ and $\boldsymbol{\tau}$ are independent variables. It follows from (5.80) that if $a<1$, and hence $\delta_1(1,0,0)=0$, that

$$W(z)=\begin{cases}\dfrac{\Pr\{\boldsymbol{w}<z\}}{\Pr\{\boldsymbol{w}<K\}}, & z\leqq K,\\[2ex] 1, & z\geqq K,\end{cases} \tag{5.81}$$

$$U(z)=\begin{cases}\dfrac{\Pr\{\boldsymbol{w}+\boldsymbol{\tau}<z\}}{\Pr\{\boldsymbol{w}<K\}}, & z<K,\\[2ex] 1, & z\geqq K,\end{cases}$$

since

$$\Pr\{\boldsymbol{w}<z\}=\frac{1}{2\pi \mathrm{i}}\int_{C_\rho}\frac{\mathrm{e}^{\rho z}}{\rho}\,\frac{(1-a)\,\rho\alpha}{\beta(\rho)+\alpha\rho-1}\,\mathrm{d}\rho,\qquad \operatorname{Re}\rho>0.$$

The result (5.81) has been found by several authors (cf. PRABHU [1965[a]], GAVER and MILLER [1962] and DALEY [1964]).

Next we derive expressions for entrance and return time distributions. From (5.59) and the relations (5.76), ..., (5.78) we obtain for $|r|\leqq 1$, $\operatorname{Re}\rho_3\geqq 0$, $\operatorname{Re}\rho_4\geqq 0$, $\operatorname{Re}\eta>\operatorname{Re}\delta_1(r,\rho_3,\rho_4)$, $0\leqq w<K$,

$$\mathrm{E}\{r^{\boldsymbol{\alpha}_{w0}}\exp(-\rho_3\boldsymbol{\gamma}_{w0}-\rho_4\boldsymbol{p}_{w0})\}$$

$$=\frac{\dfrac{r\beta(\rho_4)}{1+\alpha\rho_3-r\beta(\rho_4)}+\dfrac{r}{2\pi\mathrm{i}}\displaystyle\int_{C_\eta}\frac{\mathrm{e}^{\eta(K-w)}}{\eta}\,\frac{\beta(\rho_4+\eta)}{r\beta(\rho_4+\eta)+(\eta-\rho_3)\,\alpha-1}\,\mathrm{d}\eta}{\dfrac{1+\alpha\rho_3}{1+\alpha\rho_3-r\beta(\rho_4)}+\dfrac{1+\alpha\rho_3}{2\pi\mathrm{i}}\displaystyle\int_{C_\eta}\frac{\mathrm{e}^{\eta K}}{\eta}\,\frac{1}{r\beta(\rho_4+\eta)+(\eta-\rho_3)\,\alpha-1}\,\mathrm{d}\eta}, \tag{5.82}$$

$$\mathrm{E}\{r^{\boldsymbol{\beta}_{KK}}\exp(-\rho_3\boldsymbol{\delta}_{KK}-\rho_4\boldsymbol{q}_{KK})\}$$

$$=1-(1+\alpha\rho_3-r\beta(\rho_4))\,\frac{\dfrac{1}{2\pi\mathrm{i}}\displaystyle\int_{C_\eta}\frac{\mathrm{e}^{\eta K}}{r\beta(\rho_4+\eta)+(\eta-\rho_3)\,\alpha-1}\,\mathrm{d}\eta}{\dfrac{1}{2\pi\mathrm{i}}\displaystyle\int_{C_\eta}\left\{\frac{\alpha\eta}{r\beta(\rho_4+\eta)+(\eta-\rho_3)\,\alpha-1}-1\right\}\mathrm{e}^{\eta K}\,\mathrm{d}\eta}. \tag{5.83}$$

The other relations which can be obtained from (5.59) will be omitted since these expressions are rather lengthy.

The above relations all refer to the finite dam. We now shall obtain explicit expressions for the taboo probabilities defined in (5.61) and (5.62).

These taboo probabilities are of interest for the finite as well as the infinite dam.

From (5.63), (5.58), (5.59) and (5.64) we obtain for $\operatorname{Re}\rho_3 \geqq 0$, $\operatorname{Re}\rho_4 \geqq 0$, $0 \leqq w < K$,

$$\sum_{n=1}^{\infty} r^n \tilde{f}^{(n)}_{K;w0}(\rho_3, \rho_4) = \frac{\sum_{n=1}^{\infty} r^n \tilde{p}^{(n)}_{K;w0}(\rho_3, \rho_4)}{1 + \sum_{n=1}^{\infty} r^n \tilde{p}^{(n)}_{K;00}(\rho_3, \rho_4)}, \qquad |r| \leqq 1, \tag{5.84}$$

$$\sum_{n=1}^{\infty} r^n \tilde{p}^{(n)}_{K;w0}(\rho_3, \rho_4) = c_x(r, \rho_3, \rho_4, w) - c_K(r, \rho_3, \rho_4, K) \frac{r c_y(r, \rho_3, \rho_4, w)}{1 + c_h(r, \rho_3, \rho_4, K)},$$
$$|r| < 1. \tag{5.85}$$

From (5.84), (5.85), (5.76) and (5.78) it is now easily found that for $|r| \leqq 1$, $\operatorname{Re}\rho_3 \geqq 0$, $\operatorname{Re}\rho_4 \geqq 0$, $0 \leqq w < K$, $\operatorname{Re}\eta > \operatorname{Re}\delta_1(r, \rho_3, \rho_4)$,

$$\mathrm{E}\{(\sup_{0<t<\boldsymbol{\gamma}_{w0}} \boldsymbol{v}_t < K) r^{\boldsymbol{a}_{w0}} \exp(-\rho_3 \boldsymbol{\gamma}_{w0} - \rho_4 \boldsymbol{p}_{w0})\} = \sum_{n=1}^{\infty} r^n \tilde{f}^{(n)}_{K;w0}(\rho_3, \rho_4)$$
$$= \frac{\dfrac{r}{2\pi \mathrm{i}} \displaystyle\int_{C_\eta} \mathrm{e}^{\eta K} \frac{\beta(\rho_4 + \eta)\, \mathrm{e}^{-\eta w}}{r\beta(\rho_4 + \eta) - (\rho_3 - \eta)\alpha - 1} \mathrm{d}\eta}{\dfrac{1 + \alpha\rho_3}{2\pi \mathrm{i}} \displaystyle\int_{C_\eta} \mathrm{e}^{\eta K} \frac{\mathrm{d}\eta}{r\beta(\rho_4 + \eta) - (\rho_3 - \eta)\alpha - 1}}, \tag{5.86}$$

here $\boldsymbol{v}_t$ denotes the virtual waiting time or the content of the dam at time t. The left-hand side of (5.86) represents the Laplace-Stieltjes transform of the joint distribution of the residual busy cycle and the residual busy period and the generating function of the distribution of the number of arrivals in this busy cycle, while the supremum of the virtual waiting time during this busy cycle is less than K, given that the initial actual waiting time equals w, with $w < K$.

Taking $w = 0$, $\rho_3 = \rho_4 = 0$, $r = 1$ we have

$$\sum_{n=1}^{\infty} f^{(n)}_{K;00} = \frac{\dfrac{1}{2\pi \mathrm{i}} \displaystyle\int_{C_\eta} \mathrm{e}^{\eta K} \frac{\beta(\eta)}{1 - \alpha\eta - \beta(\eta)} \mathrm{d}\eta}{\dfrac{1}{2\pi \mathrm{i}} \displaystyle\int_{C_\eta} \mathrm{e}^{\eta K} \frac{\mathrm{d}\eta}{1 - \alpha\eta - \beta(\eta)}}, \qquad \operatorname{Re}\eta > \delta_1(1, 0, 0), \tag{5.87}$$

$$= \frac{\Pr\{\boldsymbol{w} + \boldsymbol{\tau} < K\}}{\Pr\{\boldsymbol{w} < K\}}, \qquad \text{if } a < 1 \quad \text{(cf. (5.81))},$$

which is the probability that for an $M/G/1$ queueing system with unbounded capacity the supremum of the virtual delay time during a busy cycle is less than K. The relation (5.87) has also been found by TAKÁCS [1965]. The distribution (5.87) will be investigated in section 7.4.

From (5.63) we obtain

$$\sum_{n=1}^{\infty} r^n \tilde{q}_{K;K0}^{(n)}(\rho_3, \rho_4) = \frac{c_k(r, \rho_3, \rho_4, K)}{1+c_h(r, \rho_3, \rho_4, K)},$$

so that from (5.64),

$$\mathrm{E}\{(\sup_{0<t<\boldsymbol{\delta}_{K0}} \boldsymbol{v}_t < K) r^{\boldsymbol{\beta}_{K0}} \exp(-\rho_3 \boldsymbol{\delta}_{K0} - \rho_4 \boldsymbol{q}_{K0})\} = \sum_{n=0}^{\infty} r^n \tilde{h}_{K;K0}^{(n)}(\rho_3, \rho_4)$$

$$= \left\{ \frac{1+\alpha\rho_3}{2\pi \mathrm{i}} \int_{C_\eta} \mathrm{e}^{\eta K} \frac{\alpha \, \mathrm{d}\eta}{r\beta(\rho_4+\eta) - (\rho_3-\eta)\,\alpha - 1} \right\}^{-1}, \tag{5.88}$$

for $|r| \leqq 1$, $\mathrm{Re}\,\rho_3 \geqq 0$, $\mathrm{Re}\,\rho_4 \geqq 0$, $\mathrm{Re}\,\eta > \mathrm{Re}\,\delta_1(r, \rho_3, \rho_4)$.

This expression is the Laplace-Stieltjes transform of the joint distribution of the residual busy cycle and the total amount of water which flowed into the dam during this busy cycle and the generating function of the distribution of the number of inputs during the busy cycle with no overflow in the meantime, given that the busy cycle starts at a moment that the dam is full.

In the same way as above we find that

$$\sum_{n=1}^{\infty} r^n \tilde{q}_{0;KK}^{(n)}(\rho_3, \rho_4) = c_h(r, \rho_3, \rho_4, K) - r c_y(r, \rho_3, \rho_4, 0) \frac{c_k(r, \rho_3, \rho_4, K)}{1+c_x(r, \rho_3, \rho_4, 0)},$$

and obtain from (5.64) for $|r| \leqq 1$, $\mathrm{Re}\,\rho_3 \geqq 0$, $\mathrm{Re}\,\rho_4 \geqq 0$, $\mathrm{Re}\,\eta > \mathrm{Re}\,\delta_1(r, \rho_3, \rho_4)$,

$$\mathrm{E}\{(\inf_{0<t<\boldsymbol{\delta}_{KK}} \boldsymbol{v}_t > 0) r^{\boldsymbol{\beta}_{KK}} \exp(-\rho_3 \boldsymbol{\delta}_{KK} - \rho_4 \boldsymbol{q}_{KK})\} = \sum_{n=1}^{\infty} r^n \tilde{h}_{0;KK}^{(n)}(\rho_3, \rho_4)$$

$$= 1 + \frac{1 - \dfrac{1+\alpha\rho_3 - r\beta(\rho_4)}{2\pi \mathrm{i}} \displaystyle\int_{C_\eta} \frac{\mathrm{e}^{\eta K}}{\eta} \frac{\mathrm{d}\eta}{1+(\rho_3-\eta)\,\alpha - r\beta(\rho_4+\eta)}}{\dfrac{1}{2\pi \mathrm{i}} \displaystyle\int_{C_\eta} \mathrm{e}^{\eta K} \frac{\alpha \, \mathrm{d}\eta}{1+(\rho_3-\eta)\,\alpha - r\beta(\rho_4+\eta)}}. \tag{5.89}$$

Clearly this expression is the Laplace-Stieltjes transform of the joint distribution of the (return) time between two successive overflows and the quantity of water which flowed into the dam during this return time and the generating function of the number of inputs during this time, while the dam has never been dry in the mean time.

Taking $r=1$, $\rho_3=\rho_4=0$, we obtain

$$\sum_{n=1}^{\infty} h_{0;KK}^{(n)} = \frac{1+\dfrac{1}{2\pi i}\displaystyle\int_{C_\eta} e^{\eta K}\frac{\alpha\, d\eta}{1-\alpha\eta-\beta(\eta)}}{\dfrac{1}{2\pi i}\displaystyle\int_{C_\eta} e^{\eta K}\frac{\alpha\, d\eta}{1-\alpha\eta-\beta(\eta)}}, \qquad \operatorname{Re}\eta>\delta_1(1,0,0), \tag{5.90}$$

$$= \frac{\Pr\{0<\boldsymbol{w}<K\}}{\Pr\{\boldsymbol{w}<K\}} \qquad \text{if} \quad a<1 \quad \text{(cf. (5.81))}.$$

It should be noted that

$$\sum_{n=1}^{\infty} h_{0;KK}^{(n)} + \sum_{n=0}^{\infty} h_{K;K0}^{(n)} = 1, \tag{5.91}$$

and similarly, we find

$$\sum_{n=1}^{\infty} f_{K;00}^{(n)} + \sum_{n=1}^{\infty} f_{0;0K}^{(n)} = 1. \tag{5.92}$$

The last relation implies that if we start in E_0 then with probability one we return to E_0 or reach state E_K, while the one but last relation shows that if the system starts in E_K we return to E_K or reach E_0 with probability one.

From (5.66) it follows that the average number of overflows during a busy cycle is given by

$$\mathrm{E}\{\boldsymbol{\Phi}_K\} = \frac{1}{2\pi i}\int_{C_\eta} e^{\eta K}\frac{(1-\beta(\eta))\,\alpha}{\beta(\eta)+\alpha\eta-1}\, d\eta, \qquad \operatorname{Re}\eta>\delta_1(1,0,0), \tag{5.93}$$

$$= \frac{1}{1-a}\Pr\{K-\boldsymbol{\tau}\leq\boldsymbol{w}<K\}, \qquad \text{if} \quad a<1.$$

Using the results of the present section it is possible to obtain, for the dam $M/G/1$ with infinite capacity, the distribution of the number of crossings from above of a given level K during a busy cycle and the distribution of the return time of such crossings (see section 5.8).

Finally, we shall prove that the results obtained above are independent of the assumption of rationality of $\beta(\rho)$ (cf. (5.67)), i.e. that they hold for general $B(.)$.

From the expressions obtained for

$$x(r, \rho, \rho_3, \rho_4, w), \qquad \frac{y(r, \rho, \rho_3, \rho_4, w)}{\beta_2(\rho_4+\rho)}, \qquad k(r, \rho, \rho_3, \rho_4, u), \qquad \frac{h(r, \rho, \rho_3, \rho_4, u)}{\beta_2(\rho_4+\rho)},$$

(cf. (5.71), (5.73), ..., (5.75)) it is seen that they involve only $\beta(\rho)$ and neither $\beta_1(\rho)$ nor $\beta_2(\rho)$. Consequently, the expressions (5.12), (5.18), (5.24) and (5.25) for $\omega(...)$, $\Pi(...)$, $\omega_1(...)$, $\Pi_1(...)$ contain only $\beta(\rho)$, if the functions just mentioned are substituted. The resulting expression for $\omega(r, \rho, \rho_3, \rho_4, w)$ can easily be shown to satisfy the integral equation (5.9) and the other conditions (cf. (5.4), ..., (5.6)). Since the integral equation (5.9) has been derived without using the assumption of rationality it is not difficult to verify that all results of the present section hold for general $\beta(\rho)$ (cf. end of section 4.5).

For the dam model considered in this section the quantity of water $\boldsymbol{k}_t$ flowing into the dam in a time interval $[0, t)$ is given by (cf. (II.4.100)),

$$\Pr\{\boldsymbol{k}_t<x\} = \sum_{n=0}^{\infty} \frac{(t/\alpha)^n}{n!} \mathrm{e}^{-t/\alpha} B^{n*}(x), \qquad t>0, \quad x>0,$$

$$\mathrm{E}\{\mathrm{e}^{-\rho \boldsymbol{k}_t}\} = \mathrm{e}^{-(1-\beta(\rho))t/\alpha}, \qquad t>0, \quad \operatorname{Re}\rho\geqq 0.$$

This type of input is a process with stationary, independent and non-negative increments, while its realisations are non-decreasing step-functions with a finite number of jumps in any finite interval with probability one.

The stochastic process $\{\boldsymbol{h}_t, t\in[0, \infty)\}$ with stationary, independent and non-negative increments of which the realisations are non-decreasing step-functions with probability one is characterized by

$$\mathrm{E}\{\mathrm{e}^{-\rho \boldsymbol{h}_t}\} = \exp\{-t \int_0^{\infty} (1-\mathrm{e}^{-\rho x})\, \mathrm{d}N(x)\}, \qquad t>0, \quad \operatorname{Re}\rho\geqq 0,$$

with $N(x)$ a non-decreasing function and

$$\lim_{x\to\infty} N(x) = 0, \qquad \int_0^1 x\, \mathrm{d}N(x) < \infty.$$

If $N(0+)>-\infty$ this process is of the type used in this section, while for $N(0+)=-\infty$ its realisations have an infinite number of jumps in any finite interval with probability one. This more general type of input process has been considered by several investigators (cf. PRABHU [1965[a]], TAKÁCS [1967]). Replacing the function $1-\beta(\rho)$ by $\int_0^{\infty} (1-\mathrm{e}^{-\rho x})\, \mathrm{d}N(x)$ in the relations (5.80) and (5.87) the results of Takàcs are obtained. It seems possible to apply the method developed in this chapter to processes with this general input process. We start from the relation between $\boldsymbol{v}_t$ and $\boldsymbol{v}_{t+\Delta t}$ and derive by a proper limit procedure, an integro-differential equation for the Laplace transform with respect to t of $\mathrm{E}\{\exp(-\rho \boldsymbol{v}_t)\}$, where $\boldsymbol{v}_t$ is the dam content at time t.

III.5.6. The finite dam model $G/M/1$

We consider the queueing model with

$$\alpha(\rho) = \frac{\alpha_1(\rho)}{\alpha_2(\rho)}, \qquad \beta(\rho) = \frac{1}{1+\beta\rho}, \tag{5.94}$$

with $\alpha_1(\rho)$ and $\alpha_2(\rho)$ polynomials in ρ, $\alpha_2(\rho)$ of degree m and $\alpha_1(\rho)$ of degree at most $m-1$. In the same way as in the preceding section it can be shown that the assumption of rationality of $\alpha(\rho)$ is superfluous, so that the results hold for general distributions $A(.)$. Since for the present case $n=1$ it follows from (5.12) and (5.18) that we may define

$$\begin{aligned} y_0(r, \rho_3, \rho_4, w) &\stackrel{\text{def}}{=} y(r, \rho, \rho_3, \rho_4, w), \\ h_0(r, \rho_3, \rho_4, u) &\stackrel{\text{def}}{=} h(r, \rho, \rho_3, \rho_4, u). \end{aligned} \tag{5.95}$$

From (5.10) we obtain for $|r|<1$, $\mathrm{Re}\,\rho_3 \geqq 0$, $\mathrm{Re}\,\rho_4 \geqq 0$, $0 \leqq w < K$,

$$y_0(r, \rho_3, \rho_4, w) \frac{\alpha(\rho_3-\varepsilon_1)}{1+(\rho_4+\varepsilon_1)\,\beta} \mathrm{e}^{-\varepsilon_1 K} + \frac{x(r, \varepsilon_1, \rho_3, \rho_4, w)}{\alpha_2(\rho_3-\varepsilon_1)} + \frac{r}{\varepsilon_1} \mathrm{e}^{-\varepsilon_1 w} = 0, \tag{5.96}$$

with

$$\varepsilon_1 \equiv \varepsilon_1(r, \rho_3, \rho_4),$$

while from (5.11) for $|r|<1$, $\mathrm{Re}\,\rho_3 \geqq 0$, $\mathrm{Re}\,\rho_4 \geqq 0$, $0 < \mathrm{Re}\,\xi < \nu_1$, $\mathrm{Re}\,\xi > \mathrm{Re}\,\rho$,

$$\begin{aligned} x(r, \rho, \rho_3, \rho_4, w) = -\alpha_2(\rho_3-\rho) & \frac{1+(\rho_4+\rho)\,\beta - r\alpha(\rho_3-\rho)}{2\pi\mathrm{i}(\rho-\varepsilon_1)} \\ & \cdot \int_{C_\xi} \left\{ \alpha(\rho_3-\xi) \frac{y_0(r, \rho_3, \rho_4, w)}{1+(\rho_4+\xi)\,\beta} \mathrm{e}^{-\xi K} + \frac{r}{\xi} \mathrm{e}^{-\xi w} \right\} \\ & \cdot \frac{\xi-\varepsilon_1}{1+(\rho_4+\xi)\,\beta - r\alpha(\rho_3-\xi)} \frac{\mathrm{d}\xi}{\xi-\rho}. \end{aligned} \tag{5.97}$$

Proceeding along the same lines as in the previous section it is found for $|r|<1$, $\mathrm{Re}\,\rho_3 \geqq 0$, $\mathrm{Re}\,\rho_4 \geqq 0$, $\mathrm{Re}\,\xi < \mathrm{Re}\,\varepsilon_1$ that

$$\frac{y_0(r,\rho_3,\rho_4,w)}{2\pi i}\int_{C_\xi}\frac{e^{-\xi K}}{1+(\rho_4+\xi)\beta-r\alpha(\rho_3-\xi)}\,d\xi$$
$$=-\frac{r^2}{1+\rho_4\beta-r\alpha(\rho_3)}-\frac{r^2}{2\pi i}\int_{C_\xi}\frac{e^{-\xi w}}{\xi}\frac{d\xi}{1+(\rho_4+\xi)\beta-r\alpha(\rho_3-\xi)},$$
$$0\leqq w<K. \qquad (5.98)$$

In the same way we derive for $|r|<1$, $\mathrm{Re}\,\rho_3\geqq 0$, $\mathrm{Re}\,\rho_4\geqq 0$, $0<\mathrm{Re}\,\xi<\nu_1$, $\mathrm{Re}\,\xi>\mathrm{Re}\,\rho$, $0<u\leqq K$,

$$k(r,\rho,\rho_3,\rho_4,u)=-r\alpha_2(\rho_3-\rho)\frac{1+(\rho_4+\rho)\beta-r\alpha(\rho_3-\rho)}{2\pi i(\rho-\varepsilon_1)}$$
$$\cdot\int_{C_\xi}\left\{\frac{h_0(r,\rho_3,\rho_4,u)}{1+(\rho_4+\xi)\beta}e^{-\xi K}+\frac{r}{\xi}e^{-\xi u}\right\}$$
$$\cdot\frac{\alpha(\rho_3-\xi)}{1+(\rho_4+\xi)\beta-r\alpha(\rho_3-\xi)}\frac{\xi-\varepsilon_1}{\xi-\rho}\,d\xi, \qquad (5.99)$$

and for $\mathrm{Re}\,\xi<\mathrm{Re}\,\varepsilon_1$,

$$\frac{h_0(r,\rho_3,\rho_4,u)}{2\pi i}\int_{C_\xi}\frac{e^{-\xi K}}{1+(\rho_4+\xi)\beta-r\alpha(\rho_3-\xi)}\,d\xi=$$
$$-\frac{r^2\alpha(\rho_3)}{1+\rho_4\beta-r\alpha(\rho_3)}-\frac{r^2}{2\pi i}\int_{C_\xi}\frac{e^{-\xi u}}{\xi}\frac{\alpha(\rho_3-\xi)}{1+(\rho_4+\xi)\beta-r\alpha(\rho_3-\xi)}\,d\xi. \qquad (5.100)$$

For the functions $c_x(\ldots)$, $c_y(\ldots)$, $c_k(\ldots)$ and $c_h(\ldots)$ (cf. (5.56)) we now obtain by a calculation similar to the one in the preceding section, for $|r|<1$, $\mathrm{Re}\,\rho_3\geqq 0$, $\mathrm{Re}\,\rho_4\geqq 0$, $\mathrm{Re}\,\xi<\mathrm{Re}\,\varepsilon_1$,

$$r^2c_x(r,\rho_3,\rho_4,w)$$
$$=\begin{cases}\dfrac{y_0(r,\rho_3,\rho_4,w)}{2\pi i}\displaystyle\int_{C_\xi}\left\{\frac{\beta\xi}{1+(\rho_4+\xi)\beta-r\alpha(\rho_3-\xi)}-1\right\}e^{-\xi K}\,d\xi \\ \qquad+\dfrac{r^2}{2\pi i}\displaystyle\int_{C_\xi}\frac{\beta\,e^{-\xi w}}{1+(\rho_4+\xi)\beta-r\alpha(\rho_3-\xi)}\,d\xi, \qquad 0<w<K, \\ -r^2+\dfrac{y_0(r,\rho_3,\rho_4,w)}{2\pi i}\displaystyle\int_{C_\xi}\left\{\frac{\beta\xi}{1+(\rho_4+\xi)\beta-r\alpha(\rho_3-\xi)}-1\right\}e^{-\xi K}\,d\xi, \\ \qquad\qquad w=0.\end{cases} \qquad (5.101)$$

$$r^2 c_y(r, \rho_3, \rho_4, w) = \frac{1}{\beta}\, y_0(r, \rho_3, \rho_4, w), \qquad 0 \leqq w < K, \tag{5.102}$$

$$r c_h(r, \rho_3, \rho_4, u) = \frac{1}{\beta}\, h_0(r, \rho_3, \rho_4, u), \qquad 0 < u \leqq K,$$

$$r^2 c_k(r, \rho_3, \rho_4, u)$$

$$= \frac{h_0(r, \rho_3, \rho_4, u)}{2\pi \mathrm{i}} \int_{C_\xi} \left\{ \frac{\beta\xi}{1+(\rho_4+\xi)\,\beta - r\alpha(\rho_3-\xi)} - 1 \right\} \mathrm{e}^{-\xi K}\, \mathrm{d}\xi$$

$$+ \frac{r^2}{2\pi \mathrm{i}} \int_{C_\xi} \mathrm{e}^{-\xi u}\, \frac{\beta\alpha(\rho_3-\xi)}{1+(\rho_4+\xi)\,\beta - r\alpha(\rho_3-\xi)}\, \mathrm{d}\xi, \qquad 0 < u \leqq K. \tag{5.103}$$

We derive expressions for the limit distributions $U(x)$ and $W(x)$. From (5.49) and (5.98) we obtain

$$\frac{y(\rho)}{\beta_2(\rho)} = -\frac{1}{1+\beta\rho} \left[\frac{1}{2\pi \mathrm{i}} \int_{C_\xi} \frac{\mathrm{e}^{-\xi K}}{1+\beta\xi-\alpha(-\xi)}\, \mathrm{d}\xi \right]^{-1},$$

for $\rho \neq -1/\beta$, Re $\xi < \varepsilon_1(1, 0, 0)$. Hence from (5.51) for Re $\xi < \varepsilon_1(1, 0, 0)$,

$$W(z) = \begin{cases} 0, & z < 0, \\[2ex] 1 - \dfrac{\dfrac{1}{2\pi \mathrm{i}} \displaystyle\int_{C_\xi} \dfrac{\alpha(-\xi)\, \mathrm{e}^{-\xi(K-z)}}{1+\beta\xi-\alpha(-\xi)}\, \mathrm{d}\xi}{\dfrac{1}{2\pi \mathrm{i}} \displaystyle\int_{C_\xi} \dfrac{\mathrm{e}^{-\xi K}}{1+\beta\xi-\alpha(-\xi)}\, \mathrm{d}\xi}, & 0 < z < K, \\[2ex] 1, & z \geqq K, \end{cases} \tag{5.104}$$

$$U(z) = \begin{cases} 0, & z < 0, \\[2ex] 1 - \dfrac{\dfrac{1}{2\pi \mathrm{i}} \displaystyle\int_{C_\xi} \dfrac{\mathrm{e}^{-\xi(K-z)}}{1+\beta\xi-\alpha(-\xi)}\, \mathrm{d}\xi}{\dfrac{1}{2\pi \mathrm{i}} \displaystyle\int_{C_\xi} \dfrac{\mathrm{e}^{-\xi K}}{1+\beta\xi-\alpha(-\xi)}\, \mathrm{d}\xi}, & 0 \leqq z < K, \\[2ex] 1, & z \geqq K, \end{cases}$$

$$W(0+) = -\frac{\dfrac{1}{2\pi i}\displaystyle\int_{C_\xi}\left\{\frac{\beta\xi}{1+\beta\xi-\alpha(-\xi)}-1\right\}e^{-\xi K}\,d\xi}{\dfrac{1}{2\pi i}\displaystyle\int_{C_\xi}\frac{e^{-\xi K}}{1+\beta\xi-\alpha(-\xi)}\,d\xi},$$

$$1-U(K-) = -\left\{\frac{1}{2\pi i}\int_{C_\xi}\frac{\beta\, e^{-\xi K}}{1+\beta\xi-\alpha(-\xi)}\,d\xi\right\}^{-1}.$$

As in the preceding section we obtain from (5.59) and from (5.101), (5.102) and (5.103) for the entrance and return time distributions

$$\mathrm{E}\{r^{\boldsymbol{\alpha}_{00}}\exp(-\rho_3\boldsymbol{\gamma}_{00}-\rho_4\boldsymbol{p}_{00})\} = 1+\{1+\rho_4\beta-r\alpha(\rho_3)\}$$

$$\cdot\frac{\dfrac{1}{2\pi i}\displaystyle\int_{C_\xi}\frac{e^{-\xi K}}{1+(\rho_4+\xi)\,\beta-r\alpha(\rho_3-\xi)}\,d\xi}{\dfrac{1}{2\pi i}\displaystyle\int_{C_\xi}\left\{\frac{\beta\xi}{1+(\rho_4+\xi)\,\beta-r\alpha(\rho_3-\xi)}-1\right\}e^{-\xi K}\,d\xi}, \tag{5.105}$$

$$\mathrm{E}\{r^{\boldsymbol{\beta}_{uK}}\exp(-\rho_3\boldsymbol{\delta}_{uK}-\rho_4\boldsymbol{q}_{uK})\}$$

$$=\frac{\dfrac{r\alpha(\rho_3)}{1+\rho_4\beta-r\alpha(\rho_3)}+\dfrac{r}{2\pi i}\displaystyle\int_{C_\xi}\frac{e^{-\xi u}}{\xi}\,\frac{\alpha(\rho_3-\xi)}{1+(\rho_4+\xi)\,\beta-r\alpha(\rho_3-\xi)}\,d\xi}{\dfrac{1+\rho_4\beta}{1+\rho_4\beta-r\alpha(\rho_3)}+\dfrac{1+\rho_4\beta}{2\pi i}\displaystyle\int_{C_\xi}\frac{e^{-\xi K}}{\xi}\,\frac{d\xi}{1+(\rho_4+\xi)\,\beta-r\alpha(\rho_3-\xi)}},$$

for $|r|\leq 1$, $\mathrm{Re}\,\rho_3\geq 0$, $\mathrm{Re}\,\rho_4\geq 0$, $\mathrm{Re}\,\xi<\mathrm{Re}\,\varepsilon_1$, $0<u\leq K$.

The expressions for the other entrance time distributions following from (5.57) will be omitted here, since they are rather lengthy.

In exactly the same way as in the preceding section we find for the taboo probabilities

$$\sum_{n=1}^{\infty} r^n \tilde{f}^{(n)}_{K;00}(\rho_3,\rho_4)$$

$$=1+\frac{1-\dfrac{1+\rho_4\beta-r\alpha(\rho_3)}{2\pi i}\displaystyle\int_{C_\xi}\frac{e^{-\xi K}}{\xi}\,\frac{d\xi}{1+(\rho_4+\xi)\,\beta-r\alpha(\rho_3-\xi)}}{\dfrac{1}{2\pi i}\displaystyle\int_{C_\xi}e^{-\xi K}\,\frac{\beta\,d\xi}{1+(\rho_4+\xi)\,\beta-r\alpha(\rho_3-\xi)}},$$

$$\sum_{n=1}^{\infty} r^n \tilde{h}_{0;uK}^{(n)}(\rho_3, \rho_4) = \frac{\dfrac{r}{2\pi i} \displaystyle\int_{C_\xi} e^{-\xi u} \frac{\alpha(\rho_3-\xi)\, d\xi}{1+(\rho_4+\xi)\,\beta - r\alpha(\rho_3-\xi)}}{\dfrac{1+\beta\rho_4}{2\pi i} \displaystyle\int_{C_\xi} e^{-\xi K} \frac{d\xi}{1+(\rho_4+\xi)\,\beta - r\alpha(\rho_3-\xi)}}, \tag{5.106}$$

$$\sum_{n=1}^{\infty} r^n \tilde{f}_{0;0K}^{(n)}(\rho_3, \rho_4) = -\frac{r}{\dfrac{1+\rho_4\beta}{2\pi i} \displaystyle\int_{C_\xi} e^{-\xi K} \frac{\beta\, d\xi}{1+(\rho_4+\xi)\,\beta - r\alpha(\rho_3-\xi)}},$$

for $|r| \leqq 1$, $\operatorname{Re} \rho_3 \geqq 0$, $\operatorname{Re} \rho_4 \geqq 0$, $\operatorname{Re} \xi < \operatorname{Re} \varepsilon_1$, $0 < u \leqq K$.
All results obtained in the present section may be deduced immediately from those of the previous section by using the duality relation between $\boldsymbol{w}_n$ and $K - \boldsymbol{u}_n$ (cf. (5.16)).

III.5.7. Application to the system $M/M/1$

For the $M/M/1$ system, i.e. for

$$\alpha(\rho) = \frac{1}{1+\alpha\rho}, \qquad \beta(\rho) = \frac{1}{1+\beta\rho},$$

we shall give here some of the particular results obtained in the previous sections. For the present case the expressions of the previous sections can easily be evaluated by contour integration. As most of them are rather lengthy, however, we shall only give the simpler relations.
We obtain

$$W(0+) = \begin{cases} \dfrac{1-a}{1-a\exp\{-(1-a)\,K/\beta\}}, & a \neq 1, \\[2ex] \dfrac{1}{1+K/\beta}, & a = 1, \end{cases}$$

$$1 - U(K-) = \begin{cases} \dfrac{(1-a)\exp\{-(1-a)\,K/\beta\}}{1-a\exp\{-(1-a)\,K/\beta\}}, & a \neq 1, \\[2ex] \dfrac{1}{1+K/\beta}, & a = 1, \end{cases}$$

$$W(z)=\begin{cases}\dfrac{1-a\exp\{-(1-a)\,z/\beta\}}{1-a\exp\{-(1-a)\,K/\beta\}}, & a\neq 1,\quad 0<z\leqq K,\\[2ex] \dfrac{1+z/\beta}{1+K/\beta}, & a=1,\quad 0<z\leqq K,\end{cases}$$

$$U(z)=\begin{cases}\dfrac{1-\exp\{-(1-a)\,z/\beta\}}{1-a\exp\{-(1-a)\,K/\beta\}}, & a\neq 1,\quad 0\leqq z<K,\\[2ex] \dfrac{z/\beta}{1+K/\beta}, & a=1,\quad 0\leqq z<K,\end{cases}$$

$$\sum_{n=1}^{\infty} f_{K;00}^{(n)}=\begin{cases}\dfrac{1-\exp\{-(1-a)\,K/\beta\}}{1-a\exp\{-(1-a)\,K/\beta\}}, & a\neq 1,\\[2ex] \dfrac{K/\beta}{1+K/\beta}, & a=1,\end{cases}$$

$$\sum_{n=0}^{\infty} h_{K;K0}^{(n)}=\begin{cases}\dfrac{1-a}{1-a\exp\{-(1-a)\,K/\beta\}}, & a\neq 1,\\[2ex] \dfrac{1}{1+K/\beta}, & a=1,\end{cases}$$

$$\sum_{n=1}^{\infty} h_{0;KK}^{(n)}=\begin{cases}a\,\dfrac{1-\exp\{-(1-a)\,K/\beta\}}{1-a\exp\{-(1-a)\,K/\beta\}}, & a\neq 1,\\[2ex] \dfrac{K/\beta}{1+K/\beta} & a=1,\end{cases}$$

$$\mathrm{E}\{\boldsymbol{\Phi}_K\}=\mathrm{e}^{-(1-a)K/\beta},$$

with $\boldsymbol{\Phi}_K$ the average number of times that overflow occurs during a busy cycle.

III.5.8. The distribution of crossings of a level K for the infinite dam $M/G/1$

In this section we shall consider the queueing system $M/G/1$ or equivalently, the dam $M/G/1$ with infinite capacity. In fig. 4 a realisation of the virtual waiting time or the dam content $\boldsymbol{v}_t$ has been drawn for a busy cycle.

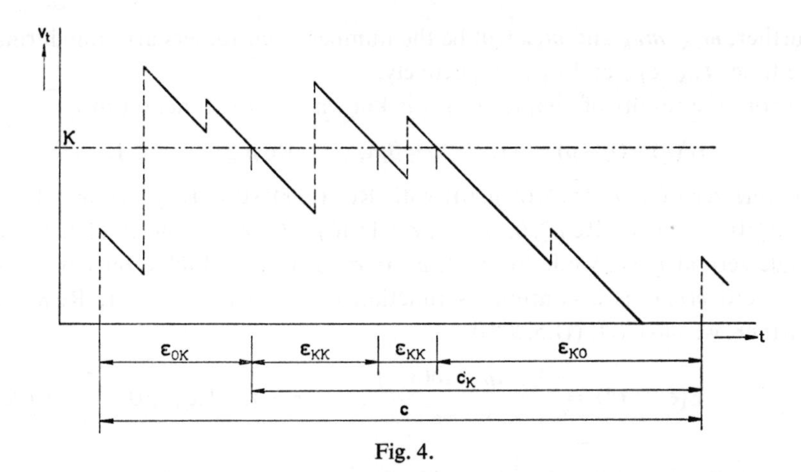

Fig. 4.

K is a fixed positive constant, and we shall study the distribution of the number of crossings of level K from above by $\boldsymbol{v}_t$ during a busy cycle. The length of a busy cycle started by an arriving customer who finds the server idle is denoted by $\boldsymbol{c}$, that of a residual busy cycle started with an initial virtual waiting time K by $\boldsymbol{c}_K$; $\boldsymbol{\varepsilon}_{0K}$ is the time between the start of the busy cycle and the moment of the first crossing of level K from above in this busy cycle if during the busy cycle $\boldsymbol{v}_t$ exceeds K; $\boldsymbol{\varepsilon}_{KK}$ is the time between the moments of two successive crossings from above of level K during a busy cycle; and finally $\boldsymbol{\varepsilon}_{K0}$ is the time between the moment of the last crossing from above and the end of the busy cycle if there is such a last crossing.

The variables $\boldsymbol{\varepsilon}_{0K}$, $\boldsymbol{\varepsilon}_{KK}$ and $\boldsymbol{\varepsilon}_{K0}$ are defined by

$$\boldsymbol{\varepsilon}_{0K} \stackrel{\text{def}}{=} \begin{cases} \inf\limits_{0<t<\boldsymbol{c}} \{t: \boldsymbol{v}_{t-} > K > \boldsymbol{v}_{t+} \mid \sup\limits_{0<\tau<\boldsymbol{c}} \boldsymbol{v}_\tau > K,\ \boldsymbol{v}_{0-} = 0,\ \boldsymbol{v}_{0+} > 0\}, \\ \infty \quad \text{if no such finite } t \text{ exists;} \end{cases} \tag{5.107}$$

$$\boldsymbol{\varepsilon}_{KK} \stackrel{\text{def}}{=} \begin{cases} \inf\limits_{0<t<\boldsymbol{c}_K} \{t: \boldsymbol{v}_{t-} > K > \boldsymbol{v}_{t+} \mid \sup\limits_{0<\tau<\boldsymbol{c}_K} \boldsymbol{v}_\tau > K,\ \boldsymbol{v}_{0+} = K\}, \\ \infty \quad \text{if no such finite } t \text{ exists;} \end{cases}$$

$$\boldsymbol{\varepsilon}_{K0} \stackrel{\text{def}}{=} \begin{cases} \inf\limits_{0<t<\boldsymbol{c}_K} \{t: \boldsymbol{v}_{t-} > 0 = \boldsymbol{v}_{t+} \mid \sup\limits_{0<\tau<\boldsymbol{c}_K} \boldsymbol{v}_\tau < K,\ \boldsymbol{v}_{0+} = K\}, \\ \infty \quad \text{if no such finite } t \text{ exists.} \end{cases}$$

The variables $\boldsymbol{n}$ and $\boldsymbol{n}_K$ will represent the number of customers arriving during a busy cycle $\boldsymbol{c}$ and during a residual busy cycle $\boldsymbol{c}_K$, respectively.

Further, $\boldsymbol{m}_{0K}$, $\boldsymbol{m}_{KK}$ and $\boldsymbol{m}_{K0}$ will be the number of customers arriving during the times $\boldsymbol{\varepsilon}_{0K}$, $\boldsymbol{\varepsilon}_{KK}$ and $\boldsymbol{\varepsilon}_{K0}$, respectively.

From the results of chapter II.4 it is known that the function of η,

$$r\beta(\eta)+(\eta-\rho)\,\alpha-1, \qquad a=\beta/\alpha, \qquad \operatorname{Re}\rho\geqq 0, \quad |r|\leqq 1,$$

has one zero $\delta(\rho, r)$ $(\equiv\delta_1(r, \rho, 0))$ with $\operatorname{Re}\delta(\rho, r)>0$ if $\operatorname{Re}\rho>0$, $|r|\leqq 1$, or $\operatorname{Re}\rho\geqq 0$, $|r|<1$, or $\operatorname{Re}\rho\geqq 0$, $|r|\leqq 1$, $a>1$; if $\rho=0$, $|r|=1$ and $a<1$ it has a single zero at $\eta=0$, while for $r=1$, $\rho=0$, $a=1$ it has a double zero at $\eta=0$. The zero $\delta(\rho, r)$ is a continuous function of r and ρ for $|r|\leqq 1$, $\operatorname{Re}\rho\geqq 0$. Further, we have (cf. (II.5.208)),

$$\mathrm{E}\{\mathrm{e}^{-\rho\boldsymbol{c}}r^{\boldsymbol{n}}\} = \frac{1+\alpha\rho-\alpha\delta(\rho, r)}{1+\alpha\rho}, \qquad |r|\leqq 1, \quad \operatorname{Re}\rho\geqq 0, \tag{5.108}$$

$$\mathrm{E}\{\mathrm{e}^{-\rho\boldsymbol{c}_K}r^{\boldsymbol{n}_K}\} = \frac{1}{1+\alpha\rho}\,\mathrm{e}^{-K\delta(\rho,r)}, \qquad |r|\leqq 1, \quad \operatorname{Re}\rho\geqq 0.$$

Note that (cf. (II.4.22) and section II.5.11),

$$\mu(\rho, r)=1+\alpha\rho-\alpha\delta(\rho, r), \qquad \delta_1(r, \rho)\equiv\delta(\rho, r).$$

Define for $z>0$, $|r|\leqq 1$, $\operatorname{Re}\rho\geqq 0$,

$$F_j(z)\overset{\text{def}}{=}\Pr\{\boldsymbol{c}<z, \boldsymbol{n}=j, \sup_{0<t<\boldsymbol{c}} \boldsymbol{v}_t<K \mid \boldsymbol{v}_{0-}=0, \boldsymbol{v}_{0+}>0\}, \quad j=1, 2, \dots, \tag{5.109}$$

$$H_j(z)\overset{\text{def}}{=}\Pr\{\boldsymbol{c}_K<z, \boldsymbol{n}_K=j, \sup_{0<\tau<\boldsymbol{c}_K} \boldsymbol{v}_\tau<K \mid \boldsymbol{v}_0=K\}, \qquad j=0, 1, \dots,$$

$$f(\rho, r)\overset{\text{def}}{=}\sum_{j=1}^{\infty} r^j \int_0^{\infty} \mathrm{e}^{-\rho z}\,\mathrm{d}F_j(z),$$

$$h(\rho, r)\overset{\text{def}}{=}\sum_{j=0}^{\infty} r^j \int_0^{\infty} \mathrm{e}^{-\rho z}\,\mathrm{d}H_j(z).$$

Expressions for $f(\rho, r)$ and $h(\rho, r)$ are given by the relations (5.86) and (5.88), i.e. for $|r|\leqq 1$, $\operatorname{Re}\rho\geqq 0$ and $\operatorname{Re}\eta>\operatorname{Re}\delta(\rho, r)$,

$$f(\rho, r) = \frac{1}{1+\alpha\rho}\,\frac{\dfrac{r}{2\pi\mathrm{i}}\displaystyle\int_{C_\eta} \mathrm{e}^{\eta K}\,\frac{\beta(\eta)}{r\beta(\eta)+(\eta-\rho)\,\alpha-1}\,\mathrm{d}\eta}{\dfrac{1}{2\pi\mathrm{i}}\displaystyle\int_{C_\eta} \mathrm{e}^{\eta K}\,\frac{1}{r\beta(\eta)+(\eta-\rho)\,\alpha-1}\,\mathrm{d}\eta}, \tag{5.110}$$

$$h(\rho, r) = \frac{1}{1+\alpha\rho}\left[\frac{1}{2\pi\mathrm{i}}\int_{C_\eta} \mathrm{e}^{\eta K}\,\frac{\alpha\,\mathrm{d}\eta}{r\beta(\eta)+(\eta-\rho)\,\alpha-1}\right]^{-1}.$$

From (5.91) we see that if $\boldsymbol{v}_{0-}=0$, $\boldsymbol{v}_{0+}>0$ then with probability one the system returns to the empty state or passes level K, and if $\boldsymbol{v}_0=K$ then with probability one the empty state is reached or the system passes level K (cf. (5.92)).

Since for the $M/G/1$ system the interarrival times are all negative exponentially distributed and independent it follows that $\boldsymbol{\varepsilon}_{0K}$, $\boldsymbol{\varepsilon}_{KK}$ and $\boldsymbol{\varepsilon}_{K0}$ are independent variables; also $\boldsymbol{m}_{0K}$, $\boldsymbol{m}_{KK}$ and $\boldsymbol{m}_{K0}$ are independent variables. Consequently, we have for $\operatorname{Re}\rho\geqq 0$, $|r|\leqq 1$,

$$\mathrm{E}\{\mathrm{e}^{-\rho\boldsymbol{c}}r^{\boldsymbol{n}}\}=f(\rho,r)+(1-f(0,1))\,\mathrm{E}\{\mathrm{e}^{-\rho\boldsymbol{\varepsilon}_{0K}}r^{\boldsymbol{m}_{0K}}\}\,\mathrm{E}\{\mathrm{e}^{-\rho\boldsymbol{c}_K}r^{\boldsymbol{n}_K}\},\tag{5.111}$$

$$\begin{aligned}\mathrm{E}\{\mathrm{e}^{-\rho\boldsymbol{c}_K}r^{\boldsymbol{n}_K}\}&=h(\rho,r)\sum_{m=0}^{\infty}[\{1-h(0,1)\}\,\mathrm{E}\{\mathrm{e}^{-\rho\boldsymbol{\varepsilon}_{KK}}r^{\boldsymbol{m}_{KK}}\}]^m\\&=\frac{h(\rho,r)}{1-(1-h(0,1))\,\mathrm{E}\{\mathrm{e}^{-\rho\boldsymbol{\varepsilon}_{KK}}r^{\boldsymbol{m}_{KK}}\}},\end{aligned}$$

since (cf. (5.91) and (5.92)),

$$\begin{aligned}1-f(0,1)&=\Pr\{\sup_{0<\tau<\boldsymbol{c}}\boldsymbol{v}_\tau>K\mid\boldsymbol{v}_{0-}=0,\ \boldsymbol{v}_{0+}>0\},\\1-h(0,1)&=\Pr\{\sup_{0<\tau<\boldsymbol{c}_K}\boldsymbol{v}_\tau>K\mid\boldsymbol{v}_0=K\}.\end{aligned}$$

From (5.108) and the above relations we obtain for $|r|\leqq 1$, $\operatorname{Re}\rho\geqq 0$,

$$\mathrm{E}\{\mathrm{e}^{-\rho\boldsymbol{\varepsilon}_{0K}}r^{\boldsymbol{m}_{0K}}\}=\{1+\alpha\rho-\alpha\delta(\rho,r)-(1+\alpha\rho)f(\rho,r)\}\frac{\mathrm{e}^{K\delta(\rho,r)}}{1-f(0,1)},\tag{5.112}$$

$$\mathrm{E}\{\mathrm{e}^{-\rho\boldsymbol{\varepsilon}_{KK}}r^{\boldsymbol{m}_{KK}}\}=\frac{1}{1-h(0,1)}\{1-(1+\alpha\rho)\,h(\rho,r)\,\mathrm{e}^{K\delta(\rho,r)}\}.$$

From these relations it follows that

$$\Psi_{0K}\overset{\text{def}}{=}\Pr\{\boldsymbol{\varepsilon}_{0K}<\infty\}=\{1-\alpha\delta(0,1)-f(0,1)\}\frac{\mathrm{e}^{K\delta(0,1)}}{1-f(0,1)},\tag{5.113}$$

$$\Psi_{KK}\overset{\text{def}}{=}\Pr\{\boldsymbol{\varepsilon}_{KK}<\infty\}=\frac{1}{1-h(0,1)}\{1-h(0,1)\,\mathrm{e}^{K\delta(0,1)}\},$$

and

$$\Psi_{0K}=\Pr\{\boldsymbol{m}_{0K}<\infty\},\qquad\Psi_{KK}=\Pr\{\boldsymbol{m}_{KK}<\infty\}.\tag{5.114}$$

Obviously

$$\Psi_{0K}=\Psi_{KK}=1\qquad\text{if and only if}\quad a\leqq 1.\tag{5.115}$$

We shall denote by $\boldsymbol{\Pi}_K$ the number of crossings of level K from above by $\boldsymbol{v}_t$ during a busy cycle. Noting the renewal property of a moment at which such a crossing from above occurs, it follows that

$$\Pr\{\boldsymbol{\Pi}_K=m, \boldsymbol{c}<\infty\}$$

$$=\begin{cases} f(0, 1), & m=0, \\ \{1-f(0, 1)\}\, \Psi_{OK}\{(1-h(0, 1))\, \Psi_{KK}\}^{m-1}\, h(0, 1), & m=1, 2, \ldots, \end{cases} \tag{5.116}$$

$$\Pr\{\boldsymbol{\Pi}_K=m, \boldsymbol{c}=\infty\}$$

$$=\begin{cases} \{1-f(0, 1)\}\{1-\Psi_{OK}\}, & m=0, \\ \{1-f(0, 1)\}\, \Psi_{OK}\{(1-h(0, 1))\, \Psi_{KK}\}^{m-1}\, (1-h(0, 1))(1-\Psi_{KK}), & \\ & m=1, 2, \ldots . \end{cases}$$

From these relations it is easily found that the number of crossings from above in a busy cycle is finite with probability 1.

From the above relations we obtain

$$\mathrm{E}\{\boldsymbol{\Pi}_K \mid \boldsymbol{c}<\infty\} = \frac{1-\alpha\delta(0, 1)-f(0, 1)}{\{1-\alpha\delta(0, 1)\}\, h(0, 1)}\, \mathrm{e}^{-K\delta(0,1)}, \tag{5.117}$$

$$\mathrm{E}\{\boldsymbol{\Pi}_K \mid \boldsymbol{c}=\infty\} = \frac{1-\alpha\delta(0, 1)-f(0, 1)}{\alpha\delta(0, 1)\, h(0, 1)}\, \{1-\mathrm{e}^{-K\delta(0,1)}\} \quad \text{for} \quad a>1,$$

$$\mathrm{E}\{\boldsymbol{\Pi}_K\} = \frac{1-\alpha\delta(0, 1)-f(0, 1)}{h(0, 1)},$$

$$\operatorname{var}\{\boldsymbol{\Pi}_K \mid \boldsymbol{c}<\infty\} = \frac{1-\alpha\delta(0, 1)-f(0, 1)}{h^2(0, 1)}\, \frac{\mathrm{e}^{-2K\delta(0,1)}}{1-\alpha\delta(0, 1)}$$

$$\cdot \left\{1 + \frac{f(0, 1)}{1-\alpha\delta(0, 1)} - h(0, 1)\, \mathrm{e}^{K\delta(0,1)}\right\},$$

$$\operatorname{var}\{\boldsymbol{\Pi}_K \mid \boldsymbol{c}=\infty\} = \frac{1-\alpha\delta(0, 1)-f(0, 1)}{h^2(0, 1)}\, \frac{\{1-\mathrm{e}^{-K\delta(0,1)}\}\, \mathrm{e}^{-K\delta(0,1)}}{\alpha\delta(0, 1)}$$

$$\cdot \left\{1 - h(0, 1)\, \mathrm{e}^{K\delta(0,1)} + \frac{1-f(0, 1)}{\alpha\delta(0, 1)} - \frac{1-\alpha\delta(0, 1)-f(0, 1)}{\alpha\delta(0, 1)}\, \mathrm{e}^{K\delta(0,1)}\right\},$$

$$\text{if} \quad a>1.$$

If $a \leqq 1$ then $\delta(0, 1)=0$. In particular, if $a<1$ we have (cf. (5.87) and (5.90)),

$$f(0, 1) = \frac{\Pr\{\boldsymbol{w}+\boldsymbol{\tau}<K\}}{\Pr\{\boldsymbol{w}<K\}}, \qquad 1-f(0, 1) = \frac{\Pr\{K-\boldsymbol{\tau}\leqq \boldsymbol{w}<K\}}{\Pr\{\boldsymbol{w}<K\}},$$

$$h(0, 1) = \frac{\Pr\{\boldsymbol{w}=0\}}{\Pr\{\boldsymbol{w}<K\}}, \qquad 1-h(0, 1) = \frac{\Pr\{0<\boldsymbol{w}<K\}}{\Pr\{\boldsymbol{w}<K\}},$$

and hence

$$\Pr\{\boldsymbol{\Pi}_K=m\} = \begin{cases} \dfrac{\Pr\{\boldsymbol{w}+\boldsymbol{\tau}<K\}}{\Pr\{\boldsymbol{w}<K\}}, & m=0, \\[2ex] \dfrac{\Pr\{K-\boldsymbol{\tau}\leqq \boldsymbol{w}<K\}}{\Pr\{\boldsymbol{w}<K\}} \left[\dfrac{\Pr\{0<\boldsymbol{w}<K\}}{\Pr\{\boldsymbol{w}<K\}}\right]^{m-1} \dfrac{\Pr\{\boldsymbol{w}=0\}}{\Pr\{\boldsymbol{w}<K\}}, & \\ & m=1, 2, \ldots, \end{cases}$$

$$\mathrm{E}\{\boldsymbol{\Pi}_K\} = \frac{1}{1-a} \Pr\{K-\boldsymbol{\tau}\leqq \boldsymbol{w}<K\},$$

$$\mathrm{var}\{\boldsymbol{\Pi}_K\} = \frac{1}{1-a} \Pr\{K-\boldsymbol{\tau}\leqq \boldsymbol{w}<K\} \frac{\Pr\{0<\boldsymbol{w}<K\}+\Pr\{\boldsymbol{w}+\boldsymbol{\tau}<K\}}{1-a}.$$

We note that if $a\leqq 1$ then the distribution of $\boldsymbol{\Pi}_K$ for the $M/G/1$ dam with infinite capacity is the same as the distribution of the number $\boldsymbol{\Phi}_K$ of overflows during a busy cycle of the $M/G/1$ dam with finite capacity (cf. (5.65) and (5.116)), and in this case this distribution is independent of the form of the tail of $B(t)$ for $t>K$.

If $\boldsymbol{\Pi}_K^{(j)}$ represents the number of crossings of level K from above in the jth busy cycle, then it is obvious that $\boldsymbol{\Pi}_K^{(j)}$, $j=1, 2, \ldots$, are independent, identically distributed variables.

If $a\leqq 1$ then the strong law of large numbers implies that (cf. (5.117)) with probability one

$$\lim_{n\to\infty} \frac{1}{n} \sum_{j=1}^{n} \boldsymbol{\Pi}_K^{(j)} = \frac{1-f(0, 1)}{h(0, 1)} = \frac{1}{2\pi \mathrm{i}} \int_{C_\eta} \mathrm{e}^{\eta K} \frac{\alpha^2 \eta \, \mathrm{d}\eta}{\beta(\eta)+\alpha\eta-1}, \qquad \mathrm{Re}\, \eta>0;$$

moreover the central limit theorem applies for the sequence $\boldsymbol{\Pi}_K^{(j)}, j=1, 2, \ldots$, since for $a\leqq 1$ the second moment of $\boldsymbol{\Pi}_K^{(j)}$ is finite.

Finally, we apply the results obtained to the $M/M/1$ queueing system. Since

$$B(t) = \begin{cases} 1-\mathrm{e}^{-t/\beta}, & t>0, \\ 0, & t<0, \end{cases}$$

we have

$$f(0,1)=\begin{cases}\dfrac{1-\exp\{-(1-a)\,K/\beta\}}{1-a\exp\{-(1-a)\,K/\beta\}}, & a\neq 1,\\[2ex] \dfrac{K/\beta}{1+K/\beta}, & a=1,\end{cases}$$

$$h(0,1)=\begin{cases}\dfrac{1-a}{1-a\exp\{-(1-a)\,K/\beta\}}, & a\neq 1,\\[2ex] \dfrac{1}{1+K/\beta}, & a=1,\end{cases}$$

$$\Psi_{0K}=\Psi_{KK}=1-\alpha\delta(0,1)=\begin{cases}a^{-1}, & a>1,\\ 1, & a\leqq 1.\end{cases}$$

Hence if $a<1$,

$$\mathrm{E}\{\boldsymbol{\Pi}_K\}=\exp\{-(1-a)\,K/\beta\},$$

$$\mathrm{var}\{\boldsymbol{\Pi}_K\}=\frac{1+a}{1-a}\exp\{-(1-a)\,K/\beta\}\{1-\exp\{-(1-a)\,K/\beta\}\};$$

while if $a=1$,

$$\mathrm{E}\{\boldsymbol{\Pi}_K\}=1,\qquad \mathrm{var}\{\boldsymbol{\Pi}_K\}=2K/\beta;$$

and if $a>1$,

$$\mathrm{E}\{\boldsymbol{\Pi}_K\mid \boldsymbol{c}<\infty\}=\exp\{(1-a)\,K/\beta\},$$

$$\mathrm{var}\{\boldsymbol{\Pi}_K\mid \boldsymbol{c}<\infty\}=\frac{1+a}{1-a}\exp\{(1-a)\,K/\beta\}\{\exp\{(1-a)\,K/\beta\}-1\},$$

$$\mathrm{E}\{\boldsymbol{\Pi}_K\mid \boldsymbol{c}=\infty\}=\frac{1-\exp\{(1-a)\,K/\beta\}}{a-1},$$

$$\mathrm{var}\{\boldsymbol{\Pi}_K\mid \boldsymbol{c}=\infty\}=\frac{\{\exp\{(1-a)\,K/\beta\}-a\}\{\exp\{(1-a)\,K/\beta\}-1\}}{(a-1)^2}.$$

III.5.9. The integral equation for the finite dam $M/G/1$

Denote the dam content at time t by $\boldsymbol{v}_t$ and define

$$V_t(\sigma,v)\overset{\text{def}}{=}\Pr\{\boldsymbol{v}_t<\sigma\mid \boldsymbol{v}_0=v\},\qquad t>0,\quad K>\sigma\geqq 0,\quad K>v\geqq 0.\tag{5.118}$$

Indicating the number of renewals in $[0,t)$ of the renewal process with renewal distribution $A(t)$ by $\boldsymbol{v}_t$, so that $\boldsymbol{v}_t$ is the number of inputs during the time interval $[0, t)$, it is easily seen that

$$\boldsymbol{v}_t = \begin{cases} [\boldsymbol{w}_{\boldsymbol{v}_t}+\boldsymbol{\tau}_{\boldsymbol{v}_t}-(t-\boldsymbol{t}_{\boldsymbol{v}_t})]^+ & \text{for } \boldsymbol{w}_{\boldsymbol{v}_t}+\boldsymbol{\tau}_{\boldsymbol{v}_t}<K, \\ [K-(t-\boldsymbol{t}_{\boldsymbol{v}_t})]^+ & \text{for } \boldsymbol{w}_{\boldsymbol{v}_t}+\boldsymbol{\tau}_{\boldsymbol{v}_t}\geqq K, \end{cases} \tag{5.119}$$

here $\boldsymbol{w}_{\boldsymbol{v}_t}$ is the content of the dam just before the last input before t.

Using the results obtained in the preceding sections of this chapter concerning the process $\{\boldsymbol{w}_n, n=1, 2, \ldots\}$, the stochastic process $\{\boldsymbol{v}_t, t\in[0, \infty)\}$ described by (5.119) may be investigated by applying a similar approach as that given in section II.5.6. By using such an approach it is possible to describe the virtual waiting time process $\boldsymbol{v}_t$ for the finite dam $K_n/K_m/1$.

For the finite dam $M/G/1$ another starting point for the investigation of the $\boldsymbol{v}_t$-process is the integro-differential equation for this process. This equation is obtained in the same way as the integro-differential equation of Takács for the infinite dam $M/G/1$ (cf. (II.4.107)).

The integro-differential equation has been used by many investigators (cf. Prabhu [1965[a]], Gaver and Miller [1962], Hasofer [1963]). It often leads rapidly to results for the stationary distribution of the process. However, a rigorous mathematical justification of the results is, although not difficult, frequently rather laborious. In the next section we shall apply the integro-differential equation to study some variants of the finite dam $M/G/1$.

From (II.4.107) it follows immediately that the integro-differential equation for the finite dam $M/G/1$ reads for $t>0$, $0<\sigma<K$, $0\leqq v\leqq K$,

$$\frac{\partial}{\partial t} V_t(\sigma, v) = \frac{\partial}{\partial \sigma} V_t(\sigma, v) - \frac{1}{\alpha} V_t(\sigma, v) + \frac{1}{\alpha} \int_{0-}^{\sigma} B(\sigma-y)\, \mathrm{d}_y V_t(y, v), \tag{5.120}$$

with

$$V_{0+}(\sigma, v) = \begin{cases} 0 & \text{for } \sigma\leqq v, \\ 1 & \text{for } \sigma>v. \end{cases}$$

If the process $\boldsymbol{v}_t$ has a stationary distribution $V(\sigma)$, then we have for every continuity point of $V(.)$,

$$\lim_{t\to\infty} V_t(\sigma, v) = V(\sigma) \qquad \text{for every } v,\ 0\leqq v\leqq K,$$

hence for $t>0$, $0<\sigma<K$,

$$\frac{\partial}{\partial \sigma} V(\sigma) - \frac{1}{\alpha} V(\sigma) + \frac{1}{\alpha} \int_{0-}^{\sigma} B(\sigma-y)\, \mathrm{d}_y V(y) = 0. \tag{5.121}$$

In section 5.3 it has been shown that the process $\{\boldsymbol{w}_n,\ n=1,2,\ldots\}$ has a stationary distribution and that the distribution function of $\boldsymbol{w}_n$ converges completely to this stationary distribution for $n\to\infty$. If $\boldsymbol{w}_1$ has the stationary distribution $W(.)$ (cf. (5.80)), then all $\boldsymbol{w}_n$ have the same distribution, so that it follows from (5.1) and (5.119) that $\boldsymbol{v}_t$ also has the distribution $W(.)$, since for the finite $M/G/1$ dam the distribution of $t-\boldsymbol{t}_{\boldsymbol{v}_t}$ is the negative exponential distribution. Consequently, the $\boldsymbol{v}_t$-process also has a stationary distribution and (cf. (5.80)),

$$V(\sigma)=\begin{cases}0, & \sigma<0,\\[2mm] \dfrac{\dfrac{1}{2\pi\mathrm{i}}\displaystyle\int_{C_\eta}\dfrac{\alpha\,\mathrm{e}^{\eta\sigma}\,\mathrm{d}\eta}{\beta(\eta)+\alpha\eta-1}}{\dfrac{1}{2\pi\mathrm{i}}\displaystyle\int_{C_\eta}\dfrac{\alpha\,\mathrm{e}^{\eta K}\,\mathrm{d}\eta}{\beta(\eta)+\alpha\eta-1}}, & 0<\sigma<K,\\[2mm] 1, & \sigma\geqq K,\end{cases}\tag{5.122}$$

$$V(0+)=\left\{\frac{1}{2\pi\mathrm{i}}\int_{C_\eta}\frac{\alpha\,\mathrm{e}^{\eta K}\,\mathrm{d}\eta}{\beta(\eta)+\alpha\eta-1}\right\}^{-1},$$

for $\operatorname{Re}\eta>\delta_1(1,0,0)$.

Substitution of the latter expression into the integro-differential equation (5.121) shows easily that $V(\sigma)$ as given by (5.122) satisfies (5.121), since

$$\int_0^\sigma\{1-B(\sigma-y)\}\,\mathrm{d}_yV(y)=\frac{1}{2\pi\mathrm{i}}\int_{C_\eta}\frac{\alpha\,\mathrm{e}^{\eta\sigma}}{\beta(\eta)+\alpha\eta-1}(1-\beta(\eta))\,\mathrm{d}\eta\,V(0+),$$

$$\frac{\partial V(\sigma)}{\partial\sigma}=\frac{1}{2\pi\mathrm{i}}\int_{C_\eta}\left\{\frac{\alpha\eta}{\beta(\eta)+\alpha\eta-1}-1\right\}\mathrm{e}^{\eta\sigma}\,\mathrm{d}\eta\,V(0+),$$

for $0<\sigma<K$, $\operatorname{Re}\eta>\delta_1(1,0,0)$. It should be noted that the Laplace-Stieltjes transform of $V(\sigma)$ is the same as that of $W(\sigma)$, viz. $\omega(\rho)$ (cf. (5.47)) with $x(\rho)$ and $y(\rho)$ to be determined from the results of section 5.5.

III.5.10. The dam $M/G/1$ with non constant release

As already stated in the introduction of the present chapter the queueing system $M/G/1$ can be considered as a dam model with constant release of water per unit of time. In this section we shall briefly discuss a dam model $M/G/1$ with a variable outflow rate. The discussion given here is based on a study of GAVER and MILLER [1962].

By $r(\sigma)$ we shall denote the outflow rate at time t if the dam content $\boldsymbol{v}_t$ at t is equal to σ, ($r(\sigma)=0$ for $\sigma=0$). With $V_t(\sigma, v)$ as defined by (5.118) the integro-differential equation for the $\boldsymbol{v}_t$-process reads

$$\frac{\partial}{\partial t} V_t(\sigma, v) = r(\sigma) \frac{\partial}{\partial \sigma} V_t(\sigma, v) - \frac{1}{\alpha} \int_{0-}^{\sigma} \{1-B(\sigma-y)\} \, \mathrm{d}_y V_t(y, v),$$

for $t>0$, $\sigma>0$, $v \geqq 0$, if it is assumed that $r(\sigma)$ is continuous for $\sigma>0$.

Obviously if the process has a stationary distribution $V(\sigma)$ then

$$r(\sigma) \frac{\partial}{\partial \sigma} V(\sigma) = \frac{1}{\alpha} \int_{0-}^{\sigma} \{1-B(\sigma-y)\} \, \mathrm{d}V(y), \qquad \sigma>0. \tag{5.123}$$

First, we shall consider the case

$$\text{(i)} \quad r(\sigma) = \begin{cases} r_1>0 & \text{for } 0<\sigma<K, \\ r_2>0 & \text{for } \sigma \geqq K. \end{cases}$$

Hence the outflow rate is r_2 if the water level exceeds K, while if it is less than K the outflow rate is r_1. Note that for r_2 tending to infinity the present model becomes that of the $M/G/1$ dam with finite capacity K. The present model is also interesting as a queueing system since if $r_2>r_1$ it may be interpreted as a model of a queueing system with the server working faster if the amount of work present in the system exceeds K.

If the $\boldsymbol{v}_t$-process has a stationary distribution $V(\sigma)$ then

$$\frac{\partial}{\partial \sigma} V(\sigma) = \begin{cases} \dfrac{1}{\alpha r_1} \displaystyle\int_{0-}^{\sigma} \{1-B(\sigma-y)\} \, \mathrm{d}V(y), & 0 \leqq \sigma < K, \\[2ex] \dfrac{1}{\alpha r_2} \displaystyle\int_{0-}^{\sigma} \{1-B(\sigma-y)\} \, \mathrm{d}V(y), & K \leqq \sigma, \end{cases} \tag{5.124}$$

the derivatives being defined as right-hand derivatives. Define

$$\varphi(\rho) \stackrel{\text{def}}{=} \int_{0-}^{\infty} \mathrm{e}^{-\rho\sigma}\, \mathrm{d}V(\sigma), \qquad \operatorname{Re} \rho \geqq 0. \tag{5.125}$$

It follows from (5.124) that

$$r_1 \int_{0+}^{K-} \mathrm{e}^{-\rho\sigma}\, \mathrm{d}V(\sigma) + r_2 \int_{K+}^{\infty} \mathrm{e}^{-\rho\sigma}\, \mathrm{d}V(\sigma) - \frac{1-\beta(\rho)}{\alpha\rho}\, \varphi(\rho) = 0, \qquad \operatorname{Re} \rho \geqq 0. \tag{5.126}$$

From (5.125) and (5.126) we obtain

$$\varphi(\rho) = \frac{V(0+) + (1 - r_2/r_1) \int_{K+}^{\infty} \mathrm{e}^{-\rho\sigma}\, \mathrm{d}V(\sigma)}{1 - \{1-\beta(\rho)\}/(\alpha r_1 \rho)}, \qquad \operatorname{Re} \rho \geqq 0, \tag{5.127}$$

and

$$\varphi(\rho) = \frac{V(0+) + (1 - r_1/r_2) \int_{0+}^{K-} \mathrm{e}^{-\rho\sigma}\, \mathrm{d}V(\sigma)}{1 - \{1-\beta(\rho)\}/(\alpha r_2 \rho)}, \qquad \operatorname{Re} \rho \geqq 0. \tag{5.128}$$

The condition

$$(\beta/\alpha)\, r_2 < 1, \tag{5.129}$$

is necessary and sufficient for the present model to have a stationary distribution. To prove this, note that it follows from the theory of the infinite dam $M/G/1$ with outflow rate r_2, that if we start in the present model with an initial dam content exceeding K the entrance time for a state with dam content less than K is finite with probability 1.

Since

$$\Big| \int_{K+}^{\infty} \mathrm{e}^{-\rho\sigma}\, \mathrm{d}V(\sigma) \Big| < \exp(-K \operatorname{Re} \rho),$$

it follows easily from (5.127) by using the inversion formula for the Laplace-Stieltjes transform that for $\operatorname{Re} \eta$ sufficiently large

$$V(\sigma) = V(0+)\, \frac{1}{2\pi \mathrm{i}} \int_{C_\eta} \frac{\alpha r_1\, \mathrm{e}^{\sigma\eta}}{\beta(\eta) + \alpha r_1 \eta - 1}\, \mathrm{d}\eta, \qquad 0 < \sigma < K. \tag{5.130}$$

Once $V(\sigma)$ for $\sigma \in (0, K)$ has been found from (5.130) we can determine $\int_{0+}^{K-} \mathrm{e}^{-\sigma\rho}\, \mathrm{d}V(\sigma)$, and then from (5.128) $V(\sigma)$ for $\sigma > K$ may be calculated by inversion; $V(0+)$ is then determined by the normalizing condition.

For $B(x)=1-\exp(-x/\beta)$, $x>0$, it is now easily found that

$$V(0+) = \left\{\frac{1}{1-a_1} + \left(\frac{-a_1}{1-a_1} + \frac{a_2}{1-a_2}\right)\exp\{-(1-a_1)\,K/\beta\}\right\}^{-1},$$

$$V(\sigma) = \begin{cases} V(0+)\,\dfrac{1}{1-a_1}\{1-a_1\exp\{-(1-a_1)\,\sigma/\beta\}\}, & 0<\sigma\leqq K, \\ V(0+)\left[\dfrac{1}{1-a_1}\{1-a_1\exp\{-(1-a_1)\,K/\beta\}\}\right. & \\ \quad + \dfrac{a_2}{1-a_2}\exp\{(a_1-a_2)\,K/\beta\}\{\exp\{-(1-a_2)\,K/\beta\} & \\ \quad \left. - \exp\{-(1-a_2)\,\sigma/\beta\}\}\right], & \text{for}\quad K<\sigma, \end{cases} \tag{5.131}$$

where

$$a_1 \overset{\text{def}}{=} \frac{\beta}{\alpha r_1}, \qquad a_2 \overset{\text{def}}{=} \frac{\beta}{\alpha r_2} < 1.$$

For $r_2\to\infty$ we again obtain the solution for the finite dam $M/M/1$ with capacity K. However, it is also of some interest to consider the distribution (5.131) for $r_1\to 0$ and also for $r_1\to\infty$.

(*ii*) $r(\sigma)=r_i>0$, for $K_{i-1}<\sigma<K_i$, $i=1, \ldots, m$,

$$K_0 \overset{\text{def}}{=} 0, \qquad K_m = \infty.$$

This model may be handled along the same lines as the preceding one. With

$$a_i \overset{\text{def}}{=} \frac{\beta}{\alpha r_i}, \qquad i=1, \ldots, m; \quad a_m<1,$$

it is found for the stationary distribution of the modified $M/M/1$ queue that

$$V(0+) = \left[\frac{1}{1-a_1} + \left(\frac{-a_1}{1-a_1} + \frac{a_2}{1-a_2}\right)\exp\{-(1-a_1)\,K_1/\beta\}\right.$$
$$+ \sum_{j=3}^{m}\left(\frac{-a_{j-1}}{1-a_{j-1}} + \frac{a_j}{1-a_j}\right)\exp\{-(1-a_{j-1})\,K_{j-1}/\beta\}$$
$$\left.\cdot\exp\{\sum_{i=2}^{j-1}(a_{i-1}-a_i)\,K_{i-1}/\beta\}\right]^{-1},$$

$$V(\sigma) = V(0+)\left\{\frac{1}{1-a_1} - \frac{a_1}{1-a_1}\exp\{-(1-a_1)\,\sigma/\beta\}\right\}, \qquad 0 \leqq \sigma < K_1,$$

$$V(\sigma) = V(0+)\Bigg[\frac{1}{1-a_1}(1-a_1\exp\{-(1-a_1)\,K_1/\beta\})$$

$$+\sum_{j=2}^{i-1}\frac{a_j}{1-a_j}\exp\{\sum_{n=2}^{j}(a_{n-1}-a_n)\,K_{n-1}/\beta\}$$

$$\cdot\{\exp\{-(1-a_j)\,K_{j-1}/\beta\}-\exp\{-(1-a_j)\,K_j/\beta\}\}$$

$$+\frac{a_i}{1-a_i}\exp\{\sum_{n=2}^{i}(a_{i-1}-a_n)\,K_{n-1}/\beta\}$$

$$\cdot\{\exp\{-(1-a_i)\,K_{i-1}/\beta\}-\exp\{-(1-a_i)\,\sigma/\beta\}\}\Bigg],$$

for $K_{i-1} < \sigma \leqq K_i, \quad i=2, \dots, m.$

(*iii*) $r(\sigma)=\sigma c, \quad \sigma>0,$

with c a positive constant. Hence for this case the outflow rate is proportional to the dam content. If the process has a stationary distribution $V(\sigma)$ then it follows from (5.123) that

$$\sigma\frac{\mathrm{d}}{\mathrm{d}\sigma}V(\sigma) = \frac{1}{\alpha c}\int_{0-}^{\sigma}\{1-B(\sigma-y)\}\,\mathrm{d}V(y), \qquad \sigma>0.$$

From this relation and from (5.125) we obtain

$$\frac{\mathrm{d}}{\mathrm{d}\rho}\varphi(\rho) = -\frac{1}{\alpha c}\varphi(\rho)\frac{1-\beta(\rho)}{\rho}, \qquad \mathrm{Re}\,\rho \geqq 0.$$

Hence, since $\varphi(0)=V(\infty)$,

$$\varphi(\rho) = \exp\left\{-\frac{1}{\alpha c}\int_{0}^{\rho}\frac{1-\beta(v)}{v}\,\mathrm{d}v\right\}. \tag{5.132}$$

Since for ρ real and $\rho\to\infty$ the integral in (5.132) diverges (the integral at the lower bound exists always if $\beta<\infty$) it follows that

$$V(0+) = \lim_{\rho\to\infty}\varphi(\rho) = 0, \qquad \rho \text{ real},$$

so that the probability of the empty state is zero. This fact is easily explained since $r(\sigma)=\sigma c$ implies that the content of the dam decreases exponentially with time between two successive inputs of the dam. The present model has been studied by KEILSON and MERMIN [1959].

III.5.11. The dam $G/G/1$ with non-instantaneous input

In the models of the dam discussed in the previous sections it has always been assumed that the input of water is instantaneous. In this section we shall consider briefly a dam model with infinite capacity and with gradual inflow. The dam content at time t will be denoted by $\boldsymbol{v}_t$. In figure 5 a realisation of $\boldsymbol{v}_t$ is shown for the dam model considered here.

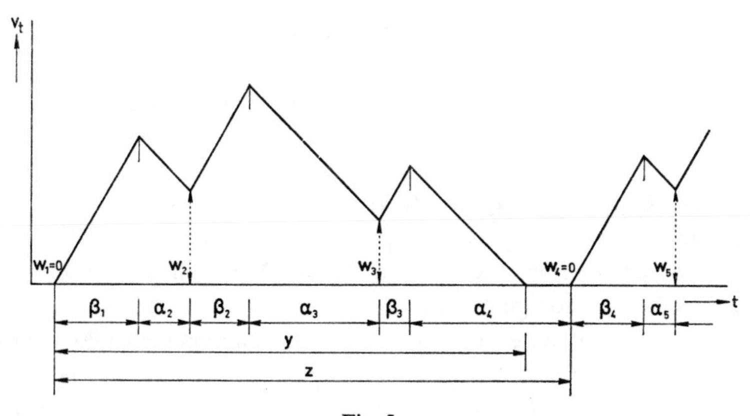

Fig. 5.

The variables $\boldsymbol{\beta}_n$, $n=1, 2, \ldots$, are independent, non-negative and identically distributed with distribution $B(.)$; similarly for $\boldsymbol{\alpha}_n$, $n=2, 3, \ldots$, with distribution $A(.)$. The families $\{\boldsymbol{\beta}_n, n=1, 2, \ldots\}$ and $\{\boldsymbol{\alpha}_n, n=2, 3, \ldots\}$ are assumed to be independent.

The release of water from the dam has a constant outflow rate r_0. The successive periods of inflow are $\boldsymbol{\beta}_1$, $\boldsymbol{\beta}_2$, … . During such periods the *net inflow rate* is constant and indicated by r_i. So that during the input period the total net inflow into the dam is $r_i\boldsymbol{\beta}_1$, while the total quantity of water which flowed into the dam during $\boldsymbol{\beta}_1$ is equal to $(r_0+r_i)\,\boldsymbol{\beta}_1$.

By $\boldsymbol{w}_n$, $n=1, 2, \ldots$, we shall denote the content of the dam just before the start of the nth inflow period.

It is easily seen that for $\boldsymbol{w}_1=0$,

$$\boldsymbol{w}_{n+1}=[\boldsymbol{w}_n+r_i\boldsymbol{\beta}_n-r_0\boldsymbol{\alpha}_{n+1}]^+, \qquad n=1,2,\dots, \qquad (5.133)$$
$$\boldsymbol{w}_1=0.$$

Obviously, this set of recurrence relations is of the same type as that for the $G/G/1$ queueing model (cf. section II.5.1).

Introducing the notation

$$\boldsymbol{a}_n \stackrel{\text{def}}{=} \begin{cases} \sum_{k=1}^{n} r_0\boldsymbol{\alpha}_{k+1}, \\ 0, \end{cases} \qquad \boldsymbol{b}_n \stackrel{\text{def}}{=} \begin{cases} \sum_{k=1}^{n} r_i\boldsymbol{\beta}_k, \\ 0, \end{cases} \qquad \boldsymbol{s}_n \stackrel{\text{def}}{=} \begin{cases} \boldsymbol{b}_n-\boldsymbol{a}_n, & n=1,2,\dots, \\ 0, & n=0, \end{cases}$$

it follows from section II.5.3 that

$$\sum_{n=1}^{\infty} r^n \,\mathrm{E}\{\exp(-\rho_1\boldsymbol{w}_n-\rho_3\boldsymbol{a}_{n-1}-\rho_4\boldsymbol{b}_{n-1}) \mid \boldsymbol{w}_1=0\}$$
$$= r\exp\left\{\sum_{n=1}^{\infty}\frac{r^n}{n}\,\mathrm{E}\{\exp(-\rho_4\boldsymbol{b}_n-\rho_3\boldsymbol{a}_n)\,((\boldsymbol{s}_n>0)\exp(-\rho_1\boldsymbol{s}_n)+(\boldsymbol{s}_n\leqq 0))\}\right\}, \qquad (5.134)$$

for $|r|<1$, $\operatorname{Re}\rho_1\geqq 0$, $\operatorname{Re}\rho_3\geqq 0$, $\operatorname{Re}\rho_4\geqq 0$.

In exactly the same way as in chapter II.5 we can now analyse the present model by starting from the relation (5.134). We shall only discuss some quantities for this storage model.

Denote the busy cycle of the dam, i.e. the time between two successive empty states of the dam at the start of input periods by $\boldsymbol{z}$, see fig. 5; $\boldsymbol{y}$ denotes the wet period of the dam. If $\boldsymbol{n}$ is the number of input periods during a busy cycle and

$$\boldsymbol{p} \stackrel{\text{def}}{=} \boldsymbol{b}_{\boldsymbol{n}}, \qquad \boldsymbol{c} \stackrel{\text{def}}{=} \boldsymbol{a}_{\boldsymbol{n}},$$

it follows that

$$\boldsymbol{z} = \frac{1}{r_i}\boldsymbol{b}_{\boldsymbol{n}} + \frac{1}{r_0}\boldsymbol{a}_{\boldsymbol{n}} = \frac{1}{r_i}\boldsymbol{p} + \frac{1}{r_0}\boldsymbol{c}, \qquad (5.135)$$

$$\boldsymbol{y} = \frac{r_i+r_0}{r_0 r_i}\boldsymbol{b}_{\boldsymbol{n}} = \left(\frac{1}{r_0}+\frac{1}{r_i}\right)\boldsymbol{p}.$$

From (5.134) we obtain

$$\mathrm{E}\{r^{\boldsymbol{n}}\exp(-\rho_3\boldsymbol{c}-\rho_4\boldsymbol{p})\} = 1-\exp\left\{-\sum_{n=1}^{\infty}\frac{r^n}{n}\,\mathrm{E}\{(\boldsymbol{s}_n\leqq 0)\exp(-\rho_3\boldsymbol{a}_n-\rho_4\boldsymbol{b}_n)\}\right\},$$

by using the same method as in section II.5.5, so that from (5.135) for $|r|\leqq 1$, Re $\rho_3\geqq 0$, Re $\rho_4\geqq 0$,

$$\mathrm{E}\{r^{\boldsymbol{n}}\exp(-\rho_3\boldsymbol{z}-\rho_4\boldsymbol{y})\} = 1-\exp\left\{-\sum_{n=1}^{\infty}\frac{r^n}{n}\right.$$

$$\left.\cdot\,\mathrm{E}\left\{\exp\left[-\frac{\rho_3}{r_0}\boldsymbol{a}_n-\left\{\frac{\rho_3}{r_i}+\left(\frac{1}{r_i}+\frac{1}{r_0}\right)\rho_4\right\}\boldsymbol{b}_n\right](\boldsymbol{s}_n\leqq 0)\right\}\right\}. \tag{5.136}$$

From the latter relation the transforms of the marginal distributions of $\boldsymbol{n}$, $\boldsymbol{z}$ and $\boldsymbol{y}$ can easily be obtained. By using Hewitt's inversion theorem (see section II.5.8) integral represention for these transforms are not difficult to derive.

From (5.136) and also from (5.133) it is readily seen that the variable $\boldsymbol{n}$ defined above has the same distribution as the number of customers served in a busy cycle in the $G/G/1$ system with service times $\boldsymbol{\tau}_n=r_i\boldsymbol{\beta}_n$ and inter-arrival times $\boldsymbol{\sigma}_n=r_0\boldsymbol{\alpha}_n$, $n=1, 2, \ldots$. Since for this process the realisations of $\boldsymbol{v}_t$, have the same number of peaks and dips in a busy cycle as the realisations of $\boldsymbol{v}_t$ for the dam with non-instantaneous input and since corresponding peaks and dips have the same height, it follows that the number of down crossings of $\boldsymbol{v}_t$ with a level K during a busy cycle (cf. section 5.8) has the same distribution for both processes. A similar conclusion applies for the probability that during a busy cycle $\boldsymbol{v}_t$ is less than K (cf. (5.87)).

Finally, we shall make some remarks on the $G/G/1$ dam with non-instantaneous input and with finite capacity K. The model is similar to that considered at the beginning of this section if the content $\boldsymbol{v}_t$ is less than K. However, if during an inflow period $\boldsymbol{\beta}_n$ the content $\boldsymbol{v}_t$ reaches K, so that overflow occurs, then for the remaining part of this inflow period the

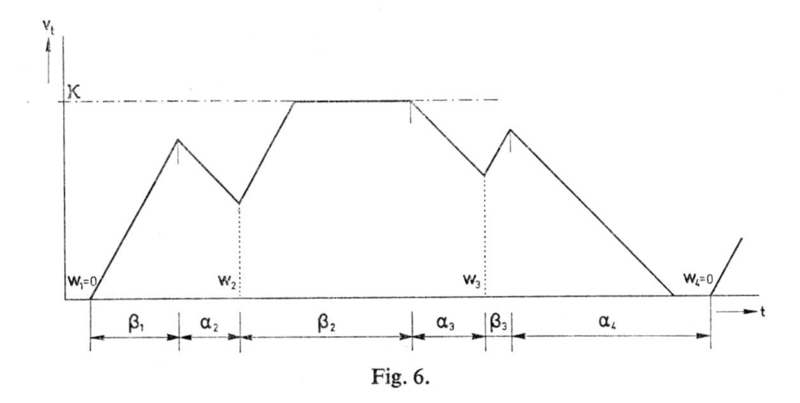

Fig. 6.

content $\boldsymbol{v}_t$ remains constant, i.e. during this time the inflow rate equals the outflow rate (see fig. 6).

Obviously, we have for the present case

$$\boldsymbol{w}_{n+1} = \begin{cases} [\boldsymbol{w}_n + r_i \boldsymbol{\beta}_n - r_0 \boldsymbol{\alpha}_{n+1}]^+ & \text{if } \boldsymbol{w}_n + r_i \boldsymbol{\beta}_n < K, \\ [K - r_0 \boldsymbol{\alpha}_{n+1}]^+ & \text{if } \boldsymbol{w}_n + r_i \boldsymbol{\beta}_n \geqq K, \end{cases}$$

or equivalently,

$$\boldsymbol{w}_{n+1} = [K - r_0 \boldsymbol{\alpha}_{n+1} + [\boldsymbol{w}_n + r_i \boldsymbol{\beta}_n - K]^-]^+.$$

From the last relation it is obvious that the process $\{\boldsymbol{w}_n,\ n=1, 2, \ldots\}$ is of exactly the same type as that described in section 5.1.

III.6. THE $M/G/1$ QUEUEING SYSTEM WITH FINITE WAITING ROOM

III.6.1. Introduction

In this chapter we shall investigate the queueing model $M/G/1$ as described in chapter II.4 but now with a limited number of waiting places. The total number of waiting places will be K. If at the moment of arrival of a customer the number of customers present in the system is less than $K+1$, the arriving customer is admitted to the system; he is served immediately if the server is idle, otherwise he occupies one of the waiting positions and waits for service. If at his arrival $K+1$ customers are present in the system, i.e. one being served and K waiting, the arriving customer is not admitted to the system. He disappears and never returns, and hence does not influence the development of the queueing process.

Denote the number of customers present in the system at time t by $\boldsymbol{x}_t$. We shall indicate the number of customers left behind in the system by the nth departing customer after $t=0$ by $\boldsymbol{z}_n$. Obviously $\{\boldsymbol{z}_n, n=1, 2, \ldots\}$ is a discrete time parameter Markov chain with stationary transition probabilities and with state space the set of integers 0, 1, 2, ..., K.

The number of customers arriving during the service between the nth and $(n+1)$th departure from the queueing system will be denoted by $\boldsymbol{g}_n$, $n=1, 2, \ldots$. Clearly

$$\mathrm{E}\{p^{\boldsymbol{g}_n}\} = \int_0^\infty \sum_{k=0}^\infty \frac{(t/\alpha)^k}{k!} \mathrm{e}^{-t/\alpha} p^k \,\mathrm{d}B(t) = \beta\{(1-p)/\alpha\}, \qquad \operatorname{Re} p \leqq 1. \tag{6.1}$$

From the description of the above model it follows that

$$\begin{aligned} \boldsymbol{z}_{n+1} &= \min(K, [\boldsymbol{z}_n - 1]^+ + \boldsymbol{g}_n), \qquad && n=1, 2, \ldots, \\ \boldsymbol{z}_1 &= z, && 0 \leqq z \leqq K, \end{aligned}$$

or equivalently,

$$
\begin{aligned}
z_{n+1} &= K+[[z_n-1]^+ + g_n - K]^-, && n=1,2,\ldots, \\
z_1 &= z, && 0 \leqq z \leqq K,
\end{aligned}
\tag{6.2}
$$

i.e. at the first departure z customers are left behind in the system.

Above we used the notation

$$[x]^+ = \max(0, x), \qquad [x]^- = \min(0, x), \qquad x \text{ real}.$$

Recurrence relations of the type (6.2) have been studied in chapters III.4 and III.5. A similar technique will be applied here. In the preceding chapters, however, the variables were continuous, whereas in (6.2) they are discrete. Hence, we shall use generating functions instead of Laplace-Stieltjes transforms for the investigation of the Markov chain $\{z_n, n=1, 2, \ldots\}$.

III.6.2. The integral equation and its solution

For x an integer and D_ξ a circle in the complex ξ-plane with centre at $\xi=0$ and with radius $|\xi|$ the following relations are easily proved by applying Cauchy's theorem:

$$p_1^{[x]^+} p_2^{[x]^-} = \frac{1}{2\pi i} \int_{D_\xi} \left(\frac{1}{\xi - p_1} - \frac{1}{\xi - p_2} \right) \xi^x \, d\xi, \qquad |p_1| < |\xi| < |p_2|, \tag{6.3}$$

$$p^{[x]^-} = \frac{1}{2\pi i} \int_{D_\xi} \left(\frac{1}{\xi - 1} - \frac{1}{\xi - p} \right) \xi^x \, d\xi, \qquad 1 < |\xi| < |p|, \tag{6.4}$$

$$p^{[x]^+} = \begin{cases} \dfrac{1}{2\pi i} \displaystyle\int_{D_\xi} \left(\dfrac{1}{\xi - p} - \dfrac{1}{\xi - 1} \right) \xi^x \, d\xi, & |p| < |\xi| < 1, \\[2ex] 1 - \dfrac{1}{2\pi i} \displaystyle\int_{D_\xi} \left(\dfrac{1}{\xi - 1} - \dfrac{1}{\xi - p} \right) \xi^x \, d\xi, & |\xi| > |p|, \ |\xi| > 1, \end{cases}$$

the positive direction of integration along D_ξ is counter clockwise.

From (6.2) and (6.4) it follows for $|p| > |\xi| > 1$, $|\eta| > |\xi|$, $|\eta| > 1$ and $n=1, 2, \ldots,$

$$p^{z_{n+1}} = \frac{p^K}{2\pi i} \int_{D_\xi} \left(\frac{1}{\xi - 1} - \frac{1}{\xi - p} \right) \xi^{g_n - K + [z_n - 1]^+} \, d\xi$$

$$= \frac{p^K}{2\pi i} \int_{D_\xi} \left(\frac{1}{\xi - 1} - \frac{1}{\xi - p} \right) \xi^{g_n - K} \, d\xi$$

$$- \frac{p^K}{2\pi i} \int_{D_\xi} \left(\frac{1}{\xi - 1} - \frac{1}{\xi - p} \right) \xi^{g_n - K} \frac{d\xi}{2\pi i} \left\{ \int_{D_\eta} \left(\frac{1}{\eta - 1} - \frac{1}{\eta - \xi} \right) \eta^{z_n - 1} \, d\eta \right\}.$$

Since the distribution of $\boldsymbol{g}_n$ is independent of n, and since $\boldsymbol{g}_n$ and z_n are independent variables it follows from (6.4) by defining

$$\boldsymbol{k}_n \stackrel{\text{def}}{=} \begin{cases} \sum_{j=1}^{n} \boldsymbol{g}_i, & n = 1, 2, \ldots, \\ 0, & n = 0, \end{cases} \tag{6.5}$$

that for $n = 1, 2, \ldots$; $|p| > |\xi| > 1$, $|\xi p_3| \leqq 1$, $|\eta| > |\xi|$, $|\eta| > 1$,

$$\mathrm{E}\{p^{z_{n+1}} p_3^{k_n} \mid z_1 = z\}$$

$$= \frac{p^K}{2\pi i} \int_{D_\xi} \left(\frac{1}{\xi - 1} - \frac{1}{\xi - p} \right) \xi^{-K} \, \mathrm{E}\{(\xi p_3)^{g_n} p_3^{k_{n-1}}\} \, d\xi$$

$$- \frac{p^K}{2\pi i} \int_{D_\xi} \left(\frac{1}{\xi - 1} - \frac{1}{\xi - p} \right) \xi^{-K} \, \mathrm{E}\{(\xi p_3)^{g_n}\} \frac{d\xi}{2\pi i}$$

$$\cdot \int_{D_\eta} \left(\frac{1}{\eta - 1} - \frac{1}{\eta - \xi} \right) \mathrm{E}\{\eta^{z_n} p_3^{k_{n-1}} \mid z_1 = z\} \frac{d\eta}{\eta}, \tag{6.6}$$

$$\mathrm{E}\{p^{z_1} p_3^{k_0} \mid z_1 = z\} = p^z.$$

In the derivation of (6.6) integration and expectation operators have been interchanged; it is easily proved that this is permitted since

$$0 \leqq z_n \leqq K \qquad \text{with probability } 1, \tag{6.7}$$

so that $\mathrm{E}\{|p|^{z_n} |p_3|^{k_{n-1}} \mid z_1 = z\}$ is bounded for $|p_3| \leqq 1$ and $|p|$ finite. Note that the sequence $\mathrm{E}\{p^{z_n} p_3^{k_{n-1}} \mid z_1 = z\}$ is uniquely determined by the set of recurrence relations (6.6).

For $|r|<1$, $|p_3|\leqq 1$, $|p|<\infty$, $0\leqq z\leqq K$, we define

$$\zeta(r,p,p_3,z)\overset{\text{def}}{=}\sum_{n=1}^{\infty} r^n\,\mathrm{E}\{p^{z_n}p_3^{k_{n-1}}\mid z_1=z\}, \tag{6.8}$$

then it follows from (6.7) that the function $\zeta(r,p,p_3,z)$ is an entire function of p for fixed r and p_3 with $|r|<1$, $|p_3|\leqq 1$; as a function of p_3 it has an analytic continuation for $\operatorname{Re} p_3\leqq 1$. Multiplying the relation (6.6) by r^n and summing over n leads to

$$\begin{aligned}\zeta(r,p,p_3,z)-rp^z &\\ &=\frac{rp^K}{2\pi\mathrm{i}}\int_{\mathrm{D}_\xi}\left(\frac{1}{\xi-1}-\frac{1}{\xi-p}\right)\xi^{-K}\frac{r\beta\{(1-\xi p_3)/\alpha\}}{1-r\beta\{(1-p_3)/\alpha\}}\,\mathrm{d}\xi\\ &\quad-\frac{rp^K}{2\pi\mathrm{i}}\int_{\mathrm{D}_\xi}\left(\frac{1}{\xi-1}-\frac{1}{\xi-p}\right)\xi^{-K}\beta\{(1-\xi p_3)/\alpha\}\\ &\quad\cdot\frac{\mathrm{d}\xi}{2\pi\mathrm{i}}\left\{\int_{\mathrm{D}_\eta}\left(\frac{1}{\eta-1}-\frac{1}{\eta-\xi}\right)\zeta(r,\eta,p_3,z)\frac{\mathrm{d}\eta}{\eta}\right\},\end{aligned} \tag{6.9}$$

for $|r|<1$, $|p|>|\xi|>1$, $|\xi p_3|<1$, $|p_3|\leqq 1$, $|\eta|>|\xi|$, $|\eta|>1$, $0\leqq z\leqq K$.

In (6.9) we replace the path of integration D_η by D_ω with $|\omega|<1$. Taking account of the residues at $\eta=\xi$ and at $\eta=1$ and noting that $\zeta(r,\eta,p_3,z)$ is an entire function of η it follows from (6.9) that

$$\begin{aligned}rp^{z-K}&=\frac{1}{2\pi\mathrm{i}}\int_{\mathrm{D}_\xi}\left(\frac{1}{\xi-1}-\frac{1}{\xi-p}\right)\xi^{-K}\left\{1-\frac{r}{\xi}\beta\{(1-\xi p_3)/\alpha\}\right\}\\ &\quad\cdot\zeta(r,\xi,p_3,z)\,\mathrm{d}\xi+\frac{r}{2\pi\mathrm{i}}\int_{\mathrm{D}_\xi}\left(\frac{1}{\xi-1}-\frac{1}{\xi-p}\right)\xi^{-K}\beta\{(1-\xi p_3)/\alpha\}\\ &\quad\cdot\frac{\mathrm{d}\xi}{2\pi\mathrm{i}}\left\{\int_{\mathrm{D}_\omega}\left(\frac{1}{\omega-1}-\frac{1}{\omega-\xi}\right)\zeta(r,\omega,p_3,z)\frac{\mathrm{d}\omega}{\omega}\right\},\end{aligned} \tag{6.10}$$

for $|p|>|\xi|>1$, $|\xi p_3|<1$, $|p_3|\leqq 1$, $|\omega|<1$, $|r|<1$, $0\leqq z\leqq K$.

To construct the solution of the integral equation (6.10) it is assumed for the present that $\beta(\rho)$ is a rational function of ρ, i.e.

$$\beta(\rho)=\frac{\beta_1(\rho)}{\beta_2(\rho)}, \tag{6.11}$$

with $\beta_2(\rho)$ a polynomial in ρ of degree n and $\beta_1(\rho)$ a polynomial in ρ of degree less than n. Our final results, however, may be shown to be independent of this assumption of rationality. It is merely introduced to keep the analysis as simple as possible. From Rouché's theorem it follows that the function

$$\xi - r\beta\{(1-\xi p_3)/\alpha\} \equiv \frac{1}{p_3}\{\xi p_3 - rp_3\beta\{(1-\xi p_3)/\alpha\}\},$$
$$|rp_3|\leqq 1, \quad p_3 \neq 0,$$

has $n+1$ zeros $\mu_i(r, p_3)$, $i=0, 1, \ldots, n$; and (cf. Takcás' lemma, app. 6),

$$\begin{aligned}
&|p_3\mu_0|<1, \ |p_3\mu_i|>1, && i=1, \ldots, n, && \text{if } |rp_3|\leqq 1, \ rp_3\neq 1, \quad (6.12)\\
&p_3\mu_0=1, \ |p_3\mu_i|>1, && i=1, \ldots, n, && \text{if } rp_3=1, \quad a<1,\\
&p_3\mu_0=p_3\mu_1=1, \ |p_3\mu_i|>1, && i=2, \ldots, n, && \text{if } rp_3=1, \quad a=1,\\
&p_3\mu_0<1, \ p_3\mu_1=1, \ |p_3\mu_i|>1, && i=2, \ldots, n, && \text{if } rp_3=1, \quad a>1,
\end{aligned}$$

with

$$a \stackrel{\text{def}}{=} \beta/\alpha, \qquad \mu_i \equiv \mu_i(r, p_3), \qquad i=0, \ldots, n.$$

For $|r|<1$, $|p_3|\leqq 1$, $0\leqq z\leqq K$ we introduce two functions $f(r, p_3, z)$ and $h(r, p, p_3, z)$ with $f(r, p_3, z)$ independent of p and with $h(r, p, p_3, z)$ a polynomial of degree $n-1$ in p. Moreover these functions should satisfy the conditions

$$rp^z + \frac{(1-p)f(r, p_3, z)}{p}\beta\{(1-pp_3)/\alpha\} + \frac{(1-p)\,h(r, p, p_3, z)}{\beta_2\{(1-pp_3)/\alpha\}}p^K = 0, \tag{6.13}$$

for $p=\mu_i(r, p_3)$, $i=0, \ldots, n$; $|r|<1$, $|p_3|\leqq 1$, $0\leqq z\leqq K$.
Clearly $f(r, p_3, z)$ and $h(r, p, p_3, z)$ are uniquely determined by these conditions.

For $|r|<1$, $|p_3|\leqq 1$, we define

$$\lambda(r, p_3) \stackrel{\text{def}}{=} \min_{i=1,\ldots,n} |\mu_i(r, p_3)|, \tag{6.14}$$

and

$$H_+(r, p, p_3) \stackrel{\text{def}}{=} \frac{p-r\beta\{(1-pp_3)/\alpha\}}{p-\mu_0(r, p_3)} \equiv \frac{pp_3-rp_3\beta\{(1-pp_3)/\alpha\}}{pp_3-p_3\mu_0(r, p_3)}; \tag{6.15}$$

obviously $H_+(r, p, p_3)$ is finite for $\operatorname{Re} pp_3 \leqq 1$, its zeros all lie outside the circle with radius $|\mu_0(r, p_3)|$.

By contour integration it is easily verified that the conditions (6.13) for $p=\mu_i(r, p_3)$, $i=1, \ldots, n$, are equivalent to

$$\frac{h(r, p, p_3, z)}{\beta_2\{(1-pp_3)/\alpha\}} = -\frac{H_+(r, p, p_3)}{2\pi \mathrm{i}} \int_{\mathrm{D}_\xi} \left\{ f(r, p_3, z)\, \xi^{-K-1} \beta\{(1-\xi p_3)/\alpha\} + \frac{r}{1-\xi}\, \xi^{z-K} \right\} \frac{1}{H_+(r, \xi, p_3)} \frac{\mathrm{d}\xi}{\xi - p}, \tag{6.16}$$

for $|\xi|>|p|$, $1<|\xi|<\lambda(r, p_3)$, $|r|<1$, $|p_3|\leqq 1$, $0\leqq z\leqq K$, and that $h(r, p, p_3, z)$ as given by (6.16) is a polynomial in p of degree $n-1$.

Next, we shall determine $f(r, p_3, z)$. We eliminate from (6.13) for $p=\mu_0$ and from (6.16) for $p=\mu_0$ the function $h(r, \mu_0, p_3, z)$. It follows for $1<|\xi| <\lambda(r, p_3)$, $|r|<1$, $|p_3|\leqq 1$, $|\omega|<|\mu_0|$ that

$$\left\{ \frac{r}{1-\mu_0}\, \mu_0^{z-K} + \frac{f(r, p_3, z)}{\mu_0}\, \mu_0^{-K} \beta\{(1-\mu_0 p_3)/\alpha\} \right\} \lim_{p\to\mu_0} \frac{p-\mu_0}{p - r\beta\{(1-pp_3)/\alpha\}}$$

$$= \frac{1}{2\pi \mathrm{i}} \int_{\mathrm{D}_\xi} \left\{ f(r, p_3, z)\, \xi^{-K-1} \beta\{(1-\xi p_3)/\alpha\} + \frac{r}{1-\xi}\, \xi^{z-K} \right\} \frac{\mathrm{d}\xi}{\xi - r\beta\{(1-\xi p_3)/\alpha\}}$$

$$= \left\{ f(r, p_3, z)\, \mu_0^{-K-1} \beta\{(1-\mu_0 p_3)/\alpha\} + \frac{r}{1-\mu_0}\, \mu_0^{z-K} \right\} \lim_{\xi\to\mu_0} \frac{\xi-\mu_0}{\xi - r\beta\{(1-\xi p_3)/\alpha\}}$$

$$- \frac{r}{1-r\beta\{(1-p_3)/\alpha\}} + \frac{1}{2\pi \mathrm{i}} \int_{\mathrm{D}_\omega} \left\{ f(r, p_3, z) \frac{\beta_2\{(1-\omega p_3)/\alpha\}}{\omega^{K+1}} + \frac{r}{1-\omega}\, \omega^{z-K} \right\} \frac{\mathrm{d}\omega}{\omega - r\beta\{(1-\omega p_3)/\alpha\}}.$$

Hence for $|r|<1$, $|p_3|\leqq 1$, $|\omega|<|\mu_0|$, $0\leqq z\leqq K$,

$$\frac{f(r, p_3, z)}{2\pi \mathrm{i}} \int_{\mathrm{D}_\omega} \frac{\mathrm{d}\omega}{\omega^{K+1}} \frac{\beta\{(1-\omega p_3)/\alpha\}}{\omega - r\beta\{(1-\omega p_3)/\alpha\}} = \frac{r}{1-r\beta\{(1-p_3)/\alpha\}} - \frac{1}{2\pi \mathrm{i}} \int_{\mathrm{D}_\omega} \frac{r}{1-\omega} \frac{\omega^{z-K}\, \mathrm{d}\omega}{\omega - r\beta\{(1-\omega p_3)/\alpha\}}, \qquad K\geqq 0, \tag{6.17}$$

and

$$\frac{f(r,p_3,z)}{2\pi i}\int_{D_\omega}\frac{\omega^{-K}\,d\omega}{\omega-r\beta\{(1-\omega p_3)/\alpha\}}=\frac{r^2}{1-r\beta\{(1-p_3)/\alpha\}}$$

$$-\frac{1}{2\pi i}\int_{D_\omega}\frac{r^2}{1-\omega}\,\frac{\omega^{z-K}\,d\omega}{\omega-r\beta\{(1-\omega p_3)/\alpha\}},\qquad K\geqq 1. \tag{6.18}$$

The relations (6.16) and (6.17) describe the functions $f(r, p_3, z)$ and $h(r, p, p_3, z)$ as defined above (cf. (6.13)).

We now prove that the function $\zeta(r, p, p_3, z)$ as defined by (6.8) is given by

$$\zeta(r,p,p_3,z)=\frac{p}{p-r\beta\{(1-pp_3)/\alpha\}}\cdot\Big\{rp^z+\frac{1-p}{p}\beta\{(1-pp_3)/\alpha\}\,f(r,p_3,z)$$

$$+\frac{(1-p)\,p^K}{\beta_2\{(1-pp_3)/\alpha\}}\,h(r,p,p_3,z)\Big\}, \tag{6.19}$$

for $|r|<1$, $|p_3|\leqq 1$, $0\leqq z\leqq K$, $|p|<\infty$.

From the conditions (6.13) it is seen that $\zeta(r, p, p_3, z)$ as given by (6.19) is an entire function of p. Substitution of the expression (6.19) into the integral equation (6.10) leads to

$$0=\frac{1}{2\pi i}\int_{D_\xi}\Big(\frac{1}{\xi-1}-\frac{1}{\xi-p}\Big)\Big\{(1-\xi)\,\xi^{-K-1}\beta\{(1-\xi p_3)/\alpha\}\,f(r,p_3,z)$$

$$+\frac{1-\xi}{\beta_2\{(1-\xi p_3)/\alpha\}}\,h(r,\xi,p_3,z)\Big\}\,d\xi$$

$$+\frac{r}{2\pi i}\int_{D_\xi}\Big(\frac{1}{\xi-1}-\frac{1}{\xi-p}\Big)\,\xi^{-K}\beta\{(1-\xi p_3)/\alpha\}$$

$$\cdot\frac{d\xi}{2\pi i}\Big\{\int_{D_\omega}\Big(\frac{1}{\omega-1}-\frac{1}{\omega-\xi}\Big)\Big[-\frac{1-\omega}{r\omega}f(r,p_3,z)$$

$$+\Big\{r\omega^{z-1}+\frac{(1-\omega)\,\omega^{K-1}}{\beta_2\{(1-\omega p_3)/\alpha\}}\,h(r,\omega,p_3,z)$$

$$+\frac{1-\omega}{r\omega}\,f(r,p_3,z)\Big\}\frac{\omega}{\omega-r\beta\{(1-\omega p_3)/\alpha\}}\Big]\,d\omega\Big\},$$

for $|p|>|\xi|>1$, $|\xi p_3|<1$, $|\omega|<1$, $|p_3|\leqq 1$, $|r|<1$, $0\leqq z\leqq K$.

The last term of the second integral in the above relation is an entire function of ω (cf. (6.13)). Hence the double integral in the relation above is equal to

$$\frac{1}{2\pi\mathrm{i}}\int_{D_\xi}\left(\frac{1}{\xi-1}-\frac{1}{\xi-p}\right)\xi^{-K}\beta\{(1-\xi p_3)/\alpha\}\frac{\xi-1}{\xi}f(r,p_3,z)\,\mathrm{d}\xi,$$

and the above relation reduces to

$$0=\frac{1}{2\pi\mathrm{i}}\int_{D_\xi}\left(\frac{1}{\xi-1}-\frac{1}{\xi-p}\right)\frac{1-\xi}{\beta_2\{(1-\xi p_3)/\alpha\}}h(r,\xi,p_3,z)\,\mathrm{d}\xi,$$

for $|p|>|\xi|>1$, $|\xi p_3|<1$, $|r|<1$, $|p_3|\leqq 1$; obviously, the latter relation is an identity. Hence, the expression (6.19) satisfies the integral equation (6.10). It remains to prove that $\zeta(r,p,p_3,z)$ as given by (6.19) is the generating function of the sequence $\mathrm{E}\{p^{z_n}p_3^{k_{n-1}}\mid z_1=z\}$, $n=1,2,\ldots$, which is uniquely determined by the set of recurrence relations (6.6). The proof of the latter statement is completely analogous to the proof of a similar statement in section 4.2 and is, therefore, omitted here.

It may now be proved that the assumption (6.11) concerning the rationality of $\beta(\rho)$ is superfluous. The statement (6.12) for $\mu_0(r,p_3)$ is still true but in general the number of zeros outside the unit circle is not finite. Note that in the expression (6.17) for $f(r,p_3,z)$ and in the expression (6.16) for $h(r,p,p_3,z)/\beta_2\{(1-pp_3)/\alpha\}$ only $\beta\{(1-p_3\xi)/\alpha\}$ occurs. The functions $\beta_1(\ldots)$ and $\beta_2(\ldots)$ do not occur here separately. Eliminating the quotient $h(r,p,p_3,z)/\beta_2\{(1-pp_3)/\alpha\}$ from (6.19) and (6.16) leads to an expression for $\zeta(r,p,p_3,z)$ with $f(r,p_3,z)$ given by (6.17) which is easily seen to be an entire function of p and which is a solution of the integral equation (6.10), representing the generating function of the sequence $\mathrm{E}\{p^{z_n}p_3^{k_{n-1}}\mid z_1=z\}$ defined by (6.6).

III.6.3. Stationary distributions

As already noted $\{z_n,\ n=1,2,\ldots\}$ is a Markov chain with stationary transition probabilities and with a finite and discrete state space. Since any two states of this Markov chain can be reached from each other in a finite number of transitions with positive probability, and since the chain is aperiodic it follows that all its states are positive recurrent, and that the chain has a stationary distribution. Hence we may define

$$z_i \overset{\text{def}}{=} \lim_{n\to\infty} \Pr\{z_n = i \mid z_1 = z\}, \qquad i=0, 1, \dots, K, \tag{6.20}$$

and the distribution $\{z_i,\ i=0, \dots, K\}$ is independent of z. Consequently, we have

$$z_j = \lim_{r\uparrow 1} (1-r) \frac{1}{2\pi\mathrm{i}} \int_{D_\omega} \frac{d\omega}{\omega^{j+1}} \zeta(r, \omega, 1, z). \tag{6.21}$$

To keep the analysis simple it will again be supposed that $\beta(\rho)$ is a rational function of ρ, although this assumption is not essential and does not influence the results. From (6.19) it follows for $|\omega| < \mu_0(r, 1)$, $|r|<1$,

$$\frac{1}{2\pi\mathrm{i}} \int_{D_\omega} \frac{d\omega}{\omega^{j+1}} \zeta(r, \omega, 1, z)$$
$$= \frac{1}{2\pi\mathrm{i}} \int_{D_\omega} \frac{d\omega}{\omega^{j+1}} \frac{r\omega^{z+1} + (1-\omega)\, \beta\{(1-\omega)/\alpha\} f(r, 1, z)}{\omega - r\beta\{(1-\omega)/\alpha\}}. \tag{6.22}$$

It is well-known (cf. app. 6) that $\mu_0(r, 1)$ is a continuous function of r for $|r| \leqq 1$,

$$\mu_0(1, 1) = \lim_{r\uparrow 1} \mu_0(r, 1),$$

that it is the zero with smallest absolute value of

$$p - \beta\{(1-p)/\alpha\};$$

$$\mu_0(1, 1) = 1 \qquad \text{if } \ a \leqq 1,$$

$$\mu_0(1, 1) < 1 \qquad \text{if } \ a > 1,$$

and that $\mu_0(1, 1)$ is the only zero inside the unit circle if $a>1$, for $a=1$ the multiplicity of the zero $\mu_0(1, 1)$ is two.
From (6.17) it follows immediately for $|\omega| < \mu_0(1, 1)$ that

$$\lim_{r\uparrow 1} (1-r) f(r, 1, z) = \left[\frac{1}{2\pi\mathrm{i}} \int_{D_\omega} \frac{d\omega}{\omega^{K+1}} \frac{\beta\{(1-\omega)/\alpha\}}{\omega - \beta\{(1-\omega)/\alpha\}} \right]^{-1}, \tag{6.23}$$

since the first integral in (6.17) converges uniformly in r for $|r| \leqq 1$ and $|\omega|$ sufficiently small. Similarly we obtain from (6.22) and (6.21) for $|\omega| < \mu_0(1, 1)$,

$$z_j = \frac{\dfrac{1}{2\pi i} \displaystyle\int_{D_\omega} \frac{d\omega}{\omega^{j+1}} \frac{(1-\omega)\,\beta\{(1-\omega)/\alpha\}}{\omega - \beta\{(1-\omega)/\alpha\}}}{\dfrac{1}{2\pi i} \displaystyle\int_{D_\omega} \frac{d\omega}{\omega^{K+1}} \frac{\beta\{(1-\omega)/\alpha\}}{\omega - \beta\{(1-\omega)/\alpha\}}}, \qquad j=0, 1, \ldots, K. \tag{6.24}$$

For $a<1$ denote by $\{v_j, j=0, 1, \ldots\}$ the probability distribution of which the generating function is given by

$$\sum_{j=0}^{\infty} v_j p^j = (1-a)(1-p) \frac{\beta\{(1-p)/\alpha\}}{\beta\{(1-p)/\alpha\}-p}, \qquad |p| \leqq 1. \tag{6.25}$$

From chapter II.4 we know that $\{v_j, j=0, 1, \ldots\}$ is the stationary distribution of the number of customers left behind in the system by a departing customer for the $M/G/1$ queueing system with an unbounded number of waiting positions.

Obviously it follows from (6.24) that

$$z_j = \frac{v_j}{v_0+v_1+\ldots+v_K}, \qquad j=0, 1, \ldots, K, \qquad \text{if} \quad a<1. \tag{6.26}$$

Denote the past service time of the customer served at time t if the server is busy at time t by $\boldsymbol{\eta}_t$. Obviously, the process $\{(\boldsymbol{x}_t, \boldsymbol{\eta}_t), t \in [0, \infty)\}$ is a Markov process with stationary transition probabilities and with state space the product space $\{0, 1, \ldots, K+1\} \times [0, \infty)$. Markov processes of this type have been described by BLANC-LAPIERRE and FORTET [1953] (see also FORTET and GRANDJEAN [1964] and GNEDENKO and KOVALENKO [1968]). From their results it follows that if $B(t)$ is absolutely continuous and if

$$\frac{1}{1-B(t)} \frac{dB(t)}{dt},$$

is bounded for $t>0$, then the transition probabilities

$$R_0(t) = \Pr\{\boldsymbol{x}_t = 0 \mid \boldsymbol{x}_0 = 0\}, \tag{6.27}$$

$$R_j(\eta, t)\, d\eta = \Pr\{\boldsymbol{x}_t = j, \eta \leqq \boldsymbol{\eta}_t < \eta + d\eta \mid \boldsymbol{x}_0 = 0\}, \qquad j=1, \ldots, K+1,$$

for $\eta>0$, $t \geqq 0$ are well defined and are the only solution of the forward Kolmogorof relations for the process $\{(\boldsymbol{x}_t, \boldsymbol{\eta}_t), t \in [0, \infty)\}$ representing transition probabilities. The forward Kolmogorof relations and the boundary conditions for $t>0$ are

$$\frac{\mathrm{d}}{\mathrm{d}t} R_0(t) + \alpha^{-1} R_0(t) = \int_{\eta=0}^{t} R_1(\eta, t) \frac{\mathrm{d}B(\eta)}{1 - B(\eta)}, \tag{6.28}$$

$$\frac{\partial}{\partial t} R_1(\eta, t) + \frac{\partial}{\partial \eta} R_1(\eta, t) = - \left\{ \alpha^{-1} + \frac{1}{1 - B(\eta)} \frac{\mathrm{d}B(\eta)}{\mathrm{d}\eta} \right\} R_1(\eta, t), \qquad \eta > 0,$$

$$\frac{\partial}{\partial t} R_j(\eta, t) + \frac{\partial}{\partial \eta} R_j(\eta, t) = - \left\{ \alpha^{-1} + \frac{1}{1 - B(\eta)} \frac{\mathrm{d}B(\eta)}{\mathrm{d}\eta} \right\} R_j(\eta, t) + \alpha^{-1} R_{j-1}(\eta, t), \qquad \eta > 0, \quad j = 2, \ldots, K,$$

$$\frac{\partial}{\partial t} R_{K+1}(\eta, t) + \frac{\partial}{\partial \eta} R_{K+1}(\eta, t) = - \frac{1}{1 - B(\eta)} \frac{\mathrm{d}B(\eta)}{\mathrm{d}\eta} R_{K+1}(\eta, t) + \alpha^{-1} R_K(\eta, t), \qquad \eta > 0,$$

$$R_1(0+, t) = \int_0^t R_2(\eta, t) \frac{\mathrm{d}B(\eta)}{1 - B(\eta)} + \alpha^{-1} R_0(t), \tag{6.29}$$

$$R_j(0+, t) = \int_0^t R_{j+1}(\eta, t) \frac{\mathrm{d}B(\eta)}{1 - B(\eta)}, \qquad j = 1, \ldots, K,$$

$$R_{K+1}(0+, t) = 0, \qquad t \geqq 0,$$

$$R_0(0+) = 1, \qquad R_j(\eta, 0+) = 0, \qquad \eta > 0; \quad j = 1, \ldots, K+1. \tag{6.30}$$

Moreover, it follows from the studies referred to above, that the Markov process has a unique stationary distribution, $\{R_0, R_j(\eta), j = 1, \ldots, K+1\}$, $\eta > 0$ and

$$R_0 = \lim_{t \to \infty} R_0(t), \tag{6.31}$$

$$R_j(\eta) = \lim_{t \to \infty} R_j(\eta, t), \qquad \eta > 0, \quad j = 1, \ldots, K+1.$$

From (6.28), ..., (6.30) it follows for R_0, $R_j(\eta)$, $j = 1, \ldots, K+1$ that,

$$\alpha^{-1}R_0 = \int_0^\infty R_1(\eta)\frac{dB(\eta)}{1-B(\eta)}, \tag{6.32}$$

$$\frac{d}{d\eta}R_1(\eta) = -\left\{\alpha^{-1} + \frac{1}{1-B(\eta)}\frac{dB(\eta)}{d\eta}\right\}R_1(\eta), \qquad \eta>0,$$

$$\frac{d}{d\eta}R_j(\eta) = -\left\{\alpha^{-1} + \frac{1}{1-B(\eta)}\frac{dB(\eta)}{d\eta}\right\}R_j(\eta) + \alpha^{-1}R_{j-1}(\eta),$$

$$j=2, \ldots, K; \quad \eta>0,$$

$$\frac{d}{d\eta}R_{K+1}(\eta) = -\frac{1}{1-B(\eta)}\frac{dB(\eta)}{d\eta}R_{K+1}(\eta) + \alpha^{-1}R_K(\eta), \qquad \eta>0,$$

$$R_1(0+) = \int_0^\infty R_2(\eta)\frac{dB(\eta)}{1-B(\eta)} + \alpha^{-1}R_0,$$

$$R_j(0+) = \int_0^\infty R_{j+1}(\eta)\frac{dB(\eta)}{1-B(\eta)}, \qquad j=2, \ldots, K,$$

$$R_{K+1}(0+) = 0.$$

The normalizing condition is given by

$$R_0 + \sum_{j=1}^{K+1}\int_0^\infty R_j(\eta)\,d\eta = 1. \tag{6.33}$$

It is not difficult to prove that the system of equations (6.32) and (6.33) has a uniquely determined solution. The solution of this system of equations, and hence the stationary distribution for the process $\{(\boldsymbol{x}_t, \boldsymbol{\eta}_t),\ t\in[0, \infty)\}$ reads:

for $|\omega|<\mu_0(1, 1)\leqq 1$, $K\geqq 1$,

$$R_0 = -\frac{\alpha}{D} = \frac{\alpha}{2\pi i D}\int_{D_\omega}\frac{d\omega}{\omega}\,\frac{(1-\omega)\,\beta\{(1-\omega)/\alpha\}}{\omega-\beta\{(1-\omega)/\alpha\}}, \tag{6.34}$$

$$R_j(\eta) = \frac{1-B(\eta)}{2\pi i D}\int_{D_\omega}\frac{d\omega}{\omega^j}\,\frac{(1-\omega)\exp\{-\eta(1-\omega)/\alpha\}}{\omega-\beta\{(1-\omega)/\alpha\}},$$

$$j=1, \ldots, K, \quad \eta>0,$$

$$R_{K+1}(\eta) = \frac{1-B(\eta)}{2\pi i D}\int_{D_\omega}\frac{1-\exp\{-\eta(1-\omega)/\alpha\}}{\omega-\beta\{(1-\omega)/\alpha\}}\,\frac{d\omega}{\omega^K}, \qquad \eta>0,$$

with

$$-\frac{\alpha}{D} \overset{\text{def}}{=} \left[1 + \frac{a}{2\pi i} \int_{D_\omega} \frac{1}{\beta\{(1-\omega)/\alpha\} - \omega} \frac{d\omega}{\omega^K}\right]^{-1}, \qquad a = \frac{\beta}{\alpha}. \tag{6.35}$$

The proof of the above statement is obtained by direct substitution of the expressions (6.34) into the equations (6.32) and (6.33) and by noting that

$$\frac{d}{d\eta} \int_{D_\omega} \frac{d\omega}{\omega^j} \frac{(1-\omega)\exp\{-\eta(1-\omega)/\alpha\}}{\omega - \beta\{(1-\omega)/\alpha\}} = -\int_{D_\omega} \frac{d\omega}{\alpha\omega^j} \frac{(1-\omega)^2 \exp\{-\eta(1-\omega)/\alpha\}}{\omega - \beta\{(1-\omega)/\alpha\}},$$

since the latter integral converges on D_ω uniformly in η for $\eta \geqq 0$. From (6.34) it follows easily that

$$R_0 = -\frac{\alpha}{D}, \tag{6.36}$$

$$R_j \overset{\text{def}}{=} \int_0^\infty R_j(\eta)\, d\eta = -\frac{\alpha}{2\pi i D} \int_{D_\omega} \frac{1-\beta\{(1-\omega)/\alpha\}}{\beta\{(1-\omega)/\alpha\} - \omega} \frac{d\omega}{\omega^j}, \qquad j = 1, \dots, K,$$

$$R_{K+1} \overset{\text{def}}{=} \int_0^\infty R_{K+1}(\eta)\, d\eta = -\frac{\alpha}{2\pi i D} \int_{D_\omega} \frac{1}{\beta\{(1-\omega)/\alpha\} - \omega} \cdot \left\{a - \frac{1-\beta\{(1-\omega)/\alpha\}}{1-\omega}\right\} \frac{d\omega}{\omega^K},$$

for $|\omega| < \mu_0(1, 1)$, $K \geqq 1$.

Here $\{R_j, j=0, \dots, K+1\}$ is the stationary distribution of the process $\{\boldsymbol{x}_t, t \in [0, \infty)\}$. It should be noted that for finite K the processes $\{\boldsymbol{z}_n, n=1, 2, \dots\}$ and $\{\boldsymbol{x}_t, t \in [0, \infty)\}$ do not have the same stationary distribution. However, from the relation

$$1 - R_{K+1} = -\frac{\alpha}{D} \frac{1}{2\pi i} \int_{D_\omega} \frac{1}{\beta\{(1-\omega)/\alpha\} - \omega} \frac{d\omega}{\omega^K}, \qquad |\omega| < \mu_0(1, 1),$$

it is easily proved that

$$z_j = \frac{R_j}{1-R_{K+1}}, \qquad j=0, \dots, K. \tag{6.37}$$

It is noted that R_{K+1} is the probability that an arriving customer is not admitted to the system.

Denote the virtual waiting time at time t by $\boldsymbol{v}_t$. Hence for $t>0$,

$$\Pr\{\boldsymbol{v}_t=0 \mid \boldsymbol{x}_0=0\} = \Pr\{\boldsymbol{x}_t=0 \mid \boldsymbol{x}_0=0\}, \tag{6.38}$$

$$\Pr\{0<\boldsymbol{v}_t<\sigma \mid \boldsymbol{x}_0=0\} =$$

$$\sum_{j=1}^{K+1} \int_{\eta=0}^{t} \{\mathrm{d}_\eta \Pr\{\boldsymbol{x}_t=j,\, \boldsymbol{\eta}_t<\eta \mid \boldsymbol{x}_0=0\}\} \frac{B(\eta+\sigma)-B(\eta)}{1-B(\eta)} * B^{(j-1)*}(\sigma).$$

From (6.27), (6.31) and (6.38) it follows by applying Helly-Bray's theorem (cf. app. 5) that $\boldsymbol{v}_t$ converges in distribution for $t\to\infty$. Hence for $\operatorname{Re}\rho\geqq 0$,

$$\int_{0-}^{\infty} \mathrm{e}^{-\rho\sigma}\, \mathrm{d}_\sigma \lim_{t\to\infty} \Pr\{\boldsymbol{v}_t<\sigma\}$$

$$= R_0 + \sum_{j=1}^{K+1} \int_0^\infty \mathrm{e}^{-\rho\sigma}\, \mathrm{d}_\sigma \left\{ \int_{\eta=0}^{\infty} R_j(\eta) \frac{B(\eta+\sigma)-B(\eta)}{1-B(\eta)} * B^{(j-1)*}(\sigma)\, \mathrm{d}\eta \right\}.$$

From (6.34) and the latter relation we obtain for $|\omega|<\mu_0(1, 1)$, $\operatorname{Re}\rho\geqq 0$,

$$\int_{0-}^{\infty} \mathrm{e}^{-\rho\sigma}\, \mathrm{d}_\sigma \lim_{t\to\infty} \Pr\{\boldsymbol{v}_t<\sigma\} = -\frac{\alpha}{D}$$

$$+\frac{\alpha}{2\pi \mathrm{i} D} \int_{\mathrm{D}_\omega} \frac{1-\omega}{\omega-\beta\{(1-\omega)/\alpha\}} \frac{\beta\{(1-\omega)/\alpha\}-\beta(\rho)}{\alpha\rho+\omega-1} \left\{1-\frac{\beta^K(\rho)}{\omega^K}\right\} \frac{\mathrm{d}\omega}{\omega-\beta(\rho)}$$

$$+\frac{\alpha}{2\pi \mathrm{i} D} \int_{\mathrm{D}_\omega} \left\{ \frac{1-\beta(\rho)}{\alpha\rho} - \frac{\beta\{(1-\omega)/\alpha\}-\beta(\rho)}{\alpha\rho+\omega-1} \right\} \frac{\beta^K(\rho)}{\omega^K} \frac{\mathrm{d}\omega}{\omega-\beta\{(1-\omega)/\alpha\}}.$$

To simplify the latter expression we take $\operatorname{Re}\rho$ so large that $|\beta(\rho)|<|\omega|$ and $\operatorname{Re}(\alpha\rho+\omega-1)>0$ for $|\omega|<\mu_0(1, 1)$. It then follows by contour integration that

$$\int_{0-}^{\infty} \mathrm{e}^{-\rho\sigma}\,\mathrm{d}_\sigma \lim_{t\to\infty} \Pr\{\boldsymbol{v}_t<\sigma\} = -\frac{\alpha}{D}\frac{\alpha\rho}{\alpha\rho+\beta(\rho)-1}$$

$$+\frac{1-\beta(\rho)}{\alpha\rho}\frac{\alpha}{2\pi\mathrm{i}D}\int_{\mathrm{D}_\omega}\frac{\omega-1}{\alpha\rho+\omega-1}\frac{\beta^K(\rho)}{\omega^K}\frac{\mathrm{d}\omega}{\omega-\beta\{(1-\omega)/\alpha\}}$$

$$-\frac{1-\beta(\rho)}{1-\alpha\rho-\beta(\rho)}\frac{\beta^K(\rho)}{(1-\alpha\rho)^K}\frac{\alpha}{D}, \tag{6.39}$$

for $|\omega|<\mu_0(1, 1)$ and Re ρ sufficiently large.

Since the distribution of the interarrival time is negative exponential

$$\frac{R_j}{1-R_{K+1}}, \qquad j=0, \ldots, K,$$

is, for the stationary situation, the probability that an arriving customer finds j customers in the system when he is admitted to the system. As above it is easily proved that for the 'first come, first served' discipline the actual waiting time of a customer arriving at time t converges in distribution for $t\to\infty$. Denote the limit distribution of the actual waiting time by $W(t)$, then for $t>0$,

$$\int_{0-}^{\infty} \mathrm{e}^{-\rho t}\,\mathrm{d}W(t) = \frac{R_0}{1-R_{K+1}}$$

$$+\int_0^\infty \mathrm{e}^{-\rho t}\,\mathrm{d}_t \sum_{j=1}^K \int_{\eta=0}^\infty \frac{R_j(\eta)}{1-R_{K+1}}\frac{B(\eta+t)-B(\eta)}{1-B(\eta)} * B^{(j-1)*}(t)\,\mathrm{d}\eta.$$

From (6.34) it is now found for $|\omega|<\mu_0(1, 1)$ and Re ρ sufficiently large that

$$\int_{0-}^{\infty} \mathrm{e}^{-\rho t}\,\mathrm{d}W(t) = \frac{1}{C}\frac{\alpha\rho}{\beta(\rho)+\alpha\rho-1}\left\{1-\frac{\beta^K(\rho)}{(1-\alpha\rho)^K}\right\}$$

$$+\frac{1}{2\pi\mathrm{i}C}\int_{\mathrm{D}_\omega}\frac{1-\omega}{\omega-\beta\{(1-\omega)/\alpha\}}\frac{\beta^K(\rho)}{\omega^K}\frac{\mathrm{d}\omega}{\alpha\rho+\omega-1}, \tag{6.40}$$

with

$$C \stackrel{\text{def}}{=} \frac{1}{2\pi\mathrm{i}}\int_{\mathrm{D}_\omega}\frac{1}{\beta\{(1-\omega)/\alpha\}-\omega}\frac{\mathrm{d}\omega}{\omega^K}, \qquad |\omega|<\mu_0(1, 1).$$

III.6.4. Entrance times and return times

We have already noted that the process $\{z_n, n=1, 2, \ldots\}$ is a discrete time parameter Markov chain with stationary transition probabilities. For this process we first define a number of stochastic variables and transition probabilities.

For $z=0, \ldots, K$, define

$$\boldsymbol{n}_{z0} \overset{\text{def}}{=} \begin{cases} 1 & \text{if} \quad z_2=0, \quad z_1=z, \\ \min\limits_{n=2,3,\ldots} \{n\colon z_{n+1}=0, z_i>0, i=2, \ldots, n\} & \text{if} \quad z_1=z, \end{cases} \tag{6.41}$$

$$\boldsymbol{n}_{zK} \overset{\text{def}}{=} \begin{cases} 1 & \text{if} \quad z_2=K, \quad z_1=z, \\ \min\limits_{n=2,3,\ldots} \{n\colon z_{n+1}=K, z_i<K, i=2, \ldots, n\} & \text{if} \quad z_1=z, \end{cases}$$

$$\boldsymbol{K}_{z0} \overset{\text{def}}{=} \boldsymbol{k}_{\boldsymbol{n}_{z0}} = \sum_{i=1}^{\boldsymbol{n}_{z0}} \boldsymbol{g}_i, \qquad \boldsymbol{K}_{zK} \overset{\text{def}}{=} \boldsymbol{k}_{\boldsymbol{n}_{zK}} = \sum_{i=1}^{\boldsymbol{n}_{zK}} \boldsymbol{g}_i.$$

Denote by E_0 the event that immediately after a departure the system is empty, and by E_K the event that a departing customer leaves K customers behind in the system, then $\boldsymbol{n}_{00}$ is the return time in discrete time of the event E_0, or equivalently, the number of customers served in a busy period. Similarly, $\boldsymbol{n}_{KK}$ is the return time in discrete time of the event E_K. Obviously, $\boldsymbol{K}_{00}$ is the number of customers arriving in a busy cycle, while $\boldsymbol{n}_{KK}$ is the number of customers arriving between two successive occurrences of E_K.

We also define for $z=0, \ldots, K$; $i=0, 1, \ldots$; $n=1, 2, \ldots$,

$$p_{z0}^{(n)}(i) \overset{\text{def}}{=} \Pr\{z_{n+1}=0, \boldsymbol{k}_n=i \mid z_1=z\}, \tag{6.42}$$

$$p_{zK}^{(n)}(i) \overset{\text{def}}{=} \Pr\{z_{n+1}=K, \boldsymbol{k}_n=i \mid z_1=z\},$$

and for $|p_3| \leqq 1$,

$$\tilde{p}_{z0}^{(n)}(p_3) \overset{\text{def}}{=} \sum_{j=0}^{\infty} p_3^j p_{z0}^{(n)}(j); \qquad \tilde{p}_{zK}^{(n)}(p_3) \overset{\text{def}}{=} \sum_{j=0}^{\infty} p_3^j p_{zK}^{(n)}(j). \tag{6.43}$$

Further for $z=0, 1, \ldots, K$; $i=0, 1, \ldots$; $n=1, 2, \ldots$,

$$f_{z0}^{(n)}(i) \overset{\text{def}}{=} \Pr\{\boldsymbol{n}_{z0}=n, \boldsymbol{K}_{z0}=i\},$$

$$f_{zK}^{(n)}(i) \overset{\text{def}}{=} \Pr\{\boldsymbol{n}_{zK}=n, \boldsymbol{K}_{zK}=i\},$$

$$p_{K;z0}^{(n)}(i) \stackrel{\text{def}}{=} \begin{cases} p_{z0}^{(1)}(i), & n=1, \\ \Pr\{z_{n+1}=0,\ z_j<K,\ j=2, \ldots, n;\ k_n=i \mid z_1=z\}, & n=2, 3, \ldots, \end{cases}$$

$$p_{0;zK}^{(n)}(i) \stackrel{\text{def}}{=} \begin{cases} p_{zK}^{(1)}(i), & n=1, \\ \Pr\{z_{n+1}=K,\ z_j>0,\ j=2, \ldots, n;\ k_n=i \mid z_1=z\}, & n=2, 3, \ldots, \end{cases}$$

$$f_{K;z0}^{(n)}(i) \stackrel{\text{def}}{=} \begin{cases} f_{z0}^{(1)}(i), & n=1, \\ \Pr\{n_{z0}=n,\ K_{z0}=i,\ z_j<K,\ j=2, \ldots, n\}, & n=2, 3, \ldots, \end{cases}$$

$$f_{0;zK}^{(n)}(i) \stackrel{\text{def}}{=} \begin{cases} f_{zK}^{(1)}(i), & n=1, \\ \Pr\{n_{zK}=n,\ K_{zK}=i,\ z_j>0,\ j=2, \ldots, n\}, & n=2, 3, \ldots. \end{cases}$$

Denoting by $\tilde{f}_{\dots}^{\dots}(p_3)$ and $\tilde{p}_{\dots}^{\dots}(p_3)$ the generating functions with respect to i of $f_{\dots}^{\dots}(i)$ and $p_{\dots}^{\dots}(i)$ (cf. (6.43)) it follows easily from the Markov property of the process $\{z_n, n=1, 2, \ldots\}$ that for $z=0, \ldots, K$; $|p_3| \leqq 1$, $n=2, 3, \ldots$,

$$\tilde{p}_{z0}^{(n)}(p_3) = \tilde{p}_{K;z0}^{(n)}(p_3) + \sum_{m=1}^{n-1} \tilde{f}_{zK}^{(m)}(p_3)\, \tilde{p}_{K0}^{(n-m)}(p_3), \tag{6.44}$$

$$\tilde{p}_{zK}^{(n)}(p_3) = \tilde{p}_{0;zK}^{(n)}(p_3) + \sum_{m=1}^{n-1} \tilde{f}_{z0}^{(m)}(p_3)\, \tilde{p}_{0K}^{(n-m)}(p_3),$$

$$\tilde{p}_{K;z0}^{(n)}(p_3) = \tilde{f}_{K;z0}^{(n)}(p_3) + \sum_{m=1}^{n-1} \tilde{f}_{K;z0}^{(m)}(p_3)\, \tilde{p}_{K;00}^{(n-m)}(p_3),$$

$$\tilde{p}_{0;zK}^{(n)}(p_3) = \tilde{f}_{0;zK}^{(n)}(p_3) + \sum_{m=1}^{n-1} \tilde{f}_{0;zK}^{(m)}(p_3)\, \tilde{p}_{0;KK}^{(n-m)}(p_3).$$

Finally, we define for $|r|<1$, $|p_3| \leqq 1$, $z=0, \ldots, K$,

$$r^2 c_f(r, p_3, z) \stackrel{\text{def}}{=} \begin{cases} -f(r, p_3, 0) - r^2, & \text{for } z=0, \\ -f(r, p_3, z), & \text{for } z>0, \end{cases} \tag{6.45}$$

$$r c_h(r, p_3, z) \stackrel{\text{def}}{=} - \lim_{|p|\to\infty} \frac{p h(r, p, p_3, z)}{\beta_2\{(1-pp_3)/\alpha\}}.$$

The function $f(r, p, z)$ is given by (6.17). From (6.16) we have

$$r c_h(r, p_3, z) = \lim_{|p|\to\infty} p \frac{p - r\beta\{(1-pp_3)/\alpha\}}{p-\mu_0}$$

$$\cdot \frac{1}{2\pi i} \int_{D_\xi} \left\{ f(r, p_3, z)\, \xi^{-K-1}\, \beta\{(1-\xi p_3)/\alpha\} + \frac{r}{1-\xi}\, \xi^{z-K} \right\}$$

$$\cdot \frac{\xi-\mu_0}{\xi - r\beta\{(1-\xi p_3)/\alpha\}} \frac{d\xi}{\xi-p},$$

for $|\xi|>|p|$, $1<|\xi|<\lambda(r,p_3)$, $|r|<1$, $|p_3|\leqq 1$. Replacing the path D_ξ by D_ω with $|\omega|<|\mu_0|$, it follows for $|\omega|<|p|$,

$$
\begin{aligned}
rc_h(r,p_3,z) = \lim_{|p|\to\infty} p\Bigg[\Big\{ & f(r,p_3,z)\,p^{-K-1}\beta\{(1-pp_3)/\alpha\} \\
&+\frac{r}{1-p}p^{z-K}\Big\}\frac{p-\mu_0}{p-r\beta\{(1-pp_3)/\alpha\}} \\
&-r\frac{1-\mu_0}{1-r\beta\{(1-p_3)/\alpha\}}\,\frac{1}{1-p}\Bigg] \\
&+\lim_{|p|\to\infty}\frac{p}{2\pi i}\int_{D_\omega}\Big\{f(r,p_3,z)\,\omega^{-K-1}\beta\{(1-\omega p_3)/\alpha\} \\
&+\frac{r}{1-\omega}\,\omega^{z-K}\Big\}\frac{\omega-\mu_0}{\omega-r\beta\{(1-\omega p_3)/\alpha\}}\,\frac{d\omega}{\omega-p}.
\end{aligned}
$$

Hence, by noting that

$$
\frac{\beta\{(1-\xi p_3)/\alpha\}}{\xi-r\beta\{(1-\xi p_3)/\alpha\}}=\frac{1}{r}\left\{-1+\frac{\xi}{\xi-r\beta\{(1-\xi p_3)/\alpha\}}\right\},
$$

and using (6.17) it follows for $|\omega|<|\mu_0|$, $|r|<1$, $|p_3|\leqq 1$ that

$$
rc_h(r,p_3,z)
$$

$$
=\begin{cases}
\dfrac{r}{1-r\beta\{(1-p_3)/\alpha\}}-\dfrac{1}{2\pi i}\displaystyle\int_{D_\omega}\frac{f(r,p_3,z)}{r\omega^{K-1}}\,\frac{d\omega}{\omega-r\beta\{(1-\omega p_3)/\alpha\}} \\
\qquad -\dfrac{r}{2\pi i}\displaystyle\int_{D_\omega}\frac{\omega^{z+1-K}}{1-\omega}\,\frac{d\omega}{\omega-r\beta\{(1-\omega p_3)/\alpha\}}, \qquad z<K, \\
\dfrac{r^2\beta\{(1-p_3)/\alpha\}}{1-r\beta\{(1-p_3)/\alpha\}}-\dfrac{1}{2\pi i}\displaystyle\int_{D_\omega}\frac{f(r,p_3,K)}{r\omega^{K-1}}\,\frac{d\omega}{\omega-r\beta\{(1-\omega p_3)/\alpha\}}, \\
\qquad\qquad z=K.
\end{cases}
\tag{6.46}
$$

Since

$$
\begin{aligned}
\lim_{|p|\to 0}\zeta(r,p,p_3,z) &= \sum_{n=1}^{\infty} r^n\,E\{p_3^{k_{n-1}}(z_n=0)\mid z_1=z\} \\
&= -\frac{1}{r}f(r,p_3,z), \qquad z=0,1,\dots,K,
\end{aligned}
$$

$$\lim_{|p|\to\infty} p^{-K}\zeta(r,p,p_3,z) = \sum_{n=1}^{\infty} r^n \,\mathrm{E}\{p_3^{k_{n-1}}(z_n=K) \mid z_1=z\}$$

$$= \begin{cases} -\lim\limits_{|p|\to\infty} \dfrac{ph(r,p,p_3,z)}{\beta_2\{(1-pp_3)/\alpha\}}, & z<K, \\[2ex] -\lim\limits_{|p|\to\infty} \dfrac{ph(r,p,p_3,K)}{\beta_2\{(1-pp_3)/\alpha\}} + r, & z=K, \end{cases}$$

where $(z_n=0)$ and similarly $(z_n=K)$ stand for the indicators of the event $\{z_n=0\}$ and the event $\{z_n=K\}$, respectively, it follows from (6.43) and (6.45) for $|r|<1$, $|p_3|\leqq 1$, $z=0, \ldots, K$ that

$$\sum_{n=1}^{\infty} r^n \tilde{p}_{z0}^{(n)}(p_3) = c_f(r,p_3,z), \tag{6.47}$$

$$\sum_{n=1}^{\infty} r^n \tilde{p}_{zK}^{(n)}(p_3) = c_h(r,p_3,z).$$

Since the successive occurrences of states E_0 and similarly of states E_K form a renewal process we have from renewal theory for $|r|\leqq 1$, $|p_3|\leqq 1$, $z=0, \ldots, K$,

$$\mathrm{E}\{r^{n_{z0}} p_3^{K_{z0}}\} = \frac{c_f(r,p_3,z)}{1+c_f(r,p_3,0)},$$

$$\mathrm{E}\{r^{n_{zK}} p^{K_{zK}}\} = \frac{c_h(r,p_3,z)}{1+c_h(r,p_3,K)}.$$

It follows from (6.18), (6.45) and (6.46) for $K\geqq 1$, $|r|\leqq 1$, $|p_3|\leqq 1$, $|\omega|<|\mu_0|$, that

$$\mathrm{E}\{r^{n_{z0}} p_3^{K_{z0}}\}$$

$$= \begin{cases} \dfrac{1 - \dfrac{1-r\beta\{(1-p_3)/\alpha\}}{2\pi\mathrm{i}} \displaystyle\int_{\mathrm{D}_\omega} \dfrac{\mathrm{d}\omega}{1-\omega}\, \dfrac{\omega^{z-K}}{\omega - r\beta\{(1-\omega p_3)/\alpha\}}}{1 - \dfrac{1-r\beta\{(1-p_3)/\alpha\}}{2\pi\mathrm{i}} \displaystyle\int_{\mathrm{D}_\omega} \dfrac{\mathrm{d}\omega}{1-\omega}\, \dfrac{\omega^{-K}}{\omega - r\beta\{(1-\omega p_3)/\alpha\}}}, & z=1, \ldots, K, \\[4ex] 1 + \dfrac{\dfrac{1-r\beta\{(1-p_3)/\alpha\}}{2\pi\mathrm{i}} \displaystyle\int_{\mathrm{D}_\omega} \dfrac{\omega^{-K}}{\omega - r\beta\{(1-\omega p_3)/\alpha\}}\,\mathrm{d}\omega}{1 - \dfrac{1-r\beta\{(1-p_3)/\alpha\}}{2\pi\mathrm{i}} \displaystyle\int_{\mathrm{D}_\omega} \dfrac{\mathrm{d}\omega}{1-\omega}\, \dfrac{\omega^{-K}}{\omega - r\beta\{(1-\omega p_3)/\alpha\}}}, & z=0, \end{cases} \tag{6.48}$$

$$\mathrm{E}\{r^{n_{KK}}p_3^{K_{KK}}\} = \frac{\dfrac{1}{2\pi\mathrm{i}}\displaystyle\int_{\mathrm{D}_\omega} \frac{\omega^{-K}\{r\beta\{(1-p_3)/\alpha\}-\omega\}}{\omega - r\beta\{(1-\omega p_3)/\alpha\}}\,\mathrm{d}\omega}{\dfrac{1}{2\pi\mathrm{i}}\displaystyle\int_{\mathrm{D}_\omega} \frac{(1-\omega)\,\omega^{-K}}{\omega - r\beta\{(1-\omega p_3)/\alpha\}}\,\mathrm{d}\omega}.$$

The expression for $\mathrm{E}\{r^{n_{zK}}p_3^{K_{zK}}\}$ is rather lengthy and therefore omitted.

For $z=0$ the first relation of (6.48) is the bivariate generating function of the joint distribution of the number of customers served in a busy cycle and of the number of customers arriving in a busy cycle. Similarly, the second relation of (6.48) is the bivariate generating function of the joint distribution of the number of customers served and of the number of customers arriving between two successive occurrences of the state E_K.

Obviously,

$$\boldsymbol{h}_{00} \overset{\text{def}}{=} \boldsymbol{K}_{00} - (\boldsymbol{n}_{00}-1)$$

is the number of customers arriving during a busy period who are not admitted to the system. Obviously $\boldsymbol{h}_{00}$ is a non-negative stochastic variable. To obtain the generating function of the distribution of $\boldsymbol{h}_{00}$ we put $q \overset{\text{def}}{=} rp_3$ and $\eta = p_3\omega$; then from (6.48) we obtain for $K \geqq 1$, $|\eta|<|p_3|$, $|\eta|<|p_3\mu_0| = |\mu_0(q,1)| \leqq 1$ (cf. (6.12)), $|q| \leqq |p_3| \leqq 1$,

$$\mathrm{E}\{q^{n_{00}}p_3^{K_{00}-n_{00}}\} = p_3^{-1}\,\mathrm{E}\{q^{n_{00}}p_3^{h_{00}}\}$$

$$= 1 + \frac{\dfrac{p_3 - q\beta\{(1-p_3)/\alpha\}}{2\pi\mathrm{i}}\,p_3^{K-1}\displaystyle\int_{\mathrm{D}_\eta} \frac{\mathrm{d}\eta}{\eta^K}\,\frac{1}{\eta - q\beta\{(1-\eta)/\alpha\}}}{1 - \dfrac{p_3 - q\beta\{(1-p_3)/\alpha\}}{2\pi\mathrm{i}}\,p_3^{K}\displaystyle\int_{\mathrm{D}_\eta} \frac{\mathrm{d}\eta}{\eta^K}\,\frac{1}{p_3-\eta}\,\frac{1}{\eta - q\beta\{(1-\eta)/\alpha\}}}.$$

From this relation it follows by contour integration that for $K \geqq 1$, $|p_3| < |\eta| < |p_3\mu_0| \leqq 1$, $|q| \leqq |p_3| < 1$,

$$\mathrm{E}\{q^{n_{00}}p_3^{h_{00}}\} = p_3 - \frac{\dfrac{1}{2\pi\mathrm{i}}\displaystyle\int_{\mathrm{D}_\eta} \frac{\mathrm{d}\eta}{\eta^K}\,\frac{1}{\eta - q\beta\{(1-\eta)/\alpha\}}}{\dfrac{1}{2\pi\mathrm{i}}\displaystyle\int_{\mathrm{D}_\eta} \frac{1}{p_3-\eta}\,\frac{\mathrm{d}\eta}{\eta^K}\,\frac{1}{\eta - q\beta\{(1-\eta)/\alpha\}}}.$$

The left-hand side is for fixed $|p_3|<1$ analytic for $|q|\leqq 1$. Hence by analytic continuation with respect to q it follows taking $q=1$ that for $K\geqq 1$ and $|p_3|<|\eta|<\mu_0(1, 1)\leqq 1$,

$$\mathrm{E}\{p_3^{h_{00}}\} = \frac{\dfrac{1}{2\pi \mathrm{i}} \displaystyle\int_{\mathrm{D}_\eta} \frac{1}{p_3-\eta} \frac{\mathrm{d}\eta}{\eta^{K-1}} \frac{1}{\eta-\beta\{(1-\eta)/\alpha\}}}{\dfrac{1}{2\pi \mathrm{i}} \displaystyle\int_{\mathrm{D}_\eta} \frac{1}{p_3-\eta} \frac{\mathrm{d}\eta}{\eta^{K}} \frac{1}{\eta-\beta\{(1-\eta)/\alpha\}}}. \tag{6.49}$$

We shall next derive expressions for some of the more important taboo return and entrance times. From (6.44) and (6.47) we have for $|r|<1$, $|p_3|\leqq 1$, $z=0, \ldots, K$,

$$\sum_{n=1}^{\infty} r^n \tilde{p}_{K;z0}^{(n)}(p_3) = c_f(r, p_3, z) - c_f(r, p_3, K) \frac{c_h(r, p_3, z)}{1+c_h(r, p_3, K)},$$

$$\sum_{n=1}^{\infty} r^n \tilde{f}_{K;00}^{(n)}(p_3) = \frac{\sum_{n=1}^{\infty} r^n \tilde{p}_{K;00}^{(n)}(p_3)}{1+\sum_{n=1}^{\infty} r^n \tilde{p}_{K;00}^{(n)}(p_3)}.$$

From these relations and from (6.18), (6.45) and (6.46) we obtain for $|r|\leqq 1$, $|\omega|<|\mu_0(r, 1)|$, $K\geqq 1$,

$$\mathrm{E}\{r^{n_{00}}(\sup_{1\leqq n\leqq n_{00}} z_n<K+1)\} = \sum_{n=1}^{\infty} r^n f_{K;00}^{(n)}$$

$$= \frac{\dfrac{1}{2\pi \mathrm{i}} \displaystyle\int_{\mathrm{D}_\omega} \frac{\mathrm{d}\omega}{\omega^{K}} \frac{\beta\{(1-\omega)/\alpha\}}{\omega-r\beta\{(1-\omega)/\alpha\}}}{\dfrac{1}{2\pi \mathrm{i}} \displaystyle\int_{\mathrm{D}_\omega} \frac{\mathrm{d}\omega}{\omega^{K+1}} \frac{\beta\{(1-\omega)/\alpha\}}{\omega-r\beta\{(1-\omega)/\alpha\}}}, \tag{6.50}$$

in particular for $|\omega|<\mu_0(1, 1)$,

$$\sum_{n=1}^{\infty} f_{K;00}^{(n)} = \frac{\dfrac{1}{2\pi \mathrm{i}} \displaystyle\int_{\mathrm{D}_\omega} \frac{\mathrm{d}\omega}{\omega^{K}} \frac{\beta\{(1-\omega)/\alpha\}}{\omega-\beta\{(1-\omega)/\alpha\}}}{\dfrac{1}{2\pi \mathrm{i}} \displaystyle\int_{\mathrm{D}_\omega} \frac{\mathrm{d}\omega}{\omega^{K+1}} \frac{\beta\{(1-\omega)/\alpha\}}{\omega-\beta\{(1-\omega)/\alpha\}}}. \tag{6.51}$$

The relation (6.50) is the generating function of the distribution of the number of customers served during a busy cycle while the number of customers present simultaneously in the system during a busy cycle is less than $K+1$. The left-hand side of (6.51) is the probability that the number of customers present simultaneously in the system during a busy cycle is less than $K+1$. The relation (6.51) has been found by TAKÁCS [1965] (cf. (II.6.77)) using combinatorial methods and has been derived in section II.4.4 by calculating taboo probabilities for the $M/G/1$ queueing system with unlimited capacity for waiting customers (cf. II.4.70). Note that for the $M/G/1$ queueing system with unlimited capacity for waiting customers the expression (6.51) represents the probability that the maximum number of customers simultaneously present in the system during a busy cycle is, at most, K.

If $a<1$ then relation (6.51) may be written as (cf. (6.25)),

$$\sum_{n=1}^{\infty} f_{K;00}^{(n)} = \frac{v_0+\ldots+v_{K-1}}{v_0+\ldots+v_K} = 1 - \frac{v_K}{v_0+\ldots+v_K}, \qquad a<1.$$

In the same way it is found for $|r|\leqq 1$, $|p_3|\leqq 1$, $K\geqq 1$, $|\omega|<|\mu_0|$ that

$$\sum_{n=1}^{\infty} r^n f_{K;K0}^{(n)} = \left[\frac{1}{2\pi i}\int_{D_\omega} \frac{d\omega}{\omega^{K+1}} \frac{r\beta\{(1-\omega)/\alpha\}}{r\beta\{(1-\omega)/\alpha\}-\omega}\right]^{-1}, \qquad |\omega|<|\mu_0(r,1)|, \tag{6.52}$$

$$\sum_{n=1}^{\infty} r^n \tilde{f}_{0;KK}^{(n)}(p_3) = 1$$

$$+\frac{\beta\{(1-p_3)/\alpha\} - \dfrac{1-r\beta\{(1-p_3)/\alpha\}}{2\pi i}\displaystyle\int_{D_\omega} \frac{d\omega}{\omega^{K+1}} \frac{\beta\{(1-\omega p_3)/\alpha\}}{\omega - r\beta\{(1-\omega p_3)/\alpha\}} \frac{1}{1-\omega}}{\dfrac{1}{2\pi i}\displaystyle\int_{D_\omega} \frac{d\omega}{\omega^{K+1}} \frac{\beta\{(1-\omega p_3)/\alpha\}}{\omega - r\beta\{(1-\omega p_3)/\alpha\}}},$$

$$\sum_{n=1}^{\infty} r^n \tilde{f}_{0;0K}^{(n)}(p_3) = \beta\{(1-p_3)/\alpha\} - \frac{\dfrac{1}{2\pi i}\displaystyle\int_{D_\omega} \frac{d\omega}{\omega^{K}} \frac{\beta\{(1-\omega p_3)/\alpha\}}{r\beta\{(1-\omega p_3)/\alpha\}-\omega}}{\dfrac{1}{2\pi i}\displaystyle\int_{D_\omega} \frac{d\omega}{\omega^{K+1}} \frac{\beta\{(1-\omega p_3)/\alpha\}}{r\beta\{(1-\omega p_3)/\alpha\}-\omega}}.$$

In particular if $a<1$ (cf. (6.25)),

$$\sum_{n=1}^{\infty} f_{K;K0}^{(n)} = \frac{1-a}{v_0+\ldots+v_K}, \tag{6.53}$$

$$\sum_{n=1}^{\infty} f_{0;KK}^{(n)} = 1 - \frac{1-a}{v_0+\ldots+v_K},$$

$$\sum_{n=1}^{\infty} f_{0;0K}^{(n)} = \frac{v_K}{v_0+\ldots+v_K}.$$

Obviously, the first relation of (6.52) is the generating function of the distribution of the entrance time in discrete time into the empty state E_0 when starting in E_K and avoiding E_K in the meantime. The second relation of (6.52) is the bivariate generating function of the joint distribution of the number of customers served and the number of customers arriving between two successive occurrences of state E_K while during this return time the system is never empty. The last relation of (6.52) is the bivariate generating function of the joint distribution of the number of customers served and the number of customers arriving during the entrance time from E_0 into E_K, avoiding E_0 in the mean time.

Denote the number of times that during a busy cycle $K+1$ customers are present in the system by $\boldsymbol{m}_{K+1}$. Hence $\boldsymbol{m}_{K+1}$ is also the number of times that during a busy cycle the state E_K occurs.

Obviously we have for the distribution of $\boldsymbol{m}_{K+1}$,

$$\Pr\{\boldsymbol{m}_{K+1}=m\} = \begin{cases} \sum_{n=1}^{\infty} f_{K;00}^{(n)}, & m=0, \\ \sum_{n=1}^{\infty} f_{0;0K}^{(n)} \{\sum_{n=1}^{\infty} f_{0;KK}^{(n)}\}^{m-1} \sum_{n=1}^{\infty} f_{K;K0}^{(n)}, & m=1, 2, \ldots, \end{cases} \tag{6.54}$$

$$\mathrm{E}\{\boldsymbol{m}_{K+1}\} = \frac{1-\sum_{n=1}^{\infty} f_{K;00}^{(n)}}{1-\sum_{n=1}^{\infty} f_{0;KK}^{(n)}}$$

$$= \frac{1}{2\pi \mathrm{i}} \int_{\mathrm{D}_\omega} \frac{\mathrm{d}\omega}{\omega^{K+1}} \frac{(1-\omega)\,\beta\{(1-\omega)/\alpha\}}{\beta\{(1-\omega)/\alpha\}-\omega}, \qquad |\omega|<\mu_0(1, 1),$$

$$= \frac{v_K}{1-a} \quad \text{if} \quad a<1 \quad \text{(cf. (6.25))}.$$

The results of this section have been derived under the assumption that $\beta(\rho)$ is a rational function of ρ. This assumption is, however, superfluous. Only minor changes are needed in (6.45), (6.46) and in the derivation of (6.47) to verify that the results of this section hold for general $\beta(\rho)$ (cf. also end of section 6.2).

In the case where

$$B(t)=1-\mathrm{e}^{-t/\beta}, \qquad t>0,$$

it follows that

$$\sum_{n=1}^{\infty} f_{K;00}^{(n)} = \begin{cases} \dfrac{1-a^K}{1-a^{K+1}}, & a\neq 1,\\[2ex] \dfrac{K}{1+K}, & a=1,\end{cases}$$

$$\sum_{n=1}^{\infty} f_{K;K0}^{(n)} = \begin{cases} \dfrac{1-a}{1-a^{K+1}}, & a\neq 1,\\[2ex] \dfrac{1}{1+K}, & a=1,\end{cases}$$

$$\sum_{n=1}^{\infty} f_{0;KK}^{(n)} = \begin{cases} a\,\dfrac{1-a^K}{1-a^{K+1}}, & a\neq 1,\\[2ex] \dfrac{K}{1+K}, & a=1,\end{cases}$$

$$\sum_{n=1}^{\infty} f_{0;0K}^{(n)} = \begin{cases} (1-a)\,\dfrac{a^K}{1-a^{K+1}}, & a\neq 1,\\[2ex] \dfrac{1}{1+K}, & a=1.\end{cases}$$

$$\mathrm{E}\{\boldsymbol{m}_{K+1}\} = a^K,$$

$$\mathrm{E}\{p^{\boldsymbol{h}_{00}}\} = \begin{cases} 1-(1-a)\,a^{K+1}\dfrac{1-p}{1-ap-a^{K+2}(1-p)}, & a\neq 1, \quad |p|\leqq 1,\\[2ex] 1-\dfrac{1-p}{K+2-(K+1)\,p}, & a=1, \quad |p|\leqq 1.\end{cases}$$

$$\mathrm{E}\{\boldsymbol{h}_{00}\} = a^{K+1}.$$

III.6.5. Return times of a state K within a busy cycle for the $M/G/1$ queue with infinite waiting room

In this section we shall again consider the queueing system $M/G/1$ with infinite waiting room. In fig. 7 a realisation of the process $\{z_n, n=1, 2, \ldots\}$ during a busy cycle is shown.
Here $\boldsymbol{n}$ denotes the number of customers served in a busy cycle; $\boldsymbol{n}_K$ is the number of customers served in a residual busy period starting with K customers. We denote the empty state at a departure of a customer by E_0, and the state with K customers left behind in the system at a departure by E_K.

If within a busy cycle z_n has reached a value of at least K, K a positive integer, then $\boldsymbol{m}_{0K}$ will represent the number of customers served during the entrance time from E_0 into state E_K; $\boldsymbol{m}_{KK}$ will be the number of customers served during the return time of E_K within a busy period; $\boldsymbol{m}_{K0}$ is the number of customers served during the entrance time into E_0 starting at the last occurrence of E_K in the busy period. We shall denote the number of times that E_K occurs in a busy cycle by $\boldsymbol{\Pi}_K$.

We shall denote by $\mu(r)$ the zero with the smallest absolute value of

$$p-r\beta\{(1-p)/\alpha\}, \qquad |r|\leqq 1.$$

From Takács' lemma it is known that $|\mu(r)|<1$ if $|r|<1$, or if $a>1$ and $|r|\leqq 1$; if $r=1$ and $a<1$ the only zero inside or on the unit circle is a single zero at $p=1$; if $r=1$, $a=1$ then there is a double zero $\mu(r)=1$; $\mu(r)$ is a continuous function of r for $|r|\leqq 1$. Moreover, we have for $|r|\leqq 1$ (cf. section II.4.4),

$$\mathrm{E}\{r^{\boldsymbol{n}}\}=\mu(r), \qquad \mathrm{E}\{r^{\boldsymbol{n}_K}\}=\mu^K(r). \tag{6.55}$$

The stochastic variables $\boldsymbol{m}_{0K}$, $\boldsymbol{m}_{KK}$ and $\boldsymbol{m}_{K0}$ are defined by

$$\boldsymbol{m}_{0K} \stackrel{\text{def}}{=} \begin{cases} \inf\limits_{1\leqq n<\boldsymbol{n}} \{n: z_n=K \mid \sup\limits_{1\leqq m<n} z_m \geqq K, z_0=0\}, \\ \infty \quad \text{if no such } n \text{ exists}, \end{cases}$$

$$\boldsymbol{m}_{KK} \stackrel{\text{def}}{=} \begin{cases} \inf\limits_{1\leqq n<\boldsymbol{n}} \{n: z_n=K \mid \sup\limits_{1\leqq m<\boldsymbol{n}_K} z_m \geqq K, z_0=K\}, \\ \infty \quad \text{if no such } n \text{ exists}, \end{cases}$$

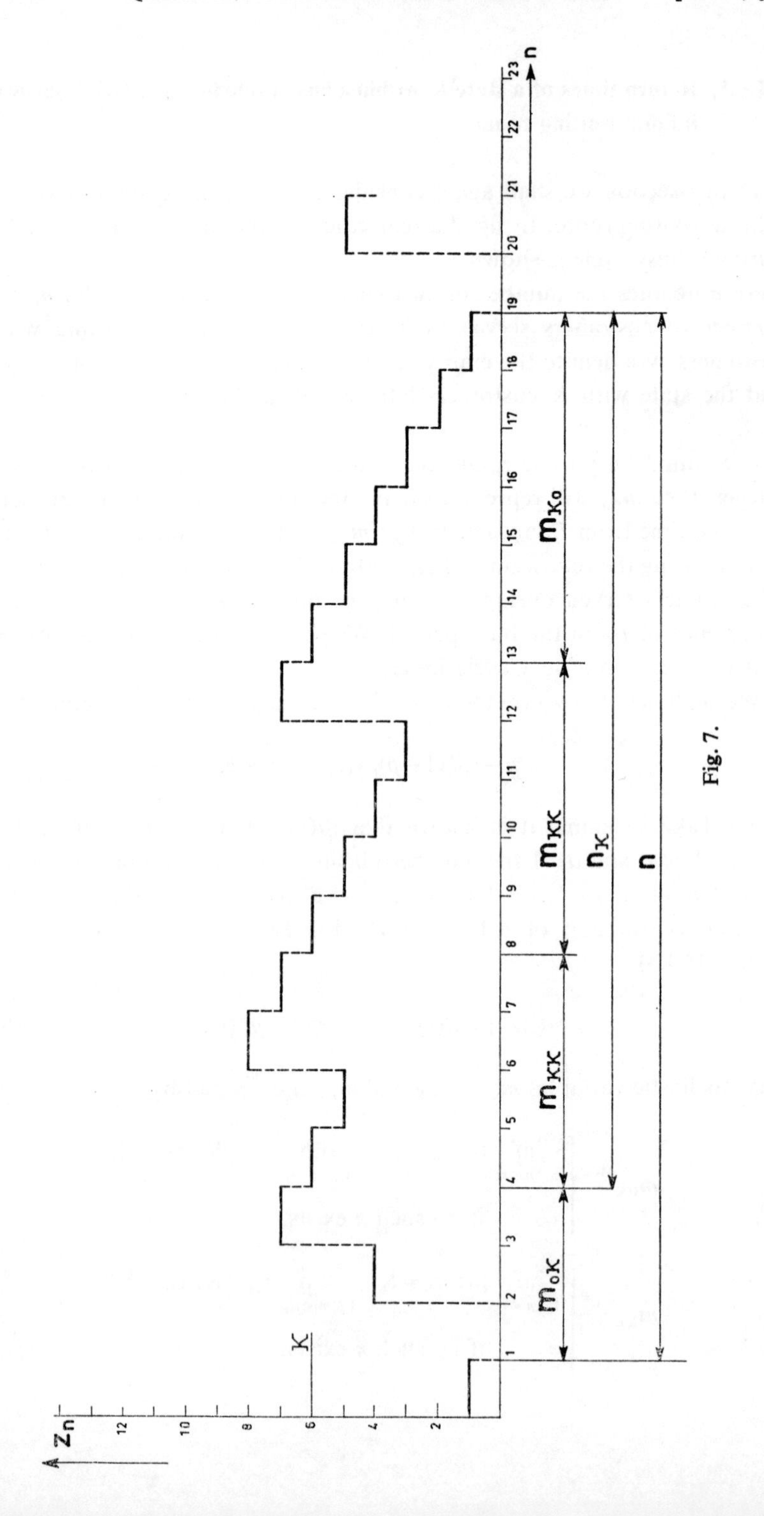
Z_n
12
10
8
6
4
2
K
1
2
3
4
5
6
7
8
9
10
11
12
13
14
15
16
17
18
19
20
21
22
23
n
m_oK
m_KK
m_KK
m_Ko
n_K
n

Fig. 7.

$$\boldsymbol{m}_{K0} \stackrel{\text{def}}{=} \begin{cases} \inf\limits_{1\leq n<\boldsymbol{n}_K} \{n: \boldsymbol{z}_n=0 \mid \sup\limits_{1\leq m<\boldsymbol{n}_K} \boldsymbol{z}_m<K, \boldsymbol{z}_0=K\}, \\ \infty \qquad \text{if no such } n \text{ exists.} \end{cases}$$

Define for $j=1, 2, \ldots,$

$$f_j \stackrel{\text{def}}{=} \Pr\{\boldsymbol{n}=j, \sup_{1\leq m<\boldsymbol{n}} \boldsymbol{z}_m<K \mid \boldsymbol{z}_0=0\},$$

$$h_j \stackrel{\text{def}}{=} \Pr\{\boldsymbol{n}_K=j, \sup_{1\leq m<\boldsymbol{n}_K} \boldsymbol{z}_m<K \mid \boldsymbol{z}_0=K\},$$

$$f(r) \stackrel{\text{def}}{=} \sum_{j=1}^{\infty} f_j r^j, \qquad h(r) \stackrel{\text{def}}{=} \sum_{j=1}^{\infty} h_j r^j, \qquad |r|\leq 1.$$

Expressions for $f(r)$ and $h(r)$ have been derived in the preceding section (cf. (6.50) and (6.52)); for $|\omega|<|\mu(r)|$, $|r|\leq 1$, $K\geq 1$,

$$f(r) = \frac{\dfrac{1}{2\pi\mathrm{i}} \displaystyle\int_{\mathrm{D}_\omega} \frac{\mathrm{d}\omega}{\omega^K} \frac{\beta\{(1-\omega)/\alpha\}}{r\beta\{(1-\omega)/\alpha\}-\omega}}{\dfrac{1}{2\pi\mathrm{i}} \displaystyle\int_{\mathrm{D}_\omega} \frac{\mathrm{d}\omega}{\omega^{K+1}} \frac{\beta\{(1-\omega)/\alpha\}}{r\beta\{(1-\omega)/\alpha\}-\omega}}, \tag{6.56}$$

$$h(r) = \left[\frac{1}{2\pi\mathrm{i}} \int_{\mathrm{D}_\omega} \frac{\mathrm{d}\omega}{\omega^{K+1}} \frac{r\beta\{(1-\omega)/\alpha\}}{r\beta\{(1-\omega)/\alpha\}-\omega}\right]^{-1}.$$

From the results of the preceding section it follows that if at a departure j customers are left behind then with probability one the system reaches the empty state or a state with at least K customers present at a departure in a finite number of steps, $j=0, 1, \ldots, K$. Consequently,

$$f(1) = \Pr\{\boldsymbol{n}<\infty, \sup_{1\leq m<\boldsymbol{n}} \boldsymbol{z}_m<K \mid \boldsymbol{z}_0=0\}, \tag{6.57}$$

$$1-f(1) = \Pr\{\sup_{1\leq m<\boldsymbol{n}} \boldsymbol{z}_m\geq K \mid \boldsymbol{z}_0=0\},$$

$$h(1) = \Pr\{\boldsymbol{n}_K<\infty, \sup_{1\leq m<\boldsymbol{n}_K} \boldsymbol{z}_m<K \mid \boldsymbol{z}_0=K\},$$

$$1-h(1) = \Pr\{\sup_{1\leq m<\boldsymbol{n}_K} \boldsymbol{z}_m\geq K \mid \boldsymbol{z}_0=K\}.$$

If $\boldsymbol{z}_m\geq K$ for at least one value of $m=1, \ldots, \boldsymbol{n}$, then $\boldsymbol{n}$ is the sum of one $\boldsymbol{m}_{0K}$, one $\boldsymbol{m}_{K0}$ and a random number of variables $\boldsymbol{m}_{KK}$. Since for the $M/G/1$ queueing system the interarrival times are independent and negative exponentially distributed it follows that these variables $\boldsymbol{m}_{0K}$, $\boldsymbol{m}_{K0}$ and $\boldsymbol{m}_{KK}$ are independent. Hence for $|r| \leqq 1$,

$$\mathrm{E}\{r^{\boldsymbol{n}}\} = f(r) + \{1-f(1)\}\, \mathrm{E}\{r^{\boldsymbol{m}_{0K}}\}\, \mathrm{E}\{r^{\boldsymbol{n}_K}\}, \tag{6.58}$$

$$\mathrm{E}\{r^{\boldsymbol{n}_K}\} = h(r) \sum_{m=0}^{\infty} [(1-h(1))\, \mathrm{E}\{r^{\boldsymbol{m}_{KK}}\}]^m$$

$$= \frac{h(r)}{1-(1-h(1))\, \mathrm{E}\{r^{\boldsymbol{m}_{KK}}\}}.$$

Hence from (6.55) and (6.58) for $|r| \leqq 1$,

$$\mathrm{E}\{r^{\boldsymbol{m}_{0K}}\} = \frac{\mu(r)-f(r)}{1-f(1)}\, \mu^{-K}(r), \tag{6.59}$$

$$\mathrm{E}\{r^{\boldsymbol{m}_{KK}}\} = \frac{1-h(r)\, \mu^{-K}(r)}{1-h(1)},$$

$$\mathrm{E}\{r^{\boldsymbol{m}_{K0}}\} = \frac{h(r)}{h(1)}.$$

It follows that

$$\Psi_{0K} \stackrel{\text{def}}{=} \Pr\{\boldsymbol{m}_{0K} < \infty\} = \frac{\mu(1)-f(1)}{1-f(1)}\, \mu^{-K}(1), \tag{6.60}$$

$$\Psi_{KK} \stackrel{\text{def}}{=} \Pr\{\boldsymbol{m}_{KK} < \infty\} = \frac{1-h(1)\, \mu^{-K}(1)}{1-h(1)}.$$

If $a \leqq 1$ then $\mu(1) = 1$ and hence

$$\Psi_{0K} = 1, \qquad \Psi_{KK} = 1 \qquad \text{for} \quad a \leqq 1. \tag{6.61}$$

Note that for $a < 1$ (cf. (6.51) and (6.53)),

$$f(1) = \frac{v_0 + \ldots + v_{K-1}}{v_0 + \ldots + v_K}, \qquad h(1) = \frac{v_0}{v_0 + \ldots + v_K}, \qquad v_0 = 1-a.$$

Next, we shall study $\boldsymbol{\Pi}_K$, the number of times that state E_K occurs in a busy period. Its distribution may now easily be determined by using the property that a moment of departure is a regeneration point of the queueing process $M/G/1$. It is necessary, however, to distinguish between finite and infinite busy cycles. It follows from (6.57) and (6.60) that

$$\Pr\{\boldsymbol{\Pi}_K=m, \boldsymbol{n}<\infty\} = \begin{cases} f(1), & m=0, \\ (1-f(1))\,\Psi_{0K}\{(1-h(1))\,\Psi_{KK}\}^{m-1}\,h(1), & m=1,2,\ldots, \end{cases} \tag{6.62}$$

$$\Pr\{\boldsymbol{\Pi}_K=m, \boldsymbol{n}=\infty\} = \begin{cases} (1-f(1))(1-\Psi_{0K}), & m=0, \\ (1-f(1))\,\Psi_{0K}\{(1-h(1))\,\Psi_{KK}\}^{m-1}(1-h(1))(1-\Psi_{KK}), & m=1,2,\ldots. \end{cases} \tag{6.63}$$

From (6.62) and (6.63) it is easily found that

$$\sum_{n=1}^{\infty} \Pr\{\boldsymbol{\Pi}_K=m\} = 1.$$

Further,

$$\mathrm{E}\{\boldsymbol{\Pi}_K \mid \boldsymbol{n}<\infty\} = \frac{\mu(1)-f(1)}{\mu(1)\,h(1)}\,\mu^K(1),$$

$$\mathrm{var}\{\boldsymbol{\Pi}_K \mid \boldsymbol{n}<\infty\} = \frac{\mu(1)-f(1)}{\mu(1)\,h^2(1)}\,\mu^{2K}(1)\left\{1+\frac{f(1)}{\mu(1)} - h(1)\,\mu^{-K}(1)\right\},$$

and if $a>1$,

$$\mathrm{E}\{\boldsymbol{\Pi}_K \mid \boldsymbol{n}=\infty\} = \frac{\mu(1)-f(1)}{1-\mu(1)}\,\frac{1-\mu^K(1)}{h(1)},$$

$$\mathrm{var}\{\boldsymbol{\Pi}_K \mid \boldsymbol{n}=\infty\} = \frac{\mu(1)-f(1)}{1-\mu(1)}\,\frac{1-\mu^K(1)}{h^2(1)}\,\mu^K(1) \cdot\left[1-h(1)\,\mu^{-K}(1)+\frac{1-f(1)}{1-\mu(1)}-\frac{\mu(1)-f(1)}{1-\mu(1)}\,\mu^{-K}(1)\right].$$

If $a<1$ then $\boldsymbol{n}$, $\boldsymbol{m}_K$, $\boldsymbol{m}_{0K}$ and $\boldsymbol{m}_{KK}$ are finite with probability one and $\Psi_{0K}=\Psi_{KK}=1$. In this case it follows from the above relations that

$$\Pr\{\boldsymbol{\Pi}_K=m\} = \begin{cases} 1-\dfrac{v_K}{v_0+\ldots+v_K}, & m=0, \\ v_0 v_K\dfrac{(v_1+\ldots+v_K)^{m-1}}{(v_0+\ldots+v_K)^{m+1}}, & m=1,2,\ldots, \end{cases}$$

$$\mathrm{E}\{\boldsymbol{\Pi}_K\} = \frac{v_K}{v_0} = \frac{1}{1-a}\,v_K,$$

$$\mathrm{var}\{\boldsymbol{\Pi}_K\} = \frac{v_K}{v_0^2}\{v_0+2(v_1+\ldots+v_{K-1})+v_K\}.$$

For $a \leqq 1$ denote the number of occurrences of state E_K in the jth busy cycle, $j=1, 2, \ldots$, supposing that the system starts in the empty state, by $\boldsymbol{\Pi}_K^{(j)}$. Since the variables $\boldsymbol{\Pi}_K^{(j)}$, $j=1, 2, \ldots$, are independent and identically distributed variables with a finite second moment it follows from the strong law of large numbers that with probability one

$$\lim_{n\to\infty} \frac{1}{n} \sum_{j=1}^{n} \boldsymbol{\Pi}_K^{(j)} = \frac{1-f(1)}{h(1)}$$

$$= \begin{cases} \dfrac{1}{2\pi \mathrm{i}} \displaystyle\int_{\mathrm{D}_\omega} \frac{\mathrm{d}\omega}{\omega^{K+1}} \frac{(1-\omega)\,\beta\{(1-\omega)/\alpha\}}{\beta\{(1-\omega)/\alpha\}-\omega}, & |\omega|<1, \quad a \leqq 1, \\[2ex] \dfrac{v_K}{v_0} = \dfrac{1}{1-a} v_K, & a<1, \end{cases}$$

and from the central limit theorem that the distribution of

$$\left\{\sum_{j=1}^{n} \boldsymbol{\Pi}_K^{(j)} - n\frac{1-f(1)}{h(1)}\right\} \Big/ \left\{\frac{1-f(1)}{h^2(1)} \{1+f(1)-h(1)\}\,\sqrt{n}\right\},$$

converges for $n\to\infty$ to the normal distribution N(0, 1) if $a \leqq 1$. Note that for $a \leqq 1$, $|\omega|<1$,

$$\frac{1-f(1)}{h^2(1)}\{1+f(1)-h(1)\}$$

$$= \frac{1-f(1)}{h(1)} \frac{1}{2\pi \mathrm{i}} \int_{\mathrm{D}_\omega} \frac{\mathrm{d}\omega}{\omega^{K+1}} (1+\omega-\omega^K) \frac{\beta\{(1-\omega)/\alpha\}}{\beta\{(1-\omega)/\alpha\}-\omega}.$$

In the preceding section the distribution of $\boldsymbol{m}_{K+1}$, the number of times that a departing customer leaves K customers behind in the system, has been derived for a $M/G/1$ system with a finite number K of waiting positions. It appears for $a \leqq 1$ that the variables $\boldsymbol{m}_{K+1}$ and $\boldsymbol{\Pi}_K$ have the same distribution.

Finally, we shall specify some of the results obtained above for the case of the $M/M/1$ queueing system, so that

$$B(t) = 1 - \mathrm{e}^{-t/\beta}, \qquad t>0, \qquad \beta(\rho) = \frac{1}{1+\beta\rho}, \qquad \mathrm{Re}\,\rho \geqq 0.$$

It follows for $|\omega|<|\mu(r)|$, $|r|\leqq 1$ that

$$\frac{1}{2\pi i}\int_{D_\omega}\frac{d\omega}{\omega^K}\,\frac{\beta\{(1-\omega)/\alpha\}}{r\beta\{(1-\omega)/\alpha\}-\omega}=\frac{1}{2\pi i}\int_{D_\omega}\frac{d\omega}{\omega^K}\,\frac{1}{a\omega^2-\omega(1+a)+r}$$

$$=\frac{1}{2\pi i}\int_{D_\omega}\frac{d\omega}{\omega^K}\,\frac{1}{a\{\omega-\mu(r)\}\{\omega-(r/a)\,\mu^{-1}(r)\}}$$

$$=\begin{cases}-\dfrac{1}{a}\left\{\mu(r)-\dfrac{r}{a\mu(r)}\right\}^{-1}\left\{\mu^{-K}(r)-\dfrac{a^K}{r^K}\,\mu^K(r)\right\}, & |r|<1,\\[2ex] \dfrac{1-a^K}{1-a}, \quad a\neq 1,\quad r=1, & \\[2ex] K, \quad a=1,\quad r=1, & \end{cases}$$

since $\mu(1)=\begin{cases}1 & \text{for } a\leqq 1,\\ a^{-1} & \text{for } a>1.\end{cases}$

From (6.56) and (6.59) we obtain for $|r|\leqq 1$,

$$\mathrm{E}\{r^{m_{KK}}\}=\frac{1-a^{K+1}}{a(1-a^K)}\left[1+\frac{a}{r}\left(\mu^2(r)-\frac{r}{a}\right)\left\{1-\frac{a^{K+1}}{r^{K+1}}\,\mu^{2K+2}(r)\right\}^{-1}\right].$$

Further

$$f(1)=\begin{cases}\dfrac{1-a^K}{1-a^{K+1}}, & a\neq 1,\\[2ex] \dfrac{K}{1+K}, & a=1,\end{cases}$$

$$h(1)=\begin{cases}\dfrac{1-a}{1-a^{K+1}}, & a\neq 1,\\[2ex] \dfrac{1}{1+K}, & a=1,\end{cases}$$

$$\Psi_{0K}=\Psi_{KK}=\begin{cases}1 & \text{if } a\leqq 1,\\ a^{-1} & \text{if } a>1.\end{cases}$$

For $a<1$,

$$\Pr\{\boldsymbol{\Pi}_K=m\} = \begin{cases} \dfrac{1-a^K}{1-a^{K+1}}, & m=0, \\[2ex] (1-a)^2 a^K \dfrac{a^{m-1}(1-a^K)^{m-1}}{(1-a^{K+1})^{m+1}}, & m=1, 2, \ldots, \end{cases}$$

$$\mathrm{E}\{\boldsymbol{\Pi}_K\} = a^K, \qquad \mathrm{var}\{\boldsymbol{\Pi}_K\} = \frac{1+a}{1-a}\, a^K(1-a^K);$$

for $a=1$,

$$\Pr\{\boldsymbol{\Pi}_K=m\} = \begin{cases} \dfrac{K}{1+K}, & m=0, \\[2ex] \dfrac{K^{m-1}}{(1+K)^{m+1}}, & m=1, 2, \ldots, \end{cases}$$

$$\mathrm{E}\{\boldsymbol{\Pi}_K\}=1, \qquad \mathrm{var}\{\boldsymbol{\Pi}_K\}=2K;$$

for $a>1$,

$$\mathrm{E}\{\boldsymbol{\Pi}_K \mid \boldsymbol{n}<\infty\} = a^{-K}, \qquad \mathrm{var}\{\boldsymbol{\Pi}_K \mid \boldsymbol{n}<\infty\} = \frac{1+a}{1-a}\, a^{-K}(a^{-K}-1),$$

$$\mathrm{E}\{\boldsymbol{\Pi}_K \mid \boldsymbol{n}=\infty\} = \frac{1-a^{-K}}{a-1}, \qquad \mathrm{var}\{\boldsymbol{\Pi}_K \mid \boldsymbol{n}=\infty\} = \frac{(1-a^{-K})(a-a^{-K})}{(1-a)^2}.$$

III.7. LIMIT THEOREMS FOR SINGLE SERVER QUEUES

III.7.1. Introduction

In this chapter we shall mainly be concerned with asymptotic properties of some characteristic quantities describing the behaviour of single server systems. These asymptotic properties divide into three classes. First we shall consider asymptotic relations for $a \to 1$. This type of asymptotics, the so-called "heavy traffic" theory, provides useful relations for the judgement of the queueing system if the system approaches its state of saturation. The second type of asymptotics is concerned with the transient behaviour of the system, i.e. it describes the relation between probabilities or expectations and absolute time t for $t \to \infty$. For instance if $a<1$ the distribution of $\boldsymbol{v}_t$ converges for $t \to \infty$ to a stationary distribution, and from the viewpoint of applicability of the model it is highly desirable to have some idea of the speed of convergence. When using a queueing model as a substitute for a practical queueing situation it may be frequently assumed that during a period of time T the parameters and structure which control the queueing process are independent of absolute time or vary only very slowly with time. Hence if $a<1$ and if there is rapid convergence to the stationary situation, this situation will be reached in a time which is small relative to T, and can be used for design purposes.

The third type of asymptotics to be considered in this chapter consists of limit distributions. In particular we shall study here extreme value distributions for the virtual waiting time, for the actual waiting time and for the number of customers simultaneously present in the system.

III.7.2. Heavy traffic theory

Heavy traffic theory describes the behaviour of queueing systems with traffic intensities close to one. The main purpose of the theory is the derivation of asymptotic expressions for $a \uparrow 1$ of those relations which describe the behaviour of queueing systems.

It is intuitively clear that the influence of the typical character of interarrival time and service time distribution on the form of the waiting time distribution will be more pronounced for small than for large traffic intensities. Consequently, if the traffic intensity a is close to one it may be expected that the behaviour of the $M/M/1$ queueing system will give a good indication of the behaviour of the $G/G/1$ queueing system if all the moments of the interarrival and service time distributions of the latter system are finite.

To illustrate this let $\boldsymbol{w}$ be a stochastic variable with distribution function the stationary distribution of the waiting time of the $G/G/1$ queueing system with $a<1$. From section II.2.3 it is known that for the $M/M/1$ system

$$\Pr\{\boldsymbol{w}<t\}=1-a\,\mathrm{e}^{-(1-a)t/\beta}, \qquad t>0;$$

hence

$$\Pr\{(1-a)\,\boldsymbol{w}<t\}=1-a\,\mathrm{e}^{-t/\beta}, \qquad t>0.$$

Consequently, it may be conjectured that under rather weak conditions the distribution of $(1-a)\,\boldsymbol{w}$ for the $G/G/1$ system with traffic a close to one is negative exponential, i.e.

$$\Pr\{(1-a)\,\boldsymbol{w}<t\}\approx 1-\mathrm{e}^{-t/d} \qquad \text{for} \quad a\uparrow 1, \quad t>0,$$

where d is a positive constant, and obviously

$$d\approx \mathrm{E}\{(1-a)\,\boldsymbol{w}\} \qquad \text{for} \quad a\uparrow 1.$$

Before proving this conjecture we shall first investigate the behaviour of $\mathrm{E}\{(1-a)\,\boldsymbol{w}\}$ for $a\uparrow 1$.

As usual we write

$$\alpha = \int_0^\infty t\,\mathrm{d}A(t), \quad \alpha_2 = \int_0^\infty t^2\,\mathrm{d}A(t), \quad \beta = \int_0^\infty t\,\mathrm{d}B(t), \quad \beta_2 = \int_0^\infty t^2\,\mathrm{d}B(t).$$

It will be assumed in this section that

$$a<1 \quad \text{and} \quad W(0+)<1.$$

In the following we shall need the queueing system $G/G_c/1$ as defined in section II.5.9. Quantities with subscript c refer to the system $G/G_c/1$, those without this subscript to the system $G/G/1$.

From the formula (II.1.1) and the theorem II.1.1 it follows that $\boldsymbol{w}_c$ and $[\boldsymbol{w}_c+\boldsymbol{\rho}_c]^+$ have the same distribution; where

$$\boldsymbol{\rho}_c \stackrel{\text{def}}{=} \boldsymbol{\rho}_{n,c} = \boldsymbol{\tau}_{n,c} - \boldsymbol{\sigma}_{n+1}.$$

Hence

$$\mathrm{E}\{\boldsymbol{w}_c\}=\mathrm{E}\{[\boldsymbol{w}_c+\boldsymbol{\rho}_c]^+\}, \qquad \mathrm{E}\{\boldsymbol{w}_c^2\}=\mathrm{E}\{([\boldsymbol{w}_c+\boldsymbol{\rho}_c]^+)^2\}; \tag{7.1}$$

note that all moments of $\boldsymbol{w}_c$ are finite (cf. (II.5.171), ff.).
Obviously

$$\boldsymbol{w}_c+\boldsymbol{\rho}_c=[\boldsymbol{w}_c+\boldsymbol{\rho}_c]^+ + [\boldsymbol{w}_c+\boldsymbol{\rho}_c]^-, \qquad (\boldsymbol{w}_c+\boldsymbol{\rho}_c)^2=([\boldsymbol{w}_c+\boldsymbol{\rho}_c]^+)^2+([\boldsymbol{w}_c+\boldsymbol{\rho}_c]^-)^2, \tag{7.2}$$

so that from (7.1) and (7.2),

$$\mathrm{E}\{\boldsymbol{\rho}_c\}=\mathrm{E}\{[\boldsymbol{w}_c+\boldsymbol{\rho}_c]^-\}. \tag{7.3}$$

Since $\boldsymbol{w}_c$ and $\boldsymbol{\rho}_c$ are independent we have from (7.1), ..., (7.3),

$$\mathrm{E}\{\boldsymbol{w}_c^2\}+2\mathrm{E}\{\boldsymbol{w}_c\}\,\mathrm{E}\{\boldsymbol{\rho}_c\}+\mathrm{E}\{\boldsymbol{\rho}_c^2\}=\mathrm{E}\{\boldsymbol{w}_c^2\}+\mathrm{E}\{([\boldsymbol{w}_c+\boldsymbol{\rho}_c]^-)^2\}$$
$$\geqq \mathrm{E}\{\boldsymbol{w}_c^2\}+\mathrm{E}^2\{[\boldsymbol{w}_c+\boldsymbol{\rho}_c]^-\}=\mathrm{E}\{\boldsymbol{w}_c^2\}+\mathrm{E}^2\{\boldsymbol{\rho}_c\}.$$

Hence, if $\alpha_2<\infty$ (note $\beta_{2,c}<\infty$), the latter inequality leads to

$$\mathrm{E}\{\boldsymbol{w}_c\} \leqq \frac{\mathrm{var}\{\boldsymbol{\rho}_c\}}{-2\mathrm{E}\{\boldsymbol{\rho}_c\}} = \frac{1}{2(1-a_c)\,\alpha}\{\beta_{2,c}-\beta_c^2+\alpha_2-\alpha^2\}. \tag{7.4}$$

In section II.5.9 it has been shown that $\beta_{2,c}\to\beta_2$, $\beta_c\to\beta$, $\mathrm{E}\{\boldsymbol{w}_c\}\to\mathrm{E}\{\boldsymbol{w}\}$ for $c\to\infty$. Consequently, if $\alpha_2<\infty$, $\beta_2<\infty$,

$$\mathrm{E}\{\boldsymbol{w}\} \leqq \frac{1}{2(1-a)\,\alpha}\{\beta_2-\beta^2+\alpha_2-\alpha^2\}. \tag{7.5}$$

This interesting inequality is due to KINGMAN [1962[a]], and gives an upperbound for $(1-a)\,\mathrm{E}\{\boldsymbol{w}\}$ for $a\uparrow 1$. To obtain the limit of $(1-a)\,\mathrm{E}\{\boldsymbol{w}\}$ for $a\uparrow 1$ we start from (II.5.169) for $G/G_c/1$,

$$\mathrm{E}\{\boldsymbol{w}_c\} = \exp\left\{-\frac{1}{2\pi\mathrm{i}}\int_{-\mathrm{i}\infty-0}^{\mathrm{i}\infty-0}\frac{\mathrm{d}\xi}{\xi^2}\log\{1-\beta_c(\xi)\,\alpha(-\xi)\}\right\},$$

or equivalently,

$$\mathrm{E}\{(1-a_c)\,w_c\}=\exp\left\{-\frac{1}{2\pi\mathrm{i}}\int_{-\mathrm{i}\infty-0}^{\mathrm{i}\infty-0}\frac{\mathrm{d}\eta}{\eta^2}\log\{1-\beta_c\{\eta(1-a_c)\}\alpha\{-\eta(1-a_c)\}\}\right\}. \tag{7.6}$$

If $\alpha_2<\infty$ we have for $|\xi|\to 0$, $-\frac{1}{2}\pi<\arg\xi<\frac{1}{2}\pi$,

$$\alpha(\xi)=1-\alpha\xi+\tfrac{1}{2}\alpha_2\xi^2+\mathrm{o}(\xi^2),$$

$$\beta_c(\xi)=1-\beta_c\xi+\tfrac{1}{2}\beta_{2,c}\xi^2+\mathrm{o}(\xi^2),$$

hence, since $\beta_c(\xi)$ is an entire function,

$$1-\beta_c(\eta(1-a_c))\,\alpha(-\eta(1-a_c))=$$

$$-\,\alpha\eta(1-a_c)^2\left\{1+\frac{\eta}{2\alpha}(\beta_{2,c}-\beta_c^2+\alpha_2-\alpha^2+(1-a_c)^2\,\alpha^2)+\frac{\mathrm{o}(\eta^2(1-a_c)^2)}{\alpha\eta(1-a_c)^2}\right\},$$

for $\eta(1-a_c)\to 0$, $-1\frac{1}{2}\pi<\arg\eta<\frac{1}{2}\pi$. (7.7)

Since

$$\frac{1}{2\pi\mathrm{i}}\int_{-\mathrm{i}\infty-0}^{\mathrm{i}\infty-0}\frac{\mathrm{d}\eta}{\eta^2}\log\eta=0,\qquad -\frac{1}{2\pi\mathrm{i}}\int_{-\mathrm{i}\infty-0}^{\mathrm{i}\infty-0}\frac{\mathrm{d}\eta}{\eta^2}\log(1+h\eta)=h\qquad \text{if}\quad h>0,$$

it follows from (7.6) and (7.7), by using the dominated convergence theorem (cf. Loève [1960]), that for $\alpha_2<\infty$,

$$\mathrm{E}\{(1-a_c)\,w_c\}=\frac{1}{2\alpha}(\beta_{2,c}-\beta_c^2+\alpha_2-\alpha^2)+\mathrm{o}(1-a_c)\qquad \text{for}\quad a_c\uparrow 1.$$

Hence from (7.5) by letting $c\to\infty$ and then $a\uparrow 1$ we have

$$\lim_{a\uparrow 1}\mathrm{E}\{(1-a)\,w\}=\frac{1}{2\alpha}(\beta_2-\beta^2+\alpha_2-\alpha^2). \tag{7.8}$$

Starting again from (II.5.150) we have for $\mathrm{Re}\,\rho>\mathrm{Re}\,\xi$,

$$\mathrm{E}\{\exp\{-\rho w_c(1-a_c)\}\} \tag{7.9}$$

$$=\exp\left\{-\frac{1}{2\pi\mathrm{i}}\int_{-\mathrm{i}\infty-0}^{\mathrm{i}\infty-0}\left(\frac{1}{\rho(1-a_c)-\xi}+\frac{1}{\xi}\right)\log\{1-\alpha(-\xi)\,\beta_c(\xi)\}\,\mathrm{d}\xi\right\}$$

$$=\exp\left\{-\frac{1}{2\pi\mathrm{i}}\int_{-\mathrm{i}\infty-0}^{\mathrm{i}\infty-0}\left(\frac{1}{\rho-\eta}+\frac{1}{\eta}\right)\log\{1-\alpha(-\eta(1-a_c))\,\beta_c(\eta(1-a_c))\}\,\mathrm{d}\eta\right\}.$$

From the latter relation and from (7.7) it follows as above that for Re $\rho \geqq 0$, $\alpha_2 < \infty$,

$$\mathrm{E}\{\exp\{-\rho(1-a_c)\,\boldsymbol{w}_c\}\} = \{1+\rho b_c\}^{-1} + \mathrm{o}(1-a_c) \qquad \text{for} \quad a_c \uparrow 1,$$

where

$$b_c \stackrel{\text{def}}{=} \frac{1}{2\alpha}\{\beta_{2,c} - \beta_c^2 + \alpha_2 - \alpha^2\}. \tag{7.10}$$

Hence from Feller's convergence theorem (cf. app. 5),

$$\lim_{a_c \uparrow 1} \Pr\{(1-a_c)\,\boldsymbol{w}_c < t\} = \begin{cases} 1-\exp(-t/b_c), & t>0, \\ 0, & t<0. \end{cases} \tag{7.11}$$

Since (cf. (II.5.155)),

$$\Pr\{\boldsymbol{w}_c < t\} \geqq \Pr\{\boldsymbol{w} < t\},$$

it follows from (7.11) if $\beta_2 < \infty$, $\alpha_2 < \infty$ that

$$\limsup_{a \uparrow 1} \Pr\{(1-a)\,\boldsymbol{w} < t\} \leqq 1 - \mathrm{e}^{-t/b}, \qquad t>0, \tag{7.12}$$

with

$$b \stackrel{\text{def}}{=} \frac{1}{2\alpha}\{\beta_2 - \beta^2 + \alpha_2 - \alpha^2\}. \tag{7.13}$$

If $\beta(\xi)$ exists for a ξ with Re $\xi < 0$ the relation (7.9) also holds for the system $G/G/1$, i.e. without subscript c, and hence if $\alpha_2 < \infty$, we obtain

$$\lim_{a \uparrow 1} \Pr\{(1-a)\,\boldsymbol{w} < t\} = 1 - \mathrm{e}^{-t/b}, \qquad t>0. \tag{7.14}$$

Further for this case all moments of $\boldsymbol{w}$ are finite and, as above, it is easily shown that lim $\mathrm{E}\{(1-a)\,\boldsymbol{w}^n\}$ also exists for $a \uparrow 1$ and every n. This limit is the nth moment of the negative exponential distribution with mean b^{-1}. The result (7.14) is due to KINGMAN [1965]. The results (7.8), (7.11) and (7.14), which confirm the conjecture stated in the beginning of this section, are basic for the description of the queueing system under heavy traffic conditions. For $a \uparrow 1$ the behaviour of many stochastic variables may be derived from these relations. We shall discuss some examples.

For the stationary situation of the $G/G_c/1$ system denote the number of customers present in the system at some moment by $\boldsymbol{x}_c$. From (II.5.140) it follows easily, assuming that $A(t)$ has no discontinuity points, that

$$\Pr\{\boldsymbol{x}_c=j\}=w_{j,c}=\begin{cases} 1-a_c, & j=0, \\ \dfrac{a_c}{2\pi\mathrm{i}}\displaystyle\int_{-\mathrm{i}\infty}^{\mathrm{i}\infty} \mathrm{E}\{\exp\{-\xi(1-a_c)\,\boldsymbol{w}_c\}\}\,\dfrac{1-\beta_c(\xi(1-a_c))}{(1-a_c)\beta_c\xi} & \\ \cdot\,\dfrac{\alpha^{j-1}(-\xi(1-a_c))-\alpha^j(-\xi(1-a_c))}{-\xi}\,\mathrm{d}\xi, & j=1,2,\dots. \end{cases}$$

Hence

$$\Pr\{\boldsymbol{x}_c\leqq j\} = 1 - a_c$$

$$+\frac{a_c}{2\pi\mathrm{i}}\int_{-\mathrm{i}\infty}^{\mathrm{i}\infty}\mathrm{E}\{\exp\{-\xi(1-a_c)\,\boldsymbol{w}_c\}\}\,\frac{1-\beta_c(\xi(1-a_c))}{(1-a_c)\xi\beta_c}\,\frac{1-\alpha^j(-\xi(1-a_c))}{-\xi}\,\mathrm{d}\xi,$$

$$j=0,1,\dots.$$

Writing $x=j/(1-a_c)$ it follows from

$$\alpha(-\xi(1-a_c))=1+\alpha\xi(1-a_c)+o(\xi(1-a_c)) \qquad \text{for} \quad \xi(1-a_c)\to 0, \quad \operatorname{Re}\xi=0,$$

that

$$\alpha^j(-\xi(1-a_c))\to \mathrm{e}^{\alpha\xi x} \qquad \text{for} \quad a_c\uparrow 1, \quad \operatorname{Re}\xi=0.$$

From (7.11) it follows by using the dominated convergence theorem that if $\alpha_2<\infty$,

$$\lim_{a_c\uparrow 1}\Pr\{(1-a_c)\,\boldsymbol{x}_c\leqq x\} = \frac{1}{2\pi\mathrm{i}}\int_{-\mathrm{i}\infty}^{\mathrm{i}\infty}\frac{1}{1+b_c\xi}\,\frac{1-\mathrm{e}^{\alpha\xi x}}{-\xi}\,\mathrm{d}\xi$$

$$= 1-\mathrm{e}^{-\alpha x/b_c}, \qquad x>0, \tag{7.15}$$

$$\lim_{a_c\uparrow 1}\mathrm{E}\{(1-a_c)\,\boldsymbol{x}_c\} = b_c/\alpha.$$

Denote by $\boldsymbol{v}_c$ a variable with distribution function the stationary distribution of the virtual waiting time of the $G/G_c/1$ system. It follows easily from (II.5.110) and (7.11) that

$$\lim_{a_c\uparrow 1}\Pr\{(1-a_c)\,\boldsymbol{v}_c<t\}=1-\exp(-t/b_c), \qquad t>0. \tag{7.16}$$

III.7.3. The relaxation time

In section II.2.1 asymptotic relations for $\mathrm{E}\{\boldsymbol{x}_t \mid \boldsymbol{x}_0=k\}$ with $t\to\infty$ have been derived for the $M/M/1$ queueing system, where $\boldsymbol{x}_t$ is the number of customers present in the system at time t. The relations obtained in that section gave rise to the concept of relaxation time. For $a<1$ this relaxation time gives an indication of the speed of approach of $\mathrm{E}\{\boldsymbol{x}_t \mid \boldsymbol{x}_0=k\}$ to its limiting value for $t\to\infty$, whereas for $a>1$ it provides a measure for the speed of the approach of $\mathrm{E}\{\boldsymbol{x}_t \mid \boldsymbol{x}_0=k\}$ to its asymptote $(a-1)\,t/\beta$. In the present section the concept of relaxation time will be investigated for more general queueing systems.

It is not difficult to prove that the relaxation time can be investigated by considering any quantity of the queueing system which is a function of continuous time. Therefore, we shall consider the behaviour of $\mathrm{E}\{\boldsymbol{v}_t\}$ and of $\Pr\{\boldsymbol{v}_t>0\}$ for $t\to\infty$, where $\boldsymbol{v}_t$ is the virtual waiting time of the system at time t. Assuming that the condition (II.5.38) is valid, it follows from (II.5.112), suppressing the condition $\boldsymbol{w}_1=0$, $\boldsymbol{t}_1=0$, that for $\operatorname{Re}\rho>0$,

$$\begin{aligned}
&\int_0^\infty \mathrm{e}^{-\rho t}\,\mathrm{d}_t\,\mathrm{E}\{\boldsymbol{v}_t\}\\
&=\frac{\beta}{1-\alpha(\rho)}-\frac{1}{\rho}\,\frac{1-\beta(\rho)}{1-\alpha(\rho)}\exp\left[-\sum_{n=1}^{\infty}\frac{1}{n}\,\mathrm{E}\{(\mathrm{e}^{-\rho\boldsymbol{a}_n}-\mathrm{e}^{-\rho\boldsymbol{b}_n})(\boldsymbol{s}_n>0)\}\right]\\
&=\frac{\beta}{1-\alpha(\rho)}-\frac{1}{\rho}\exp\left[-\sum_{n=1}^{\infty}\frac{1}{n}\,\mathrm{E}\{(\mathrm{e}^{-\rho\boldsymbol{b}_n}-\mathrm{e}^{-\rho\boldsymbol{a}_n})(\boldsymbol{s}_n\leqq 0)\}\right]\\
&=\frac{\beta}{1-\alpha(\rho)}-\frac{1}{\rho}\int_0^\infty \mathrm{e}^{-\rho t}\,\mathrm{d}_t\,\Pr\{\boldsymbol{v}_t>0\}.
\end{aligned}\tag{7.17}$$

It is easily found for $0<\operatorname{Re}\xi<\operatorname{Re}\rho$ and $0<\operatorname{Re}\eta<\operatorname{Re}\rho$ that

$$\begin{aligned}
&\int_0^\infty \mathrm{e}^{-\rho t}\,\mathrm{d}_t\,\Pr\{\boldsymbol{v}_t>0\}\\
&=\{1-\beta(\rho)\}\exp\left\{-\frac{1}{2\pi\mathrm{i}}\int_{C_\xi}\left(\frac{1}{\rho-\xi}+\frac{1}{\xi}\right)\log\{1-\beta(\xi)\,\alpha(\rho-\xi)\}\,\mathrm{d}\xi\right\}\\
&=\{1-\beta(\rho)\}\exp\left\{-\frac{1}{2\pi\mathrm{i}}\int_{C_\eta}\left(\frac{1}{\rho-\eta}+\frac{1}{\eta}\right)\log\{1-\beta(\rho-\eta)\,\alpha(\eta)\}\,\mathrm{d}\eta\right\}.
\end{aligned}\tag{7.18}$$

It will be assumed that $\alpha(\rho)$ and $\beta(\rho)$ are analytic in a halfplane containing in its interior the axis Re $\rho=0$; denote the abcissa of convergence of $\beta(\rho)$ and $\alpha(\rho)$ by ζ_β and ζ_α. Since $B(t)$ and $A(t)$ are non-decreasing functions, ζ_β and ζ_α are both negative. Moreover, it will be assumed that

$$\alpha(\rho)\to\infty \quad \text{for} \quad \rho\to\zeta_\alpha, \quad \text{Im}\,\rho=0; \quad \beta(\rho)\to\infty \quad \text{for} \quad \rho\to\zeta_\beta, \quad \text{Im}\,\rho=0, \tag{7.19}$$

and that $A(t)$ is not a lattice distribution.

For real values of ξ the function $\log \beta(\xi)$ is for $\xi>\zeta_\beta$ a monotonic decreasing function which is convex (DOETSCH [1943]). For fixed and real ρ the function $\log \alpha^{-1}(\rho-\xi)$, $\rho-\xi>\zeta_\alpha$ is a monotonic decreasing function which is concave. Since $\alpha(\rho-\xi)\to\infty$ for $\xi\to\rho-\zeta_\alpha$, $\alpha(\rho-\xi)\to 0$ for $\xi\to-\infty$ and $\beta(\rho)\to\infty$ for $\rho\to\zeta_\beta$, it is easily verified that for ρ not too small the function

$$1-\beta(\xi)\,\alpha(\rho-\xi), \tag{7.20}$$

has two real simple zeros in the interval $(\zeta_\beta, \rho-\zeta_\alpha)$. Denote the larger one by $\xi_0(\rho)$ and the smaller one by $\xi_1(\rho)$. Moreover, it is easily verified that a unique real value ρ_0 of ρ exists for which the zeros $\xi_0(\rho)$ and $\xi_1(\rho)$ coincide, i.e. the function (7.20) has a real zero of multiplicity two,

$$\xi_0(\rho_0)=\xi_1(\rho_0).$$

Obviously, for $\rho=\rho_0$, $\xi=\xi_0(\rho_0)$ we have

$$0=\frac{\mathrm{d}}{\mathrm{d}\xi}\{\beta(\xi)\,\alpha(\rho-\xi)-1\} = \beta'(\xi)\,\alpha(\rho-\xi) - \beta(\xi)\,\alpha'(\rho-\xi). \tag{7.21}$$

It is also easily found that for real $\rho>\rho_0$:
if $a<1$ then $\rho_0>0$, and

$$\begin{aligned} &\zeta_\beta<\xi_1(\rho)<\xi_0(\rho)\leqq\rho\leqq 0,\\ \text{or}\quad &\\ &\zeta_\beta<\xi_1(\rho)<0\leqq\rho\leqq\xi_0(\rho)<\rho-\zeta_\alpha; \end{aligned} \tag{7.22}$$

if $a=1$ then $\rho_0=0$ and

$$\zeta_\beta<\xi_1(\rho)\leqq\rho\leqq\xi_0(\rho)<\rho-\zeta_\alpha;$$

if $a>1$ then $\rho_0<0$, and

$$\begin{aligned} &\zeta_\beta<\xi_1(\rho)\leqq 0\leqq\rho<\xi_0(\rho)<\rho-\zeta_\alpha,\\ \text{or}\quad &\\ &\zeta_\beta<\rho\leqq 0\leqq\xi_1(\rho)<\xi_0(\rho)<\rho-\zeta_\alpha. \end{aligned}$$

For complex values of η and ρ such that

$$\operatorname{Re}\rho>\rho_0, \qquad \xi_1(\operatorname{Re}\rho)<\operatorname{Re}\eta<\xi_0(\operatorname{Re}\rho),$$

we have

$$|\beta(\eta)| \leqq \beta(\operatorname{Re}\eta) < \frac{1}{\alpha(\operatorname{Re}(\rho-\eta))} \leqq \frac{1}{|\alpha(\rho-\eta)|}. \tag{7.23}$$

From the latter inequality it follows that for $\varepsilon>0$ but sufficiently small the inequality

$$|\beta(\eta)-\varepsilon| < \left|\frac{1}{\alpha(\rho-\eta)} - \varepsilon\right|,$$

holds. Since by assumption $A(t)$ is not a lattice distribution $\alpha(\rho-\eta)-\varepsilon^{-1}$ has for every fixed ρ exactly one zero and it follows from the last inequality by applying Rouché's theorem that the function (7.20) has exactly one zero $\xi_0(\rho)$ in the halfplane $\operatorname{Re}\xi>\xi_1(\operatorname{Re}\rho)$ if $\operatorname{Re}\rho>\rho_0$. Since $\beta(\rho)$ and $\alpha(\rho)$ are analytic functions in their domain of definition this zero is an analytic function of ρ for $\operatorname{Re}\rho>\rho_0$ and if ρ is real it is equal to the zero $\xi_0(\rho)$ defined above. Henceforth it will be denoted by $\xi_0(\rho)$.
From (7.21) and

$$1-\beta(\xi_0(\rho))\,\alpha(\rho-\xi_0(\rho))\equiv 0, \tag{7.24}$$

it is easily found that

$$\left|\frac{\mathrm{d}\xi_0(\rho)}{\mathrm{d}\rho}\right| \to \infty \qquad \text{for} \quad \rho\to\rho_0, \tag{7.25}$$

so that $\xi_0(\rho)$ has a singularity at $\rho=\rho_0$. We shall prove that

$$\left|\lim_{\rho\to\rho_0} \frac{\xi_0(\rho)-\xi_0(\rho_0)}{\sqrt{(\rho-\rho_0)}}\right| < \infty, \qquad \operatorname{Re}\rho>\rho_0, \tag{7.26}$$

so that $\xi_0(\rho)$ has a branch point of order two at $\rho=\rho_0$.
With $u \overset{\text{def}}{=} \rho-\rho_0$, $\operatorname{Re}(\rho-\rho_0)\geqq 0$ we may write

$$\begin{aligned} \xi_0(\rho)&=\xi_0(\rho_0)+c(u)\sqrt{u}, \\ \rho-\xi_0(\rho)&=\rho-\xi_0(\rho_0)+u-c(u)\sqrt{u}, \end{aligned} \tag{7.27}$$

where $c(u)$ is a function of $\sqrt{u}$ and $c(u)\sqrt{u}$ is obviously analytic for $\operatorname{Re}u>0$. Since $\beta(\xi)$ and $\alpha(\rho-\xi)$ are analytic at $\rho=\rho_0$, $\xi=\xi_0(\rho_0)$ we have

$$\begin{aligned} \beta(\xi)&=\beta(\xi_0)+(\xi-\xi_0)\,\beta'(\xi_0)+\tfrac{1}{2}(\xi-\xi_0)^2\,\beta''(\xi_0)+o((\xi-\xi_0)^2), \quad \xi\to\xi_0, \\ \alpha(\eta)&=\alpha(\eta_0)+(\eta-\eta_0)\,\alpha'(\eta_0)+\tfrac{1}{2}(\eta-\eta_0)^2\,\alpha''(\eta_0)+o((\eta-\eta_0)^2), \quad \eta\to\eta_0, \end{aligned} \tag{7.28}$$

with $\xi_0=\xi_0(\rho_0)$, $\eta_0=\rho_0-\xi_0(\rho_0)$. Taking $\xi=\xi_0(\rho)$, $\eta=\rho-\xi_0(\rho)$ and inserting the relations (7.28) and (7.27) in (7.24) gives

$$\begin{aligned} c^2(0+) &= -2\beta(\xi)\,\alpha'(\rho-\xi)\,\{\beta''(\xi)\,\alpha(\rho-\xi)-2\beta'(\xi)\,\alpha'(\rho-\xi)+\beta(\xi)\,\alpha''(\rho-\xi)\}^{-1} \\ &= -2\beta(\xi)\,\alpha'(\rho-\xi)\,\{\mathrm{E}\{(\boldsymbol{\tau}-\boldsymbol{\sigma})^2\,\mathrm{e}^{-\boldsymbol{\tau}\xi-\boldsymbol{\sigma}(\rho-\xi)}\}\}^{-1}, \end{aligned} \tag{7.29}$$

for $\rho=\rho_0$, and where $\boldsymbol{\tau}$ and $\boldsymbol{\sigma}$ stand for a service time and interarrival time, respectively. Obviously $c^2(0+)$ is finite and hence (7.26) is proved, so that $\xi_0(\rho)$ has a branch point of order two at $\rho=\rho_0$. From (7.29) it is seen that $c^2(0+)$ is positive. The choice of the sign of $c(0+)$ determines the branch of $\xi_0(\rho)$. Since $\xi_0(\rho)$ is real for ρ real and $\rho>\rho_0$, and $\xi_0(\rho)$ is increasing for increasing ρ we have to take

$$c(0+)>0.$$

Obviously, the other branch of $\xi_0(\rho)$ with branch point at $\rho=\rho_0$ is also a zero of the function (7.20) and is obtained by taking $c(0+)<0$. This branch yields for ρ real and $\rho>\rho_0$ the other real zero $\xi_1(\rho)$, defined above. For $\mathrm{Re}\,\rho>\rho_0$ this other branch of $\xi_0(\rho)$ will from now on be denoted by $\xi_1(\rho)$. Since (cf. above) the function (7.20) has only one zero in the halfplane $\mathrm{Re}\,\xi>\xi_1(\mathrm{Re}\,\rho)$ if $\mathrm{Re}\,\rho>\rho_0$ it follows that

$$\mathrm{Re}\,\xi_1(\rho)\leqq\xi_1(\mathrm{Re}\,\rho), \qquad \mathrm{Re}\,\rho>\rho_0. \tag{7.30}$$

It is easily verified that the relations (7.22) remain valid for complex ρ if in these relations $\xi_1(\rho)$, $\xi_0(\rho)$ and ρ are replaced by their real parts respectively.

If $B(t)$ is a lattice distribution with period τ_0 so that

$$\beta(\xi) = \beta(\xi + 2\pi\mathrm{i}\,h/\tau_0), \qquad h=0, \pm1, \pm2, \dots,$$

then it is easily proved that if $\xi(\rho)$ is a zero of the function (7.20) then

$$\xi(\rho + 2\pi\mathrm{i}\,h/\tau_0) = \xi(\rho) + 2\pi\mathrm{i}\,\frac{h}{\tau_0},$$

so that

$$\mathrm{Re}\,\xi(\rho + 2\pi\mathrm{i}\,h/\tau_0) = \mathrm{Re}\,\xi(\rho).$$

Taking ρ real it follows from the latter relation by the same argument as above that if $B(t)$ is a lattice distribution with period τ_0 then, besides the branchpoint $\rho=\rho_0$, $\xi_0(\rho)$ also has branchpoints at $\rho=\rho_0+2\pi\mathrm{i}\,h/\tau_0$, $h=\pm1$, $\pm2, \dots$.

For any zero $\xi(\rho)$ of the function (7.20) except $\xi_0(\rho)$ we have for $\mathrm{Re}\,\rho>\rho_0$,

$$\operatorname{Re}\xi(\rho) \leqq \xi_1(\operatorname{Re}\rho).$$

Suppose that a ρ exists such that

$$\operatorname{Re}\xi(\rho) = \xi_1(\rho_1), \qquad \rho_1 = \operatorname{Re}\rho, \qquad \operatorname{Im}\rho \neq 0,$$

then from

$$\beta(\xi_1(\rho_1)) = \beta(\operatorname{Re}\xi(\rho)) \geqq |\beta(\xi(\rho))| = \frac{1}{|\alpha(\rho-\xi(\rho))|} \geqq \frac{1}{\alpha(\operatorname{Re}(\rho-\xi(\rho)))}$$

$$= \frac{1}{\alpha(\rho_1-\xi_1(\rho_1))} = \beta(\xi_1(\rho_1)),$$

we obtain

$$|\alpha(\rho-\xi(\rho))| = \alpha(\operatorname{Re}\rho - \operatorname{Re}\xi(\rho)),$$

$$\beta(\operatorname{Re}\xi(\rho)) = |\beta(\xi(\rho))| = |\beta(\operatorname{Re}\xi(\rho) + \mathrm{i}\operatorname{Im}\xi(\rho))|.$$

Since $A(t)$ is not a lattice distribution the first relation implies

$$\operatorname{Im}\{\rho-\xi(\rho)\} = 0,$$

whereas the second relation implies that $B(t)$ is a lattice distribution, say with period τ_0, such that

$$\operatorname{Im}\rho = \operatorname{Im}\xi(\rho) = 2\pi\, h/\tau_0, \qquad h = 0, \pm 1, \pm 2, \dots .$$

From these results we see that $\xi_0(\rho)$ for $\operatorname{Re}\rho \geqq \rho_0$ has more than one branch point for $\rho = \rho_0$ if and only if $\beta(\xi)$ is periodic.

From (7.18) we obtain for $0 < \operatorname{Re}\rho < \operatorname{Re}\xi < \operatorname{Re}\xi_0(\rho)$,

$$\int_0^\infty \mathrm{e}^{-\rho t}\,\mathrm{d}_t \Pr\{\boldsymbol{v}_t > 0\} = \exp\left\{-\frac{1}{2\pi\mathrm{i}} \int_{C_\xi} \left(\frac{1}{\rho-\xi} + \frac{1}{\xi}\right) \log\{1-\beta(\xi)\,\alpha(\rho-\xi)\}\,\mathrm{d}\xi\right\}. \tag{7.31}$$

From the properties of the zeros of (7.20) it follows by partial integration that for $0 < \operatorname{Re}\rho < \operatorname{Re}\xi < \operatorname{Re}\xi_0(\rho) < \operatorname{Re}\zeta < \operatorname{Re}(\rho-\zeta_\alpha)$,

$$-\frac{1}{2\pi\mathrm{i}} \int_{C_\xi} \left(\frac{1}{\rho-\xi} + \frac{1}{\xi}\right) \log\{1-\beta(\xi)\,\alpha(\rho-\xi)\}\,\mathrm{d}\xi$$

$$= \frac{1}{2\pi\mathrm{i}} \int_{C_\xi} \frac{\mathrm{d}\{1-\beta(\xi)\,\alpha(\rho-\xi)\}/\mathrm{d}\xi}{1-\beta(\xi)\,\alpha(\rho-\xi)} \log\frac{\xi}{\xi-\rho}\,\mathrm{d}\xi$$

$$= \log\frac{\xi_0(\rho)-\rho}{\xi_0(\rho)} - \frac{1}{2\pi\mathrm{i}} \int_{C_\zeta} \left(\frac{1}{\rho-\zeta} + \frac{1}{\zeta}\right) \log\{1-\beta(\zeta)\,\alpha(\rho-\zeta)\}\,\mathrm{d}\zeta. \tag{7.32}$$

Define for $0<\text{Re}\,\rho<\text{Re}\,\xi_0(\rho)<\text{Re}\,\zeta<\text{Re}(\rho-\zeta_\alpha)$,

$$H(\rho) \stackrel{\text{def}}{=} \exp\left\{-\frac{1}{2\pi i}\int_{C_\zeta}\left(\frac{1}{\rho-\zeta}+\frac{1}{\zeta}\right)\log\{1-\beta(\zeta)\,\alpha(\rho-\zeta)\}\,d\zeta\right\}. \tag{7.33}$$

It is readily verified that $H(\rho)$ is an analytic function of ρ for $\text{Re}\,\rho>\rho_0$, and $\text{Re}(\rho-\zeta_\alpha)>\text{Re}\,\zeta>\max(0, \text{Re}\,\xi_0(\rho))$, so that by analytic continuation $H(\rho)$ is also defined for $\text{Re}\,\rho>\rho_0$. From the above relations and by using the inversion formula for the Laplace-Stieltjes transform we obtain

$$\Pr\{\boldsymbol{v}_t>0\} = \frac{1}{2\pi i}\int_{C_\rho}\frac{e^{\rho t}}{\rho}\,\frac{\xi_0(\rho)-\rho}{\xi_0(\rho)}\,H(\rho)\,d\rho, \qquad \text{Re}\,\rho>0, \quad t>0. \tag{7.34}$$

The integrand in (7.34) has a singularity at $\rho=0$. If $a\neq 1$ it is a simple pole, while for $a=1$ it is a pole combined with a branch point (cf. (7.26)). For $a<1$ we have $\xi_0(0)=0$, and from (7.24),

$$\frac{d\xi_0(\rho)}{d\rho} = \frac{1}{1-a} \qquad \text{for} \quad \rho=0, \quad a<1.$$

If $a>1$ then $\xi_0(0)>0$ (cf. (7.22)). Hence, since $H(0)=1$,

$$\Pr\{\boldsymbol{v}_t>0\} = \begin{cases} a+\dfrac{1}{2\pi i}\displaystyle\int_{C_\rho}\frac{e^{\rho t}}{\rho}\,\frac{\xi_0(\rho)-\rho}{\xi_0(\rho)}\,H(\rho)\,d\rho, & a<1, \quad t>0, \quad \text{Re}\,\rho=0-, \\[2ex] 1+\dfrac{1}{2\pi i}\displaystyle\int_{C_\rho}\frac{e^{\rho t}}{\rho}\,\frac{\xi_0(\rho)-\rho}{\xi_0(\rho)}\,H(\rho)\,d\rho, & a>1, \quad t>0, \quad \text{Re}\,\rho=0-. \end{cases} \tag{7.35}$$

To obtain an asymptotic expression for $\Pr\{\boldsymbol{v}_t>0\}$, $t\to\infty$ from the above expression we may use a well-known technique based on the fact that $\xi_0(\rho)$ has a branch point at $\rho=\rho_0$, and (if $\beta(\xi)$ is periodic) also at $\rho=\rho_0+2\pi i\,h/\tau_0$, $h=\pm1, \pm2, \ldots$. To apply this method (cf. DOETSCH [1943] ch. 13) we must use the expansions of $\xi_0(\rho)$ in the neighbourhood of its branch points (cf. (7.26)). However, the saddle point method is somewhat easier to use here (cf. WISHART [1966]).

Define for $\text{Re}\,\rho\geq\rho_0$,

$$s \stackrel{\text{def}}{=} \rho - \xi_0(\rho), \qquad s_0 \stackrel{\text{def}}{=} \rho_0 - \xi_0(\rho_0), \tag{7.36}$$

so that

$$1-\beta(\rho-s)\,\alpha(s)=0. \tag{7.37}$$

Since $\beta(\xi)$ is a monotonic and continuous function of ξ for ξ real and $\xi>\zeta_\beta$, it has a uniquely defined inverse function $\beta^*(.)$ and

$$\rho = h(s) \stackrel{\text{def}}{=} s + \beta^*(1/\alpha(s)) \qquad \text{for } s \text{ real and } s \geqq s_0. \tag{7.38}$$

Obviously, the function $h(s)$ can be extended in the complex plane for $\operatorname{Re} s \geqq s_0$, and this extension is uniquely defined if we require (cf. (7.36)),

$$s = h(s) - \xi_0(h(s)), \qquad \operatorname{Re} s \geqq s_0, \tag{7.39}$$

and if $\beta(\xi)$ is not periodic. In the following the analytic extension of $h(s)$ will be supposed to satisfy (7.39). Moreover, it will be assumed for the present that $B(t)$ is not a lattice distribution. It is easily found from (7.36), (7.37), (7.29) and (7.21) that for $s = s_0$,

$$\frac{\mathrm{d}h(s)}{\mathrm{d}s} = 0, \tag{7.40}$$

$$\frac{\mathrm{d}^2 h(s)}{\mathrm{d}s^2} = -\frac{1}{\alpha(s_0)\,\beta'(\rho_0 - s_0)}\,\mathrm{E}\{(\boldsymbol{\tau}-\boldsymbol{\sigma})^2 \exp\{-(\rho_0 - s_0)\,\boldsymbol{\tau} - s_0\boldsymbol{\sigma}\}\} > 0.$$

Consider first the case $a>1$. From (7.35) we obtain for $t>0$, $\operatorname{Re}\rho = \rho_0 +$,

$$\begin{aligned}
&\Pr\{\boldsymbol{v}_t > 0\}\\
&= 1 + \frac{1}{2\pi\mathrm{i}} \int_{C_\rho} \frac{\mathrm{e}^{\rho t}}{\rho}\, \frac{\xi_0(\rho) - \rho}{\xi_0(\rho)}\, H(\rho)\,\mathrm{d}\rho\\
&= 1 + \frac{1}{2\pi\mathrm{i}} \int_{C_s} \mathrm{e}^{th(s)}\, \frac{s}{s-\rho}\, \frac{H(h(s))}{h(s)}\, h'(s)\,\mathrm{d}s\\
&= 1 + \frac{1}{2\pi\mathrm{i}t} \int_{C_s} \mathrm{e}^{th(s)} \left\{ \frac{H(\rho)}{(\rho - s)^2} + \frac{s}{\rho - s}\, \frac{\mathrm{d}\rho}{\mathrm{d}s} \left\{ \frac{\mathrm{d}}{\mathrm{d}\rho} \frac{H(\rho)}{\rho} - \frac{H(\rho)}{\rho(\rho - s)} \right\}\right\} \mathrm{d}s,\\
&\qquad \rho = h(s),
\end{aligned} \tag{7.41}$$

where C_s is the contour in the s-plane defined by $s = \rho - \xi_0(\rho)$, $\operatorname{Re}\rho = \rho_0 +$; the last equality sign in (7.41) is based on partial integration. From (7.40) it is seen that $s = s_0$ is a saddle point of $\exp\{th(s)\}$, the contour of the steepest descent being the line $\operatorname{Re} s = s_0$. Since $B(t)$ is supposed not to be a lattice distribution $s = s_0$ is the only saddle point on $\operatorname{Re} s = s_0$. From the properties of $\xi_0(\rho)$ and from (7.38) it is seen that in the last integral of

(7.41) the contour C_s may be replaced by the line Re $s=s_0$. Since for $a>1$, $\xi_0(\rho_0)>0$ we may write for Re $\rho=\rho_0+$, Re $\zeta=\xi_0(\rho_0)+$ (cf. (7.33)),

$$H(\rho)=\exp\left\{-\frac{1}{2\pi\mathrm{i}}\int_{C_\zeta}\left(\frac{1}{\rho-\zeta}+\frac{1}{\zeta}\right)\log\{1-\beta(\zeta)\,\alpha(\rho-\zeta)\}\,\mathrm{d}\zeta\right\}. \tag{7.42}$$

Hence, since $\mathrm{d}\rho/\mathrm{d}s=0$ for $s=s_0$ (cf. (7.40)) we obtain from saddle point asymptotics (cf. DE BRUIJN [1958]),

$$\Pr\{\boldsymbol{v}_t>0\}$$

$$=1+\frac{1}{2\pi\mathrm{i}t}\,\frac{H(\rho_0+)}{(\rho_0-s_0)^2}\int_{s_0-\mathrm{i}\infty}^{s_0+\mathrm{i}\infty}\exp[t\{h(s_0)+\tfrac{1}{2}v^2h''(s_0)\}]\,\mathrm{d}v\left\{1+\mathrm{O}\left(\frac{1}{t}\right)\right\}$$

$$=1-\frac{H(\rho_0+)}{\xi_0^2(\rho_0)}\,\frac{1}{\{h''(s_0)\}^{\frac{1}{2}}}\,\frac{\mathrm{e}^{t\rho_0}}{t^{\frac{3}{2}}\sqrt{2\pi}}\left\{1+\mathrm{O}\left(\frac{1}{t}\right)\right\},\qquad t\to\infty,\quad a>1. \tag{7.43}$$

If $a<1$ we obtain by the same technique from (7.35),

$$\Pr\{\boldsymbol{v}_t>0\}=a-\frac{H(\rho_0+)}{\xi_0^2(\rho_0)}\,\frac{1}{\{h''(s_0)\}^{\frac{1}{2}}}\,\frac{\mathrm{e}^{t\rho_0}}{t^{\frac{3}{2}}\sqrt{2\pi}}\left\{1+\mathrm{O}\left(\frac{1}{t}\right)\right\},$$

$$t\to\infty,\quad a<1, \tag{7.44}$$

with

$$H(\rho)=\exp\left\{-\frac{1}{2\pi\mathrm{i}}\int_{C_\zeta}\left(\frac{1}{\rho-\zeta}+\frac{1}{\zeta}\right)\log\{1-\beta(\zeta)\,\alpha(\rho-\zeta)\}\,\mathrm{d}\zeta\right\},$$

$$\mathrm{Re}\,\rho=\rho_0+,\qquad 0<\mathrm{Re}\,\zeta<\rho_0-\zeta_\alpha,\qquad \text{since for}\quad a<1,\quad \xi_0(\rho_0)<0.$$

For $a=1$ we have $\rho_0=0$, $\xi_0(\rho_0)=0$, so that from (7.27) and (7.36),

$$\xi_0(\rho)=c(\rho)\sqrt{\rho}=\rho-s\qquad\text{for}\quad \mathrm{Re}\,\rho>0,\quad |\rho|\ \text{small}. \tag{7.45}$$

Hence

$$h'(s)=\frac{\mathrm{d}\rho}{\mathrm{d}s}=\frac{2\sqrt{\rho}}{2\sqrt{\rho}-c(\rho)-2\rho c'(\rho)},$$

and

$$\frac{\mathrm{e}^{\rho t}}{\rho}H(\rho)\,\frac{\xi_0(\rho)-\rho}{\xi_0(\rho)}\,\frac{\mathrm{d}\rho}{\mathrm{d}s}=\frac{\mathrm{e}^{th(s)}}{\sqrt{\rho}}\,\frac{2H(h(s))}{2\sqrt{\rho}-c(\rho)-2\rho c'(\rho)}$$

$$-\frac{\mathrm{e}^{th(s)}}{c(\rho)}\,\frac{2H(h(s))}{2\sqrt{\rho}-c(\rho)-2\rho c'(\rho)}\qquad\text{with}\quad \rho=h(s).$$

Write (cf. (7.34)) for $\mathrm{Re}\,\rho = 0+$,

$$\Pr\{\boldsymbol{v}_t > 0\} = \frac{1}{2\pi \mathrm{i}} \int_{C_\rho} \frac{\mathrm{e}^{\rho t}}{\rho} \frac{\xi_0(\rho) - \rho}{\xi_0(\rho)} H(\rho)\, \mathrm{d}\rho$$

$$= \frac{1}{2\pi \mathrm{i}} \int_{C_{s_1}} \frac{\mathrm{e}^{th(s)}}{\sqrt{\rho}} \frac{2H(h(s))\, \mathrm{d}s}{2\sqrt{\rho} - c(\rho) - 2\rho c'(\rho)}$$

$$- \frac{1}{2\pi \mathrm{i}} \int_{C_s} \frac{\mathrm{e}^{th(s)}}{c(\rho)} \frac{2H(h(s))\, \mathrm{d}s}{2\sqrt{\rho} - c(\rho) - 2\rho c'(\rho)}, \tag{7.46}$$

with $\rho = h(s)$ and C_{s_1} the contour consisting of the half-lines $\mathrm{Re}\, s = 0$, $|\mathrm{Im}\, s| \geqq \delta$ and the semi-circle $|s| = \delta$, $|\arg s| \leqq \frac{1}{2}\pi$, $\delta > 0$ and small while C_s is the line $\mathrm{Re}\, s = 0$. It is easily seen from (7.45) and

$$\sqrt{\rho} = \sqrt{h(s)} = \sqrt{\tfrac{1}{2} s^2 h''(s)}, \qquad |s| \to 0,$$

that the first integral in (7.46) does not contribute to the leading term of the asymptotic expansion of $\Pr\{\boldsymbol{v}_t > 0\}$ for $t \to \infty$. Hence, from (7.46), (7.29) and (7.40) since $H(0+) = 1$,

$$\Pr\{\boldsymbol{v}_t > 0\} \approx \frac{2}{2\pi} \frac{1}{c^2(0+)} \int_{-\infty}^{\infty} \mathrm{e}^{-\frac{1}{2} t v^2 h''(0)}\, \mathrm{d}v = \frac{2}{\sqrt{2\pi}} \frac{1}{c^2(0+)\{th''(0+)\}^{\frac{1}{2}}}$$

$$= \frac{1}{\sqrt{2\pi}} \frac{\mathrm{E}^{\frac{1}{2}}\{(\boldsymbol{\tau} - \boldsymbol{\sigma})^2\}}{\sqrt{\alpha t}} \left\{1 + \mathrm{O}\left(\frac{1}{t}\right)\right\}, \qquad t \to \infty, \quad a = 1. \tag{7.47}$$

From the results obtained above we may now derive the asymptotic relations for $\mathrm{E}\{\boldsymbol{v}_t\}$ by starting from (7.17) and (7.34); for $\mathrm{Re}\,\rho > 0$,

$$\int_0^\infty \mathrm{e}^{-\rho t}\, \mathrm{d}\mathrm{E}\{\boldsymbol{v}_t\} = \frac{\beta}{1 - \alpha(\rho)} - \frac{\xi_0(\rho) - \rho}{\rho \xi_0(\rho)} H(\rho). \tag{7.48}$$

The same technique as above may be used; there are, however, two minor differences. First, we have to account for the residues at $\rho = 0$ of $(\mathrm{e}^{\rho t}/\rho)$ times the expression (7.48) (cf. (7.34)). These residues are (cf. (II.5.113)),

$$a\left\{\frac{\beta_2}{2\beta} + \int_0^\infty t\, \mathrm{d}W(t)\right\} \qquad \text{if} \quad a < 1,$$

$$(a - 1)\, t \qquad \text{if} \quad a > 1.$$

Secondly, the term $\beta/(1-\alpha(\rho))$ may have poles in the region $\rho_0<\operatorname{Re}\rho<0$. For the present we shall exclude this possibility. The asymptotic relations for $\mathrm{E}\{\boldsymbol{v}_t\}$ are given below.

If: (i) condition (II.5.38) *holds, (ii)* $\alpha(\rho)$ *and* $\beta(\rho)$ *are analytic in a region containing in its interior the axis* $\operatorname{Re}\rho=0$, *(iii)* $A(t)$ *is not a lattice distribution, (iv)* $1-\alpha(\rho)$ *has no zeros in the region* $\operatorname{Re}\rho_0<\operatorname{Re}\rho<0$, *(v)* $B(t)$ *is not a lattice distribution, then for* $t\to\infty$,

$$\Pr\{\boldsymbol{v}_t>0 \mid \boldsymbol{w}_1=0,\ \boldsymbol{t}_1=0\}$$

$$=\begin{cases} a-\dfrac{H(\rho_0+)}{\xi_0^2(\rho_0)}\{h''(s_0)\}^{-\frac{1}{2}}\dfrac{\mathrm{e}^{t\rho_0}}{t^{\frac{3}{2}}\sqrt{2\pi}}\left\{1+\mathrm{O}\left(\dfrac{1}{t}\right)\right\}, & a<1,\\[2ex] \dfrac{\mathrm{E}^{\frac{1}{2}}\{(\boldsymbol{\tau}-\boldsymbol{\sigma})^2\}}{\sqrt{2\pi\alpha t}}\left\{1+\mathrm{O}\left(\dfrac{1}{t}\right)\right\}, & a=1,\\[2ex] 1-\dfrac{H(\rho_0+)}{\xi_0^2(\rho_0)}\{h''(s_0)\}^{-\frac{1}{2}}\dfrac{\mathrm{e}^{t\rho_0}}{t^{\frac{3}{2}}\sqrt{2\pi}}\left\{1+\mathrm{O}\left(\dfrac{1}{t}\right)\right\}, & a>1,\end{cases} \tag{7.49}$$

$$\mathrm{E}\{\boldsymbol{v}_t \mid \boldsymbol{w}_1=0,\ \boldsymbol{t}_1=0\}$$

$$\begin{cases} =a\left\{\dfrac{\beta_2}{2\beta}+\displaystyle\int_0^\infty t\,\mathrm{d}W(t)\right\}+\dfrac{H(\rho_0+)}{\rho_0\xi_0^2(\rho_0)}\{h''(s_0)\}^{-\frac{1}{2}}\dfrac{\mathrm{e}^{t\rho_0}}{t^{\frac{3}{2}}\sqrt{2\pi}}\left\{1+\mathrm{O}\left(\dfrac{1}{t}\right)\right\}, & a<1,\\[2ex] \approx\left[\dfrac{2t}{\pi\alpha}\mathrm{E}\{(\boldsymbol{\tau}-\boldsymbol{\sigma})^2\}\right]^{\frac{1}{2}}, & a=1,\\[2ex] =(a-1)\,t+\dfrac{H(\rho_0+)}{\rho_0\xi_0^2(\rho_0)}\{h''(s_0)\}^{-\frac{1}{2}}\dfrac{\mathrm{e}^{t\rho_0}}{t^{\frac{3}{2}}\sqrt{2\pi}}\left\{1+\mathrm{O}\left(\dfrac{1}{t}\right)\right\}, & a>1,\end{cases}$$

where ρ_0 *is that real value of* ρ *for which* $1-\beta(\xi)\,\alpha(\rho-\xi)$ *has a real zero* $\xi_0(\rho_0)$ *of multiplicity two (cf.* (7.20) *and* (7.21)), *and where* $s_0=\rho_0-\xi_0(\rho_0)$,

$$H(\rho)=\exp\left\{-\frac{1}{2\pi\mathrm{i}}\int_{C_\xi}\left(\frac{1}{\rho-\zeta}+\frac{1}{\zeta}\right)\log\{1-\beta(\zeta)\,\alpha(\rho-\zeta)\}\,\mathrm{d}\zeta\right\},$$

with $\operatorname{Re}\rho>\rho_0$, $\rho_0-\zeta_\alpha>\operatorname{Re}\zeta>\max(0,\ \xi_0(\rho_0))$;

$$h''(s_0)=-\{\alpha(s_0)\,\beta'(\rho_0-s_0)\}^{-1}\,\mathrm{E}\{(\boldsymbol{\tau}-\boldsymbol{\sigma})^2\exp\{-(\rho_0-s_0)\,\boldsymbol{\tau}-s_0\boldsymbol{\sigma}\}\}.$$

The results show that the approach of $\Pr\{\boldsymbol{v}_t>0\}$ and $\mathrm{E}\{\boldsymbol{v}_t\}$ to their limiting values for $t\to\infty$ behave like

$$t^{-\frac{3}{2}}\,\mathrm{e}^{t\rho_0} \qquad \text{if} \quad a\neq 1.$$

Hence the decay has a negative exponential character, and for this reason $|\rho_0^{-1}|$ is denoted as the relaxation time of the $G/G/1$ queueing system. Due to the factor $t^{-\frac{3}{2}}$ the decay is, however, somewhat stronger than exponential.

We have the following remarks to make about the above conditions. The condition (II.5.38) has been discussed in section II.5.4. If $a>1$ the condition that $\beta(\xi)$ should exist for a ξ with $\mathrm{Re}\,\xi<0$ is not necessary, similarly, if $a<1$ the same condition for $\alpha(\xi)$ is superfluous. If $B(t)$ is a lattice distribution then $\xi_0(\rho)$, besides the branch point $\rho=\rho_0$, also has branch points with $\mathrm{Re}\,\rho=\rho_0$, the latter also contribute to the term with

$$\mathrm{e}^{t\rho_0}$$

in the asymptotic relation. If $A(t)$ is a lattice distribution then $\Pr\{\boldsymbol{v}_t>0\}$ and $\mathrm{E}\{\boldsymbol{v}_t\}$ have no limits for $t\to\infty$ (cf. (II.5.112)). If $1-\alpha(\rho)$ has a zero ρ_1 with $\rho_0<\mathrm{Re}\,\rho_1<0$, then the asymptotic expression for $\mathrm{E}\{\boldsymbol{v}_t\}$ will contain a term with factor

$$\exp(t\,\mathrm{Re}\,\rho_1);$$

such a term has a slower decay than $\exp(t\rho_0)$. The presence of such a term is due to the transient phenomenon of the arrival process of the customers (cf. sections I.6.1 and I.6.3), and obviously, the interarrival process approaches its stationary situation slower than the queueing process. It should also be noted that the initial conditions influence the coefficient of the factor

$$t^{-\frac{3}{2}}\,\mathrm{e}^{t\rho_0}.$$

For "heavy traffic" conditions the relaxation time $|\rho_0^{-1}|$ is easily determined. Since ρ_0 and $\xi_0(\rho_0)$ both tend to zero for $a\uparrow 1$ we write

$$\rho_0=(1-a)\,r_0, \qquad \xi_0(\rho_0)=(1-a)\,\eta_0.$$

Hence

$$\begin{aligned}0 &= 1-\beta(\eta_0(1-a))\,\alpha((1-a)\,r_0-(1-a)\,\eta_0)\\ &= (1-a)\Big\{(r_0-\eta_0)\,\alpha+\eta_0\beta-\tfrac{1}{2}(1-a)\,\eta_0^2\beta_2-(1-a)\,\eta_0(r_0-\eta_0)\,\alpha\beta\\ &\quad -\tfrac{1}{2}(1-a)(r_0-\eta_0)^2\,\alpha_2+\frac{1}{1-a}\,\mathrm{o}((1-a)^2)\Big\}, \qquad a\uparrow 1.\end{aligned}$$

This equation in η_0 should have a double root. Neglecting $1-\sqrt{a}$ with respect to 1 the latter condition yields (cf. (7.13)),

$$\rho_0=(1-a)\,r_0\approx-(1-\sqrt{a})^2/b,\qquad \xi_0(\rho_0)=(1-a)\,\eta_0\approx-(1-\sqrt{a})/b,\qquad a\approx 1. \tag{7.50}$$

From (7.49) it now follows that for $a<1$, $1-a\approx 0$,

$$\Pr\{\boldsymbol{v}_t>0\}\approx a-\frac{1-\sqrt{a}}{2\sqrt{\pi}}\,\frac{\exp\{-t(1-\sqrt{a})^2/b\}}{\{t(1-\sqrt{a})^2/b\}^{\frac{3}{2}}},\qquad t\to\infty, \tag{7.51}$$

with

$$b=\frac{1}{2\alpha}\,\{\beta_2-\beta^2+\alpha_2-\alpha^2\}.$$

Finally, we shall specialize some of the relations obtained above for various queueing systems. Firstly for the system $M/G/1$, the relations for determining ρ_0 and $\xi_0(\rho_0)$ are

$$1+(\rho_0-\xi_0(\rho_0))\,\alpha-\beta(\xi_0(\rho_0))=0,$$
$$-\alpha-\beta'(\xi_0(\rho_0))=0.$$

From these relations, it is found that

$$h''(s_0)=\frac{1}{\alpha}\,\beta''(\xi_0(\rho_0)).$$

Further, for $\rho_0+1/\alpha>\operatorname{Re}\zeta>\max(0,\xi_0(\rho_0))$,

$$H(\rho_0+)=\exp\left\{-\frac{1}{2\pi\mathrm{i}}\int_{C_\zeta}\left(\frac{1}{\rho-\zeta}+\frac{1}{\zeta}\right)\log\frac{1+(\rho_0-\zeta)\,\alpha-\beta(\zeta)}{1+(\rho_0-\zeta)\,\alpha}\,\mathrm{d}\zeta\right\}$$

$$=\exp\left\{\frac{1}{2\pi\mathrm{i}}\int_{C_\zeta}\left[\frac{\mathrm{d}\{1+(\rho_0-\zeta)\,\alpha-\beta(\zeta)\}/\mathrm{d}\zeta}{1+(\rho_0-\zeta)\,\alpha-\beta(\zeta)}-\frac{\mathrm{d}\{1+(\rho_0-\zeta)\,\alpha\}/\mathrm{d}\zeta}{1+(\rho_0-\zeta)\,\alpha}\right]\right.$$

$$\left.\cdot\log\frac{\zeta}{\zeta-\rho_0}\,\mathrm{d}\zeta\right\}=1+\alpha\rho_0.$$

The integrals are evaluated by applying Cauchy's theorem to a contour consisting of C_ζ and a sufficiently large semi-circle to the right of C_ζ and by noting that $1+(\rho_0-\zeta)\,\alpha-\beta(\zeta)$ has no zeros and that $1+(\rho_0-\zeta)\,\alpha$ has exactly one zero in the right half plane bounded to the left by C_ζ. Hence, we obtain from (7.49) for $a<1$,

$$\Pr\{\boldsymbol{v}_t>0 \mid \boldsymbol{w}_1=0, \boldsymbol{t}_1=0\}$$

$$= a - \frac{1+\alpha\rho_0}{\xi_0^2(\rho_0)} \frac{\alpha^{\frac{1}{2}}}{\{\beta''(\xi_0(\rho_0))\}^{\frac{1}{2}}} \frac{\mathrm{e}^{t\rho_0}}{t^{\frac{3}{2}}\sqrt{2\pi}} \left\{1 + \mathrm{O}\left(\frac{1}{t}\right)\right\}, \qquad t\to\infty, \tag{7.52}$$

if $B(t)$ is not a lattice distribution. It should be noted that if at time $t=0$ the system is empty, i.e. the number $\boldsymbol{x}_0$ of customers present in the system is zero then $\Pr\{\boldsymbol{v}_t>0 \mid \boldsymbol{x}_0=0\}$ for $t\to\infty$ is given by (7.52) with the factor $1+\alpha\rho_0$ replaced by 1.

For the queueing system $E_m/E_n/1$ (cf. sections I.6.5 and II.1.2) supposing $n\geqq 1$, $m\geqq 1$, we have

$$\alpha(\rho) = \left\{1 + \frac{\alpha\rho}{m}\right\}^{-m}, \qquad \operatorname{Re}\rho > -\frac{m}{\alpha},$$

$$\beta(\rho) = \left\{1 + \frac{\beta\rho}{n}\right\}^{-n}, \qquad \operatorname{Re}\rho > -\frac{n}{\beta}.$$

The equations to be satisfied by ρ_0 and $\xi_0(\rho_0)$ are

$$\left\{1 + \frac{\rho-\xi}{m}\alpha\right\}^m \left\{1 + \frac{\beta\xi}{n}\right\}^n - 1 = 0,$$

$$\frac{\mathrm{d}}{\mathrm{d}\xi}\left[\left\{1 + \frac{(\rho-\xi)\,\alpha}{m}\right\}^m \left\{1 + \frac{\beta\xi}{n}\right\}^n\right] = 0.$$

From these equations it is readily found that

$$\beta\rho_0 = -(n+ma) + (m+n)\, a^{m/(n+m)},$$

$$\beta\xi_0(\rho_0) = -n + na^{m/(n+m)}.$$

A simple calculation yields

$$h''(s_0) = \beta\,\frac{n+m}{mn}\, a^{-m/(n+m)},$$

for $H(\rho_0+)$ it is found that

$$H(\rho_0+) = \left\{1 + \frac{\alpha\rho_0}{m}\right\}^m.$$

Therefore, for $a<1$ and $t\to\infty$,

$$\Pr\{\boldsymbol{v}_t>0 \mid \boldsymbol{w}_1=0, \boldsymbol{t}_1=0\}$$

$$= a - \frac{\{1+\alpha\rho_0/m\}^m}{\{\beta\xi_0(\rho_0)\}^2}\left\{\frac{mn}{m+n}\right\}^{\frac{1}{2}} a^{\frac{1}{2}m/(n+m)} \frac{e^{t\rho_0}}{(t/\beta)^{\frac{3}{2}}\sqrt{2\pi}}\left\{1+\mathrm{O}\left(\frac{1}{t}\right)\right\}.$$

It is of some interest to consider the behaviour of ρ_0 for $n\to\infty$, and $m\to\infty$, respectively. It is easily found that

$$\lim_{n\to\infty} \beta\rho_0 = m(1-a) + m\log a,$$

$$\lim_{m\to\infty} \alpha\rho_0 = n\left(1-\frac{1}{a}\right) + n\log\frac{1}{a}.$$

Since

$$\frac{\mathrm{d}}{\mathrm{d}n}\,|\beta\rho_0| = 1 - a^{m/(n+m)} + a^{m/(n+m)}\log a^{m/(n+m)} > 0, \qquad a\neq 1,$$

it is seen that for increasing n, and similarly for increasing m, the relaxation time $|\rho_0^{-1}|$ decreases. Note that for the system $E_m/D/1$ we have

$$\beta\rho_0 = m(1-a) + m\log a.$$

III.7.4. Limit theorems for the $M/G/1$ queueing system

As usual we denote the virtual waiting time at time t by $\boldsymbol{v}_t$, the number of customers present in the system at time t by $\boldsymbol{x}_t$ and the actual waiting time of the nth arriving customer by $\boldsymbol{w}_n$ with $\boldsymbol{w}_1=0$. With $\boldsymbol{c}$ the duration of a busy cycle and $\boldsymbol{n}$ the number of customers served in a busy cycle we define

$$\boldsymbol{v}_{\max} \overset{\text{def}}{=} \sup_{0<t<\boldsymbol{c}} \boldsymbol{v}_t, \tag{7.53}$$

$$\boldsymbol{w}_{\max} \overset{\text{def}}{=} \sup_{1\leq n\leq \boldsymbol{n}} \boldsymbol{w}_n,$$

$$\boldsymbol{x}_{\max} \overset{\text{def}}{=} \sup_{0<t<\boldsymbol{c}} \boldsymbol{x}_t,$$

so that $\boldsymbol{v}_{\max}$ is the supremum of the virtual waiting time during a busy cycle, $\boldsymbol{w}_{\max}$ is the supremum of all actual waiting times of a busy cycle and $\boldsymbol{x}_{\max}$ is the supremum of the number of customers simultaneously present in a busy cycle.

Expressions for the distributions of the quantities defined in (7.53) have been derived in preceding sections. Denoting by δ the zero of $\beta(\eta)+\alpha\eta-1$ with $\operatorname{Re}\eta>0$ and by μ the zero of $\beta\{(1-\omega)/\alpha\}-\omega$ with $|\omega|<1$ for $a=\beta/\alpha>1$, and noting that $\delta=0$, $\mu=1$ if $a\leqq 1$ we have from (5.87),

$$\Pr\{\boldsymbol{v}_{\max}<v\} = \begin{cases} \dfrac{\dfrac{1}{2\pi i}\displaystyle\int_{C_\eta} e^{\eta v}\frac{\beta(\eta)}{\beta(\eta)+\alpha\eta-1}\,d\eta}{\dfrac{1}{2\pi i}\displaystyle\int_{C_\eta} e^{\eta v}\frac{d\eta}{\beta(\eta)+\alpha\eta-1}}, & \operatorname{Re}\eta>\delta,\quad v>0, \\ 0, & v<0. \end{cases} \tag{7.54}$$

From (4.73) it follows that

$$\Pr\{\boldsymbol{w}_{\max}<w\} = \begin{cases} \dfrac{\dfrac{1}{2\pi i}\displaystyle\int_{C_\eta} e^{\eta w}\frac{d\eta}{\beta(\eta)+\alpha\eta-1}}{\dfrac{1}{2\pi i}\displaystyle\int_{C_\eta} \frac{e^{\eta w}}{1-\alpha\eta}\frac{d\eta}{\beta(\eta)+\alpha\eta-1}}, & \dfrac{1}{\alpha}>\operatorname{Re}\eta>\delta, \\ & w>0, \\ 0, & w<0; \end{cases} \tag{7.55}$$

and from (6.51) (cf. also (II.4.70)) that

$$\Pr\{\boldsymbol{x}_{\max}\leqq x\} = \frac{\dfrac{1}{2\pi i}\displaystyle\int_{D_\omega} \frac{d\omega}{\omega^x}\frac{\beta\{(1-\omega)/\alpha\}}{\beta\{(1-\omega)/\alpha\}-\omega}}{\dfrac{1}{2\pi i}\displaystyle\int_{D_\omega} \frac{d\omega}{\omega^{x+1}}\frac{\beta\{(1-\omega)/\alpha\}}{\beta\{(1-\omega)/\alpha\}-\omega}}, \qquad |\omega|<\mu,\quad x=1,2,\dots, \tag{7.56}$$

with D_ω a circle in the complex ω-plane with centre at $\omega=0$ and radius $|\omega|$, the positive direction of integration being counter clockwise. From the above relations it is readily found by contour integration that for $a\leqq 1$,

$$1-\Pr\{\boldsymbol{v}_{\max}<v\} = \frac{\dfrac{1}{2\pi i}\displaystyle\int_{C_\eta} e^{\eta v}\left\{\frac{\alpha\eta}{\beta(\eta)+\alpha\eta-1}-1\right\}d\eta}{\dfrac{1}{2\pi i}\displaystyle\int_{C_\eta} e^{\eta v}\frac{d\eta}{\beta(\eta)+\alpha\eta-1}}, \qquad \operatorname{Re}\eta>0,\quad v>0, \tag{7.57}$$

$$1-\Pr\{\boldsymbol{w}_{\max}<w\}$$

$$=\frac{\dfrac{\alpha}{2\pi\mathrm{i}}\displaystyle\int_{C_\eta}\frac{\mathrm{e}^{\eta w}}{1-\alpha\eta}\,\frac{\eta\,\mathrm{d}\eta}{\beta(\eta)+\alpha\eta-1}}{\dfrac{1}{2\pi\mathrm{i}}\displaystyle\int_{C_\eta}\frac{\mathrm{e}^{\eta w}}{1-\alpha\eta}\,\frac{\mathrm{d}\eta}{\beta(\eta)+\alpha\eta-1}},\qquad \frac{1}{\alpha}>\operatorname{Re}\eta>0,\quad w>0,$$

$$=\frac{\dfrac{1}{2\pi\mathrm{i}}\displaystyle\int_{C_\xi}\frac{\mathrm{e}^{\xi w}}{1-\alpha\xi}\left\{\frac{\alpha\xi}{\beta(\xi)+\alpha\xi-1}-\frac{1}{\beta(1/\alpha)}\right\}\mathrm{d}\xi}{\dfrac{1}{2\pi\mathrm{i}}\displaystyle\int_{C_\xi}\frac{\mathrm{e}^{\xi w}}{1-\alpha\xi}\left\{\frac{1}{\beta(\xi)+\alpha\xi-1}-\frac{1}{\beta(1/\alpha)}\right\}\mathrm{d}\xi},\qquad \operatorname{Re}\xi>0,\quad w>0, \tag{7.58}$$

$$1-\Pr\{\boldsymbol{x}_{\max}\leqq x\}$$

$$=\frac{\dfrac{1}{2\pi\mathrm{i}}\displaystyle\int_{D_\omega}\frac{\mathrm{d}\omega}{\omega^x}\,\frac{1-\omega}{\beta\{(1-\omega)/\alpha\}-\omega}}{\dfrac{1}{2\pi\mathrm{i}}\displaystyle\int_{D_\omega}\frac{\mathrm{d}\omega}{\omega^x}\,\frac{1}{\beta\{(1-\omega)/\alpha\}-\omega}},\qquad |\omega|<1,\quad x=2,3,\dots. \tag{7.59}$$

Define

$$H(t)\overset{\mathrm{def}}{=}\begin{cases}\dfrac{1}{\beta}\displaystyle\int_0^t\{1-B(\tau)\}\,\mathrm{d}\tau,\\[2ex] 0,\end{cases}\qquad h(t)\overset{\mathrm{def}}{=}\begin{cases}\dfrac{1}{\beta}\{1-B(t)\}, & t>0,\\[2ex] 0, & t<0,\end{cases} \tag{7.60}$$

so that $H(t)$ is a distribution function having a bounded and monotone density function; and write for $a\leqq 1$,

$$K(t,a)\overset{\mathrm{def}}{=}\sum_{n=0}^{\infty}a^nH^{n*}(t). \tag{7.61}$$

It follows that

$$\int_{0-}^{\infty}\mathrm{e}^{-\eta t}\,\mathrm{d}_tK(t,a)=\frac{\alpha\eta}{\beta(\eta)+\alpha\eta-1},\qquad \operatorname{Re}\eta>0. \tag{7.62}$$

Obviously, $K(t, 1)$ is the renewal function of a renewal process with $H(t)$ as renewal distribution. Since $H(t)$ has a bounded and monotone density function, $K(t, a)$ has for $a \leqq 1$ a bounded derivative $k(t, a)$ with respect to t (cf. FELLER [1966] p. 358), so that with

$$k(t, a) \stackrel{\text{def}}{=} \frac{\mathrm{d}}{\mathrm{d}t} K(t, a), \qquad t>0, \tag{7.63}$$

we have

$$\int_0^\infty \mathrm{e}^{-\eta t} k(t, a)\, \mathrm{d}t = \frac{\alpha\eta}{\beta(\eta)+\alpha\eta-1} - 1, \qquad \operatorname{Re} \eta>0.$$

Since for $a \leqq 1$, $w>0$,

$$\int_0^\infty K(w+\tau, a)\, \mathrm{e}^{-\tau/\alpha} \frac{\mathrm{d}\tau}{\alpha} = \mathrm{e}^{w/\alpha} \int_{t=w}^\infty \mathrm{e}^{-t/\alpha} K(t, a) \frac{\mathrm{d}t}{\alpha}$$

$$= K(w, a) + \mathrm{e}^{w/\alpha} \int_{t=w}^\infty \mathrm{e}^{-t/\alpha} k(t, a)\, \mathrm{d}t,$$

we have for $a \leqq 1$, $\operatorname{Re} \xi>0$,

$$\int_0^\infty \mathrm{e}^{-\xi w} \int_0^\infty K(w+t, a)\, \mathrm{e}^{-t/\alpha}\, \mathrm{d}t\, \mathrm{d}w = \frac{\alpha^2}{1-\alpha\xi} \left\{ \frac{1}{\beta(\xi)+\alpha\xi-1} - \frac{1}{\beta(1/\alpha)} \right\},$$
$$\tag{7.64}$$
$$\int_0^\infty \mathrm{e}^{-\xi w} \int_0^\infty k(w+t, a)\, \mathrm{e}^{-t/\alpha}\, \mathrm{d}t\, \mathrm{d}w = \frac{\alpha}{1-\alpha\xi} \left\{ \frac{\alpha\xi}{\beta(\xi)+\alpha\xi-1} - \frac{1}{\beta(1/\alpha)} \right\}.$$

Using generating functions it is easily found that for $|\omega|<1$, $x=0, 1, 2, \dots$; $a \leqq 1$,

$$\frac{1}{2\pi \mathrm{i}} \int_{D_\omega} \frac{\mathrm{d}\omega}{\omega^{x+1}} \frac{1-\omega}{\beta\{(1-\omega)/\alpha\}-\omega} = \int_{0-}^\infty \frac{(t/\alpha)^x}{x!} \mathrm{e}^{-t/\alpha}\, \mathrm{d}_t K(t, a), \tag{7.65}$$

$$\frac{1}{2\pi \mathrm{i}} \int_{D_\omega} \frac{\mathrm{d}\omega}{\omega^{x+1}} \frac{1}{\beta\{(1-\omega)/\alpha\}-\omega} = \int_{0-}^\infty \frac{(t/\alpha)^x}{x!} \mathrm{e}^{-t/\alpha} K(t, a) \frac{\mathrm{d}t}{\alpha}.$$

Let $\boldsymbol{w}$ and $\boldsymbol{\sigma}$ be two independent non-negative variables, with distribution functions defined by

$$\begin{aligned} \mathrm{E}\{\mathrm{e}^{-\rho\boldsymbol{\sigma}}\} &= \frac{1}{1+\alpha\rho}, && \operatorname{Re}\rho > -\frac{1}{\alpha}, \\ \mathrm{E}\{\mathrm{e}^{-\rho\boldsymbol{w}}\} &= \frac{(1-a)\,\alpha\rho}{\beta(\rho)+\alpha\rho-1}, && \operatorname{Re}\rho \geqq 0, \quad a<1, \end{aligned} \tag{7.66}$$

obviously $\boldsymbol{w}$ has the stationary distribution of the (virtual or actual) waiting time of the $M/G/1$ queueing system.

Using the inversion formulas for the Laplace-Stieltjes transform and for the Laplace transform we obtain from (7.57), ..., (7.59) and (7.64), ..., (7.66),

$$1-\Pr\{\boldsymbol{v}_{\max}<v\}$$

$$= \begin{cases} \alpha\dfrac{k(v,a)}{K(v,a)} = \alpha\dfrac{\mathrm{d}}{\mathrm{d}v}\log K(v,a), & v>0, \quad a\leqq 1, \\[2ex] \alpha\dfrac{\mathrm{d}}{\mathrm{d}v}\log\Pr\{\boldsymbol{w}<v\}, & v>0, \quad a<1; \end{cases} \tag{7.67}$$

$$1-\Pr\{\boldsymbol{w}_{\max}<w\}$$

$$= \begin{cases} \dfrac{\alpha\int_0^\infty k(w+t,a)\,\mathrm{e}^{-t/\alpha}\,\mathrm{d}t}{\int_0^\infty K(w+t,a)\,\mathrm{e}^{-t/\alpha}\,\mathrm{d}t} = \alpha\dfrac{\mathrm{d}}{\mathrm{d}w}\log\displaystyle\int_0^\infty K(w+t,a)\,\mathrm{e}^{-t/\alpha}\,\mathrm{d}t, \\ \qquad\qquad\qquad\qquad\qquad \text{for} \quad w>0, \quad a\leqq 1, \\[2ex] \alpha\dfrac{\mathrm{d}}{\mathrm{d}w}\log\Pr\{\boldsymbol{w}<w+\boldsymbol{\sigma}\} \qquad \text{for} \quad w>0, \quad a<1; \end{cases} \tag{7.68}$$

$$1-\Pr\{\boldsymbol{x}_{\max}\leqq x\}$$

$$= \frac{\displaystyle\int_0^\infty \frac{(t/\alpha)^{x-1}}{(x-1)!}\,\mathrm{e}^{-t/\alpha}\,\mathrm{d}_t K(t,a)}{\displaystyle\int_0^\infty \frac{(t/\alpha)^{x-1}}{(x-1)!}\,\mathrm{e}^{-t/\alpha}\,K(t,a)\,\frac{\mathrm{d}t}{\alpha}}, \qquad x=2,3,\ldots, \quad a\leqq 1. \tag{7.69}$$

From the relations

$$\mathrm{E}\{\boldsymbol{v}_{\max}\} = \int_0^\infty \{1-\Pr\{\boldsymbol{v}_{\max}<v\}\}\,\mathrm{d}v,$$

$$\mathrm{E}\{\boldsymbol{v}_{\max}^2\} = 2\int_0^\infty v\{1-\Pr\{\boldsymbol{v}_{\max}<v\}\}\,\mathrm{d}v,$$

$$\Pr\{\boldsymbol{w}<0+\} = 1-a \quad \text{if} \quad a<1,$$

it is found from (7.68) and (7.69) that for $a<1$,

$$\mathrm{E}\{\boldsymbol{v}_{\max}\} = \alpha \log \frac{1}{1-a}, \tag{7.70}$$

$$\mathrm{E}\{\boldsymbol{v}_{\max}^2\} = -2\alpha \int_0^\infty \log\{1-\Pr\{\boldsymbol{w}\geqq v\}\}\,\mathrm{d}v,$$

$$\mathrm{E}\{\boldsymbol{w}_{\max}\} = \alpha \log \frac{\beta(1/\alpha)}{1-a},$$

$$\mathrm{E}\{\boldsymbol{w}_{\max}^2\} = -2\alpha \int_0^\infty \log \Pr\{\boldsymbol{w}<w+\boldsymbol{\sigma}\}\,\mathrm{d}w.$$

Since for $a<1$,

$$\log\{1-\Pr\{\boldsymbol{w}\geqq v\}\} = -\sum_{n=1}^\infty \frac{1}{n}[\Pr\{\boldsymbol{w}\geqq v\}]^n, \qquad v>0,$$

$$\Pr\{\boldsymbol{w}\geqq w\} < a, \qquad w>0,$$

we have for $a<1$,

$$2\alpha \int_0^\infty \Pr\{\boldsymbol{w}\geqq w\}\,\mathrm{d}w < \mathrm{E}\{\boldsymbol{v}_{\max}^2\} < 2\alpha \sum_{n=1}^\infty \frac{a^{n-1}}{n} \int_0^\infty \Pr\{\boldsymbol{w}\geqq w\}\,\mathrm{d}w,$$

so that, since

$$\mathrm{E}\{\boldsymbol{w}\} = \int_0^\infty \Pr\{\boldsymbol{w}\geqq w\}\,\mathrm{d}w = \frac{1}{2}\frac{a\beta}{1-a}\frac{\beta_2}{\beta^2},$$

with β_2 the second moment of $B(t)$,

$$\frac{\beta_2}{1-a} < \mathrm{E}\{\boldsymbol{v}_{\max}^2\} < \frac{\beta_2}{1-a}\,\frac{1}{a}\log\frac{1}{1-a}, \qquad a<1. \tag{7.71}$$

Consequently, the second moment of $\boldsymbol{v}_{\max}$ is finite if and only if $\beta_2<\infty$. A similar conclusion holds for $\boldsymbol{w}_{\max}$. It should be noted that $\mathrm{E}\{\boldsymbol{v}_{\max}\}$ and $\mathrm{E}\{\boldsymbol{w}_{\max}\}$ are finite if and only if $a<1$, while $\mathrm{E}\{\boldsymbol{w}\}$ is finite if and only if $a<1$, $\beta_2<\infty$. Obviously, for $a<1$ but a not too small, $\mathrm{E}\{\boldsymbol{v}_{\max}\}$ will be smaller than $\mathrm{E}\{\boldsymbol{w}\}$. Although this result seems rather paradoxical it is easily explained by noting that $\boldsymbol{v}_{\max}$ has only one realisation in a busy cycle, while the actual waiting time may have any number of realisations in a busy cycle.

From now on it will be assumed that the server is idle at time $t=0$. Denote the supremum of $\boldsymbol{v}_t$, of $\boldsymbol{w}_n$ and of $\boldsymbol{x}_t$, in the jth busy cycle by $\boldsymbol{v}_{\max}^{(j)}$, $\boldsymbol{w}_{\max}^{(j)}$ and $\boldsymbol{x}_{\max}^{(j)}$ respectively. Obviously, $\boldsymbol{v}_{\max}^{(j)}$, $j=1, 2, \ldots$, are independent, identically distributed variables with finite first moment if $a<1$ and with finite second moment if $\beta_2<\infty$. Hence, if $a<1$ the strong law of large numbers applies for the sequence $\boldsymbol{v}_{\max}^{(j)}$, $j=1, 2, \ldots$; whereas for $\beta_2<\infty$ the central limit theorem also applies for this sequence.

Define for $n=1, 2, \ldots$,

$$\boldsymbol{V}_n \stackrel{\text{def}}{=} \max_{1\leq j\leq n} \boldsymbol{v}_{\max}^{(j)}, \qquad \boldsymbol{W}_n \stackrel{\text{def}}{=} \max_{1\leq j\leq n} \boldsymbol{w}_{\max}^{(j)}, \tag{7.72}$$

$$\boldsymbol{X}_n \stackrel{\text{def}}{=} \max_{1\leq j\leq n} \boldsymbol{x}_{\max}^{(j)},$$

so that $\boldsymbol{V}_n$ is the supremum of the virtual waiting time in n busy cycles, $\boldsymbol{W}_n$ that of the actual waiting time in n busy cycles, and $\boldsymbol{X}_n$ is the supremum of the number of customers present simultaneously in the system in n busy cycles. For these variables we shall derive some limit theorems.

THEOREM 7.1. *If $a=1$ and β_2, the second moment of $B(t)$, is finite then the distributions of*

$$\frac{1}{n\beta}\boldsymbol{V}_n, \text{ of } \frac{1}{n\beta}\boldsymbol{W}_n \text{ and of } \frac{1}{n}\boldsymbol{X}_n \text{ all converge} \quad \text{for } n\to\infty$$

to the distribution $G(x)$ with

$$G(x)=\begin{cases} \mathrm{e}^{-1/x} & \text{for } x>0,\\ 0 & \text{for } x<0.\end{cases}$$

Proof. Since $\beta_2/2\beta$ is the first moment of $H(t)$ (cf. (7.60)) and since $h(t)$ is monotonic we have (cf. section I.6.2, theorems 6.1 and 6.5) from (7.61),

$$\lim_{t\to\infty}\frac{K(t,1)}{t}=\frac{2\beta}{\beta_2}, \qquad \lim_{t\to\infty} k(t,1)=\frac{2\beta}{\beta_2}. \tag{7.73}$$

Hence from (7.67) since $a=1$,

$$\lim_{v\to\infty} v\{1-\Pr\{\boldsymbol{v}_{\max}^{(J)}<v\}\}=\alpha=\beta.$$

From the last relation and from

$$\Pr\left\{\frac{1}{n\beta}\,\boldsymbol{V}_n<x\right\}=[\Pr\{\boldsymbol{v}_{\max}^{(J)}<n\beta x\}]^n=\left\{1-\frac{\beta}{n\beta x}+\mathrm{o}\left(\frac{1}{n}\right)\right\}^n,$$

$$x>0,\quad n\to\infty,$$

it follows immediately that

$$\lim_{n\to\infty}\Pr\left\{\frac{1}{n\beta}\,\boldsymbol{V}_n<x\right\}=\begin{cases}\mathrm{e}^{-1/x}, & x>0,\\ 0, & x<0.\end{cases}$$

The statement for $\boldsymbol{V}_n$ has been proved.

From (7.68) for $a=1$,

$$1-\Pr\{\boldsymbol{w}_{\max}<w\}=\alpha\frac{\int_0^\infty k(w+\tau,1)\,\mathrm{e}^{-\tau/\alpha}\,\mathrm{d}\tau/\alpha}{\int_0^\infty K(w+\tau,1)\,\mathrm{e}^{-\tau/\alpha}\,\mathrm{d}\tau/\alpha},\qquad w>0. \tag{7.74}$$

For given $\varepsilon>0$ a finite number $W(\varepsilon)>0$ exists such that

$$\left|k(w,1)-\frac{2\beta}{\beta_2}\right|<\varepsilon\qquad\text{for all } w>W(\varepsilon),$$

so that

$$\left|k(w+t,1)-\frac{2\beta}{\beta_2}\right|<\varepsilon\qquad\text{for all } w>W(\varepsilon),\ t\geqq 0.$$

Consequently, since $k(t, 1)$ is bounded

$$\lim_{w\to\infty}\int_0^\infty k(t+w,1)\,\mathrm{e}^{-t/\alpha}\,\frac{\mathrm{d}t}{\alpha}=\frac{2\beta}{\beta_2}\int_0^\infty \mathrm{e}^{-t/\alpha}\,\frac{\mathrm{d}t}{\alpha}=\frac{2\beta}{\beta_2}. \tag{7.75}$$

Using (7.73) the same argument yields

$$\lim_{w\to\infty}\frac{1}{w}\int_0^\infty K(w+t,1)\,\mathrm{e}^{-t/\alpha}\,\frac{\mathrm{d}t}{\alpha}$$

$$=\lim_{w\to\infty}\int_0^\infty \frac{K(w+t,1)}{w+t}\,\mathrm{e}^{-t/\alpha}\left\{1+\frac{t}{w}\right\}\frac{\mathrm{d}t}{\alpha}=\frac{2\beta}{\beta_2}.$$

Hence from (7.74),

$$\lim_{w\to\infty} w\{1-\Pr\{\boldsymbol{w}_{\max}^{(j)}<w\}\}=\alpha=\beta,$$

so that, as above, the statement for $\boldsymbol{W}_n$ follows.

For $x=1, 2, \ldots$,

$$\int_{W(\varepsilon)}^{\infty} \frac{(t/\alpha)^{x-1}}{(x-1)!} e^{-t/\alpha} \left| k(t,1) - \frac{2\beta}{\beta_2} \right| \frac{dt}{\alpha} \leqq \varepsilon \int_{W(\varepsilon)}^{\infty} \frac{(t/\alpha)^{x-1}}{(x-1)!} e^{-t/\alpha} \frac{dt}{\alpha} \leqq \varepsilon,$$

and

$$\lim_{x\to\infty} \int_0^{W(\varepsilon)} \frac{(t/\alpha)^{x-1}}{(x-1)!} e^{-t/\alpha} k(t,1) \frac{dt}{\alpha}$$

$$\leqq \max_{0\leqq t\leqq W(\varepsilon)} k(t,1) \lim_{x\to\infty} \int_0^{W(\varepsilon)} \frac{(t/\alpha)^{x-1}}{(x-1)!} e^{-t/\alpha} \frac{dt}{\alpha} = 0.$$

It follows that

$$\lim_{x\to\infty} \int_0^{\infty} \frac{(t/\alpha)^{x-1}}{(x-1)!} e^{-t/\alpha} \left(k(t,1) - \frac{2\beta}{\beta_2} \right) \frac{dt}{\alpha} = 0,$$

or that

$$\lim_{x\to\infty} \int_0^{\infty} \frac{(t/\alpha)^{x-1}}{(x-1)!} e^{-t/\alpha} k(t,1) \frac{dt}{\alpha} = \frac{2\beta}{\beta_2}.$$

In the same way it is shown that

$$\lim_{x\to\infty} \int_0^{\infty} \frac{(t/\alpha)^{x}}{x!} e^{-t/\alpha} \frac{K(t,1)}{t} \frac{dt}{\alpha} = \frac{2\beta}{\beta_2}.$$

Consequently from (7.69),

$$\lim_{x\to\infty} x\{1-\Pr\{\boldsymbol{x}_{\max}^{(j)}\leqq x\}\}=1,$$

and, as above, this relation leads to the statement for $\boldsymbol{X}_n$. The theorem is proved.

THEOREM 7.2. *If* $a<1$, $\rho_0<0$, ρ_0 *being the abscissa of convergence of* $\beta(\rho)$, *and if* $\beta(\rho)\to\infty$ *for* $\rho\to\rho_0$, $\operatorname{Im}\rho=0$, *then for* $-\infty<x<\infty$,

$$\lim_{n\to\infty} \Pr\left\{\frac{1}{\beta} V_n < \frac{x+\log(nb_1)}{-\varepsilon\beta}\right\} = \mathrm{e}^{-\mathrm{e}^{-x}},$$

$$\lim_{n\to\infty} \Pr\left\{\frac{1}{\beta} W_n < \frac{x+\log(nb_2)}{-\varepsilon\beta}\right\} = \mathrm{e}^{-\mathrm{e}^{-x}},$$

$$\mathrm{e}^{-(1-\alpha\varepsilon)\mathrm{e}^{-x}} \leqq \liminf_{n\to\infty} \Pr\left\{X_n < \frac{x+\log(nb_1)}{\log(1-\alpha\varepsilon)}\right\}$$

$$\leqq \limsup_{n\to\infty} \Pr\left\{X_n < \frac{x+\log(nb_1)}{\log(1-\alpha\varepsilon)}\right\} \leqq \mathrm{e}^{-\mathrm{e}^{-x}},$$

with

$$b_1 = \frac{\alpha-\beta}{\alpha+\beta'(\varepsilon)}\,\alpha\varepsilon, \qquad b_2 = \frac{\alpha-\beta}{\alpha+\beta'(\varepsilon)}\,\frac{\alpha\varepsilon}{1-\alpha\varepsilon},$$

$$\beta'(\rho) = -\int_0^\infty t\,\mathrm{e}^{-\rho t}\,\mathrm{d}B(t), \qquad \operatorname{Re}\rho>\rho_0,$$

and ε *is the zero of* $\beta(\eta)+\alpha\eta-1$, $\operatorname{Re}\eta<0$, *which is nearest to the imaginary axis* $\operatorname{Re}\eta=0$.

Proof. Since $\rho_0<0$ and $a<1$ the function $\beta(\eta)+\alpha\eta-1$ has, for $\operatorname{Re}\eta<0$, a real zero. Denote its real zero nearest to the axis $\operatorname{Re}\eta=0$ by ε. Clearly, $\varepsilon>\rho_0$. From

$$|\beta(\eta)|\leqq\beta(\operatorname{Re}\eta)=1-\alpha\varepsilon<|1-\alpha\eta| \qquad \text{for} \quad \operatorname{Re}\eta=\varepsilon, \quad \eta\neq\varepsilon,$$

it follows that ε is the only zero with $\operatorname{Re}\eta=\varepsilon$. From

$$|\beta(\eta)|\leqq\beta(\operatorname{Re}\eta)<1-\alpha\operatorname{Re}\eta<|1-\alpha\eta| \qquad \text{for} \quad \varepsilon<\operatorname{Re}\eta<0,$$

and from Rouché's theorem it is seen that $\beta(\eta)+\alpha\eta-1$ has only one zero with $\operatorname{Re}\eta>\varepsilon$; this zero is $\eta=0$. Hence ε is the zero with $\operatorname{Re}\eta<0$ nearest to the axis $\operatorname{Re}\eta=0$. Moreover ε is a simple zero since

$$\beta'(\varepsilon)+\alpha=\beta'(\varepsilon)+\frac{1-\beta(\varepsilon)}{\varepsilon}=-\sum_{n=1}^{\infty}\int_0^\infty (n-1)\,\frac{(-\varepsilon)^{n-1}}{n!}\,t^n\,\mathrm{d}B(t)<0,$$

the series being convergent. If $\beta(\eta)+\alpha\eta-1$ has a second zero ε_1 with $\operatorname{Re}\varepsilon_1<0$ and $\rho_0<\operatorname{Re}\varepsilon_1<\varepsilon$, let C_ξ be a line parallel to the imaginary axis with $\operatorname{Re} \varepsilon_1 < \operatorname{Re} \xi < \varepsilon$ if ε_1 exists, and $\rho_0 < \operatorname{Re} \xi < \varepsilon$ otherwise. The function $\beta(\eta) + \alpha\eta - 1$ is analytic for $\operatorname{Re} \eta > \operatorname{Re} \xi$ and has simple zeros at $\eta = \varepsilon$ and $\eta = 0$. From Cauchy's theorem it follows for $\operatorname{Re} \eta > 0 > \varepsilon > \operatorname{Re} \xi > \operatorname{Re} \varepsilon_1 > \rho_0$,

$$\frac{1}{2\pi i} \int_{C_\eta} e^{\eta v} \left\{ \frac{\alpha\eta}{\beta(\eta) + \alpha\eta - 1} - 1 \right\} d\eta$$

$$= \frac{\alpha\varepsilon}{\alpha + \beta'(\varepsilon)} e^{\varepsilon v} + \frac{1}{2\pi i} \int_{C_\xi} e^{\xi v} \left\{ \frac{\alpha\xi}{\beta(\xi) + \alpha\xi - 1} - 1 \right\} d\xi, \qquad v > 0,$$

$$\frac{1}{2\pi i} \int_{C_\eta} e^{\eta v} \frac{d\eta}{\beta(\eta) + \alpha\eta - 1}$$

$$= \frac{1}{\alpha - \beta} + \frac{1}{\alpha + \beta'(\varepsilon)} e^{\varepsilon v} + \frac{1}{2\pi i} \int_{C_\xi} e^{\xi v} \frac{d\xi}{\beta(\xi) + \alpha\xi - 1}, \qquad v > 0.$$

It is easily verified that

$$\lim_{v \to \infty} \frac{e^{-\varepsilon v}}{2\pi i} \int_{C_\xi} e^{\xi v} \left\{ \frac{\alpha\xi}{\beta(\xi) + \alpha\xi - 1} - 1 \right\} d\xi = 0,$$

$$\lim_{v \to \infty} \frac{1}{2\pi i} \int_{C_\xi} e^{\xi v} \frac{d\xi}{\beta(\xi) + \alpha\xi - 1} = 0.$$

Hence from (7.57) we obtain

$$\lim_{v \to \infty} e^{-\varepsilon v} \{1 - \Pr\{\boldsymbol{v}_{\max}^{(J)} < v\}\} = \frac{\alpha - \beta}{\alpha + \beta'(\varepsilon)} \alpha\varepsilon = b_1 > 0.$$

Hence

$$\Pr\left\{\frac{1}{\beta} \boldsymbol{V}_n < \frac{x + \log(nb_1)}{-\varepsilon\beta}\right\} = \left[\Pr\left\{\boldsymbol{v}_{\max}^{(J)} < \frac{x + \log(nb_1)}{-\varepsilon}\right\}\right]^n,$$

so that for $n \to \infty$,

$$\Pr\left\{\frac{1}{\beta} \boldsymbol{V}_n < \frac{x + \log(nb_1)}{-\varepsilon\beta}\right\} = \left[1 - b_1 \exp\{-x - \log(nb_1)\} + o\left(\frac{1}{n}\right)\right]^n,$$

i.e.

$$\lim_{n \to \infty} \Pr\left\{\frac{1}{\beta} \boldsymbol{V}_n < \frac{x + \log(nb_1)}{-\varepsilon\beta}\right\} = e^{-e^{-x}}, \qquad -\infty < x < \infty.$$

This proves the statement for $\boldsymbol{V}_n$; that for $\boldsymbol{W}_n$ is proved in the same way. To prove the statement for $\boldsymbol{X}_n$ start from (7.59) and deform the path of integration D_ω to a circle with radius $|\omega|>1$ and such that the first zero, other than 1, of $\beta\{(1-\omega)/\alpha\}-\omega$ is an interior point of this circle. It is easily found that for integer values of z,

$$\lim_{z\to\infty} (1-\alpha\varepsilon)^z \{1-\Pr\{\boldsymbol{x}_{\max}\leqq z\}\}=b_1.$$

For y a positive number and z an integer such that $y-1\leqq z<y$ it follows that with $0<\gamma<1$ and y sufficiently large

$$1-\frac{b_1}{(1-\alpha\varepsilon)^{y-1}}(1+O(\gamma^y))\leqq \Pr\{\boldsymbol{x}_{\max}<y\}<1-\frac{b_1}{(1-\alpha\varepsilon)^y}(1+O(\gamma^y)).$$

Taking

$$y=\frac{x+\log(nb_1)}{\log(1-\alpha\varepsilon)},$$

yields, for $n\to\infty$, the statement for $\boldsymbol{X}_n$.

COROLLARY TO THEOREM 7.2. *For $a<1$ the variables $\boldsymbol{V}_n/\beta\log n$, $\boldsymbol{W}_n/\beta\log n$ and $\boldsymbol{X}_n/\log n$ converge for $n\to\infty$ in probability to $-1/\varepsilon\beta$, $-1/\varepsilon\beta$ and $1/\log(1-\alpha\varepsilon)$, respectively.*

Proof. For every fixed $x>0$ it follows from the theorem above that for $n\to\infty$,

$$\Pr\left\{\left|\frac{1}{\beta}\frac{\boldsymbol{V}_n}{\log n}+\frac{1}{\varepsilon\beta}+\frac{\log b_1}{\varepsilon\beta\log n}\right|>\frac{x}{-\varepsilon\beta\log n}\right\}\to \mathrm{e}^{-\mathrm{e}^{x}}+1-\mathrm{e}^{-\mathrm{e}^{-x}}.$$

Hence with $x=z\log N$, $z>0$, we have for every fixed $N>1$,

$$\begin{aligned}\lim_{n\to\infty}\Pr&\left\{\left|\frac{1}{\beta}\frac{\boldsymbol{V}_n}{\log n}+\frac{1}{\varepsilon\beta}+\frac{\log b_1}{\varepsilon\beta\log n}\right|\geqq\frac{z}{-\varepsilon\beta}\right\}\\ &\leqq\lim_{n\to\infty}\Pr\left\{\left|\frac{1}{\beta}\frac{\boldsymbol{V}_n}{\log n}+\frac{1}{\varepsilon\beta}+\frac{\log b_1}{\varepsilon\beta\log n}\right|\geqq\frac{z\log N}{-\varepsilon\beta\log n}\right\}\\ &=\exp(-N^z)+1-\exp(-N^{-z}).\end{aligned}$$

Letting $N\to\infty$ we obtain for every $z>0$,

$$\lim_{n\to\infty}\Pr\left\{\left|\frac{1}{\beta}\frac{\boldsymbol{V}_n}{\log n}+\frac{1}{\varepsilon\beta}\right|\geqq\frac{z}{-\varepsilon\beta}\right\}=0,$$

from which the statement for $\boldsymbol{V}_n$ follows; the other ones are proved similarly.

Within a busy cycle a realisation of $\boldsymbol{v}_t$ may have a number of intersections with level K. There are no intersections at all if during the busy cycle the virtual delay time is always less than K. Denote by $\boldsymbol{\Pi}_K^{(j)}$ the number of intersections of level K from above with $\boldsymbol{v}_t$ in the jth busy cycle, $j=1, 2, \ldots$. The distribution of $\boldsymbol{\Pi}_K^{(j)}$ is given by (5.116).

Denote the state with K customers left behind in the system at a departure by E_K. Let $\boldsymbol{\Lambda}_K^{(j)}$ represent the number of times that state E_K occurs in the jth busy cycle. The distribution of $\boldsymbol{\Lambda}_K^{(j)}$ is given by (6.62).

Obviously, the variables $\boldsymbol{\Pi}_K^{(j)}, j=1, 2, \ldots$, are independent, and identically distributed. The same statement applies for $\boldsymbol{\Lambda}_K^{(j)}, j=1, 2, \ldots$.

Define

$$\boldsymbol{P}_{K,n} \stackrel{\text{def}}{=} \max_{1\leq j\leq n} \boldsymbol{\Pi}_K^{(j)}, \qquad \boldsymbol{L}_{K,n} \stackrel{\text{def}}{=} \max_{1\leq j\leq n} \boldsymbol{\Lambda}_K^{(j)},$$

with

$$f(0) \stackrel{\text{def}}{=} \Pr\{\boldsymbol{v}_{\max}<K\}, \qquad h(0) \stackrel{\text{def}}{=} \left\{\frac{1}{2\pi\mathrm{i}} \int_{C_\eta} \mathrm{e}^{\eta K} \frac{\alpha\, \mathrm{d}\eta}{\beta(\eta)+\alpha\eta-1}\right\}^{-1},$$

$$\operatorname{Re}\eta>0, \quad a\leqq 1,$$

and

$$f(1) \stackrel{\text{def}}{=} \Pr\{\boldsymbol{x}_{\max}\leqq K\}, \qquad h(1) \stackrel{\text{def}}{=} \left\{\frac{1}{2\pi\mathrm{i}} \int_{D_\omega} \frac{\mathrm{d}\omega}{\omega^{K+1}} \frac{\beta\{(1-\omega)/\alpha\}}{\beta\{(1-\omega)/\alpha\}-\omega}\right\}^{-1},$$

$$|\omega|<1, \quad a\leqq 1,$$

we then have:

THEOREM 7.3. *If $a\leqq 1$ then for $-\infty<x<\infty$,*

$$\exp\left\{-\frac{\mathrm{e}^{-x}}{1-h(0)}\right\} \leqq \liminf_{n\to\infty} \Pr\left\{\boldsymbol{P}_{K,n} < \frac{x+\log(n(1-f(0)))}{-\log(1-h(0))}\right\}$$

$$\leqq \limsup_{n\to\infty} \Pr\left\{\boldsymbol{P}_{K,n} < \frac{x+\log(n(1-f(0)))}{-\log(1-h(0))}\right\} \leqq \mathrm{e}^{-\mathrm{e}^{-x}},$$

$$\exp\left\{-\frac{\mathrm{e}^{-x}}{1-h(1)}\right\} \leqq \liminf_{n\to\infty} \Pr\left\{\boldsymbol{L}_{K,n} < \frac{x+\log(n(1-f(1)))}{-\log(1-h(1))}\right\}$$

$$\leqq \limsup_{n\to\infty} \Pr\left\{\boldsymbol{L}_{K,n} < \frac{x+\log(n(1-f(1)))}{-\log(1-h(1))}\right\} \leqq \mathrm{e}^{-\mathrm{e}^{-x}}.$$

Proof. It is easily verified that for $a\leqq 1$,

$$\Pr\{\boldsymbol{\Pi}_K^{(j)}\geqq m\} = \frac{1-f(0)}{1-h(0)} \exp\{m \log(1-h(0))\},$$

from which the statement of the theorem follows as in the preceding theorem. Similarly for $\boldsymbol{L}_{K,n}$. As before we obtain

COROLLARY TO THEOREM 7.3. *For $a \leqq 1$ the variables $\boldsymbol{P}_{K,n}/\log n$ and $\boldsymbol{L}_{K,n}/\log n$ converge for $n \to \infty$ in probability to*

$$(-\log\{1-h(0)\})^{-1} \quad \textit{and} \quad (-\log\{1-h(1)\})^{-1},$$

respectively.

Note that for

$$B(t) = \begin{cases} 1-\mathrm{e}^{-t/\beta}, & t>0, \\ 0, & t<0, \end{cases}$$

we have

$$\varepsilon\beta = -(1-a), \qquad b_1 = 1-a, \qquad b_2 = a(1-a), \qquad 1-\alpha\varepsilon = a^{-1};$$

for the values of $f(0)$ and $h(0)$ see end of section 5.8, while $f(1)$ and $h(1)$ are given at the end of section 6.5.

III.7.5. Limit theorems for the $G/M/1$ queueing system

For the system $G/M/1$ the variables $\boldsymbol{v}_{\max}$, $\boldsymbol{w}_{\max}$, $\boldsymbol{x}_{\max}$, $\boldsymbol{V}_n$, $\boldsymbol{W}_n$ and $\boldsymbol{X}_n$ will have the same meaning as those for the system $M/G/1$ (cf. (7.53) and (7.72)).

For $a \leqq 1$ we have from (5.106),

$$1 - \Pr\{\boldsymbol{v}_{\max} < v\} = \begin{cases} \left\{\dfrac{1}{2\pi\mathrm{i}} \displaystyle\int_{C_\xi} \mathrm{e}^{\xi v} \dfrac{\beta\, \mathrm{d}\xi}{\alpha(\xi)+\beta\xi-1}\right\}^{-1}, & \operatorname{Re}\xi > \Psi, \quad v>0, \\ 0, \quad v<0; & \end{cases} \tag{7.76}$$

from (4.85),

$$1 - \Pr\{\boldsymbol{w}_{\max} < w\}$$
$$= \begin{cases} \left\{\dfrac{1}{2\pi\mathrm{i}} \displaystyle\int_{C_\xi} \dfrac{\mathrm{e}^{\xi v}}{1-\beta\xi} \dfrac{\beta\, \mathrm{d}\xi}{\alpha(\xi)+\beta\xi-1}\right\}^{-1}, & \dfrac{1}{\beta} > \operatorname{Re}\xi > \Psi, \quad w>0, \\ 0, \quad w<0; & \end{cases} \tag{7.77}$$

and from (II.3.93),

$$1-\Pr\{\boldsymbol{x}_{\max}\leqq x\}$$
$$=\left\{\frac{1}{2\pi i}\int_{D_\omega}\frac{d\omega}{\omega^x}\,\frac{1}{\alpha\{(1-\omega)/\beta\}-\omega}\right\}^{-1},\qquad |\omega|<\varphi,\quad x=1,2,\ldots;\tag{7.78}$$

here Ψ represents the zero of $\alpha(\xi)+\beta\xi-1$ with Re $\xi>0$, and φ the zero of $\alpha\{(1-\omega)/\beta\}-\omega$ with $|\omega|<1$ if $a<1$, while $\Psi=0$, $\varphi=1$ if $a=1$.

Write

$$N(t)\overset{\text{def}}{=}\begin{cases}0, & t<0,\\ \dfrac{1}{\alpha}\displaystyle\int_0^t\{1-A(u)\}\,du, & t>0,\end{cases}$$

and

$$M(t)\overset{\text{def}}{=}\sum_{n=0}^{\infty}\{N(t)\}^{n*},\qquad t>0,$$

so that $M(t)$ is the renewal function of the renewal process with $N(t)$ as the renewal distribution. As in the preceding section (cf. the derivation of (7.67), ..., (7.69)) we obtain from (7.76), ..., (7.78), for $a=1$,

$$1-\Pr\{\boldsymbol{v}_{\max}<v\}=\{M(v)\}^{-1},\qquad v>0,\tag{7.79}$$

$$1-\Pr\{\boldsymbol{w}_{\max}<w\}=\left\{\int_0^\infty M(w+t)\,e^{-t/\beta}\,\frac{dt}{\beta}\right\}^{-1},\qquad w>0,$$

$$1-\Pr\{\boldsymbol{x}_{\max}\leqq x\}=\left\{\int_0^\infty\frac{(t/\beta)^{x-1}}{(x-1)!}\,e^{-t/\beta}\,M(t)\,\frac{dt}{\beta}\right\}^{-1},\qquad x=1,2,\ldots.$$

If the second moment α_2 of $A(t)$ is finite then from renewal theory (cf. section I.6.2),

$$\lim_{t\to\infty}\frac{M(t)}{t}=\frac{2\alpha}{\alpha_2}.$$

The same argument as used in theorem 7.1 of the preceding section leads immediately to

THEOREM 7.4. *If $a=1$ and $\alpha_2<\infty$ then the distribution functions of $2\alpha V_n/n\alpha_2$, of $2\alpha W_n/n\alpha_2$ and of $2\alpha^2X_n/n\alpha_2$ all converge to $G(x)$ for $n\to\infty$.*

Further we have

THEOREM 7.5. *If $a<1$ then for $-\infty<x<\infty$,*

$$\lim_{n\to\infty} \Pr\left\{\frac{1}{\beta} V_n < \frac{x+\log(nc_1)}{\Psi\beta}\right\} = e^{-e^{-x}},$$

$$\lim_{n\to\infty} \Pr\left\{\frac{1}{\beta} W_n < \frac{x+\log(nc_2)}{\Psi\beta}\right\} = e^{-e^{-x}},$$

$$\exp\left(-\frac{e^{-x}}{1-\beta\Psi}\right) \leqq \liminf_{n\to\infty} \Pr\left\{X_n < \frac{x+\log(nc_1)}{-\log(1-\beta\Psi)}\right\}$$

$$\leqq \limsup_{n\to\infty} \Pr\left\{X_n < \frac{x+\log(nc_1)}{-\log(1-\beta\Psi)}\right\} \leqq e^{-e^{-x}},$$

with

$$c_1 = \frac{\alpha'(\Psi)+\beta}{\beta}, \qquad c_2 = \frac{\alpha'(\Psi)+\beta}{\beta}(1-\beta\Psi),$$

$$\alpha'(\rho) = -\int_0^\infty t\, e^{-\rho t}\, dA(t), \qquad \operatorname{Re}\rho \geqq 0.$$

Proof. If $a<1$ then $\alpha(\xi)+\beta\xi-1$, $\operatorname{Re}\xi>0$ has exactly one zero Ψ, which is real and simple; hence for $\operatorname{Re}\xi>\Psi>\operatorname{Re}\eta>0$, $v>0$,

$$\frac{1}{2\pi i}\int_{C_\xi} e^{\xi v} \frac{\beta\, d\xi}{\alpha(\xi)+\beta\xi-1} = \frac{\beta\, e^{v\Psi}}{\alpha'(\Psi)+\beta} + \frac{1}{2\pi i}\int_{C_\eta} e^{\eta v} \frac{\beta\, d\eta}{\alpha(\eta)+\beta\eta-1}.$$

From this relation and from (7.76) it follows immediately that

$$\lim_{v\to\infty} e^{v\Psi}\{1-\Pr\{\boldsymbol{v}_{\max}^{(J)}<v\}\} = \frac{\alpha'(\Psi)+\beta}{\beta} = \frac{1}{\beta}\left\{\alpha'(\Psi)+\frac{1-\alpha(\Psi)}{\Psi}\right\} > 0,$$

and the proof of the statement for V_n follows as in the proof of theorem 7.2. The proof for the statement of W_n is similar. To prove the statement for X_n move the path of integration D_ω to a circle with radius $|\zeta|$ and so that $\varphi<|\zeta|<1$. The statement for X_n is now easily derived.

COROLLARY TO THEOREM 7.5. *For $a<1$ the variables $V_n/\beta\log n$, $W_n/\beta\log n$ and $X_n/\log n$ converge for $n\to\infty$ in probability to $1/\beta\Psi$, $1/\beta\Psi$ and $\{-\log(1-\beta\Psi)\}^{-1}$, respectively.*

The proof is analogous to that of the corollary in the preceding section.

In theorem 7.4 we needed the condition $\alpha_2<\infty$ as well as the condition $a=1$. Although for practical applications the former condition is hardly restrictive, it is of some interest to have information about the behaviour of V_n if $A(t)$ does not have a finite second moment. We shall, therefore, study $\boldsymbol{v}_{\max}$ and V_n for the case that $A(t)$ has the property

$$1 - A(t) \approx \frac{k}{\Gamma(1-k)} \frac{\alpha^{k+1}}{t^{k+1}} L(t), \qquad t \to \infty, \quad 0<k<1, \tag{7.80}$$

where $L(t)$ is a slowly decreasing function i.e. for every $x>0$,

$$\frac{L(tx)}{L(t)} \to 1 \qquad \text{for} \quad t \to \infty.$$

Obviously, $A(t)$ has a finite first moment. It follows (cf. FELLER [1966] p. 273) that for $t \to \infty$,

$$\frac{t\{1-A(t)\}}{\int_t^\infty \{1-A(\tau)\}\,\mathrm{d}\tau} \to |-(k+1)+1| = k,$$

so that since

$$N(t) = \frac{1}{\alpha} \int_0^t \{1-A(\tau)\}\,\mathrm{d}\tau = 1 - \frac{1}{\alpha} \int_t^\infty \{1-A(\tau)\}\,\mathrm{d}\tau,$$

we have

$$1 - N(t) \approx \frac{1}{\Gamma(1-k)} \frac{\alpha^k}{t^k} L(t) \qquad \text{for} \quad t \to \infty.$$

Consequently (cf. FELLER [1966] p. 423),

$$1 - \frac{1-\alpha(\rho)}{\alpha\rho} \approx \alpha^k \rho^k L(1/\rho), \qquad \rho \to 0 \quad (\rho \text{ real}), \tag{7.81}$$

since

$$\int_0^\infty \mathrm{e}^{-\rho t}\,\mathrm{d}N(t) = \frac{1-\alpha(\rho)}{\alpha\rho}, \qquad \operatorname{Re}\rho>0.$$

Because for $a=1$, i.e. $\alpha=\beta$,

$$\int_0^\infty \mathrm{e}^{-\rho t}\,\mathrm{d}M(t) = \frac{\alpha\rho}{\alpha(\rho)+\alpha\rho-1}, \qquad \operatorname{Re}\rho>0,$$

we obtain from (7.81), with $L^* = L^{-1}$, a slowly decreasing function,

$$\int_0^\infty e^{-\rho t}\, dM(t) \approx \alpha^{-k}\rho^{-k}L^*(1/\rho) \qquad \text{for} \quad \rho \to 0.$$

Since $M(t)$ is non-decreasing it follows from the last relation and a well-known Tauberian theorem (cf. app. 5) that

$$M(t) \approx \frac{\alpha^{-k}t^k}{\Gamma(1+k)}\, L^*(t) \qquad \text{for} \quad t \to \infty. \tag{7.82}$$

Hence, from (7.79) we obtain the existence of the limit

$$d \overset{\text{def}}{=} \lim_{v \to \infty} \left(\frac{v}{\alpha}\right)^k \{1 - \Pr\{v_{\max} < v\}\} > 0. \tag{7.83}$$

As before it follows from (7.83) that

$$\lim_{n \to \infty} \Pr\left\{\frac{V_n}{\alpha n^{1/k}} < x\right\} = \begin{cases} \exp(-dx^{-k}) & \text{for} \quad x > 0. \\ 0, & \qquad\quad x < 0. \end{cases} \tag{7.84}$$

III.7.6. Behaviour of v_t for $t \to \infty$ if $a > 1$

In this section we shall consider for the $G/G/1$ system the behaviour of the virtual waiting time v_t for $t \to \infty$. It is assumed here that α and β are both finite. In fig. 8 a realisation of v_t starting with $w_1 = 0$ is shown.

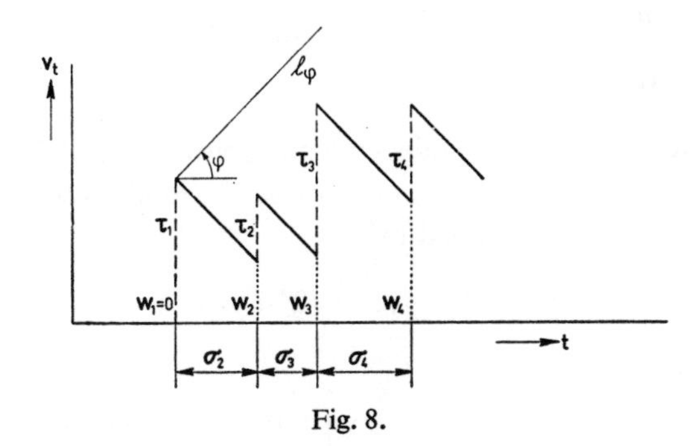

Fig. 8.

Consider fig. 9. Here we constructed the broken line A11′22′33′... . The projection of A1 on the horizontal axis is equal to σ_2, that of 1′2 is σ_3, that of 2′3 is σ_4, and so on; further $11' = \tau_2$, $22' = \tau_3$, $33' = \tau_4$, and so on.

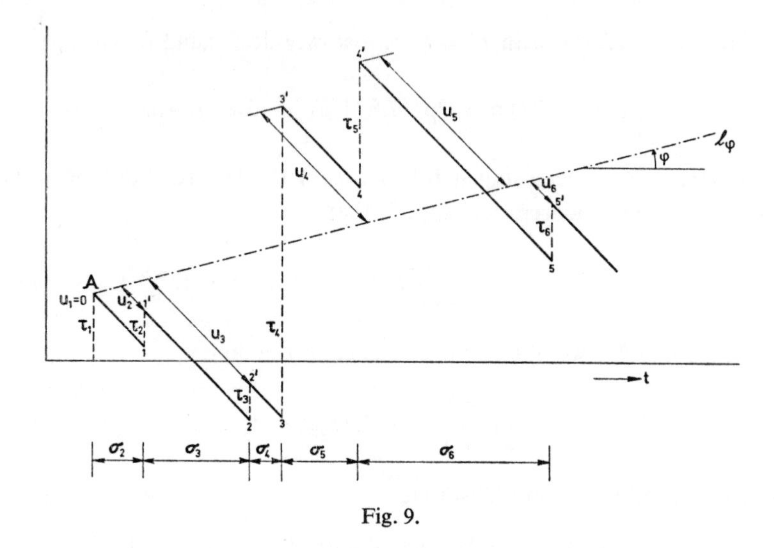

Fig. 9.

This broken line is in general not a realisation of the $\boldsymbol{v}_t$-process for the $G/G/1$ system. Only if this broken line A11'22'... never intersects the horizontal axis it may represent such a realisation.

We shall denote the distance from point n' of the broken line to the line l_φ by $\boldsymbol{u}_{n+1}$, this distance being measured along a line of slope 135° with the horizontal axis, and $\boldsymbol{u}_{n+1}$ is positive if point n' lies below the line l_φ. Clearly we have for the stochastic process $\{\boldsymbol{u}_n,\ n=1, 2, \ldots\}$,

$$\boldsymbol{u}_{n+1} = \boldsymbol{u}_n + \boldsymbol{\sigma}_{n+1}\sqrt{2} - \boldsymbol{\tau}_{n+1}\frac{\sqrt{2}}{1+\tan\varphi}, \qquad n=1, 2, \ldots, \tag{7.85}$$
$$\boldsymbol{u}_1 = 0.$$

Since $\boldsymbol{u}_n$ is the sum of $n-1$ independent, identically distributed variables this process $\{\boldsymbol{u}_n,\ n=1, 2, \ldots\}$ is an example of a random walk.

For $a>1$ let φ_0 be determined by

$$a-1=\tan\varphi_0, \qquad 0°<\varphi_0<90°. \tag{7.86}$$

It follows that

$$\mathrm{E}\{\boldsymbol{u}_{n+1}-\boldsymbol{u}_n\} = \alpha\sqrt{2}\left(1-\frac{a}{1+\tan\varphi}\right)\begin{cases} >0 & \text{for } 90°>\varphi>\varphi_0, \\ <0 & \text{for } 0°<\varphi<\varphi_0. \end{cases} \tag{7.87}$$

Suppose $a>1$ and take in (7.85) $\varphi=\varphi_0$. It then follows from fluctuation theory (cf. theorem I.6.8 and (I.6.100)) that the process $\{\boldsymbol{u}_n,\ n=1, 2, \ldots\}$

oscillates between $-\infty$ and $+\infty$ with probability one, and consequently the line l_{φ_0} has, with probability one an infinite number of intersections with the broken line A11′22′33′... . For $0° < \varphi < 90°$ but $\varphi \neq \varphi_0$ it is seen from (7.85), (7.87) and (I.6.100) that, with probability one, the broken line A11′22′33′... has only a finite number of intersections with l_φ.

Since $a > 1$ the probability that a busy cycle is infinite is positive, and hence with probability one a realisation of the $\boldsymbol{v}_t$-process for the $G/G/1$ queueing system will only contain a finite number of finite busy cycles. From the above results for the process $\{\boldsymbol{u}_n, n=1, 2, \ldots\}$ it is seen that any line l_φ which intersects a realisation of the $\boldsymbol{v}_t$-process has with probability one an infinite number of intersections with this realisation if $\varphi = \varphi_0$ and a finite number of intersections if $\varphi \neq \varphi_0$. Consequently, from (7.86) with probability one for every $\varepsilon > 0$,

$$-\varepsilon < \liminf_{t\to\infty} \frac{\boldsymbol{v}_t}{t} - (a-1) \leqq \limsup_{t\to\infty} \frac{\boldsymbol{v}_t}{t} - (a-1) < \varepsilon,$$

i.e. with probability one

$$\lim_{t\to\infty} \frac{\boldsymbol{v}_t}{t} = a - 1.$$

APPENDIX

1. Abelian theorems for power series (cf. Titchmarsh [1952])

If all a_n are real and $\sum_{n=0}^{\infty} a_n$ is convergent then $\sum_{n=0}^{\infty} a_n s^n$ is uniformly convergent for $0 \leqq s \leqq 1$ and

$$\lim_{s \uparrow 1} \sum_{n=0}^{\infty} a_n s^n = \sum_{n=0}^{\infty} a_n.$$

If $a_n \geqq 0$, $n=0, 1, \ldots$, and $\sum_{n=0}^{\infty} a_n s^n$ converges for $|s|<1$ then

$$\lim_{s \uparrow 1} \sum_{n=0}^{\infty} a_n s^n = \sum_{n=0}^{\infty} a_n,$$

in the sense that both members are infinite or both are finite and equal.
If $\sum_{n=0}^{\infty} a_n s^n$ converges for $|s|<1$ and $\lim_{n\to\infty} a_n = a$ then

$$\lim_{s \uparrow 1} (1-s) \sum_{n=0}^{\infty} a_n s^n = a.$$

2. Abelian theorems for Laplace transforms (cf. Doetsch [1943])

If $\int_0^\infty a(t)\,\mathrm{d}t$ is convergent then $\int_0^\infty \mathrm{e}^{-st}\, a(t)\,\mathrm{d}t$ is uniformly convergent for $\mathrm{Re}\, s \geqq 0$ *and*

$$\lim_{s \to 0} \int_0^\infty \mathrm{e}^{-st}\, a(t)\,\mathrm{d}t = \int_0^\infty a(t)\,\mathrm{d}t, \qquad |\arg s| \leqq \theta < \tfrac{1}{2}\pi.$$

If $a(t) \geqq 0$ for $t \geqq 0$ and $\int_0^\infty \mathrm{e}^{-st}\, a(t)\,\mathrm{d}t$ converges for $\mathrm{Re}\, s > 0$ *then*

$$\lim_{s\to 0}\int_0^\infty e^{-st}\, a(t)\, dt = \int_0^\infty a(t)\, dt, \qquad |\arg s| \leqq \theta < \tfrac{1}{2}\pi,$$

in the sense that both members are infinite or both are finite and equal.
If $\int_0^\infty e^{-st}\, a(t)\, dt$ *is convergent for* Re $s>0$ *and* $a(t)$ *has a limit for* $t\to\infty$ *then*

$$\lim_{t\to\infty} a(t) = \lim_{s\to 0} s\int_0^\infty e^{-st}\, a(t)\, dt, \qquad |\arg s| \leqq \theta < \tfrac{1}{2}\pi.$$

3. Abelian theorem for Laplace-Stieltjes transforms (cf. Widder [1946])

If $a(t)$, $t \geqq 0$ *is of bounded variation in every finite interval, if* $\int_0^\infty e^{-st}\, da(t)$ *is convergent for* Re $s>0$ *and if* $a(t)$ *has a limit for* $t\to\infty$ *then*

$$\lim_{t\to\infty} a(t) = \lim_{s\to 0}\int_0^\infty e^{-st}\, da(t), \qquad |\arg s| \leqq \theta < \tfrac{1}{2}\pi.$$

4. Tauberian theorem for Laplace-Stieltjes transforms (cf. Widder [1946])

If $a(t)$, $t \geqq 0$, *is non-decreasing and* $\alpha(s) = \int_0^\infty e^{-st}\, da(t)$ *exists for* $s>0$, *and if for* $\gamma \geqq 0$,

$$\alpha(s) \approx \frac{A}{s^\gamma} \qquad \text{for} \quad s\downarrow 0,$$

then

$$a(t) \approx \frac{A t^\gamma}{\Gamma(\gamma+1)} \qquad \text{for} \quad t\to\infty.$$

5. Convergence theorems

Helly-Bray theorem for probability distributions (Loève [1960])

If the sequence of probability distributions $F_n(x)$, $n=1, 2, \ldots$, *converges for* $n\to\infty$ *to a (proper) distribution* $F(x)$ *for every finite* x *which is a continuity point of* $F(x)$ *and if* $g(x)$, $-\infty<x<\infty$ *is a bounded and continuous function then*

$$\lim_{n\to\infty}\int_{-\infty}^{\infty} g(x)\, dF_n(x) = \int_{-\infty}^{\infty} g(x)\, dF(x).$$

If $g(x)\to 0$ for $x\to\infty$ and for $x\to-\infty$ then the theorem is still true if $F(x)$ is an improper probability distribution, i.e. if $\lim_{x\to\infty} F(x)<1$.

CONVERGENCE THEOREM OF FELLER FOR LAPLACE-STIELTJES TRANSFORMS (cf. FELLER [1966])

Let $F_n(x)$, $n=1,2,\ldots$, be a sequence of distributions with $F_n(x)=0$ for $x<0$. If for every $\rho>0$,

$$\varphi_n(\rho) \stackrel{\text{def}}{=} \int_{0-}^{\infty} \mathrm{e}^{-\rho t}\,\mathrm{d}F_n(t) \to \varphi(\rho) \qquad \textit{for} \quad n\to\infty,$$

then $\varphi(\rho)$ is the Laplace-Stieltjes transform of a possibly improper distribution $F(x)$ and $F_n(x)\to F(x)$ at every continuity point of $F(x)$. $F(x)$ is a proper probability distribution if and only if $\varphi(\rho)\to 1$ for $\rho\downarrow 0$.

CONVERGENCE THEOREM OF FELLER FOR GENERATING FUNCTIONS (cf. FELLER [1957])

Let $\{a_k^{(n)},\ k=0,1,\ldots\}$, $n=1,2,\ldots$, be a sequence of discrete probability distributions with $\sum_{k=0}^{\infty} a_k^{(n)}=1$. If for every r with $0\leqq r<1$,

$$A_n(r) \stackrel{\text{def}}{=} \sum_{k=0}^{\infty} a_k^{(n)} r^k \to A(r) \qquad \textit{for} \quad n\to\infty,$$

then $A(r)$ is the generating function of a possibly improper discrete probability distribution $\{a_k,\ k=0,1,\ldots\}$. The converse statement is also true. $\{a_k,\ k=0,1,\ldots\}$ is a proper discrete distribution if and only if $A(r)\to 1$ for $r\uparrow 1$.

6. Zero's of analytic functions

ROUCHÉ'S THEOREM (cf. TITCHMARSH [1952])

If $f(z)$ and $g(z)$ are analytic inside and on a closed contour D *and $|g(z)|<|f(z)|$ on* D *then $f(z)$ and $f(z)+g(z)$ have the same number of zeros inside* D.

LAGRANGE'S EXPANSION THEOREM (cf. WHITTAKER and WATSON [1946])

For $f(z)$ and $g(z)$ functions of z analytic on and inside a contour D *surrounding a point a, and for r satisfying the inequality*

$$|rg(z)| < |z-a|,$$

for every z on D, *the function* $z-a-rg(z)$ *of z has exactly one zero* η *inside* D, *and*

$$f(\eta) = f(a) + \sum_{n=1}^{\infty} \frac{r^n}{n!} \frac{\mathrm{d}^{n-1}}{\mathrm{d}x^{n-1}} [f'(x)\{g(x)\}^n]\Big|_{x=a}.$$

Let $Z(t)$ be a probability distribution of a non-negative stochastic variable with $Z(0+)=0$ and with finite first moment ζ. Denote the n-fold convolution of $Z(t)$ with itself by $Z^{n*}(t)$,

$$Z^{0*}(t) = \begin{cases} 0 & \text{for } t \leqq 0, \\ 1 & \text{for } t > 0, \end{cases}$$

$$Z^{n*}(t) = \int_0^t Z^{(n-1)*}(t-u)\,\mathrm{d}Z(u), \qquad t>0, \quad n=1, 2, \dots .$$

Define

$$\zeta(\rho) \stackrel{\text{def}}{=} \int_0^\infty \mathrm{e}^{-\rho t}\,\mathrm{d}Z(t), \qquad \operatorname{Re}\rho \geqq 0,$$

$$f(z) \stackrel{\text{def}}{=} z - r\zeta\{\rho+(1-z)\,\delta\}, \qquad |r| \leqq 1, \quad \operatorname{Re}\rho \geqq 0, \quad \delta>0, \quad \operatorname{Re} z \leqq 1.$$

For $f(z)$ we shall now prove a lemma slightly different from the original formulation given by Takács [1962]. Suppose $Z(t)$ is not a lattice distribution.

Lemma of Takács

(*i*) *The zero* $z(\rho, r)$ *of* $f(z)$ *which has the smallest absolute value is unique, it is a continuous function of r and* ρ, *for* $|r| \leqq 1$, $\operatorname{Re}\rho \geqq 0$, *and*

$$z(\rho, r) = r \sum_{n=1}^{\infty} \frac{(-\delta r)^{n-1}}{n!} \frac{\mathrm{d}^{n-1}}{\mathrm{d}\rho^{n-1}} \{\zeta(\rho+\delta)\}^n$$

$$= \sum_{n=1}^{\infty} \frac{r^n \delta^{n-1}}{n!} \int_0^\infty \mathrm{e}^{-(\rho+\delta)t}\, t^{n-1}\,\mathrm{d}Z^{n*}(t),$$

$$z^k(\rho, r) = k \sum_{n=k}^{\infty} \frac{r^n \delta^{n-k}}{n(n-k)!} \int_0^\infty \mathrm{e}^{-(\rho+\delta)t}\, t^{n-k}\,\mathrm{d}Z^{n*}(t), \qquad k=1, 2, \dots;$$

(*ii*) $z(\rho, r)$ *is the generating function and Laplace-Stieltjes transform of a possibly improper joint distribution of two positive stochastic variables, of which one is discrete, the other continuous;*

(*iii*) $|z(\rho, r)| < 1$ *for* $|r| \leqq 1$, $\mathrm{Re}\,\rho \geqq 0$, *except for* $r=1$, $\rho=0$, $\delta\zeta \leqq 1$, *in the latter case* $z(\rho, r)=1$;

(*iv*) *if* $f(z)$ *has a zero with* $|z|=1$ *then* $r=1$, $\rho=0$ *and this zero is* $z=1$, *it is a simple zero if* $\delta\zeta \neq 1$, *and a double zero if* $\delta\zeta=1$;

(*v*) *if* $r=1$, $\rho=0$, $\delta\zeta>1$ *then* $0<z(0, 1)<1$.

Remark. *If* $Z(t)$ *is a lattice distribution with period* τ *then the statements above remain true but in (iii), (iv) and (v),* $\rho=0$ *should be replaced by* $\rho=2\pi\mathrm{i}n/\tau$, $n=\ldots, -1, 0, 1, \ldots$.

Proof. Denote by $\Delta \arg f(z)$ the increase of $\arg f(z)$ if z describes the circle $|z|=1$ once. For $\rho=0$, $|r|=1$, $r\neq 1$, we have

$$\Delta \arg f(z) = \Delta \arg z + \Delta \arg\left\{1 - \frac{r}{z}\,\zeta\{\delta(1-z)\}\right\}.$$

For $|z|=1$, $z\neq 1$ the point

$$q = 1 - \frac{r}{z}\,\zeta\{\delta(1-z)\},$$

is always an interior point of the circle with radius 1 and centre at $q=1$ in the q-plane, while for $z=1$ the point $q=1-r\neq 0$ is on the boundary of this circle. Hence $\Delta \arg q=0$, i.e. $\Delta \arg f(z)=2\pi$ if z describes $|z|=1$. Since $f(z)$ is analytic for $\mathrm{Re}\, z\leqq 1$, it follows from a well-known theorem on zeros of analytic functions (cf. TITCHMARSH [1962]) that $f(z)$ has exactly one zero inside $|z|=1$ if $\rho=0$, $|r|=1$, $r\neq 1$. For $r=1$, $\rho=0$, $\delta\zeta<1$ and $|z|<1$ we have

$$\left|\frac{\mathrm{d}}{\mathrm{d}z}\,\zeta\{\delta(1-z)\}\right| < 1,$$

hence

$$|\zeta\{\delta(1-z)\}-1| < |z-1| \qquad \text{for} \quad |z|<1,$$

so that $f(z)$ has no zero inside $|z|=1$, but $z=1$ is a zero. If $|r|=1$ and z_0 with $|z_0|=1$ is a zero of $f(z)$ then $1=|z_0|=|r|\,|\zeta\{\rho+\delta(1-z_0)\}|$ so that $\delta(1-z_0)+\rho=0$, since $Z(t)$ is not a lattice distribution. Hence $\rho=0$, $z_0=1$, $r=1$, since $\mathrm{Re}\,\rho\geqq 0$, $|z_0|=1$ and $|r|=1$. If $z=1$ is a zero then

$$\frac{\mathrm{d}}{\mathrm{d}z}\,\zeta\{\delta(1-z)\} \neq 1 \qquad \text{for} \quad z=1, \quad \delta\zeta\neq 1,$$

and

$$\frac{d}{dz}\zeta\{\delta(1-z)\} = 1, \qquad \frac{d^2}{dz^2}\zeta\{\delta(1-z)\} \neq 0 \qquad \text{for} \quad z=1, \quad \delta\zeta=1,$$

so that $z=1$ is a simple zero if $\delta\zeta\neq 1$, and a double zero if $\delta\zeta=1$. For $\varepsilon>0$, but sufficiently small we have with $|z|=1-\varepsilon$ for $|r|<1$, $\operatorname{Re}\rho\geqq 0$, or $|r|\leqq 1$, $\operatorname{Re}\rho>0$,

$$|r\zeta\{\rho+(1-z)\,\delta\}|<1-\varepsilon=|z|;$$

and for $r\leqq 1$, $\operatorname{Re}\rho\geqq 0$, $\delta\zeta>1$,

$$|r\zeta\{\rho+\delta(1-z)\}|\leqq\zeta\{\delta\operatorname{Re}(1-z)\}\leqq\zeta(\varepsilon\delta)=1-\varepsilon\delta\zeta+o(\varepsilon)<1-\varepsilon=|z|.$$

For $|r|=1$, $\operatorname{Re}\rho=0$, $\rho\neq 0$ and $|z|=1$ we have

$$|r\zeta\{\rho+\delta(1-z)\}|<1=|z|,$$

since $Z(t)$ is not a lattice distribution. From Rouché's theorem and the inequalities above it follows that $f(z)$ has exactly one zero inside $|z|=1$ for those conditions for which the inequalities hold. Since $\zeta\{\delta(1-z)\}$ is an increasing function of z it is readily seen that $0<z(0, 1)<1$ if $\delta\zeta>1$. The statements (*iii*), ..., (*v*) have now been proved, and it follows that $z(\rho, r)$ is unique. From Lagrange's expansion theorem and the inequalities above the series representations of $z(\rho, r)$ in (*i*) follow for those conditions for which the inequalities hold, i.e. for $|r|\leqq 1$, $\operatorname{Re}\rho\geqq 0$, except $r=1$, $\rho=0$, $\delta\zeta\leqq 1$ and $|r|=1$, $r\neq 1$, $\rho=0$. To prove the series representation of $z(\rho, r)$ for the latter cases it is noted that for the $M/G/1$ queueing system with service time distribution $Z(t)$ the function $\mathrm{E}\{r^{\boldsymbol{n}}\exp(-\rho\boldsymbol{p})\}$, $|r|\leqq 1$, $\operatorname{Re}\rho\geqq 0$, with $\boldsymbol{p}$ the duration of the busy cycle and $\boldsymbol{n}$ the number of customers served during $\boldsymbol{p}$ is a zero of $f(z)$ (cf. section II.4.4). Hence $z(\rho, r)=\mathrm{E}\{r^{\boldsymbol{n}}\exp(-\rho\boldsymbol{p})\}$ for all those r and ρ for which $f(z)$ has exactly one zero with $|z|\leqq 1$, i.e. for all $|r|\leqq 1$, $\operatorname{Re}\rho\geqq 0$, except $r=1$, $\rho=0$, $\delta\zeta>1$. Since $\mathrm{E}\{r^{\boldsymbol{n}}\exp(-\rho\boldsymbol{p})\}$ is continuous for $|r|\leqq 1$, $\operatorname{Re}\rho\geqq 0$, and since the series representation of $z(\rho, r)$ holds for $r=1$, $\rho=0$, $\delta\zeta>1$, it follows that $z(\rho, r)$ is continuous for $|r|\leqq 1$, $\operatorname{Re}\rho\geqq 0$, and that the series representation of (*i*) is valid for $|r|\leqq 1$, $\operatorname{Re}\rho\geqq 0$, the proof is complete.

If $Z(t)$ is a lattice distribution with period τ then by noting that $\zeta(\rho)=0$ implies $\rho=2\pi i n/\tau$, $n=\ldots, -1, 0, 1, \ldots$, and conversely, the lemma is proved as above.

Note: since for $\operatorname{Re}\rho>0$, $|r|<1$,

$$z(\rho, 1)=\zeta\{\rho+(1-z(\rho, 1))\,\delta\}, \qquad z(0, r)=r\zeta\{(1-z(0, r))\,\delta\},$$

we have

$$\frac{\mathrm{d}z(\rho,1)}{\mathrm{d}\rho} = \zeta'\{\rho+(1-z(\rho,1))\,\delta\}\left\{1-\delta\frac{\mathrm{d}z(\rho,1)}{\mathrm{d}\rho}\right\}, \qquad \mathrm{Re}\,\rho>0,$$

$$\frac{\mathrm{d}z(0,r)}{\mathrm{d}r} = \frac{z(0,r)}{r} - r\delta\zeta'\{(1-z(0,r))\,\delta\}\frac{\mathrm{d}z(0,r)}{\mathrm{d}r}, \qquad |r|<1.$$

It follows if $\delta\zeta \leqq 1$ that

$$\left\{\lim_{\rho\downarrow 0}\frac{\mathrm{d}z(\rho,1)}{\mathrm{d}\rho}\right\}^{-1} = -\frac{1-\delta\zeta}{\zeta}, \qquad \left\{\lim_{r\uparrow 1}\frac{\mathrm{d}z(0,r)}{\mathrm{d}r}\right\}^{-1} = 1-\delta\zeta.$$

7. Wald's theorem (cf. FELLER [1966])

Let $\boldsymbol{x}_n$, $n=1, 2, \ldots$, be a sequence of independent, identically distributed variables with finite mean $x=\mathrm{E}\{\boldsymbol{x}_n\}$, and $\boldsymbol{\nu}$ a variable with state space the set of positive integers and with finite mean. If the event $\{\boldsymbol{\nu}=n\}$ is independent of the variables $\boldsymbol{x}_{n+1}, \boldsymbol{x}_{n+2}, \ldots$, for every $n=1, 2, \ldots$, then

$$\mathrm{E}\{\boldsymbol{x}_1+\ldots+\boldsymbol{x}_{\boldsymbol{\nu}}\}=x\,\mathrm{E}\{\boldsymbol{\nu}\}.$$

NOTES ON LITERATURE

Chapter I.2

The treatment of discrete time parameter Markov chains with denumerable state space and stationary transition probabilities given in chapter I.2 is mainly along the lines as given by FELLER [1957]. A very profound study of these Markov chains may be found in CHUNG [1960]. The interested reader is also referred to KEMENY and SNELL [1960], PARZEN [1962], FRÉCHET [1952], DOOB [1953], BLANC-LAPIERRE and FORTET [1953], KEMENY, SNELL and KNAPP [1966].

Chapter I.3

For a very fundamental treatment of the theory of continuous time parameter Markov chains the reader is again referred to CHUNG [1960]. DOOB [1953] discusses only chains with a finite state space. BLANC-LAPIERRE and FORTET [1953] treat the general case. The discussion given in the present chapter is far too restricted for a general theory. For practical purposes, however, it is effective. See in this respect also the discussion in FELLER [1957].

Chapter I.4

For a fundamental paper concerning the solutions of the backward and forward differential equations of birth and death processes see LEDERMANN and REUTER [1953] and also KEMPERMAN [1961[a]]. We have already mentioned the studies of KARLIN and MCGREGOR [1957, 1959], and the book of BAILEY [1964], which gives an excellent treatment of stochastic processes from an applied point of view. We should also mention here the book of HARRIS [1963] and that of SPITZER [1964]. An excellent chapter on the Poisson process is available in PARZEN [1962], see also BLANC-LAPIERRE and FORTET [1953]. The birth and death process of section I.4.4 has been treated by various authors of whom we mention here HEATHCOTE and MOYAL [1959] and BAILEY [1957]; for further references on this subject see chapter II.2.

Chapter I.6

In chapter I.6 we have discussed mainly those parts of renewal theory which are important for queueing theory. An excellent review paper on renewal theory has been written by SMITH [1958]; here practically all references on renewal theory up to 1958 are given. This paper which is fundamental also discusses processes closely related to renewal theory. A very interesting booklet on renewal theory has been written by COX [1962]. Here the reader will find much detailed information about renewal theory and related processes, and also applications to reliability theory. A very readable chapter on renewal theory can be found in the book by ARROW, KARLIN and SCARF [1958]. For more recent developments and extensions of renewal theory the reader is referred to FELLER [1966] and SMITH [1960]. We mention here only the recent work of FELLER and OREY [1961] concerning a generalisation of Blackwell's theorem for extended renewal processes, i.e. renewal processes for which the renewal time may be negative with positive probability. The chapter on regenerative processes is new, but partly related to semi-Markov processes and Markov-renewal processes as studied by PYKE [1961[a], 1961[b]] and to the study of SMITH [1955]. The regenerative processes studied here are of a somewhat more general type than those of Smith.

Fluctuation theory has received a new impetus by the studies of SPARRE ANDERSEN [1953] and SPITZER [1956, 1957, 1960]. The latter author has recognized the relation between fluctuation theory and queueing theory. Many studies on this topic have been published, we refer the reader to KEMPERMAN [1961[b]], FELLER [1966], SPITZER [1964], PRABHU [1965[a], 1965[b]] and KINGMAN [1966]. The latter paper contains an extensive list of references.

Chapter II.2

The first investigators of the $M/M/1$ queueing system seem to be Erlang (cf. BROCKMEYER, HALSTRØM and JENSEN [1948]) and MOLINA [1927]. The explicit time-dependent solution appears in literature around 1954 (cf. CLARKE [1953], BAILEY [1954[a] and 1957], LEDERMANN and REUTER [1954], CONOLLY [1958], CHAMPERNOWNE [1956]). MORSE [1955] and TAKÁCS [1962] present the time dependent solution by means of trigonometric functions. MORSE [1955] introduces the notion of relaxation time for the $M/M/1$ system. He takes as relaxation time $2a\beta/(1-a)^2$ which differs by a factor $2a/(1+\sqrt{a})^2$ from the relaxation time as defined in (2.11). His derivation is based on an approximating expression for the correlation function of the x_t-process. Our definition for the relaxation time is based on the behaviour of $p_{ij}(t)$ and $\mathrm{E}\{x_t \mid x_0=k\}$ for large values of t and is identical with that of KINGMAN [1962[c]]. However, it is always difficult to point out which definition of relaxation time should be taken; what is essential is that it is proportional to

$$\frac{\beta}{(1-\sqrt{a})^2}\left(=\frac{(1+\sqrt{a})^2\beta}{(1-a)^2}\right).$$

The busy period has been the subject of many papers. Here we mention only the studies by PALM [1943 and 1957] and BAILEY [1954]. NEUTS [1964] investigates the joint distribution of the length of the busy period and the maximum number of customers present during this busy period. He applies the birth and death technique; the method used here is based on taboo probabilities and seems to be a more general approach (cf. sections II.3.5 and II.4.4). Considering the discrete time parameter Markov chain with the jump matrix of the x_t-process as the one-step transition matrix, we see that $F_{h;10}(\infty)$ is the ruin probability for the classical ruin problem (cf. FELLER [1957]) where the gambler has a probability $a(1+a)^{-1}$ to gain and $(1+a)^{-1}$ to loose one unit and starts with one unit, while he and his adversary together own h units. $F_{h;10}(\infty)$ can be immediately obtained from this relation.

The departure process has also been studied by REICH [1957 and 1965].

Chapter II.3

The queueing system $G/M/1$ has been treated by KENDALL [1953] in an important paper on the method of regeneration points, which in our notation means the study of the process $\{y_n, n=0, 1, ...\}$. An extensive study has been given by TAKÁCS [1960, 1962]. We further mention here the work of SMITH [1953], CONOLLY [1959], PRABHU [1964], HEATHCOTE [1965] and COHEN [1967[a]].

Chapter II.4

The queueing process $M/G/1$ received a good deal of attention in the literature. The approach to the investigation of the $M/G/1$ process as discussed above started with Kendall's investigation in 1953. TAKÁCS [1961, 1962, 1964[a]] obtained many fundamental results. Here we mention further the work of GAVER [1959] and BENEŠ [1957, 1963]. PRABHU [1965[a]] studied the virtual waiting time process extensively and gives a detailed account of it in his book. For Pollaczek's studies we refer to chapter 5, see also section 6.6 where TakÁcs' combinatorial method is applied to the queueing system $M/G/1$.

The queueing system with constant service time, i.e. the system $M/D/1$, already studied by Erlang in 1909 (cf. BROCKMEYER, HALSTRØM and JENSEN [1948]) received much attention in the literature. In particular we mention the study of CROMMELIN [1932, 1933] whose method is also applicable to the many server queue with constant service time and a Poisson arrival process. For the stationary distribution of the waiting time of the $M/D/1$ system a number of graphs is available (cf. LE GALL [1962], KAUFMANN and CRUON [1961] and especially MOLNAR [1952]).

Chapter II.5

Pollaczek considered the single server queue $G/G/1$ as early as 1930. His theory for the single server queue was published in 1957 (POLLACZEK [1957], see also LE GALL [1962] and SYSKI [1967]). In his investigations Pollaczek assumed that the Laplace-Stieltjes transforms $\beta(\rho)$ and $\alpha(\rho)$ are analytic for $\mathrm{Re}\,\rho>\delta$, with $\delta<0$. The method used in this chapter, sections 1 and 2, is essentially based on Pollaczek's ideas. Due to the fact that we have started by investigating the joint distribution of $\boldsymbol{w}_n$, $\boldsymbol{d}_n$, $\boldsymbol{a}_{n-1}$ and $\boldsymbol{b}_{n-1}$ no assumption about analyticity of $\alpha(\rho)$ and $\beta(\rho)$ for $\mathrm{Re}\,\rho<0$ is needed. From the studies of KEMPERMAN [1961[b]] and KINGMAN [1962[b]] the idea of the present approach arose. The duality of $\boldsymbol{w}_n$ and $\boldsymbol{d}_n$ has already been mentioned by

KINGMAN [1965]. The results of section 5.4 can also be obtained from fluctuation theory, in fact SPITZER [1956] obtained these results in his basic papers on fluctuation theory (cf. also SPITZER [1960] and PRABHU [1965[a]]). Keilson and Runnenburg (cf. discussion of KINGMAN's paper [1965])pointed out that the limit distribution for the actual waiting time is infinitely divisible; this property is also mentioned in a remark by SPITZER [1956]. The fact that a queueing system can have a non-trivial stationary waiting time distribution if $A(t)$ and $B(t)$ have no finite first moment has been noted by SPITZER [1960]. Formula (5.58) for the transform of the joint distribution of busy period, busy cycle, idle period and number of customers served during a busy period is due to KINGMAN [1962] (see also FINCH [1961] and POLLACZEK [1952]). For derivations based on fluctuation theory see PRABHU [1965[a]]. The concept of strong busy period and weak busy period is of importance for queueing situations in computers. The formulas for the residual busy period and the residual busy cycle seem to be new. It is rather difficult to obtain these results by using fluctuation theory and it was for this reason that we have preferred the Pollaczek-integral-equation approach. The relation (5.107) for the joint distribution of the virtual delay time at time t, of the accumulated empty time at time t and of the total amount of service time brought in by the customers having arrived in $[0, t)$ is new. This formula is of special importance in the theory of dams. The limit distribution for the virtual delay time has been obtained by TAKÁCS [1963[a]]. The important relation (5.131) between average delay and average queue length has been proved by LITTLE [1961] and JEWELL [1967] (cf. also TAKÁCS [1963]). The distribution function for the queue length has been investigated by LE GALL [1962] and TAKÁCS [1962, 1963]. The $G/G/1$ system with truncated service time distribution is presented here for the first time. The systems $G/K_n/1$ and $K_m/G/1$ have been studied by various authors, in particular if K_n is an Erlang distribution E_n, see KINGMAN [1962], CONOLLY [1960], PRABHU [1965[a]], POLLACZEK [1957], LE GALL [1962] and JACKSON [1954, 1956].

An interesting aspect is that for the system $G/G/1$ the converse of Smith's theorem is true.

Chapter II.6

The method of inclusion of a supplementary variable has been applied by many investigators. We mention here KOSTEN [1942, 1948], COX [1955] and particularly the recent book of GNEDENKO and KOVALENKO [1968], where an extensive analysis of the method is given; see also COX and MILLER [1965]. We also mention here TAKÁCS [1963b], COHEN [1957], SYSKI [1962], KEILSON and KOOHARIAN [1960]. The method of collective marks has also been applied by KEMPERMAN [1961], for further references see RUNNENBURG [1965]. The method of phases has been extensively used by MORSE [1958]. In his book a detailed account of its applications to the single server queue is given. He illustrates the influence of the various types of interarrival and service time distributions on the average waiting time and queue length. We have already mentioned the work of JACKSON and NICKOLS [1956]; see also BROCKMEYER, HALSTRØM and JENSEN [1948], WISHART [1956] and GUPTA and GOYAL [1964]. Combinatorial methods have been applied in queueing theory by several authors. The most extensive study is that of TAKÁCS [1965 and 1967]; see also TAKÁCS [1964a], BENEŠ [1963] and PYKE [1959].

Chapter III.2

BAILEY [1954b], DOWNTON [1955] and TAKÁCS [1962] considered the $M/G/1$ queue with delayed service and batches of constant size. These authors apply the method of the imbedded Markov chain (cf. chapter II.4), see also FOSTER [1961], FOSTER and NYUNT [1961] and FOSTER and PERERA [1964]. The same model for the $G/G/1$ queue has been studied by LE GALL [1962], whose investigation is based on the solution of the $G/G/1$ queueing model (cf. section 2.9, where the same idea has been used). CONOLLY [1960] discusses the $M/G/1$ queue with single service and group arrivals, groups being of constant size. MILLER [1959] seems to be the first author to consider group arrivals and group service for the $M/G/1$ queue. After setting up the general relations he restricts the analysis mainly to group arrivals and single service. KEILSON [1962a] studies the $M/G/1$ queue with group arrivals and group service for the general case. He also investigates the transportation problem. The method of inclusion of a supplementary variable is applied here for the derivation of the relations between the system variables. The resulting partial differential equations are transformed into a Hilbert problem and he formally discusses its solution. LE GALL [1962] discusses the $G/G/1$ queue with group arrivals and single service; also this queue with individual arrivals and group service is dealt with for the case of constant batches (p. 267). The latter case is especially interesting. The problem is reduced to an integral equation of Pollaczek's type, the kernel being more complicated, however, than that in chapter II.5. The solution is rather intricate and the interested reader is referred to Le Gall's book for further details. The method used in section 2.7 for $G/M/1$ with single arrivals and batch service was introduced in the paper of COHEN [1963]. Applying fluctuation theory BHAT [1964] considers the $G/G/1$ queue with group arrivals and batch services. His studies (cf. also PRABHU [1965b]) centre on those bulk queues which can be described by relations of the type (2.64) and (2.99). For these cases he discusses $\Pr\{y_n=j \mid y_1=i\}$ and its limit for $n\to\infty$. A number of interesting examples is given.

Chapter III.3

The "last come, first served" discipline for $M/G/1$ has been studied by VAULOT [1954]; WISHART [1960] considered this discipline for $M/G/1$ and $G/M/1$, see also RIORDAN [1962] for $M/G/1$.

The "random service" discipline for $M/M/1$ has been studied by PALM [1938; see 1957], VAULOT [1946] and POLLACZEK [1946; see also 1959]. For $M/D/1$ see BURKE [1959]. The solution for $M/G/1$ has been obtained by LE GALL [1962] and by KINGMAN [1962c]. Le Gall applies asymptotic methods to calculate the waiting time distribution. Kingman (see also TAKÁCS [1963b]) derives a set of recurrence relations for the determination of the moments of the waiting time distribution. Graphs for the waiting time distribution for $M/M/1$ with random service are given by LE GALL [1962], WILKINSON [1953], RIORDAN [1953], KAUFMANN and CRUON [1961] and MORSE [1958].

Random service for $G/M/1$ has been discussed by LE GALL [1962]. This author also treats the $G/G/1$ system but a workable solution is not obtained.

The important concept of "completion times" for queueing with interrupted service has been introduced by GAVER [1962] and KEILSON [1962b].

Queueing with priority disciplines appeared for the first time in the literature in a study of COBHAM [1954] treating for $M/G/1$ the "non break-in" priority discipline with k levels. COHEN [1956] considered for $M/M/n$ the priority discipline with two levels, all customers having the same service time distribution. KESTEN and RUNNENBURG [1957] obtained waiting time distributions for $M/G/1$. PHIPPS [1956] introduced continuous priorities and considered the discipline: customers with shorter service time have priority (cf. (3.121)). HEATHCOTE [1959, 1960 and 1961] initiated the research of pre-emptive resume priorities. Pre-emptive repeat seems to have been introduced by GAVER [1962]. A very interesting and unifying study has been given by TAKÁCS [1964[b]]. MILLER [1960] studied the imbedded Markov chain for $M/G/1$ with non break-in priorities for two levels. JAISWAL [1962] applied the method of included variable and obtains the generating function of the joint distribution of queue lengths for first and second priority customers. See further papers by YEO [1962] and WELCH [1964]. A very readable survey paper on priority disciplines has been given by GAVER [1965]. The recent monograph by JAISWAL [1968] contains an interesting approach to and review of priority systems.

An extensive comparative study of priority systems has been published by AVI-ITZHAK and NAOR [1963]. These authors introduced "discretionary priorities", see also AVI-ITZHAK, BROSH and NAOR [1964]. This is a priority discipline which is between non break-in and pre-emptive resume. The discipline is pre-emptive resume if the past service time of a second priority customer does not exceed a given time, while it is non break-in otherwise. BROSH and NAOR [1963] discuss the determination of the optimum priority distribution $P(y)$ for minimizing total cost given the cost per unit waiting time for the various types of arriving customers. Their study starts from an idea developed by COX and SMITH [1961].

Chapter III.5

The model for the dam with finite capacity and instantaneous input has been investigated by PRABHU [1958] for the $E_k/D/1$ system, by KOVALENKO [1961] for the $M/M/1$ model and by GAVER and MILLER [1962] for the $M/G/1$ model. DALEY [1964] has studied the $G/G/1$ model, see also GHOSAL [1960]. The greater part of these studies are centred on the investigation of the stationary distribution of the dam content. Except for Daley, who starts from an integral equation which is the analogue of Lindley's for the general queueing system $G/G/1$, and for Kovalenko who applies a birth and death method, most authors start from the integro-differential equation for the dam content, see sections 5.9 and 5.10. The method used in this chapter has been developed by the author in his papers COHEN [1967[b], 1968[a]], see also COHEN and GREENBERG [1968] for down crossings. For the dam model with non-instantaneous inflow see also GAVER and MILLER [1962]. A comprehensive discussion of dam models is given by PRABHU [1965[a]]. For dam models with an input process with stationary, independent increments, see TAKÁCS [1966, 1967].

Chapter III.6

Queueing systems with limited waiting room have obtained relatively little attention in the literature. Some results can be found in RIORDAN [1962] p. 70. The ergodic distributions for the $M/G/1$ and the $G/M/1$ systems have been studied by KEILSON [1966]. For the $M/M/1$ queue the problem is much simpler, see SAATY [1961] p. 127, FINCH [1958] and SYSKI [1960]. Many studies of systems with finite waiting room are for models in storage and dam theory, see PRABHU [1965[a]], WEESAKUL [1961], KINNEY [1962] and BHAT [1965]. The discussion given in the present chapter is new.

For finite dam theory, PRABHU [1965[a]] and WEESAKUL [1961] consider a relation of the type

$$z_{n+1}=[\min(z_n+g_n, K)-m]^+, \qquad K>m, \quad n=1, 2, \dots,$$

whereas Kinney investigates a process described by

$$z_{n+1}=\min[K, [z_n+g_n-K_n]^+], \qquad n=1, 2, \dots .$$

Here, $\boldsymbol{g}_n$ is the number of customers arriving in the nth service interval, while $\boldsymbol{K}_n$ is the number of departures in this interval. The service intervals are of constant length, whereas $\boldsymbol{g}_n$, $\boldsymbol{K}_n$, $n=1, 2, \dots$, are independent variables, all $\boldsymbol{g}_n$ and also all $\boldsymbol{K}_n$ are identically distributed. Recurrence relations of the type just mentioned are of the same structure as those discussed in chapters III.4, III.5 and the present chapter and may be analysed by the methods described in these chapters.

The queueing system discussed in this chapter is a typical example of queueing with restricted accessibility, i.e. not every customer will be admitted to the system. The admittance of a customer to the system depends on the state of the queueing system at the moment of his arrival. Two interesting examples of systems with restricted accessibility are described by

$$\boldsymbol{w}_{n+1}=\begin{cases}[\boldsymbol{w}_n+\boldsymbol{\tau}_n-\boldsymbol{\sigma}_{n+1}]^+ & \text{if} \quad \boldsymbol{w}_n<H,\\ [\boldsymbol{w}_n-\boldsymbol{\sigma}_{n+1}]^+ & \text{if} \quad \boldsymbol{w}_n\geqq H,\end{cases}$$

and

$$\boldsymbol{w}_{n+1}=\begin{cases}[\boldsymbol{w}_n+\boldsymbol{\tau}_n-\boldsymbol{\sigma}_{n+1}]^+ & \text{if} \quad \boldsymbol{w}_n+\boldsymbol{\tau}_n<H,\\ [\boldsymbol{w}_n-\boldsymbol{\sigma}_{n+1}]^+ & \text{if} \quad \boldsymbol{w}_n+\boldsymbol{\tau}_n\geqq H,\end{cases}$$

where H is a positive constant. In these models the customer is refused if and only if his actual waiting time or the sum of his actual waiting time and service time exceeds H. Both models have been analyzed by applying the method applied in chapters III.4 and III.5.

Chapter III.7

The first researches on heavy traffic theory were made by Kingman, who obtained a number of important results, see KINGMAN [1961, 1962[a], 1962[b] and 1965]. Section 7.1 is based mainly on Kingman's ideas. A detailed study on heavy traffic theory is also given by BOROVKOV [1964], see also SAMANDAROV [1963].

Asymptotic relations of the type discussed in section 7.3 appeared already in the work of POLLACZEK [1957]. The concept of relaxation time in queueing has been introduced by MORSE [1955], (see also section II.2.1), who found the expression for the $M/M/1$ queue. KINGMAN [1962[c]] obtained expressions for $M/G/1$ and $M/D/1$. The method used in the present section is based on an idea of WISHART [1966] who applied this method to the $M/G/1$ system. CRAVEN [1963] studied asymptotic behaviour for bulk queues; very detailed results have also been obtained by NAWIJN [1967]. Limit distributions of the type discussed in section 7.4 have been obtained by COHEN [1968[b]]. This paper contains an error; C. W. Anderson noted that $\boldsymbol{X}_n$ for $a<1$ and $\boldsymbol{P}_{K,n}, \boldsymbol{L}_{K,n}$ for $a\leqq 1$ have no limiting distributions for $n\rightarrow\infty$ (private communication, cf. also GNEDENKO [1943]). Distributions of maxima for birth and death processes have been studied by KARLIN and MCGREGOR [1960].

REFERENCES

ANDERSEN, E. Sparre, On the fluctuations of sums of random variables, I and II, Math. Scand. **1** (1953) 263–285, **2** (1954) 195–223.

ARROW, K. J., KARLIN, S. and SCARF, H., Studies in Mathematical Theory of Inventory and Production (Stanford University Press, Stanford, Calif., 1958).

AVI-ITZHAK, B. and NAOR, P., Multipurpose service stations in queueing problems, Proc. 3rd Intern. Conf. Operations Res., Oslo, 1963, eds. G. Kreweras and G. Morlat (Dunod, Paris, English Universities Press, London, 1964).

AVI-ITZHAK, B., BROSH, I. and NAOR, P., On discretionary priority queueing, Z. Angew. Math. Mech. **44** (1964) 235–242.

BAILEY, N. T. J., A continuous time treatment of a simple queue using generating functions, J. Roy. Statist. Soc. Ser. B **16** (1954[a]) 288–291.

BAILEY, N. T. J., On queueing processes with bulk service, J. Roy. Statist. Soc. Ser. B **16** (1954[b]) 80–87.

BAILEY, N. T. J., Some further results in the non-equilibrium theory of a simple queue, J. Roy. Statist. Soc. Ser. B **19** (1957) 326–333.

BAILEY, N. T. J., The Elements of Stochastic Processes with Applications to the Natural Sciences (Wiley, New York, 1964).

BENEŠ, V. E., On queues with Poisson arrivals, Ann. Math. Statist. **28** (1957) 670–677.

BENEŠ, V. E., General Stochastic Processes in the Theory of Queues (Addison-Wesley, Reading, Mass., 1963).

BHAT, U. N., Imbedded Markov chain analysis of single server bulk queues, J. Austral. Math. Soc. **4** (1962) 244–263.

BHAT, U. N., Customer overflow in queues with finite waiting place, Austral. J. Statist. **7** (1965) 15–19.

BLACKWELL, D., A renewal theorem, Duke Math. J. **15** (1948) 145–150.

BLANC-LAPIERRE, A. and FORTET, R., Théorie des Fonctions Aléatoires (Masson, Paris, 1953).

BLUM, J. R. and ROSENBLATT, R., On the structure of infinitely divisible distributions, Pacific J. Math. **9** (1957) 1–7.

BOCHNER, S., Harmonic Analysis and the Theory of Probability (University of California Press, Berkely and Los Angeles, 1955).

BOROVKOV, A. A., Some limit theorems in the theory of mass service, Theor. Probability Appl. **9** (1964) 550–565.

BROCKMEYER, E., HALSTRØM, H. L. and JENSEN, A., The Life and Works of A. K. Erlang, Trans. Danish Acad. Tech. Sci. **2** (1948).

BROSH, I. and NAOR, P., On optimal disciplines in priority queueing, Bull. Intern. Statist. Inst. Proc. 34th session, Ottawa, 1963.

BRUYN, N. G. DE, Asymptotic Methods in Analysis (North-Holland Publishing Co., Amsterdam, 1958).

BURKE, P. J., The output of a queueing system, Operations Res. **4** (1956) 699–704.

BURKE, P. J., Equilibrium delay distribution for one channel with constant holding time, Poisson input and random service, Bell System Tech. J. **38** (1959) 1021–1031.

CHAMPERNOWNE, D. G., An elementary method of the solution of the queueing problem with a single server and a constant parameter, J. Roy. Statist. Soc. Ser. B **18** (1956) 125–128.

CHUNG, K. L., Markov Chains with Stationary Transition Probabilities (Springer Verlag, Berlin, 1960).

CHUNG, K. L. and FUCHS, W. H. J., On the distributions of sums of random variables, Four papers on probability, Mem. Amer. Math. Soc. **6** (1950) 1–12.

CLARKE, A. B., The time dependent waiting line problem, Univ. Michigan report M 720-1R39, 1953.

COBHAM, A., Priority assignment in waiting line problems, Operations Res. **2** (1954) 70–76.

COHEN, J. W., Certain delay problems for a full availability trunk group loaded by two traffic sources, Communication News (Philips, Holland) **16** (1956) 105–113.

COHEN, J. W., The generalized Engset formulae, Philips Telecommunication Review **18** (1957) 158–170.

COHEN, J. W., On derived and nonstationary Markov chains, Teor. Verojatnost. i Primenen **7** (1962[a]) 410–432.

COHEN, J. W., Derived Markov chains, I, II, III, Nederl. Akad. Wetensch. Indagationes Math. **24** (1962[b]) 55–92.

COHEN, J. W., Applications of derived Markov chains in queueing theory, Appl. Sci. Res. Ser. B **10** (1963) 269–302.

COHEN, J. W., Distribution of the maximum number of customers present simultaneously during a busy period for the system $M/G/1$ and for the system $G/M/1$, J. Appl. Probability **4** (1967[a]) 162–179.

COHEN, J. W., On two integral equations of queueing theory, J. Appl. Probability **4** (1967[b]) 343–355.

COHEN, J. W., Single server queue with uniformly bounded virtual waiting time, J. Appl. Probability **5** (1968[a]) 93–122.

COHEN, J. W., Extreme value distributions for the $M/G/1$ and the $G/M/1$ queueing systems, Ann. Inst. H. Poincaré Sect. B **4** (1968[b]) 83–98.

COHEN, J. W. and GREENBERG, I., Distributions of crossings of level K in a busy cycle of the $M/G/1$ queue, Ann. Inst. H. Poincaré Sect. B **4** (1968) 75–81.

CONOLLY, B. W., A difference equation technique applied to the simple queue, J. Roy. Statist. Soc. Ser. B **20** (1958) 165–167.

CONOLLY, B. W., The busy period in relation to the queueing process $G/M/1$, Biometrika **46** (1959) 246–251.

CONOLLY, B. W., The busy period in relation to the single server queueing system with general independent input and Erlangian service time, J. Roy. Statist. Soc. Ser. B **22** (1960) 89–96.
COX, D. R., The analysis of non-Markovian stochastic processes by the inclusion of supplementary variables, Proc. Cambridge Phil. Soc. **51** (1955) 433–441.
COX, D. R., Renewal Theory (Methuen, London, 1962).
COX, D. R. and MILLER, H. D., The Theory of Stochastic Processes (Methuen, London, 1965).
COX, D. R. and SMITH, W. L., Queues (Methuen, London, 1961).
CRAVEN, B. D., Asymptotic behaviour of the bulk service queue, J. Austral. Math. Soc. **3** (1963) 503–512.
CROMMELIN, C. D., Delay probability formulae when the holding times are constant, Post Office Electr. Engrs. J. **25** (1932) 41–50.
CROMMELIN, C. D., Delay probability formulae, Post Office Electr. Engrs. J. **26** (1933) 266–274.
DALEY, D. J., Single server queueing systems with uniformly limited queueing time, J. Austral. Math. Soc. **4** (1964) 489–505.
DOBRUSHIN, R. L., Lemma on the limit of compound random functions (in Russian), Usp. Math. Nauk. **10** (1955) 157–159.
DOETSCH, G., Theorie und Anwendung der Laplace Transformation (Dover, New York, 1943).
DOOB, J. L., Stochastic Processes (Wiley, New York, 1953).
DOWNTON, F., Waiting times in bulk service queues, J. Roy. Statist. Soc. Ser. B **17** (1955) 256–261.
DYNKIN, E. B., Markov Processes, I, II (Springer Verlag, Berlin, 1965).
ERDÉLYI, A. et al., Tables of integral transforms (McGraw-Hill, New York, 1954).
ERDÉLYI, A. et al., Higher transcendental functions (McGraw-Hill, New York, 1953).
FABENS, A. J., The solution of queueing and inventory models by semi-Markov processes, J. Roy. Statist. Soc. Ser. B **23** (1961) 113–127.
FELLER, W., On the integral equation of renewal theory, Ann. Math. Statist. **12** (1941) 243–267.
FELLER, W., Probability Theory and its Applications, Vol. I (Wiley, New York, 1957).
FELLER, W., Probability Theory and its Applications, Vol. II (Wiley, New York, 1966).
FELLER, W. and OREY, S., A renewal theorem, J. Math. Mech. **10** (1961) 619–624.
FINCH, P. D., The effect of the size of the waiting room on a simple queue, J. Roy. Statist. Soc. Ser. B **20** (1958) 182–186.
FINCH, P. D., A probability limit theorem with application to a generalization of queueing theory, Acta Math. Acad. Sci. Hungar. **10** (1959[a]) 317–325.
FINCH, P. D., The output process of the queueing system $M/G/1$, J. Roy. Statist. Soc. Ser. B **21** (1959[b]) 375–380.
FINCH, P. D., On the busy period in the queueing system $G/G/1$, J. Austral. Math. Soc. **2** (1961) 217–228.
FINCH, P. D., On partial sums of Lagrange's series with application to the theory of queues, J. Austral. Math. Soc. **3** (1963) 488–490.
FORTET, R. and GRANDJEAN, Ch., Study of congestion in a loss system, Fourth Intern. Teletraffic Congress, London 1964, Post Office Telecommunications, special issue, 1964.

FOSTER, F. G., On stochastic matrices associated with certain queueing processes, Ann. Math. Statist. **24** (1953) 355–360.

FOSTER, F. G., Queues with batch arrivals I, Acta Math. Acad. Sci. Hungar. **12** (1961) 1–10.

FOSTER, F. G. and NYUNT, K. M., Queues with batch departures I, Ann. Math. Statist. **32** (1961) 1324–1332.

FOSTER, F. G. and PERERA, A. G. A. D., Queues with batch departures II, Ann. Math. Statist. **35** (1964) 1147–1156.

FRÉCHET, M., Méthodes des Fonctions Arbitraires, Théorie des Événements en Chaine dans le Cas d'un Nombre Fini d'États Possibles (Gauthier Villars, Paris, 1952).

GAVER, D. P., Imbedded Markov chain analysis of a waiting line process in continuous time, Ann. Math. Statist. **30** (1959) 698–720.

GAVER, D. P., A waiting line with interrupted service, including priorities, J. Roy. Statist. Soc. Ser. B **24** (1962) 73–90.

GAVER, D. P., Competitive queueing: idleness probabilities under priority disciplines, J. Roy. Statist. Soc. Ser. B **25** (1963) 489–499.

GAVER, D. P., On priority type disciplines in queueing, Proc. Symp. on Congestion Theory, eds. W. L. Smith and W. E. Wilkinson (University of North-Carolina Press, Chapel Hill, N.C., 1965).

GAVER, D. P. and MILLER, R. G., Limiting distributions for some storage problems, Studies in Applied Probability and Management Science, eds. K. J. Arrow, S. Karlin and H. Scarf (Stanford University Press, Stanford, Calif., 1962).

GHOSAL, A., Emptiness in the finite dam, Ann. Math. Statist. **31** (1960) 803–808.

GNEDENKO, B., Sur la distribution limite du terme maximum d'une série aléatoire, Ann. Math. **44** (1943) 423–453.

GNEDENKO, B. and KOVALENKO, I. N., Introduction to Queueing Theory (Israel Program for Scientific Translations, Jerusalem, 1968).

GUPTA, S. K. and GOYAL, J. K., Queues with Poisson input and hyper-exponential output with finite waiting space, Operations Res. **12** (1964) 75–81.

HARDY, G. H., LITTLEWOOD, J. E. and PÓLYA, G., Inequalities (Cambridge University Press, Cambridge, 1962).

HARRIS, Th. E., The Theory of Branching Processes (Springer Verlag, Berlin, 1963).

HASOFER, A. M., On the integrability, continuity and differentiability of a family of functions introduced by L. Takács, Ann. Math. Statist. **34** (1963) 1045–1049.

HEATHCOTE, C. R., The time-dependent problem for a queue with preemptive priorities, Operations Res. **7** (1959) 670–680.

HEATHCOTE, C. R., A simple queue with several preemptive priority classes, Operations Res. **8** (1960) 630–638.

HEATHCOTE, C. R., Preemptive priority queueing, Biometrika **48** (1961) 57–63.

HEATHCOTE, C. R., On the maximum of the queue $GI/M/1$, J. Appl. Probability **2** (1965) 206–214.

HEATHCOTE, C. R. and MOYAL, J. E., The random walk in continuous time and its application to the theory of queues, Biometrika **46** (1959) 400–411.

HEWITT, E., Remarks on the Fourier-Stieltjes transforms, Ann. Math. **57** (1953) 458–474.

HILLE, E. and PHILLIPS, R. S., Functional Analysis of Semi-groups (Am. Math. Soc., Providence, R.I., 1957).

JACKSON, R. R. P., Queueing system with phase type service, Operational Res. Quart. **5** (1954) 109–120.

JACKSON, R. R. P., Queueing processes with phase type service, J. Roy. Statist. Soc. Ser. B **18** (1956) 129–132.

JACKSON, R. R. P. and NICKOLS, D. G., Some equilibrium results for the queueing process $E/M/1$, J. Roy. Statist. Soc. Ser. B **18** (1956) 275–279.

JAISWAL, N. K., Time dependent solution of the 'Head-of-the Line' priority queue, J. Roy. Statist. Soc. Ser. B **24** (1962) 91–101.

JAISWAL, N. K., Priority Queues (Academic Press, New York, 1968).

JEWELL, S. W., A simple proof of: $L=\lambda W$, Operations Res. **15** (1967) 1109–1116.

KANTERS, P. J. A., Foster's criteria for periodic chains (in Dutch), Report Mathematical Institute Technological Univ. Delft, 1965.

KARLIN, S. and MCGREGOR, J., The classification of birth and death processes, Trans. Amer. Math. Soc. **86** (1957) 366–400.

KARLIN, S. and MCGREGOR, J., Many server queueing processes with Poisson input and exponential service times, Pacific J. Math. **8** (1959) 87–118.

KARLIN, S. and MCGREGOR, J., Occupation time laws for birth and death processes, Proc. Fourth Berkeley Symp. on Math. Statist. and Probability II, ed. J. Neyman (University of California Press, Berkeley, Calif., 1960).

KAUFMANN, A. and CRUON, R., Les Phénomènes d'Attente (Dunod, Paris, 1961).

KEILSON, J., The general bulk queue as a Hilbert problem, J. Roy. Statist. Soc. Ser. B **24** (1962[a]) 344–358.

KEILSON, J., Queues subject to service interruptions, Ann. Math. Statist. **33** (1962[b]) 1314–1322.

KEILSON, J., Green's Function Methods in Probability Theory (Griffin, London, 1965).

KEILSON, J., The ergodic queue length distribution for queueing systems with finite capacity, J. Roy. Statist. Soc. Ser. B **28** (1966) 190–201.

KEILSON, J. and KOOHARIAN, A., On time dependent queueing processes, Ann. Math. Statist. **31** (1960) 104–112.

KEILSON, J. and MERMIN, N. D., The second order distribution of integrated shot noise, Trans. I.R.E. **5** (1959) 75–77.

KEMENY, J. G. and SNELL, J. L., Finite Markov Chains (Van Nostrand, New York, 1960).

KEMENY, J. G., SNELL, J. L. and KNAPP, A. W., Denumerable Markov Chains (Van Nostrand, New York, 1966).

KEMPERMAN, J. H. B., An Analytical Approach to the Differential Equations of the Birth and Death Process, Mathematical Research Center, Univ. of Wisconsin, 1961[a].

KEMPERMAN, J. H. B., The Passage Problem for a Stationary Markov Chain (University of Chicago Press, Chicago, 1961[b]).

KENDALL, D. G., Some problems in the theory of queues, J. Roy. Statist. Soc. Ser. B **13** (1951) 151–185.

KENDALL, D. G., Stochastic processes occurring in the theory of queues and their analysis by the method of the imbedded Markov chain, Ann. Math. Statist. **24** (1953) 338–354.

KENDALL, D. G., Geometric ergodicity and the theory of queues, Mathematical Methods in the Social Sciences (Stanford University Press, Stanford, Calif., 1960).

KESTEN, H. and RUNNENBURG, J. Th., Priority in waiting line problems, Nederl. Akad. Wetensch. Indagationes Math. **60** (1957) 312–336.

KINGMAN, J. F. C., The single server queue in heavy traffic, Proc. Cambridge Phil. Soc. **57** (1961) 902–904.

KINGMAN, J. F. C., Some inequalities for the queue $G/G/1$, Biometrika **49** (1962[a]) 315–324.

KINGMAN, J. F. C., On queues in heavy traffic, J. Roy. Statist. Soc. Ser. B **24** (1962[b]) 383–392.

KINGMAN, J. F. C., On queues in which customers are served in random order, Proc. Cambridge Phil. Soc. **58** (1962[c]) 79–91.

KINGMAN, J. F. C., The use of Spitzer's identity in the investigation of the busy period and other quantities in the queue $GI/G/1$, Austral. J. Math. **2** (1962[d]) 345–356.

KINGMAN, J. F. C., On continuous time models in the theory of dams, Austral. J. Math. **3** (1963) 480–487.

KINGMAN, J. F. C., The heavy traffic approximation in the theory of queues, Proc. Symp. on Congestion Theory, eds. W. L. Smith and W. E. Wilkinson (University of North-Carolina Press, Chapel Hill, N.C., 1965).

KINGMAN, J. F. C., On the algebra of queues. J. Appl. Probability **3** (1966) 285–326.

KINNEY, J. R., A transient discrete time queue with finite storage, Ann. Math. Statist. **33** (1962) 130–136.

KHINTCHINE, A. Y., Mathematical Methods in the Theory of Queueing (Griffin, London, 1960).

KNOPP, K., Theory of Functions I, II (Dover, New York, 1947).

KOSTEN, L., On Loss and Queueing Problems (in Dutch), Thesis, Technological Univ. Delft, 1942.

KOSTEN, L., On the validity of the Erlang and Engset loss formulae, Het P.T.T. Bedrijf (Dutch Post Office journal) **2** (1948) 42–45.

KOVALENKO, I. N., Some queueing problems with restrictions, Theor. Probability Appl. **6** (1961) 222–228.

LAMBOTTE, J. P. and TEGHEM, J., Utilisation de la théorie des processus semi-Markoviens dans l'étude des problèmes de files d'attente, Proc. Nato Conference on Queueing Theory, Lisbon, 1966, ed. R. Cruon (English University Press, London, 1967).

LEDERMANN, W. and REUTER, G. E. H., Spectral theory for the differential equations of simple birth and death processes, Phil. Trans. Roy. Soc. London Ser. A **246** (1954) 321–369.

LE GALL, P., Les Systèmes avec ou sans Attente et les Processus Stochastiques (Dunod, Paris, 1962).

LINDLEY, D. V., The theory of queues with a single server, Proc. Cambridge Phil. Soc. **48** (1952) 277–289.

LITTLE, J. D. C., A proof for the queueing formula: $L=\lambda W$, Operations Res. **9** (1961) 383–387.

LOÈVE, M., Probability Theory (Van Nostrand, New York, 1960).

LOYNES, R. M., Stationary waiting time distributions for single server queues, Ann. Math. Statist. **33** (1962) 1323–1348.

LUKACS, E., Characteristic Functions (Griffin, London, 1960).

MILLER, R. G., A contribution to the theory of bulk queues, J. Roy. Statist. Soc. Ser. B **21** (1959) 320–337.

MILLER, R. G., Priority queues, Ann. Math. Statist. **31** (1960) 86–103.

MOLINA, E. C., Application of the theory of probability to telephone trunking problems, Bell System Tech. J. **6** (1927) 461–494.

MOLNAR, I., Delay probability charts for telephone traffic where the holding times are constant, Engineering Notes 2032, Automatic Electric Chicago, 1952.

MORSE, Ph. M., Stochastic processes of waiting lines, Operations Res. **3** (1955) 255–261.
MORSE, Ph. M., Queues, Inventories and Maintenance (Wiley, New York, 1958).
NAWIJN, W. M., Relaxation times for queueing processes (in Dutch), Report Mathematical Institute Technological Univ. Delft, 1967.
NEWELL, G. F., Queues for a fixed cycle traffic light, Ann. Math. Statist. **31** (1960) 589–597.
NEWELL, G. F., Approximation methods for queues with application to the fixed-cycle traffic light, SIAM Rev. **7** (1965) 223–239.
NEUTS, M. F., The distribution of the maximum length of a Poisson queue during a busy period, Operations Res. **12** (1964) 281–285.
NEUTS, M. F., The single server queue with Poisson input and semi-Markov service times, J. Appl. Probability **3** (1966[a]) 202–230.
NEUTS, M. F., Two Markov chains arising from examples of queues with state-dependent service times, Purdue Univ. Statistical Department, Mimeographed series **96**, 1966[b].
NEUTS, M. F., A general class of bulk queues with Poisson input, Ann. Math. Statist. **38** (1967) 759–770.
PALM, C., Intensitätsschwankungen im Fernsprechverkehr, Ericsson Technics **44** (1943) 1–89.
PALM, C., Waiting times with random served queue, Tele **1** (1957) 1–107 (English ed.).
PARZEN, E., Stochastic Processes (Holden-Day, San Francisco, 1962).
PHIBBS, T. E., Machine repair as a priority waiting line problem, Operations Res. **4** (1956) 76–85.
POLLACZEK, F., La loi d'attente des appels téléphoniques, C. R. Acad. Sci. Paris **222** (1946) 353–355.
POLLACZEK, F., Sur la répartition des périodes d'occupation ininterrompue d'un guichet, C. R. Acad. Sci. Paris **234** (1952) 2042–2044.
POLLACZEK, F., Problèmes Stochastiques Posés par le Phénomène de Formation d'une Queue d'Attente à un Guichet et par des Phénomènes Apparentés (Gauthiers Villars, Paris, 1957).
POLLACZEK, F., Application de la théorie des probabilités à des problèmes posés par l'encombrement des reseaux téléphoniques, Ann. Télécommunications **14** (1959) 165–183.
POLLACZEK, F., Théorie Analytique des Problèmes Stochastiques Relatifs à un Groupe de Lignes Téléphoniques avec Dispositif d'Attente (Gauthiers Villars, Paris, 1961).
POLLACZEK, F., Concerning an analytical method for the treatment of queueing problems, Proc. Symp. on Congestion Theory, eds. W. L. Smith and W. E. Wilkinson (University of North-Carolina Press, Chapel Hill, N.C., 1965).
PRABHU, N. U., On the integral equation for the finite dam, Quart. J. Math. Oxford Ser. **9** (1958) 183–188.
PRABHU, N. U., A waiting time process in the queue $G/M/1$, Acta Math. Acad. Sci. Hungar. **15** (1964) 363–371.
PRABHU, N. U., Queues and Inventories (Wiley, New York, 1965[a]).
PRABHU, N. U., Stochastic Processes (Macmillan, New York, 1965[b]).
PYKE, R., The supremum and infinum of a Poisson process, Ann. Math. Statist. **30** (1959) 568–576.
PYKE, R., Markov renewal processes; definitions and preliminary properties, Ann. Math. Statist. **32** (1961[a]) 1231–1242.

PYKE, R., Markov renewal processes with finitely many states, Ann. Math. Statist. **32** (1961[b]) 1243–1259.

REICH, E., Waiting times when queues are in tandem, Ann. Math. Statist. **28** (1957) 768–773.

REICH, E., On the integro-differential equation of Takács, I, Ann. Math. Statist. **29** (1958) 563–570.

REICH, E., Departure processes, Proc. Symp. on Congestion Theory, eds. W. L. Smith and W. E. Wilkinson (University of North-Carolina Press, Chapel Hill, N.C., 1965).

RIORDAN, J., Delay curves for calls served in random order, Bell System Tech. J. **32** (1953) 100–119.

RIORDAN, J., Stochastic Service Systems (Wiley, New York, 1962).

ROES, P. B. M., A many server bulk queue, Operations Res. **14** (1966) 1037–1044.

RUNNENBURG, J. Th., On the Use of Markov Processes in One Server Waiting Time Problems and Renewal Theory, Thesis, Univ. Amsterdam (Poortpress, 1960).

RUNNENBURG, J. Th., On the use of the method of collective marks in queueing theory, Proc. Symp. on Congestion Theory, eds. W. L. Smith and W. E. Wilkinson (University of North-Carolina Press, Chapel Hill, N.C., 1965).

SAATY, T. L., Elements of Queueing Theory (McGraw-Hill, New York, 1961).

SAMANDAROV, E. G., Service systems in heavy traffic, Theor. Probability Appl. **8** (1963) 307–309.

SMITH, W. L., On the distribution of queueing times, Proc. Cambridge Phil. Soc. **49** (1953) 449–461.

SMITH, W. L., Asymptotic renewal theorems, Proc. Roy. Soc. Edinburgh Sect. A **64** (1954) 9–48.

SMITH, W. L., Regenerative stochastic processes, Proc. Roy. Soc. Ser. A **232** (1955) 6–31.

SMITH, W. L., Renewal theory and its ramifications, J. Roy. Statist. Soc. Ser. B **20** (1958) 243–302.

SMITH, W. L., On some general renewal theorems for not-identically distributed variables, Proc. Fourth Berkeley Symp. on Math. Statistics and Probability II, ed. J. Neyman (University of California Press, Berkeley, Calif., 1960).

SPITZER, F., A combinatorial lemma and its application to probability theory, Trans. Am. Math. Soc. **82** (1956) 323–339.

SPITZER, F., The Wiener-Hopf equation whose kernel is a probability density, Duke Math. J. **24** (1957) 327–343.

SPITZER, F., A Tauberian theorem and its probability interpretation, Trans. Am. Math. Soc. **94** (1960) 150–160.

SPITZER, F., Principles of Random Walk (Van Nostrand, New York, 1964).

STAM, A. J., On the Q-matrix of a derived Markov chain, Nederl. Akad. Wetensch. Indagationes Math. **24** (1962) 578–582.

SYSKI, R., Introduction to Congestion Theory in Telephone Systems (Oliver and Boyd, London, 1962).

SYSKI, R., Markovian queues, Proc. Symp. on Congestion Theory, eds. W. L. Smith and W. E. Wilkinson (University of North-Carolina Press, Chapel Hill, N.C., 1965).

SYSKI, R., Pollaczek's method in queueing theory, Proc. Nato Conference on queueing theory, Lisbon, 1966, ed. R. Cruon (English University Press, London, 1967).

TAKÁCS, L., Transient behaviour of single server queueing processes with recurrent input and exponentially distributed service times, Operations Res. 8 (1960) 231–245.

TAKÁCS, L., The transient behaviour of a single server queueing process with a Poisson input, Proc. Fourth Berkely Symp. on Math. Statistics and Probability II, ed. J. Neyman (University of California Press, Berkely, Calif., 1961).

TAKÁCS, L., Introduction to the Theory of Queues (Oxford University Press, New York, 1962).

TAKÁCS, L., The limiting distribution of the virtual waiting time and the queue size for a single server queue with recurrent input and general service times, Sankhyā Ser. A **25** (1963[a]) 91–100.

TAKÁCS, L., Delay distributions for one line with Poisson input, general holding times and various orders of service, Bell System Tech. J. **42** (1963[b]) 487–504.

TAKÁCS, L., Occupation time problems in the theory of queues, Operations Res. **12** (1964[a]) 753–767.

TAKÁCS, L., Priority queues, Operations Res. **12** (1964[b]) 63–74.

TAKÁCS, L., Applications of ballot theorems in the theory of queues, Proc. Symp. on Congestion Theory, eds. W. L. Smith and W. E. Wilkinson (University of North-Carolina Press, Chapel Hill, N.C., 1965).

TAKÁCS, L., The distribution of the content of a dam when the input has stationary independent increments, J. Math. Mech. **15** (1966) 101–112.

TAKÁCS, L., Combinatorial Methods in the Theory of Stochastic Processes (Wiley, New York, 1967).

TITCHMARSH, E. C., Theory of Functions (Oxford University Press, London, 1952).

VAULOT, E., Delais d'attente des appels téléphoniques traités au hasard, C. R. Acad. Sci. Paris **222** (1946) 268–269.

VAULOT, E., Delais d'attente des appels téléphoniques dans l'ordre inverse de leur arrivé, C. R. Acad. Sci. Paris **238** (1954) 1188–1189.

WEESAKUL, B., First emptiness in a finite dam, J. Roy. Statist Soc. Ser. B **23** (1961) 343–351.

WELCH, P. D., On pre-emptive resume priority queues, Ann. Math. Statist. **35** (1964) 600–612.

WENDEL, J. G., Spitzer's formula: a short proof, Proc. Am. Math. Soc. **9** (1958) 905–908.

WENDEL, J. G., Order statistics of partial sums, Ann. Math. Statist. **31** (1960) 1034–1044.

WHITTAKER, E. T. and WATSON, G. N., A Course of Modern Analysis (Cambridge University Press, London, 1946).

WIDDER, D. V., The Laplace Transform (Princeton University Press, Princeton, 1946).

WILKINSON, R. I., Working curves for delayed exponential calls served in random order, Bell System Tech. J. **32** (1953) 360–383.

WISHART, D. M. G., A queueing system with χ^2-service time distribution, Ann. Math. Statist. **27** (1956) 768–779.

WISHART, D. M. G., Queueing systems in which the discipline is "last come, first served", Operations Res. **8** (1960) 591–599.

WISHART, D. M. G., Discussion on Kingman's paper, J. Roy. Statist. Soc. Ser. B **28** (1966) 417–447.

YEO, G. F., Single server queue with modified service mechanism, J. Austral. Math. Soc. **2** (1962) 499–507.

YEO, G. F., Pre-emptive priority queues, J. Austral. Math. Soc. **3** (1963) 491–502.

AUTHOR INDEX

INDEX OF NOTATIONS

Only the symbols which are used *frequently throughout* the book are listed below.

(A) $\overset{\text{def}}{=} \begin{cases} 1 \text{ if } A \text{ occurs,} \\ 0 \text{ if } A \text{ does not occur,} \end{cases}$ (indicator function)

$A(.)$ interarrival time distribution

a load or traffic

$\boldsymbol{a}_n$ $\overset{\text{def}}{=} \sum_{i=1}^{n} \boldsymbol{\sigma}_{i+1}$

$B(.)$ service time distribution

$\boldsymbol{b}_n$ $\overset{\text{def}}{=} \sum_{i=1}^{n} \boldsymbol{\tau}_i$

C_ξ contour parallel to the imaginary axis running from $\varepsilon - \mathrm{i}\delta$ to $\varepsilon + \mathrm{i}\delta$, with $\varepsilon = \mathrm{Re}\ \xi$ and $\delta \to \infty$

$\boldsymbol{c}$ busy cycle

$\boldsymbol{c}_1$ residual busy cycle

$\boldsymbol{c}_\mathrm{i}$ completion time

$\boldsymbol{c}_\mathrm{s}$ strong busy cycle

$D(.)$ distribution of the busy period

D_ξ circular contour with centre at $\xi = 0$

$\boldsymbol{d}_n$*) total idle time during $[0, \boldsymbol{t}_n)$

$\mathrm{E}\{.\}$ expectation

$\boldsymbol{e}_t$ total idle time during $[0, t)$

$\boldsymbol{f}_t$ total busy time during $[0, t)$

$\boldsymbol{g}_{1,n}$ size of the nth group

*) In chapter III. 2 $\boldsymbol{d}_n$ has a different meaning.

$\boldsymbol{g}_{2,n}$	capacity for service of the nth service
$\boldsymbol{h}_t$	number of departures during $[0,t)$
I	unit matrix
$\boldsymbol{i}$	idle period
$\boldsymbol{k}_t$	amount of work brought in during $[0,t)$
$m(t)$	renewal function
$\boldsymbol{n}$	number of customers in a busy period
P	matrix of one-step transition probabilities
$P^{(m)}$	matrix of m-step transition probabilities
$P(.)$	distribution of the busy cycle or priority distribution
p_{ij}	one-step transition probability from state E_i to state E_j
$p_{ij}^{(m)}$	m-step transition probability from state E_i to state E_j
$\boldsymbol{p}$	busy period
$\boldsymbol{p}_1$	residual busy period
$\boldsymbol{p}_s$	strong busy period
$\boldsymbol{r}_n$	moment of departure of the nth (arriving) customer
$\boldsymbol{r}_n'$	moment of the nth departure after $t=0$
$\boldsymbol{t}_n$	arrival moment of the nth customer
$U(.)$ *)	normalized unit step-function; $U(x)=\begin{cases}0, & x<0\\ \frac{1}{2}, & x=0\\ 1, & x>0\end{cases}$
$U_1(.)$	unit step-function; $U_1(x)=\begin{cases}0, & x\leqq 0\\ 1, & x>0\end{cases}$
$V(.)$	limit distribution of the virtual waiting time
$\boldsymbol{v}_t$	virtual waiting time or content of a dam at time t
$W(.)$	limit distribution of the actual waiting time
$\boldsymbol{w}_n$	actual waiting time of the nth arriving customer
$[x]^+$	$\overset{\text{def}}{=}$ max $(0, x)$
$[x]^-$	$\overset{\text{def}}{=}$ min $(0, x)$
$\boldsymbol{x}_t$	queue length at time t
$\boldsymbol{y}_n$	queue length at time $\boldsymbol{t}_n-$
$\boldsymbol{z}_n$	queue length at time $\boldsymbol{r}_n'+$
α	first moment of the interarrival time distribution
α_n	nth moment of the interarrival time distribution
$\alpha(.)$	Laplace-Stieltjes transform of the interarrival time distribution
$\alpha_1(.)$	numerator of $\alpha(.)$ if $\alpha(.)$ is rational
$\alpha_2(.)$	denominator of $\alpha(.)$ if $\alpha(.)$ is rational

*) On page 599 ff. $U(.)$ has a different meaning.

$\boldsymbol{\alpha}_{ij}$	entrance time from state E_i into state E_j
β	first moment of the service time distribution
β_n	nth moment of the service time distribution
$\beta(.)$	Laplace-Stieltjes transform of the service time distribution
$\beta_1(.)$	numerator of $\beta(.)$ if $\beta(.)$ is rational
$\beta_2(.)$	denominator of $\beta(.)$ if $\beta(.)$ is rational
δ_{ij}	$\stackrel{\text{def}}{=}\begin{cases}0 \text{ if } i\neq j,\\ 1 \text{ if } i=j,\end{cases}$ (Kronecker's symbol)
$\boldsymbol{\xi}_t$	residual lifetime at time t
$\boldsymbol{\eta}_t$	past lifetime at time t
$\boldsymbol{\theta}_v$	initial busy period if $\boldsymbol{v}_0=v$
λ_i	birth rate in state E_i
μ	average renewal time
μ_i	death rate in state E_i
$\boldsymbol{v}_n$	number of arrivals during nth service time
$\boldsymbol{v}_t$	number of arrivals (renewals) during $[0,t)$
$\boldsymbol{\rho}_n$	$\stackrel{\text{def}}{=}\boldsymbol{\tau}_n-\boldsymbol{\sigma}_{n+1}$
$\boldsymbol{\sigma}_n$	interarrival time between the $(n-1)$th and nth arrival
$\boldsymbol{\tau}_n$	service time of the nth arriving customer

SUBJECT INDEX